Student Solutions Manual
for
MULTIVARIABLE CALCULUS
SEVENTH EDITION

DAN CLEGG
Palomar College

BARBARA FRANK
Cape Fear Community College

BROOKS/COLE
CENGAGE Learning

Australia · Brazil · Japan · Korea · Mexico · Singapore · Spain · United Kingdom · United States

© 2012 Brooks/Cole, Cengage Learning

ALL RIGHTS RESERVED. No part of this work covered by the copyright herein may be reproduced, transmitted, stored, or used in any form or by any means graphic, electronic, or mechanical, including but not limited to photocopying, recording, scanning, digitizing, taping, Web distribution, information networks, or information storage and retrieval systems, except as permitted under Section 107 or 108 of the 1976 United States Copyright Act, without the prior written permission of the publisher.

For product information and technology assistance, contact us at
**Cengage Learning Customer & Sales Support,
1-800-354-9706**

For permission to use material from this text or product, submit all requests online at
www.cengage.com/permissions
Further permissions questions can be emailed to
permissionrequest@cengage.com

ISBN-13: 987-0-8400-4945-2
ISBN-10: 0-8400-4945-5

Brooks/Cole
20 Davis Drive
Belmont, CA 94002-3098
USA

Cengage Learning is a leading provider of customized learning solutions with office locations around the globe, including Singapore, the United Kingdom, Australia, Mexico, Brazil, and Japan. Locate your local office at:
www.cengage.com/global

Cengage Learning products are represented in Canada by Nelson Education, Ltd.

To learn more about Brooks/Cole, visit
www.cengage.com/brookscole

Purchase any of our products at your local college store or at our preferred online store **www.cengagebrain.com**

Printed in the United States of America
2 3 4 5 6 7   15 14 13 12 11

© 2012 Cengage Learning. All Rights Reserved. May not be scanned, copied, or duplicated, or posted to a publicly accessible website, in whole or in part.

□ PREFACE

This *Student Solutions Manual* contains detailed solutions to selected exercises in the text *Multivariable Calculus,* Seventh Edition (Chapters 10–17 of *Calculus,* Seventh Edition, and *Calculus: Early Transcendentals,* Seventh Edition) by James Stewart. Specifically, it includes solutions to the odd-numbered exercises in each chapter section, review section, True-False Quiz, and Problems Plus section. Also included are all solutions to the Concept Check questions.

Because of differences between the regular version and the *Early Transcendentals* version of the text, some references are given in a dual format. In these cases, readers of the *Early Transcendentals* text should use the references denoted by "ET."

Each solution is presented in the context of the corresponding section of the text. In general, solutions to the initial exercises involving a new concept illustrate that concept in more detail; this knowledge is then utilized in subsequent solutions. Thus, while the intermediate steps of a solution are given, you may need to refer back to earlier exercises in the section or prior sections for additional explanation of the concepts involved. Note that, in many cases, different routes to an answer may exist which are equally valid; also, answers can be expressed in different but equivalent forms. Thus, the goal of this manual is not to give the definitive solution to each exercise, but rather to assist you as a student in understanding the concepts of the text and learning how to apply them to the challenge of solving a problem.

We would like to thank James Stewart for entrusting us with the writing of this manual and offering suggestions and Kathi Townes of TECH-arts for typesetting and producing this manual as well as creating the illustrations. We also thank Richard Stratton, Liz Covello, and Elizabeth Neustaetter of Brooks/Cole, Cengage Learning, for their trust, assistance, and patience.

DAN CLEGG
Palomar College

BARBARA FRANK
Cape Fear Community College

© 2012 Cengage Learning. All Rights Reserved. May not be scanned, copied, or duplicated, or posted to a publicly accessible website, in whole or in part.

ABBREVIATIONS AND SYMBOLS

CD	concave downward
CU	concave upward
D	the domain of f
FDT	First Derivative Test
HA	horizontal asymptote(s)
I	interval of convergence
I/D	Increasing/Decreasing Test
IP	inflection point(s)
R	radius of convergence
VA	vertical asymptote(s)
$\overset{CAS}{=}$	indicates the use of a computer algebra system.
$\overset{H}{=}$	indicates the use of l'Hospital's Rule.
$\overset{j}{=}$	indicates the use of Formula j in the Table of Integrals in the back endpapers.
$\overset{s}{=}$	indicates the use of the substitution $\{u = \sin x, du = \cos x\, dx\}$.
$\overset{c}{=}$	indicates the use of the substitution $\{u = \cos x, du = -\sin x\, dx\}$.

© 2012 Cengage Learning. All Rights Reserved. May not be scanned, copied, or duplicated, or posted to a publicly accessible website, in whole or in part.

CONTENTS

© 2012 Cengage Learning. All Rights Reserved. May not be scanned, copied, or duplicated, or posted to a publicly accessible website, in whole or in part.

© 2012 Cengage Learning. All Rights Reserved. May not be scanned, copied, or duplicated, or posted to a publicly accessible website, in whole or in part.

□ 16 VECTOR CALCULUS   303

□ 17 SECOND-ORDER DIFFERENTIAL EQUATIONS   345

□ APPENDIX   359

© 2012 Cengage Learning. All Rights Reserved. May not be scanned, copied, or duplicated, or posted to a publicly accessible website, in whole or in part.

10 ☐ PARAMETRIC EQUATIONS AND POLAR COORDINATES

10.1 Curves Defined by Parametric Equations

1. $x = t^2 + t$, $y = t^2 - t$, $-2 \le t \le 2$

t	-2	-1	0	1	2
x	2	0	0	2	6
y	6	2	0	0	2

3. $x = \cos^2 t$, $y = 1 - \sin t$, $0 \le t \le \pi/2$

t	0	$\pi/6$	$\pi/3$	$\pi/2$
x	1	$3/4$	$1/4$	0
y	1	$1/2$	$1 - \frac{\sqrt{3}}{2} \approx 0.13$	0

5. $x = 3 - 4t$, $y = 2 - 3t$

(a)

t	-1	0	1	2
x	7	3	-1	-5
y	5	2	-1	-4

(b) $x = 3 - 4t \Rightarrow 4t - -x + 3 \Rightarrow t = -\frac{1}{4}x + \frac{3}{4}$, so

$$y = 2 - 3t = 2 - 3\left(-\frac{1}{4}x + \frac{3}{4}\right) = 2 + \frac{3}{4}x - \frac{9}{4} \Rightarrow y = \frac{3}{4}x - \frac{1}{4}$$

7. $x = 1 - t^2$, $y = t - 2$, $-2 \le t \le 2$

(a)

t	-2	-1	0	1	2
x	-3	0	1	0	-3
y	-4	-3	-2	-1	0

(b) $y = t - 2 \Rightarrow t = y + 2$, so $x = 1 - t^2 = 1 - (y + 2)^2 \Rightarrow$

$x = -(y + 2)^2 + 1$, or $x = -y^2 - 4y - 3$, with $-4 \le y \le 0$

© 2012 Cengage Learning. All Rights Reserved. May not be scanned, copied, or duplicated, or posted to a publicly accessible website, in whole or in part.

9. $x = \sqrt{t}$, $y = 1 - t$

(a)

t	0	1	2	3	4
x	0	1	1.414	1.732	2
y	1	0	−1	−2	−3

(b) $x = \sqrt{t}$ $\Rightarrow$ $t = x^2$ $\Rightarrow$ $y = 1 - t = 1 - x^2$. Since $t \geq 0$, $x \geq 0$.

So the curve is the right half of the parabola $y = 1 - x^2$.

11. (a) $x = \sin \frac{1}{2}\theta$, $y = \cos \frac{1}{2}\theta$, $-\pi \leq \theta \leq \pi$.

$x^2 + y^2 = \sin^2 \frac{1}{2}\theta + \cos^2 \frac{1}{2}\theta = 1$. For $-\pi \leq \theta \leq 0$, we have

$-1 \leq x \leq 0$ and $0 \leq y \leq 1$. For $0 < \theta \leq \pi$, we have $0 < x \leq 1$

and $1 > y \geq 0$. The graph is a semicircle.

(b)

13. (a) $x = \sin t$, $y = \csc t$, $0 < t < \frac{\pi}{2}$. $y = \csc t = \dfrac{1}{\sin t} = \dfrac{1}{x}$.

For $0 < t < \frac{\pi}{2}$, we have $0 < x < 1$ and $y > 1$. Thus, the curve is the

portion of the hyperbola $y = 1/x$ with $y > 1$.

(b)

15. (a) $x = e^{2t}$ $\Rightarrow$ $2t = \ln x$ $\Rightarrow$ $t = \frac{1}{2} \ln x$.

$y = t + 1 = \frac{1}{2} \ln x + 1$.

(b)

17. (a) $x = \sinh t$, $y = \cosh t$ $\Rightarrow$ $y^2 - x^2 = \cosh^2 t - \sinh^2 t = 1$. Since

$y = \cosh t \geq 1$, we have the upper branch of the hyperbola $y^2 - x^2 = 1$.

(b)

19. $x = 3 + 2\cos t$, $y = 1 + 2\sin t$, $\pi/2 \leq t \leq 3\pi/2$.   By Example 4 with $r = 2$, $h = 3$, and $k = 1$, the motion of the particle

takes place on a circle centered at $(3, 1)$ with a radius of 2. As t goes from $\frac{\pi}{2}$ to $\frac{3\pi}{2}$, the particle starts at the point $(3, 3)$ and

moves counterclockwise along the circle $(x - 3)^2 + (y - 1)^2 = 4$ to $(3, -1)$ [one-half of a circle].

21. $x = 5\sin t$, $y = 2\cos t$ $\Rightarrow$ $\sin t = \dfrac{x}{5}$, $\cos t = \dfrac{y}{2}$. $\sin^2 t + \cos^2 t = 1$ $\Rightarrow$ $\left(\dfrac{x}{5}\right)^2 + \left(\dfrac{y}{2}\right)^2 = 1$. The motion of the

particle takes place on an ellipse centered at $(0, 0)$. As t goes from $-\pi$ to 5π, the particle starts at the point $(0, -2)$ and moves

clockwise around the ellipse 3 times.

23. We must have $1 \leq x \leq 4$ and $2 \leq y \leq 3$. So the graph of the curve must be contained in the rectangle $[1, 4]$ by $[2, 3]$.

© 2012 Cengage Learning. All Rights Reserved. May not be scanned, copied, or duplicated, or posted to a publicly accessible website, in whole or in part.

25. When $t = -1$, $(x, y) = (0, -1)$. As t increases to 0, x decreases to -1 and y increases to 0. As t increases from 0 to 1, x increases to 0 and y increases to 1. As t increases beyond 1, both x and y increase. For $t < -1$, x is positive and decreasing and y is negative and increasing. We could achieve greater accuracy by estimating x- and y-values for selected values of t from the given graphs and plotting the corresponding points.

27. When $t = 0$ we see that $x = 0$ and $y = 0$, so the curve starts at the origin. As t increases from 0 to $\frac{1}{2}$, the graphs show that y increases from 0 to 1 while x increases from 0 to 1, decreases to 0 and to -1, then increases back to 0, so we arrive at the point $(0, 1)$. Similarly, as t increases from $\frac{1}{2}$ to 1, y decreases from 1 to 0 while x repeats its pattern, and we arrive back at the origin. We could achieve greater accuracy by estimating x- and y-values for selected values of t from the given graphs and plotting the corresponding points.

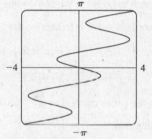

29. Use $y = t$ and $x = t - 2\sin \pi t$ with a t-interval of $[-\pi, \pi]$.

31. (a) $x = x_1 + (x_2 - x_1)t$, $y = y_1 + (y_2 - y_1)t$, $0 \le t \le 1$. Clearly the curve passes through $P_1(x_1, y_1)$ when $t = 0$ and through $P_2(x_2, y_2)$ when $t = 1$. For $0 < t < 1$, x is strictly between x_1 and x_2 and y is strictly between y_1 and y_2. For every value of t, x and y satisfy the relation $y - y_1 = \dfrac{y_2 - y_1}{x_2 - x_1}(x - x_1)$, which is the equation of the line through $P_1(x_1, y_1)$ and $P_2(x_2, y_2)$.

    Finally, any point (x, y) on that line satisfies $\dfrac{y - y_1}{y_2 - y_1} = \dfrac{x - x_1}{x_2 - x_1}$; if we call that common value t, then the given parametric equations yield the point (x, y); and any (x, y) on the line between $P_1(x_1, y_1)$ and $P_2(x_2, y_2)$ yields a value of t in $[0, 1]$. So the given parametric equations exactly specify the line segment from $P_1(x_1, y_1)$ to $P_2(x_2, y_2)$.

(b) $x = -2 + [3 - (-2)]t = -2 + 5t$ and $y = 7 + (-1 - 7)t = 7 - 8t$ for $0 \le t \le 1$.

33. The circle $x^2 + (y - 1)^2 = 4$ has center $(0, 1)$ and radius 2, so by Example 4 it can be represented by $x = 2\cos t$, $y = 1 + 2\sin t$, $0 \le t \le 2\pi$. This representation gives us the circle with a counterclockwise orientation starting at $(2, 1)$.

(a) To get a clockwise orientation, we could change the equations to $x = 2\cos t$, $y = 1 - 2\sin t$, $0 \le t \le 2\pi$.

(b) To get three times around in the counterclockwise direction, we use the original equations $x = 2\cos t$, $y = 1 + 2\sin t$ with the domain expanded to $0 \le t \le 6\pi$.

© 2012 Cengage Learning. All Rights Reserved. May not be scanned, copied, or duplicated, or posted to a publicly accessible website, in whole or in part.

(c) To start at $(0, 3)$ using the original equations, we must have $x_1 = 0$; that is, $2 \cos t = 0$. Hence, $t = \frac{\pi}{2}$. So we use

$x = 2 \cos t$, $y = 1 + 2 \sin t$, $\frac{\pi}{2} \leq t \leq \frac{3\pi}{2}$.

Alternatively, if we want t to start at 0, we could change the equations of the curve. For example, we could use

$x = -2 \sin t$, $y = 1 + 2 \cos t$, $0 \leq t \leq \pi$.

35. *Big circle:* It's centered at $(2, 2)$ with a radius of 2, so by Example 4, parametric equations are

$$x = 2 + 2 \cos t, \qquad y = 2 + 2 \sin t, \qquad 0 \leq t \leq 2\pi$$

Small circles: They are centered at $(1, 3)$ and $(3, 3)$ with a radius of 0.1. By Example 4, parametric equations are

$$\text{(left)} \qquad x = 1 + 0.1 \cos t, \qquad y = 3 + 0.1 \sin t, \qquad 0 \leq t \leq 2\pi$$

and $\qquad\qquad \text{(right)} \qquad x = 3 + 0.1 \cos t, \qquad y = 3 + 0.1 \sin t, \qquad 0 \leq t \leq 2\pi$

Semicircle: It's the lower half of a circle centered at $(2, 2)$ with radius 1. By Example 4, parametric equations are

$$x = 2 + 1 \cos t, \qquad y = 2 + 1 \sin t, \qquad \pi \leq t \leq 2\pi$$

To get all four graphs on the same screen with a typical graphing calculator, we need to change the last t-interval to $[0, 2\pi]$ in order to match the others. We can do this by changing t to $0.5t$. This change gives us the upper half. There are several ways to get the lower half—one is to change the "+" to a "−" in the y-assignment, giving us

$$x = 2 + 1 \cos(0.5t), \qquad y = 2 - 1 \sin(0.5t), \qquad 0 \leq t \leq 2\pi$$

37. (a) $x = t^3 \;\Rightarrow\; t = x^{1/3}$, so $y = t^2 = x^{2/3}$.

We get the entire curve $y = x^{2/3}$ traversed in a left to right direction.

(b) $x = t^6 \;\Rightarrow\; t = x^{1/6}$, so $y = t^4 = x^{4/6} = x^{2/3}$.

Since $x = t^6 \geq 0$, we only get the right half of the curve $y = x^{2/3}$.

(c) $x = e^{-3t} = (e^{-t})^3$ [so $e^{-t} = x^{1/3}$],

$y = e^{-2t} = (e^{-t})^2 = (x^{1/3})^2 = x^{2/3}$.

If $t < 0$, then x and y are both larger than 1. If $t > 0$, then x and y are between 0 and 1. Since $x > 0$ and $y > 0$, the curve never quite reaches the origin.

39. The case $\frac{\pi}{2} < \theta < \pi$ is illustrated. C has coordinates $(r\theta, r)$ as in Example 7, and Q has coordinates $(r\theta, r + r \cos(\pi - \theta)) = (r\theta, r(1 - \cos\theta))$ [since $\cos(\pi - \alpha) = \cos\pi \cos\alpha + \sin\pi \sin\alpha = -\cos\alpha$], so P has coordinates $(r\theta - r \sin(\pi - \theta), r(1 - \cos\theta)) = (r(\theta - \sin\theta), r(1 - \cos\theta))$ [since $\sin(\pi - \alpha) = \sin\pi \cos\alpha - \cos\pi \sin\alpha = \sin\alpha$]. Again we have the parametric equations $x = r(\theta - \sin\theta)$, $y = r(1 - \cos\theta)$.

© 2012 Cengage Learning. All Rights Reserved. May not be scanned, copied, or duplicated, or posted to a publicly accessible website, in whole or in part.

41. It is apparent that $x = |OQ|$ and $y = |QP| = |ST|$. From the diagram,

$x = |OQ| = a\cos\theta$ and $y = |ST| = b\sin\theta$. Thus, the parametric equations are

$x = a\cos\theta$ and $y = b\sin\theta$. To eliminate θ we rearrange: $\sin\theta = y/b$ $\Rightarrow$

$\sin^2\theta = (y/b)^2$ and $\cos\theta = x/a$ $\Rightarrow$ $\cos^2\theta = (x/a)^2$. Adding the two

equations: $\sin^2\theta + \cos^2\theta = 1 = x^2/a^2 + y^2/b^2$. Thus, we have an ellipse.

43. $C = (2a\cot\theta, 2a)$, so the x-coordinate of P is $x = 2a\cot\theta$. Let $B = (0, 2a)$.

Then $\angle OAB$ is a right angle and $\angle OBA = \theta$, so $|OA| = 2a\sin\theta$ and

$A = ((2a\sin\theta)\cos\theta, (2a\sin\theta)\sin\theta)$. Thus, the y-coordinate of P

is $y = 2a\sin^2\theta$.

45. (a)

There are 2 points of intersection:

$(-3, 0)$ and approximately $(-2.1, 1.4)$.

(b) A collision point occurs when $x_1 = x_2$ and $y_1 = y_2$ for the same t. So solve the equations:

$$3\sin t = -3 + \cos t \quad \textbf{(1)}$$

$$2\cos t = 1 + \sin t \quad \textbf{(2)}$$

From **(2)**, $\sin t = 2\cos t - 1$. Substituting into **(1)**, we get $3(2\cos t - 1) = -3 + \cos t$ $\Rightarrow$ $5\cos t = 0$ $(\star)$ $\Rightarrow$

$\cos t = 0$ $\Rightarrow$ $t = \frac{\pi}{2}$ or $\frac{3\pi}{2}$. We check that $t = \frac{3\pi}{2}$ satisfies **(1)** and **(2)** but $t = \frac{\pi}{2}$ does not. So the only collision point

occurs when $t = \frac{3\pi}{2}$, and this gives the point $(-3, 0)$. [We could check our work by graphing x_1 and x_2 together as

functions of t and, on another plot, y_1 and y_2 as functions of t. If we do so, we see that the only value of t for which *both*

pairs of graphs intersect is $t = \frac{3\pi}{2}$.]

(c) The circle is centered at $(3, 1)$ instead of $(-3, 1)$. There are still 2 intersection points: $(3, 0)$ and $(2.1, 1.4)$, but there are

no collision points, since $(\star)$ in part (b) becomes $5\cos t = 6$ $\Rightarrow$ $\cos t = \frac{6}{5} > 1$.

47. $x = t^2$, $y = t^3 - ct$. We use a graphing device to produce the graphs for various values of c with $-\pi \leq t \leq \pi$. Note that all

the members of the family are symmetric about the x-axis. For $c < 0$, the graph does not cross itself, but for $c = 0$ it has a

cusp at $(0, 0)$ and for $c > 0$ the graph crosses itself at $x = c$, so the loop grows larger as c increases.

© 2012 Cengage Learning. All Rights Reserved. May not be scanned, copied, or duplicated, or posted to a publicly accessible website, in whole or in part.

49. $x = t + a\cos t$, $y = t + a\sin t$, $a > 0$. From the first figure, we see that

curves roughly follow the line $y = x$, and they start having loops when a

is between 1.4 and 1.6. The loops increase in size as a increases.

While not required, the following is a solution to determine the *exact* values for which the curve has a loop,

that is, we seek the values of a for which there exist parameter values t and u such that $t < u$ and

$$(t + a\cos t, t + a\sin t) = (u + a\cos u, u + a\sin u).$$

In the diagram at the left, T denotes the point (t, t), U the point (u, u),

and P the point $(t + a\cos t, t + a\sin t) = (u + a\cos u, u + a\sin u)$.

Since $\overline{PT} = \overline{PU} = a$, the triangle PTU is isosceles. Therefore its base

angles, $\alpha = \angle PTU$ and $\beta = \angle PUT$ are equal. Since $\alpha = t - \frac{\pi}{4}$ and

$\beta = 2\pi - \frac{3\pi}{4} - u = \frac{5\pi}{4} - u$, the relation $\alpha = \beta$ implies that

$u + t = \frac{3\pi}{2}$ **(1)**.

Since $\overline{TU} = \text{distance}((t, t), (u, u)) = \sqrt{2(u - t)^2} = \sqrt{2}\,(u - t)$, we see that

$\cos\alpha = \dfrac{\frac{1}{2}\overline{TU}}{\overline{PT}} = \dfrac{(u - t)/\sqrt{2}}{a}$, so $u - t = \sqrt{2}\,a\cos\alpha$, that is,

$u - t = \sqrt{2}\,a\cos\left(t - \frac{\pi}{4}\right)$ **(2)**. Now $\cos\left(t - \frac{\pi}{4}\right) = \sin\left[\frac{\pi}{2} - \left(t - \frac{\pi}{4}\right)\right] = \sin\left(\frac{3\pi}{4} - t\right)$,

so we can rewrite **(2)** as $u - t = \sqrt{2}\,a\sin\left(\frac{3\pi}{4} - t\right)$ **(2′)**. Subtracting **(2′)** from **(1)** and

dividing by 2, we obtain $t = \frac{3\pi}{4} - \frac{\sqrt{2}}{2}a\sin\left(\frac{3\pi}{4} - t\right)$, or $\frac{3\pi}{4} - t = \frac{a}{\sqrt{2}}\sin\left(\frac{3\pi}{4} - t\right)$ **(3)**.

Since $a > 0$ and $t < u$, it follows from **(2′)** that $\sin\left(\frac{3\pi}{4} - t\right) > 0$. Thus from **(3)** we see that $t < \frac{3\pi}{4}$. [We have

implicitly assumed that $0 < t < \pi$ by the way we drew our diagram, but we lost no generality by doing so since replacing t

by $t + 2\pi$ merely increases x and y by 2π. The curve's basic shape repeats every time we change t by 2π.] Solving for a in

(3), we get $a = \dfrac{\sqrt{2}\left(\frac{3\pi}{4} - t\right)}{\sin\left(\frac{3\pi}{4} - t\right)}$. Write $z = \frac{3\pi}{4} - t$. Then $a = \dfrac{\sqrt{2}\,z}{\sin z}$, where $z > 0$. Now $\sin z < z$ for $z > 0$, so $a > \sqrt{2}$.

$\left[\text{As } z \to 0^+, \text{ that is, as } t \to \left(\frac{3\pi}{4}\right)^-, a \to \sqrt{2}\right]$.

51. Note that all the Lissajous figures are symmetric about the x-axis. The parameters a and b simply stretch the graph in the

x- and y-directions respectively. For $a = b = n = 1$ the graph is simply a circle with radius 1. For $n = 2$ the graph crosses

© 2012 Cengage Learning. All Rights Reserved. May not be scanned, copied, or duplicated, or posted to a publicly accessible website, in whole or in part.

itself at the origin and there are loops above and below the x-axis. In general, the figures have $n - 1$ points of intersection, all of which are on the y-axis, and a total of n closed loops.

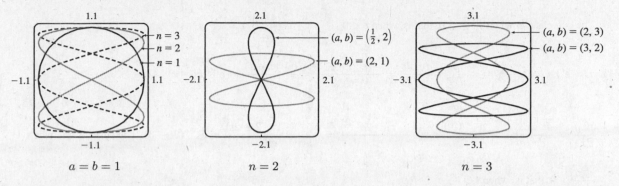

$$a = b = 1 \qquad\qquad n = 2 \qquad\qquad n = 3$$

10.2   Calculus with Parametric Curves

1. $x = t \sin t$, $y = t^2 + t$ $\Rightarrow$ $\dfrac{dy}{dt} = 2t + 1$, $\dfrac{dx}{dt} = t \cos t + \sin t$, and $\dfrac{dy}{dx} = \dfrac{dy/dt}{dx/dt} = \dfrac{2t + 1}{t \cos t + \sin t}$.

3. $x = 1 + 4t - t^2$, $y = 2 - t^3$; $t = 1$. $\dfrac{dy}{dt} = -3t^2$, $\dfrac{dx}{dt} = 4 - 2t$, and $\dfrac{dy}{dx} = \dfrac{dy/dt}{dx/dt} = \dfrac{-3t^2}{4 - 2t}$. When $t = 1$,

$(x, y) = (4, 1)$ and $dy/dx = -\frac{3}{2}$, so an equation of the tangent to the curve at the point corresponding to $t = 1$ is

$y - 1 = -\frac{3}{2}(x - 4)$, or $y = -\frac{3}{2}x + 7$.

5. $x = t \cos t$, $y = t \sin t$; $t = \pi$. $\dfrac{dy}{dt} = t \cos t + \sin t$, $\dfrac{dx}{dt} = t(-\sin t) + \cos t$, and $\dfrac{dy}{dx} = \dfrac{dy/dt}{dx/dt} = \dfrac{t \cos t + \sin t}{-t \sin t + \cos t}$.

When $t = \pi$, $(x, y) = (-\pi, 0)$ and $dy/dx = -\pi/(-1) = \pi$, so an equation of the tangent to the curve at the point

corresponding to $t = \pi$ is $y - 0 = \pi[x - (-\pi)]$, or $y = \pi x + \pi^2$.

7. (a) $x = 1 + \ln t$, $y = t^2 + 2$; $(1, 3)$. $\dfrac{dy}{dt} = 2t$, $\dfrac{dx}{dt} = \dfrac{1}{t}$, and $\dfrac{dy}{dx} = \dfrac{dy/dt}{dx/dt} = \dfrac{2t}{1/t} = 2t^2$. At $(1, 3)$,

$x = 1 + \ln t = 1$ $\Rightarrow$ $\ln t = 0$ $\Rightarrow$ $t = 1$ and $\dfrac{dy}{dx} = 2$, so an equation of the tangent is $y - 3 = 2(x - 1)$,

or $y = 2x + 1$.

(b) $x = 1 + \ln t$ $\Rightarrow$ $\ln t = x - 1$ $\Rightarrow$ $t = e^{x-1}$, so $y = t^2 + 2 = (e^{x-1})^2 + 2 = e^{2x-2} + 2$, and $y' = e^{2x-2} \cdot 2$.

At $(1, 3)$, $y' = e^{2(1)-2} \cdot 2 = 2$, so an equation of the tangent is $y - 3 = 2(x - 1)$, or $y = 2x + 1$.

9. $x = 6 \sin t$, $y = t^2 + t$; $(0, 0)$.

$\dfrac{dy}{dx} = \dfrac{dy/dt}{dx/dt} = \dfrac{2t + 1}{6 \cos t}$. The point $(0, 0)$ corresponds to $t = 0$, so the

slope of the tangent at that point is $\frac{1}{6}$. An equation of the tangent is therefore

$y - 0 = \frac{1}{6}(x - 0)$, or $y = \frac{1}{6}x$.

© 2012 Cengage Learning. All Rights Reserved. May not be scanned, copied, or duplicated, or posted to a publicly accessible website, in whole or in part.

11. $x = t^2 + 1$, $y = t^2 + t$ $\Rightarrow$ $\dfrac{dy}{dx} = \dfrac{dy/dt}{dx/dt} = \dfrac{2t+1}{2t} = 1 + \dfrac{1}{2t}$ $\Rightarrow$ $\dfrac{d^2y}{dx^2} = \dfrac{\frac{d}{dt}\left(\frac{dy}{dx}\right)}{dx/dt} = \dfrac{-1/(2t^2)}{2t} = -\dfrac{1}{4t^3}$.

The curve is CU when $\dfrac{d^2y}{dx^2} > 0$, that is, when $t < 0$.

13. $x = e^t$, $y = te^{-t}$ $\Rightarrow$ $\dfrac{dy}{dx} = \dfrac{dy/dt}{dx/dt} = \dfrac{-te^{-t} + e^{-t}}{e^t} = \dfrac{e^{-t}(1-t)}{e^t} = e^{-2t}(1-t)$ $\Rightarrow$

$\dfrac{d^2y}{dx^2} = \dfrac{\frac{d}{dt}\left(\frac{dy}{dx}\right)}{dx/dt} = \dfrac{e^{-2t}(-1) + (1-t)(-2e^{-2t})}{e^t} = \dfrac{e^{-2t}(-1-2+2t)}{e^t} = e^{-3t}(2t-3)$. The curve is CU when

$\dfrac{d^2y}{dx^2} > 0$, that is, when $t > \frac{3}{2}$.

15. $x = 2\sin t$, $y = 3\cos t$, $0 < t < 2\pi$.

$\dfrac{dy}{dx} = \dfrac{dy/dt}{dx/dt} = \dfrac{-3\sin t}{2\cos t} = -\dfrac{3}{2}\tan t$, so $\dfrac{d^2y}{dx^2} = \dfrac{\frac{d}{dt}\left(\frac{dy}{dx}\right)}{dx/dt} = \dfrac{-\frac{3}{2}\sec^2 t}{2\cos t} = -\dfrac{3}{4}\sec^3 t$.

The curve is CU when $\sec^3 t < 0$ $\Rightarrow$ $\sec t < 0$ $\Rightarrow$ $\cos t < 0$ $\Rightarrow$ $\frac{\pi}{2} < t < \frac{3\pi}{2}$.

17. $x = t^3 - 3t$, $y = t^2 - 3$. $\dfrac{dy}{dt} = 2t$, so $\dfrac{dy}{dt} = 0$ $\Leftrightarrow$ $t = 0$ $\Leftrightarrow$

$(x, y) = (0, -3)$. $\dfrac{dx}{dt} = 3t^2 - 3 = 3(t+1)(t-1)$, so $\dfrac{dx}{dt} = 0$ $\Leftrightarrow$

$t = -1$ or 1 $\Leftrightarrow$ $(x, y) = (2, -2)$ or $(-2, -2)$. The curve has a horizontal

tangent at $(0, -3)$ and vertical tangents at $(2, -2)$ and $(-2, -2)$.

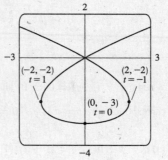

19. $x = \cos\theta$, $y = \cos 3\theta$. The whole curve is traced out for $0 \le \theta \le \pi$.

$\dfrac{dy}{d\theta} = -3\sin 3\theta$, so $\dfrac{dy}{d\theta} = 0$ $\Leftrightarrow$ $\sin 3\theta = 0$ $\Leftrightarrow$ $3\theta = 0, \pi, 2\pi,$ or 3π $\Leftrightarrow$

$\theta = 0, \frac{\pi}{3}, \frac{2\pi}{3},$ or π $\Leftrightarrow$ $(x, y) = (1, 1), (\frac{1}{2}, -1), (-\frac{1}{2}, 1),$ or $(-1, -1)$.

$\dfrac{dx}{d\theta} = -\sin\theta$, so $\dfrac{dx}{d\theta} = 0$ $\Leftrightarrow$ $\sin\theta = 0$ $\Leftrightarrow$ $\theta = 0$ or π $\Leftrightarrow$

$(x, y) = (1, 1)$ or $(-1, -1)$. Both $\dfrac{dy}{d\theta}$ and $\dfrac{dx}{d\theta}$ equal 0 when $\theta = 0$ and π.

To find the slope when $\theta = 0$, we find $\lim\limits_{\theta \to 0} \dfrac{dy}{dx} = \lim\limits_{\theta \to 0} \dfrac{-3\sin 3\theta}{-\sin\theta} \overset{\text{H}}{=} \lim\limits_{\theta \to 0} \dfrac{-9\cos 3\theta}{-\cos\theta} = 9$, which is the same slope when $\theta = \pi$.

Thus, the curve has horizontal tangents at $(\frac{1}{2}, -1)$ and $(-\frac{1}{2}, 1)$, and there are no vertical tangents.

© 2012 Cengage Learning. All Rights Reserved. May not be scanned, copied, or duplicated, or posted to a publicly accessible website, in whole or in part.

21. From the graph, it appears that the rightmost point on the curve $x = t - t^6$, $y = e^t$

    is about $(0.6, 2)$. To find the exact coordinates, we find the value of t for which the

    graph has a vertical tangent, that is, $0 = dx/dt = 1 - 6t^5 \iff t = 1/\sqrt[5]{6}$.

    Hence, the rightmost point is

    $$\left(1/\sqrt[5]{6} - 1/\left(6\sqrt[5]{6}\right), e^{1/\sqrt[5]{6}}\right) = \left(5 \cdot 6^{-6/5}, e^{6^{-1/5}}\right) \approx (0.58, 2.01).$$

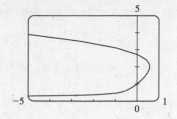

23. We graph the curve $x = t^4 - 2t^3 - 2t^2$, $y = t^3 - t$ in the viewing rectangle $[-2, 1.1]$ by $[-0.5, 0.5]$. This rectangle

    corresponds approximately to $t \in [-1, 0.8]$.

    We estimate that the curve has horizontal tangents at about $(-1, -0.4)$ and $(-0.17, 0.39)$ and vertical tangents at

    about $(0, 0)$ and $(-0.19, 0.37)$. We calculate $\dfrac{dy}{dx} = \dfrac{dy/dt}{dx/dt} = \dfrac{3t^2 - 1}{4t^3 - 6t^2 - 4t}$. The horizontal tangents occur when

    $dy/dt = 3t^2 - 1 = 0 \iff t = \pm\frac{1}{\sqrt{3}}$, so both horizontal tangents are shown in our graph. The vertical tangents occur when

    $dx/dt = 2t(2t^2 - 3t - 2) = 0 \iff 2t(2t + 1)(t - 2) = 0 \iff t = 0, -\frac{1}{2}$ or 2. It seems that we have missed one vertical

    tangent, and indeed if we plot the curve on the t-interval $[-1.2, 2.2]$ we see that there is another vertical tangent at $(-8, 6)$.

25. $x = \cos t$, $y = \sin t \cos t$. $dx/dt = -\sin t$, $dy/dt = -\sin^2 t + \cos^2 t = \cos 2t$.

    $(x, y) = (0, 0) \iff \cos t = 0 \iff t$ is an odd multiple of $\frac{\pi}{2}$. When $t = \frac{\pi}{2}$,

    $dx/dt = -1$ and $dy/dt = -1$, so $dy/dx = 1$. When $t = \frac{3\pi}{2}$, $dx/dt = 1$ and

    $dy/dt = -1$. So $dy/dx = -1$. Thus, $y = x$ and $y = -x$ are both tangent to the

    curve at $(0, 0)$.

27. $x = r\theta - d\sin\theta$, $y = r - d\cos\theta$.

    (a) $\dfrac{dx}{d\theta} = r - d\cos\theta$, $\dfrac{dy}{d\theta} = d\sin\theta$, so $\dfrac{dy}{dx} = \dfrac{d\sin\theta}{r - d\cos\theta}$.

    (b) If $0 < d < r$, then $|d\cos\theta| \le d < r$, so $r - d\cos\theta \ge r - d > 0$. This shows that $dx/d\theta$ never vanishes,

    so the trochoid can have no vertical tangent if $d < r$.

29. $x = 2t^3$, $y = 1 + 4t - t^2 \implies \dfrac{dy}{dx} = \dfrac{dy/dt}{dx/dt} = \dfrac{4 - 2t}{6t^2}$. Now solve $\dfrac{dy}{dx} = 1 \iff \dfrac{4 - 2t}{6t^2} = 1 \iff$

    $6t^2 + 2t - 4 = 0 \iff 2(3t - 2)(t + 1) = 0 \iff t = \frac{2}{3}$ or $t = -1$. If $t = \frac{2}{3}$, the point is $\left(\frac{16}{27}, \frac{29}{9}\right)$, and if $t = -1$,

    the point is $(-2, -4)$.

© 2012 Cengage Learning. All Rights Reserved. May not be scanned, copied, or duplicated, or posted to a publicly accessible website, in whole or in part.

31. By symmetry of the ellipse about the x- and y-axes,

$$A = 4 \int_0^a y\,dx = 4 \int_{\pi/2}^0 b \sin\theta\,(-a\sin\theta)\,d\theta = 4ab \int_0^{\pi/2} \sin^2\theta\,d\theta = 4ab \int_0^{\pi/2} \tfrac{1}{2}(1 - \cos 2\theta)\,d\theta$$

$$= 2ab\left[\theta - \tfrac{1}{2}\sin 2\theta\right]_0^{\pi/2} = 2ab\left(\tfrac{\pi}{2}\right) = \pi ab$$

33. The curve $x = 1 + e^t$, $y = t - t^2 = t(1 - t)$ intersects the x-axis when $y = 0$,

that is, when $t = 0$ and $t = 1$. The corresponding values of x are 2 and $1 + e$.

The shaded area is given by

$$\int_{x=2}^{x=1+e} (y_T - y_B)\,dx = \int_{t=0}^{t=1} [y(t) - 0]\,x'(t)\,dt = \int_0^1 (t - t^2)e^t\,dt$$

$$= \int_0^1 te^t\,dt - \int_0^1 t^2 e^t\,dt = \int_0^1 te^t\,dt - \left[t^2 e^t\right]_0^1 + 2\int_0^1 te^t\,dt \qquad \text{[Formula 97 or parts]}$$

$$= 3\int_0^1 te^t\,dt - (e - 0) = 3\left[(t - 1)e^t\right]_0^1 - e \qquad \text{[Formula 96 or parts]}$$

$$= 3[0 - (-1)] - e = 3 - e$$

35. $x = r\theta - d\sin\theta$, $y = r - d\cos\theta$.

$$A = \int_0^{2\pi r} y\,dx = \int_0^{2\pi} (r - d\cos\theta)(r - d\cos\theta)\,d\theta = \int_0^{2\pi} (r^2 - 2dr\cos\theta + d^2\cos^2\theta)\,d\theta$$

$$= \left[r^2\theta - 2dr\sin\theta + \tfrac{1}{2}d^2\left(\theta + \tfrac{1}{2}\sin 2\theta\right)\right]_0^{2\pi} = 2\pi r^2 + \pi d^2$$

37. $x = t + e^{-t}$, $y = t - e^{-t}$, $0 \le t \le 2$. $dx/dt = 1 - e^{-t}$ and $dy/dt = 1 + e^{-t}$, so

$$(dx/dt)^2 + (dy/dt)^2 = (1 - e^{-t})^2 + (1 + e^{-t})^2 = 1 - 2e^{-t} + e^{-2t} + 1 + 2e^{-t} + e^{-2t} = 2 + 2e^{-2t}.$$

Thus, $L = \int_a^b \sqrt{(dx/dt)^2 + (dy/dt)^2}\,dt = \int_0^2 \sqrt{2 + 2e^{-2t}}\,dt \approx 3.1416$.

39. $x = t - 2\sin t$, $y = 1 - 2\cos t$, $0 \le t \le 4\pi$. $dx/dt = 1 - 2\cos t$ and $dy/dt = 2\sin t$, so

$$(dx/dt)^2 + (dy/dt)^2 = (1 - 2\cos t)^2 + (2\sin t)^2 = 1 - 4\cos t + 4\cos^2 t + 4\sin^2 t = 5 - 4\cos t.$$

Thus, $L = \int_a^b \sqrt{(dx/dt)^2 + (dy/dt)^2}\,dt = \int_0^{4\pi} \sqrt{5 - 4\cos t}\,dt \approx 26.7298$.

41. $x = 1 + 3t^2$, $y = 4 + 2t^3$, $0 \le t \le 1$. $dx/dt = 6t$ and $dy/dt = 6t^2$, so $(dx/dt)^2 + (dy/dt)^2 = 36t^2 + 36t^4$

Thus, $\quad L = \int_0^1 \sqrt{36t^2 + 36t^4}\,dt = \int_0^1 6t\sqrt{1 + t^2}\,dt = 6\int_1^2 \sqrt{u}\left(\tfrac{1}{2}du\right) \quad [u = 1 + t^2,\, du = 2t\,dt]$

$$= 3\left[\tfrac{2}{3}u^{3/2}\right]_1^2 = 2(2^{3/2} - 1) = 2(2\sqrt{2} - 1)$$

43. $x = t\sin t$, $y = t\cos t$, $0 \le t \le 1$. $\dfrac{dx}{dt} = t\cos t + \sin t$ and $\dfrac{dy}{dt} = -t\sin t + \cos t$, so

$$\left(\frac{dx}{dt}\right)^2 + \left(\frac{dy}{dt}\right)^2 = t^2\cos^2 t + 2t\sin t\cos t + \sin^2 t + t^2\sin^2 t - 2t\sin t\cos t + \cos^2 t$$

$$= t^2(\cos^2 t + \sin^2 t) + \sin^2 t + \cos^2 t = t^2 + 1.$$

Thus, $L = \int_0^1 \sqrt{t^2 + 1}\,dt \overset{21}{=} \left[\tfrac{1}{2}t\sqrt{t^2 + 1} + \tfrac{1}{2}\ln\left(t + \sqrt{t^2 + 1}\right)\right]_0^1 = \tfrac{1}{2}\sqrt{2} + \tfrac{1}{2}\ln\left(1 + \sqrt{2}\right)$.

© 2012 Cengage Learning. All Rights Reserved. May not be scanned, copied, or duplicated, or posted to a publicly accessible website, in whole or in part.

45.

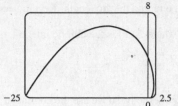

$x = e^t \cos t,\ y = e^t \sin t,\ 0 \le t \le \pi.$

$$\left(\tfrac{dx}{dt}\right)^2 + \left(\tfrac{dy}{dt}\right)^2 = [e^t(\cos t - \sin t)]^2 + [e^t(\sin t + \cos t)]^2$$

$$= (e^t)^2(\cos^2 t - 2\cos t \sin t + \sin^2 t)$$

$$+ (e^t)^2(\sin^2 t + 2\sin t \cos t + \cos^2 t$$

$$= e^{2t}(2\cos^2 t + 2\sin^2 t) = 2e^{2t}$$

Thus, $L = \int_0^\pi \sqrt{2e^{2t}}\,dt = \int_0^\pi \sqrt{2}\,e^t\,dt = \sqrt{2}\left[e^t\right]_0^\pi = \sqrt{2}\,(e^\pi - 1).$

47.

The figure shows the curve $x = \sin t + \sin 1.5t,\ y = \cos t$ for $0 \le t \le 4\pi.$

$dx/dt = \cos t + 1.5\cos 1.5t$ and $dy/dt = -\sin t$, so

$(dx/dt)^2 + (dy/dt)^2 = \cos^2 t + 3\cos t \cos 1.5t + 2.25\cos^2 1.5t + \sin^2 t.$

Thus, $L = \int_0^{4\pi} \sqrt{1 + 3\cos t \cos 1.5t + 2.25\cos^2 1.5t}\,dt \approx 16.7102.$

49. $x = t - e^t,\ y = t + e^t,\ -6 \le t \le 6.$

$\left(\tfrac{dx}{dt}\right)^2 + \left(\tfrac{dy}{dt}\right)^2 = (1 - e^t)^2 + (1 + e^t)^2 = (1 - 2e^t + e^{2t}) + (1 + 2e^t + e^{2t}) = 2 + 2e^{2t}$, so $L = \int_{-6}^6 \sqrt{2 + 2e^{2t}}\,dt.$

Set $f(t) = \sqrt{2 + 2e^{2t}}$. Then by Simpson's Rule with $n = 6$ and $\Delta t = \frac{6-(-6)}{6} = 2$, we get

$L \approx \tfrac{2}{3}[f(-6) + 4f(-4) + 2f(-2) + 4f(0) + 2f(2) + 4f(4) + f(6)] \approx 612.3053.$

51. $x = \sin^2 t,\ y = \cos^2 t,\ 0 \le t \le 3\pi.$

$(dx/dt)^2 + (dy/dt)^2 = (2\sin t \cos t)^2 + (-2\cos t \sin t)^2 = 8\sin^2 t \cos^2 t = 2\sin^2 2t \quad \Rightarrow$

Distance $= \int_0^{3\pi} \sqrt{2}\,|\sin 2t|\,dt = 6\sqrt{2} \int_0^{\pi/2} \sin 2t\,dt$ [by symmetry] $= -3\sqrt{2}\left[\cos 2t\right]_0^{\pi/2} = -3\sqrt{2}\,(-1 - 1) = 6\sqrt{2}.$

The full curve is traversed as t goes from 0 to $\frac{\pi}{2}$, because the curve is the segment of $x + y = 1$ that lies in the first quadrant

(since $x, y \ge 0$), and this segment is completely traversed as t goes from 0 to $\frac{\pi}{2}$. Thus, $L = \int_0^{\pi/2} \sin 2t\,dt = \sqrt{2}$, as above.

53. $x = a\sin\theta,\ y = b\cos\theta,\ 0 \le \theta \le 2\pi.$

$$\left(\tfrac{dx}{dt}\right)^2 + \left(\tfrac{dy}{dt}\right)^2 = (a\cos\theta)^2 + (-b\sin\theta)^2 = a^2\cos^2\theta + b^2\sin^2\theta = a^2(1 - \sin^2\theta) + b^2\sin^2\theta$$

$$= a^2 - (a^2 - b^2)\sin^2\theta = a^2 - c^2\sin^2\theta = a^2\left(1 - \tfrac{c^2}{a^2}\sin^2\theta\right) = a^2(1 - e^2\sin^2\theta)$$

So $L = 4\int_0^{\pi/2} \sqrt{a^2\left(1 - e^2\sin^2\theta\right)}\,d\theta$ [by symmetry] $= 4a\int_0^{\pi/2} \sqrt{1 - e^2\sin^2\theta}\,d\theta.$

55. (a) $x = 11\cos t - 4\cos(11t/2),\ y = 11\sin t - 4\sin(11t/2).$

Notice that $0 \le t \le 2\pi$ does not give the complete curve because

$x(0) \ne x(2\pi)$. In fact, we must take $t \in [0, 4\pi]$ in order to obtain the

complete curve, since the first term in each of the parametric equations has

period 2π and the second has period $\frac{2\pi}{11/2} = \frac{4\pi}{11}$, and the least common

integer multiple of these two numbers is $4\pi.$

© 2012 Cengage Learning. All Rights Reserved. May not be scanned, copied, or duplicated, or posted to a publicly accessible website, in whole or in part.

(b) We use the CAS to find the derivatives dx/dt and dy/dt, and then use Theorem 6 to find the arc length. Recent versions of Maple express the integral $\int_0^{4\pi} \sqrt{(dx/dt)^2 + (dy/dt)^2}\, dt$ as $88E(2\sqrt{2}\,i)$, where $E(x)$ is the elliptic integral

$\int_0^1 \dfrac{\sqrt{1 - x^2 t^2}}{\sqrt{1 - t^2}}\, dt$ and i is the imaginary number $\sqrt{-1}$.

Some earlier versions of Maple (as well as Mathematica) cannot do the integral exactly, so we use the command `evalf(Int(sqrt(diff(x,t)^2+diff(y,t)^2),t=0..4*Pi));` to estimate the length, and find that the arc length is approximately 294.03. Derive's `Para_arc_length` function in the utility file `Int_apps` simplifies the integral to $11\int_0^{4\pi} \sqrt{-4\cos t \, \cos\left(\frac{11t}{2}\right) - 4\sin t \, \sin\left(\frac{11t}{2}\right) + 5}\, dt$.

57. $x = t\sin t$, $y = t\cos t$, $0 \le t \le \pi/2$. $dx/dt = t\cos t + \sin t$ and $dy/dt = -t\sin t + \cos t$, so

$(dx/dt)^2 + (dy/dt)^2 = t^2\cos^2 t + 2t\sin t \, \cos t + \sin^2 t + t^2\sin^2 t - 2t\sin t \, \cos t + \cos^2 t$

$= t^2(\cos^2 t + \sin^2 t) + \sin^2 t + \cos^2 t = t^2 + 1$

$S = \int 2\pi y \, ds = \int_0^{\pi/2} 2\pi t \cos t \sqrt{t^2 + 1}\, dt \approx 4.7394.$

59. $x = 1 + te^t$, $y = (t^2 + 1)e^t$, $0 \le t \le 1$.

$\left(\frac{dx}{dt}\right)^2 + \left(\frac{dy}{dt}\right)^2 = (te^t + e^t)^2 + [(t^2 + 1)e^t + e^t(2t)]^2 = [e^t(t + 1)]^2 + [e^t(t^2 + 2t + 1)]^2$

$= e^{2t}(t + 1)^2 + e^{2t}(t + 1)^4 = e^{2t}(t + 1)^2[1 + (t + 1)^2]$, so

$S = \int 2\pi y \, ds = \int_0^1 2\pi(t^2 + 1)e^t \sqrt{e^{2t}(t + 1)^2(t^2 + 2t + 2)}\, dt = \int_0^1 2\pi(t^2 + 1)e^{2t}(t + 1)\sqrt{t^2 + 2t + 2}\, dt \approx 103.5999.$

61. $x = t^3$, $y = t^2$, $0 \le t \le 1$. $\left(\frac{dx}{dt}\right)^2 + \left(\frac{dy}{dt}\right)^2 = (3t^2)^2 + (2t)^2 = 9t^4 + 4t^2.$

$S = \int_0^1 2\pi y \sqrt{\left(\frac{dx}{dt}\right)^2 + \left(\frac{dy}{dt}\right)^2}\, dt = \int_0^1 2\pi t^2 \sqrt{9t^4 + 4t^2}\, dt = 2\pi \int_0^1 t^2 \sqrt{t^2(9t^2 + 4)}\, dt$

$= 2\pi \int_4^{13} \left(\frac{u - 4}{9}\right)\sqrt{u}\left(\frac{1}{18}\, du\right) \quad \begin{bmatrix} u = 9t^2 + 4,\ t^2 = (u - 4)/9, \\ du = 18t\, dt, \text{ so } t\, dt = \frac{1}{18}\, du \end{bmatrix} = \frac{2\pi}{9 \cdot 18}\int_4^{13}(u^{3/2} - 4u^{1/2})\, du$

$= \frac{\pi}{81}\left[\frac{2}{5}u^{5/2} - \frac{8}{3}u^{3/2}\right]_4^{13} = \frac{\pi}{81} \cdot \frac{2}{15}\left[3u^{5/2} - 20u^{3/2}\right]_4^{13}$

$= \frac{2\pi}{1215}\left[(3 \cdot 13^2 \sqrt{13} - 20 \cdot 13\sqrt{13}) - (3 \cdot 32 - 20 \cdot 8)\right] = \frac{2\pi}{1215}(247\sqrt{13} + 64)$

63. $x = a\cos^3 \theta$, $y = a\sin^3 \theta$, $0 \le \theta \le \frac{\pi}{2}$. $\left(\frac{dx}{d\theta}\right)^2 + \left(\frac{dy}{d\theta}\right)^2 = (-3a\cos^2 \theta \, \sin \theta)^2 + (3a\sin^2 \theta \, \cos \theta)^2 = 9a^2\sin^2 \theta \, \cos^2 \theta.$

$S = \int_0^{\pi/2} 2\pi \cdot a\sin^3 \theta \cdot 3a\sin \theta \, \cos \theta \, d\theta = 6\pi a^2 \int_0^{\pi/2} \sin^4 \theta \, \cos \theta \, d\theta = \frac{6}{5}\pi a^2 \left[\sin^5 \theta\right]_0^{\pi/2} = \frac{6}{5}\pi a^2$

65. $x = 3t^2$, $y = 2t^3$, $0 \le t \le 5$ $\Rightarrow$ $\left(\frac{dx}{dt}\right)^2 + \left(\frac{dy}{dt}\right)^2 = (6t)^2 + (6t^2)^2 = 36t^2(1 + t^2)$ $\Rightarrow$

$S = \int_0^5 2\pi x \sqrt{(dx/dt)^2 + (dy/dt)^2}\, dt = \int_0^5 2\pi(3t^2)6t\sqrt{1 + t^2}\, dt = 18\pi \int_0^5 t^2 \sqrt{1 + t^2}\, 2t\, dt$

$= 18\pi \int_1^{26}(u - 1)\sqrt{u}\, du \quad \begin{bmatrix} u = 1 + t^2, \\ du = 2t\, dt \end{bmatrix} = 18\pi \int_1^{26}(u^{3/2} - u^{1/2})\, du = 18\pi\left[\frac{2}{5}u^{5/2} - \frac{2}{3}u^{3/2}\right]_1^{26}$

$= 18\pi\left[\left(\frac{2}{5} \cdot 676\sqrt{26} - \frac{2}{3} \cdot 26\sqrt{26}\right) - \left(\frac{2}{5} - \frac{2}{3}\right)\right] = \frac{24}{5}\pi(949\sqrt{26} + 1)$

© 2012 Cengage Learning. All Rights Reserved. May not be scanned, copied, or duplicated, or posted to a publicly accessible website, in whole or in part.

67. If f' is continuous and $f'(t) \neq 0$ for $a \leq t \leq b$, then either $f'(t) > 0$ for all t in $[a, b]$ or $f'(t) < 0$ for all t in $[a, b]$. Thus, f is monotonic (in fact, strictly increasing or strictly decreasing) on $[a, b]$. It follows that f has an inverse. Set $F = g \circ f^{-1}$, that is, define F by $F(x) = g(f^{-1}(x))$. Then $x = f(t) \Rightarrow f^{-1}(x) = t$, so $y = g(t) = g(f^{-1}(x)) = F(x)$.

69. (a) $\phi = \tan^{-1}\left(\dfrac{dy}{dx}\right) \Rightarrow \dfrac{d\phi}{dt} = \dfrac{d}{dt}\tan^{-1}\left(\dfrac{dy}{dx}\right) = \dfrac{1}{1 + (dy/dx)^2}\left[\dfrac{d}{dt}\left(\dfrac{dy}{dx}\right)\right]$. But $\dfrac{dy}{dx} = \dfrac{dy/dt}{dx/dt} = \dfrac{\dot{y}}{\dot{x}} \Rightarrow$

$\dfrac{d}{dt}\left(\dfrac{dy}{dx}\right) = \dfrac{d}{dt}\left(\dfrac{\dot{y}}{\dot{x}}\right) = \dfrac{\ddot{y}\dot{x} - \ddot{x}\dot{y}}{\dot{x}^2} \Rightarrow \dfrac{d\phi}{dt} = \dfrac{1}{1 + (\dot{y}/\dot{x})^2}\left(\dfrac{\ddot{y}\dot{x} - \ddot{x}\dot{y}}{\dot{x}^2}\right) = \dfrac{\dot{x}\ddot{y} - \ddot{x}\dot{y}}{\dot{x}^2 + \dot{y}^2}$. Using the Chain Rule, and the

fact that $s = \displaystyle\int_0^t \sqrt{\left(\dfrac{dx}{dt}\right)^2 + \left(\dfrac{dy}{dt}\right)^2}\, dt \Rightarrow \dfrac{ds}{dt} = \sqrt{\left(\dfrac{dx}{dt}\right)^2 + \left(\dfrac{dy}{dt}\right)^2} = (\dot{x}^2 + \dot{y}^2)^{1/2}$, we have that

$\dfrac{d\phi}{ds} = \dfrac{d\phi/dt}{ds/dt} = \left(\dfrac{\dot{x}\ddot{y} - \ddot{x}\dot{y}}{\dot{x}^2 + \dot{y}^2}\right)\dfrac{1}{(\dot{x}^2 + \dot{y}^2)^{1/2}} = \dfrac{\dot{x}\ddot{y} - \ddot{x}\dot{y}}{(\dot{x}^2 + \dot{y}^2)^{3/2}}$. So $\kappa = \left|\dfrac{d\phi}{ds}\right| = \left|\dfrac{\dot{x}\ddot{y} - \ddot{x}\dot{y}}{(\dot{x}^2 + \dot{y}^2)^{3/2}}\right| = \dfrac{|\dot{x}\ddot{y} - \ddot{x}\dot{y}|}{(\dot{x}^2 + \dot{y}^2)^{3/2}}$.

(b) $x = x$ and $y = f(x) \Rightarrow \dot{x} = 1, \ddot{x} = 0$ and $\dot{y} = \dfrac{dy}{dx}, \ddot{y} = \dfrac{d^2y}{dx^2}$.

So $\kappa = \dfrac{\left|1 \cdot (d^2y/dx^2) - 0 \cdot (dy/dx)\right|}{[1 + (dy/dx)^2]^{3/2}} = \dfrac{\left|d^2y/dx^2\right|}{[1 + (dy/dx)^2]^{3/2}}$.

71. $x = \theta - \sin\theta \Rightarrow \dot{x} = 1 - \cos\theta \Rightarrow \ddot{x} = \sin\theta$, and $y = 1 - \cos\theta \Rightarrow \dot{y} = \sin\theta \Rightarrow \ddot{y} = \cos\theta$. Therefore,

$\kappa = \dfrac{\left|\cos\theta - \cos^2\theta - \sin^2\theta\right|}{[(1 - \cos\theta)^2 + \sin^2\theta]^{3/2}} = \dfrac{\left|\cos\theta - (\cos^2\theta + \sin^2\theta)\right|}{(1 - 2\cos\theta + \cos^2\theta + \sin^2\theta)^{3/2}} = \dfrac{|\cos\theta - 1|}{(2 - 2\cos\theta)^{3/2}}$. The top of the arch is

characterized by a horizontal tangent, and from Example 2(b) in Section 10.2, the tangent is horizontal when $\theta = (2n - 1)\pi$,

so take $n = 1$ and substitute $\theta = \pi$ into the expression for κ: $\kappa = \dfrac{|\cos\pi - 1|}{(2 - 2\cos\pi)^{3/2}} = \dfrac{|-1 - 1|}{[2 - 2(-1)]^{3/2}} = \dfrac{1}{4}$.

73. The coordinates of T are $(r\cos\theta, r\sin\theta)$. Since TP was unwound from

arc TA, TP has length $r\theta$. Also $\angle PTQ = \angle PTR - \angle QTR = \frac{1}{2}\pi - \theta$,

so P has coordinates $x = r\cos\theta + r\theta\cos\left(\frac{1}{2}\pi - \theta\right) = r(\cos\theta + \theta\sin\theta)$,

$y = r\sin\theta - r\theta\sin\left(\frac{1}{2}\pi - \theta\right) = r(\sin\theta - \theta\cos\theta)$.

10.3 Polar Coordinates

1. (a) $\left(2, \frac{\pi}{3}\right)$

By adding 2π to $\frac{\pi}{3}$, we obtain the point $\left(2, \frac{7\pi}{3}\right)$. The direction opposite $\frac{\pi}{3}$ is $\frac{4\pi}{3}$, so $\left(-2, \frac{4\pi}{3}\right)$ is a point that satisfies the $r < 0$ requirement.

© 2012 Cengage Learning. All Rights Reserved. May not be scanned, copied, or duplicated, or posted to a publicly accessible website, in whole or in part.

(b) $\left(1, -\frac{3\pi}{4}\right)$

$r > 0$: $\left(1, -\frac{3\pi}{4} + 2\pi\right) = \left(1, \frac{5\pi}{4}\right)$

$r < 0$: $\left(-1, -\frac{3\pi}{4} + \pi\right) = \left(-1, \frac{\pi}{4}\right)$

(c) $\left(-1, \frac{\pi}{2}\right)$

$r > 0$: $\left(-(-1), \frac{\pi}{2} + \pi\right) = \left(1, \frac{3\pi}{2}\right)$

$r < 0$: $\left(-1, \frac{\pi}{2} + 2\pi\right) = \left(-1, \frac{5\pi}{2}\right)$

3. (a)

$x = 1 \cos \pi = 1(-1) = -1$ and

$y = 1 \sin \pi = 1(0) = 0$ give us

the Cartesian coordinates $(-1, 0)$.

(b)

$x = 2 \cos\left(-\frac{2\pi}{3}\right) = 2\left(-\frac{1}{2}\right) = -1$ and

$y = 2 \sin\left(-\frac{2\pi}{3}\right) = 2\left(-\frac{\sqrt{3}}{2}\right) = -\sqrt{3}$

give us $\left(-1, -\sqrt{3}\right)$.

(c)

$x = -2 \cos \frac{3\pi}{4} = -2\left(-\frac{\sqrt{2}}{2}\right) = \sqrt{2}$ and

$y = -2 \sin \frac{3\pi}{4} = -2\left(\frac{\sqrt{2}}{2}\right) = -\sqrt{2}$

gives us $\left(\sqrt{2}, -\sqrt{2}\right)$.

5. (a) $x = 2$ and $y = -2 \;\Rightarrow\; r = \sqrt{2^2 + (-2)^2} = 2\sqrt{2}$ and $\theta = \tan^{-1}\left(\frac{-2}{2}\right) = -\frac{\pi}{4}$. Since $(2, -2)$ is in the fourth

quadrant, the polar coordinates are (i) $\left(2\sqrt{2}, \frac{7\pi}{4}\right)$ and (ii) $\left(-2\sqrt{2}, \frac{3\pi}{4}\right)$.

(b) $x = -1$ and $y = \sqrt{3} \;\Rightarrow\; r = \sqrt{(-1)^2 + \left(\sqrt{3}\right)^2} = 2$ and $\theta = \tan^{-1}\left(\frac{\sqrt{3}}{-1}\right) = \frac{2\pi}{3}$. Since $\left(-1, \sqrt{3}\right)$ is in the second

quadrant, the polar coordinates are (i) $\left(2, \frac{2\pi}{3}\right)$ and (ii) $\left(-2, \frac{5\pi}{3}\right)$.

© 2012 Cengage Learning. All Rights Reserved. May not be scanned, copied, or duplicated, or posted to a publicly accessible website, in whole or in part.

7. $r \geq 1$. The curve $r = 1$ represents a circle with center
O and radius 1. So $r \geq 1$ represents the region on or
outside the circle. Note that θ can take on any value.

9. $r \geq 0$, $\pi/4 \leq \theta \leq 3\pi/4$.

$\theta = k$ represents a line through O.

11. $2 < r < 3$, $\frac{5\pi}{3} \leq \theta \leq \frac{7\pi}{3}$

13. Converting the polar coordinates $(2, \pi/3)$ and $(4, 2\pi/3)$ to Cartesian coordinates gives us $\left(2\cos\frac{\pi}{3}, 2\sin\frac{\pi}{3}\right) = \left(1, \sqrt{3}\right)$ and

$\left(4\cos\frac{2\pi}{3}, 4\sin\frac{2\pi}{3}\right) = \left(-2, 2\sqrt{3}\right)$. Now use the distance formula.

$$d = \sqrt{(x_2 - x_1)^2 + (y_2 - y_1)^2} = \sqrt{(-2-1)^2 + \left(2\sqrt{3} - \sqrt{3}\right)^2} = \sqrt{9+3} = \sqrt{12} = 2\sqrt{3}$$

15. $r^2 = 5 \iff x^2 + y^2 = 5$, a circle of radius $\sqrt{5}$ centered at the origin.

17. $r = 2\cos\theta \implies r^2 = 2r\cos\theta \iff x^2 + y^2 = 2x \iff x^2 - 2x + 1 + y^2 = 1 \iff (x-1)^2 + y^2 = 1$, a circle of
radius 1 centered at $(1, 0)$. The first two equations are actually equivalent since $r^2 - 2r\cos\theta \implies r(r - 2\cos\theta) = 0 \implies$
$r = 0$ or $r = 2\cos\theta$. But $r = 2\cos\theta$ gives the point $r = 0$ (the pole) when $\theta = 0$. Thus, the equation $r = 2\cos\theta$ is
equivalent to the compound condition $(r = 0$ or $r = 2\cos\theta)$.

19. $r^2 \cos 2\theta = 1 \iff r^2(\cos^2\theta - \sin^2\theta) = 1 \iff (r\cos\theta)^2 - (r\sin\theta)^2 = 1 \iff x^2 - y^2 = 1$, a hyperbola centered at
the origin with foci on the x-axis.

21. $y = 2 \iff r\sin\theta = 2 \iff r = \dfrac{2}{\sin\theta} \iff r = 2\csc\theta$

23. $y = 1 + 3x \iff r\sin\theta = 1 + 3r\cos\theta \iff r\sin\theta - 3r\cos\theta = 1 \iff r(\sin\theta - 3\cos\theta) = 1 \iff$

$r = \dfrac{1}{\sin\theta - 3\cos\theta}$

25. $x^2 + y^2 = 2cx \iff r^2 = 2cr\cos\theta \iff r^2 - 2cr\cos\theta = 0 \iff r(r - 2c\cos\theta) = 0 \iff r = 0$ or $r = 2c\cos\theta$.
$r = 0$ is included in $r = 2c\cos\theta$ when $\theta = \frac{\pi}{2} + n\pi$, so the curve is represented by the single equation $r = 2c\cos\theta$.

© 2012 Cengage Learning. All Rights Reserved. May not be scanned, copied, or duplicated, or posted to a publicly accessible website, in whole or in part.

27. (a) The description leads immediately to the polar equation $\theta = \frac{\pi}{6}$, and the Cartesian equation $y = \tan\left(\frac{\pi}{6}\right) x = \frac{1}{\sqrt{3}} x$ is slightly more difficult to derive.

(b) The easier description here is the Cartesian equation $x = 3$.

29. $r = -2\sin\theta$

31. $r = 2(1 + \cos\theta)$

33. $r = \theta, \quad \theta \geq 0$

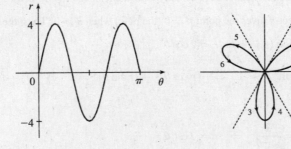

35. $r = 4\sin 3\theta$

37. $r = 2\cos 4\theta$

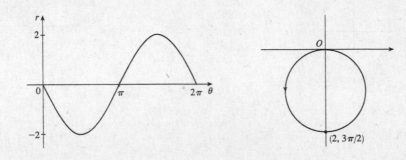

© 2012 Cengage Learning. All Rights Reserved. May not be scanned, copied, or duplicated, or posted to a publicly accessible website, in whole or in part.

39. $r = 1 - 2\sin\theta$

41. $r^2 = 9\sin 2\theta$

43. $r = 2 + \sin 3\theta$

45. $r = 1 + 2\cos 2\theta$

47. For $\theta = 0$, π, and 2π, r has its minimum value of about 0.5. For $\theta = \frac{\pi}{2}$ and $\frac{3\pi}{2}$, r attains its maximum value of 2.

We see that the graph has a similar shape for $0 \le \theta \le \pi$ and $\pi \le \theta \le 2\pi$.

© 2012 Cengage Learning. All Rights Reserved. May not be scanned, copied, or duplicated, or posted to a publicly accessible website, in whole or in part.

49. $x = r \cos \theta = (4 + 2 \sec \theta) \cos \theta = 4 \cos \theta + 2$. Now, $r \to \infty$ $\Rightarrow$

$(4 + 2 \sec \theta) \to \infty$ $\Rightarrow$ $\theta \to \left(\frac{\pi}{2}\right)^-$ or $\theta \to \left(\frac{3\pi}{2}\right)^+$ [since we need only

consider $0 \le \theta < 2\pi$], so $\lim\limits_{r \to \infty} x = \lim\limits_{\theta \to \pi/2^-} (4 \cos \theta + 2) = 2$. Also,

$r \to -\infty$ $\Rightarrow$ $(4 + 2 \sec \theta) \to -\infty$ $\Rightarrow$ $\theta \to \left(\frac{\pi}{2}\right)^+$ or $\theta \to \left(\frac{3\pi}{2}\right)^-$, so

$\lim\limits_{r \to -\infty} x = \lim\limits_{\theta \to \pi/2^+} (4 \cos \theta + 2) = 2$. Therefore, $\lim\limits_{r \to \pm\infty} x = 2$ $\Rightarrow$ $x = 2$ is a vertical asymptote.

51. To show that $x = 1$ is an asymptote we must prove $\lim\limits_{r \to \pm\infty} x = 1$.

$x = (r) \cos \theta = (\sin \theta \tan \theta) \cos \theta = \sin^2 \theta$. Now, $r \to \infty$ $\Rightarrow$ $\sin \theta \tan \theta \to \infty$ $\Rightarrow$

$\theta \to \left(\frac{\pi}{2}\right)^-$, so $\lim\limits_{r \to \infty} x = \lim\limits_{\theta \to \pi/2^-} \sin^2 \theta = 1$. Also, $r \to -\infty$ $\Rightarrow$ $\sin \theta \tan \theta \to -\infty$ $\Rightarrow$

$\theta \to \left(\frac{\pi}{2}\right)^+$, so $\lim\limits_{r \to -\infty} x = \lim\limits_{\theta \to \pi/2^+} \sin^2 \theta = 1$. Therefore, $\lim\limits_{r \to \pm\infty} x = 1$ $\Rightarrow$ $x = 1$ is

a vertical asymptote. Also notice that $x = \sin^2 \theta \ge 0$ for all θ, and $x = \sin^2 \theta \le 1$ for all θ. And $x \ne 1$, since the curve is not

defined at odd multiples of $\frac{\pi}{2}$. Therefore, the curve lies entirely within the vertical strip $0 \le x < 1$.

53. (a) We see that the curve $r = 1 + c \sin \theta$ crosses itself at the origin, where $r = 0$ (in fact the inner loop corresponds to

negative r-values,) so we solve the equation of the limaçon for $r = 0$ $\Leftrightarrow$ $c \sin \theta = -1$ $\Leftrightarrow$ $\sin \theta = -1/c$. Now if

$|c| < 1$, then this equation has no solution and hence there is no inner loop. But if $c < -1$, then on the interval $(0, 2\pi)$

the equation has the two solutions $\theta = \sin^{-1}(-1/c)$ and $\theta = \pi - \sin^{-1}(-1/c)$, and if $c > 1$, the solutions are

$\theta = \pi + \sin^{-1}(1/c)$ and $\theta = 2\pi - \sin^{-1}(1/c)$. In each case, $r < 0$ for θ between the two solutions, indicating a loop.

(b) For $0 < c < 1$, the dimple (if it exists) is characterized by the fact that y has a local maximum at $\theta = \frac{3\pi}{2}$. So we

determine for what c-values $\dfrac{d^2 y}{d\theta^2}$ is negative at $\theta = \frac{3\pi}{2}$, since by the Second Derivative Test this indicates a maximum:

$y = r \sin \theta = \sin \theta + c \sin^2 \theta$ $\Rightarrow$ $\dfrac{dy}{d\theta} = \cos \theta + 2c \sin \theta \cos \theta = \cos \theta + c \sin 2\theta$ $\Rightarrow$ $\dfrac{d^2 y}{d\theta^2} = -\sin \theta + 2c \cos 2\theta$.

At $\theta = \frac{3\pi}{2}$, this is equal to $-(-1) + 2c(-1) = 1 - 2c$, which is negative only for $c > \frac{1}{2}$. A similar argument shows that

for $-1 < c < 0$, y only has a local minimum at $\theta = \frac{\pi}{2}$ (indicating a dimple) for $c < -\frac{1}{2}$.

55. $r = 2 \sin \theta$ $\Rightarrow$ $x = r \cos \theta = 2 \sin \theta \cos \theta = \sin 2\theta$, $y = r \sin \theta = 2 \sin^2 \theta$ $\Rightarrow$

$$\frac{dy}{dx} = \frac{dy/d\theta}{dx/d\theta} = \frac{2 \cdot 2 \sin \theta \cos \theta}{\cos 2\theta \cdot 2} = \frac{\sin 2\theta}{\cos 2\theta} = \tan 2\theta$$

When $\theta = \dfrac{\pi}{6}$, $\dfrac{dy}{dx} = \tan\left(2 \cdot \dfrac{\pi}{6}\right) = \tan \dfrac{\pi}{3} = \sqrt{3}$.   [*Another method:* Use Equation 3.]

57. $r = 1/\theta$ $\Rightarrow$ $x = r \cos \theta = (\cos \theta)/\theta$, $y = r \sin \theta = (\sin \theta)/\theta$ $\Rightarrow$

$$\frac{dy}{dx} = \frac{dy/d\theta}{dx/d\theta} = \frac{\sin \theta(-1/\theta^2) + (1/\theta) \cos \theta}{\cos \theta(-1/\theta^2) - (1/\theta) \sin \theta} \cdot \frac{\theta^2}{\theta^2} = \frac{-\sin \theta + \theta \cos \theta}{-\cos \theta - \theta \sin \theta}$$

When $\theta = \pi$, $\dfrac{dy}{dx} = \dfrac{-0 + \pi(-1)}{-(-1) - \pi(0)} = \dfrac{-\pi}{1} = -\pi$.

© 2012 Cengage Learning. All Rights Reserved. May not be scanned, copied, or duplicated, or posted to a publicly accessible website, in whole or in part.

59. $r = \cos 2\theta \;\Rightarrow\; x = r\cos\theta = \cos 2\theta \cos\theta, \; y = r\sin\theta = \cos 2\theta \sin\theta \;\Rightarrow$

$$\frac{dy}{dx} = \frac{dy/d\theta}{dx/d\theta} = \frac{\cos 2\theta \cos\theta + \sin\theta\,(-2\sin 2\theta)}{\cos 2\theta\,(-\sin\theta) + \cos\theta\,(-2\sin 2\theta)}$$

When $\theta = \frac{\pi}{4}$, $\dfrac{dy}{dx} = \dfrac{0(\sqrt{2}/2) + (\sqrt{2}/2)(-2)}{0(-\sqrt{2}/2) + (\sqrt{2}/2)(-2)} = \dfrac{-\sqrt{2}}{-\sqrt{2}} = 1$.

61. $r = 3\cos\theta \;\Rightarrow\; x = r\cos\theta = 3\cos\theta\cos\theta, \; y = r\sin\theta = 3\cos\theta\sin\theta \;\Rightarrow$

$\frac{dy}{d\theta} = -3\sin^2\theta + 3\cos^2\theta = 3\cos 2\theta = 0 \;\Rightarrow\; 2\theta = \frac{\pi}{2}$ or $\frac{3\pi}{2} \;\Leftrightarrow\; \theta = \frac{\pi}{4}$ or $\frac{3\pi}{4}$.

So the tangent is horizontal at $\left(\frac{3}{\sqrt{2}}, \frac{\pi}{4}\right)$ and $\left(-\frac{3}{\sqrt{2}}, \frac{3\pi}{4}\right)$ $\left[\text{same as }\left(\frac{3}{\sqrt{2}}, -\frac{\pi}{4}\right)\right]$.

$\frac{dx}{d\theta} = -6\sin\theta\cos\theta = -3\sin 2\theta = 0 \;\Rightarrow\; 2\theta = 0$ or $\pi \;\Leftrightarrow\; \theta = 0$ or $\frac{\pi}{2}$. So the tangent is vertical at $(3,0)$ and $\left(0, \frac{\pi}{2}\right)$.

63. $r = 1 + \cos\theta \;\Rightarrow\; x = r\cos\theta = \cos\theta\,(1 + \cos\theta), \; y = r\sin\theta = \sin\theta\,(1 + \cos\theta) \;\Rightarrow$

$\frac{dy}{d\theta} = (1 + \cos\theta)\cos\theta - \sin^2\theta = 2\cos^2\theta + \cos\theta - 1 = (2\cos\theta - 1)(\cos\theta + 1) = 0 \;\Rightarrow\; \cos\theta = \frac{1}{2}$ or $-1 \;\Rightarrow$

$\theta = \frac{\pi}{3}, \pi,$ or $\frac{5\pi}{3} \;\Rightarrow\;$ horizontal tangent at $\left(\frac{3}{2}, \frac{\pi}{3}\right)$, $(0, \pi)$, and $\left(\frac{3}{2}, \frac{5\pi}{3}\right)$.

$\frac{dx}{d\theta} = -(1 + \cos\theta)\sin\theta - \cos\theta\sin\theta = -\sin\theta\,(1 + 2\cos\theta) = 0 \;\Rightarrow\; \sin\theta = 0$ or $\cos\theta = -\frac{1}{2} \;\Rightarrow$

$\theta = 0, \pi, \frac{2\pi}{3},$ or $\frac{4\pi}{3} \;\Rightarrow\;$ vertical tangent at $(2, 0)$, $\left(\frac{1}{2}, \frac{2\pi}{3}\right)$, and $\left(\frac{1}{2}, \frac{4\pi}{3}\right)$.

Note that the tangent is horizontal, not vertical when $\theta = \pi$, since $\lim\limits_{\theta \to \pi} \dfrac{dy/d\theta}{dx/d\theta} = 0$.

65. $r = a\sin\theta + b\cos\theta \;\Rightarrow\; r^2 = ar\sin\theta + br\cos\theta \;\Rightarrow\; x^2 + y^2 = ay + bx \;\Rightarrow$

$x^2 - bx + \left(\frac{1}{2}b\right)^2 + y^2 - ay + \left(\frac{1}{2}a\right)^2 = \left(\frac{1}{2}b\right)^2 + \left(\frac{1}{2}a\right)^2 \;\Rightarrow\; \left(x - \frac{1}{2}b\right)^2 + \left(y - \frac{1}{2}a\right)^2 = \frac{1}{4}(a^2 + b^2)$, and this is a circle

with center $\left(\frac{1}{2}b, \frac{1}{2}a\right)$ and radius $\frac{1}{2}\sqrt{a^2 + b^2}$.

67. $r = 1 + 2\sin(\theta/2)$. The parameter interval is $[0, 4\pi]$.

69. $r = e^{\sin\theta} - 2\cos(4\theta)$.

The parameter interval is $[0, 2\pi]$.

71. $r = 1 + \cos^{999}\theta$. The parameter interval is $[0, 2\pi]$.

© 2012 Cengage Learning. All Rights Reserved. May not be scanned, copied, or duplicated, or posted to a publicly accessible website, in whole or in part.

73. It appears that the graph of $r = 1 + \sin\left(\theta - \frac{\pi}{6}\right)$ is the same shape as

the graph of $r = 1 + \sin\theta$, but rotated counterclockwise about the

origin by $\frac{\pi}{6}$. Similarly, the graph of $r = 1 + \sin\left(\theta - \frac{\pi}{3}\right)$ is rotated by

$\frac{\pi}{3}$. In general, the graph of $r = f(\theta - \alpha)$ is the same shape as that of

$r = f(\theta)$, but rotated counterclockwise through α about the origin.

That is, for any point (r_0, θ_0) on the curve $r = f(\theta)$, the point

$(r_0, \theta_0 + \alpha)$ is on the curve $r = f(\theta - \alpha)$, since $r_0 = f(\theta_0) = f((\theta_0 + \alpha) - \alpha)$.

75. Consider curves with polar equation $r = 1 + c\cos\theta$, where c is a real number. If $c = 0$, we get a circle of radius 1 centered at

the pole. For $0 < c \le 0.5$, the curve gets slightly larger, moves right, and flattens out a bit on the left side. For $0.5 < c < 1$,

the left side has a dimple shape. For $c = 1$, the dimple becomes a cusp. For $c > 1$, there is an internal loop. For $c \ge 0$, the

rightmost point on the curve is $(1 + c, 0)$. For $c < 0$, the curves are reflections through the vertical axis of the curves

with $c > 0$.

$c = 0.25$ $\qquad\qquad$ $c = 0.75$ $\qquad\qquad$ $c = 1$ $\qquad\qquad$ $c = 2$

77. $\tan\psi = \tan(\phi - \theta) = \dfrac{\tan\phi - \tan\theta}{1 + \tan\phi\,\tan\theta} = \dfrac{\dfrac{dy}{dx} - \tan\theta}{1 + \dfrac{dy}{dx}\tan\theta} = \dfrac{\dfrac{dy/d\theta}{dx/d\theta} - \tan\theta}{1 + \dfrac{dy/d\theta}{dx/d\theta}\tan\theta}$

$= \dfrac{\dfrac{dy}{d\theta} - \dfrac{dx}{d\theta}\tan\theta}{\dfrac{dx}{d\theta} + \dfrac{dy}{d\theta}\tan\theta} = \dfrac{\left(\dfrac{dr}{d\theta}\sin\theta + r\cos\theta\right) - \tan\theta\left(\dfrac{dr}{d\theta}\cos\theta - r\sin\theta\right)}{\left(\dfrac{dr}{d\theta}\cos\theta - r\sin\theta\right) + \tan\theta\left(\dfrac{dr}{d\theta}\sin\theta + r\cos\theta\right)} = \dfrac{r\cos\theta + r \cdot \dfrac{\sin^2\theta}{\cos\theta}}{\dfrac{dr}{d\theta}\cos\theta + \dfrac{dr}{d\theta}\cdot\dfrac{\sin^2\theta}{\cos\theta}}$

$= \dfrac{r\cos^2\theta + r\sin^2\theta}{\dfrac{dr}{d\theta}\cos^2\theta + \dfrac{dr}{d\theta}\sin^2\theta} = \dfrac{r}{dr/d\theta}$

10.4 Areas and Lengths in Polar Coordinates

1. $r = e^{-\theta/4}$, $\pi/2 \le \theta \le \pi$.

$A = \displaystyle\int_{\pi/2}^{\pi} \tfrac{1}{2}r^2\,d\theta = \int_{\pi/2}^{\pi} \tfrac{1}{2}(e^{-\theta/4})^2\,d\theta = \int_{\pi/2}^{\pi} \tfrac{1}{2}e^{-\theta/2}\,d\theta = \tfrac{1}{2}\left[-2e^{-\theta/2}\right]_{\pi/2}^{\pi} = -1(e^{-\pi/2} - e^{-\pi/4}) = e^{-\pi/4} - e^{-\pi/2}$

© 2012 Cengage Learning. All Rights Reserved. May not be scanned, copied, or duplicated, or posted to a publicly accessible website, in whole or in part.

3. $r^2 = 9\sin 2\theta,\ r \geq 0,\ 0 \leq \theta \leq \pi/2.$

$$A = \int_0^{\pi/2} \tfrac{1}{2} r^2\, d\theta = \int_0^{\pi/2} \tfrac{1}{2}(9\sin 2\theta)\, d\theta = \tfrac{9}{2}\left[-\tfrac{1}{2}\cos 2\theta\right]_0^{\pi/2} = -\tfrac{9}{4}(-1-1) = \tfrac{9}{2}$$

5. $r = \sqrt{\theta},\ 0 \leq \theta \leq 2\pi.$ $A = \int_0^{2\pi} \tfrac{1}{2} r^2\, d\theta = \int_0^{2\pi} \tfrac{1}{2}\left(\sqrt{\theta}\right)^2 d\theta = \int_0^{2\pi} \tfrac{1}{2}\theta\, d\theta = \left[\tfrac{1}{4}\theta^2\right]_0^{2\pi} = \pi^2$

7. $r = 4 + 3\sin\theta,\ -\tfrac{\pi}{2} \leq \theta \leq \tfrac{\pi}{2}.$

$$A = \int_{-\pi/2}^{\pi/2} \tfrac{1}{2}\left((4+3\sin\theta)^2\right) d\theta = \tfrac{1}{2}\int_{-\pi/2}^{\pi/2} (16 + 24\sin\theta + 9\sin^2\theta)\, d\theta$$

$$= \tfrac{1}{2}\int_{-\pi/2}^{\pi/2} (16 + 9\sin^2\theta)\, d\theta \qquad \text{[by Theorem 4.5.6(b) [ET 5.5.7(b)]]}$$

$$= \tfrac{1}{2}\cdot 2\int_0^{\pi/2} \left[16 + 9\cdot\tfrac{1}{2}(1-\cos 2\theta)\right] d\theta \qquad \text{[by Theorem 4.5.6(a) [ET 5.5.7(a)]]}$$

$$= \int_0^{\pi/2} \left(\tfrac{41}{2} - \tfrac{9}{2}\cos 2\theta\right) d\theta = \left[\tfrac{41}{2}\theta - \tfrac{9}{4}\sin 2\theta\right]_0^{\pi/2} = \left(\tfrac{41\pi}{4} - 0\right) - (0-0) = \tfrac{41\pi}{4}$$

9. The area is bounded by $r = 2\sin\theta$ for $\theta = 0$ to $\theta = \pi$.

$$A = \int_0^{\pi} \tfrac{1}{2} r^2\, d\theta = \tfrac{1}{2}\int_0^{\pi} (2\sin\theta)^2\, d\theta = \tfrac{1}{2}\int_0^{\pi} 4\sin^2\theta\, d\theta$$

$$= 2\int_0^{\pi} \tfrac{1}{2}(1-\cos 2\theta)\, d\theta = \left[\theta - \tfrac{1}{2}\sin 2\theta\right]_0^{\pi} = \pi$$

Also, note that this is a circle with radius 1, so its area is $\pi(1)^2 = \pi$.

11. $A = \int_0^{2\pi} \tfrac{1}{2} r^2\, d\theta = \int_0^{2\pi} \tfrac{1}{2}(3 + 2\cos\theta)^2\, d\theta = \tfrac{1}{2}\int_0^{2\pi} (9 + 12\cos\theta + 4\cos^2\theta)\, d\theta$

$$= \tfrac{1}{2}\int_0^{2\pi} \left[9 + 12\cos\theta + 4\cdot\tfrac{1}{2}(1+\cos 2\theta)\right] d\theta$$

$$= \tfrac{1}{2}\int_0^{2\pi} (11 + 12\cos\theta + 2\cos 2\theta)\, d\theta = \tfrac{1}{2}\left[11\theta + 12\sin\theta + \sin 2\theta\right]_0^{2\pi}$$

$$= \tfrac{1}{2}(22\pi) = 11\pi$$

13. $A = \int_0^{2\pi} \tfrac{1}{2} r^2\, d\theta = \int_0^{2\pi} \tfrac{1}{2}(2 + \sin 4\theta)^2\, d\theta = \tfrac{1}{2}\int_0^{2\pi} (4 + 4\sin 4\theta + \sin^2 4\theta)\, d\theta$

$$= \tfrac{1}{2}\int_0^{2\pi} \left[4 + 4\sin 4\theta + \tfrac{1}{2}(1 - \cos 8\theta)\right] d\theta$$

$$= \tfrac{1}{2}\int_0^{2\pi} \left(\tfrac{9}{2} + 4\sin 4\theta - \tfrac{1}{2}\cos 8\theta\right) d\theta = \tfrac{1}{2}\left[\tfrac{9}{2}\theta - \cos 4\theta - \tfrac{1}{16}\sin 8\theta\right]_0^{2\pi}$$

$$= \tfrac{1}{2}\left[(9\pi - 1) - (-1)\right] = \tfrac{9}{2}\pi$$

© 2012 Cengage Learning. All Rights Reserved. May not be scanned, copied, or duplicated, or posted to a publicly accessible website, in whole or in part.

15. $A = \int_0^{2\pi} \frac{1}{2} r^2 \, d\theta = \int_0^{2\pi} \frac{1}{2} \left(\sqrt{1 + \cos^2 5\theta} \right)^2 d\theta$

$= \frac{1}{2} \int_0^{2\pi} (1 + \cos^2 5\theta) \, d\theta = \frac{1}{2} \int_0^{2\pi} \left[1 + \frac{1}{2}(1 + \cos 10\theta) \right] d\theta$

$= \frac{1}{2} \left[\frac{3}{2}\theta + \frac{1}{20} \sin 10\theta \right]_0^{2\pi} = \frac{1}{2}(3\pi) = \frac{3}{2}\pi$

17. The curve passes through the pole when $r = 0 \Rightarrow 4\cos 3\theta = 0 \Rightarrow \cos 3\theta = 0 \Rightarrow 3\theta = \frac{\pi}{2} + \pi n \Rightarrow$

$\theta = \frac{\pi}{6} + \frac{\pi}{3}n$. The part of the shaded loop above the polar axis is traced out for

$\theta = 0$ to $\theta = \pi/6$, so we'll use $-\pi/6$ and $\pi/6$ as our limits of integration.

$A = \int_{-\pi/6}^{\pi/6} \frac{1}{2}(4\cos 3\theta)^2 \, d\theta = 2 \int_0^{\pi/6} \frac{1}{2}(16\cos^2 3\theta) \, d\theta$

$= 16 \int_0^{\pi/6} \frac{1}{2}(1 + \cos 6\theta) \, d\theta = 8 \left[\theta + \frac{1}{6}\sin 6\theta \right]_0^{\pi/6} = 8\left(\frac{\pi}{6} \right) = \frac{4}{3}\pi$

$r = 4\cos 3\theta$

19. $r = 0 \Rightarrow \sin 4\theta = 0 \Rightarrow 4\theta = \pi n \Rightarrow \theta = \frac{\pi}{4}n$.

$A = \int_0^{\pi/4} \frac{1}{2}(\sin 4\theta)^2 \, d\theta = \frac{1}{2} \int_0^{\pi/4} \sin^2 4\theta \, d\theta = \frac{1}{2} \int_0^{\pi/4} \frac{1}{2}(1 - \cos 8\theta) \, d\theta$

$= \frac{1}{4} \left[\theta - \frac{1}{8}\sin 8\theta \right]_0^{\pi/4} = \frac{1}{4}\left(\frac{\pi}{4} \right) = \frac{1}{16}\pi$

$r = \sin 4\theta$

21.

This is a limaçon, with inner loop traced out between $\theta = \frac{7\pi}{6}$ and $\frac{11\pi}{6}$ [found by solving $r = 0$].

$A = 2 \int_{7\pi/6}^{3\pi/2} \frac{1}{2}(1 + 2\sin\theta)^2 \, d\theta = \int_{7\pi/6}^{3\pi/2} (1 + 4\sin\theta + 4\sin^2\theta) \, d\theta = \int_{7\pi/6}^{3\pi/2} \left[1 + 4\sin\theta + 4 \cdot \frac{1}{2}(1 - \cos 2\theta) \right] d\theta$

$= \left[\theta - 4\cos\theta + 2\theta - \sin 2\theta \right]_{7\pi/6}^{3\pi/2} = \left(\frac{9\pi}{2} \right) - \left(\frac{7\pi}{2} + 2\sqrt{3} - \frac{\sqrt{3}}{2} \right) = \pi - \frac{3\sqrt{3}}{2}$

23. $2\cos\theta = 1 \Rightarrow \cos\theta = \frac{1}{2} \Rightarrow \theta = \frac{\pi}{3}$ or $\frac{5\pi}{3}$.

$A = 2 \int_0^{\pi/3} \frac{1}{2}[(2\cos\theta)^2 - 1^2] \, d\theta = \int_0^{\pi/3} (4\cos^2\theta - 1) \, d\theta$

$= \int_0^{\pi/3} \left\{ 4 \left[\frac{1}{2}(1 + \cos 2\theta) \right] - 1 \right\} d\theta = \int_0^{\pi/3} (1 + 2\cos 2\theta) \, d\theta$

$= \left[\theta + \sin 2\theta \right]_0^{\pi/3} = \frac{\pi}{3} + \frac{\sqrt{3}}{2}$

© 2012 Cengage Learning. All Rights Reserved. May not be scanned, copied, or duplicated, or posted to a publicly accessible website, in whole or in part.

25. To find the area inside the leminiscate $r^2 = 8\cos 2\theta$ and outside the circle $r = 2$, we first note that the two curves intersect when $r^2 = 8\cos 2\theta$ and $r = 2$, that is, when $\cos 2\theta = \frac{1}{2}$. For $-\pi < \theta \le \pi$, $\cos 2\theta = \frac{1}{2} \iff 2\theta = \pm\pi/3$ or $\pm 5\pi/3 \iff \theta = \pm\pi/6$ or $\pm 5\pi/6$. The figure shows that the desired area is 4 times the area between the curves from 0 to $\pi/6$. Thus,

$$A = 4\int_0^{\pi/6}\left[\tfrac{1}{2}(8\cos 2\theta) - \tfrac{1}{2}(2)^2\right]d\theta = 8\int_0^{\pi/6}(2\cos 2\theta - 1)\,d\theta$$

$$= 8\left[\sin 2\theta - \theta\right]_0^{\pi/6} = 8(\sqrt{3}/2 - \pi/6) = 4\sqrt{3} - 4\pi/3$$

27. $3\cos\theta = 1 + \cos\theta \iff \cos\theta = \frac{1}{2} \Rightarrow \theta = \frac{\pi}{3}$ or $-\frac{\pi}{3}$.

$$A = 2\int_0^{\pi/3}\tfrac{1}{2}[(3\cos\theta)^2 - (1 + \cos\theta)^2]\,d\theta$$

$$= \int_0^{\pi/3}(8\cos^2\theta - 2\cos\theta - 1)\,d\theta = \int_0^{\pi/3}[4(1 + \cos 2\theta) - 2\cos\theta - 1]\,d\theta$$

$$= \int_0^{\pi/3}(3 + 4\cos 2\theta - 2\cos\theta)\,d\theta = \left[3\theta + 2\sin 2\theta - 2\sin\theta\right]_0^{\pi/3}$$

$$= \pi + \sqrt{3} - \sqrt{3} = \pi$$

29. $\sqrt{3}\cos\theta = \sin\theta \Rightarrow \sqrt{3} = \dfrac{\sin\theta}{\cos\theta} \Rightarrow \tan\theta = \sqrt{3} \Rightarrow \theta = \frac{\pi}{3}$.

$$A = \int_0^{\pi/3}\tfrac{1}{2}(\sin\theta)^2\,d\theta + \int_{\pi/3}^{\pi/2}\tfrac{1}{2}\left(\sqrt{3}\cos\theta\right)^2\,d\theta$$

$$= \int_0^{\pi/3}\tfrac{1}{2}\cdot\tfrac{1}{2}(1 - \cos 2\theta)\,d\theta + \int_{\pi/3}^{\pi/2}\tfrac{1}{2}\cdot 3\cdot\tfrac{1}{2}(1 + \cos 2\theta)\,d\theta$$

$$= \tfrac{1}{4}\left[\theta - \tfrac{1}{2}\sin 2\theta\right]_0^{\pi/3} + \tfrac{3}{4}\left[\theta + \tfrac{1}{2}\sin 2\theta\right]_{\pi/3}^{\pi/2}$$

$$= \tfrac{1}{4}\left[\left(\tfrac{\pi}{3} - \tfrac{\sqrt{3}}{4}\right) - 0\right] + \tfrac{3}{4}\left[\left(\tfrac{\pi}{2} + 0\right) - \left(\tfrac{\pi}{3} + \tfrac{\sqrt{3}}{4}\right)\right]$$

$$= \tfrac{\pi}{12} - \tfrac{\sqrt{3}}{16} + \tfrac{\pi}{8} - \tfrac{3\sqrt{3}}{16} = \tfrac{5\pi}{24} - \tfrac{\sqrt{3}}{4}$$

31. $\sin 2\theta = \cos 2\theta \Rightarrow \dfrac{\sin 2\theta}{\cos 2\theta} = 1 \Rightarrow \tan 2\theta = 1 \to 2\theta = \frac{\pi}{4} \Rightarrow \theta = \frac{\pi}{8} \Rightarrow$

$$A = 8\cdot 2\int_0^{\pi/8}\tfrac{1}{2}\sin^2 2\theta\,d\theta = 8\int_0^{\pi/8}\tfrac{1}{2}(1 - \cos 4\theta)\,d\theta$$

$$= 4\left[\theta - \tfrac{1}{4}\sin 4\theta\right]_0^{\pi/8} = 4\left(\tfrac{\pi}{8} - \tfrac{1}{4}\cdot 1\right) = \tfrac{\pi}{2} - 1$$

33. $\sin 2\theta = \cos 2\theta \Rightarrow \tan 2\theta = 1 \Rightarrow 2\theta = \frac{\pi}{4} \Rightarrow \theta = \frac{\pi}{8}$

$$A = 4\int_0^{\pi/8}\tfrac{1}{2}\sin 2\theta\,d\theta \quad [\text{since } r^2 = \sin 2\theta]$$

$$= \int_0^{\pi/8}2\sin 2\theta\,d\theta = \left[-\cos 2\theta\right]_0^{\pi/8}$$

$$= -\tfrac{1}{2}\sqrt{2} - (-1) = 1 - \tfrac{1}{2}\sqrt{2}$$

© 2012 Cengage Learning. All Rights Reserved. May not be scanned, copied, or duplicated, or posted to a publicly accessible website, in whole or in part.

35. The darker shaded region (from $\theta = 0$ to $\theta = 2\pi/3$) represents $\frac{1}{2}$ of the desired area plus $\frac{1}{2}$ of the area of the inner loop.

From this area, we'll subtract $\frac{1}{2}$ of the area of the inner loop (the lighter shaded region from $\theta = 2\pi/3$ to $\theta = \pi$), and then

double that difference to obtain the desired area.

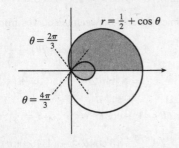

$$A = 2\left[\int_0^{2\pi/3} \frac{1}{2}\left(\frac{1}{2} + \cos\theta\right)^2 d\theta - \int_{2\pi/3}^{\pi} \frac{1}{2}\left(\frac{1}{2} + \cos\theta\right)^2 d\theta\right]$$

$$= \int_0^{2\pi/3}\left(\frac{1}{4} + \cos\theta + \cos^2\theta\right)d\theta - \int_{2\pi/3}^{\pi}\left(\frac{1}{4} + \cos\theta + \cos^2\theta\right)d\theta$$

$$= \int_0^{2\pi/3}\left[\frac{1}{4} + \cos\theta + \frac{1}{2}(1 + \cos 2\theta)\right]d\theta$$

$$\qquad - \int_{2\pi/3}^{\pi}\left[\frac{1}{4} + \cos\theta + \frac{1}{2}(1 + \cos 2\theta)\right]d\theta$$

$$= \left[\frac{\theta}{4} + \sin\theta + \frac{\theta}{2} + \frac{\sin 2\theta}{4}\right]_0^{2\pi/3} - \left[\frac{\theta}{4} + \sin\theta + \frac{\theta}{2} + \frac{\sin 2\theta}{4}\right]_{2\pi/3}^{\pi}$$

$$= \left(\frac{\pi}{6} + \frac{\sqrt{3}}{2} + \frac{\pi}{3} - \frac{\sqrt{3}}{8}\right) - \left(\frac{\pi}{4} + \frac{\pi}{2}\right) + \left(\frac{\pi}{6} + \frac{\sqrt{3}}{2} + \frac{\pi}{3} - \frac{\sqrt{3}}{8}\right)$$

$$= \frac{\pi}{4} + \frac{3}{4}\sqrt{3} = \frac{1}{4}\left(\pi + 3\sqrt{3}\right)$$

37. The pole is a point of intersection.

$$1 + \sin\theta = 3\sin\theta \quad\Rightarrow\quad 1 = 2\sin\theta \quad\Rightarrow\quad \sin\theta = \frac{1}{2} \quad\Rightarrow$$

$\theta = \frac{\pi}{6}$ or $\frac{5\pi}{6}$.

The other two points of intersection are $\left(\frac{3}{2}, \frac{\pi}{6}\right)$ and $\left(\frac{3}{2}, \frac{5\pi}{6}\right)$.

39. $2\sin 2\theta = 1 \quad\Rightarrow\quad \sin 2\theta = \frac{1}{2} \quad\Rightarrow\quad 2\theta = \frac{\pi}{6}, \frac{5\pi}{6}, \frac{13\pi}{6},$ or $\frac{17\pi}{6}$.

By symmetry, the eight points of intersection are given by

$(1, \theta)$, where $\theta = \frac{\pi}{12}, \frac{5\pi}{12}, \frac{13\pi}{12},$ and $\frac{17\pi}{12}$, and

$(-1, \theta)$, where $\theta = \frac{7\pi}{12}, \frac{11\pi}{12}, \frac{19\pi}{12},$ and $\frac{23\pi}{12}$.

[There are many ways to describe these points.]

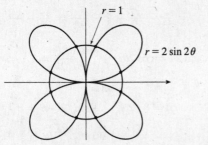

41. The pole is a point of intersection. $\sin\theta = \sin 2\theta = 2\sin\theta\cos\theta \quad\Leftrightarrow$

$\sin\theta\,(1 - 2\cos\theta) = 0 \quad\Leftrightarrow\quad \sin\theta = 0$ or $\cos\theta = \frac{1}{2} \quad\Rightarrow$

$\theta = 0, \pi, \frac{\pi}{3},$ or $-\frac{\pi}{3} \quad\Rightarrow\quad$ the other intersection points are $\left(\frac{\sqrt{3}}{2}, \frac{\pi}{3}\right)$

and $\left(\frac{\sqrt{3}}{2}, \frac{2\pi}{3}\right)$ [by symmetry].

© 2012 Cengage Learning. All Rights Reserved. May not be scanned, copied, or duplicated, or posted to a publicly accessible website, in whole or in part.

43.

From the first graph, we see that the pole is one point of intersection. By zooming in or using the cursor, we find the θ-values of the intersection points to be $\alpha \approx 0.88786 \approx 0.89$ and $\pi - \alpha \approx 2.25$. (The first of these values may be more easily estimated by plotting $y = 1 + \sin x$ and $y = 2x$ in rectangular coordinates; see the second graph.) By symmetry, the total area contained is twice the area contained in the first quadrant, that is,

$$A = 2 \int_0^\alpha \tfrac{1}{2}(2\theta)^2 \, d\theta + 2 \int_\alpha^{\pi/2} \tfrac{1}{2}(1 + \sin \theta)^2 \, d\theta = \int_0^\alpha 4\theta^2 \, d\theta + \int_\alpha^{\pi/2} \left[1 + 2\sin \theta + \tfrac{1}{2}(1 - \cos 2\theta) \right] d\theta$$

$$= \left[\tfrac{4}{3}\theta^3 \right]_0^\alpha + \left[\theta - 2\cos \theta + \left(\tfrac{1}{2}\theta - \tfrac{1}{4}\sin 2\theta \right) \right]_\alpha^{\pi/2} = \tfrac{4}{3}\alpha^3 + \left[\left(\tfrac{\pi}{2} + \tfrac{\pi}{4} \right) - \left(\alpha - 2\cos \alpha + \tfrac{1}{2}\alpha - \tfrac{1}{4}\sin 2\alpha \right) \right] \approx 3.4645$$

45. $L = \displaystyle\int_a^b \sqrt{r^2 + (dr/d\theta)^2} \, d\theta = \int_0^\pi \sqrt{(2\cos\theta)^2 + (-2\sin\theta)^2} \, d\theta$

$$= \int_0^\pi \sqrt{4(\cos^2\theta + \sin^2\theta)} \, d\theta = \int_0^\pi \sqrt{4} \, d\theta = \left[2\theta \right]_0^\pi = 2\pi$$

As a check, note that the curve is a circle of radius 1, so its circumference is $2\pi(1) = 2\pi$.

47. $L = \displaystyle\int_a^b \sqrt{r^2 + (dr/d\theta)^2} \, d\theta = \int_0^{2\pi} \sqrt{(\theta^2)^2 + (2\theta)^2} \, d\theta = \int_0^{2\pi} \sqrt{\theta^4 + 4\theta^2} \, d\theta$

$$= \int_0^{2\pi} \sqrt{\theta^2(\theta^2 + 4)} \, d\theta = \int_0^{2\pi} \theta\sqrt{\theta^2 + 4} \, d\theta$$

Now let $u = \theta^2 + 4$, so that $du = 2\theta \, d\theta$ $\left[\theta \, d\theta = \tfrac{1}{2}\,du \right]$ and

$$\int_0^{2\pi} \theta\sqrt{\theta^2 + 4} \, d\theta = \int_4^{4\pi^2 + 4} \tfrac{1}{2}\sqrt{u} \, du = \tfrac{1}{2} \cdot \tfrac{2}{3} \left[u^{3/2} \right]_4^{4(\pi^2+1)} = \tfrac{1}{3} \left[4^{3/2}(\pi^2 + 1)^{3/2} - 4^{3/2} \right] = \tfrac{8}{3} \left[(\pi^2 + 1)^{3/2} - 1 \right]$$

49. The curve $r = \cos^4(\theta/4)$ is completely traced with $0 \le \theta \le 4\pi$.

$$r^2 + (dr/d\theta)^2 = \left[\cos^4(\theta/4) \right]^2 + \left[4\cos^3(\theta/4) \cdot (-\sin(\theta/4)) \cdot \tfrac{1}{4} \right]^2$$

$$= \cos^8(\theta/4) + \cos^6(\theta/4) \sin^2(\theta/4)$$

$$= \cos^6(\theta/4)[\cos^2(\theta/4) + \sin^2(\theta/4)] = \cos^6(\theta/4)$$

$$L = \int_0^{4\pi} \sqrt{\cos^6(\theta/4)} \, d\theta = \int_0^{4\pi} \left| \cos^3(\theta/4) \right| \, d\theta$$

$$= 2 \int_0^{2\pi} \cos^3(\theta/4) \, d\theta \quad [\text{since } \cos^3(\theta/4) \ge 0 \text{ for } 0 \le \theta \le 2\pi] \quad = 8 \int_0^{\pi/2} \cos^3 u \, du \quad \left[u = \tfrac{1}{4}\theta \right]$$

$$= 8 \int_0^{\pi/2} (1 - \sin^2 u) \cos u \, du = 8 \int_0^1 (1 - x^2) \, dx \quad \begin{bmatrix} x = \sin u, \\ dx = \cos u \, du \end{bmatrix}$$

$$= 8 \left[x - \tfrac{1}{3}x^3 \right]_0^1 = 8 \left(1 - \tfrac{1}{3} \right) = \tfrac{16}{3}$$

© 2012 Cengage Learning. All Rights Reserved. May not be scanned, copied, or duplicated, or posted to a publicly accessible website, in whole or in part.

51. One loop of the curve $r = \cos 2\theta$ is traced with $-\pi/4 \le \theta \le \pi/4$.

$$r^2 + \left(\frac{dr}{d\theta}\right)^2 = \cos^2 2\theta + (-2\sin 2\theta)^2 = \cos^2 2\theta + 4\sin^2 2\theta = 1 + 3\sin^2 2\theta \quad \Rightarrow \quad L \int_{-\pi/4}^{\pi/4} \sqrt{1 + 3\sin^2 2\theta}\, d\theta \approx 2.4221.$$

53. The curve $r = \sin(6\sin\theta)$ is completely traced with $0 \le \theta \le \pi$. $\quad r = \sin(6\sin\theta) \quad \Rightarrow \quad \frac{dr}{d\theta} = \cos(6\sin\theta) \cdot 6\cos\theta$, so

$$r^2 + \left(\frac{dr}{d\theta}\right)^2 = \sin^2(6\sin\theta) + 36\cos^2\theta\cos^2(6\sin\theta) \quad \Rightarrow \quad L \int_0^\pi \sqrt{\sin^2(6\sin\theta) + 36\cos^2\theta\,\cos^2(6\sin\theta)}\, d\theta \approx 8.0091.$$

55. (a) From (10.2.6),

$$S = \int_a^b 2\pi y \sqrt{(dx/d\theta)^2 + (dy/d\theta)^2}\, d\theta$$

$$= \int_a^b 2\pi y \sqrt{r^2 + (dr/d\theta)^2}\, d\theta \qquad \text{[from the derivation of Equation 10.4.5]}$$

$$= \int_a^b 2\pi r \sin\theta \sqrt{r^2 + (dr/d\theta)^2}\, d\theta$$

(b) The curve $r^2 = \cos 2\theta$ goes through the pole when $\cos 2\theta = 0 \quad \Rightarrow$

$2\theta = \frac{\pi}{2} \quad \Rightarrow \quad \theta = \frac{\pi}{4}$. We'll rotate the curve from $\theta = 0$ to $\theta = \frac{\pi}{4}$ and double

this value to obtain the total surface area generated.

$r^2 = \cos 2\theta \quad \Rightarrow \quad 2r\dfrac{dr}{d\theta} = -2\sin 2\theta \quad \Rightarrow \quad \left(\dfrac{dr}{d\theta}\right)^2 = \dfrac{\sin^2 2\theta}{r^2} = \dfrac{\sin^2 2\theta}{\cos 2\theta}$.

$$S = 2\int_0^{\pi/4} 2\pi\sqrt{\cos 2\theta}\,\sin\theta\,\sqrt{\cos 2\theta + (\sin^2 2\theta)/\cos 2\theta}\, d\theta = 4\pi\int_0^{\pi/4}\sqrt{\cos 2\theta}\,\sin\theta\,\sqrt{\frac{\cos^2 2\theta + \sin^2 2\theta}{\cos 2\theta}}\, d\theta$$

$$= 4\pi\int_0^{\pi/4}\sqrt{\cos 2\theta}\,\sin\theta\,\frac{1}{\sqrt{\cos 2\theta}}\, d\theta = 4\pi\int_0^{\pi/4}\sin\theta\, d\theta = 4\pi\big[-\cos\theta\big]_0^{\pi/4} = -4\pi\left(\frac{\sqrt{2}}{2} - 1\right) = 2\pi\left(2 - \sqrt{2}\right)$$

10.5 Conic Sections

1. $x^2 = 6y$ and $x^2 = 4py \quad \Rightarrow \quad 4p = 6 \quad \Rightarrow \quad p = \frac{3}{2}$.

The vertex is $(0,0)$, the focus is $\left(0, \frac{3}{2}\right)$, and the directrix

is $y = -\frac{3}{2}$.

3. $2x = -y^2 \quad \Rightarrow \quad y^2 = -2x. \ 4p = -2 \quad \Rightarrow \quad p = -\frac{1}{2}$.

The vertex is $(0,0)$, the focus is $\left(-\frac{1}{2}, 0\right)$, and the

directrix is $x = \frac{1}{2}$.

© 2012 Cengage Learning. All Rights Reserved. May not be scanned, copied, or duplicated, or posted to a publicly accessible website, in whole or in part.

5. $(x+2)^2 = 8(y-3)$. $4p = 8$, so $p = 2$. The vertex is $(-2,3)$, the focus is $(-2,5)$, and the directrix is $y = 1$.

7. $y^2 + 2y + 12x + 25 = 0 \Rightarrow$

$y^2 + 2y + 1 = -12x - 24 \Rightarrow$

$(y+1)^2 = -12(x+2)$. $4p = -12$, so $p = -3$.

The vertex is $(-2,-1)$, the focus is $(-5,-1)$, and the directrix is $x = 1$.

9. The equation has the form $y^2 = 4px$, where $p < 0$. Since the parabola passes through $(-1,1)$, we have $1^2 = 4p(-1)$, so $4p = -1$ and an equation is $y^2 = -x$ or $x = -y^2$. $4p = -1$, so $p = -\frac{1}{4}$ and the focus is $\left(-\frac{1}{4},0\right)$ while the directrix is $x = \frac{1}{4}$.

11. $\dfrac{x^2}{2} + \dfrac{y^2}{4} = 1 \Rightarrow a = \sqrt{4} = 2, b = \sqrt{2}, c = \sqrt{a^2 - b^2} = \sqrt{4-2} = \sqrt{2}$. The

ellipse is centered at $(0,0)$, with vertices at $(0,\pm 2)$. The foci are $\left(0,\pm\sqrt{2}\right)$.

13. $x^2 + 9y^2 = 9 \Leftrightarrow \dfrac{x^2}{9} + \dfrac{y^2}{1} = 1 \Rightarrow a = \sqrt{9} = 3$,

$b = \sqrt{1} = 1, c = \sqrt{a^2 - b^2} = \sqrt{9-1} = \sqrt{8} = 2\sqrt{2}$.

The ellipse is centered at $(0,0)$, with vertices $(\pm 3, 0)$.

The foci are $\left(\pm 2\sqrt{2}, 0\right)$.

15. $9x^2 - 18x + 4y^2 = 27 \Leftrightarrow$

$9(x^2 - 2x + 1) + 4y^2 = 27 + 9 \Leftrightarrow$

$9(x-1)^2 + 4y^2 = 36 \Leftrightarrow \dfrac{(x-1)^2}{4} + \dfrac{y^2}{9} = 1 \Rightarrow$

$a = 3, b = 2, c = \sqrt{5} \Rightarrow$ center $(1,0)$,

vertices $(1,\pm 3)$, foci $\left(1,\pm\sqrt{5}\right)$

17. The center is $(0,0)$, $a = 3$, and $b = 2$, so an equation is $\dfrac{x^2}{4} + \dfrac{y^2}{9} = 1$. $c = \sqrt{a^2 - b^2} = \sqrt{5}$, so the foci are $\left(0,\pm\sqrt{5}\right)$.

© 2012 Cengage Learning. All Rights Reserved. May not be scanned, copied, or duplicated, or posted to a publicly accessible website, in whole or in part.

19. $\dfrac{y^2}{25} - \dfrac{x^2}{9} = 1 \quad \Rightarrow \quad a = 5, b = 3, c = \sqrt{25 + 9} = \sqrt{34} \quad \Rightarrow$

center $(0, 0)$, vertices $(0, \pm 5)$, foci $\left(0, \pm\sqrt{34}\right)$, asymptotes $y = \pm\frac{5}{3}x$.

Note: It is helpful to draw a $2a$-by-$2b$ rectangle whose center is the center of

the hyperbola. The asymptotes are the extended diagonals of the rectangle.

21. $x^2 - y^2 = 100 \quad \Leftrightarrow \quad \dfrac{x^2}{100} - \dfrac{y^2}{100} = 1 \quad \Rightarrow \quad a = b = 10,$

$c = \sqrt{100 + 100} = 10\sqrt{2} \quad \Rightarrow \quad$ center $(0, 0)$, vertices $(\pm 10, 0)$,

foci $\left(\pm 10\sqrt{2}, 0\right)$, asymptotes $y = \pm\frac{10}{10}x = \pm x$

23. $4x^2 - y^2 - 24x - 4y + 28 = 0 \quad \Leftrightarrow$

$4(x^2 - 6x + 9) - (y^2 + 4y + 4) = -28 + 36 - 4 \quad \Leftrightarrow$

$4(x - 3)^2 - (y + 2)^2 = 4 \quad \Leftrightarrow \quad \dfrac{(x - 3)^2}{1} - \dfrac{(y + 2)^2}{4} = 1 \quad \Rightarrow$

$a = \sqrt{1} = 1, b = \sqrt{4} = 2, c = \sqrt{1 + 4} = \sqrt{5} \quad \Rightarrow$

center $(3, -2)$, vertices $(4, -2)$ and $(2, -2)$, foci $\left(3 \pm \sqrt{5}, -2\right)$,

asymptotes $y + 2 = \pm 2(x - 3)$.

25. $x^2 = y + 1 \quad \Leftrightarrow \quad x^2 = 1(y + 1)$. This is an equation of a *parabola* with $4p = 1$, so $p = \frac{1}{4}$. The vertex is $(0, -1)$ and the

focus is $\left(0, -\frac{3}{4}\right)$.

27. $x^2 = 4y - 2y^2 \quad \Leftrightarrow \quad x^2 + 2y^2 - 4y = 0 \quad \Leftrightarrow \quad x^2 + 2(y^2 - 2y + 1) = 2 \quad \Leftrightarrow \quad x^2 + 2(y - 1)^2 = 2 \quad \Leftrightarrow$

$\dfrac{x^2}{2} + \dfrac{(y - 1)^2}{1} = 1$. This is an equation of an *ellipse* with vertices at $\left(\pm\sqrt{2}, 1\right)$. The foci are at $\left(\pm\sqrt{2 - 1}, 1\right) = (\pm 1, 1)$.

29. $y^2 + 2y = 4x^2 + 3 \quad \Leftrightarrow \quad y^2 + 2y + 1 = 4x^2 + 4 \quad \Leftrightarrow \quad (y + 1)^2 - 4x^2 = 4 \quad \Leftrightarrow \quad \dfrac{(y + 1)^2}{4} - x^2 = 1$. This is an equation

of a *hyperbola* with vertices $(0, -1 \pm 2) = (0, 1)$ and $(0, -3)$. The foci are at $\left(0, -1 \pm \sqrt{4 + 1}\right) = \left(0, -1 \pm \sqrt{5}\right)$.

31. The parabola with vertex $(0, 0)$ and focus $(1, 0)$ opens to the right and has $p = 1$, so its equation is $y^2 = 4px$, or $y^2 = 4x$.

33. The distance from the focus $(-4, 0)$ to the directrix $x = 2$ is $2 - (-4) = 6$, so the distance from the focus to the vertex is

$\frac{1}{2}(6) = 3$ and the vertex is $(-1, 0)$. Since the focus is to the left of the vertex, $p = -3$. An equation is $y^2 = 4p(x + 1) \quad \Rightarrow$

$y^2 = -12(x + 1)$.

© 2012 Cengage Learning. All Rights Reserved. May not be scanned, copied, or duplicated, or posted to a publicly accessible website, in whole or in part.

35. A parabola with vertical axis and vertex $(2, 3)$ has equation $y - 3 = a(x - 2)^2$. Since it passes through $(1, 5)$, we have

$5 - 3 = a(1 - 2)^2 \quad \Rightarrow \quad a = 2$, so an equation is $y - 3 = 2(x - 2)^2$.

37. The ellipse with foci $(\pm 2, 0)$ and vertices $(\pm 5, 0)$ has center $(0, 0)$ and a horizontal major axis, with $a = 5$ and $c = 2$,

so $b^2 = a^2 - c^2 = 25 - 4 = 21$. An equation is $\dfrac{x^2}{25} + \dfrac{y^2}{21} = 1$.

39. Since the vertices are $(0, 0)$ and $(0, 8)$, the ellipse has center $(0, 4)$ with a vertical axis and $a = 4$. The foci at $(0, 2)$ and $(0, 6)$

are 2 units from the center, so $c = 2$ and $b = \sqrt{a^2 - c^2} = \sqrt{4^2 - 2^2} = \sqrt{12}$. An equation is $\dfrac{(x - 0)^2}{b^2} + \dfrac{(y - 4)^2}{a^2} = 1 \quad \Rightarrow$

$\dfrac{x^2}{12} + \dfrac{(y - 4)^2}{16} = 1$.

41. An equation of an ellipse with center $(-1, 4)$ and vertex $(-1, 0)$ is $\dfrac{(x + 1)^2}{b^2} + \dfrac{(y - 4)^2}{4^2} = 1$. The focus $(-1, 6)$ is 2 units

from the center, so $c = 2$. Thus, $b^2 + 2^2 = 4^2 \quad \Rightarrow \quad b^2 = 12$, and the equation is $\dfrac{(x + 1)^2}{12} + \dfrac{(y - 4)^2}{16} = 1$.

43. An equation of a hyperbola with vertices $(\pm 3, 0)$ is $\dfrac{x^2}{3^2} - \dfrac{y^2}{b^2} = 1$. Foci $(\pm 5, 0) \quad \Rightarrow \quad c = 5$ and $3^2 + b^2 = 5^2 \quad \Rightarrow$

$b^2 = 25 - 9 = 16$, so the equation is $\dfrac{x^2}{9} - \dfrac{y^2}{16} = 1$.

45. The center of a hyperbola with vertices $(-3, -4)$ and $(-3, 6)$ is $(-3, 1)$, so $a = 5$ and an equation is

$\dfrac{(y - 1)^2}{5^2} - \dfrac{(x + 3)^2}{b^2} = 1$. Foci $(-3, -7)$ and $(-3, 9) \quad \Rightarrow \quad c = 8$, so $5^2 + b^2 = 8^2 \quad \Rightarrow \quad b^2 = 64 - 25 = 39$ and the

equation is $\dfrac{(y - 1)^2}{25} - \dfrac{(x + 3)^2}{39} = 1$.

47. The center of a hyperbola with vertices $(\pm 3, 0)$ is $(0, 0)$, so $a = 3$ and an equation is $\dfrac{x^2}{3^2} - \dfrac{y^2}{b^2} = 1$.

Asymptotes $y = \pm 2x \quad \Rightarrow \quad \dfrac{b}{a} = 2 \quad \Rightarrow \quad b = 2(3) = 6$ and the equation is $\dfrac{x^2}{9} - \dfrac{y^2}{36} = 1$.

49. In Figure 8, we see that the point on the ellipse closest to a focus is the closer vertex (which is a distance

$a - c$ from it) while the farthest point is the other vertex (at a distance of $a + c$). So for this lunar orbit,

$(a - c) + (a + c) = 2a = (1728 + 110) + (1728 + 314)$, or $a = 1940$; and $(a + c) - (a - c) = 2c = 314 - 110$,

or $c = 102$. Thus, $b^2 = a^2 - c^2 = 3{,}753{,}196$, and the equation is $\dfrac{x^2}{3{,}763{,}600} + \dfrac{y^2}{3{,}753{,}196} = 1$.

51. (a) Set up the coordinate system so that A is $(-200, 0)$ and B is $(200, 0)$.

$|PA| - |PB| = (1200)(980) = 1{,}176{,}000 \text{ ft} = \frac{2450}{11} \text{ mi} = 2a \quad \Rightarrow \quad a = \frac{1225}{11}$, and $c = 200$ so

$b^2 = c^2 - a^2 = \dfrac{3{,}339{,}375}{121} \quad \Rightarrow \quad \dfrac{121x^2}{1{,}500{,}625} - \dfrac{121y^2}{3{,}339{,}375} = 1$.

© 2012 Cengage Learning. All Rights Reserved. May not be scanned, copied, or duplicated, or posted to a publicly accessible website, in whole or in part.

(b) Due north of B $\Rightarrow$ $x = 200$ $\Rightarrow$ $\dfrac{(121)(200)^2}{1{,}500{,}625} - \dfrac{121y^2}{3{,}339{,}375} = 1$ $\Rightarrow$ $y = \dfrac{133{,}575}{539} \approx 248$ mi

53. The function whose graph is the upper branch of this hyperbola is concave upward. The function is

$$y = f(x) = a\sqrt{1 + \dfrac{x^2}{b^2}} = \dfrac{a}{b}\sqrt{b^2 + x^2}, \text{ so } y' = \dfrac{a}{b}x(b^2 + x^2)^{-1/2} \text{ and}$$

$$y'' = \dfrac{a}{b}\left[(b^2 + x^2)^{-1/2} - x^2(b^2 + x^2)^{-3/2}\right] = ab(b^2 + x^2)^{-3/2} > 0 \text{ for all } x, \text{ and so } f \text{ is concave upward.}$$

55. (a) If $k > 16$, then $k - 16 > 0$, and $\dfrac{x^2}{k} + \dfrac{y^2}{k - 16} = 1$ is an *ellipse* since it is the sum of two squares on the left side.

(b) If $0 < k < 16$, then $k - 16 < 0$, and $\dfrac{x^2}{k} + \dfrac{y^2}{k - 16} = 1$ is a *hyperbola* since it is the difference of two squares on the left side.

(c) If $k < 0$, then $k - 16 < 0$, and there is *no curve* since the left side is the sum of two negative terms, which cannot equal 1.

(d) In case (a), $a^2 = k$, $b^2 = k - 16$, and $c^2 = a^2 - b^2 = 16$, so the foci are at $(\pm 4, 0)$. In case (b), $k - 16 < 0$, so $a^2 = k$, $b^2 = 16 - k$, and $c^2 = a^2 + b^2 = 16$, and so again the foci are at $(\pm 4, 0)$.

57. $x^2 = 4py$ $\Rightarrow$ $2x = 4py'$ $\Rightarrow$ $y' = \dfrac{x}{2p}$, so the tangent line at (x_0, y_0) is

$y - \dfrac{x_0^2}{4p} = \dfrac{x_0}{2p}(x - x_0)$. This line passes through the point $(a, -p)$ on the

directrix, so $-p - \dfrac{x_0^2}{4p} = \dfrac{x_0}{2p}(a - x_0)$ $\Rightarrow$ $-4p^2 - x_0^2 = 2ax_0 - 2x_0^2$ $\Leftrightarrow$

$x_0^2 - 2ax_0 - 4p^2 = 0$ $\Leftrightarrow$ $x_0^2 - 2ax_0 + a^2 = a^2 + 4p^2$ $\Leftrightarrow$

$(x_0 - a)^2 = a^2 + 4p^2$ $\Leftrightarrow$ $x_0 = a \pm \sqrt{a^2 + 4p^2}$. The slopes of the tangent lines at $x = a \pm \sqrt{a^2 + 4p^2}$

are $\dfrac{a \pm \sqrt{a^2 + 4p^2}}{2p}$, so the product of the two slopes is

$$\frac{a + \sqrt{a^2 + 4p^2}}{2p} \cdot \frac{a - \sqrt{a^2 + 4p^2}}{2p} = \frac{a^2 - (a^2 + 4p^2)}{4p^2} = \frac{-4p^2}{4p^2} = -1,$$

showing that the tangent lines are perpendicular.

59. $9x^2 + 4y^2 = 36$ $\Leftrightarrow$ $\dfrac{x^2}{4} + \dfrac{y^2}{9} = 1$. We use the parametrization $x = 2\cos t$, $y = 3\sin t$, $0 \le t \le 2\pi$. The circumference

is given by

$$L = \int_0^{2\pi} \sqrt{(dx/dt)^2 + (dy/dt)^2}\, dt = \int_0^{2\pi} \sqrt{(-2\sin t)^2 + (3\cos t)^2}\, dt = \int_0^{2\pi} \sqrt{4\sin^2 t + 9\cos^2 t}\, dt$$

$$= \int_0^{2\pi} \sqrt{4 + 5\cos^2 t}\, dt$$

Now use Simpson's Rule with $n = 8$, $\Delta t = \dfrac{2\pi - 0}{8} = \dfrac{\pi}{4}$, and $f(t) = \sqrt{4 + 5\cos^2 t}$ to get

$$L \approx S_8 = \tfrac{\pi/4}{3}\left[f(0) + 4f\!\left(\tfrac{\pi}{4}\right) + 2f\!\left(\tfrac{\pi}{2}\right) + 4f\!\left(\tfrac{3\pi}{4}\right) + 2f(\pi) + 4f\!\left(\tfrac{5\pi}{4}\right) + 2f\!\left(\tfrac{3\pi}{2}\right) + 4f\!\left(\tfrac{7\pi}{4}\right) + f(2\pi)\right] \approx 15.9.$$

© 2012 Cengage Learning. All Rights Reserved. May not be scanned, copied, or duplicated, or posted to a publicly accessible website, in whole or in part.

61. $\dfrac{x^2}{a^2} - \dfrac{y^2}{b^2} = 1 \Rightarrow \dfrac{y^2}{b^2} = \dfrac{x^2 - a^2}{a^2} \Rightarrow y = \pm \dfrac{b}{a} \sqrt{x^2 - a^2}$.

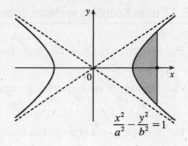

$A = 2 \displaystyle\int_a^c \dfrac{b}{a} \sqrt{x^2 - a^2}\, dx \overset{39}{=} \dfrac{2b}{a}\left[\dfrac{x}{2}\sqrt{x^2 - a^2} - \dfrac{a^2}{2}\ln\left|x + \sqrt{x^2 - a^2}\right|\right]_a^c$

$= \dfrac{b}{a}\left[c\sqrt{c^2 - a^2} - a^2 \ln\left|c + \sqrt{c^2 - a^2}\right| + a^2 \ln|a|\right]$

Since $a^2 + b^2 = c^2, c^2 - a^2 = b^2$, and $\sqrt{c^2 - a^2} = b$.

$= \dfrac{b}{a}\left[cb - a^2 \ln(c + b) + a^2 \ln a\right] = \dfrac{b}{a}\left[cb + a^2(\ln a - \ln(b+c))\right]$

$= b^2 c/a + ab\ln[a/(b+c)]$, where $c^2 = a^2 + b^2$.

63. $9x^2 + 4y^2 = 36 \Leftrightarrow \dfrac{x^2}{4} + \dfrac{y^2}{9} = 1 \Rightarrow a = 3, b = 2$. By symmetry, $\overline{x} = 0$. By Example 2 in Section 7.3, the area of the

top half of the ellipse is $\frac{1}{2}(\pi ab) = 3\pi$. Solve $9x^2 + 4y^2 = 36$ for y to get an equation for the top half of the ellipse:

$9x^2 + 4y^2 = 36 \Leftrightarrow 4y^2 = 36 - 9x^2 \Leftrightarrow y^2 = \frac{9}{4}(4 - x^2) \Rightarrow y = \frac{3}{2}\sqrt{4 - x^2}$. Now

$\overline{y} = \dfrac{1}{A}\displaystyle\int_a^b \dfrac{1}{2}[f(x)]^2\, dx = \dfrac{1}{3\pi}\int_{-2}^2 \dfrac{1}{2}\left(\dfrac{3}{2}\sqrt{4 - x^2}\right)^2 dx = \dfrac{3}{8\pi}\int_{-2}^2 (4 - x^2)\, dx$

$= \dfrac{3}{8\pi} \cdot 2 \displaystyle\int_0^2 (4 - x^2)\, dx = \dfrac{3}{4\pi}\left[4x - \dfrac{1}{3}x^3\right]_0^2 = \dfrac{3}{4\pi}\left(\dfrac{16}{3}\right) = \dfrac{4}{\pi}$

so the centroid is $(0, 4/\pi)$.

65. Differentiating implicitly, $\dfrac{x^2}{a^2} + \dfrac{y^2}{b^2} = 1 \Rightarrow \dfrac{2x}{a^2} + \dfrac{2yy'}{b^2} = 0 \Rightarrow y' = -\dfrac{b^2 x}{a^2 y}$ $[y \neq 0]$. Thus, the slope of the tangent

line at P is $-\dfrac{b^2 x_1}{a^2 y_1}$. The slope of $F_1 P$ is $\dfrac{y_1}{x_1 + c}$ and of $F_2 P$ is $\dfrac{y_1}{x_1 - c}$. By the formula from Problems Plus, we have

$$\tan \alpha = \dfrac{\dfrac{y_1}{x_1 + c} + \dfrac{b^2 x_1}{a^2 y_1}}{1 - \dfrac{b^2 x_1 y_1}{a^2 y_1(x_1 + c)}} = \dfrac{a^2 y_1^2 + b^2 x_1(x_1 + c)}{a^2 y_1(x_1 + c) - b^2 x_1 y_1} = \dfrac{a^2 b^2 + b^2 c x_1}{c^2 x_1 y_1 + a^2 c y_1} \qquad \left[\begin{array}{l}\text{using } b^2 x_1^2 + a^2 y_1^2 = a^2 b^2, \\ \text{and } a^2 - b^2 = c^2\end{array}\right]$$

$$= \dfrac{b^2(c x_1 + a^2)}{c y_1(c x_1 + a^2)} = \dfrac{b^2}{c y_1}$$

and

$$\tan \beta = \dfrac{-\dfrac{b^2 x_1}{a^2 y_1} - \dfrac{y_1}{x_1 - c}}{1 - \dfrac{b^2 x_1 y_1}{a^2 y_1(x_1 - c)}} = \dfrac{-a^2 y_1^2 - b^2 x_1(x_1 - c)}{a^2 y_1(x_1 - c) - b^2 x_1 y_1} = \dfrac{-a^2 b^2 + b^2 c x_1}{c^2 x_1 y_1 - a^2 c y_1} = \dfrac{b^2(c x_1 - a^2)}{c y_1(c x_1 - a^2)} = \dfrac{b^2}{c y_1}$$

Thus, $\alpha = \beta$.

© 2012 Cengage Learning. All Rights Reserved. May not be scanned, copied, or duplicated, or posted to a publicly accessible website, in whole or in part.

10.6 Conic Sections in Polar Coordinates

1. The directrix $x = 4$ is to the right of the focus at the origin, so we use the form with "$+ e\cos\theta$" in the denominator.

(See Theorem 6 and Figure 2.) An equation is $r = \dfrac{ed}{1 + e\cos\theta} = \dfrac{\frac{1}{2}\cdot 4}{1 + \frac{1}{2}\cos\theta} = \dfrac{4}{2 + \cos\theta}$.

3. The directrix $y = 2$ is above the focus at the origin, so we use the form with "$+ e\sin\theta$" in the denominator. An equation is

$$r = \frac{ed}{1 + e\sin\theta} = \frac{1.5(2)}{1 + 1.5\sin\theta} = \frac{6}{2 + 3\sin\theta}.$$

5. The vertex $(4, 3\pi/2)$ is 4 units below the focus at the origin, so the directrix is 8 units below the focus $(d = 8)$, and we

use the form with "$- e\sin\theta$" in the denominator. $e = 1$ for a parabola, so an equation is

$$r = \frac{ed}{1 - e\sin\theta} = \frac{1(8)}{1 - 1\sin\theta} = \frac{8}{1 - \sin\theta}.$$

7. The directrix $r = 4\sec\theta$ (equivalent to $r\cos\theta = 4$ or $x = 4$) is to the right of the focus at the origin, so we will use the form

with "$+ e\cos\theta$" in the denominator. The distance from the focus to the directrix is $d = 4$, so an equation is

$$r = \frac{ed}{1 + e\cos\theta} = \frac{\frac{1}{2}(4)}{1 + \frac{1}{2}\cos\theta}\cdot\frac{2}{2} = \frac{4}{2 + \cos\theta}.$$

9. $r = \dfrac{4}{5 - 4\sin\theta}\cdot\dfrac{1/5}{1/5} = \dfrac{4/5}{1 - \frac{4}{5}\sin\theta}$, where $e = \frac{4}{5}$ and $ed = \frac{4}{5}$ $\Rightarrow$ $d = 1$.

(a) Eccentricity $= e = \frac{4}{5}$

(b) Since $e = \frac{4}{5} < 1$, the conic is an ellipse.

(c) Since "$- e\sin\theta$" appears in the denominator, the directrix is below the focus

at the origin, $d = |Fl| = 1$, so an equation of the directrix is $y = -1$.

(d) The vertices are $\left(4, \frac{\pi}{2}\right)$ and $\left(\frac{4}{9}, \frac{3\pi}{2}\right)$.

11. $r = \dfrac{2}{3 + 3\sin\theta}\cdot\dfrac{1/3}{1/3} = \dfrac{2/3}{1 + 1\sin\theta}$, where $e = 1$ and $ed = \frac{2}{3}$ $\Rightarrow$ $d = \frac{2}{3}$.

(a) Eccentricity $= e = 1$

(b) Since $e = 1$, the conic is a parabola.

(c) Since "$+ e\sin\theta$" appears in the denominator, the directrix is above the focus

at the origin. $d = |Fl| = \frac{2}{3}$, so an equation of the directrix is $y = \frac{2}{3}$.

(d) The vertex is at $\left(\frac{1}{3}, \frac{\pi}{2}\right)$, midway between the focus and directrix.

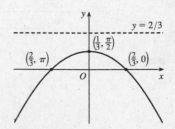

© 2012 Cengage Learning. All Rights Reserved. May not be scanned, copied, or duplicated, or posted to a publicly accessible website, in whole or in part.

13. $r = \dfrac{9}{6 + 2\cos\theta} \cdot \dfrac{1/6}{1/6} = \dfrac{3/2}{1 + \frac{1}{3}\cos\theta}$, where $e = \frac{1}{3}$ and $ed = \frac{3}{2}$ $\Rightarrow$ $d = \frac{9}{2}$.

(a) Eccentricity $= e = \frac{1}{3}$

(b) Since $e = \frac{1}{3} < 1$, the conic is an ellipse.

(c) Since "$+e\cos\theta$" appears in the denominator, the directrix is to the right of

the focus at the origin. $d = |Fl| = \frac{9}{2}$, so an equation of the directrix is

$x = \frac{9}{2}$.

(d) The vertices are $\left(\frac{9}{8}, 0\right)$ and $\left(\frac{9}{4}, \pi\right)$, so the center is midway between them,

that is, $\left(\frac{9}{16}, \pi\right)$.

15. $r = \dfrac{3}{4 - 8\cos\theta} \cdot \dfrac{1/4}{1/4} = \dfrac{3/4}{1 - 2\cos\theta}$, where $e = 2$ and $ed = \frac{3}{4}$ $\Rightarrow$ $d = \frac{3}{8}$.

(a) Eccentricity $= e = 2$

(b) Since $e = 2 > 1$, the conic is a hyperbola.

(c) Since "$-e\cos\theta$" appears in the denominator, the directrix is to the left of

the focus at the origin. $d = |Fl| = \frac{3}{8}$, so an equation of the directrix is

$x = -\frac{3}{8}$.

(d) The vertices are $\left(-\frac{3}{4}, 0\right)$ and $\left(\frac{1}{4}, \pi\right)$, so the center is midway between them,

that is, $\left(\frac{1}{2}, \pi\right)$.

17. (a) $r = \dfrac{1}{1 - 2\sin\theta}$, where $e = 2$ and $ed = 1$ $\Rightarrow$ $d = \frac{1}{2}$. The eccentricity

$e = 2 > 1$, so the conic is a hyperbola. Since "$-e\sin\theta$" appears in the

denominator, the directrix is below the focus at the origin. $d = |Fl| = \frac{1}{2}$,

so an equation of the directrix is $y = -\frac{1}{2}$. The vertices are $\left(-1, \frac{\pi}{2}\right)$ and

$\left(\frac{1}{3}, \frac{3\pi}{2}\right)$, so the center is midway between them, that is, $\left(\frac{2}{3}, \frac{3\pi}{2}\right)$.

(b) By the discussion that precedes Example 4, the equation

is $r = \dfrac{1}{1 - 2\sin\left(\theta - \frac{3\pi}{4}\right)}$.

© 2012 Cengage Learning. All Rights Reserved. May not be scanned, copied, or duplicated, or posted to a publicly accessible website, in whole or in part.

19. For $e < 1$ the curve is an ellipse. It is nearly circular when e is close to 0. As e
increases, the graph is stretched out to the right, and grows larger (that is, its
right-hand focus moves to the right while its left-hand focus remains at the
origin.) At $e = 1$, the curve becomes a parabola with focus at the origin.

21. $|PF| = e|Pl| \Rightarrow r = e[d - r\cos(\pi - \theta)] = e(d + r\cos\theta) \Rightarrow$
$r(1 - e\cos\theta) = ed \Rightarrow r = \dfrac{ed}{1 - e\cos\theta}$

23. $|PF| = e|Pl| \Rightarrow r = e[d - r\sin(\theta - \pi)] = e(d + r\sin\theta) \Rightarrow$
$r(1 - e\sin\theta) = ed \Rightarrow r = \dfrac{ed}{1 - e\sin\theta}$

25. We are given $e = 0.093$ and $a = 2.28 \times 10^8$. By (7), we have

$$r = \frac{a(1 - e^2)}{1 + e\cos\theta} = \frac{2.28 \times 10^8[1 - (0.093)^2]}{1 + 0.093\cos\theta} \approx \frac{2.26 \times 10^8}{1 + 0.093\cos\theta}$$

27. Here $2a$ = length of major axis = 36.18 AU $\Rightarrow a = 18.09$ AU and $e = 0.97$. By (7), the equation of the orbit is

$r = \dfrac{18.09[1 - (0.97)^2]}{1 + 0.97\cos\theta} \approx \dfrac{1.07}{1 + 0.97\cos\theta}$. By (8), the maximum distance from the comet to the sun is

$18.09(1 + 0.97) \approx 35.64$ AU or about 3.314 billion miles.

29. The minimum distance is at perihelion, where $4.6 \times 10^7 = r = a(1 - e) = a(1 - 0.206) = a(0.794) \Rightarrow$

$a = 4.6 \times 10^7/0.794$. So the maximum distance, which is at aphelion, is

$r = a(1 + e) = (4.6 \times 10^7/0.794)(1.206) \approx 7.0 \times 10^7$ km.

31. From Exercise 29, we have $e = 0.206$ and $a(1 - e) = 4.6 \times 10^7$ km. Thus, $a = 4.6 \times 10^7/0.794$. From (7), we can write the

equation of Mercury's orbit as $r = a\dfrac{1 - e^2}{1 + e\cos\theta}$. So since

$\dfrac{dr}{d\theta} = \dfrac{a(1 - e^2)e\sin\theta}{(1 + e\cos\theta)^2} \Rightarrow$

$$r^2 + \left(\frac{dr}{d\theta}\right)^2 = \frac{a^2(1 - e^2)^2}{(1 + e\cos\theta)^2} + \frac{a^2(1 - e^2)^2 e^2\sin^2\theta}{(1 + e\cos\theta)^4} = \frac{a^2(1 - e^2)^2}{(1 + e\cos\theta)^4}(1 + 2e\cos\theta + e^2)$$

© 2012 Cengage Learning. All Rights Reserved. May not be scanned, copied, or duplicated, or posted to a publicly accessible website, in whole or in part.

the length of the orbit is

$$L = \int_0^{2\pi} \sqrt{r^2 + (dr/d\theta)^2}\, d\theta = a(1 - e^2) \int_0^{2\pi} \frac{\sqrt{1 + e^2 + 2e\cos\theta}}{(1 + e\cos\theta)^2}\, d\theta \approx 3.6 \times 10^8 \text{ km}$$

This seems reasonable, since Mercury's orbit is nearly circular, and the circumference of a circle of radius a is $2\pi a \approx 3.6 \times 10^8$ km.

10   Review

CONCEPT CHECK

1. (a) A parametric curve is a set of points of the form $(x, y) = (f(t), g(t))$, where f and g are continuous functions of a variable t.

   (b) Sketching a parametric curve, like sketching the graph of a function, is difficult to do in general. We can plot points on the curve by finding $f(t)$ and $g(t)$ for various values of t, either by hand or with a calculator or computer. Sometimes, when f and g are given by formulas, we can eliminate t from the equations $x = f(t)$ and $y = g(t)$ to get a Cartesian equation relating x and y. It may be easier to graph that equation than to work with the original formulas for x and y in terms of t.

2. (a) You can find $\dfrac{dy}{dx}$ as a function of t by calculating $\dfrac{dy}{dx} = \dfrac{dy/dt}{dx/dt}$ [if $dx/dt \neq 0$].

   (b) Calculate the area as $\int_a^b y\, dx = \int_\alpha^\beta g(t)\, f'(t)\, dt$ [or $\int_\beta^\alpha g(t)\, f'(t)\, dt$ if the leftmost point is $(f(\beta), g(\beta))$ rather than $(f(\alpha), g(\alpha))$].

3. (a) $L = \int_\alpha^\beta \sqrt{(dx/dt)^2 + (dy/dt)^2}\, dt = \int_\alpha^\beta \sqrt{[f'(t)]^2 + [g'(t)]^2}\, dt$

   (b) $S = \int_\alpha^\beta 2\pi y \sqrt{(dx/dt)^2 + (dy/dt)^2}\, dt = \int_\alpha^\beta 2\pi g(t) \sqrt{[f'(t)]^2 + [g'(t)]^2}\, dt$

4. (a) See Figure 5 in Section 10.3.

   (b) $x = r\cos\theta,\ y = r\sin\theta$

   (c) To find a polar representation (r, θ) with $r \geq 0$ and $0 \leq \theta < 2\pi$, first calculate $r = \sqrt{x^2 + y^2}$. Then θ is specified by $\cos\theta = x/r$ and $\sin\theta = y/r$.

5. (a) Calculate $\dfrac{dy}{dx} = \dfrac{\dfrac{dy}{d\theta}}{\dfrac{dx}{d\theta}} = \dfrac{\dfrac{d}{d\theta}(y)}{\dfrac{d}{d\theta}(x)} = \dfrac{\dfrac{d}{d\theta}(r\sin\theta)}{\dfrac{d}{d\theta}(r\cos\theta)} = \dfrac{\left(\dfrac{dr}{d\theta}\right)\sin\theta + r\cos\theta}{\left(\dfrac{dr}{d\theta}\right)\cos\theta - r\sin\theta}$, where $r = f(\theta)$.

   (b) Calculate $A = \int_a^b \tfrac{1}{2}r^2\, d\theta = \int_a^b \tfrac{1}{2}[f(\theta)]^2\, d\theta$

   (c) $L = \int_a^b \sqrt{(dx/d\theta)^2 + (dy/d\theta)^2}\, d\theta = \int_a^b \sqrt{r^2 + (dr/d\theta)^2}\, d\theta = \int_a^b \sqrt{[f(\theta)]^2 + [f'(\theta)]^2}\, d\theta$

6. (a) A parabola is a set of points in a plane whose distances from a fixed point F (the focus) and a fixed line l (the directrix) are equal.

   (b) $x^2 = 4py;\ y^2 = 4px$

© 2012 Cengage Learning. All Rights Reserved. May not be scanned, copied, or duplicated, or posted to a publicly accessible website, in whole or in part.

7. (a) An ellipse is a set of points in a plane the sum of whose distances from two fixed points (the foci) is a constant.

(b) $\dfrac{x^2}{a^2} + \dfrac{y^2}{a^2 - c^2} = 1$.

8. (a) A hyperbola is a set of points in a plane the difference of whose distances from two fixed points (the foci) is a constant. This difference should be interpreted as the larger distance minus the smaller distance.

(b) $\dfrac{x^2}{a^2} - \dfrac{y^2}{c^2 - a^2} = 1$

(c) $y = \pm \dfrac{\sqrt{c^2 - a^2}}{a} x$

9. (a) If a conic section has focus F and corresponding directrix l, then the eccentricity e is the fixed ratio $|PF| / |Pl|$ for points P of the conic section.

(b) $e < 1$ for an ellipse; $e > 1$ for a hyperbola; $e = 1$ for a parabola.

(c) $x = d$: $r = \dfrac{ed}{1 + e \cos\theta}$. $x = -d$: $r = \dfrac{ed}{1 - e\cos\theta}$. $y = d$: $r = \dfrac{ed}{1 + e\sin\theta}$. $y = -d$: $r = \dfrac{ed}{1 - e\sin\theta}$.

TRUE-FALSE QUIZ

1. False. Consider the curve defined by $x = f(t) = (t - 1)^3$ and $y = g(t) = (t - 1)^2$. Then $g'(t) = 2(t - 1)$, so $g'(1) = 0$, but its graph has a *vertical* tangent when $t = 1$. *Note:* The statement is true if $f'(1) \neq 0$ when $g'(1) = 0$.

3. False. For example, if $f(t) = \cos t$ and $g(t) = \sin t$ for $0 \leq t \leq 4\pi$, then the curve is a circle of radius 1, hence its length is 2π, but $\int_0^{4\pi} \sqrt{[f'(t)]^2 + [g'(t)]^2}\, dt = \int_0^{4\pi} \sqrt{(-\sin t)^2 + (\cos t)^2}\, dt = \int_0^{4\pi} 1\, dt = 4\pi$, since as t increases from 0 to 4π, the circle is traversed twice.

5. True. The curve $r = 1 - \sin 2\theta$ is unchanged if we rotate it through $180°$ about O because $1 - \sin 2(\theta + \pi) = 1 - \sin(2\theta + 2\pi) = 1 - \sin 2\theta$. So it's unchanged if we replace r by $-r$. (See the discussion after Example 8 in Section 10.3.) In other words, it's the same curve as $r = -(1 - \sin 2\theta) = \sin 2\theta - 1$.

7. False. The first pair of equations gives the portion of the parabola $y = x^2$ with $x \geq 0$, whereas the second pair of equations traces out the whole parabola $y = x^2$.

9. True. By rotating and translating the parabola, we can assume it has an equation of the form $y = cx^2$, where $c > 0$. The tangent at the point (a, ca^2) is the line $y - ca^2 = 2ca(x - a)$; i.e., $y = 2cax - ca^2$. This tangent meets the parabola at the points (x, cx^2) where $cx^2 = 2cax - ca^2$. This equation is equivalent to $x^2 = 2ax - a^2$ [since $c > 0$]. But $x^2 = 2ax - a^2 \Leftrightarrow x^2 - 2ax + a^2 = 0 \Leftrightarrow (x - a)^2 = 0 \Leftrightarrow x = a \Leftrightarrow (x, cx^2) = (a, ca^2)$. This shows that each tangent meets the parabola at exactly one point.

© 2012 Cengage Learning. All Rights Reserved. May not be scanned, copied, or duplicated, or posted to a publicly accessible website, in whole or in part.

EXERCISES

1. $x = t^2 + 4t$, $y = 2 - t$, $-4 \le t \le 1$. $t = 2 - y$, so

$x = (2 - y)^2 + 4(2 - y) = 4 - 4y + y^2 + 8 - 4y = y^2 - 8y + 12 \quad\Leftrightarrow$

$x + 4 = y^2 - 8y + 16 = (y - 4)^2$. This is part of a parabola with vertex

$(-4, 4)$, opening to the right.

3. $y = \sec\theta = \dfrac{1}{\cos\theta} = \dfrac{1}{x}$. Since $0 \le \theta \le \pi/2$, $0 < x \le 1$ and $y \ge 1$.

This is part of the hyperbola $y = 1/x$.

5. Three different sets of parametric equations for the curve $y = \sqrt{x}$ are

(i) $x = t, \quad y = \sqrt{t}$

(ii) $x = t^4, \quad y = t^2$

(iii) $x = \tan^2 t, \quad y = \tan t, \quad 0 \le t < \pi/2$

There are many other sets of equations that also give this curve.

7. (a)   The Cartesian coordinates are $x = 4\cos\dfrac{2\pi}{3} = 4\left(-\dfrac{1}{2}\right) = -2$ and

$y = 4\sin\dfrac{2\pi}{3} = 4\left(\dfrac{\sqrt{3}}{2}\right) = 2\sqrt{3}$, that is, the point $\left(-2, 2\sqrt{3}\right)$.

(b) Given $x = -3$ and $y = 3$, we have $r = \sqrt{(-3)^2 + 3^2} = \sqrt{18} = 3\sqrt{2}$. Also, $\tan\theta = \dfrac{y}{x} \implies \tan\theta = \dfrac{3}{-3}$, and since

$(-3, 3)$ is in the second quadrant, $\theta = \dfrac{3\pi}{4}$. Thus, one set of polar coordinates for $(-3, 3)$ is $\left(3\sqrt{2}, \dfrac{3\pi}{4}\right)$, and two others are

$\left(3\sqrt{2}, \dfrac{11\pi}{4}\right)$ and $\left(-3\sqrt{2}, \dfrac{7\pi}{4}\right)$.

9. $r = 1 - \cos\theta$. This cardioid is
symmetric about the polar axis.

© 2012 Cengage Learning. All Rights Reserved. May not be scanned, copied, or duplicated, or posted to a publicly accessible website, in whole or in part.

11. $r = \cos 3\theta$. This is a three-leaved rose. The curve is traced twice.

13. $r = 1 + \cos 2\theta$. The curve is symmetric about the pole and both the horizontal and vertical axes.

15. $r = \dfrac{3}{1 + 2\sin\theta}$ $\Rightarrow$ $e = 2 > 1$, so the conic is a hyperbola. $de = 3$ $\Rightarrow$

$d = \frac{3}{2}$ and the form "$+2\sin\theta$" imply that the directrix is above the focus at the origin and has equation $y = \frac{3}{2}$. The vertices are $\left(1, \frac{\pi}{2}\right)$ and $\left(-3, \frac{3\pi}{2}\right)$.

17. $x + y = 2$ $\Leftrightarrow$ $r\cos\theta + r\sin\theta = 2$ $\Leftrightarrow$ $r(\cos\theta + \sin\theta) = 2$ $\Leftrightarrow$ $r = \dfrac{2}{\cos\theta + \sin\theta}$

19. $r = (\sin\theta)/\theta$. As $\theta \to \pm\infty$, $r \to 0$. As $\theta \to 0$, $r \to 1$. In the first figure, there are an infinite number of x-intercepts at $x = \pi n$, n a nonzero integer. These correspond to pole points in the second figure.

 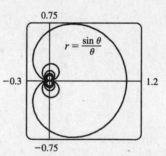

21. $x = \ln t$, $y = 1 + t^2$; $t = 1$. $\dfrac{dy}{dt} = 2t$ and $\dfrac{dx}{dt} = \dfrac{1}{t}$, so $\dfrac{dy}{dx} = \dfrac{dy/dt}{dx/dt} = \dfrac{2t}{1/t} = 2t^2$.

When $t = 1$, $(x, y) = (0, 2)$ and $dy/dx = 2$.

23. $r = e^{-\theta}$ $\Rightarrow$ $y = r\sin\theta = e^{-\theta}\sin\theta$ and $x = r\cos\theta = e^{-\theta}\cos\theta$ $\Rightarrow$

$\dfrac{dy}{dx} = \dfrac{dy/d\theta}{dx/d\theta} = \dfrac{\frac{dr}{d\theta}\sin\theta + r\cos\theta}{\frac{dr}{d\theta}\cos\theta - r\sin\theta} = \dfrac{-e^{-\theta}\sin\theta + e^{-\theta}\cos\theta}{-e^{-\theta}\cos\theta - e^{-\theta}\sin\theta} \cdot \dfrac{-e^{\theta}}{-e^{\theta}} = \dfrac{\sin\theta - \cos\theta}{\cos\theta + \sin\theta}$.

When $\theta = \pi$, $\dfrac{dy}{dx} = \dfrac{0 - (-1)}{-1 + 0} = \dfrac{1}{-1} = -1$.

© 2012 Cengage Learning. All Rights Reserved. May not be scanned, copied, or duplicated, or posted to a publicly accessible website, in whole or in part.

25. $x = t + \sin t$, $y = t - \cos t$ $\Rightarrow$ $\dfrac{dy}{dx} = \dfrac{dy/dt}{dx/dt} = \dfrac{1 + \sin t}{1 + \cos t}$ $\Rightarrow$

$$\frac{d^2y}{dx^2} = \frac{\dfrac{d}{dt}\left(\dfrac{dy}{dx}\right)}{dx/dt} = \frac{\dfrac{(1 + \cos t)\cos t - (1 + \sin t)(-\sin t)}{(1 + \cos t)^2}}{1 + \cos t} = \frac{\cos t + \cos^2 t + \sin t + \sin^2 t}{(1 + \cos t)^3} = \frac{1 + \cos t + \sin t}{(1 + \cos t)^3}$$

27. We graph the curve $x = t^3 - 3t$, $y = t^2 + t + 1$ for $-2.2 \le t \le 1.2$.

By zooming in or using a cursor, we find that the lowest point is about

$(1.4, 0.75)$. To find the exact values, we find the t-value at which

$dy/dt = 2t + 1 = 0$ $\Leftrightarrow$ $t = -\frac{1}{2}$ $\Leftrightarrow$ $(x, y) = \left(\frac{11}{8}, \frac{3}{4}\right)$.

29. $x = 2a\cos t - a\cos 2t$ $\Rightarrow$ $\dfrac{dx}{dt} = -2a\sin t + 2a\sin 2t = 2a\sin t(2\cos t - 1) = 0$ $\Leftrightarrow$

$\sin t = 0$ or $\cos t = \frac{1}{2}$ $\Rightarrow$ $t = 0$, $\frac{\pi}{3}$, π, or $\frac{5\pi}{3}$.

$y = 2a\sin t - a\sin 2t$ $\Rightarrow$ $\dfrac{dy}{dt} = 2a\cos t - 2a\cos 2t = 2a\left(1 + \cos t - 2\cos^2 t\right) = 2a(1 - \cos t)(1 + 2\cos t) = 0$ $\Rightarrow$

$t = 0$, $\frac{2\pi}{3}$, or $\frac{4\pi}{3}$.

Thus the graph has vertical tangents where $t = \frac{\pi}{3}$, π and $\frac{5\pi}{3}$, and horizontal tangents where $t = \frac{2\pi}{3}$ and $\frac{4\pi}{3}$. To determine

what the slope is where $t = 0$, we use l'Hospital's Rule to evaluate $\lim\limits_{t \to 0} \dfrac{dy/dt}{dx/dt} = 0$, so there is a horizontal tangent there.

t	x	y
0	a	0
$\frac{\pi}{3}$	$\frac{3}{2}a$	$\frac{\sqrt{3}}{2}a$
$\frac{2\pi}{3}$	$-\frac{1}{2}a$	$\frac{3\sqrt{3}}{2}a$
π	$-3a$	0
$\frac{4\pi}{3}$	$-\frac{1}{2}a$	$-\frac{3\sqrt{3}}{2}a$
$\frac{5\pi}{3}$	$\frac{3}{2}a$	$-\frac{\sqrt{3}}{2}a$

31. The curve $r^2 = 9\cos 5\theta$ has 10 "petals." For instance, for $-\frac{\pi}{10} \le \theta \le \frac{\pi}{10}$, there are two petals, one with $r > 0$ and one

with $r < 0$.

$$A = 10 \int_{-\pi/10}^{\pi/10} \tfrac{1}{2}r^2\, d\theta = 5\int_{-\pi/10}^{\pi/10} 9\cos 5\theta\, d\theta = 5 \cdot 9 \cdot 2\int_0^{\pi/10} \cos 5\theta\, d\theta = 18\big[\sin 5\theta\big]_0^{\pi/10} = 18$$

33. The curves intersect when $4\cos\theta = 2$ $\Rightarrow$ $\cos\theta = \frac{1}{2}$ $\Rightarrow$ $\theta = \pm\frac{\pi}{3}$

for $-\pi \le \theta \le \pi$. The points of intersection are $\left(2, \frac{\pi}{3}\right)$ and $\left(2, -\frac{\pi}{3}\right)$.

© 2012 Cengage Learning. All Rights Reserved. May not be scanned, copied, or duplicated, or posted to a publicly accessible website, in whole or in part.

35. The curves intersect where $2\sin\theta = \sin\theta + \cos\theta \Rightarrow$

$\sin\theta = \cos\theta \Rightarrow \theta = \frac{\pi}{4}$, and also at the origin (at which $\theta = \frac{3\pi}{4}$

on the second curve).

$\theta = \frac{\pi}{4}$

$r = 2\sin\theta$ $r = \sin\theta + \cos\theta$

$$A = \int_0^{\pi/4} \tfrac{1}{2}(2\sin\theta)^2 \, d\theta + \int_{\pi/4}^{3\pi/4} \tfrac{1}{2}(\sin\theta + \cos\theta)^2 \, d\theta$$

$$= \int_0^{\pi/4} (1 - \cos 2\theta) \, d\theta + \tfrac{1}{2}\int_{\pi/4}^{3\pi/4}(1 + \sin 2\theta) \, d\theta$$

$$= \left[\theta - \tfrac{1}{2}\sin 2\theta\right]_0^{\pi/4} + \left[\tfrac{1}{2}\theta - \tfrac{1}{4}\cos 2\theta\right]_{\pi/4}^{3\pi/4} = \tfrac{1}{2}(\pi - 1)$$

37. $x = 3t^2$, $y = 2t^3$.

$$L = \int_0^2 \sqrt{(dx/dt)^2 + (dy/dt)^2}\, dt = \int_0^2 \sqrt{(6t)^2 + (6t^2)^2}\, dt = \int_0^2 \sqrt{36t^2 + 36t^4}\, dt = \int_0^2 \sqrt{36t^2}\sqrt{1 + t^2}\, dt$$

$$= \int_0^2 6\,|t|\,\sqrt{1 + t^2}\, dt = 6\int_0^2 t\sqrt{1 + t^2}\, dt = 6\int_1^5 u^{1/2}\left(\tfrac{1}{2}du\right) \qquad \left[u = 1 + t^2,\, du = 2t\,dt\right]$$

$$= 6 \cdot \tfrac{1}{2} \cdot \tfrac{2}{3}\left[u^{3/2}\right]_1^5 = 2(5^{3/2} - 1) = 2\left(5\sqrt{5} - 1\right)$$

39. $L = \int_\pi^{2\pi} \sqrt{r^2 + (dr/d\theta)^2}\, d\theta = \int_\pi^{2\pi}\sqrt{(1/\theta)^2 + (-1/\theta^2)^2}\, d\theta = \int_\pi^{2\pi} \dfrac{\sqrt{\theta^2 + 1}}{\theta^2}\, d\theta$

$$\overset{24}{=} \left[-\dfrac{\sqrt{\theta^2 + 1}}{\theta} + \ln\left(\theta + \sqrt{\theta^2 + 1}\right)\right]_\pi^{2\pi} = \dfrac{\sqrt{\pi^2 + 1}}{\pi} - \dfrac{\sqrt{4\pi^2 + 1}}{2\pi} + \ln\left(\dfrac{2\pi + \sqrt{4\pi^2 + 1}}{\pi + \sqrt{\pi^2 + 1}}\right)$$

$$= \dfrac{2\sqrt{\pi^2 + 1} - \sqrt{4\pi^2 + 1}}{2\pi} + \ln\left(\dfrac{2\pi + \sqrt{4\pi^2 + 1}}{\pi + \sqrt{\pi^2 + 1}}\right)$$

41. $x = 4\sqrt{t}$, $y = \dfrac{t^3}{3} + \dfrac{1}{2t^2}$, $1 \le t \le 4 \Rightarrow$

$$S = \int_1^4 2\pi y\sqrt{(dx/dt)^2 + (dy/dt)^2}\, dt = \int_1^4 2\pi\left(\tfrac{1}{3}t^3 + \tfrac{1}{2}t^{-2}\right)\sqrt{\left(2/\sqrt{t}\right)^2 + (t^2 - t^{-3})^2}\, dt$$

$$= 2\pi\int_1^4\left(\tfrac{1}{3}t^3 + \tfrac{1}{2}t^{-2}\right)\sqrt{(t^2 + t^{-3})^2}\, dt = 2\pi\int_1^4\left(\tfrac{1}{3}t^5 + \tfrac{5}{6} + \tfrac{1}{2}t^{-5}\right)dt = 2\pi\left[\tfrac{1}{18}t^6 + \tfrac{5}{6}t - \tfrac{1}{8}t^{-4}\right]_1^4 = \dfrac{471{,}295}{1024}\pi$$

43. For all c except -1, the curve is asymptotic to the line $x = 1$. For

$c < -1$, the curve bulges to the right near $y = 0$. As c increases, the

bulge becomes smaller, until at $c = -1$ the curve is the straight line $x = 1$.

As c continues to increase, the curve bulges to the left, until at $c = 0$ there

is a cusp at the origin. For $c > 0$, there is a loop to the left of the origin,

whose size and roundness increase as c increases. Note that the x-intercept

of the curve is always $-c$.

© 2012 Cengage Learning. All Rights Reserved. May not be scanned, copied, or duplicated, or posted to a publicly accessible website, in whole or in part.

45. $\dfrac{x^2}{9} + \dfrac{y^2}{8} = 1$ is an ellipse with center $(0,0)$.

$a = 3, b = 2\sqrt{2}, c = 1 \;\Rightarrow$

foci $(\pm 1, 0)$, vertices $(\pm 3, 0)$.

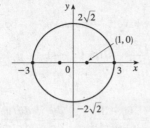

47. $6y^2 + x - 36y + 55 = 0 \;\Leftrightarrow$

$6(y^2 - 6y + 9) = -(x+1) \;\Leftrightarrow$

$(y-3)^2 = -\tfrac{1}{6}(x+1)$, a parabola with vertex $(-1, 3)$,

opening to the left, $p = -\tfrac{1}{24} \;\Rightarrow$ focus $\left(-\tfrac{25}{24}, 3\right)$ and

directrix $x = -\tfrac{23}{24}$.

49. The ellipse with foci $(\pm 4, 0)$ and vertices $(\pm 5, 0)$ has center $(0,0)$ and a horizontal major axis, with $a = 5$ and $c = 4$,

so $b^2 = a^2 - c^2 = 5^2 - 4^2 = 9$. An equation is $\dfrac{x^2}{25} + \dfrac{y^2}{9} = 1$.

51. The center of a hyperbola with foci $(0, \pm 4)$ is $(0,0)$, so $c = 4$ and an equation is $\dfrac{y^2}{a^2} - \dfrac{x^2}{b^2} = 1$.

The asymptote $y = 3x$ has slope 3, so $\dfrac{a}{b} - \dfrac{3}{1} \;\Rightarrow\; a = 3b$ and $a^2 + b^2 = c^2 \;\Rightarrow\; (3b)^2 + b^2 = 4^2 \;\Rightarrow$

$10b^2 = 16 \;\Rightarrow\; b^2 = \tfrac{8}{5}$ and so $a^2 = 16 - \tfrac{8}{5} = \tfrac{72}{5}$. Thus, an equation is $\dfrac{y^2}{72/5} - \dfrac{x^2}{8/5} = 1$, or $\dfrac{5y^2}{72} - \dfrac{5x^2}{8} = 1$.

53. $x^2 = -(y - 100)$ has its vertex at $(0, 100)$, so one of the vertices of the ellipse is $(0, 100)$. Another form of the equation of a

parabola is $x^2 = 4p(y - 100)$ so $4p(y - 100) = -(y - 100) \;\Rightarrow\; 4p = -1 \;\Rightarrow\; p = -\tfrac{1}{4}$. Therefore the shared focus is

found at $\left(0, \tfrac{300}{4}\right)$ so $2c = \tfrac{399}{4} - 0 \;\Rightarrow\; c = \tfrac{399}{8}$ and the center of the ellipse is $\left(0, \tfrac{399}{8}\right)$. So $a = 100 - \tfrac{399}{8} = \tfrac{401}{8}$ and

$b^2 = a^2 - c^2 = \dfrac{401^2 - 399^2}{8^2} = 25$. So the equation of the ellipse is $\dfrac{x^2}{b^2} + \dfrac{\left(y - \tfrac{399}{8}\right)^2}{a^2} = 1 \;\Rightarrow\; \dfrac{x^2}{25} + \dfrac{\left(y - \tfrac{399}{8}\right)^2}{\left(\tfrac{401}{8}\right)^2} = 1$,

or $\dfrac{x^2}{25} + \dfrac{(8y - 399)^2}{160{,}801} = 1$.

55. Directrix $x = 4 \;\Rightarrow\; d = 4$, so $e = \tfrac{1}{3} \;\Rightarrow\; r = \dfrac{ed}{1 + e\cos\theta} = \dfrac{4}{3 + \cos\theta}$.

57. (a) If (a, b) lies on the curve, then there is some parameter value t_1 such that $\dfrac{3t_1}{1 + t_1^3} = a$ and $\dfrac{3t_1^2}{1 + t_1^3} = b$. If $t_1 = 0$,

the point is $(0,0)$, which lies on the line $y = x$. If $t_1 \neq 0$, then the point corresponding to $t = \dfrac{1}{t_1}$ is given by

$x = \dfrac{3(1/t_1)}{1 + (1/t_1)^3} = \dfrac{3t_1^2}{t_1^3 + 1} = b, \; y = \dfrac{3(1/t_1)^2}{1 + (1/t_1)^3} = \dfrac{3t_1}{t_1^3 + 1} = a$. So (b, a) also lies on the curve. [Another way to see

this is to do part (e) first; the result is immediate.] The curve intersects the line $y = x$ when $\dfrac{3t}{1 + t^3} = \dfrac{3t^2}{1 + t^3} \;\Rightarrow$

$t = t^2 \;\Rightarrow\; t = 0$ or 1, so the points are $(0,0)$ and $\left(\tfrac{3}{2}, \tfrac{3}{2}\right)$.

© 2012 Cengage Learning. All Rights Reserved. May not be scanned, copied, or duplicated, or posted to a publicly accessible website, in whole or in part.

(b) $\dfrac{dy}{dt} = \dfrac{(1+t^3)(6t) - 3t^2(3t^2)}{(1+t^3)^2} = \dfrac{6t - 3t^4}{(1+t^3)^2} = 0$ when $6t - 3t^4 = 3t(2 - t^3) = 0 \;\Rightarrow\; t = 0$ or $t = \sqrt[3]{2}$, so there are

horizontal tangents at $(0,0)$ and $\left(\sqrt[3]{2}, \sqrt[3]{4}\right)$. Using the symmetry from part (a), we see that there are vertical tangents at

$(0,0)$ and $\left(\sqrt[3]{4}, \sqrt[3]{2}\right)$.

(c) Notice that as $t \to -1^+$, we have $x \to -\infty$ and $y \to \infty$. As $t \to -1^-$, we have $x \to \infty$ and $y \to -\infty$. Also

$y - (-x - 1) = y + x + 1 = \dfrac{3t + 3t^2 + (1+t^3)}{1 + t^3} = \dfrac{(t+1)^3}{1+t^3} = \dfrac{(t+1)^2}{t^2 - t + 1} \to 0$ as $t \to -1$. So $y = -x - 1$ is a

slant asymptote.

(d) $\dfrac{dx}{dt} = \dfrac{(1+t^3)(3) - 3t(3t^2)}{(1+t^3)^2} = \dfrac{3 - 6t^3}{(1+t^3)^2}$ and from part (b) we have $\dfrac{dy}{dt} = \dfrac{6t - 3t^4}{(1+t^3)^2}$. So $\dfrac{dy}{dx} = \dfrac{dy/dt}{dx/dt} = \dfrac{t(2 - t^3)}{1 - 2t^3}$.

Also $\dfrac{d^2y}{dx^2} = \dfrac{\dfrac{d}{dt}\left(\dfrac{dy}{dx}\right)}{dx/dt} = \dfrac{2(1+t^3)^4}{3(1 - 2t^3)^3} > 0 \;\Leftrightarrow\; t < \dfrac{1}{\sqrt[3]{2}}$.

So the curve is concave upward there and has a minimum point at $(0,0)$

and a maximum point at $\left(\sqrt[3]{2}, \sqrt[3]{4}\right)$. Using this together with the

information from parts (a), (b), and (c), we sketch the curve.

(e) $x^3 + y^3 = \left(\dfrac{3t}{1+t^3}\right)^3 + \left(\dfrac{3t^2}{1+t^3}\right)^3 = \dfrac{27t^3 + 27t^6}{(1+t^3)^3} = \dfrac{27t^3(1+t^3)}{(1+t^3)^3} = \dfrac{27t^3}{(1+t^3)^2}$ and

$3xy = 3\left(\dfrac{3t}{1+t^3}\right)\left(\dfrac{3t^2}{1+t^3}\right) = \dfrac{27t^3}{(1+t^3)^2}$, so $x^3 + y^3 = 3xy$.

(f) We start with the equation from part (e) and substitute $x = r\cos\theta$, $y = r\sin\theta$. Then $x^3 + y^3 = 3xy \;\Rightarrow$

$r^3 \cos^3\theta + r^3 \sin^3\theta = 3r^2 \cos\theta\sin\theta$. For $r \neq 0$, this gives $r = \dfrac{3\cos\theta\sin\theta}{\cos^3\theta + \sin^3\theta}$. Dividing numerator and denominator

by $\cos^3\theta$, we obtain $r = \dfrac{3\left(\dfrac{1}{\cos\theta}\right)\dfrac{\sin\theta}{\cos\theta}}{1 + \dfrac{\sin^3\theta}{\cos^3\theta}} = \dfrac{3\sec\theta\tan\theta}{1 + \tan^3\theta}$.

(g) The loop corresponds to $\theta \in \left(0, \dfrac{\pi}{2}\right)$, so its area is

$A = \displaystyle\int_0^{\pi/2} \dfrac{r^2}{2}\, d\theta = \dfrac{1}{2}\int_0^{\pi/2} \left(\dfrac{3\sec\theta\tan\theta}{1 + \tan^3\theta}\right)^2 d\theta = \dfrac{9}{2}\int_0^{\pi/2} \dfrac{\sec^2\theta\tan^2\theta}{(1 + \tan^3\theta)^2}\, d\theta = \dfrac{9}{2}\int_0^{\infty} \dfrac{u^2\, du}{(1 + u^3)^2}$ [let $u = \tan\theta$]

$\qquad = \displaystyle\lim_{b\to\infty} \dfrac{9}{2}\left[-\dfrac{1}{3}(1 + u^3)^{-1}\right]_0^b = \dfrac{3}{2}$

(h) By symmetry, the area between the folium and the line $y = -x - 1$ is equal to the enclosed area in the third quadrant,

plus twice the enclosed area in the fourth quadrant. The area in the third quadrant is $\dfrac{1}{2}$, and since $y = -x - 1 \;\Rightarrow$

$r\sin\theta = -r\cos\theta - 1 \;\Rightarrow\; r = -\dfrac{1}{\sin\theta + \cos\theta}$, the area in the fourth quadrant is

$\dfrac{1}{2}\displaystyle\int_{-\pi/2}^{-\pi/4} \left[\left(-\dfrac{1}{\sin\theta + \cos\theta}\right)^2 - \left(\dfrac{3\sec\theta\tan\theta}{1 + \tan^3\theta}\right)^2\right] d\theta \overset{\text{CAS}}{=} \dfrac{1}{2}$. Therefore, the total area is $\dfrac{1}{2} + 2\left(\dfrac{1}{2}\right) = \dfrac{3}{2}$.

© 2012 Cengage Learning. All Rights Reserved. May not be scanned, copied, or duplicated, or posted to a publicly accessible website, in whole or in part.

□ PROBLEMS PLUS

1. $x = \int_1^t \frac{\cos u}{u}\,du$, $y = \int_1^t \frac{\sin u}{u}\,du$, so by FTC1, we have $\frac{dx}{dt} = \frac{\cos t}{t}$ and $\frac{dy}{dt} = \frac{\sin t}{t}$. Vertical tangent lines occur when

$\frac{dx}{dt} = 0 \iff \cos t = 0$. The parameter value corresponding to $(x, y) = (0, 0)$ is $t = 1$, so the nearest vertical tangent

occurs when $t = \frac{\pi}{2}$. Therefore, the arc length between these points is

$$L = \int_1^{\pi/2} \sqrt{\left(\frac{dx}{dt}\right)^2 + \left(\frac{dy}{dt}\right)^2}\,dt = \int_1^{\pi/2} \sqrt{\frac{\cos^2 t}{t^2} + \frac{\sin^2 t}{t^2}}\,dt = \int_1^{\pi/2} \frac{dt}{t} = \big[\ln t\big]_1^{\pi/2} = \ln\frac{\pi}{2}$$

3. In terms of x and y, we have $x = r\cos\theta = (1 + c\sin\theta)\cos\theta = \cos\theta + c\sin\theta\cos\theta = \cos\theta + \frac{1}{2}c\sin 2\theta$ and

$y = r\sin\theta = (1 + c\sin\theta)\sin\theta = \sin\theta + c\sin^2\theta$. Now $-1 \le \sin\theta \le 1 \implies -1 \le \sin\theta + c\sin^2\theta \le 1 + c \le 2$, so

$-1 \le y \le 2$. Furthermore, $y = 2$ when $c = 1$ and $\theta = \frac{\pi}{2}$, while $y = -1$ for $c = 0$ and $\theta = \frac{3\pi}{2}$. Therefore, we need a viewing

rectangle with $-1 \le y \le 2$.

To find the x-values, look at the equation $x = \cos\theta + \frac{1}{2}c\sin 2\theta$ and use the fact that $\sin 2\theta \ge 0$ for $0 \le \theta \le \frac{\pi}{2}$ and

$\sin 2\theta \le 0$ for $-\frac{\pi}{2} \le \theta \le 0$. [Because $r = 1 + c\sin\theta$ is symmetric about the y-axis, we only need to consider

$-\frac{\pi}{2} \le \theta \le \frac{\pi}{2}$.] So for $-\frac{\pi}{2} \le \theta \le 0$, x has a maximum value when $c = 0$ and then $x = \cos\theta$ has a maximum value

of 1 at $\theta = 0$. Thus, the maximum value of x must occur on $\left[0, \frac{\pi}{2}\right]$ with $c = 1$. Then $x = \cos\theta + \frac{1}{2}\sin 2\theta \implies$

$\frac{dx}{d\theta} = -\sin\theta + \cos 2\theta = -\sin\theta + 1 - 2\sin^2\theta \implies \frac{dx}{d\theta} = -(2\sin\theta - 1)(\sin\theta + 1) = 0$ when $\sin\theta = -1$ or $\frac{1}{2}$

[but $\sin\theta \ne -1$ for $0 \le \theta \le \frac{\pi}{2}$]. If $\sin\theta = \frac{1}{2}$, then $\theta = \frac{\pi}{6}$ and

$x = \cos\frac{\pi}{6} + \frac{1}{2}\sin\frac{\pi}{3} = \frac{3}{4}\sqrt{3}$. Thus, the maximum value of x is $\frac{3}{4}\sqrt{3}$, and,

by symmetry, the minimum value is $-\frac{3}{4}\sqrt{3}$. Therefore, the smallest

viewing rectangle that contains every member of the family of polar curves

$r = 1 + c\sin\theta$, where $0 \le c \le 1$, is $\left[-\frac{3}{4}\sqrt{3}, \frac{3}{4}\sqrt{3}\right] \times [-1, 2]$.

5. Without loss of generality, assume the hyperbola has equation $\frac{x^2}{a^2} - \frac{y^2}{b^2} = 1$. Use implicit differentiation to get

$\frac{2x}{a^2} - \frac{2yy'}{b^2} = 0$, so $y' = \frac{b^2 x}{a^2 y}$. The tangent line at the point (c, d) on the hyperbola has equation $y - d = \frac{b^2 c}{a^2 d}(x - c)$.

The tangent line intersects the asymptote $y = \frac{b}{a}x$ when $\frac{b}{a}x - d = \frac{b^2 c}{a^2 d}(x - c) \implies abdx - a^2 d^2 = b^2 cx - b^2 c^2 \implies$

© 2012 Cengage Learning. All Rights Reserved. May not be scanned, copied, or duplicated, or posted to a publicly accessible website, in whole or in part.

$abdx - b^2cx = a^2d^2 - b^2c^2 \quad \Rightarrow \quad x = \dfrac{a^2d^2 - b^2c^2}{b(ad - bc)} = \dfrac{ad + bc}{b}$ and the y-value is $\dfrac{b}{a}\dfrac{ad + bc}{b} = \dfrac{ad + bc}{a}$.

Similarly, the tangent line intersects $y = -\dfrac{b}{a}x$ at $\left(\dfrac{bc - ad}{b}, \dfrac{ad - bc}{a}\right)$. The midpoint of these intersection points is

$\left(\dfrac{1}{2}\left(\dfrac{ad + bc}{b} + \dfrac{bc - ad}{b}\right), \dfrac{1}{2}\left(\dfrac{ad + bc}{a} + \dfrac{ad - bc}{a}\right)\right) = \left(\dfrac{1}{2}\dfrac{2bc}{b}, \dfrac{1}{2}\dfrac{2ad}{a}\right) = (c, d)$, the point of tangency.

Note: If $y = 0$, then at $(\pm a, 0)$, the tangent line is $x = \pm a$, and the points of intersection are clearly equidistant from the point of tangency.

© 2012 Cengage Learning. All Rights Reserved. May not be scanned, copied, or duplicated, or posted to a publicly accessible website, in whole or in part.

11 □ INFINITE SEQUENCES AND SERIES

11.1 Sequences

1. (a) A sequence is an ordered list of numbers. It can also be defined as a function whose domain is the set of positive integers.

(b) The terms a_n approach 8 as n becomes large. In fact, we can make a_n as close to 8 as we like by taking n sufficiently large.

(c) The terms a_n become large as n becomes large. In fact, we can make a_n as large as we like by taking n sufficiently large.

3. $a_n = \dfrac{2n}{n^2+1}$, so the sequence is $\left\{ \dfrac{2}{1+1}, \dfrac{4}{4+1}, \dfrac{6}{9+1}, \dfrac{8}{16+1}, \dfrac{10}{25+1}, \cdots \right\} = \left\{ 1, \dfrac{4}{5}, \dfrac{3}{5}, \dfrac{8}{17}, \dfrac{5}{13}, \cdots \right\}$.

5. $a_n = \dfrac{(-1)^{n-1}}{5^n}$, so the sequence is $\left\{ \dfrac{1}{5^1}, \dfrac{-1}{5^2}, \dfrac{1}{5^3}, \dfrac{-1}{5^4}, \dfrac{1}{5^5}, \cdots \right\} = \left\{ \dfrac{1}{5}, -\dfrac{1}{25}, \dfrac{1}{125}, -\dfrac{1}{625}, \dfrac{1}{3125}, \cdots \right\}$.

7. $a_n = \dfrac{1}{(n+1)!}$, so the sequence is $\left\{ \dfrac{1}{2!}, \dfrac{1}{3!}, \dfrac{1}{4!}, \dfrac{1}{5!}, \dfrac{1}{6!}, \cdots \right\} = \left\{ \dfrac{1}{2}, \dfrac{1}{6}, \dfrac{1}{24}, \dfrac{1}{120}, \dfrac{1}{720}, \cdots \right\}$.

9. $a_1 = 1$, $a_{n+1} = 5a_n - 3$. Each term is defined in terms of the preceding term. $a_2 = 5a_1 - 3 = 5(1) - 3 = 2$.

$a_3 = 5a_2 - 3 = 5(2) - 3 = 7$. $a_4 = 5a_3 - 3 = 5(7) - 3 = 32$. $a_5 = 5a_4 - 3 = 5(32) - 3 = 157$.

The sequence is $\{1, 2, 7, 32, 157, \ldots\}$.

11. $a_1 = 2$, $a_{n+1} = \dfrac{a_n}{1+a_n}$. $a_2 = \dfrac{u_1}{1+a_1} = \dfrac{2}{1+2} = \dfrac{2}{3}$. $a_3 = \dfrac{a_2}{1+a_2} = \dfrac{2/3}{1+2/3} = \dfrac{2}{5}$. $a_4 = \dfrac{a_3}{1+a_3} = \dfrac{2/5}{1+2/5} = \dfrac{2}{7}$.

$a_5 = \dfrac{a_4}{1+a_4} = \dfrac{2/7}{1+2/7} = \dfrac{2}{9}$. The sequence is $\left\{ 2, \dfrac{2}{3}, \dfrac{2}{5}, \dfrac{2}{7}, \dfrac{2}{9}, \cdots \right\}$.

13. $\left\{ 1, \dfrac{1}{3}, \dfrac{1}{5}, \dfrac{1}{7}, \dfrac{1}{9}, \cdots \right\}$. The denominator of the nth term is the nth positive odd integer, so $a_n = \dfrac{1}{2n-1}$.

15. $\left\{ -3, 2, -\dfrac{4}{3}, \dfrac{8}{9}, -\dfrac{16}{27}, \cdots \right\}$. The first term is -3 and each term is $-\dfrac{2}{3}$ times the preceding one, so $a_n = -3\left(-\dfrac{2}{3}\right)^{n-1}$.

17. $\left\{ \dfrac{1}{2}, -\dfrac{4}{3}, \dfrac{9}{4}, -\dfrac{16}{5}, \dfrac{25}{6}, \cdots \right\}$. The numerator of the nth term is n^2 and its denominator is $n + 1$. Including the alternating signs,

we get $a_n = (-1)^{n+1} \dfrac{n^2}{n+1}$.

© 2012 Cengage Learning. All Rights Reserved. May not be scanned, copied, or duplicated, or posted to a publicly accessible website, in whole or in part.

19.

n	$a_n = \dfrac{3n}{1+6n}$
1	0.4286
2	0.4615
3	0.4737
4	0.4800
5	0.4839
6	0.4865
7	0.4884
8	0.4898
9	0.4909
10	0.4918

It appears that $\lim\limits_{n\to\infty} a_n = 0.5$.

$$\lim_{n\to\infty} \frac{3n}{1+6n} = \lim_{n\to\infty} \frac{(3n)/n}{(1+6n)/n} = \lim_{n\to\infty} \frac{3}{1/n + 6} = \frac{3}{6} = \frac{1}{2}$$

21.

n	$a_n = 1 + \left(-\frac{1}{2}\right)^n$
1	0.5000
2	1.2500
3	0.8750
4	1.0625
5	0.9688
6	1.0156
7	0.9922
8	1.0039
9	0.9980
10	1.0010

It appears that $\lim\limits_{n\to\infty} a_n = 1$.

$$\lim_{n\to\infty} \left(1 + \left(-\tfrac{1}{2}\right)^n\right) = \lim_{n\to\infty} 1 + \lim_{n\to\infty} \left(-\tfrac{1}{2}\right)^n = 1 + 0 = 1 \text{ since}$$

$$\lim_{n\to\infty} \left(-\tfrac{1}{2}\right)^n = 0 \text{ by (9).}$$

23. $a_n = 1 - (0.2)^n$, so $\lim\limits_{n\to\infty} a_n = 1 - 0 = 1$ by (9).   Converges

25. $a_n = \dfrac{3+5n^2}{n+n^2} = \dfrac{(3+5n^2)/n^2}{(n+n^2)/n^2} = \dfrac{5+3/n^2}{1+1/n}$, so $a_n \to \dfrac{5+0}{1+0} = 5$ as $n \to \infty$.   Converges

27. Because the natural exponential function is continuous at 0, Theorem 7 enables us to write

$$\lim_{n\to\infty} a_n = \lim_{n\to\infty} e^{1/n} = e^{\lim_{n\to\infty}(1/n)} = e^0 = 1. \quad \text{Converges}$$

29. If $b_n = \dfrac{2n\pi}{1+8n}$, then $\lim\limits_{n\to\infty} b_n = \lim\limits_{n\to\infty} \dfrac{(2n\pi)/n}{(1+8n)/n} = \lim\limits_{n\to\infty} \dfrac{2\pi}{1/n+8} = \dfrac{2\pi}{8} = \dfrac{\pi}{4}$. Since $\tan$ is continuous at $\frac{\pi}{4}$, by

Theorem 7, $\lim\limits_{n\to\infty} \tan\!\left(\dfrac{2n\pi}{1+8n}\right) = \tan\!\left(\lim\limits_{n\to\infty}\dfrac{2n\pi}{1+8n}\right) = \tan\dfrac{\pi}{4} = 1$.   Converges

31. $a_n = \dfrac{n^2}{\sqrt{n^3+4n}} = \dfrac{n^2/\sqrt{n^3}}{\sqrt{n^3+4n}/\sqrt{n^3}} = \dfrac{\sqrt{n}}{\sqrt{1+4/n^2}}$, so $a_n \to \infty$ as $n \to \infty$ since $\lim\limits_{n\to\infty} \sqrt{n} = \infty$ and

$\lim\limits_{n\to\infty} \sqrt{1+4/n^2} = 1$.   Diverges

33. $\lim\limits_{n\to\infty} |a_n| = \lim\limits_{n\to\infty} \left|\dfrac{(-1)^n}{2\sqrt{n}}\right| = \dfrac{1}{2} \lim\limits_{n\to\infty} \dfrac{1}{n^{1/2}} = \dfrac{1}{2}(0) = 0$, so $\lim\limits_{n\to\infty} a_n = 0$ by (6).   Converges

© 2012 Cengage Learning. All Rights Reserved. May not be scanned, copied, or duplicated, or posted to a publicly accessible website, in whole or in part.

35. $a_n = \cos(n/2)$.   This sequence diverges since the terms don't approach any particular real number as $n \to \infty$. The terms take on values between -1 and 1.

37. $a_n = \dfrac{(2n-1)!}{(2n+1)!} = \dfrac{(2n-1)!}{(2n+1)(2n)(2n-1)!} = \dfrac{1}{(2n+1)(2n)} \to 0$ as $n \to \infty$.   Converges

39. $a_n = \dfrac{e^n + e^{-n}}{e^{2n} - 1} \cdot \dfrac{e^{-n}}{e^{-n}} = \dfrac{1 + e^{-2n}}{e^n - e^{-n}} \to 0$ as $n \to \infty$ because $1 + e^{-2n} \to 1$ and $e^n - e^{-n} \to \infty$.   Converges

41. $a_n = n^2 e^{-n} = \dfrac{n^2}{e^n}$. Since $\lim\limits_{x \to \infty} \dfrac{x^2}{e^x} \overset{\text{H}}{=} \lim\limits_{x \to \infty} \dfrac{2x}{e^x} \overset{\text{H}}{=} \lim\limits_{x \to \infty} \dfrac{2}{e^x} = 0$, it follows from Theorem 3 that $\lim\limits_{n \to \infty} a_n = 0$.   Converges

43. $0 \le \dfrac{\cos^2 n}{2^n} \le \dfrac{1}{2^n}$ [since $0 \le \cos^2 n \le 1$],   so since $\lim\limits_{n \to \infty} \dfrac{1}{2^n} = 0$, $\left\{ \dfrac{\cos^2 n}{2^n} \right\}$ converges to 0 by the Squeeze Theorem.

45. $a_n = n \sin(1/n) = \dfrac{\sin(1/n)}{1/n}$.   Since $\lim\limits_{x \to \infty} \dfrac{\sin(1/x)}{1/x} = \lim\limits_{t \to 0+} \dfrac{\sin t}{t}$ [where $t = 1/x$] $= 1$, it follows from Theorem 3 that $\{a_n\}$ converges to 1.

47. $y = \left(1 + \dfrac{2}{x}\right)^x \;\Rightarrow\; \ln y = x \ln\left(1 + \dfrac{2}{x}\right)$, so

$$\lim_{x \to \infty} \ln y = \lim_{x \to \infty} \dfrac{\ln(1 + 2/x)}{1/x} \overset{\text{H}}{=} \lim_{x \to \infty} \dfrac{\left(\dfrac{1}{1 + 2/x}\right)\left(-\dfrac{2}{x^2}\right)}{-1/x^2} = \lim_{x \to \infty} \dfrac{2}{1 + 2/x} = 2 \;\Rightarrow$$

$$\lim_{x \to \infty} \left(1 + \dfrac{2}{x}\right)^x = \lim_{x \to \infty} e^{\ln y} = e^2, \text{ so by Theorem 3, } \lim_{n \to \infty} \left(1 + \dfrac{2}{n}\right)^n = e^2. \quad \text{Converges}$$

49. $a_n = \ln(2n^2 + 1) - \ln(n^2 + 1) = \ln\left(\dfrac{2n^2 + 1}{n^2 + 1}\right) = \ln\left(\dfrac{2 + 1/n^2}{1 + 1/n^2}\right) \to \ln 2$ as $n \to \infty$.   Converges

51. $a_n = \arctan(\ln n)$. Let $f(x) = \arctan(\ln x)$. Then $\lim\limits_{x \to \infty} f(x) = \dfrac{\pi}{2}$ since $\ln x \to \infty$ as $x \to \infty$ and $\arctan$ is continuous.

Thus, $\lim\limits_{n \to \infty} a_n = \lim\limits_{n \to \infty} f(n) = \dfrac{\pi}{2}$.   Converges

53. $\{0, 1, 0, 0, 1, 0, 0, 0, 1, \ldots\}$ diverges since the sequence takes on only two values, 0 and 1, and never stays arbitrarily close to either one (or any other value) for n sufficiently large.

55. $a_n = \dfrac{n!}{2^n} = \dfrac{1}{2} \cdot \dfrac{2}{2} \cdot \dfrac{3}{2} \cdots \cdot \dfrac{(n-1)}{2} \cdot \dfrac{n}{2} \ge \dfrac{1}{2} \cdot \dfrac{n}{2}$ [for $n > 1$] $= \dfrac{n}{4} \to \infty$ as $n \to \infty$, so $\{a_n\}$ diverges.

57.

From the graph, it appears that the sequence converges to 1.

$\{(-2/e)^n\}$ converges to 0 by (7), and hence $\{1 + (-2/e)^n\}$

converges to $1 + 0 = 1$.

© 2012 Cengage Learning. All Rights Reserved. May not be scanned, copied, or duplicated, or posted to a publicly accessible website, in whole or in part.

59.

From the graph, it appears that the sequence converges to $\frac{1}{2}$.

As $n \to \infty$,

$$a_n = \sqrt{\frac{3 + 2n^2}{8n^2 + n}} = \sqrt{\frac{3/n^2 + 2}{8 + 1/n}} \quad \Rightarrow \quad \sqrt{\frac{0 + 2}{8 + 0}} = \sqrt{\frac{1}{4}} = \frac{1}{2},$$

so $\lim\limits_{n \to \infty} a_n = \frac{1}{2}$.

61.

From the graph, it appears that the sequence $\{a_n\} = \left\{ \dfrac{n^2 \cos n}{1 + n^2} \right\}$ is

divergent, since it oscillates between 1 and -1 (approximately). To

prove this, suppose that $\{a_n\}$ converges to L. If $b_n = \dfrac{n^2}{1 + n^2}$, then

$\{b_n\}$ converges to 1, and $\lim\limits_{n \to \infty} \dfrac{a_n}{b_n} = \dfrac{L}{1} = L$. But $\dfrac{a_n}{b_n} = \cos n$, so

$\lim\limits_{n \to \infty} \dfrac{a_n}{b_n}$ does not exist. This contradiction shows that $\{a_n\}$ diverges.

63.

From the graph, it appears that the sequence approaches 0.

$$0 < a_n = \frac{1 \cdot 3 \cdot 5 \cdots (2n-1)}{(2n)^n} = \frac{1}{2n} \cdot \frac{3}{2n} \cdot \frac{5}{2n} \cdots \cdot \frac{2n-1}{2n}$$

$$\le \frac{1}{2n} \cdot (1) \cdot (1) \cdots \cdot (1) = \frac{1}{2n} \to 0 \text{ as } n \to \infty$$

So by the Squeeze Theorem, $\left\{ \dfrac{1 \cdot 3 \cdot 5 \cdots (2n-1)}{(2n)^n} \right\}$ converges to 0.

65. (a) $a_n = 1000(1.06)^n \quad \Rightarrow \quad a_1 = 1060, a_2 = 1123.60, a_3 = 1191.02, a_4 = 1262.48,$ and $a_5 = 1338.23.$

(b) $\lim\limits_{n \to \infty} a_n = 1000 \lim\limits_{n \to \infty} (1.06)^n$, so the sequence diverges by (9) with $r = 1.06 > 1$.

67. (a) We are given that the initial population is 5000, so $P_0 = 5000$. The number of catfish increases by 8% per month and is

decreased by 300 per month, so $P_1 = P_0 + 8\% P_0 - 300 = 1.08 P_0 - 300$, $P_2 = 1.08 P_1 - 300$, and so on. Thus,

$P_n = 1.08 P_{n-1} - 300$.

(b) Using the recursive formula with $P_0 = 5000$, we get $P_1 = 5100$, $P_2 = 5208$, $P_3 = 5325$ (rounding any portion of a

catfish), $P_4 = 5451$, $P_5 = 5587$, and $P_6 = 5734$, which is the number of catfish in the pond after six months.

69. If $|r| \ge 1$, then $\{r^n\}$ diverges by (9), so $\{nr^n\}$ diverges also, since $|nr^n| = n |r^n| \ge |r^n|$. If $|r| < 1$ then

$$\lim_{x \to \infty} xr^x = \lim_{x \to \infty} \frac{x}{r^{-x}} \overset{\text{H}}{=} \lim_{x \to \infty} \frac{1}{(-\ln r) r^{-x}} = \lim_{x \to \infty} \frac{r^x}{-\ln r} = 0, \text{ so } \lim_{n \to \infty} nr^n = 0, \text{ and hence } \{nr^n\} \text{ converges}$$

whenever $|r| < 1$.

© 2012 Cengage Learning. All Rights Reserved. May not be scanned, copied, or duplicated, or posted to a publicly accessible website, in whole or in part.

71. Since $\{a_n\}$ is a decreasing sequence, $a_n > a_{n+1}$ for all $n \geq 1$. Because all of its terms lie between 5 and 8, $\{a_n\}$ is a bounded sequence. By the Monotonic Sequence Theorem, $\{a_n\}$ is convergent; that is, $\{a_n\}$ has a limit L. L must be less than 8 since $\{a_n\}$ is decreasing, so $5 \leq L < 8$.

73. $a_n = \dfrac{1}{2n+3}$ is decreasing since $a_{n+1} = \dfrac{1}{2(n+1)+3} = \dfrac{1}{2n+5} < \dfrac{1}{2n+3} = a_n$ for each $n \geq 1$. The sequence is bounded since $0 < a_n \leq \frac{1}{5}$ for all $n \geq 1$. Note that $a_1 = \frac{1}{5}$.

75. The terms of $a_n = n(-1)^n$ alternate in sign, so the sequence is not monotonic. The first five terms are $-1, 2, -3, 4$, and -5. Since $\lim\limits_{n\to\infty} |a_n| = \lim\limits_{n\to\infty} n = \infty$, the sequence is not bounded.

77. $a_n = \dfrac{n}{n^2+1}$ defines a decreasing sequence since for $f(x) = \dfrac{x}{x^2+1}$, $f'(x) = \dfrac{(x^2+1)(1) - x(2x)}{(x^2+1)^2} = \dfrac{1-x^2}{(x^2+1)^2} \leq 0$ for $x \geq 1$. The sequence is bounded since $0 < a_n \leq \frac{1}{2}$ for all $n \geq 1$.

79. For $\left\{ \sqrt{2}, \sqrt{2\sqrt{2}}, \sqrt{2\sqrt{2\sqrt{2}}}, \ldots \right\}$, $a_1 = 2^{1/2}$, $a_2 = 2^{3/4}$, $a_3 = 2^{7/8}, \ldots$, so $a_n = 2^{(2^n-1)/2^n} = 2^{1-(1/2^n)}$.

$\lim\limits_{n\to\infty} a_n = \lim\limits_{n\to\infty} 2^{1-(1/2^n)} = 2^1 = 2$.

Alternate solution: Let $L = \lim\limits_{n\to\infty} a_n$. (We could show the limit exists by showing that $\{a_n\}$ is bounded and increasing.)

Then L must satisfy $L = \sqrt{2 \cdot L} \Rightarrow L^2 = 2L \Rightarrow L(L-2) = 0$. $L \neq 0$ since the sequence increases, so $L = 2$.

81. $a_1 = 1$, $a_{n+1} = 3 - \dfrac{1}{a_n}$. We show by induction that $\{a_n\}$ is increasing and bounded above by 3. Let P_n be the proposition that $a_{n+1} > a_n$ and $0 < a_n < 3$. Clearly P_1 is true. Assume that P_n is true. Then $a_{n+1} > a_n \Rightarrow \dfrac{1}{a_{n+1}} < \dfrac{1}{a_n} \Rightarrow$

$-\dfrac{1}{a_{n+1}} > -\dfrac{1}{a_n}$. Now $a_{n+2} = 3 - \dfrac{1}{a_{n+1}} > 3 - \dfrac{1}{a_n} = a_{n+1} \Leftrightarrow P_{n+1}$. This proves that $\{a_n\}$ is increasing and bounded above by 3, so $1 = a_1 < a_n < 3$, that is, $\{a_n\}$ is bounded, and hence convergent by the Monotonic Sequence Theorem.

If $L = \lim\limits_{n\to\infty} a_n$, then $\lim\limits_{n\to\infty} a_{n+1} = L$ also, so L must satisfy $L = 3 - 1/L \Rightarrow L^2 - 3L + 1 = 0 \Rightarrow L = \frac{3 \pm \sqrt{5}}{2}$.

But $L > 1$, so $L = \frac{3 + \sqrt{5}}{2}$.

83. (a) Let a_n be the number of rabbit pairs in the nth month. Clearly $a_1 = 1 = a_2$. In the nth month, each pair that is 2 or more months old (that is, a_{n-2} pairs) will produce a new pair to add to the a_{n-1} pairs already present. Thus, $a_n = a_{n-1} + a_{n-2}$, so that $\{a_n\} = \{f_n\}$, the Fibonacci sequence.

(b) $a_n = \dfrac{f_{n+1}}{f_n} \Rightarrow a_{n-1} = \dfrac{f_n}{f_{n-1}} = \dfrac{f_{n-1} + f_{n-2}}{f_{n-1}} = 1 + \dfrac{f_{n-2}}{f_{n-1}} = 1 + \dfrac{1}{f_{n-1}/f_{n-2}} = 1 + \dfrac{1}{a_{n-2}}$. If $L = \lim\limits_{n\to\infty} a_n$,

then $L = \lim\limits_{n\to\infty} a_{n-1}$ and $L = \lim\limits_{n\to\infty} a_{n-2}$, so L must satisfy $L = 1 + \dfrac{1}{L} \Rightarrow L^2 - L - 1 = 0 \Rightarrow L = \frac{1+\sqrt{5}}{2}$

[since L must be positive].

© 2012 Cengage Learning. All Rights Reserved. May not be scanned, copied, or duplicated, or posted to a publicly accessible website, in whole or in part.

85. (a)

From the graph, it appears that the sequence $\left\{ \dfrac{n^5}{n!} \right\}$

converges to 0, that is, $\displaystyle\lim_{n\to\infty} \dfrac{n^5}{n!} = 0.$

(b)

From the first graph, it seems that the smallest possible value of N corresponding to $\varepsilon = 0.1$ is 9, since $n^5/n! < 0.1$

whenever $n \geq 10$, but $9^5/9! > 0.1$. From the second graph, it seems that for $\varepsilon = 0.001$, the smallest possible value for N

is 11 since $n^5/n! < 0.001$ whenever $n \geq 12$.

87. Theorem 6: If $\displaystyle\lim_{n\to\infty} |a_n| = 0$ then $\displaystyle\lim_{n\to\infty} -|a_n| = 0$, and since $-|a_n| \leq a_n \leq |a_n|$, we have that $\displaystyle\lim_{n\to\infty} a_n = 0$ by the

Squeeze Theorem.

89. To Prove: If $\displaystyle\lim_{n\to\infty} a_n = 0$ and $\{b_n\}$ is bounded, then $\displaystyle\lim_{n\to\infty} (a_n b_n) = 0.$

Proof: Since $\{b_n\}$ is bounded, there is a positive number M such that $|b_n| \leq M$ and hence, $|a_n||b_n| \leq |a_n| M$ for

all $n \geq 1$. Let $\varepsilon > 0$ be given. Since $\displaystyle\lim_{n\to\infty} a_n = 0$, there is an integer N such that $|a_n - 0| < \dfrac{\varepsilon}{M}$ if $n > N$. Then

$|a_n b_n - 0| = |a_n b_n| = |a_n||b_n| \leq |a_n| M = |a_n - 0| M < \dfrac{\varepsilon}{M} \cdot M = \varepsilon$ for all $n > N$. Since ε was arbitrary,

$\displaystyle\lim_{n\to\infty} (a_n b_n) = 0.$

91. (a) First we show that $a > a_1 > b_1 > b$.

$a_1 - b_1 = \frac{a+b}{2} - \sqrt{ab} = \frac{1}{2}\left(a - 2\sqrt{ab} + b\right) = \frac{1}{2}\left(\sqrt{a} - \sqrt{b}\right)^2 > 0$ [since $a > b$] $\Rightarrow$ $a_1 > b_1$. Also

$a - a_1 = a - \frac{1}{2}(a+b) = \frac{1}{2}(a-b) > 0$ and $b - b_1 = b - \sqrt{ab} = \sqrt{b}\left(\sqrt{b} - \sqrt{a}\right) < 0$, so $a > a_1 > b_1 > b$. In the same

way we can show that $a_1 > a_2 > b_2 > b_1$ and so the given assertion is true for $n = 1$. Suppose it is true for $n = k$, that is,

$a_k > a_{k+1} > b_{k+1} > b_k$. Then

© 2012 Cengage Learning. All Rights Reserved. May not be scanned, copied, or duplicated, or posted to a publicly accessible website, in whole or in part.

$$a_{k+2} - b_{k+2} = \tfrac{1}{2}(a_{k+1} + b_{k+1}) - \sqrt{a_{k+1}b_{k+1}} = \tfrac{1}{2}\left(a_{k+1} - 2\sqrt{a_{k+1}b_{k+1}} + b_{k+1}\right) = \tfrac{1}{2}\left(\sqrt{a_{k+1}} - \sqrt{b_{k+1}}\right)^2 > 0,$$

$$a_{k+1} - a_{k+2} = a_{k+1} - \tfrac{1}{2}(a_{k+1} + b_{k+1}) = \tfrac{1}{2}(a_{k+1} - b_{k+1}) > 0, \text{ and}$$

$$b_{k+1} - b_{k+2} = b_{k+1} - \sqrt{a_{k+1}b_{k+1}} = \sqrt{b_{k+1}}\left(\sqrt{b_{k+1}} - \sqrt{a_{k+1}}\right) < 0 \quad \Rightarrow \quad a_{k+1} > a_{k+2} > b_{k+2} > b_{k+1},$$

so the assertion is true for $n = k + 1$. Thus, it is true for all n by mathematical induction.

(b) From part (a) we have $a > a_n > a_{n+1} > b_{n+1} > b_n > b$, which shows that both sequences, $\{a_n\}$ and $\{b_n\}$, are monotonic and bounded. So they are both convergent by the Monotonic Sequence Theorem.

(c) Let $\lim\limits_{n \to \infty} a_n = \alpha$ and $\lim\limits_{n \to \infty} b_n = \beta$. Then $\lim\limits_{n \to \infty} a_{n+1} = \lim\limits_{n \to \infty} \dfrac{a_n + b_n}{2} \quad \Rightarrow \quad \alpha = \dfrac{\alpha + \beta}{2} \quad \Rightarrow$

$2\alpha = \alpha + \beta \quad \Rightarrow \quad \alpha = \beta.$

93. (a) Suppose $\{p_n\}$ converges to p. Then $p_{n+1} = \dfrac{bp_n}{a + p_n} \quad \Rightarrow \quad \lim\limits_{n \to \infty} p_{n+1} = \dfrac{b \lim\limits_{n \to \infty} p_n}{a + \lim\limits_{n \to \infty} p_n} \quad \Rightarrow \quad p = \dfrac{bp}{a + p} \quad \Rightarrow$

$p^2 + ap = bp \quad \Rightarrow \quad p(p + a - b) = 0 \quad \Rightarrow \quad p = 0 \text{ or } p = b - a.$

(b) $p_{n+1} = \dfrac{bp_n}{a + p_n} = \dfrac{\left(\dfrac{b}{a}\right)p_n}{1 + \dfrac{p_n}{a}} < \left(\dfrac{b}{a}\right)p_n$ since $1 + \dfrac{p_n}{a} > 1$.

(c) By part (b), $p_1 < \left(\dfrac{b}{a}\right)p_0$, $p_2 < \left(\dfrac{b}{a}\right)p_1 < \left(\dfrac{b}{a}\right)^2 p_0$, $p_3 < \left(\dfrac{b}{a}\right)p_2 < \left(\dfrac{b}{a}\right)^3 p_0$, etc. In general, $p_n < \left(\dfrac{b}{a}\right)^n p_0$,

so $\lim\limits_{n \to \infty} p_n \leq \lim\limits_{n \to \infty} \left(\dfrac{b}{a}\right)^n \cdot p_0 = 0$ since $b < a$. $\left[\text{By (7), } \lim\limits_{n \to \infty} r^n = 0 \text{ if } -1 < r < 1. \text{ Here } r = \dfrac{b}{a} \in (0, 1).\right]$

(d) Let $a < b$. We first show, by induction, that if $p_0 < b - a$, then $p_n < b - a$ and $p_{n+1} > p_n$.

For $n = 0$, we have $p_1 - p_0 = \dfrac{bp_0}{a + p_0} - p_0 = \dfrac{p_0(b - a - p_0)}{a + p_0} > 0$ since $p_0 < b - a$. So $p_1 > p_0$.

Now we suppose the assertion is true for $n = k$, that is, $p_k < b - a$ and $p_{k+1} > p_k$. Then

$b - a - p_{k+1} = b - a - \dfrac{bp_k}{a + p_k} = \dfrac{a(b - a) + bp_k - ap_k - bp_k}{a + p_k} = \dfrac{a(b - a - p_k)}{a + p_k} > 0$ because $p_k < b - a$. So

$p_{k+1} < b - a$. And $p_{k+2} - p_{k+1} = \dfrac{bp_{k+1}}{a + p_{k+1}} - p_{k+1} = \dfrac{p_{k+1}(b - a - p_{k+1})}{a + p_{k+1}} > 0$ since $p_{k+1} < b - a$. Therefore,

$p_{k+2} > p_{k+1}$. Thus, the assertion is true for $n = k + 1$. It is therefore true for all n by mathematical induction.

A similar proof by induction shows that if $p_0 > b - a$, then $p_n > b - a$ and $\{p_n\}$ is decreasing.

In either case the sequence $\{p_n\}$ is bounded and monotonic, so it is convergent by the Monotonic Sequence Theorem.

It then follows from part (a) that $\lim\limits_{n \to \infty} p_n = b - a.$

11.2   Series

1. (a) A sequence is an ordered list of numbers whereas a series is the *sum* of a list of numbers.

(b) A series is convergent if the sequence of partial sums is a convergent sequence. A series is divergent if it is not convergent.

© 2012 Cengage Learning. All Rights Reserved. May not be scanned, copied, or duplicated, or posted to a publicly accessible website, in whole or in part.

3. $\sum\limits_{n=1}^{\infty} a_n = \lim\limits_{n\to\infty} s_n = \lim\limits_{n\to\infty} [2 - 3(0.8)^n] = \lim\limits_{n\to\infty} 2 - 3 \lim\limits_{n\to\infty} (0.8)^n = 2 - 3(0) = 2$

5. For $\sum\limits_{n=1}^{\infty} \dfrac{1}{n^3}, a_n = \dfrac{1}{n^3}$. $s_1 = a_1 = \dfrac{1}{1^3} = 1$, $s_2 = s_1 + a_2 = 1 + \dfrac{1}{2^3} = 1.125$, $s_3 = s_2 + a_3 \approx 1.1620$,

$s_4 = s_3 + a_4 \approx 1.1777$, $s_5 = s_4 + a_5 \approx 1.1857$, $s_6 = s_5 + a_6 \approx 1.1903$, $s_7 = s_6 + a_7 \approx 1.1932$, and

$s_8 = s_7 + a_8 \approx 1.1952$. It appears that the series is convergent.

7. For $\sum\limits_{n=1}^{\infty} \dfrac{n}{1 + \sqrt{n}}, a_n = \dfrac{n}{1 + \sqrt{n}}$. $s_1 = a_1 = \dfrac{1}{1 + \sqrt{1}} = 0.5$, $s_2 = s_1 + a_2 = 0.5 + \dfrac{2}{1 + \sqrt{2}} \approx 1.3284$,

$s_3 = s_2 + a_3 \approx 2.4265$, $s_4 = s_3 + a_4 \approx 3.7598$, $s_5 = s_4 + a_5 \approx 5.3049$, $s_6 = s_5 + a_6 \approx 7.0443$,

$s_7 = s_6 + a_7 \approx 8.9644$, $s_8 = s_7 + a_8 \approx 11.0540$. It appears that the series is divergent.

9.

n	s_n
1	-2.40000
2	-1.92000
3	-2.01600
4	-1.99680
5	-2.00064
6	-1.99987
7	-2.00003
8	-1.99999
9	-2.00000
10	-2.00000

From the graph and the table, it seems that the series converges to -2. In fact, it is a geometric

series with $a = -2.4$ and $r = -\frac{1}{5}$, so its sum is $\sum\limits_{n=1}^{\infty} \dfrac{12}{(-5)^n} = \dfrac{-2.4}{1 - \left(-\frac{1}{5}\right)} = \dfrac{-2.4}{1.2} = -2$.

Note that the dot corresponding to $n = 1$ is part of both $\{a_n\}$ and $\{s_n\}$.

TI-86 Note: To graph $\{a_n\}$ and $\{s_n\}$, set your calculator to Param mode and DrawDot mode. (DrawDot is under

GRAPH, MORE, FORMT (F3).) Now under E (t) = make the assignments: xt1=t, yt1=12/(-5)^t, xt2=t,

yt2=sum seq(yt1,t,1,t,1). (sum and seq are under LIST, OPS (F5), MORE.) Under WIND use

1,10,1,0,10,1,-3,1,1 to obtain a graph similar to the one above. Then use TRACE (F4) to see the values.

11.

n	s_n
1	0.44721
2	1.15432
3	1.98637
4	2.88080
5	3.80927
6	4.75796
7	5.71948
8	6.68962
9	7.66581
10	8.64639

The series $\sum\limits_{n=1}^{\infty} \dfrac{n}{\sqrt{n^2 + 4}}$ diverges, since its terms do not approach 0.

© 2012 Cengage Learning. All Rights Reserved. May not be scanned, copied, or duplicated, or posted to a publicly accessible website, in whole or in part.

13.

n	s_n
1	0.29289
2	0.42265
3	0.50000
4	0.55279
5	0.59175
6	0.62204
7	0.64645
8	0.66667
9	0.68377
10	0.69849

From the graph and the table, it seems that the series converges.

$$\sum_{n=1}^{k}\left(\frac{1}{\sqrt{n}}-\frac{1}{\sqrt{n+1}}\right)=\left(\frac{1}{\sqrt{1}}-\frac{1}{\sqrt{2}}\right)+\left(\frac{1}{\sqrt{2}}-\frac{1}{\sqrt{3}}\right)+\cdots+\left(\frac{1}{\sqrt{k}}-\frac{1}{\sqrt{k+1}}\right)$$

$$=1-\frac{1}{\sqrt{k+1}},$$

so $\displaystyle\sum_{n=1}^{\infty}\left(\frac{1}{\sqrt{n}}-\frac{1}{\sqrt{n+1}}\right)=\lim_{k\to\infty}\left(1-\frac{1}{\sqrt{k+1}}\right)=1.$

15. (a) $\displaystyle\lim_{n\to\infty}a_n=\lim_{n\to\infty}\frac{2n}{3n+1}=\frac{2}{3}$, so the *sequence* $\{a_n\}$ is convergent by (11.1.1).

(b) Since $\displaystyle\lim_{n\to\infty}a_n=\frac{2}{3}\neq 0$, the *series* $\displaystyle\sum_{n=1}^{\infty}a_n$ is divergent by the Test for Divergence.

17. $3-4+\frac{16}{3}-\frac{64}{9}+\cdots$ is a geometric series with ratio $r=-\frac{4}{3}$. Since $|r|=\frac{4}{3}>1$, the series diverges.

19. $10-2+0.4-0.08+\cdots$ is a geometric series with ratio $-\frac{2}{10}=-\frac{1}{5}$. Since $|r|=\frac{1}{5}<1$, the series converges to

$$\frac{a}{1-r}=\frac{10}{1-(-1/5)}=\frac{10}{6/5}=\frac{50}{6}=\frac{25}{3}.$$

21. $\displaystyle\sum_{n=1}^{\infty}6(0.9)^{n-1}$ is a geometric series with first term $a=6$ and ratio $r=0.9$. Since $|r|=0.9<1$, the series converges to

$$\frac{a}{1-r}=\frac{6}{1-0.9}=\frac{6}{0.1}=60.$$

23. $\displaystyle\sum_{n=1}^{\infty}\frac{(-3)^{n-1}}{4^n}=\frac{1}{4}\sum_{n=1}^{\infty}\left(-\frac{3}{4}\right)^{n-1}$. The latter series is geometric with $a=1$ and ratio $r=-\frac{3}{4}$. Since $|r|=\frac{3}{4}<1$, it

converges to $\dfrac{1}{1-(-3/4)}=\frac{4}{7}$. Thus, the given series converges to $\left(\frac{1}{4}\right)\left(\frac{4}{7}\right)=\frac{1}{7}$.

25. $\displaystyle\sum_{n=0}^{\infty}\frac{\pi^n}{3^{n+1}}=\frac{1}{3}\sum_{n=0}^{\infty}\left(\frac{\pi}{3}\right)^n$ is a geometric series with ratio $r=\frac{\pi}{3}$. Since $|r|>1$, the series diverges.

27. $\frac{1}{3}+\frac{1}{6}+\frac{1}{9}+\frac{1}{12}+\frac{1}{15}+\cdots=\displaystyle\sum_{n=1}^{\infty}\frac{1}{3n}=\frac{1}{3}\sum_{n=1}^{\infty}\frac{1}{n}$. This is a constant multiple of the divergent harmonic series, so

it diverges.

29. $\displaystyle\sum_{n=1}^{\infty}\frac{n-1}{3n-1}$ diverges by the Test for Divergence since $\displaystyle\lim_{n\to\infty}a_n=\lim_{n\to\infty}\frac{n-1}{3n-1}=\frac{1}{3}\neq 0.$

© 2012 Cengage Learning. All Rights Reserved. May not be scanned, copied, or duplicated, or posted to a publicly accessible website, in whole or in part.

31. Converges.

$$\sum_{n=1}^{\infty} \frac{1+2^n}{3^n} = \sum_{n=1}^{\infty} \left(\frac{1}{3^n} + \frac{2^n}{3^n} \right) = \sum_{n=1}^{\infty} \left[\left(\frac{1}{3} \right)^n + \left(\frac{2}{3} \right)^n \right] \qquad \text{[sum of two convergent geometric series]}$$

$$= \frac{1/3}{1-1/3} + \frac{2/3}{1-2/3} = \frac{1}{2} + 2 = \frac{5}{2}$$

33. $\sum_{n=1}^{\infty} \sqrt[n]{2} = 2 + \sqrt{2} + \sqrt[3]{2} + \sqrt[4]{2} + \cdots$ diverges by the Test for Divergence since

$$\lim_{n \to \infty} a_n = \lim_{n \to \infty} \sqrt[n]{2} = \lim_{n \to \infty} 2^{1/n} = 2^0 = 1 \neq 0.$$

35. $\sum_{n=1}^{\infty} \ln \left(\frac{n^2+1}{2n^2+1} \right)$ diverges by the Test for Divergence since

$$\lim_{n \to \infty} a_n = \lim_{n \to \infty} \ln \left(\frac{n^2+1}{2n^2+1} \right) = \ln \left(\lim_{n \to \infty} \frac{n^2+1}{2n^2+1} \right) = \ln \tfrac{1}{2} \neq 0.$$

37. $\sum_{k=0}^{\infty} \left(\frac{\pi}{3} \right)^k$ is a geometric series with ratio $r = \frac{\pi}{3} \approx 1.047$. It diverges because $|r| \geq 1$.

39. $\sum_{n=1}^{\infty} \arctan n$ diverges by the Test for Divergence since $\lim_{n \to \infty} a_n = \lim_{n \to \infty} \arctan n = \frac{\pi}{2} \neq 0.$

41. $\sum_{n=1}^{\infty} \frac{1}{e^n} = \sum_{n=1}^{\infty} \left(\frac{1}{e} \right)^n$ is a geometric series with first term $a = \frac{1}{e}$ and ratio $r = \frac{1}{e}$. Since $|r| = \frac{1}{e} < 1$, the series converges

to $\dfrac{1/e}{1-1/e} = \dfrac{1/e}{1-1/e} \cdot \dfrac{e}{e} = \dfrac{1}{e-1}$. By Example 7, $\sum_{n=1}^{\infty} \dfrac{1}{n(n+1)} = 1$. Thus, by Theorem 8(ii),

$$\sum_{n=1}^{\infty} \left(\frac{1}{e^n} + \frac{1}{n(n+1)} \right) = \sum_{n=1}^{\infty} \frac{1}{e^n} + \sum_{n=1}^{\infty} \frac{1}{n(n+1)} = \frac{1}{e-1} + 1 = \frac{1}{e-1} + \frac{e-1}{e-1} = \frac{e}{e-1}.$$

43. Using partial fractions, the partial sums of the series $\sum_{n=2}^{\infty} \dfrac{2}{n^2-1}$ are

$$s_n = \sum_{i=2}^{n} \frac{2}{(i-1)(i+1)} = \sum_{i=2}^{n} \left(\frac{1}{i-1} - \frac{1}{i+1} \right)$$

$$= \left(1 - \frac{1}{3} \right) + \left(\frac{1}{2} - \frac{1}{4} \right) + \left(\frac{1}{3} - \frac{1}{5} \right) + \cdots + \left(\frac{1}{n-3} - \frac{1}{n-1} \right) + \left(\frac{1}{n-2} - \frac{1}{n} \right)$$

This sum is a telescoping series and $s_n = 1 + \dfrac{1}{2} - \dfrac{1}{n-1} - \dfrac{1}{n}.$

Thus, $\sum_{n=2}^{\infty} \dfrac{2}{n^2-1} = \lim_{n \to \infty} s_n = \lim_{n \to \infty} \left(1 + \dfrac{1}{2} - \dfrac{1}{n-1} - \dfrac{1}{n} \right) = \dfrac{3}{2}.$

45. For the series $\sum_{n=1}^{\infty} \dfrac{3}{n(n+3)}$, $s_n = \sum_{i=1}^{n} \dfrac{3}{i(i+3)} = \sum_{i=1}^{n} \left(\dfrac{1}{i} - \dfrac{1}{i+3} \right)$ [using partial fractions]. The latter sum is

$$\left(1 - \tfrac{1}{4}\right) + \left(\tfrac{1}{2} - \tfrac{1}{5}\right) + \left(\tfrac{1}{3} - \tfrac{1}{6}\right) + \left(\tfrac{1}{4} - \tfrac{1}{7}\right) + \cdots + \left(\tfrac{1}{n-3} - \tfrac{1}{n}\right) + \left(\tfrac{1}{n-2} - \tfrac{1}{n+1}\right) + \left(\tfrac{1}{n-1} - \tfrac{1}{n+2}\right) + \left(\tfrac{1}{n} - \tfrac{1}{n+3}\right)$$

$$= 1 + \tfrac{1}{2} + \tfrac{1}{3} - \tfrac{1}{n+1} - \tfrac{1}{n+2} - \tfrac{1}{n+3} \qquad \text{[telescoping series]}$$

Thus, $\sum_{n=1}^{\infty} \dfrac{3}{n(n+3)} = \lim_{n \to \infty} s_n = \lim_{n \to \infty} \left(1 + \tfrac{1}{2} + \tfrac{1}{3} - \tfrac{1}{n+1} - \tfrac{1}{n+2} - \tfrac{1}{n+3} \right) = 1 + \tfrac{1}{2} + \tfrac{1}{3} = \tfrac{11}{6}.$ Converges

© 2012 Cengage Learning. All Rights Reserved. May not be scanned, copied, or duplicated, or posted to a publicly accessible website, in whole or in part.

47. For the series $\sum_{n=1}^{\infty}\left(e^{1/n}-e^{1/(n+1)}\right)$,

$$s_n = \sum_{i=1}^{n}\left(e^{1/i}-e^{1/(i+1)}\right) = \left(e^1 - e^{1/2}\right) + \left(e^{1/2}-e^{1/3}\right) + \cdots + \left(e^{1/n}-e^{1/(n+1)}\right) = e - e^{1/(n+1)}$$

[telescoping series]

Thus, $\sum_{n=1}^{\infty}\left(e^{1/n}-e^{1/(n+1)}\right) = \lim_{n\to\infty} s_n = \lim_{n\to\infty}\left(e - e^{1/(n+1)}\right) = e - e^0 = e - 1$.   Converges

49. (a) Many people would guess that $x < 1$, but note that x consists of an infinite number of 9s.

(b) $x = 0.99999\ldots = \dfrac{9}{10} + \dfrac{9}{100} + \dfrac{9}{1000} + \dfrac{9}{10,000} + \cdots = \sum_{n=1}^{\infty}\dfrac{9}{10^n}$, which is a geometric series with $a_1 = 0.9$ and

$r = 0.1$. Its sum is $\dfrac{0.9}{1-0.1} = \dfrac{0.9}{0.9} = 1$, that is, $x = 1$.

(c) The number 1 has two decimal representations, $1.00000\ldots$ and $0.99999\ldots$.

(d) Except for 0, all rational numbers that have a terminating decimal representation can be written in more than one way. For

example, 0.5 can be written as $0.49999\ldots$ as well as $0.50000\ldots$.

51. $0.\overline{8} = \dfrac{8}{10} + \dfrac{8}{10^2} + \cdots$ is a geometric series with $a = \dfrac{8}{10}$ and $r = \dfrac{1}{10}$. It converges to $\dfrac{a}{1-r} = \dfrac{8/10}{1-1/10} = \dfrac{8}{9}$.

53. $2.\overline{516} = 2 + \dfrac{516}{10^3} + \dfrac{516}{10^6} + \cdots$. Now $\dfrac{516}{10^3} + \dfrac{516}{10^6} + \cdots$ is a geometric series with $a = \dfrac{516}{10^3}$ and $r = \dfrac{1}{10^3}$. It converges to

$\dfrac{a}{1-r} = \dfrac{516/10^3}{1-1/10^3} = \dfrac{516/10^3}{999/10^3} = \dfrac{516}{999}$. Thus, $2.\overline{516} = 2 + \dfrac{516}{999} = \dfrac{2514}{999} = \dfrac{838}{333}$.

55. $1.5\overline{342} = 1.53 + \dfrac{42}{10^4} + \dfrac{42}{10^6} + \cdots$. Now $\dfrac{42}{10^4} + \dfrac{42}{10^6} + \cdots$ is a geometric series with $a = \dfrac{42}{10^4}$ and $r = \dfrac{1}{10^2}$.

It converges to $\dfrac{a}{1-r} = \dfrac{42/10^4}{1-1/10^2} = \dfrac{42/10^4}{99/10^2} = \dfrac{42}{9900}$.

Thus, $1.5\overline{342} = 1.53 + \dfrac{42}{9900} = \dfrac{153}{100} + \dfrac{42}{9900} = \dfrac{15,147}{9900} + \dfrac{42}{9900} = \dfrac{15,189}{9900}$ or $\dfrac{5063}{3300}$.

57. $\sum_{n=1}^{\infty}(-5)^n x^n = \sum_{n=1}^{\infty}(-5x)^n$ is a geometric series with $r = -5x$, so the series converges $\Leftrightarrow$ $|r| < 1$ $\Leftrightarrow$

$|-5x| < 1$ $\Leftrightarrow$ $|x| < \frac{1}{5}$, that is, $-\frac{1}{5} < x < \frac{1}{5}$. In that case, the sum of the series is $\dfrac{a}{1-r} = \dfrac{-5x}{1-(-5x)} = \dfrac{-5x}{1+5x}$.

59. $\sum_{n=0}^{\infty}\dfrac{(x-2)^n}{3^n} = \sum_{n=0}^{\infty}\left(\dfrac{x-2}{3}\right)^n$ is a geometric series with $r = \dfrac{x-2}{3}$, so the series converges $\Leftrightarrow$ $|r| < 1$ $\Leftrightarrow$

$\left|\dfrac{x-2}{3}\right| < 1$ $\Leftrightarrow$ $-1 < \dfrac{x-2}{3} < 1$ $\Leftrightarrow$ $-3 < x - 2 < 3$ $\Leftrightarrow$ $-1 < x < 5$. In that case, the sum of the series is

$\dfrac{a}{1-r} = \dfrac{1}{1-\dfrac{x-2}{3}} = \dfrac{1}{\dfrac{3-(x-2)}{3}} = \dfrac{3}{5-x}$.

61. $\sum_{n=0}^{\infty}\dfrac{2^n}{x^n} = \sum_{n=0}^{\infty}\left(\dfrac{2}{x}\right)^n$ is a geometric series with $r = \dfrac{2}{x}$, so the series converges $\Leftrightarrow$ $|r| < 1$ $\Leftrightarrow$ $\left|\dfrac{2}{x}\right| < 1$ $\Leftrightarrow$

$2 < |x|$ $\Leftrightarrow$ $x > 2$ or $x < -2$. In that case, the sum of the series is $\dfrac{a}{1-r} = \dfrac{1}{1-2/x} = \dfrac{x}{x-2}$.

© 2012 Cengage Learning. All Rights Reserved. May not be scanned, copied, or duplicated, or posted to a publicly accessible website, in whole or in part.

63. $\sum_{n=0}^{\infty} e^{nx} = \sum_{n=0}^{\infty} (e^x)^n$ is a geometric series with $r = e^x$, so the series converges $\Leftrightarrow |r| < 1 \Leftrightarrow |e^x| < 1 \Leftrightarrow$

$-1 < e^x < 1 \Leftrightarrow 0 < e^x < 1 \Leftrightarrow x < 0$. In that case, the sum of the series is $\dfrac{a}{1-r} = \dfrac{1}{1-e^x}$.

65. After defining f, We use `convert(f,parfrac);` in Maple, `Apart` in Mathematica, or `Expand Rational` and

`Simplify` in Derive to find that the general term is $\dfrac{3n^2 + 3n + 1}{(n^2 + n)^3} = \dfrac{1}{n^3} - \dfrac{1}{(n+1)^3}$. So the nth partial sum is

$$s_n = \sum_{k=1}^{n} \left(\frac{1}{k^3} - \frac{1}{(k+1)^3} \right) = \left(1 - \frac{1}{2^3} \right) + \left(\frac{1}{2^3} - \frac{1}{3^3} \right) + \cdots + \left(\frac{1}{n^3} - \frac{1}{(n+1)^3} \right) = 1 - \frac{1}{(n+1)^3}$$

The series converges to $\lim\limits_{n\to\infty} s_n = 1$. This can be confirmed by directly computing the sum using

`sum(f,n=1..infinity);` (in Maple), `Sum[f,{n,1,Infinity}]` (in Mathematica), or `Calculus Sum`

(from 1 to ∞) and `Simplify` (in Derive).

67. For $n = 1$, $a_1 = 0$ since $s_1 = 0$. For $n > 1$,

$$a_n = s_n - s_{n-1} = \frac{n-1}{n+1} - \frac{(n-1)-1}{(n-1)+1} = \frac{(n-1)n - (n+1)(n-2)}{(n+1)n} = \frac{2}{n(n+1)}$$

Also, $\sum\limits_{n=1}^{\infty} a_n = \lim\limits_{n\to\infty} s_n = \lim\limits_{n\to\infty} \dfrac{1 - 1/n}{1 + 1/n} = 1$.

69. (a) The quantity of the drug in the body after the first tablet is 150 mg. After the second tablet, there is 150 mg plus 5%

of the first 150- mg tablet, that is, $[150 + 150(0.05)]$ mg. After the third tablet, the quantity is

$[150 + 150(0.05) + 150(0.05)^2] = 157.875$ mg. After n tablets, the quantity (in mg) is

$150 + 150(0.05) + \cdots + 150(0.05)^{n-1}$. We can use Formula 3 to write this as $\dfrac{150(1 - 0.05^n)}{1 - 0.05} = \dfrac{3000}{19}(1 - 0.05^n)$.

(b) The number of milligrams remaining in the body in the long run is $\lim\limits_{n\to\infty} \left[\frac{3000}{19}(1 - 0.05^n) \right] = \frac{3000}{19}(1 - 0) \approx 157.895$,

only 0.02 mg more than the amount after 3 tablets.

71. (a) The first step in the chain occurs when the local government spends D dollars. The people who receive it spend a

fraction c of those D dollars, that is, Dc dollars. Those who receive the Dc dollars spend a fraction c of it, that is,

Dc^2 dollars. Continuing in this way, we see that the total spending after n transactions is

$S_n = D + Dc + Dc^2 + \cdots + Dc^{n-1} = \dfrac{D(1-c^n)}{1-c}$ by (3).

(b) $\lim\limits_{n\to\infty} S_n = \lim\limits_{n\to\infty} \dfrac{D(1-c^n)}{1-c} = \dfrac{D}{1-c} \lim\limits_{n\to\infty} (1 - c^n) = \dfrac{D}{1-c}$ $\left[\text{since } 0 < c < 1 \Rightarrow \lim\limits_{n\to\infty} c^n = 0 \right]$

$= \dfrac{D}{s}$ [since $c + s = 1$] $= kD$ [since $k = 1/s$]

If $c = 0.8$, then $s = 1 - c = 0.2$ and the multiplier is $k = 1/s = 5$.

73. $\sum\limits_{n=2}^{\infty} (1+c)^{-n}$ is a geometric series with $a = (1+c)^{-2}$ and $r = (1+c)^{-1}$, so the series converges when

$|(1+c)^{-1}| < 1 \Leftrightarrow |1+c| > 1 \Leftrightarrow 1+c > 1$ or $1 + c < -1 \Leftrightarrow c > 0$ or $c < -2$. We calculate the sum of the

© 2012 Cengage Learning. All Rights Reserved. May not be scanned, copied, or duplicated, or posted to a publicly accessible website, in whole or in part.

series and set it equal to 2: $\dfrac{(1+c)^{-2}}{1-(1+c)^{-1}} = 2 \ \Leftrightarrow \ \left(\dfrac{1}{1+c}\right)^2 = 2 - 2\left(\dfrac{1}{1+c}\right) \ \Leftrightarrow \ 1 = 2(1+c)^2 - 2(1+c) \ \Leftrightarrow$

$2c^2 + 2c - 1 = 0 \ \Leftrightarrow \ c = \dfrac{-2 \pm \sqrt{12}}{4} = \dfrac{\pm\sqrt{3} - 1}{2}$. However, the negative root is inadmissible because $-2 < \dfrac{-\sqrt{3}-1}{2} < 0$.

So $c = \dfrac{\sqrt{3}-1}{2}$.

75. $e^{s_n} = e^{1+\frac{1}{2}+\frac{1}{3}+\cdots+\frac{1}{n}} = e^1 e^{1/2} e^{1/3} \cdots e^{1/n} > (1+1)\left(1+\tfrac{1}{2}\right)\left(1+\tfrac{1}{3}\right)\cdots\left(1+\tfrac{1}{n}\right) \qquad [e^x > 1 + x]$

$\qquad = \dfrac{2}{1}\dfrac{3}{2}\dfrac{4}{3}\cdots\dfrac{n+1}{n} = n+1$

Thus, $e^{s_n} > n+1$ and $\displaystyle\lim_{n\to\infty} e^{s_n} = \infty$. Since $\{s_n\}$ is increasing, $\displaystyle\lim_{n\to\infty} s_n = \infty$, implying that the harmonic series is

divergent.

77. Let d_n be the diameter of C_n. We draw lines from the centers of the C_i to

the center of D (or C), and using the Pythagorean Theorem, we can write

$1^2 + \left(1 - \tfrac{1}{2}d_1\right)^2 = \left(1 + \tfrac{1}{2}d_1\right)^2 \ \Leftrightarrow$

$1 = \left(1 + \tfrac{1}{2}d_1\right)^2 - \left(1 - \tfrac{1}{2}d_1\right)^2 = 2d_1$ [difference of squares] $\Rightarrow \ d_1 = \tfrac{1}{2}$.

Similarly,

$1 = \left(1 + \tfrac{1}{2}d_2\right)^2 - \left(1 - d_1 - \tfrac{1}{2}d_2\right)^2 = 2d_2 + 2d_1 - d_1^2 - d_1 d_2$

$\quad = (2 - d_1)(d_1 + d_2) \ \Leftrightarrow$

$d_2 = \dfrac{1}{2 - d_1} - d_1 = \dfrac{(1 - d_1)^2}{2 - d_1}$, $1 = \left(1 + \tfrac{1}{2}d_3\right)^2 - \left(1 - d_1 - d_2 - \tfrac{1}{2}d_3\right)^2 \ \Leftrightarrow \ d_3 = \dfrac{[1 - (d_1 + d_2)]^2}{2 - (d_1 + d_2)}$, and in general,

$d_{n+1} = \dfrac{\left(1 - \sum_{i=1}^{n} d_i\right)^2}{2 - \sum_{i=1}^{n} d_i}$. If we actually calculate d_2 and d_3 from the formulas above, we find that they are $\dfrac{1}{6} = \dfrac{1}{2\cdot 3}$ and

$\dfrac{1}{12} = \dfrac{1}{3\cdot 4}$ respectively, so we suspect that in general, $d_n = \dfrac{1}{n(n+1)}$. To prove this, we use induction: Assume that for all

$k \leq n,\ d_k = \dfrac{1}{k(k+1)} = \dfrac{1}{k} - \dfrac{1}{k+1}$. Then $\displaystyle\sum_{i=1}^{n} d_i = 1 - \dfrac{1}{n+1} = \dfrac{n}{n+1}$ [telescoping sum]. Substituting this into our

formula for d_{n+1}, we get $d_{n+1} = \dfrac{\left[1 - \dfrac{n}{n+1}\right]^2}{2 - \left(\dfrac{n}{n+1}\right)} = \dfrac{\dfrac{1}{(n+1)^2}}{\dfrac{n+2}{n+1}} = \dfrac{1}{(n+1)(n+2)}$, and the induction is complete.

Now, we observe that the partial sums $\sum_{i=1}^{n} d_i$ of the diameters of the circles approach 1 as $n \to \infty$; that is,

$\displaystyle\sum_{n=1}^{\infty} a_n = \sum_{n=1}^{\infty} \dfrac{1}{n(n+1)} = 1$, which is what we wanted to prove.

79. The series $1 - 1 + 1 - 1 + 1 - 1 + \cdots$ diverges (geometric series with $r = -1$) so we cannot say that

$0 = 1 - 1 + 1 - 1 + 1 - 1 + \cdots$.

81. $\sum_{n=1}^{\infty} ca_n = \displaystyle\lim_{n\to\infty} \sum_{i=1}^{n} ca_i = \lim_{n\to\infty} c\sum_{i=1}^{n} a_i = c\lim_{n\to\infty} \sum_{i=1}^{n} a_i = c\sum_{n=1}^{\infty} a_n$, which exists by hypothesis.

© 2012 Cengage Learning. All Rights Reserved. May not be scanned, copied, or duplicated, or posted to a publicly accessible website, in whole or in part.

83. Suppose on the contrary that $\sum(a_n + b_n)$ converges. Then $\sum(a_n + b_n)$ and $\sum a_n$ are convergent series. So by Theorem 8(iii), $\sum[(a_n + b_n) - a_n]$ would also be convergent. But $\sum[(a_n + b_n) - a_n] = \sum b_n$, a contradiction, since $\sum b_n$ is given to be divergent.

85. The partial sums $\{s_n\}$ form an increasing sequence, since $s_n - s_{n-1} = a_n > 0$ for all n. Also, the sequence $\{s_n\}$ is bounded since $s_n \le 1000$ for all n. So by the Monotonic Sequence Theorem, the sequence of partial sums converges, that is, the series $\sum a_n$ is convergent.

87. (a) At the first step, only the interval $\left(\frac{1}{3}, \frac{2}{3}\right)$ (length $\frac{1}{3}$) is removed. At the second step, we remove the intervals $\left(\frac{1}{9}, \frac{2}{9}\right)$ and $\left(\frac{7}{9}, \frac{8}{9}\right)$, which have a total length of $2 \cdot \left(\frac{1}{3}\right)^2$. At the third step, we remove 2^2 intervals, each of length $\left(\frac{1}{3}\right)^3$. In general, at the nth step we remove 2^{n-1} intervals, each of length $\left(\frac{1}{3}\right)^n$, for a length of $2^{n-1} \cdot \left(\frac{1}{3}\right)^n = \frac{1}{3}\left(\frac{2}{3}\right)^{n-1}$. Thus, the total length of all removed intervals is $\sum_{n=1}^{\infty} \frac{1}{3}\left(\frac{2}{3}\right)^{n-1} = \frac{1/3}{1 - 2/3} = 1$ [geometric series with $a = \frac{1}{3}$ and $r = \frac{2}{3}$]. Notice that at the nth step, the leftmost interval that is removed is $\left(\left(\frac{1}{3}\right)^n, \left(\frac{2}{3}\right)^n\right)$, so we never remove 0, and 0 is in the Cantor set. Also, the rightmost interval removed is $\left(1 - \left(\frac{2}{3}\right)^n, 1 - \left(\frac{1}{3}\right)^n\right)$, so 1 is never removed. Some other numbers in the Cantor set are $\frac{1}{3}, \frac{2}{3}, \frac{1}{9}, \frac{2}{9}, \frac{7}{9},$ and $\frac{8}{9}$.

(b) The area removed at the first step is $\frac{1}{9}$; at the second step, $8 \cdot \left(\frac{1}{9}\right)^2$; at the third step, $(8)^2 \cdot \left(\frac{1}{9}\right)^3$. In general, the area removed at the nth step is $(8)^{n-1}\left(\frac{1}{9}\right)^n = \frac{1}{9}\left(\frac{8}{9}\right)^{n-1}$, so the total area of all removed squares is

$$\sum_{n=1}^{\infty} \frac{1}{9}\left(\frac{8}{9}\right)^{n-1} = \frac{1/9}{1 - 8/9} = 1.$$

89. (a) For $\sum_{n=1}^{\infty} \frac{n}{(n+1)!}$, $s_1 = \frac{1}{1 \cdot 2} = \frac{1}{2}$, $s_2 = \frac{1}{2} + \frac{2}{1 \cdot 2 \cdot 3} = \frac{5}{6}$, $s_3 = \frac{5}{6} + \frac{3}{1 \cdot 2 \cdot 3 \cdot 4} = \frac{23}{24}$,

$s_4 = \frac{23}{24} + \frac{4}{1 \cdot 2 \cdot 3 \cdot 4 \cdot 5} = \frac{119}{120}$. The denominators are $(n+1)!$, so a guess would be $s_n = \frac{(n+1)! - 1}{(n+1)!}$.

(b) For $n = 1$, $s_1 = \frac{1}{2} = \frac{2! - 1}{2!}$, so the formula holds for $n = 1$. Assume $s_k = \frac{(k+1)! - 1}{(k+1)!}$. Then

$$s_{k+1} = \frac{(k+1)! - 1}{(k+1)!} + \frac{k+1}{(k+2)!} = \frac{(k+1)! - 1}{(k+1)!} + \frac{k+1}{(k+1)!(k+2)} = \frac{(k+2)! - (k+2) + k + 1}{(k+2)!}$$
$$= \frac{(k+2)! - 1}{(k+2)!}$$

Thus, the formula is true for $n = k + 1$. So by induction, the guess is correct.

(c) $\lim_{n \to \infty} s_n = \lim_{n \to \infty} \frac{(n+1)! - 1}{(n+1)!} = \lim_{n \to \infty}\left[1 - \frac{1}{(n+1)!}\right] = 1$ and so $\sum_{n=1}^{\infty} \frac{n}{(n+1)!} = 1$.

© 2012 Cengage Learning. All Rights Reserved. May not be scanned, copied, or duplicated, or posted to a publicly accessible website, in whole or in part.

11.3 The Integral Test and Estimates of Sums

1. The picture shows that $a_2 = \dfrac{1}{2^{1.3}} < \displaystyle\int_1^2 \dfrac{1}{x^{1.3}}\,dx$,

$a_3 = \dfrac{1}{3^{1.3}} < \displaystyle\int_2^3 \dfrac{1}{x^{1.3}}\,dx$, and so on, so $\displaystyle\sum_{n=2}^{\infty} \dfrac{1}{n^{1.3}} < \int_1^{\infty} \dfrac{1}{x^{1.3}}\,dx$. The

integral converges by (7.8.2) with $p = 1.3 > 1$, so the series converges.

3. The function $f(x) = 1/\sqrt[5]{x} = x^{-1/5}$ is continuous, positive, and decreasing on $[1, \infty)$, so the Integral Test applies.

$\displaystyle\int_1^{\infty} x^{-1/5}\,dx = \lim_{t\to\infty} \int_1^t x^{-1/5}\,dx = \lim_{t\to\infty} \left[\tfrac{5}{4}x^{4/5}\right]_1^t = \lim_{t\to\infty} \left(\tfrac{5}{4}t^{4/5} - \tfrac{5}{4}\right) = \infty$, so $\displaystyle\sum_{n=1}^{\infty} 1/\sqrt[5]{n}$ diverges.

5. The function $f(x) = \dfrac{1}{(2x+1)^3}$ is continuous, positive, and decreasing on $[1, \infty)$, so the Integral Test applies.

$\displaystyle\int_1^{\infty} \dfrac{1}{(2x+1)^3}\,dx = \lim_{t\to\infty} \int_1^t \dfrac{1}{(2x+1)^3}\,dx = \lim_{t\to\infty} \left[-\dfrac{1}{4}\dfrac{1}{(2x+1)^2}\right]_1^t = \lim_{t\to\infty}\left(-\dfrac{1}{4(2t+1)^2} + \dfrac{1}{36}\right) = \dfrac{1}{36}$.

Since this improper integral is convergent, the series $\displaystyle\sum_{n=1}^{\infty} \dfrac{1}{(2n+1)^3}$ is also convergent by the Integral Test.

7. The function $f(x) - \dfrac{x}{x^2+1}$ is continuous, positive, and decreasing on $[1, \infty)$, so the Integral Test applies.

$\displaystyle\int_1^{\infty} \dfrac{x}{x^2+1}\,dx = \lim_{t\to\infty} \int_1^t \dfrac{x}{x^2+1}\,dx = \lim_{t\to\infty}\left[\tfrac{1}{2}\ln(x^2+1)\right]_1^t = \tfrac{1}{2}\lim_{t\to\infty}\left[\ln(t^2+1) - \ln 2\right] = \infty$. Since this improper

integral is divergent, the series $\displaystyle\sum_{n=1}^{\infty} \dfrac{n}{n^2+1}$ is also divergent by the Integral Test.

9. $\displaystyle\sum_{n=1}^{\infty} \dfrac{1}{n^{\sqrt{2}}}$ is a p-series with $p = \sqrt{2} > 1$, so it converges by (1).

11. $1 + \dfrac{1}{8} + \dfrac{1}{27} + \dfrac{1}{64} + \dfrac{1}{125} + \cdots = \displaystyle\sum_{n=1}^{\infty} \dfrac{1}{n^3}$. This is a p-series with $p = 3 > 1$, so it converges by (1).

13. $1 + \dfrac{1}{3} + \dfrac{1}{5} + \dfrac{1}{7} + \dfrac{1}{9} + \cdots = \displaystyle\sum_{n=1}^{\infty} \dfrac{1}{2n-1}$. The function $f(x) = \dfrac{1}{2x-1}$ is

continuous, positive, and decreasing on $[1, \infty)$, so the Integral Test applies.

$\displaystyle\int_1^{\infty} \dfrac{1}{2x-1}\,dx = \lim_{t\to\infty} \int_1^t \dfrac{1}{2x-1}\,dx = \lim_{t\to\infty}\left[\tfrac{1}{2}\ln|2x-1|\right]_1^t = \tfrac{1}{2}\lim_{t\to\infty}\left(\ln(2t-1) - 0\right) = \infty$, so the series $\displaystyle\sum_{n=1}^{\infty} \dfrac{1}{2n-1}$

diverges.

15. $\displaystyle\sum_{n=1}^{\infty} \dfrac{\sqrt{n}+4}{n^2} = \sum_{n=1}^{\infty}\left(\dfrac{\sqrt{n}}{n^2} + \dfrac{4}{n^2}\right) = \sum_{n=1}^{\infty} \dfrac{1}{n^{3/2}} + \sum_{n=1}^{\infty} \dfrac{4}{n^2}$. $\displaystyle\sum_{n=1}^{\infty} \dfrac{1}{n^{3/2}}$ is a convergent p-series with $p = \tfrac{3}{2} > 1$.

$\displaystyle\sum_{n=1}^{\infty} \dfrac{4}{n^2} = 4\sum_{n=1}^{\infty} \dfrac{1}{n^2}$ is a constant multiple of a convergent p-series with $p = 2 > 1$, so it converges. The sum of two

convergent series is convergent, so the original series is convergent.

© 2012 Cengage Learning. All Rights Reserved. May not be scanned, copied, or duplicated, or posted to a publicly accessible website, in whole or in part.

17. The function $f(x) = \dfrac{1}{x^2 + 4}$ is continuous, positive, and decreasing on $[1, \infty)$, so we can apply the Integral Test.

$$\int_1^\infty \frac{1}{x^2 + 4}\, dx = \lim_{t \to \infty} \int_1^t \frac{1}{x^2 + 4}\, dx = \lim_{t \to \infty} \left[\frac{1}{2} \tan^{-1} \frac{x}{2} \right]_1^t = \frac{1}{2} \lim_{t \to \infty} \left[\tan^{-1}\left(\frac{t}{2}\right) - \tan^{-1}\left(\frac{1}{2}\right) \right]$$

$$= \frac{1}{2} \left[\frac{\pi}{2} - \tan^{-1}\left(\frac{1}{2}\right) \right]$$

Therefore, the series $\sum\limits_{n=1}^\infty \dfrac{1}{n^2 + 4}$ converges.

19. $\sum\limits_{n=1}^\infty \dfrac{\ln n}{n^3} = \sum\limits_{n=2}^\infty \dfrac{\ln n}{n^3}$ since $\dfrac{\ln 1}{1} = 0$. The function $f(x) = \dfrac{\ln x}{x^3}$ is continuous and positive on $[2, \infty)$.

$$f'(x) = \frac{x^3(1/x) - (\ln x)(3x^2)}{(x^3)^2} = \frac{x^2 - 3x^2 \ln x}{x^6} = \frac{1 - 3\ln x}{x^4} < 0 \iff 1 - 3\ln x < 0 \iff \ln x > \tfrac{1}{3} \iff$$

$x > e^{1/3} \approx 1.4$, so f is decreasing on $[2, \infty)$, and the Integral Test applies.

$$\int_2^\infty \frac{\ln x}{x^3}\, dx = \lim_{t \to \infty} \int_2^t \frac{\ln x}{x^3}\, dx \overset{(\star)}{=} \lim_{t \to \infty} \left[-\frac{\ln x}{2x^2} - \frac{1}{4x^2} \right]_1^t = \lim_{t \to \infty} \left[-\frac{1}{4t^2}(2\ln t + 1) + \frac{1}{4} \right] \overset{(\star\star)}{=} \frac{1}{4},$$ so the series $\sum\limits_{n=2}^\infty \dfrac{\ln n}{n^3}$

converges.

$(\star)$: $u = \ln x,\ dv = x^{-3}\, dx \implies du = (1/x)\, dx,\ v = -\tfrac{1}{2}x^{-2}$, so

$$\int \frac{\ln x}{x^3}\, dx = -\tfrac{1}{2} x^{-2} \ln x - \int -\tfrac{1}{2} x^{-2}(1/x)\, dx = -\tfrac{1}{2} x^{-2} \ln x + \tfrac{1}{2} \int x^{-3}\, dx = -\tfrac{1}{2} x^{-2} \ln x - \tfrac{1}{4} x^{-2} + C.$$

$(\star\star)$: $\lim\limits_{t \to \infty} \left(-\dfrac{2\ln t + 1}{4t^2} \right) \overset{\text{H}}{=} -\lim\limits_{t \to \infty} \dfrac{2/t}{8t} = -\tfrac{1}{4} \lim\limits_{t \to \infty} \dfrac{1}{t^2} = 0.$

21. $f(x) = \dfrac{1}{x \ln x}$ is continuous and positive on $[2, \infty)$, and also decreasing since $f'(x) = -\dfrac{1 + \ln x}{x^2(\ln x)^2} < 0$ for $x > 2$, so we can

use the Integral Test. $\displaystyle\int_2^\infty \frac{1}{x \ln x}\, dx = \lim_{t \to \infty} \left[\ln(\ln x) \right]_2^t = \lim_{t \to \infty} \left[\ln(\ln t) - \ln(\ln 2) \right] = \infty$, so the series $\sum\limits_{n=2}^\infty \dfrac{1}{n \ln n}$ diverges.

23. The function $f(x) = e^{1/x}/x^2$ is continuous, positive, and decreasing on $[1, \infty)$, so the Integral Test applies.

$[g(x) = e^{1/x}$ is decreasing and dividing by x^2 doesn't change that fact.$]$

$$\int_1^\infty f(x)\, dx = \lim_{t \to \infty} \int_1^t \frac{e^{1/x}}{x^2}\, dx = \lim_{t \to \infty} \left[-e^{1/x} \right]_1^t = -\lim_{t \to \infty} (e^{1/t} - e) = -(1 - e) = e - 1,$$ so the series $\sum\limits_{n=1}^\infty \dfrac{e^{1/n}}{n^2}$

converges.

25. The function $f(x) = \dfrac{1}{x^2 + x^3} = \dfrac{1}{x^2} - \dfrac{1}{x} + \dfrac{1}{x + 1}$ [by partial fractions] is continuous, positive and decreasing on $[1, \infty)$,

so the Integral Test applies.

$$\int_1^\infty f(x)\, dx = \lim_{t \to \infty} \int_1^t \left(\frac{1}{x^2} - \frac{1}{x} + \frac{1}{x + 1} \right) dx = \lim_{t \to \infty} \left[-\frac{1}{x} - \ln x + \ln(x + 1) \right]_1^t$$

$$= \lim_{t \to \infty} \left[-\frac{1}{t} + \ln \frac{t + 1}{t} + 1 - \ln 2 \right] = 0 + 0 + 1 - \ln 2$$

The integral converges, so the series $\sum\limits_{n=1}^\infty \dfrac{1}{n^2 + n^3}$ converges.

© 2012 Cengage Learning. All Rights Reserved. May not be scanned, copied, or duplicated, or posted to a publicly accessible website, in whole or in part.

27. The function $f(x) = \dfrac{\cos \pi x}{\sqrt{x}}$ is neither positive nor decreasing on $[1, \infty)$, so the hypotheses of the Integral Test are not

satisfied for the series $\displaystyle\sum_{n=1}^{\infty} \dfrac{\cos \pi n}{\sqrt{n}}$.

29. We have already shown (in Exercise 21) that when $p = 1$ the series $\displaystyle\sum_{n=2}^{\infty} \dfrac{1}{n(\ln n)^p}$ diverges, so assume that $p \neq 1$.

$f(x) = \dfrac{1}{x(\ln x)^p}$ is continuous and positive on $[2, \infty)$, and $f'(x) = -\dfrac{p + \ln x}{x^2 (\ln x)^{p+1}} < 0$ if $x > e^{-p}$, so that f is eventually

decreasing and we can use the Integral Test.

$$\int_2^{\infty} \frac{1}{x(\ln x)^p}\, dx = \lim_{t \to \infty} \left[\frac{(\ln x)^{1-p}}{1-p} \right]_2^t \quad [\text{for } p \neq 1] = \lim_{t \to \infty} \left[\frac{(\ln t)^{1-p}}{1-p} - \frac{(\ln 2)^{1-p}}{1-p} \right]$$

This limit exists whenever $1 - p < 0 \;\Leftrightarrow\; p > 1$, so the series converges for $p > 1$.

31. Clearly the series cannot converge if $p \geq -\tfrac{1}{2}$, because then $\displaystyle\lim_{n \to \infty} n(1 + n^2)^p \neq 0$. So assume $p < -\tfrac{1}{2}$. Then

$f(x) = x(1 + x^2)^p$ is continuous, positive, and eventually decreasing on $[1, \infty)$, and we can use the Integral Test.

$$\int_1^{\infty} x(1 + x^2)^p\, dx = \lim_{t \to \infty} \left[\frac{1}{2} \cdot \frac{(1 + x^2)^{p+1}}{p+1} \right]_1^t = \frac{1}{2(p+1)} \lim_{t \to \infty} [(1 + t^2)^{p+1} - 2^{p+1}].$$

This limit exists and is finite $\;\Leftrightarrow\; p + 1 < 0 \;\Leftrightarrow\; p < -1$, so the series converges whenever $p < -1$.

33. Since this is a p-series with $p = x$, $\zeta(x)$ is defined when $x > 1$. Unless specified otherwise, the domain of a function f is the

set of real numbers x such that the expression for $f(x)$ makes sense and defines a real number. So, in the case of a series, it's

the set of real numbers x such that the series is convergent.

35. (a) $\displaystyle\sum_{n=1}^{\infty} \left(\frac{3}{n} \right)^4 = \sum_{n=1}^{\infty} \frac{81}{n^4} = 81 \sum_{n=1}^{\infty} \frac{1}{n^4} = 81 \left(\frac{\pi^4}{90} \right) = \frac{9\pi^4}{10}$

(b) $\displaystyle\sum_{k=5}^{\infty} \frac{1}{(k-2)^4} = \frac{1}{3^4} + \frac{1}{4^4} + \frac{1}{5^4} + \cdots = \sum_{k=3}^{\infty} \frac{1}{k^4} = \frac{\pi^4}{90} - \left(\frac{1}{1^4} + \frac{1}{2^4} \right)$ [subtract a_1 and a_2] $= \frac{\pi^4}{90} - \frac{17}{16}$

37. (a) $f(x) = \dfrac{1}{x^2}$ is positive and continuous and $f'(x) = -\dfrac{2}{x^3}$ is negative for $x > 0$, and so the Integral Test applies.

$$\sum_{n=1}^{\infty} \frac{1}{n^2} \approx s_{10} = \frac{1}{1^2} + \frac{1}{2^2} + \frac{1}{3^2} + \cdots + \frac{1}{10^2} \approx 1.549768.$$

$$R_{10} \leq \int_{10}^{\infty} \frac{1}{x^2}\, dx = \lim_{t \to \infty} \left[\frac{-1}{x} \right]_{10}^t = \lim_{t \to \infty} \left(-\frac{1}{t} + \frac{1}{10} \right) = \frac{1}{10}, \text{ so the error is at most } 0.1.$$

(b) $s_{10} + \displaystyle\int_{11}^{\infty} \frac{1}{x^2}\, dx \leq s \leq s_{10} + \int_{10}^{\infty} \frac{1}{x^2}\, dx \;\Rightarrow\; s_{10} + \frac{1}{11} \leq s \leq s_{10} + \frac{1}{10} \;\Rightarrow$

$1.549768 + 0.090909 = 1.640677 \leq s \leq 1.549768 + 0.1 = 1.649768$, so we get $s \approx 1.64522$ (the average of 1.640677

and 1.649768) with error ≤ 0.005 (the maximum of $1.649768 - 1.64522$ and $1.64522 - 1.640677$, rounded up).

© 2012 Cengage Learning. All Rights Reserved. May not be scanned, copied, or duplicated, or posted to a publicly accessible website, in whole or in part.

(c) The estimate in part (b) is $s \approx 1.64522$ with error ≤ 0.005. The exact value given in Exercise 34 is $\pi^2/6 \approx 1.644934$. The difference is less than 0.0003.

(d) $R_n \leq \int_n^\infty \frac{1}{x^2}\,dx = \frac{1}{n}$. So $R_n < 0.001$ if $\frac{1}{n} < \frac{1}{1000}$ $\Leftrightarrow$ $n > 1000$.

39. $f(x) = 1/(2x+1)^6$ is continuous, positive, and decreasing on $[1, \infty)$, so the Integral Test applies. Using (3),

$$R_n \leq \int_n^\infty (2x+1)^{-6}\,dx = \lim_{t \to \infty} \left[\frac{-1}{10(2x+1)^5} \right]_n^t = \frac{1}{10(2n+1)^5}. \text{ To be correct to five decimal places, we want}$$

$$\frac{1}{10(2n+1)^5} \leq \frac{5}{10^6} \quad \Leftrightarrow \quad (2n+1)^5 \geq 20{,}000 \quad \Leftrightarrow \quad n \geq \tfrac{1}{2}\left(\sqrt[5]{20{,}000} - 1 \right) \approx 3.12, \text{ so use } n = 4.$$

$$s_4 = \sum_{n=1}^{4} \frac{1}{(2n+1)^6} = \frac{1}{3^6} + \frac{1}{5^6} + \frac{1}{7^6} + \frac{1}{9^6} \approx 0.001\,446 \approx 0.00145.$$

41. $\sum_{n=1}^{\infty} n^{-1.001} = \sum_{n=1}^{\infty} \frac{1}{n^{1.001}}$ is a convergent p-series with $p = 1.001 > 1$. Using (2), we get

$$R_n \leq \int_n^\infty x^{-1.001}\,dx = \lim_{t \to \infty} \left[\frac{x^{-0.001}}{-0.001} \right]_n^t = -1000 \lim_{t \to \infty} \left[\frac{1}{x^{0.001}} \right]_n^t = -1000 \left(-\frac{1}{n^{0.001}} \right) = \frac{1000}{n^{0.001}}.$$

We want $R_n < 0.000\,000\,005$ $\Leftrightarrow$ $\frac{1000}{n^{0.001}} < 5 \times 10^{-9}$ $\Leftrightarrow$ $n^{0.001} > \frac{1000}{5 \times 10^{-9}}$ $\Leftrightarrow$

$$n > \left(2 \times 10^{11} \right)^{1000} = 2^{1000} \times 10^{11{,}000} \approx 1.07 \times 10^{301} \times 10^{11{,}000} = 1.07 \times 10^{11{,}301}.$$

43. (a) From the figure, $a_2 + a_3 + \cdots + a_n \leq \int_1^n f(x)\,dx$, so with

$$f(x) = \frac{1}{x}, \frac{1}{2} + \frac{1}{3} + \frac{1}{4} + \cdots + \frac{1}{n} \leq \int_1^n \frac{1}{x}\,dx = \ln n.$$

Thus, $s_n = 1 + \frac{1}{2} + \frac{1}{3} + \frac{1}{4} + \cdots + \frac{1}{n} \leq 1 + \ln n$.

(b) By part (a), $s_{10^6} \leq 1 + \ln 10^6 \approx 14.82 < 15$ and

$$s_{10^9} \leq 1 + \ln 10^9 \approx 21.72 < 22.$$

45. $b^{\ln n} = \left(e^{\ln b} \right)^{\ln n} = \left(e^{\ln n} \right)^{\ln b} = n^{\ln b} = \frac{1}{n^{-\ln b}}$. This is a p-series, which converges for all b such that $-\ln b > 1$ $\Leftrightarrow$

$$\ln b < -1 \quad \Leftrightarrow \quad b < e^{-1} \quad \Leftrightarrow \quad b < 1/e \text{ [with } b > 0\text{]}.$$

11.4 The Comparison Tests

1. (a) We cannot say anything about $\sum a_n$. If $a_n > b_n$ for all n and $\sum b_n$ is convergent, then $\sum a_n$ could be convergent or divergent. (See the note after Example 2.)

(b) If $a_n < b_n$ for all n, then $\sum a_n$ is convergent. [This is part (i) of the Comparison Test.]

3. $\dfrac{n}{2n^3+1} < \dfrac{n}{2n^3} = \dfrac{1}{2n^2} < \dfrac{1}{n^2}$ for all $n \geq 1$, so $\sum_{n=1}^{\infty} \dfrac{n}{2n^3+1}$ converges by comparison with $\sum_{n=1}^{\infty} \dfrac{1}{n^2}$, which converges because it is a p-series with $p = 2 > 1$.

© 2012 Cengage Learning. All Rights Reserved. May not be scanned, copied, or duplicated, or posted to a publicly accessible website, in whole or in part.

5. $\dfrac{n+1}{n\sqrt{n}} > \dfrac{n}{n\sqrt{n}} = \dfrac{1}{\sqrt{n}}$ for all $n \geq 1$, so $\displaystyle\sum_{n=1}^{\infty} \dfrac{n+1}{n\sqrt{n}}$ diverges by comparison with $\displaystyle\sum_{n=1}^{\infty} \dfrac{1}{\sqrt{n}}$, which diverges because it is a

p-series with $p = \frac{1}{2} \leq 1$.

7. $\dfrac{9^n}{3+10^n} < \dfrac{9^n}{10^n} = \left(\dfrac{9}{10}\right)^n$ for all $n \geq 1$. $\displaystyle\sum_{n=1}^{\infty} \left(\frac{9}{10}\right)^n$ is a convergent geometric series $\left(|r| = \frac{9}{10} < 1\right)$, so $\displaystyle\sum_{n=1}^{\infty} \dfrac{9^n}{3+10^n}$

converges by the Comparison Test.

9. $\dfrac{\ln k}{k} > \dfrac{1}{k}$ for all $k \geq 3$ [since $\ln k > 1$ for $k \geq 3$], so $\displaystyle\sum_{k=3}^{\infty} \dfrac{\ln k}{k}$ diverges by comparison with $\displaystyle\sum_{k=3}^{\infty} \dfrac{1}{k}$, which diverges because it

is a p-series with $p = 1 \leq 1$ (the harmonic series). Thus, $\displaystyle\sum_{k=1}^{\infty} \dfrac{\ln k}{k}$ diverges since a finite number of terms doesn't affect the

convergence or divergence of a series.

11. $\dfrac{\sqrt[3]{k}}{\sqrt{k^3 + 4k + 3}} < \dfrac{\sqrt[3]{k}}{\sqrt{k^3}} = \dfrac{k^{1/3}}{k^{3/2}} = \dfrac{1}{k^{7/6}}$ for all $k \geq 1$, so $\displaystyle\sum_{k=1}^{\infty} \dfrac{\sqrt[3]{k}}{\sqrt{k^3 + 4k + 3}}$ converges by comparison with $\displaystyle\sum_{k=1}^{\infty} \dfrac{1}{k^{7/6}}$,

which converges because it is a p-series with $p = \frac{7}{6} > 1$.

13. $\dfrac{\arctan n}{n^{1.2}} < \dfrac{\pi/2}{n^{1.2}}$ for all $n \geq 1$, so $\displaystyle\sum_{n=1}^{\infty} \dfrac{\arctan n}{n^{1.2}}$ converges by comparison with $\dfrac{\pi}{2} \displaystyle\sum_{n=1}^{\infty} \dfrac{1}{n^{1.2}}$, which converges because it is a

constant times a p-series with $p = 1.2 > 1$.

15. $\dfrac{4^{n+1}}{3^n - 2} > \dfrac{4 \cdot 4^n}{3^n} = 4\left(\dfrac{4}{3}\right)^n$ for all $n \geq 1$. $\displaystyle\sum_{n=1}^{\infty} 4\left(\frac{4}{3}\right)^n = 4\displaystyle\sum_{n=1}^{\infty}\left(\frac{4}{3}\right)^n$ is a divergent geometric series $\left(|r| = \frac{4}{3} > 1\right)$, so

$\displaystyle\sum_{n=1}^{\infty} \dfrac{4^{n+1}}{3^n - 2}$ diverges by the Comparison Test.

17. Use the Limit Comparison Test with $a_n = \dfrac{1}{\sqrt{n^2 + 1}}$ and $b_n = \dfrac{1}{n}$:

$$\lim_{n\to\infty} \frac{a_n}{b_n} = \lim_{n\to\infty} \frac{n}{\sqrt{n^2+1}} = \lim_{n\to\infty} \frac{1}{\sqrt{1+(1/n^2)}} = 1 > 0. \text{ Since the harmonic series } \sum_{n=1}^{\infty} \frac{1}{n} \text{ diverges, so does}$$

$\displaystyle\sum_{n=1}^{\infty} \dfrac{1}{\sqrt{n^2 + 1}}$.

19. Use the Limit Comparison Test with $a_n = \dfrac{1 + 4^n}{1 + 3^n}$ and $b_n = \dfrac{4^n}{3^n}$:

$$\lim_{n\to\infty} \frac{a_n}{b_n} = \lim_{n\to\infty} \frac{\dfrac{1+4^n}{1+3^n}}{\dfrac{4^n}{3^n}} = \lim_{n\to\infty} \frac{1+4^n}{1+3^n} \cdot \frac{3^n}{4^n} = \lim_{n\to\infty} \frac{1+4^n}{4^n} \cdot \frac{3^n}{1+3^n} = \lim_{n\to\infty} \left(\frac{1}{4^n} + 1\right) \cdot \frac{1}{\dfrac{1}{3^n}+1} = 1 > 0$$

Since the geometric series $\sum b_n = \sum \left(\frac{4}{3}\right)^n$ diverges, so does $\displaystyle\sum_{n=1}^{\infty} \dfrac{1+4^n}{1+3^n}$. Alternatively, use the Comparison Test with

$\dfrac{1+4^n}{1+3^n} > \dfrac{1+4^n}{3^n+3^n} > \dfrac{4^n}{2(3^n)} = \dfrac{1}{2}\left(\dfrac{4}{3}\right)^n$ or use the Test for Divergence.

© 2012 Cengage Learning. All Rights Reserved. May not be scanned, copied, or duplicated, or posted to a publicly accessible website, in whole or in part.

21. Use the Limit Comparison Test with $a_n = \dfrac{\sqrt{n+2}}{2n^2+n+1}$ and $b_n = \dfrac{1}{n^{3/2}}$:

$$\lim_{n\to\infty} \frac{a_n}{b_n} = \lim_{n\to\infty} \frac{n^{3/2}\sqrt{n+2}}{2n^2+n+1} = \lim_{n\to\infty} \frac{(n^{3/2}\sqrt{n+2})/(n^{3/2}\sqrt{n})}{(2n^2+n+1)/n^2} = \lim_{n\to\infty} \frac{\sqrt{1+2/n}}{2+1/n+1/n^2} = \frac{\sqrt{1}}{2} = \frac{1}{2} > 0.$$

Since $\displaystyle\sum_{n=1}^{\infty} \frac{1}{n^{3/2}}$ is a convergent p-series $\left[p = \frac{3}{2} > 1\right]$, the series $\displaystyle\sum_{n=1}^{\infty} \frac{\sqrt{n+2}}{2n^2+n+1}$ also converges.

23. Use the Limit Comparison Test with $a_n = \dfrac{5+2n}{(1+n^2)^2}$ and $b_n = \dfrac{1}{n^3}$:

$$\lim_{n\to\infty} \frac{a_n}{b_n} = \lim_{n\to\infty} \frac{n^3(5+2n)}{(1+n^2)^2} = \lim_{n\to\infty} \frac{5n^3+2n^4}{(1+n^2)^2} \cdot \frac{1/n^4}{1/(n^2)^2} = \lim_{n\to\infty} \frac{\frac{5}{n}+2}{\left(\frac{1}{n^2}+1\right)^2} = 2 > 0.$$ Since $\displaystyle\sum_{n=1}^{\infty} \frac{1}{n^3}$ is a convergent

p-series $[p = 3 > 1]$, the series $\displaystyle\sum_{n=1}^{\infty} \frac{5+2n}{(1+n^2)^2}$ also converges.

25. $\dfrac{\sqrt{n^4+1}}{n^3+n^2} > \dfrac{\sqrt{n^4}}{n^2(n+1)} = \dfrac{n^2}{n^2(n+1)} = \dfrac{1}{n+1}$ for all $n \geq 1$, so $\displaystyle\sum_{n=1}^{\infty} \frac{\sqrt{n^4+1}}{n^3+n^2}$ diverges by comparison with

$\displaystyle\sum_{n=1}^{\infty} \frac{1}{n+1} = \sum_{n=2}^{\infty} \frac{1}{n}$, which diverges because it is a p-series with $p = 1 \leq 1$.

27. Use the Limit Comparison Test with $a_n = \left(1+\dfrac{1}{n}\right)^2 e^{-n}$ and $b_n = e^{-n}$: $\displaystyle\lim_{n\to\infty} \frac{a_n}{b_n} = \lim_{n\to\infty} \left(1+\frac{1}{n}\right)^2 = 1 > 0.$ Since

$\displaystyle\sum_{n=1}^{\infty} e^{-n} = \sum_{n=1}^{\infty} \frac{1}{e^n}$ is a convergent geometric series $\left[|r| = \frac{1}{e} < 1\right]$, the series $\displaystyle\sum_{n=1}^{\infty} \left(1+\frac{1}{n}\right)^2 e^{-n}$ also converges.

29. Clearly $n! = n(n-1)(n-2)\cdots(3)(2) \geq 2\cdot2\cdot2\cdots\cdots2\cdot2 = 2^{n-1}$, so $\dfrac{1}{n!} \leq \dfrac{1}{2^{n-1}}$. $\displaystyle\sum_{n=1}^{\infty} \frac{1}{2^{n-1}}$ is a convergent geometric

series $\left[|r| = \frac{1}{2} < 1\right]$, so $\displaystyle\sum_{n=1}^{\infty} \frac{1}{n!}$ converges by the Comparison Test.

31. Use the Limit Comparison Test with $a_n = \sin\left(\dfrac{1}{n}\right)$ and $b_n = \dfrac{1}{n}$. Then $\sum a_n$ and $\sum b_n$ are series with positive terms and

$$\lim_{n\to\infty} \frac{a_n}{b_n} = \lim_{n\to\infty} \frac{\sin(1/n)}{1/n} = \lim_{\theta\to0} \frac{\sin\theta}{\theta} = 1 > 0. \text{ Since } \sum_{n=1}^{\infty} b_n \text{ is the divergent harmonic series,}$$

$\displaystyle\sum_{n=1}^{\infty} \sin(1/n)$ also diverges.   [Note that we could also use l'Hospital's Rule to evaluate the limit:

$$\lim_{x\to\infty} \frac{\sin(1/x)}{1/x} \overset{\text{H}}{=} \lim_{x\to\infty} \frac{\cos(1/x)\cdot(-1/x^2)}{-1/x^2} = \lim_{x\to\infty} \cos\frac{1}{x} = \cos 0 = 1.]$$

33. $\displaystyle\sum_{n=1}^{10} \frac{1}{\sqrt{n^4+1}} = \frac{1}{\sqrt{2}} + \frac{1}{\sqrt{17}} + \frac{1}{\sqrt{82}} + \cdots + \frac{1}{\sqrt{10,001}} \approx 1.24856.$ Now $\dfrac{1}{\sqrt{n^4+1}} < \dfrac{1}{\sqrt{n^4}} = \dfrac{1}{n^2}$, so the error is

$$R_{10} \leq T_{10} \leq \int_{10}^{\infty} \frac{1}{x^2}\,dx = \lim_{t\to\infty} \left[-\frac{1}{x}\right]_{10}^{t} = \lim_{t\to\infty} \left(-\frac{1}{t} + \frac{1}{10}\right) = \frac{1}{10} = 0.1.$$

© 2012 Cengage Learning. All Rights Reserved. May not be scanned, copied, or duplicated, or posted to a publicly accessible website, in whole or in part.

35. $\sum_{n=1}^{10} 5^{-n} \cos^2 n = \dfrac{\cos^2 1}{5} + \dfrac{\cos^2 2}{5^2} + \dfrac{\cos^2 3}{5^3} + \cdots + \dfrac{\cos^2 10}{5^{10}} \approx 0.07393$. Now $\dfrac{\cos^2 n}{5^n} \le \dfrac{1}{5^n}$, so the error is

$$R_{10} \le T_{10} \le \int_{10}^{\infty} \frac{1}{5^x}\, dx = \lim_{t \to \infty} \int_{10}^{t} 5^{-x}\, dx = \lim_{t \to \infty} \left[-\frac{5^{-x}}{\ln 5} \right]_{10}^{t} = \lim_{t \to \infty} \left(-\frac{5^{-t}}{\ln 5} + \frac{5^{-10}}{\ln 5} \right) = \frac{1}{5^{10} \ln 5} < 6.4 \times 10^{-8}.$$

37. Since $\dfrac{d_n}{10^n} \le \dfrac{9}{10^n}$ for each n, and since $\sum_{n=1}^{\infty} \dfrac{9}{10^n}$ is a convergent geometric series $\left(|r| = \frac{1}{10} < 1 \right)$, $0.d_1 d_2 d_3 \ldots = \sum_{n=1}^{\infty} \dfrac{d_n}{10^n}$

will always converge by the Comparison Test.

39. Since $\sum a_n$ converges, $\displaystyle\lim_{n \to \infty} a_n = 0$, so there exists N such that $|a_n - 0| < 1$ for all $n > N \;\Rightarrow\; 0 \le a_n < 1$ for

all $n > N \;\Rightarrow\; 0 \le a_n^2 \le a_n$. Since $\sum a_n$ converges, so does $\sum a_n^2$ by the Comparison Test.

41. (a) Since $\displaystyle\lim_{n \to \infty} \frac{a_n}{b_n} = \infty$, there is an integer N such that $\dfrac{a_n}{b_n} > 1$ whenever $n > N$. (Take $M = 1$ in Definition 11.1.5.)

Then $a_n > b_n$ whenever $n > N$ and since $\sum b_n$ is divergent, $\sum a_n$ is also divergent by the Comparison Test.

(b) (i) If $a_n = \dfrac{1}{\ln n}$ and $b_n = \dfrac{1}{n}$ for $n \ge 2$, then $\displaystyle\lim_{n \to \infty} \frac{a_n}{b_n} = \lim_{n \to \infty} \frac{n}{\ln n} = \lim_{x \to \infty} \frac{x}{\ln x} \overset{\text{H}}{=} \lim_{x \to \infty} \frac{1}{1/x} = \lim_{x \to \infty} x = \infty$,

so by part (a), $\displaystyle\sum_{n=2}^{\infty} \frac{1}{\ln n}$ is divergent.

(ii) If $a_n = \dfrac{\ln n}{n}$ and $b_n = \dfrac{1}{n}$, then $\displaystyle\sum_{n=1}^{\infty} b_n$ is the divergent harmonic series and $\displaystyle\lim_{n \to \infty} \frac{a_n}{b_n} = \lim_{n \to \infty} \ln n = \lim_{x \to \infty} \ln x = \infty$,

so $\displaystyle\sum_{n=1}^{\infty} a_n$ diverges by part (a).

43. $\displaystyle\lim_{n \to \infty} n a_n = \lim_{n \to \infty} \frac{a_n}{1/n}$, so we apply the Limit Comparison Test with $b_n = \dfrac{1}{n}$. Since $\displaystyle\lim_{n \to \infty} n a_n > 0$ we know that either both

series converge or both series diverge, and we also know that $\displaystyle\sum_{n=1}^{\infty} \frac{1}{n}$ diverges [p-series with $p = 1$]. Therefore, $\sum a_n$ must be

divergent.

45. Yes. Since $\sum a_n$ is a convergent series with positive terms, $\displaystyle\lim_{n \to \infty} a_n = 0$ by Theorem 11.2.6, and $\sum b_n = \sum \sin(a_n)$ is a

series with positive terms (for large enough n). We have $\displaystyle\lim_{n \to \infty} \frac{b_n}{a_n} = \lim_{n \to \infty} \frac{\sin(a_n)}{a_n} = 1 > 0$ by Theorem 2.4.2

[ET Theorem 3.3.2]. Thus, $\sum b_n$ is also convergent by the Limit Comparison Test.

11.5 Alternating Series

1. (a) An alternating series is a series whose terms are alternately positive and negative.

(b) An alternating series $\displaystyle\sum_{n=1}^{\infty} a_n = \sum_{n=1}^{\infty} (-1)^{n-1} b_n$, where $b_n = |a_n|$, converges if $0 < b_{n+1} \le b_n$ for all n and $\displaystyle\lim_{n \to \infty} b_n = 0$.

(This is the Alternating Series Test.)

(c) The error involved in using the partial sum s_n as an approximation to the total sum s is the remainder $R_n = s - s_n$ and the

size of the error is smaller than b_{n+1}; that is, $|R_n| \le b_{n+1}$. (This is the Alternating Series Estimation Theorem.)

© 2012 Cengage Learning. All Rights Reserved. May not be scanned, copied, or duplicated, or posted to a publicly accessible website, in whole or in part.

3. $-\dfrac{2}{5} + \dfrac{4}{6} - \dfrac{6}{7} + \dfrac{8}{8} - \dfrac{10}{9} + \cdots = \displaystyle\sum_{n=1}^{\infty} (-1)^n \dfrac{2n}{n+4}$. Now $\displaystyle\lim_{n\to\infty} b_n = \lim_{n\to\infty} \dfrac{2n}{n+4} = \lim_{n\to\infty} \dfrac{2}{1+4/n} = \dfrac{2}{1} \neq 0$. Since

$\displaystyle\lim_{n\to\infty} a_n \neq 0$ (in fact the limit does not exist), the series diverges by the Test for Divergence.

5. $\displaystyle\sum_{n=1}^{\infty} a_n = \sum_{n=1}^{\infty} (-1)^{n-1} \dfrac{1}{2n+1} = \sum_{n=1}^{\infty} (-1)^{n-1} b_n$. Now $b_n = \dfrac{1}{2n+1} > 0$, $\{b_n\}$ is decreasing, and $\displaystyle\lim_{n\to\infty} b_n = 0$, so the

series converges by the Alternating Series Test.

7. $\displaystyle\sum_{n=1}^{\infty} a_n = \sum_{n=1}^{\infty} (-1)^n \dfrac{3n-1}{2n+1} = \sum_{n=1}^{\infty} (-1)^n b_n$. Now $\displaystyle\lim_{n\to\infty} b_n = \lim_{n\to\infty} \dfrac{3-1/n}{2+1/n} = \dfrac{3}{2} \neq 0$. Since $\displaystyle\lim_{n\to\infty} a_n \neq 0$

(in fact the limit does not exist), the series diverges by the Test for Divergence.

9. $\displaystyle\sum_{n=1}^{\infty} a_n = \sum_{n=1}^{\infty} (-1)^n e^{-n} = \sum_{n=1}^{\infty} (-1)^n b_n$. Now $b_n = \dfrac{1}{e^n} > 0$, $\{b_n\}$ is decreasing, and $\displaystyle\lim_{n\to\infty} b_n = 0$, so the series converges

by the Alternating Series Test.

11. $b_n = \dfrac{n^2}{n^3+4} > 0$ for $n \geq 1$. $\{b_n\}$ is decreasing for $n \geq 2$ since

$\left(\dfrac{x^2}{x^3+4}\right)' = \dfrac{(x^3+4)(2x) - x^2(3x^2)}{(x^3+4)^2} = \dfrac{x(2x^3+8-3x^3)}{(x^3+4)^2} = \dfrac{x(8-x^3)}{(x^3+4)^2} < 0$ for $x > 2$. Also,

$\displaystyle\lim_{n\to\infty} b_n = \lim_{n\to\infty} \dfrac{1/n}{1+4/n^3} = 0$. Thus, the series $\displaystyle\sum_{n=1}^{\infty} (-1)^{n+1} \dfrac{n^2}{n^3+4}$ converges by the Alternating Series Test.

13. $\displaystyle\lim_{n\to\infty} b_n = \lim_{n\to\infty} e^{2/n} = e^0 = 1$, so $\displaystyle\lim_{n\to\infty} (-1)^{n-1} e^{2/n}$ does not exist. Thus, the series $\displaystyle\sum_{n=1}^{\infty} (-1)^{n-1} e^{2/n}$ diverges by the

Test for Divergence.

15. $a_n = \dfrac{\sin\left(n+\frac{1}{2}\right)\pi}{1+\sqrt{n}} = \dfrac{(-1)^n}{1+\sqrt{n}}$.   Now $b_n = \dfrac{1}{1+\sqrt{n}} > 0$ for $n \geq 0$, $\{b_n\}$ is decreasing, and $\displaystyle\lim_{n\to\infty} b_n = 0$, so the series

$\displaystyle\sum_{n=0}^{\infty} \dfrac{\sin\left(n+\frac{1}{2}\right)\pi}{1+\sqrt{n}}$ converges by the Alternating Series Test.

17. $\displaystyle\sum_{n=1}^{\infty} (-1)^n \sin\left(\dfrac{\pi}{n}\right)$. $b_n = \sin\left(\dfrac{\pi}{n}\right) > 0$ for $n \geq 2$ and $\sin\left(\dfrac{\pi}{n}\right) \geq \sin\left(\dfrac{\pi}{n+1}\right)$, and $\displaystyle\lim_{n\to\infty} \sin\left(\dfrac{\pi}{n}\right) = \sin 0 = 0$, so the

series converges by the Alternating Series Test.

19. $\dfrac{n^n}{n!} = \dfrac{n \cdot n \cdots \cdot n}{1 \cdot 2 \cdots \cdot n} \geq n \;\Rightarrow\; \displaystyle\lim_{n\to\infty} \dfrac{n^n}{n!} = \infty \;\Rightarrow\; \lim_{n\to\infty} \dfrac{(-1)^n n^n}{n!}$ does not exist. So the series $\displaystyle\sum_{n=1}^{\infty} (-1)^n \dfrac{n^n}{n!}$ diverges

by the Test for Divergence.

21.

The graph gives us an estimate for the sum of the series

$\displaystyle\sum_{n=1}^{\infty} \dfrac{(-0.8)^n}{n!}$ of -0.55.

$b_8 = \dfrac{(0.8)^8}{8!} \approx 0.000\,004$, so

© 2012 Cengage Learning. All Rights Reserved. May not be scanned, copied, or duplicated, or posted to a publicly accessible website, in whole or in part.

$$\sum_{n=1}^{\infty} \frac{(-0.8)^n}{n!} \approx s_7 = \sum_{n=1}^{7} \frac{(-0.8)^n}{n!}$$

$$\approx -0.8 + 0.32 - 0.085\overline{3} + 0.0170\overline{6} - 0.002\,731 + 0.000\,364 - 0.000\,042 \approx -0.5507$$

Adding b_8 to s_7 does not change the fourth decimal place of s_7, so the sum of the series, correct to four decimal places, is -0.5507.

23. The series $\sum_{n=1}^{\infty} \frac{(-1)^{n+1}}{n^6}$ satisfies (i) of the Alternating Series Test because $\frac{1}{(n+1)^6} < \frac{1}{n^6}$ and (ii) $\lim\limits_{n \to \infty} \frac{1}{n^6} = 0$, so the

series is convergent. Now $b_5 = \frac{1}{5^6} = 0.000064 > 0.00005$ and $b_6 = \frac{1}{6^6} \approx 0.00002 < 0.00005$, so by the Alternating Series

Estimation Theorem, $n = 5$. (That is, since the 6th term is less than the desired error, we need to add the first 5 terms to get the

sum to the desired accuracy.)

25. The series $\sum_{n=0}^{\infty} \frac{(-1)^n}{10^n\, n!}$ satisfies (i) of the Alternating Series Test because $\frac{1}{10^{n+1}(n+1)!} < \frac{1}{10^n\, n!}$ and (ii) $\lim\limits_{n \to \infty} \frac{1}{10^n\, n!} = 0$,

so the series is convergent. Now $b_3 = \frac{1}{10^3\, 3!} \approx 0.000\,167 > 0.000\,005$ and $b_4 = \frac{1}{10^4\, 4!} = 0.000\,004 < 0.000\,005$, so by

the Alternating Series Estimation Theorem, $n = 4$ (since the series starts with $n = 0$, not $n = 1$). (That is, since the 5th term

is less than the desired error, we need to add the first 4 terms to get the sum to the desired accuracy.)

27. $b_4 = \frac{1}{8!} = \frac{1}{40{,}320} \approx 0.000\,025$, so

$$\sum_{n=1}^{\infty} \frac{(-1)^n}{(2n)!} \approx s_3 = \sum_{n=1}^{3} \frac{(-1)^n}{(2n)!} = -\frac{1}{2} + \frac{1}{24} - \frac{1}{720} \approx -0.459\,722$$

Adding b_4 to s_3 does not change the fourth decimal place of s_3, so by the Alternating Series Estimation Theorem, the sum of

the series, correct to four decimal places, is -0.4597.

29. $b_7 = \frac{7^2}{10^7} = 0.000\,004\,9$, so

$$\sum_{n=1}^{\infty} \frac{(-1)^{n-1} n^2}{10^n} \approx s_6 = \sum_{n=1}^{6} \frac{(-1)^{n-1} n^2}{10^n} = \frac{1}{10} - \frac{4}{100} + \frac{9}{1000} - \frac{16}{10{,}000} + \frac{25}{100{,}000} - \frac{36}{1{,}000{,}000} = 0.067\,614$$

Adding b_7 to s_6 does not change the fourth decimal place of s_6, so by the Alternating Series Estimation Theorem, the sum of

the series, correct to four decimal places, is 0.0676.

31. $\sum_{n=1}^{\infty} \frac{(-1)^{n-1}}{n} = 1 - \frac{1}{2} + \frac{1}{3} - \frac{1}{4} + \cdots + \frac{1}{49} - \frac{1}{50} + \frac{1}{51} - \frac{1}{52} + \cdots$. The 50th partial sum of this series is an

underestimate, since $\sum_{n=1}^{\infty} \frac{(-1)^{n-1}}{n} = s_{50} + \left(\frac{1}{51} - \frac{1}{52} \right) + \left(\frac{1}{53} - \frac{1}{54} \right) + \cdots$, and the terms in parentheses are all positive.

The result can be seen geometrically in Figure 1.

33. Clearly $b_n = \frac{1}{n+p}$ is decreasing and eventually positive and $\lim\limits_{n \to \infty} b_n = 0$ for any p. So the series converges (by the

Alternating Series Test) for any p for which every b_n is defined, that is, $n + p \neq 0$ for $n \geq 1$, or p is not a negative integer.

© 2012 Cengage Learning. All Rights Reserved. May not be scanned, copied, or duplicated, or posted to a publicly accessible website, in whole or in part.

35. $\sum b_{2n} = \sum 1/(2n)^2$ clearly converges (by comparison with the p-series for $p = 2$). So suppose that $\sum (-1)^{n-1} b_n$

converges. Then by Theorem 11.2.8(ii), so does $\sum \left[(-1)^{n-1} b_n + b_n\right] = 2\left(1 + \frac{1}{3} + \frac{1}{5} + \cdots\right) = 2\sum \dfrac{1}{2n-1}$. But this

diverges by comparison with the harmonic series, a contradiction. Therefore, $\sum (-1)^{n-1} b_n$ must diverge. The Alternating

Series Test does not apply since $\{b_n\}$ is not decreasing.

11.6 Absolute Convergence and the Ratio and Root Tests

1. (a) Since $\lim\limits_{n\to\infty} \left|\dfrac{a_{n+1}}{a_n}\right| = 8 > 1$, part (b) of the Ratio Test tells us that the series $\sum a_n$ is divergent.

(b) Since $\lim\limits_{n\to\infty} \left|\dfrac{a_{n+1}}{a_n}\right| = 0.8 < 1$, part (a) of the Ratio Test tells us that the series $\sum a_n$ is absolutely convergent (and

   therefore convergent).

(c) Since $\lim\limits_{n\to\infty} \left|\dfrac{a_{n+1}}{a_n}\right| = 1$, the Ratio Test fails and the series $\sum a_n$ might converge or it might diverge.

3. $\lim\limits_{n\to\infty} \left|\dfrac{a_{n+1}}{a_n}\right| = \lim\limits_{n\to\infty} \left|\dfrac{n+1}{5^{n+1}} \cdot \dfrac{5^n}{n}\right| = \lim\limits_{n\to\infty} \left|\dfrac{1}{5} \cdot \dfrac{n+1}{n}\right| = \dfrac{1}{5} \lim\limits_{n\to\infty} \dfrac{1 + 1/n}{1} = \dfrac{1}{5}(1) = \dfrac{1}{5} < 1$, so the series $\sum\limits_{n=1}^{\infty} \dfrac{n}{5^n}$ is

absolutely convergent by the Ratio Test.

5. $b_n = \dfrac{1}{5n+1} > 0$ for $n \geq 0$, $\{b_n\}$ is decreasing for $n \geq 0$, and $\lim\limits_{n\to\infty} b_n = 0$, so $\sum\limits_{n=0}^{\infty} \dfrac{(-1)^n}{5n+1}$ converges by the Alternating

Series Test. To determine absolute convergence, choose $a_n = \dfrac{1}{n}$ to get

$$\lim\limits_{n\to\infty} \dfrac{a_n}{b_n} = \lim\limits_{n\to\infty} \dfrac{1/n}{1/(5n+1)} = \lim\limits_{n\to\infty} \dfrac{5n+1}{n} = 5 > 0, \text{ so } \sum\limits_{n=1}^{\infty} \dfrac{1}{5n+1} \text{ diverges by the Limit Comparison Test with the}$$

harmonic series. Thus, the series $\sum\limits_{n=0}^{\infty} \dfrac{(-1)^n}{5n+1}$ is conditionally convergent.

7. $\lim\limits_{k\to\infty} \left|\dfrac{a_{k+1}}{a_k}\right| = \lim\limits_{k\to\infty} \left[\dfrac{(k+1)\left(\frac{2}{3}\right)^{k+1}}{k\left(\frac{2}{3}\right)^k}\right] = \lim\limits_{k\to\infty} \dfrac{k+1}{k}\left(\dfrac{2}{3}\right)^1 = \dfrac{2}{3} \lim\limits_{k\to\infty} \left(1 + \dfrac{1}{k}\right) = \dfrac{2}{3}(1) = \dfrac{2}{3} < 1$, so the series

$\sum\limits_{n=1}^{\infty} k\left(\frac{2}{3}\right)^k$ is absolutely convergent by the Ratio Test. Since the terms of this series are positive, absolute convergence is the

same as convergence.

9. $\lim\limits_{n\to\infty} \left|\dfrac{a_{n+1}}{a_n}\right| = \lim\limits_{n\to\infty} \left[\dfrac{(1.1)^{n+1}}{(n+1)^4} \cdot \dfrac{n^4}{(1.1)^n}\right] = \lim\limits_{n\to\infty} \dfrac{(1.1)n^4}{(n+1)^4} = (1.1) \lim\limits_{n\to\infty} \dfrac{1}{\dfrac{(n+1)^4}{n^4}} = (1.1) \lim\limits_{n\to\infty} \dfrac{1}{(1 + 1/n)^4}$

$$= (1.1)(1) = 1.1 > 1,$$

so the series $\sum\limits_{n=1}^{\infty} (-1)^n \dfrac{(1.1)^n}{n^4}$ diverges by the Ratio Test.

11. Since $0 \leq \dfrac{e^{1/n}}{n^3} \leq \dfrac{e}{n^3} = e\left(\dfrac{1}{n^3}\right)$ and $\sum\limits_{n=1}^{\infty} \dfrac{1}{n^3}$ is a convergent p-series $[p = 3 > 1]$, $\sum\limits_{n=1}^{\infty} \dfrac{e^{1/n}}{n^3}$ converges, and so

$\sum\limits_{n=1}^{\infty} \dfrac{(-1)^n e^{1/n}}{n^3}$ is absolutely convergent.

© 2012 Cengage Learning. All Rights Reserved. May not be scanned, copied, or duplicated, or posted to a publicly accessible website, in whole or in part.

13. $\lim\limits_{n\to\infty} \left| \dfrac{a_{n+1}}{a_n} \right| = \lim\limits_{n\to\infty} \left[\dfrac{10^{n+1}}{(n+2)\,4^{2n+3}} \cdot \dfrac{(n+1)\,4^{2n+1}}{10^n} \right] = \lim\limits_{n\to\infty} \left(\dfrac{10}{4^2} \cdot \dfrac{n+1}{n+2} \right) = \dfrac{5}{8} < 1$, so the series $\sum\limits_{n=1}^{\infty} \dfrac{10^n}{(n+1)4^{2n+1}}$

is absolutely convergent by the Ratio Test. Since the terms of this series are positive, absolute convergence is the same as convergence.

15. $\left| \dfrac{(-1)^n \arctan n}{n^2} \right| < \dfrac{\pi/2}{n^2}$, so since $\sum\limits_{n=1}^{\infty} \dfrac{\pi/2}{n^2} = \dfrac{\pi}{2} \sum\limits_{n=1}^{\infty} \dfrac{1}{n^2}$ converges ($p = 2 > 1$), the given series $\sum\limits_{n=1}^{\infty} \dfrac{(-1)^n \arctan n}{n^2}$

converges absolutely by the Comparison Test.

17. $\sum\limits_{n=2}^{\infty} \dfrac{(-1)^n}{\ln n}$ converges by the Alternating Series Test since $\lim\limits_{n\to\infty} \dfrac{1}{\ln n} = 0$ and $\left\{ \dfrac{1}{\ln n} \right\}$ is decreasing. Now $\ln n < n$, so

$\dfrac{1}{\ln n} > \dfrac{1}{n}$, and since $\sum\limits_{n=2}^{\infty} \dfrac{1}{n}$ is the divergent (partial) harmonic series, $\sum\limits_{n=2}^{\infty} \dfrac{1}{\ln n}$ diverges by the Comparison Test. Thus,

$\sum\limits_{n=2}^{\infty} \dfrac{(-1)^n}{\ln n}$ is conditionally convergent.

19. $\dfrac{|\cos(n\pi/3)|}{n!} \leq \dfrac{1}{n!}$ and $\sum\limits_{n=1}^{\infty} \dfrac{1}{n!}$ converges (use the Ratio Test), so the series $\sum\limits_{n=1}^{\infty} \dfrac{\cos(n\pi/3)}{n!}$ converges absolutely by the

Comparison Test.

21. $\lim\limits_{n\to\infty} \sqrt[n]{|a_n|} = \lim\limits_{n\to\infty} \dfrac{n^2+1}{2n^2+1} = \lim\limits_{n\to\infty} \dfrac{1+1/n^2}{2+1/n^2} = \dfrac{1}{2} < 1$, so the series $\sum\limits_{n=1}^{\infty} \left(\dfrac{n^2+1}{2n^2+1} \right)^n$ is absolutely convergent by the

Root Test.

23. $\lim\limits_{n\to\infty} \sqrt[n]{|a_n|} = \lim\limits_{n\to\infty} \sqrt[n]{\left(1+\dfrac{1}{n}\right)^{n^2}} = \lim\limits_{n\to\infty} \left(1+\dfrac{1}{n}\right)^n = e > 1$ [by Equation 7.4.9 (or 7.4*.9) [ET 3.6.6]],

so the series $\sum\limits_{n=1}^{\infty} \left(1+\dfrac{1}{n}\right)^{n^2}$ diverges by the Root Test.

25. $\lim\limits_{n\to\infty} \left| \dfrac{a_{n+1}}{a_n} \right| = \lim\limits_{n\to\infty} \left| \dfrac{(n+1)^{100}100^{n+1}}{(n+1)!} \cdot \dfrac{n!}{n^{100}100^n} \right| = \lim\limits_{n\to\infty} \dfrac{100}{n+1} \left(\dfrac{n+1}{n} \right)^{100} = \lim\limits_{n\to\infty} \dfrac{100}{n+1} \left(1+\dfrac{1}{n}\right)^{100}$

$= 0 \cdot 1 = 0 < 1$

so the series $\sum\limits_{n=1}^{\infty} \dfrac{n^{100}100^n}{n!}$ is absolutely convergent by the Ratio Test.

27. Use the Ratio Test with the series

$1 - \dfrac{1 \cdot 3}{3!} + \dfrac{1 \cdot 3 \cdot 5}{5!} - \dfrac{1 \cdot 3 \cdot 5 \cdot 7}{7!} + \cdots + (-1)^{n-1}\dfrac{1 \cdot 3 \cdot 5 \cdot \cdots \cdot (2n-1)}{(2n-1)!} + \cdots = \sum\limits_{n=1}^{\infty} (-1)^{n-1}\dfrac{1 \cdot 3 \cdot 5 \cdot \cdots \cdot (2n-1)}{(2n-1)!}.$

$\lim\limits_{n\to\infty} \left| \dfrac{a_{n+1}}{a_n} \right| = \lim\limits_{n\to\infty} \left| \dfrac{(-1)^n \cdot 1 \cdot 3 \cdot 5 \cdot \cdots \cdot (2n-1)[2(n+1)-1]}{[2(n+1)-1]!} \cdot \dfrac{(2n-1)!}{(-1)^{n-1} \cdot 1 \cdot 3 \cdot 5 \cdot \cdots \cdot (2n-1)} \right|$

$= \lim\limits_{n\to\infty} \left| \dfrac{(-1)(2n+1)(2n-1)!}{(2n+1)(2n)(2n-1)!} \right| = \lim\limits_{n\to\infty} \dfrac{1}{2n} = 0 < 1,$

so the given series is absolutely convergent and therefore convergent.

© 2012 Cengage Learning. All Rights Reserved. May not be scanned, copied, or duplicated, or posted to a publicly accessible website, in whole or in part.

29. $\sum_{n=1}^{\infty} \dfrac{2 \cdot 4 \cdot 6 \cdot \cdots \cdot (2n)}{n!} = \sum_{n=1}^{\infty} \dfrac{(2 \cdot 1) \cdot (2 \cdot 2) \cdot (2 \cdot 3) \cdot \cdots \cdot (2 \cdot n)}{n!} = \sum_{n=1}^{\infty} \dfrac{2^n n!}{n!} = \sum_{n=1}^{\infty} 2^n$, which diverges by the Test for

Divergence since $\lim\limits_{n \to \infty} 2^n = \infty$.

31. By the recursive definition, $\lim\limits_{n \to \infty} \left| \dfrac{a_{n+1}}{a_n} \right| = \lim\limits_{n \to \infty} \left| \dfrac{5n+1}{4n+3} \right| = \dfrac{5}{4} > 1$, so the series diverges by the Ratio Test.

33. The series $\sum\limits_{n=1}^{\infty} \dfrac{b_n^n \cos n\pi}{n} = \sum\limits_{n=1}^{\infty} (-1)^n \dfrac{b_n^n}{n}$, where $b_n > 0$ for $n \geq 1$ and $\lim\limits_{n \to \infty} b_n = \dfrac{1}{2}$.

$\lim\limits_{n \to \infty} \left| \dfrac{a_{n+1}}{a_n} \right| = \lim\limits_{n \to \infty} \left| \dfrac{(-1)^{n+1} b_n^{n+1}}{n+1} \cdot \dfrac{n}{(-1)^n b_n^n} \right| = \lim\limits_{n \to \infty} b_n \dfrac{n}{n+1} = \dfrac{1}{2}(1) = \dfrac{1}{2} < 1$, so the series $\sum\limits_{n=1}^{\infty} \dfrac{b_n^n \cos n\pi}{n}$ is

absolutely convergent by the Ratio Test.

35. (a) $\lim\limits_{n \to \infty} \left| \dfrac{1/(n+1)^3}{1/n^3} \right| = \lim\limits_{n \to \infty} \dfrac{n^3}{(n+1)^3} = \lim\limits_{n \to \infty} \dfrac{1}{(1+1/n)^3} = 1$. Inconclusive

(b) $\lim\limits_{n \to \infty} \left| \dfrac{(n+1)}{2^{n+1}} \cdot \dfrac{2^n}{n} \right| = \lim\limits_{n \to \infty} \dfrac{n+1}{2n} = \lim\limits_{n \to \infty} \left(\dfrac{1}{2} + \dfrac{1}{2n} \right) = \dfrac{1}{2}$. Conclusive (convergent)

(c) $\lim\limits_{n \to \infty} \left| \dfrac{(-3)^n}{\sqrt{n+1}} \cdot \dfrac{\sqrt{n}}{(-3)^{n-1}} \right| = 3 \lim\limits_{n \to \infty} \sqrt{\dfrac{n}{n+1}} = 3 \lim\limits_{n \to \infty} \sqrt{\dfrac{1}{1+1/n}} = 3$. Conclusive (divergent)

(d) $\lim\limits_{n \to \infty} \left| \dfrac{\sqrt{n+1}}{1+(n+1)^2} \cdot \dfrac{1+n^2}{\sqrt{n}} \right| = \lim\limits_{n \to \infty} \left[\sqrt{1 + \dfrac{1}{n}} \cdot \dfrac{1/n^2 + 1}{1/n^2 + (1+1/n)^2} \right] = 1$. Inconclusive

37. (a) $\lim\limits_{n \to \infty} \left| \dfrac{a_{n+1}}{a_n} \right| = \lim\limits_{n \to \infty} \left| \dfrac{x^{n+1}}{(n+1)!} \cdot \dfrac{n!}{x^n} \right| = \lim\limits_{n \to \infty} \left| \dfrac{x}{n+1} \right| = |x| \lim\limits_{n \to \infty} \dfrac{1}{n+1} = |x| \cdot 0 = 0 < 1$, so by the Ratio Test the

series $\sum\limits_{n=0}^{\infty} \dfrac{x^n}{n!}$ converges for all x.

(b) Since the series of part (a) always converges, we must have $\lim\limits_{n \to \infty} \dfrac{x^n}{n!} = 0$ by Theorem 11.2.6.

39. (a) $s_5 = \sum\limits_{n=1}^{5} \dfrac{1}{n2^n} = \dfrac{1}{2} + \dfrac{1}{8} + \dfrac{1}{24} + \dfrac{1}{64} + \dfrac{1}{160} = \dfrac{661}{960} \approx 0.68854$. Now the ratios

$r_n = \dfrac{a_{n+1}}{a_n} = \dfrac{n2^n}{(n+1)2^{n+1}} = \dfrac{n}{2(n+1)}$ form an increasing sequence, since

$r_{n+1} - r_n = \dfrac{n+1}{2(n+2)} - \dfrac{n}{2(n+1)} = \dfrac{(n+1)^2 - n(n+2)}{2(n+1)(n+2)} = \dfrac{1}{2(n+1)(n+2)} > 0$. So by Exercise 34(b), the error

in using s_5 is $R_5 \leq \dfrac{a_6}{1 - \lim\limits_{n \to \infty} r_n} = \dfrac{1/(6 \cdot 2^6)}{1 - 1/2} = \dfrac{1}{192} \approx 0.00521$.

(b) The error in using s_n as an approximation to the sum is $R_n = \dfrac{a_{n+1}}{1 - \frac{1}{2}} = \dfrac{2}{(n+1)2^{n+1}}$. We want $R_n < 0.00005 \quad \Leftrightarrow$

$\dfrac{1}{(n+1)2^n} < 0.00005 \quad \Leftrightarrow \quad (n+1)2^n > 20{,}000$. To find such an n we can use trial and error or a graph. We calculate

$(11+1)2^{11} = 24{,}576$, so $s_{11} = \sum\limits_{n=1}^{11} \dfrac{1}{n2^n} \approx 0.693109$ is within 0.00005 of the actual sum.

© 2012 Cengage Learning. All Rights Reserved. May not be scanned, copied, or duplicated, or posted to a publicly accessible website, in whole or in part.

41. (i) Following the hint, we get that $|a_n| < r^n$ for $n \geq N$, and so since the geometric series $\sum_{n=1}^{\infty} r^n$ converges $[0 < r < 1]$, the series $\sum_{n=N}^{\infty} |a_n|$ converges as well by the Comparison Test, and hence so does $\sum_{n=1}^{\infty} |a_n|$, so $\sum_{n=1}^{\infty} a_n$ is absolutely convergent.

(ii) If $\lim_{n \to \infty} \sqrt[n]{|a_n|} = L > 1$, then there is an integer N such that $\sqrt[n]{|a_n|} > 1$ for all $n \geq N$, so $|a_n| > 1$ for $n \geq N$. Thus, $\lim_{n \to \infty} a_n \neq 0$, so $\sum_{n=1}^{\infty} a_n$ diverges by the Test for Divergence.

(iii) Consider $\sum_{n=1}^{\infty} \frac{1}{n}$ [diverges] and $\sum_{n=1}^{\infty} \frac{1}{n^2}$ [converges]. For each sum, $\lim_{n \to \infty} \sqrt[n]{|a_n|} = 1$, so the Root Test is inconclusive.

43. (a) Since $\sum a_n$ is absolutely convergent, and since $|a_n^+| \leq |a_n|$ and $|a_n^-| \leq |a_n|$ (because a_n^+ and a_n^- each equal either a_n or 0), we conclude by the Comparison Test that both $\sum a_n^+$ and $\sum a_n^-$ must be absolutely convergent.

Or: Use Theorem 11.2.8.

(b) We will show by contradiction that both $\sum a_n^+$ and $\sum a_n^-$ must diverge. For suppose that $\sum a_n^+$ converged. Then so would $\sum \left(a_n^+ - \frac{1}{2} a_n \right)$ by Theorem 11.2.8. But $\sum \left(a_n^+ - \frac{1}{2} a_n \right) = \sum \left[\frac{1}{2} \left(a_n + |a_n| \right) - \frac{1}{2} a_n \right] = \frac{1}{2} \sum |a_n|$, which diverges because $\sum a_n$ is only conditionally convergent. Hence, $\sum a_n^+$ can't converge. Similarly, neither can $\sum a_n^-$.

45. Suppose that $\sum a_n$ is conditionally convergent.

(a) $\sum n^2 a_n$ is divergent: Suppose $\sum n^2 a_n$ converges. Then $\lim_{n \to \infty} n^2 a_n = 0$ by Theorem 6 in Section 11.2, so there is an integer $N > 0$ such that $n > N \Rightarrow n^2 |a_n| < 1$. For $n > N$, we have $|a_n| < \frac{1}{n^2}$, so $\sum_{n > N} |a_n|$ converges by comparison with the convergent p-series $\sum_{n > N} \frac{1}{n^2}$. In other words, $\sum a_n$ converges absolutely, contradicting the assumption that $\sum a_n$ is conditionally convergent. This contradiction shows that $\sum n^2 a_n$ diverges.

Remark: The same argument shows that $\sum n^p a_n$ diverges for any $p > 1$.

(b) $\sum_{n=2}^{\infty} \frac{(-1)^n}{n \ln n}$ is conditionally convergent. It converges by the Alternating Series Test, but does not converge absolutely $\left[\text{by the Integral Test, since the function } f(x) = \frac{1}{x \ln x} \text{ is continuous, positive, and decreasing on } [2, \infty) \text{ and} \right.$
$\int_2^{\infty} \frac{dx}{x \ln x} = \lim_{t \to \infty} \int_2^t \frac{dx}{x \ln x} = \lim_{t \to \infty} \left[\ln(\ln x) \right]_2^t = \infty \Big]$. Setting $a_n = \frac{(-1)^n}{n \ln n}$ for $n \geq 2$, we find that $\sum_{n=2}^{\infty} n a_n = \sum_{n=2}^{\infty} \frac{(-1)^n}{\ln n}$ converges by the Alternating Series Test.

It is easy to find conditionally convergent series $\sum a_n$ such that $\sum n a_n$ diverges. Two examples are $\sum_{n=1}^{\infty} \frac{(-1)^{n-1}}{n}$ and $\sum_{n=1}^{\infty} \frac{(-1)^{n-1}}{\sqrt{n}}$, both of which converge by the Alternating Series Test and fail to converge absolutely because $\sum |a_n|$ is a p-series with $p \leq 1$. In both cases, $\sum n a_n$ diverges by the Test for Divergence.

© 2012 Cengage Learning. All Rights Reserved. May not be scanned, copied, or duplicated, or posted to a publicly accessible website, in whole or in part.

11.7 Strategy for Testing Series

1. $\dfrac{1}{n+3^n} < \dfrac{1}{3^n} = \left(\dfrac{1}{3}\right)^n$ for all $n \geq 1$. $\displaystyle\sum_{n=1}^{\infty} \left(\dfrac{1}{3}\right)^n$ is a convergent geometric series $\left[|r| = \tfrac{1}{3} < 1\right]$, so $\displaystyle\sum_{n=1}^{\infty} \dfrac{1}{n+3^n}$

converges by the Comparison Test.

3. $\displaystyle\lim_{n\to\infty} |a_n| = \lim_{n\to\infty} \dfrac{n}{n+2} = 1$, so $\displaystyle\lim_{n\to\infty} a_n = \lim_{n\to\infty} (-1)^n \dfrac{n}{n+2}$ does not exist. Thus, the series $\displaystyle\sum_{n=1}^{\infty} (-1)^n \dfrac{n}{n+2}$ diverges by

the Test for Divergence.

5. $\displaystyle\lim_{n\to\infty} \left|\dfrac{a_{n+1}}{a_n}\right| = \lim_{n\to\infty} \left|\dfrac{(n+1)^2 \, 2^n}{(-5)^{n+1}} \cdot \dfrac{(-5)^n}{n^2 \, 2^{n-1}}\right| = \lim_{n\to\infty} \dfrac{2(n+1)^2}{5n^2} = \dfrac{2}{5} \lim_{n\to\infty} \left(1+\dfrac{1}{n}\right)^2 = \dfrac{2}{5}(1) = \dfrac{2}{5} < 1$, so the series

$\displaystyle\sum_{n=1}^{\infty} \dfrac{n^2 \, 2^{n-1}}{(-5)^n}$ converges by the Ratio Test.

7. Let $f(x) = \dfrac{1}{x\sqrt{\ln x}}$. Then f is positive, continuous, and decreasing on $[2, \infty)$, so we can apply the Integral Test.

Since $\displaystyle\int \dfrac{1}{x\sqrt{\ln x}}\, dx \; \begin{bmatrix} u = \ln x, \\ du = dx/x \end{bmatrix} = \int u^{-1/2}\, du = 2u^{1/2} + C = 2\sqrt{\ln x} + C$, we find

$\displaystyle\int_2^{\infty} \dfrac{dx}{x\sqrt{\ln x}} = \lim_{t\to\infty} \int_2^t \dfrac{dx}{x\sqrt{\ln x}} = \lim_{t\to\infty} \left[2\sqrt{\ln x}\right]_2^t = \lim_{t\to\infty} \left(2\sqrt{\ln t} - 2\sqrt{\ln 2}\right) = \infty$. Since the integral diverges, the

given series $\displaystyle\sum_{n=2}^{\infty} \dfrac{1}{n\sqrt{\ln n}}$ diverges.

9. $\displaystyle\sum_{k=1}^{\infty} k^2 e^{-k} = \sum_{k=1}^{\infty} \dfrac{k^2}{e^k}$. Using the Ratio Test, we get

$\displaystyle\lim_{k\to\infty} \left|\dfrac{a_{k+1}}{a_k}\right| = \lim_{k\to\infty} \left|\dfrac{(k+1)^2}{e^{k+1}} \cdot \dfrac{e^k}{k^2}\right| = \lim_{k\to\infty} \left[\left(\dfrac{k+1}{k}\right)^2 \cdot \dfrac{1}{e}\right] = 1^2 \cdot \dfrac{1}{e} = \dfrac{1}{e} < 1$, so the series converges.

11. $\displaystyle\sum_{n=1}^{\infty} \left(\dfrac{1}{n^3} + \dfrac{1}{3^n}\right) = \sum_{n=1}^{\infty} \dfrac{1}{n^3} + \sum_{n=1}^{\infty} \left(\dfrac{1}{3}\right)^n$. The first series converges since it is a p-series with $p = 3 > 1$ and the second

series converges since it is geometric with $|r| = \tfrac{1}{3} < 1$. The sum of two convergent series is convergent.

13. $\displaystyle\lim_{n\to\infty} \left|\dfrac{a_{n+1}}{a_n}\right| = \lim_{n\to\infty} \left|\dfrac{3^{n+1}(n+1)^2}{(n+1)!} \cdot \dfrac{n!}{3^n n^2}\right| = \lim_{n\to\infty} \dfrac{3(n+1)^2}{(n+1)n^2} = 3\lim_{n\to\infty} \dfrac{n+1}{n^2} = 0 < 1$, so the series $\displaystyle\sum_{n=1}^{\infty} \dfrac{3^n n^2}{n!}$

converges by the Ratio Test.

15. $a_k = \dfrac{2^{k-1}3^{k+1}}{k^k} = \dfrac{2^k 2^{-1} 3^k 3^1}{k^k} = \dfrac{3}{2}\left(\dfrac{2 \cdot 3}{k}\right)^k$. By the Root Test, $\displaystyle\lim_{k\to\infty} \sqrt[k]{\left(\dfrac{6}{k}\right)^k} = \lim_{k\to\infty} \dfrac{6}{k} = 0 < 1$, so the series

$\displaystyle\sum_{k=1}^{\infty} \left(\dfrac{6}{k}\right)^k$ converges. It follows from Theorem 8(i) in Section 11.2 that the given series, $\displaystyle\sum_{k=1}^{\infty} \dfrac{2^{k-1}3^{k+1}}{k^k} = \sum_{k=1}^{\infty} \dfrac{3}{2}\left(\dfrac{6}{k}\right)^k$,

also converges.

© 2012 Cengage Learning. All Rights Reserved. May not be scanned, copied, or duplicated, or posted to a publicly accessible website, in whole or in part.

17. $\lim\limits_{n\to\infty}\left|\dfrac{a_{n+1}}{a_n}\right| = \lim\limits_{n\to\infty}\left|\dfrac{1\cdot 3\cdot 5\cdot\cdots\cdot(2n-1)(2n+1)}{2\cdot 5\cdot 8\cdot\cdots\cdot(3n-1)(3n+2)}\cdot\dfrac{2\cdot 5\cdot 8\cdot\cdots\cdot(3n-1)}{1\cdot 3\cdot 5\cdot\cdots\cdot(2n-1)}\right| = \lim\limits_{n\to\infty}\dfrac{2n+1}{3n+2}$

$\qquad = \lim\limits_{n\to\infty}\dfrac{2+1/n}{3+2/n} = \dfrac{2}{3} < 1,$

so the series $\sum\limits_{n=1}^{\infty}\dfrac{1\cdot 3\cdot 5\cdot\cdots\cdot(2n-1)}{2\cdot 5\cdot 8\cdot\cdots\cdot(3n-1)}$ converges by the Ratio Test.

19. Let $f(x) = \dfrac{\ln x}{\sqrt{x}}$. Then $f'(x) = \dfrac{2-\ln x}{2x^{3/2}} < 0$ when $\ln x > 2$ or $x > e^2$, so $\dfrac{\ln n}{\sqrt{n}}$ is decreasing for $n > e^2$.

By l'Hospital's Rule, $\lim\limits_{n\to\infty}\dfrac{\ln n}{\sqrt{n}} = \lim\limits_{n\to\infty}\dfrac{1/n}{1/\left(2\sqrt{n}\right)} = \lim\limits_{n\to\infty}\dfrac{2}{\sqrt{n}} = 0$, so the series $\sum\limits_{n=1}^{\infty}(-1)^n\dfrac{\ln n}{\sqrt{n}}$ converges by the

Alternating Series Test.

21. $\lim\limits_{n\to\infty}|a_n| = \lim\limits_{n\to\infty}\left|(-1)^n\cos(1/n^2)\right| = \lim\limits_{n\to\infty}\left|\cos(1/n^2)\right| = \cos 0 = 1$, so the series $\sum\limits_{n=1}^{\infty}(-1)^n\cos(1/n^2)$ diverges by the

Test for Divergence.

23. Using the Limit Comparison Test with $a_n = \tan\left(\dfrac{1}{n}\right)$ and $b_n = \dfrac{1}{n}$, we have

$\lim\limits_{n\to\infty}\dfrac{a_n}{b_n} = \lim\limits_{n\to\infty}\dfrac{\tan(1/n)}{1/n} = \lim\limits_{x\to\infty}\dfrac{\tan(1/x)}{1/x} \overset{\text{H}}{=} \lim\limits_{x\to\infty}\dfrac{\sec^2(1/x)\cdot(-1/x^2)}{-1/x^2} = \lim\limits_{x\to\infty}\sec^2(1/x) = 1^2 = 1 > 0$. Since

$\sum\limits_{n=1}^{\infty} b_n$ is the divergent harmonic series, $\sum\limits_{n=1}^{\infty} a_n$ is also divergent.

25. Use the Ratio Test. $\lim\limits_{n\to\infty}\left|\dfrac{a_{n+1}}{a_n}\right| = \lim\limits_{n\to\infty}\left|\dfrac{(n+1)!}{e^{(n+1)^2}}\cdot\dfrac{e^{n^2}}{n!}\right| = \lim\limits_{n\to\infty}\dfrac{(n+1)n!\cdot e^{n^2}}{e^{n^2+2n+1}n!} = \lim\limits_{n\to\infty}\dfrac{n+1}{e^{2n+1}} = 0 < 1$, so $\sum\limits_{n=1}^{\infty}\dfrac{n!}{e^{n^2}}$

converges.

27. $\displaystyle\int_2^{\infty}\dfrac{\ln x}{x^2}\,dx = \lim\limits_{t\to\infty}\left[-\dfrac{\ln x}{x} - \dfrac{1}{x}\right]_1^t$ [using integration by parts] $\overset{\text{H}}{=} 1$. So $\sum\limits_{n=1}^{\infty}\dfrac{\ln n}{n^2}$ converges by the Integral Test, and since

$\dfrac{k\ln k}{(k+1)^3} < \dfrac{k\ln k}{k^3} = \dfrac{\ln k}{k^2}$, the given series $\sum\limits_{k=1}^{\infty}\dfrac{k\ln k}{(k+1)^3}$ converges by the Comparison Test.

29. $\sum\limits_{n=1}^{\infty} a_n = \sum\limits_{n=1}^{\infty}(-1)^n\dfrac{1}{\cosh n} = \sum\limits_{n=1}^{\infty}(-1)^n b_n$. Now $b_n = \dfrac{1}{\cosh n} > 0$, $\{b_n\}$ is decreasing, and $\lim\limits_{n\to\infty} b_n = 0$, so the series

converges by the Alternating Series Test.

Or: Write $\dfrac{1}{\cosh n} = \dfrac{2}{e^n+e^{-n}} < \dfrac{2}{e^n}$ and $\sum\limits_{n=1}^{\infty}\dfrac{1}{e^n}$ is a convergent geometric series, so $\sum\limits_{n=1}^{\infty}\dfrac{1}{\cosh n}$ is convergent by the

Comparison Test. So $\sum\limits_{n=1}^{\infty}(-1)^n\dfrac{1}{\cosh n}$ is absolutely convergent and therefore convergent.

31. $\lim\limits_{k\to\infty} a_k = \lim\limits_{k\to\infty}\dfrac{5^k}{3^k+4^k} = $ [divide by 4^k] $\lim\limits_{k\to\infty}\dfrac{(5/4)^k}{(3/4)^k+1} = \infty$ since $\lim\limits_{k\to\infty}\left(\dfrac{3}{4}\right)^k = 0$ and $\lim\limits_{k\to\infty}\left(\dfrac{5}{4}\right)^k = \infty$.

Thus, $\sum\limits_{k=1}^{\infty}\dfrac{5^k}{3^k+4^k}$ diverges by the Test for Divergence.

© 2012 Cengage Learning. All Rights Reserved. May not be scanned, copied, or duplicated, or posted to a publicly accessible website, in whole or in part.

33. $\lim\limits_{n\to\infty} \sqrt[n]{|a_n|} = \lim\limits_{n\to\infty} \left(\dfrac{n}{n+1}\right)^{n^2/n} = \lim\limits_{n\to\infty} \dfrac{1}{[(n+1)/n]^n} = \dfrac{1}{\lim\limits_{n\to\infty}(1+1/n)^n} = \dfrac{1}{e} < 1$, so the series $\sum\limits_{n=1}^{\infty} \left(\dfrac{n}{n+1}\right)^{n^2}$

converges by the Root Test.

35. $a_n = \dfrac{1}{n^{1+1/n}} = \dfrac{1}{n \cdot n^{1/n}}$, so let $b_n = \dfrac{1}{n}$ and use the Limit Comparison Test. $\lim\limits_{n\to\infty}\dfrac{a_n}{b_n} = \lim\limits_{n\to\infty}\dfrac{1}{n^{1/n}} = 1 > 0$

[see Exercise 4.4.61], so the series $\sum\limits_{n=1}^{\infty} \dfrac{1}{n^{1+1/n}}$ diverges by comparison with the divergent harmonic series.

37. $\lim\limits_{n\to\infty} \sqrt[n]{|a_n|} = \lim\limits_{n\to\infty} (2^{1/n} - 1) = 1 - 1 = 0 < 1$, so the series $\sum\limits_{n=1}^{\infty} \left(\sqrt[n]{2} - 1\right)^n$ converges by the Root Test.

11.8 Power Series

1. A power series is a series of the form $\sum_{n=0}^{\infty} c_n x^n = c_0 + c_1 x + c_2 x^2 + c_3 x^3 + \cdots$, where x is a variable and the c_n's are

constants called the coefficients of the series.

More generally, a series of the form $\sum_{n=0}^{\infty} c_n(x-a)^n = c_0 + c_1(x-a) + c_2(x-a)^2 + \cdots$ is called a power series in

$(x-a)$ or a power series centered at a or a power series about a, where a is a constant.

3. If $a_n = (-1)^n n x^n$, then

$$\lim_{n\to\infty} \left|\frac{a_{n+1}}{a_n}\right| = \lim_{n\to\infty} \left|\frac{(-1)^{n+1}(n+1)x^{n+1}}{(-1)^n nx^n}\right| = \lim_{n\to\infty} \left|(-1)\frac{n+1}{n}x\right| = \lim_{n\to\infty} \left[\left(1+\frac{1}{n}\right)|x|\right] = |x|. \text{ By the Ratio Test, the}$$

series $\sum\limits_{n=1}^{\infty}(-1)^n nx^n$ converges when $|x| < 1$, so the radius of convergence $R = 1$. Now we'll check the endpoints, that is,

$x = \pm 1$. Both series $\sum\limits_{n=1}^{\infty}(-1)^n n(\pm 1)^n = \sum\limits_{n=1}^{\infty}(\mp 1)^n n$ diverge by the Test for Divergence since $\lim\limits_{n\to\infty}|(\mp 1)^n n| = \infty$. Thus,

the interval of convergence is $I = (-1, 1)$.

5. If $a_n = \dfrac{x^n}{2n-1}$, then $\lim\limits_{n\to\infty}\left|\dfrac{a_{n+1}}{a_n}\right| = \lim\limits_{n\to\infty}\left|\dfrac{x^{n+1}}{2n+1} \cdot \dfrac{2n-1}{x^n}\right| = \lim\limits_{n\to\infty}\left(\dfrac{2n-1}{2n+1}|x|\right) = \lim\limits_{n\to\infty}\left(\dfrac{2-1/n}{2+1/n}|x|\right) = |x|$. By

the Ratio Test, the series $\sum\limits_{n=1}^{\infty}\dfrac{x^n}{2n-1}$ converges when $|x| < 1$, so $R = 1$. When $x = 1$, the series $\sum\limits_{n=1}^{\infty}\dfrac{1}{2n-1}$ diverges by

comparison with $\sum\limits_{n=1}^{\infty}\dfrac{1}{2n}$ since $\dfrac{1}{2n-1} > \dfrac{1}{2n}$ and $\dfrac{1}{2}\sum\limits_{n=1}^{\infty}\dfrac{1}{n}$ diverges since it is a constant multiple of the harmonic series.

When $x = -1$, the series $\sum\limits_{n=1}^{\infty}\dfrac{(-1)^n}{2n-1}$ converges by the Alternating Series Test. Thus, the interval of convergence is $[-1, 1)$.

7. If $a_n = \dfrac{x^n}{n!}$, then $\lim\limits_{n\to\infty}\left|\dfrac{a_{n+1}}{a_n}\right| = \lim\limits_{n\to\infty}\left|\dfrac{x^{n+1}}{(n+1)!} \cdot \dfrac{n!}{x^n}\right| = \lim\limits_{n\to\infty}\left|\dfrac{x}{n+1}\right| = |x|\lim\limits_{n\to\infty}\dfrac{1}{n+1} = |x| \cdot 0 = 0 < 1$ for *all* real x.

So, by the Ratio Test, $R = \infty$ and $I = (-\infty, \infty)$.

9. If $a_n = (-1)^n \dfrac{n^2 x^n}{2^n}$, then

$$\lim_{n\to\infty}\left|\frac{a_{n+1}}{a_n}\right| = \lim_{n\to\infty}\left|\frac{(n+1)^2 x^{n+1}}{2^{n+1}} \cdot \frac{2^n}{n^2 x^n}\right| = \lim_{n\to\infty}\left|\frac{x(n+1)^2}{2n^2}\right| = \lim_{n\to\infty}\left[\frac{|x|}{2}\left(1+\frac{1}{n}\right)^2\right] = \frac{|x|}{2}(1)^2 = \tfrac{1}{2}|x|. \text{ By the}$$

© 2012 Cengage Learning. All Rights Reserved. May not be scanned, copied, or duplicated, or posted to a publicly accessible website, in whole or in part.

Ratio Test, the series $\sum\limits_{n=1}^{\infty} (-1)^n \dfrac{n^2 x^n}{2^n}$ converges when $\frac{1}{2}|x| < 1 \;\Leftrightarrow\; |x| < 2$, so the radius of convergence is $R = 2$.

When $x = \pm 2$, both series $\sum\limits_{n=1}^{\infty} (-1)^n \dfrac{n^2 (\pm 2)^n}{2^n} = \sum\limits_{n=1}^{\infty} (\mp 1)^n n^2$ diverge by the Test for Divergence since

$\lim\limits_{n \to \infty} \left| (\mp 1)^n n^2 \right| = \infty$. Thus, the interval of convergence is $I = (-2, 2)$.

11. If $a_n = \dfrac{(-3)^n x^n}{n^{3/2}}$, then

$$\lim_{n \to \infty} \left| \frac{a_{n+1}}{a_n} \right| = \lim_{n \to \infty} \left| \frac{(-3)^{n+1} x^{n+1}}{(n+1)^{3/2}} \cdot \frac{n^{3/2}}{(-3)^n x^n} \right| = \lim_{n \to \infty} \left| -3x \left(\frac{n}{n+1} \right)^{3/2} \right| = 3|x| \lim_{n \to \infty} \left(\frac{1}{1+1/n} \right)^{3/2}$$

$$= 3|x|\,(1) = 3|x|$$

By the Ratio Test, the series $\sum\limits_{n=1}^{\infty} \dfrac{(-3)^n}{n\sqrt{n}} x^n$ converges when $3|x| < 1 \;\Leftrightarrow\; |x| < \frac{1}{3}$, so $R = \frac{1}{3}$. When $x = \frac{1}{3}$, the series

$\sum\limits_{n=1}^{\infty} \dfrac{(-1)^n}{n^{3/2}}$ converges by the Alternating Series Test. When $x = -\frac{1}{3}$, the series $\sum\limits_{n=1}^{\infty} \dfrac{1}{n^{3/2}}$ is a convergent p-series

$\left(p = \frac{3}{2} > 1 \right)$. Thus, the interval of convergence is $\left[-\frac{1}{3}, \frac{1}{3} \right]$.

13. If $a_n = (-1)^n \dfrac{x^n}{4^n \ln n}$, then $\lim\limits_{n \to \infty} \left| \dfrac{a_{n+1}}{a_n} \right| = \lim\limits_{n \to \infty} \left| \dfrac{x^{n+1}}{4^{n+1} \ln(n+1)} \cdot \dfrac{4^n \ln n}{x^n} \right| = \dfrac{|x|}{4} \lim\limits_{n \to \infty} \dfrac{\ln n}{\ln(n+1)} = \dfrac{|x|}{4} \cdot 1$

[by l'Hospital's Rule] $= \dfrac{|x|}{4}$. By the Ratio Test, the series converges when $\dfrac{|x|}{4} < 1 \;\Leftrightarrow\; |x| < 4$, so $R = 4$. When

$x = -4$, $\sum\limits_{n=2}^{\infty} (-1)^n \dfrac{x^n}{4^n \ln n} = \sum\limits_{n=2}^{\infty} \dfrac{[(-1)(-4)]^n}{4^n \ln n} = \sum\limits_{n=2}^{\infty} \dfrac{1}{\ln n}$. Since $\ln n < n$ for $n \geq 2$, $\dfrac{1}{\ln n} > \dfrac{1}{n}$ and $\sum\limits_{n=2}^{\infty} \dfrac{1}{n}$ is the

divergent harmonic series (without the $n = 1$ term), $\sum\limits_{n=2}^{\infty} \dfrac{1}{\ln n}$ is divergent by the Comparison Test. When $x = 4$,

$\sum\limits_{n=2}^{\infty} (-1)^n \dfrac{x^n}{4^n \ln n} = \sum\limits_{n=2}^{\infty} (-1)^n \dfrac{1}{\ln n}$, which converges by the Alternating Series Test. Thus, $I = (-4, 4]$.

15. If $a_n = \dfrac{(x-2)^n}{n^2+1}$, then $\lim\limits_{n \to \infty} \left| \dfrac{a_{n+1}}{a_n} \right| = \lim\limits_{n \to \infty} \left| \dfrac{(x-2)^{n+1}}{(n+1)^2+1} \cdot \dfrac{n^2+1}{(x-2)^n} \right| = |x-2| \lim\limits_{n \to \infty} \dfrac{n^2+1}{(n+1)^2+1} = |x-2|$. By the

Ratio Test, the series $\sum\limits_{n=0}^{\infty} \dfrac{(x-2)^n}{n^2+1}$ converges when $|x-2| < 1 \;\;[R = 1] \;\Leftrightarrow\; -1 < x - 2 < 1 \;\Leftrightarrow\; 1 < x < 3$. When

$x = 1$, the series $\sum\limits_{n=0}^{\infty} (-1)^n \dfrac{1}{n^2+1}$ converges by the Alternating Series Test; when $x = 3$, the series $\sum\limits_{n=0}^{\infty} \dfrac{1}{n^2+1}$ converges by

comparison with the p-series $\sum\limits_{n=1}^{\infty} \dfrac{1}{n^2} \;\;[p = 2 > 1]$. Thus, the interval of convergence is $I = [1, 3]$.

17. If $a_n = \dfrac{3^n (x+4)^n}{\sqrt{n}}$, then $\lim\limits_{n \to \infty} \left| \dfrac{a_{n+1}}{a_n} \right| = \lim\limits_{n \to \infty} \left| \dfrac{3^{n+1} (x+4)^{n+1}}{\sqrt{n+1}} \cdot \dfrac{\sqrt{n}}{3^n (x+4)^n} \right| = 3|x+4| \lim\limits_{n \to \infty} \dfrac{\sqrt{n}}{\sqrt{n+1}} = 3|x+4|$.

By the Ratio Test, the series $\sum\limits_{n=1}^{\infty} \dfrac{3^n (x+4)^n}{\sqrt{n}}$ converges when $3|x+4| < 1 \;\Leftrightarrow\; |x+4| < \frac{1}{3} \;\;\left[R = \frac{1}{3} \right] \;\Leftrightarrow\;$

© 2012 Cengage Learning. All Rights Reserved. May not be scanned, copied, or duplicated, or posted to a publicly accessible website, in whole or in part.

$-\frac{1}{3} < x + 4 < \frac{1}{3}$ ⇔ $-\frac{13}{3} < x < -\frac{11}{3}$. When $x = -\frac{13}{3}$, the series $\sum\limits_{n=1}^{\infty} (-1)^n \dfrac{1}{\sqrt{n}}$ converges by the Alternating Series

Test; when $x = -\frac{11}{3}$, the series $\sum\limits_{n=1}^{\infty} \dfrac{1}{\sqrt{n}}$ diverges $\left[p = \frac{1}{2} \leq 1 \right]$. Thus, the interval of convergence is $I = \left[-\frac{13}{3}, -\frac{11}{3} \right)$.

19. If $a_n = \dfrac{(x-2)^n}{n^n}$, then $\lim\limits_{n \to \infty} \sqrt[n]{|a_n|} = \lim\limits_{n \to \infty} \dfrac{|x-2|}{n} = 0$, so the series converges for all x (by the Root Test).

$R = \infty$ and $I = (-\infty, \infty)$.

21. $a_n = \dfrac{n}{b^n} (x-a)^n$, where $b > 0$.

$\lim\limits_{n \to \infty} \left| \dfrac{a_{n+1}}{a_n} \right| = \lim\limits_{n \to \infty} \dfrac{(n+1)|x-a|^{n+1}}{b^{n+1}} \cdot \dfrac{b^n}{n|x-a|^n} = \lim\limits_{n \to \infty} \left(1 + \dfrac{1}{n} \right) \dfrac{|x-a|}{b} = \dfrac{|x-a|}{b}$.

By the Ratio Test, the series converges when $\dfrac{|x-a|}{b} < 1$ ⇔ $|x-a| < b$ [so $R = b$] ⇔ $-b < x - a < b$ ⇔

$a - b < x < a + b$. When $|x - a| = b$, $\lim\limits_{n \to \infty} |a_n| = \lim\limits_{n \to \infty} n = \infty$, so the series diverges. Thus, $I = (a - b, a + b)$.

23. If $a_n = n! (2x-1)^n$, then $\lim\limits_{n \to \infty} \left| \dfrac{a_{n+1}}{a_n} \right| = \lim\limits_{n \to \infty} \left| \dfrac{(n+1)! (2x-1)^{n+1}}{n!(2x-1)^n} \right| = \lim\limits_{n \to \infty} (n+1)|2x-1| \to \infty$ as $n \to \infty$

for all $x \neq \frac{1}{2}$. Since the series diverges for all $x \neq \frac{1}{2}$, $R = 0$ and $I = \left\{ \frac{1}{2} \right\}$.

25. If $a_n = \dfrac{(5x-4)^n}{n^3}$, then

$$\lim_{n \to \infty} \left| \dfrac{a_{n+1}}{a_n} \right| = \lim_{n \to \infty} \left| \dfrac{(5x-4)^{n+1}}{(n+1)^3} \cdot \dfrac{n^3}{(5x-4)^n} \right| = \lim_{n \to \infty} |5x-4| \left(\dfrac{n}{n+1} \right)^3 = \lim_{n \to \infty} |5x-4| \left(\dfrac{1}{1 + 1/n} \right)^3$$

$$= |5x-4| \cdot 1 = |5x-4|$$

By the Ratio Test, $\sum\limits_{n=1}^{\infty} \dfrac{(5x-4)^n}{n^3}$ converges when $|5x-4| < 1$ ⇔ $\left| x - \frac{4}{5} \right| < \frac{1}{5}$ ⇔ $-\frac{1}{5} < x - \frac{4}{5} < \frac{1}{5}$ ⇔

$\frac{3}{5} < x < 1$, so $R = \frac{1}{5}$. When $x = 1$, the series $\sum\limits_{n=1}^{\infty} \dfrac{1}{n^3}$ is a convergent p-series ($p = 3 > 1$). When $x = \frac{3}{5}$, the series

$\sum\limits_{n=1}^{\infty} \dfrac{(-1)^n}{n^3}$ converges by the Alternating Series Test. Thus, the interval of convergence is $I = \left[\frac{3}{5}, 1 \right]$.

27. If $a_n = \dfrac{x^n}{1 \cdot 3 \cdot 5 \cdot \,\cdots\, \cdot (2n-1)}$, then

$\lim\limits_{n \to \infty} \left| \dfrac{a_{n+1}}{a_n} \right| = \lim\limits_{n \to \infty} \left| \dfrac{x^{n+1}}{1 \cdot 3 \cdot 5 \cdot \,\cdots\, \cdot (2n-1)(2n+1)} \cdot \dfrac{1 \cdot 3 \cdot 5 \cdot \,\cdots\, \cdot (2n-1)}{x^n} \right| = \lim\limits_{n \to \infty} \dfrac{|x|}{2n+1} = 0 < 1$. Thus, by

the Ratio Test, the series $\sum\limits_{n=1}^{\infty} \dfrac{x^n}{1 \cdot 3 \cdot 5 \cdot \,\cdots\, \cdot (2n-1)}$ converges for *all* real x and we have $R = \infty$ and $I = (-\infty, \infty)$.

29. (a) We are given that the power series $\sum_{n=0}^{\infty} c_n x^n$ is convergent for $x = 4$. So by Theorem 3, it must converge for at least

$-4 < x \leq 4$. In particular, it converges when $x = -2$; that is, $\sum_{n=0}^{\infty} c_n (-2)^n$ is convergent.

© 2012 Cengage Learning. All Rights Reserved. May not be scanned, copied, or duplicated, or posted to a publicly accessible website, in whole or in part.

(b) It does not follow that $\sum_{n=0}^{\infty} c_n(-4)^n$ is necessarily convergent. [See the comments after Theorem 3 about convergence at the endpoint of an interval. An example is $c_n = (-1)^n/(n4^n)$.]

31. If $a_n = \dfrac{(n!)^k}{(kn)!} x^n$, then

$$\lim_{n \to \infty} \left| \frac{a_{n+1}}{a_n} \right| = \lim_{n \to \infty} \frac{[(n+1)!]^k (kn)!}{(n!)^k [k(n+1)]!} |x| = \lim_{n \to \infty} \frac{(n+1)^k}{(kn+k)(kn+k-1)\cdots(kn+2)(kn+1)} |x|$$

$$= \lim_{n \to \infty} \left[\frac{(n+1)}{(kn+1)} \frac{(n+1)}{(kn+2)} \cdots \frac{(n+1)}{(kn+k)} \right] |x|$$

$$= \lim_{n \to \infty} \left[\frac{n+1}{kn+1} \right] \lim_{n \to \infty} \left[\frac{n+1}{kn+2} \right] \cdots \lim_{n \to \infty} \left[\frac{n+1}{kn+k} \right] |x|$$

$$= \left(\frac{1}{k} \right)^k |x| < 1 \quad \Leftrightarrow \quad |x| < k^k \text{ for convergence, and the radius of convergence is } R = k^k.$$

33. No. If a power series is centered at a, its interval of convergence is symmetric about a. If a power series has an infinite radius of convergence, then its interval of convergence must be $(-\infty, \infty)$, not $[0, \infty)$.

35. (a) If $a_n = \dfrac{(-1)^n x^{2n+1}}{n!(n+1)! 2^{2n+1}}$, then

$$\lim_{n \to \infty} \left| \frac{a_{n+1}}{a_n} \right| = \lim_{n \to \infty} \left| \frac{x^{2n+3}}{(n+1)!(n+2)! 2^{2n+3}} \cdot \frac{n!(n+1)! 2^{2n+1}}{x^{2n+1}} \right| = \left(\frac{x}{2} \right)^2 \lim_{n \to \infty} \frac{1}{(n+1)(n+2)} = 0 \text{ for all } x.$$

So $J_1(x)$ converges for all x and its domain is $(-\infty, \infty)$.

(b), (c) The initial terms of $J_1(x)$ up to $n = 5$ are $a_0 = \dfrac{x}{2}$,

$a_1 = -\dfrac{x^3}{16}, a_2 = \dfrac{x^5}{384}, a_3 = -\dfrac{x^7}{18,432}, a_4 = \dfrac{x^9}{1,474,560}$,

and $a_5 = -\dfrac{x^{11}}{176,947,200}$. The partial sums seem to approximate $J_1(x)$ well near the origin, but as $|x|$ increases, we need to take a large number of terms to get a good approximation.

37. $s_{2n-1} = 1 + 2x + x^2 + 2x^3 + x^4 + 2x^5 + \cdots + x^{2n-2} + 2x^{2n-1}$

$$= 1(1+2x) + x^2(1+2x) + x^4(1+2x) + \cdots + x^{2n-2}(1+2x) = (1+2x)(1+x^2+x^4+\cdots+x^{2n-2})$$

$$= (1+2x)\frac{1-x^{2n}}{1-x^2} \text{ [by (11.2.3) with } r = x^2] \to \frac{1+2x}{1-x^2} \text{ as } n \to \infty \text{ by (11.2.4), when } |x| < 1.$$

Also $s_{2n} = s_{2n-1} + x^{2n} \to \dfrac{1+2x}{1-x^2}$ since $x^{2n} \to 0$ for $|x| < 1$. Therefore, $s_n \to \dfrac{1+2x}{1-x^2}$ since s_{2n} and s_{2n-1} both

approach $\dfrac{1+2x}{1-x^2}$ as $n \to \infty$. Thus, the interval of convergence is $(-1, 1)$ and $f(x) = \dfrac{1+2x}{1-x^2}$.

© 2012 Cengage Learning. All Rights Reserved. May not be scanned, copied, or duplicated, or posted to a publicly accessible website, in whole or in part.

39. We use the Root Test on the series $\sum c_n x^n$. We need $\lim\limits_{n\to\infty} \sqrt[n]{|c_n x^n|} = |x| \lim\limits_{n\to\infty} \sqrt[n]{|c_n|} = c\,|x| < 1$ for convergence, or $|x| < 1/c$, so $R = 1/c$.

41. For $2 < x < 3$, $\sum c_n x^n$ diverges and $\sum d_n x^n$ converges. By Exercise 11.2.69, $\sum (c_n + d_n)\,x^n$ diverges. Since both series converge for $|x| < 2$, the radius of convergence of $\sum (c_n + d_n)\,x^n$ is 2.

11.9 Representations of Functions as Power Series

1. If $f(x) = \sum\limits_{n=0}^{\infty} c_n x^n$ has radius of convergence 10, then $f'(x) = \sum\limits_{n=1}^{\infty} n c_n x^{n-1}$ also has radius of convergence 10 by Theorem 2.

3. Our goal is to write the function in the form $\dfrac{1}{1-r}$, and then use Equation (1) to represent the function as a sum of a power series. $f(x) = \dfrac{1}{1+x} = \dfrac{1}{1-(-x)} = \sum\limits_{n=0}^{\infty}(-x)^n = \sum\limits_{n=0}^{\infty}(-1)^n x^n$ with $|-x| < 1 \;\Leftrightarrow\; |x| < 1$, so $R = 1$ and $I = (-1, 1)$.

5. $f(x) = \dfrac{2}{3-x} = \dfrac{2}{3}\left(\dfrac{1}{1-x/3}\right) = \dfrac{2}{3}\sum\limits_{n=0}^{\infty}\left(\dfrac{x}{3}\right)^n$ or, equivalently, $2\sum\limits_{n=0}^{\infty}\dfrac{1}{3^{n+1}}\,x^n$. The series converges when $\left|\dfrac{x}{3}\right| < 1$, that is, when $|x| < 3$, so $R = 3$ and $I = (-3, 3)$.

7. $f(x) = \dfrac{x}{9+x^2} = \dfrac{x}{9}\left[\dfrac{1}{1+(x/3)^2}\right] = \dfrac{x}{9}\left[\dfrac{1}{1-\{-(x/3)^2\}}\right] = \dfrac{x}{9}\sum\limits_{n=0}^{\infty}\left[-\left(\dfrac{x}{3}\right)^2\right]^n = \dfrac{x}{9}\sum\limits_{n=0}^{\infty}(-1)^n\dfrac{x^{2n}}{9^n} = \sum\limits_{n=0}^{\infty}(-1)^n\dfrac{x^{2n+1}}{9^{n+1}}$

The geometric series $\sum\limits_{n=0}^{\infty}\left[-\left(\dfrac{x}{3}\right)^2\right]^n$ converges when $\left|-\left(\dfrac{x}{3}\right)^2\right| < 1 \;\Leftrightarrow\; \dfrac{|x^2|}{9} < 1 \;\Leftrightarrow\; |x|^2 < 9 \;\Leftrightarrow\; |x| < 3$, so $R = 3$ and $I = (-3, 3)$.

9. $f(x) = \dfrac{1+x}{1-x} = (1+x)\left(\dfrac{1}{1-x}\right) = (1+x)\sum\limits_{n=0}^{\infty} x^n = \sum\limits_{n=0}^{\infty} x^n + \sum\limits_{n=0}^{\infty} x^{n+1} = 1 + \sum\limits_{n=1}^{\infty} x^n + \sum\limits_{n=1}^{\infty} x^n = 1 + 2\sum\limits_{n=1}^{\infty} x^n$.

The series converges when $|x| < 1$, so $R = 1$ and $I = (-1, 1)$.

A second approach: $f(x) = \dfrac{1+x}{1-x} = \dfrac{-(1-x)+2}{1-x} = -1 + 2\left(\dfrac{1}{1-x}\right) = -1 + 2\sum\limits_{n=0}^{\infty} x^n = 1 + 2\sum\limits_{n=1}^{\infty} x^n$.

A third approach:

$$f(x) = \dfrac{1+x}{1-x} = (1+x)\left(\dfrac{1}{1-x}\right) = (1+x)(1 + x + x^2 + x^3 + \cdots)$$

$$= (1 + x + x^2 + x^3 + \cdots) + (x + x^2 + x^3 + x^4 + \cdots) = 1 + 2x + 2x^2 + 2x^3 + \cdots = 1 + 2\sum\limits_{n=1}^{\infty} x^n.$$

© 2012 Cengage Learning. All Rights Reserved. May not be scanned, copied, or duplicated, or posted to a publicly accessible website, in whole or in part.

11. $f(x) = \dfrac{3}{x^2 - x - 2} = \dfrac{3}{(x-2)(x+1)} = \dfrac{A}{x-2} + \dfrac{B}{x+1}$ $\Rightarrow$ $3 = A(x+1) + B(x-2)$. Let $x = 2$ to get $A = 1$ and

$x = -1$ to get $B = -1$. Thus

$$\dfrac{3}{x^2 - x - 2} = \dfrac{1}{x-2} - \dfrac{1}{x+1} = \dfrac{1}{-2}\left(\dfrac{1}{1-(x/2)}\right) - \dfrac{1}{1-(-x)} = -\dfrac{1}{2}\sum_{n=0}^{\infty}\left(\dfrac{x}{2}\right)^n - \sum_{n=0}^{\infty}(-x)^n$$

$$= \sum_{n=0}^{\infty}\left[-\dfrac{1}{2}\left(\dfrac{1}{2}\right)^n - 1(-1)^n\right]x^n = \sum_{n=0}^{\infty}\left[(-1)^{n+1} - \dfrac{1}{2^{n+1}}\right]x^n$$

We represented f as the sum of two geometric series; the first converges for $x \in (-2, 2)$ and the second converges for $(-1, 1)$.

Thus, the sum converges for $x \in (-1, 1) = I$.

13. (a) $f(x) = \dfrac{1}{(1+x)^2} = \dfrac{d}{dx}\left(\dfrac{-1}{1+x}\right) = -\dfrac{d}{dx}\left[\sum_{n=0}^{\infty}(-1)^n x^n\right]$ [from Exercise 3]

$= \sum_{n=1}^{\infty}(-1)^{n+1} nx^{n-1}$ [from Theorem 2(i)] $= \sum_{n=0}^{\infty}(-1)^n(n+1)x^n$ with $R = 1$.

In the last step, note that we *decreased* the initial value of the summation variable n by 1, and then *increased* each

occurrence of n in the term by 1 [also note that $(-1)^{n+2} = (-1)^n$].

(b) $f(x) = \dfrac{1}{(1+x)^3} = -\dfrac{1}{2}\dfrac{d}{dx}\left[\dfrac{1}{(1+x)^2}\right] = -\dfrac{1}{2}\dfrac{d}{dx}\left[\sum_{n=0}^{\infty}(-1)^n(n+1)x^n\right]$ [from part (a)]

$= -\dfrac{1}{2}\sum_{n=1}^{\infty}(-1)^n(n+1)nx^{n-1} = \dfrac{1}{2}\sum_{n=0}^{\infty}(-1)^n(n+2)(n+1)x^n$ with $R = 1$.

(c) $f(x) = \dfrac{x^2}{(1+x)^3} = x^2 \cdot \dfrac{1}{(1+x)^3} = x^2 \cdot \dfrac{1}{2}\sum_{n=0}^{\infty}(-1)^n(n+2)(n+1)x^n$ [from part (b)]

$= \dfrac{1}{2}\sum_{n=0}^{\infty}(-1)^n(n+2)(n+1)x^{n+2}$

To write the power series with x^n rather than x^{n+2}, we will *decrease* each occurrence of n in the term by 2 and *increase*

the initial value of the summation variable by 2. This gives us $\dfrac{1}{2}\sum_{n=2}^{\infty}(-1)^n(n)(n-1)x^n$ with $R = 1$.

15. $f(x) = \ln(5 - x) = -\displaystyle\int \dfrac{dx}{5-x} = -\dfrac{1}{5}\int \dfrac{dx}{1 - x/5} = -\dfrac{1}{5}\int\left[\sum_{n=0}^{\infty}\left(\dfrac{x}{5}\right)^n\right]dx = C - \dfrac{1}{5}\sum_{n=0}^{\infty}\dfrac{x^{n+1}}{5^n(n+1)} = C - \sum_{n=1}^{\infty}\dfrac{x^n}{n\,5^n}$

Putting $x = 0$, we get $C = \ln 5$. The series converges for $|x/5| < 1$ $\Leftrightarrow$ $|x| < 5$, so $R = 5$.

17. We know that $\dfrac{1}{1+4x} = \dfrac{1}{1-(-4x)} = \sum_{n=0}^{\infty}(-4x)^n$. Differentiating, we get

$\dfrac{-4}{(1+4x)^2} = \sum_{n=1}^{\infty}(-4)^n nx^{n-1} = \sum_{n=0}^{\infty}(-4)^{n+1}(n+1)x^n$, so

$f(x) = \dfrac{x}{(1+4x)^2} = \dfrac{-x}{4} \cdot \dfrac{-4}{(1+4x)^2} = \dfrac{-x}{4}\sum_{n=0}^{\infty}(-4)^{n+1}(n+1)x^n = \sum_{n=0}^{\infty}(-1)^n 4^n(n+1)x^{n+1}$

for $|-4x| < 1$ $\Leftrightarrow$ $|x| < \dfrac{1}{4}$, so $R = \dfrac{1}{4}$.

© 2012 Cengage Learning. All Rights Reserved. May not be scanned, copied, or duplicated, or posted to a publicly accessible website, in whole or in part.

19. By Example 5, $\dfrac{1}{(1-x)^2} = \sum\limits_{n=0}^{\infty} (n+1)x^n$. Thus,

$$f(x) = \frac{1+x}{(1-x)^2} = \frac{1}{(1-x)^2} + \frac{x}{(1-x)^2} = \sum_{n=0}^{\infty} (n+1)x^n + \sum_{n=0}^{\infty} (n+1)x^{n+1}$$

$$= \sum_{n=0}^{\infty} (n+1)x^n + \sum_{n=1}^{\infty} nx^n \qquad \text{[make the starting values equal]}$$

$$= 1 + \sum_{n=1}^{\infty} [(n+1)+n]x^n = 1 + \sum_{n=1}^{\infty} (2n+1)x^n = \sum_{n=0}^{\infty} (2n+1)x^n \ \text{ with } R = 1.$$

21. $f(x) = \dfrac{x}{x^2+16} = \dfrac{x}{16}\left(\dfrac{1}{1-(-x^2/16)}\right) = \dfrac{x}{16}\sum\limits_{n=0}^{\infty}\left(-\dfrac{x^2}{16}\right)^n = \dfrac{x}{16}\sum\limits_{n=0}^{\infty}(-1)^n\dfrac{1}{16^n}x^{2n} = \sum\limits_{n=0}^{\infty}(-1)^n\dfrac{1}{16^{n+1}}x^{2n+1}.$

The series converges when $\left|-x^2/16\right| < 1 \ \Leftrightarrow \ x^2 < 16 \ \Leftrightarrow \ |x| < 4$, so $R = 4$. The partial sums are $s_1 = \dfrac{x}{16}$,

$s_2 = s_1 - \dfrac{x^3}{16^2}, s_3 = s_2 + \dfrac{x^5}{16^3}, s_4 = s_3 - \dfrac{x^7}{16^4}, s_5 = s_4 + \dfrac{x^9}{16^5}, \ldots$. Note that s_1 corresponds to the first term of the infinite

sum, regardless of the value of the summation variable and the value of the exponent.

As n increases, $s_n(x)$ approximates f better on the interval of convergence, which is $(-4, 4)$.

23. $f(x) = \ln\left(\dfrac{1+x}{1-x}\right) = \ln(1+x) - \ln(1-x) = \displaystyle\int \dfrac{dx}{1+x} + \int \dfrac{dx}{1-x} = \int \dfrac{dx}{1-(-x)} + \int \dfrac{dx}{1-x}$

$= \displaystyle\int\left[\sum_{n=0}^{\infty}(-1)^n x^n + \sum_{n=0}^{\infty} x^n\right] dx = \int [(1 - x + x^2 - x^3 + x^4 - \cdots) + (1 + x + x^2 + x^3 + x^4 + \cdots)]\, dx$

$= \displaystyle\int (2 + 2x^2 + 2x^4 + \cdots)\, dx = \int \sum_{n=0}^{\infty} 2x^{2n}\, dx = C + \sum_{n=0}^{\infty} \dfrac{2x^{2n+1}}{2n+1}$

But $f(0) = \ln \frac{1}{1} = 0$, so $C = 0$ and we have $f(x) = \displaystyle\sum_{n=0}^{\infty} \dfrac{2x^{2n+1}}{2n+1}$ with $R = 1$. If $x = \pm 1$, then $f(x) = \pm 2\displaystyle\sum_{n=0}^{\infty}\dfrac{1}{2n+1}$,

which both diverge by the Limit Comparison Test with $b_n = \dfrac{1}{n}$.

© 2012 Cengage Learning. All Rights Reserved. May not be scanned, copied, or duplicated, or posted to a publicly accessible website, in whole or in part.

The partial sums are $s_1 = \dfrac{2x}{1}$, $s_2 = s_1 + \dfrac{2x^3}{3}$, $s_3 = s_2 + \dfrac{2x^5}{5}$, ...

As n increases, $s_n(x)$ approximates f better on the interval of convergence, which is $(-1, 1)$.

25. $\dfrac{t}{1-t^8} = t \cdot \dfrac{1}{1-t^8} = t \displaystyle\sum_{n=0}^{\infty} (t^8)^n = \displaystyle\sum_{n=0}^{\infty} t^{8n+1}$ $\Rightarrow$ $\displaystyle\int \dfrac{t}{1-t^8}\, dt = C + \displaystyle\sum_{n=0}^{\infty} \dfrac{t^{8n+2}}{8n+2}$. The series for $\dfrac{1}{1-t^8}$ converges

when $\left| t^8 \right| < 1$ $\Leftrightarrow$ $|t| < 1$, so $R = 1$ for that series and also the series for $t/(1-t^8)$. By Theorem 2, the series for

$\displaystyle\int \dfrac{t}{1-t^8}\, dt$ also has $R = 1$.

27. From Example 6, $\ln(1+x) = \displaystyle\sum_{n=1}^{\infty} (-1)^{n-1} \dfrac{x^n}{n}$ for $|x| < 1$, so $x^2 \ln(1+x) = \displaystyle\sum_{n=1}^{\infty} (-1)^{n-1} \dfrac{x^{n+2}}{n}$ and

$\displaystyle\int x^2 \ln(1+x)\, dx = C + \displaystyle\sum_{n=1}^{\infty} (-1)^n \dfrac{x^{n+3}}{n(n+3)}$. $R = 1$ for the series for $\ln(1+x)$, so $R = 1$ for the series representing

$x^2 \ln(1+x)$ as well. By Theorem 2, the series for $\displaystyle\int x^2 \ln(1+x)\, dx$ also has $R = 1$.

29. $\dfrac{1}{1+x^5} = \dfrac{1}{1-(-x^5)} = \displaystyle\sum_{n=0}^{\infty} (-x^5)^n = \displaystyle\sum_{n=0}^{\infty} (-1)^n x^{5n}$ $\Rightarrow$

$\displaystyle\int \dfrac{1}{1+x^5}\, dx = \displaystyle\int \sum_{n=0}^{\infty} (-1)^n x^{5n}\, dx = C + \displaystyle\sum_{n=0}^{\infty} (-1)^n \dfrac{x^{5n+1}}{5n+1}$. Thus,

$I = \displaystyle\int_0^{0.2} \dfrac{1}{1+x^5}\, dx = \left[x - \dfrac{x^6}{6} + \dfrac{x^{11}}{11} - \cdots \right]_0^{0.2} = 0.2 - \dfrac{(0.2)^6}{6} + \dfrac{(0.2)^{11}}{11} - \cdots$. The series is alternating, so if we use

the first two terms, the error is at most $(0.2)^{11}/11 \approx 1.9 \times 10^{-9}$. So $I \approx 0.2 - (0.2)^6/6 \approx 0.199\,989$ to six decimal places.

31. We substitute $3x$ for x in Example 7, and find that

$\displaystyle\int x \arctan(3x)\, dx = \displaystyle\int x \sum_{n=0}^{\infty} (-1)^n \dfrac{(3x)^{2n+1}}{2n+1}\, dx = \displaystyle\int \sum_{n=0}^{\infty} (-1)^n \dfrac{3^{2n+1} x^{2n+2}}{2n+1}\, dx = C + \displaystyle\sum_{n=0}^{\infty} (-1)^n \dfrac{3^{2n+1} x^{2n+3}}{(2n+1)(2n+3)}$

So $\displaystyle\int_0^{0.1} x \arctan(3x)\, dx = \left[\dfrac{3x^3}{1 \cdot 3} - \dfrac{3^3 x^5}{3 \cdot 5} + \dfrac{3^5 x^7}{5 \cdot 7} - \dfrac{3^7 x^9}{7 \cdot 9} + \cdots \right]_0^{0.1}$

$= \dfrac{1}{10^3} - \dfrac{9}{5 \times 10^5} + \dfrac{243}{35 \times 10^7} - \dfrac{2187}{63 \times 10^9} + \cdots$.

The series is alternating, so if we use three terms, the error is at most $\dfrac{2187}{63 \times 10^9} \approx 3.5 \times 10^{-8}$. So

$\displaystyle\int_0^{0.1} x \arctan(3x)\, dx \approx \dfrac{1}{10^3} - \dfrac{9}{5 \times 10^5} + \dfrac{243}{35 \times 10^7} \approx 0.000\,983$ to six decimal places.

© 2012 Cengage Learning. All Rights Reserved. May not be scanned, copied, or duplicated, or posted to a publicly accessible website, in whole or in part.

33. By Example 7, $\arctan x = x - \dfrac{x^3}{3} + \dfrac{x^5}{5} - \dfrac{x^7}{7} + \cdots$, so $\arctan 0.2 = 0.2 - \dfrac{(0.2)^3}{3} + \dfrac{(0.2)^5}{5} - \dfrac{(0.2)^7}{7} + \cdots$.

The series is alternating, so if we use three terms, the error is at most $\dfrac{(0.2)^7}{7} \approx 0.000\,002$.

Thus, to five decimal places, $\arctan 0.2 \approx 0.2 - \dfrac{(0.2)^3}{3} + \dfrac{(0.2)^5}{5} \approx 0.197\,40$.

35. (a) $J_0(x) = \displaystyle\sum_{n=0}^{\infty} \dfrac{(-1)^n x^{2n}}{2^{2n}(n!)^2}$, $J_0'(x) = \displaystyle\sum_{n=1}^{\infty} \dfrac{(-1)^n 2n x^{2n-1}}{2^{2n}(n!)^2}$, and $J_0''(x) = \displaystyle\sum_{n=1}^{\infty} \dfrac{(-1)^n 2n(2n-1)x^{2n-2}}{2^{2n}(n!)^2}$, so

$$x^2 J_0''(x) + x J_0'(x) + x^2 J_0(x) = \sum_{n=1}^{\infty} \dfrac{(-1)^n 2n(2n-1)x^{2n}}{2^{2n}(n!)^2} + \sum_{n=1}^{\infty} \dfrac{(-1)^n 2n x^{2n}}{2^{2n}(n!)^2} + \sum_{n=0}^{\infty} \dfrac{(-1)^n x^{2n+2}}{2^{2n}(n!)^2}$$

$$= \sum_{n=1}^{\infty} \dfrac{(-1)^n 2n(2n-1)x^{2n}}{2^{2n}(n!)^2} + \sum_{n=1}^{\infty} \dfrac{(-1)^n 2n x^{2n}}{2^{2n}(n!)^2} + \sum_{n=1}^{\infty} \dfrac{(-1)^{n-1} x^{2n}}{2^{2n-2}[(n-1)!]^2}$$

$$= \sum_{n=1}^{\infty} \dfrac{(-1)^n 2n(2n-1)x^{2n}}{2^{2n}(n!)^2} + \sum_{n=1}^{\infty} \dfrac{(-1)^n 2n x^{2n}}{2^{2n}(n!)^2} + \sum_{n=1}^{\infty} \dfrac{(-1)^n (-1)^{-1} 2^2 n^2 x^{2n}}{2^{2n}(n!)^2}$$

$$= \sum_{n=1}^{\infty} (-1)^n \left[\dfrac{2n(2n-1) + 2n - 2^2 n^2}{2^{2n}(n!)^2} \right] x^{2n}$$

$$= \sum_{n=1}^{\infty} (-1)^n \left[\dfrac{4n^2 - 2n + 2n - 4n^2}{2^{2n}(n!)^2} \right] x^{2n} = 0$$

(b) $\displaystyle\int_0^1 J_0(x)\,dx = \int_0^1 \left[\sum_{n=0}^{\infty} \dfrac{(-1)^n x^{2n}}{2^{2n}(n!)^2} \right] dx = \int_0^1 \left(1 - \dfrac{x^2}{4} + \dfrac{x^4}{64} - \dfrac{x^6}{2304} + \cdots \right) dx$

$$= \left[x - \dfrac{x^3}{3 \cdot 4} + \dfrac{x^5}{5 \cdot 64} - \dfrac{x^7}{7 \cdot 2304} + \cdots \right]_0^1 = 1 - \dfrac{1}{12} + \dfrac{1}{320} - \dfrac{1}{16,128} + \cdots$$

Since $\frac{1}{16,128} \approx 0.000062$, it follows from The Alternating Series Estimation Theorem that, correct to three decimal places,

$\int_0^1 J_0(x)\,dx \approx 1 - \frac{1}{12} + \frac{1}{320} \approx 0.920$.

37. (a) $f(x) = \displaystyle\sum_{n=0}^{\infty} \dfrac{x^n}{n!} \ \Rightarrow\ f'(x) = \sum_{n=1}^{\infty} \dfrac{n x^{n-1}}{n!} = \sum_{n=1}^{\infty} \dfrac{x^{n-1}}{(n-1)!} = \sum_{n=0}^{\infty} \dfrac{x^n}{n!} = f(x)$

(b) By Theorem 9.4.2, the only solution to the differential equation $df(x)/dx = f(x)$ is $f(x) = Ke^x$, but $f(0) = 1$, so $K = 1$ and $f(x) = e^x$.

Or: We could solve the equation $df(x)/dx = f(x)$ as a separable differential equation.

39. If $a_n = \dfrac{x^n}{n^2}$, then by the Ratio Test, $\displaystyle\lim_{n \to \infty} \left| \dfrac{a_{n+1}}{a_n} \right| = \lim_{n \to \infty} \left| \dfrac{x^{n+1}}{(n+1)^2} \cdot \dfrac{n^2}{x^n} \right| = |x| \lim_{n \to \infty} \left(\dfrac{n}{n+1} \right)^2 = |x| < 1$ for

convergence, so $R = 1$. When $x = \pm 1$, $\displaystyle\sum_{n=1}^{\infty} \left| \dfrac{x^n}{n^2} \right| = \sum_{n=1}^{\infty} \dfrac{1}{n^2}$ which is a convergent p-series ($p = 2 > 1$), so the interval of

convergence for f is $[-1, 1]$. By Theorem 2, the radii of convergence of f' and f'' are both 1, so we need only check the

endpoints. $f(x) = \displaystyle\sum_{n=1}^{\infty} \dfrac{x^n}{n^2} \ \Rightarrow\ f'(x) = \sum_{n=1}^{\infty} \dfrac{n x^{n-1}}{n^2} = \sum_{n=0}^{\infty} \dfrac{x^n}{n+1}$, and this series diverges for $x = 1$ (harmonic series)

© 2012 Cengage Learning. All Rights Reserved. May not be scanned, copied, or duplicated, or posted to a publicly accessible website, in whole or in part.

and converges for $x = -1$ (Alternating Series Test), so the interval of convergence is $[-1, 1)$. $f''(x) = \sum_{n=1}^{\infty} \frac{nx^{n-1}}{n+1}$ diverges

at both 1 and -1 (Test for Divergence) since $\lim\limits_{n \to \infty} \frac{n}{n+1} = 1 \neq 0$, so its interval of convergence is $(-1, 1)$.

41. By Example 7, $\tan^{-1} x = \sum_{n=0}^{\infty} (-1)^n \frac{x^{2n+1}}{2n+1}$ for $|x| < 1$. In particular, for $x = \frac{1}{\sqrt{3}}$, we

have $\frac{\pi}{6} = \tan^{-1}\left(\frac{1}{\sqrt{3}}\right) = \sum_{n=0}^{\infty} (-1)^n \frac{(1/\sqrt{3})^{2n+1}}{2n+1} = \sum_{n=0}^{\infty} (-1)^n \left(\frac{1}{3}\right)^n \frac{1}{\sqrt{3}} \frac{1}{2n+1}$, so

$\pi = \frac{6}{\sqrt{3}} \sum_{n=0}^{\infty} \frac{(-1)^n}{(2n+1)3^n} = 2\sqrt{3} \sum_{n=0}^{\infty} \frac{(-1)^n}{(2n+1)3^n}$.

11.10 Taylor and Maclaurin Series

1. Using Theorem 5 with $\sum_{n=0}^{\infty} b_n (x-5)^n$, $b_n = \frac{f^{(n)}(a)}{n!}$, so $b_8 = \frac{f^{(8)}(5)}{8!}$.

3. Since $f^{(n)}(0) = (n+1)!$, Equation 7 gives the Maclaurin series

$\sum_{n=0}^{\infty} \frac{f^{(n)}(0)}{n!} x^n = \sum_{n=0}^{\infty} \frac{(n+1)!}{n!} x^n = \sum_{n=0}^{\infty} (n+1)x^n$. Applying the Ratio Test with $a_n = (n+1)x^n$ gives us

$\lim\limits_{n \to \infty} \left| \frac{a_{n+1}}{a_n} \right| = \lim\limits_{n \to \infty} \left| \frac{(n+2)x^{n+1}}{(n+1)x^n} \right| = |x| \lim\limits_{n \to \infty} \frac{n+2}{n+1} = |x| \cdot 1 = |x|$. For convergence, we must have $|x| < 1$, so the

radius of convergence $R = 1$.

5.

n	$f^{(n)}(x)$	$f^{(n)}(0)$
0	$(1-x)^{-2}$	1
1	$2(1-x)^{-3}$	2
2	$6(1-x)^{-4}$	6
3	$24(1-x)^{-5}$	24
4	$120(1-x)^{-6}$	120
⋮	⋮	⋮

$(1-x)^{-2} = f(0) + f'(0)x + \frac{f''(0)}{2!}x^2 + \frac{f'''(0)}{3!}x^3 + \frac{f^{(4)}(0)}{4!}x^4 + \cdots$

$= 1 + 2x + \frac{6}{2}x^2 + \frac{24}{6}x^3 + \frac{120}{24}x^4 + \cdots$

$= 1 + 2x + 3x^2 + 4x^3 + 5x^4 + \cdots = \sum_{n=0}^{\infty} (n+1)x^n$

$\lim\limits_{n \to \infty} \left| \frac{a_{n+1}}{a_n} \right| = \lim\limits_{n \to \infty} \left| \frac{(n+2)x^{n+1}}{(n+1)x^n} \right| = |x| \lim\limits_{n \to \infty} \frac{n+2}{n+1} = |x|(1) = |x| < 1$

for convergence, so $R = 1$.

7.

n	$f^{(n)}(x)$	$f^{(n)}(0)$
0	$\sin \pi x$	0
1	$\pi \cos \pi x$	π
2	$-\pi^2 \sin \pi x$	0
3	$-\pi^3 \cos \pi x$	$-\pi^3$
4	$\pi^4 \sin \pi x$	0
5	$\pi^5 \cos \pi x$	π^5
⋮	⋮	⋮

$\sin \pi x = f(0) + f'(0)x + \frac{f''(0)}{2!}x^2 + \frac{f'''(0)}{3!}x^3$

$\qquad + \frac{f^{(4)}(0)}{4!}x^4 + \frac{f^{(5)}(0)}{5!}x^5 + \cdots$

$= 0 + \pi x + 0 - \frac{\pi^3}{3!}x^3 + 0 + \frac{\pi^5}{5!}x^5 + \cdots$

$= \pi x - \frac{\pi^3}{3!}x^3 + \frac{\pi^5}{5!}x^5 - \frac{\pi^7}{7!}x^7 + \cdots$

$= \sum_{n=0}^{\infty} (-1)^n \frac{\pi^{2n+1}}{(2n+1)!} x^{2n+1}$

$\lim\limits_{n \to \infty} \left| \frac{a_{n+1}}{a_n} \right| = \lim\limits_{n \to \infty} \left| \frac{\pi^{2n+3} x^{2n+3}}{(2n+3)!} \cdot \frac{(2n+1)!}{\pi^{2n+1} x^{2n+1}} \right| = \lim\limits_{n \to \infty} \frac{\pi^2 x^2}{(2n+3)(2n+2)} = 0 < 1$ for all x, so $R = \infty$.

© 2012 Cengage Learning. All Rights Reserved. May not be scanned, copied, or duplicated, or posted to a publicly accessible website, in whole or in part.

9.

n	$f^{(n)}(x)$	$f^{(n)}(0)$
0	2^x	1
1	$2^x(\ln 2)$	$\ln 2$
2	$2^x(\ln 2)^2$	$(\ln 2)^2$
3	$2^x(\ln 2)^3$	$(\ln 2)^3$
4	$2^x(\ln 2)^4$	$(\ln 2)^4$
⋮	⋮	⋮

$$2^x = \sum_{n=0}^{\infty} \frac{f^{(n)}(0)}{n!}\, x^n = \sum_{n=0}^{\infty} \frac{(\ln 2)^n}{n!}\, x^n.$$

$$\lim_{n\to\infty}\left|\frac{a_{n+1}}{a_n}\right| = \lim_{n\to\infty}\left|\frac{(\ln 2)^{n+1} x^{n+1}}{(n+1)!} \cdot \frac{n!}{(\ln 2)^n x^n}\right|$$

$$= \lim_{n\to\infty} \frac{(\ln 2)\,|x|}{n+1} = 0 < 1 \quad \text{for all } x, \text{ so } R = \infty.$$

11.

n	$f^{(n)}(x)$	$f^{(n)}(0)$
0	$\sinh x$	0
1	$\cosh x$	1
2	$\sinh x$	0
3	$\cosh x$	1
4	$\sinh x$	0
⋮	⋮	⋮

$$f^{(n)}(0) = \begin{cases} 0 & \text{if } n \text{ is even} \\ 1 & \text{if } n \text{ is odd} \end{cases} \quad \text{so } \sinh x = \sum_{n=0}^{\infty} \frac{x^{2n+1}}{(2n+1)!}.$$

Use the Ratio Test to find R. If $a_n = \dfrac{x^{2n+1}}{(2n+1)!}$, then

$$\lim_{n\to\infty}\left|\frac{a_{n+1}}{a_n}\right| = \lim_{n\to\infty}\left|\frac{x^{2n+3}}{(2n+3)!} \cdot \frac{(2n+1)!}{x^{2n+1}}\right| = x^2 \cdot \lim_{n\to\infty} \frac{1}{(2n+3)(2n+2)}$$

$$= 0 < 1 \quad \text{for all } x, \text{ so } R = \infty.$$

13.

n	$f^{(n)}(x)$	$f^{(n)}(1)$
0	$x^4 - 3x^2 + 1$	-1
1	$4x^3 - 6x$	-2
2	$12x^2 - 6$	6
3	$24x$	24
4	24	24
5	0	0
6	0	0
⋮	⋮	⋮

$f^{(n)}(x) = 0$ for $n \geq 5$, so f has a finite series expansion about $a = 1$.

$$f(x) = x^4 - 3x^2 + 1 = \sum_{n=0}^{4} \frac{f^{(n)}(1)}{n!}(x-1)^n$$

$$= \frac{-1}{0!}(x-1)^0 + \frac{-2}{1!}(x-1)^1 + \frac{6}{2!}(x-1)^2$$

$$+ \frac{24}{3!}(x-1)^3 + \frac{24}{4!}(x-1)^4$$

$$= -1 - 2(x-1) + 3(x-1)^2 + 4(x-1)^3 + (x-1)^4$$

A finite series converges for all x, so $R = \infty$.

15.

n	$f^{(n)}(x)$	$f^{(n)}(2)$
0	$\ln x$	$\ln 2$
1	$1/x$	$1/2$
2	$-1/x^2$	$-1/2^2$
3	$2/x^3$	$2/2^3$
4	$-6/x^4$	$-6/2^4$
5	$24/x^5$	$24/2^5$
⋮	⋮	⋮

$$f(x) = \ln x = \sum_{n=0}^{\infty} \frac{f^{(n)}(2)}{n!}(x-2)^n$$

$$= \frac{\ln 2}{0!}(x-2)^0 + \frac{1}{1!\,2^1}(x-2)^1 + \frac{-1}{2!\,2^2}(x-2)^2 + \frac{2}{3!\,2^3}(x-2)^3$$

$$+ \frac{-6}{4!\,2^4}(x-2)^4 + \frac{24}{5!\,2^5}(x-2)^5 + \cdots$$

$$= \ln 2 + \sum_{n=1}^{\infty} (-1)^{n+1} \frac{(n-1)!}{n!\,2^n}(x-2)^n$$

$$= \ln 2 + \sum_{n=1}^{\infty} (-1)^{n+1} \frac{1}{n\,2^n}(x-2)^n$$

© 2012 Cengage Learning. All Rights Reserved. May not be scanned, copied, or duplicated, or posted to a publicly accessible website, in whole or in part.

$$\lim_{n\to\infty}\left|\frac{a_{n+1}}{a_n}\right| = \lim_{n\to\infty}\left|\frac{(-1)^{n+2}(x-2)^{n+1}}{(n+1)\,2^{n+1}}\cdot\frac{n\,2^n}{(-1)^{n+1}(x-2)^n}\right| = \lim_{n\to\infty}\left|\frac{(-1)(x-2)n}{(n+1)2}\right| = \lim_{n\to\infty}\left(\frac{n}{n+1}\right)\frac{|x-2|}{2}$$

$$= \frac{|x-2|}{2} < 1 \quad\text{for convergence, so } |x-2| < 2 \text{ and } R = 2.$$

17.

n	$f^{(n)}(x)$	$f^{(n)}(3)$
0	e^{2x}	e^6
1	$2e^{2x}$	$2e^6$
2	$2^2 e^{2x}$	$4e^6$
3	$2^3 e^{2x}$	$8e^6$
4	$2^4 e^{2x}$	$16e^6$
⋮	⋮	⋮

$$f(x) = e^{2x} = \sum_{n=0}^{\infty}\frac{f^{(n)}(3)}{n!}(x-3)^n$$

$$= \frac{e^6}{0!}(x-3)^0 + \frac{2e^6}{1!}(x-3)^1 + \frac{4e^6}{2!}(x-3)^2$$

$$+ \frac{8e^6}{3!}(x-3)^3 + \frac{16e^6}{4!}(x-3)^4 + \cdots$$

$$= \sum_{n=0}^{\infty}\frac{2^n e^6}{n!}(x-3)^n$$

$$\lim_{n\to\infty}\left|\frac{a_{n+1}}{a_n}\right| = \lim_{n\to\infty}\left|\frac{2^{n+1}e^6(x-3)^{n+1}}{(n+1)!}\cdot\frac{n!}{2^n e^6(x-3)^n}\right| = \lim_{n\to\infty}\frac{2\,|x-3|}{n+1} = 0 < 1 \quad\text{for all } x, \text{ so } R = \infty.$$

19.

n	$f^{(n)}(x)$	$f^{(n)}(\pi)$
0	$\cos x$	-1
1	$-\sin x$	0
2	$-\cos x$	1
3	$\sin x$	0
4	$\cos x$	-1
⋮	⋮	⋮

$$f(x) = \cos x = \sum_{k=0}^{\infty}\frac{f^{(k)}(\pi)}{k!}(x-\pi)^k$$

$$= -1 + \frac{(x-\pi)^2}{2!} - \frac{(x-\pi)^4}{4!} + \frac{(x-\pi)^6}{6!} -$$

$$= \sum_{n=0}^{\infty}(-1)^{n+1}\frac{(x-\pi)^{2n}}{(2n)!}$$

$$\lim_{n\to\infty}\left|\frac{a_{n+1}}{a_n}\right| = \lim_{n\to\infty}\left[\frac{|x-\pi|^{2n+2}}{(2n+2)!}\cdot\frac{(2n)!}{|x-\pi|^{2n}}\right]$$

$$= \lim_{n\to\infty}\frac{|x-\pi|^2}{(2n+2)(2n+1)} = 0 < 1 \quad\text{for all } x, \text{ so } R = \infty.$$

21. If $f(x) = \sin \pi x$, then $f^{(n+1)}(x) = \pm\pi^{n+1}\sin \pi x$ or $\pm\pi^{n+1}\cos \pi x$. In each case, $\left|f^{(n+1)}(x)\right| \leq \pi^{n+1}$, so by Formula 9

with $a = 0$ and $M = \pi^{n+1}$, $|R_n(x)| \leq \dfrac{\pi^{n+1}}{(n+1)!}\,|x|^{n+1} = \dfrac{|\pi x|^{n+1}}{(n+1)!}$. Thus, $|R_n(x)| \to 0$ as $n \to \infty$ by Equation 10.

So $\lim\limits_{n\to\infty} R_n(x) = 0$ and, by Theorem 8, the series in Exercise 7 represents $\sin \pi x$ for all x.

23. If $f(x) = \sinh x$, then for all n, $f^{(n+1)}(x) = \cosh x$ or $\sinh x$. Since $|\sinh x| < |\cosh x| = \cosh x$ for all x, we have

$\left|f^{(n+1)}(x)\right| \leq \cosh x$ for all n. If d is any positive number and $|x| \leq d$, then $\left|f^{(n+1)}(x)\right| \leq \cosh x \leq \cosh d$, so by

Formula 9 with $a = 0$ and $M = \cosh d$, we have $|R_n(x)| \leq \dfrac{\cosh d}{(n+1)!}\,|x|^{n+1}$. It follows that $|R_n(x)| \to 0$ as $n \to \infty$ for

$|x| \leq d$ (by Equation 10). But d was an arbitrary positive number. So by Theorem 8, the series represents $\sinh x$ for all x.

© 2012 Cengage Learning. All Rights Reserved. May not be scanned, copied, or duplicated, or posted to a publicly accessible website, in whole or in part.

25. $\sqrt[4]{1-x} = [1+(-x)]^{1/4} = \sum\limits_{n=0}^{\infty}\binom{1/4}{n}(-x)^n = 1+\frac{1}{4}(-x)+\frac{\frac{1}{4}\left(-\frac{3}{4}\right)}{2!}(-x)^2+\frac{\frac{1}{4}\left(-\frac{3}{4}\right)\left(-\frac{7}{4}\right)}{3!}(-x)^3+\cdots$

$$= 1-\frac{1}{4}x+\sum\limits_{n=2}^{\infty}\frac{(-1)^{n-1}(-1)^n\cdot[3\cdot7\cdot\cdots\cdot(4n-5)]}{4^n\cdot n!}x^n$$

$$= 1-\frac{1}{4}x-\sum\limits_{n=2}^{\infty}\frac{3\cdot7\cdot\cdots\cdot(4n-5)}{4^n\cdot n!}x^n$$

and $|-x|<1 \quad\Leftrightarrow\quad |x|<1$, so $R=1$.

27. $\dfrac{1}{(2+x)^3} = \dfrac{1}{[2(1+x/2)]^3} = \dfrac{1}{8}\left(1+\dfrac{x}{2}\right)^{-3} = \dfrac{1}{8}\sum\limits_{n=0}^{\infty}\binom{-3}{n}\left(\dfrac{x}{2}\right)^n$. The binomial coefficient is

$$\binom{-3}{n} = \frac{(-3)(-4)(-5)\cdot\cdots\cdot(-3-n+1)}{n!} = \frac{(-3)(-4)(-5)\cdot\cdots\cdot[-(n+2)]}{n!}$$

$$= \frac{(-1)^n\cdot2\cdot3\cdot4\cdot5\cdot\cdots\cdot(n+1)(n+2)}{2\cdot n!} = \frac{(-1)^n(n+1)(n+2)}{2}$$

Thus, $\dfrac{1}{(2+x)^3} = \dfrac{1}{8}\sum\limits_{n=0}^{\infty}\dfrac{(-1)^n(n+1)(n+2)}{2}\dfrac{x^n}{2^n} = \sum\limits_{n=0}^{\infty}\dfrac{(-1)^n(n+1)(n+2)x^n}{2^{n+4}}$ for $\left|\dfrac{x}{2}\right|<1 \Leftrightarrow |x|<2$, so $R=2$.

29. $\sin x = \sum\limits_{n=0}^{\infty}(-1)^n\dfrac{x^{2n+1}}{(2n+1)!} \quad\Rightarrow\quad f(x) = \sin(\pi x) = \sum\limits_{n=0}^{\infty}(-1)^n\dfrac{(\pi x)^{2n+1}}{(2n+1)!} = \sum\limits_{n=0}^{\infty}(-1)^n\dfrac{\pi^{2n+1}}{(2n+1)!}x^{2n+1}$, $R=\infty$.

31. $e^x = \sum\limits_{n=0}^{\infty}\dfrac{x^n}{n!} \quad\Rightarrow\quad e^{2x} = \sum\limits_{n=0}^{\infty}\dfrac{(2x)^n}{n!} = \sum\limits_{n=0}^{\infty}\dfrac{2^n\,x^n}{n!}$, so $f(x) = e^x+e^{2x} = \sum\limits_{n=0}^{\infty}\dfrac{1}{n!}x^n + \sum\limits_{n=0}^{\infty}\dfrac{2^n}{n!}x^n = \sum\limits_{n=0}^{\infty}\dfrac{2^n+1}{n!}x^n$,

$R=\infty$.

33. $\cos x = \sum\limits_{n=0}^{\infty}(-1)^n\dfrac{x^{2n}}{(2n)!} \quad\Rightarrow\quad \cos\left(\tfrac{1}{2}x^2\right) = \sum\limits_{n=0}^{\infty}(-1)^n\dfrac{\left(\tfrac{1}{2}x^2\right)^{2n}}{(2n)!} = \sum\limits_{n=0}^{\infty}(-1)^n\dfrac{x^{4n}}{2^{2n}(2n)!}$, so

$$f(x) = x\cos\left(\tfrac{1}{2}x^2\right) = \sum\limits_{n=0}^{\infty}(-1)^n\dfrac{1}{2^{2n}(2n)!}x^{4n+1}, \quad R=\infty.$$

35. We must write the binomial in the form $(1+$ expression$)$, so we'll factor out a 4.

$$\frac{x}{\sqrt{4+x^2}} = \frac{x}{\sqrt{4(1+x^2/4)}} = \frac{x}{2\sqrt{1+x^2/4}} = \frac{x}{2}\left(1+\frac{x^2}{4}\right)^{-1/2} = \frac{x}{2}\sum\limits_{n=0}^{\infty}\binom{-\frac{1}{2}}{n}\left(\frac{x^2}{4}\right)^n$$

$$= \frac{x}{2}\left[1+\left(-\tfrac{1}{2}\right)\frac{x^2}{4}+\frac{\left(-\frac{1}{2}\right)\left(-\frac{3}{2}\right)}{2!}\left(\frac{x^2}{4}\right)^2+\frac{\left(-\frac{1}{2}\right)\left(-\frac{3}{2}\right)\left(-\frac{5}{2}\right)}{3!}\left(\frac{x^2}{4}\right)^3+\cdots\right]$$

$$= \frac{x}{2}+\frac{x}{2}\sum\limits_{n=1}^{\infty}(-1)^n\frac{1\cdot3\cdot5\cdot\cdots\cdot(2n-1)}{2^n\cdot4^n\cdot n!}x^{2n}$$

$$= \frac{x}{2}+\sum\limits_{n=1}^{\infty}(-1)^n\frac{1\cdot3\cdot5\cdot\cdots\cdot(2n-1)}{n!\,2^{3n+1}}x^{2n+1} \text{ and } \frac{x^2}{4}<1 \Leftrightarrow \frac{|x|}{2}<1 \Leftrightarrow |x|<2, \text{ so } R=2.$$

37. $\sin^2 x = \dfrac{1}{2}(1-\cos2x) = \dfrac{1}{2}\left[1-\sum\limits_{n=0}^{\infty}\dfrac{(-1)^n(2x)^{2n}}{(2n)!}\right] = \dfrac{1}{2}\left[1-1-\sum\limits_{n=1}^{\infty}\dfrac{(-1)^n(2x)^{2n}}{(2n)!}\right] = \sum\limits_{n=1}^{\infty}\dfrac{(-1)^{n+1}2^{2n-1}x^{2n}}{(2n)!}$,

$R=\infty$

© 2012 Cengage Learning. All Rights Reserved. May not be scanned, copied, or duplicated, or posted to a publicly accessible website, in whole or in part.

39. $\cos x \overset{(16)}{=} \sum_{n=0}^{\infty} (-1)^n \dfrac{x^{2n}}{(2n)!} \quad \Rightarrow$

$$f(x) = \cos(x^2) = \sum_{n=0}^{\infty} \dfrac{(-1)^n \, (x^2)^{2n}}{(2n)!} = \sum_{n=0}^{\infty} \dfrac{(-1)^n x^{4n}}{(2n)!}$$

$$= 1 - \tfrac{1}{2}x^4 + \tfrac{1}{24}x^8 - \tfrac{1}{720}x^{12} + \cdots$$

The series for $\cos x$ converges for all x, so the same is true of the series for $f(x)$, that is, $R = \infty$. Notice that, as n increases, $T_n(x)$ becomes a better approximation to $f(x)$.

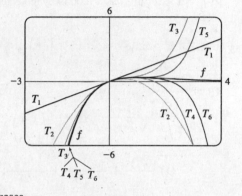

$T_0 = T_1 = T_2 = T_3$

$T_8 = T_9 = T_{10} = T_{11}$

$T_4 = T_5 = T_6 = T_7$

41. $e^x \overset{(11)}{=} \sum_{n=0}^{\infty} \dfrac{x^n}{n!}$, so $e^{-x} = \sum_{n=0}^{\infty} \dfrac{(-x)^n}{n!} = \sum_{n=0}^{\infty} (-1)^n \dfrac{x^n}{n!}$, so

$$f(x) = xe^{-x} = \sum_{n=0}^{\infty} (-1)^n \dfrac{1}{n!} x^{n+1}$$

$$= x - x^2 + \tfrac{1}{2}x^3 - \tfrac{1}{6}x^4 + \tfrac{1}{24}x^5 - \tfrac{1}{120}x^6 + \cdots$$

$$= \sum_{n=1}^{\infty} (-1)^{n-1} \dfrac{x^n}{(n-1)!}$$

The series for e^x converges for all x, so the same is true of the series for $f(x)$; that is, $R = \infty$. From the graphs of f and the first few Taylor polynomials, we see that $T_n(x)$ provides a closer fit to $f(x)$ near 0 as n increases.

43. $5° = 5° \left(\dfrac{\pi}{180°} \right) = \dfrac{\pi}{36}$ radians and $\cos x = \sum_{n=0}^{\infty} (-1)^n \dfrac{x^{2n}}{(2n)!} = 1 - \dfrac{x^2}{2!} + \dfrac{x^4}{4!} - \dfrac{x^6}{6!} + \cdots$, so

$\cos \dfrac{\pi}{36} = 1 - \dfrac{(\pi/36)^2}{2!} + \dfrac{(\pi/36)^4}{4!} - \dfrac{(\pi/36)^6}{6!} + \cdots$. Now $1 - \dfrac{(\pi/36)^2}{2!} \approx 0.99619$ and adding $\dfrac{(\pi/36)^4}{4!} \approx 2.4 \times 10^{-6}$

does not affect the fifth decimal place, so $\cos 5° \approx 0.99619$ by the Alternating Series Estimation Theorem.

45. (a) $1/\sqrt{1-x^2} = \left[1 + (-x^2) \right]^{-1/2} = 1 + \left(-\tfrac{1}{2} \right)(-x^2) + \dfrac{\left(-\frac{1}{2} \right)\left(-\frac{3}{2} \right)}{2!}(-x^2)^2 + \dfrac{\left(-\frac{1}{2} \right)\left(-\frac{3}{2} \right)\left(-\frac{5}{2} \right)}{3!}(-x^2)^3 + \cdots$

$$= 1 + \sum_{n=1}^{\infty} \dfrac{1 \cdot 3 \cdot 5 \cdot \cdots \cdot (2n-1)}{2^n \cdot n!} x^{2n}$$

(b) $\sin^{-1} x = \int \dfrac{1}{\sqrt{1-x^2}} \, dx = C + x + \sum_{n=1}^{\infty} \dfrac{1 \cdot 3 \cdot 5 \cdot \cdots \cdot (2n-1)}{(2n+1)2^n \cdot n!} x^{2n+1}$

$$= x + \sum_{n=1}^{\infty} \dfrac{1 \cdot 3 \cdot 5 \cdot \cdots \cdot (2n-1)}{(2n+1)2^n \cdot n!} x^{2n+1} \quad \text{since } 0 = \sin^{-1} 0 = C.$$

47. $\cos x \overset{(16)}{=} \sum_{n=0}^{\infty} (-1)^n \dfrac{x^{2n}}{(2n)!} \quad \Rightarrow \quad \cos(x^3) = \sum_{n=0}^{\infty} (-1)^n \dfrac{(x^3)^{2n}}{(2n)!} = \sum_{n=0}^{\infty} (-1)^n \dfrac{x^{6n}}{(2n)!} \quad \Rightarrow$

$x \cos(x^3) = \sum_{n=0}^{\infty} (-1)^n \dfrac{x^{6n+1}}{(2n)!} \quad \Rightarrow \quad \int x \cos(x^3) \, dx = C + \sum_{n=0}^{\infty} (-1)^n \dfrac{x^{6n+2}}{(6n+2)(2n)!}$, with $R = \infty$.

49. $\cos x \overset{(16)}{=} \sum_{n=0}^{\infty} (-1)^n \dfrac{x^{2n}}{(2n)!} \quad \Rightarrow \quad \cos x - 1 = \sum_{n=1}^{\infty} (-1)^n \dfrac{x^{2n}}{(2n)!} \quad \Rightarrow \quad \dfrac{\cos x - 1}{x} = \sum_{n=1}^{\infty} (-1)^n \dfrac{x^{2n-1}}{(2n)!} \quad \Rightarrow$

$\int \dfrac{\cos x - 1}{x} \, dx = C + \sum_{n=1}^{\infty} (-1)^n \dfrac{x^{2n}}{2n \cdot (2n)!}$, with $R = \infty$.

© 2012 Cengage Learning. All Rights Reserved. May not be scanned, copied, or duplicated, or posted to a publicly accessible website, in whole or in part.

51. $\arctan x = \sum_{n=0}^{\infty} (-1)^n \dfrac{x^{2n+1}}{2n+1}$ for $|x| < 1$, so $x^3 \arctan x = \sum_{n=0}^{\infty} (-1)^n \dfrac{x^{2n+4}}{2n+1}$ for $|x| < 1$ and

$$\int x^3 \arctan x \, dx = C + \sum_{n=0}^{\infty} (-1)^n \frac{x^{2n+5}}{(2n+1)(2n+5)}.$$ Since $\frac{1}{2} < 1$, we have

$$\int_0^{1/2} x^3 \arctan x \, dx = \sum_{n=0}^{\infty} (-1)^n \frac{(1/2)^{2n+5}}{(2n+1)(2n+5)} = \frac{(1/2)^5}{1 \cdot 5} - \frac{(1/2)^7}{3 \cdot 7} + \frac{(1/2)^9}{5 \cdot 9} - \frac{(1/2)^{11}}{7 \cdot 11} + \cdots .$$ Now

$\dfrac{(1/2)^5}{1 \cdot 5} - \dfrac{(1/2)^7}{3 \cdot 7} + \dfrac{(1/2)^9}{5 \cdot 9} \approx 0.0059$ and subtracting $\dfrac{(1/2)^{11}}{7 \cdot 11} \approx 6.3 \times 10^{-6}$ does not affect the fourth decimal place,

so $\int_0^{1/2} x^3 \arctan x \, dx \approx 0.0059$ by the Alternating Series Estimation Theorem.

53. $\sqrt{1 + x^4} = (1 + x^4)^{1/2} = \sum_{n=0}^{\infty} \binom{1/2}{n} (x^4)^n$, so $\int \sqrt{1 + x^4} \, dx = C + \sum_{n=0}^{\infty} \binom{1/2}{n} \dfrac{x^{4n+1}}{4n+1}$ and hence, since $0.4 < 1$,

we have

$$I = \int_0^{0.4} \sqrt{1 + x^4} \, dx = \sum_{n=0}^{\infty} \binom{1/2}{n} \frac{(0.4)^{4n+1}}{4n+1}$$

$$= (1)\frac{(0.4)^1}{0!} + \frac{\frac{1}{2}}{1!} \frac{(0.4)^5}{5} + \frac{\frac{1}{2}(-\frac{1}{2})}{2!} \frac{(0.4)^9}{9} + \frac{\frac{1}{2}(-\frac{1}{2})(-\frac{3}{2})}{3!} \frac{(0.4)^{13}}{13} + \frac{\frac{1}{2}(-\frac{1}{2})(-\frac{3}{2})(-\frac{5}{2})}{4!} \frac{(0.4)^{17}}{17} + \cdots$$

$$= 0.4 + \frac{(0.4)^5}{10} - \frac{(0.4)^9}{72} + \frac{(0.4)^{13}}{208} - \frac{5(0.4)^{17}}{2176} + \cdots$$

Now $\dfrac{(0.4)^9}{72} \approx 3.6 \times 10^{-6} < 5 \times 10^{-6}$, so by the Alternating Series Estimation Theorem, $I \approx 0.4 + \dfrac{(0.4)^5}{10} \approx 0.40102$

(correct to five decimal places).

55. $\lim\limits_{x \to 0} \dfrac{x - \ln(1 + x)}{x^2} = \lim\limits_{x \to 0} \dfrac{x - \left(x - \frac{1}{2}x^2 + \frac{1}{3}x^3 - \frac{1}{4}x^4 + \frac{1}{5}x^5 - \cdots\right)}{x^2} = \lim\limits_{x \to 0} \dfrac{\frac{1}{2}x^2 - \frac{1}{3}x^3 + \frac{1}{4}x^4 - \frac{1}{5}x^5 + \cdots}{x^2}$

$$= \lim_{x \to 0} \left(\tfrac{1}{2} - \tfrac{1}{3}x + \tfrac{1}{4}x^2 - \tfrac{1}{5}x^3 + \cdots\right) = \tfrac{1}{2}$$

since power series are continuous functions.

57. $\lim\limits_{x \to 0} \dfrac{\sin x - x + \frac{1}{6}x^3}{x^5} = \lim\limits_{x \to 0} \dfrac{\left(x - \frac{1}{3!}x^3 + \frac{1}{5!}x^5 - \frac{1}{7!}x^7 + \cdots\right) - x + \frac{1}{6}x^3}{x^5}$

$$= \lim_{x \to 0} \frac{\frac{1}{5!}x^5 - \frac{1}{7!}x^7 + \cdots}{x^5} = \lim_{x \to 0} \left(\frac{1}{5!} - \frac{x^2}{7!} + \frac{x^4}{9!} - \cdots\right) = \frac{1}{5!} = \frac{1}{120}$$

since power series are continuous functions.

59. From Equation 11, we have $e^{-x^2} = 1 - \dfrac{x^2}{1!} + \dfrac{x^4}{2!} - \dfrac{x^6}{3!} + \cdots$ and we know that $\cos x = 1 - \dfrac{x^2}{2!} + \dfrac{x^4}{4!} - \cdots$ from

Equation 16. Therefore, $e^{-x^2} \cos x = \left(1 - x^2 + \frac{1}{2}x^4 - \cdots\right)\left(1 - \frac{1}{2}x^2 + \frac{1}{24}x^4 - \cdots\right)$. Writing only the terms with

degree ≤ 4, we get $e^{-x^2} \cos x = 1 - \frac{1}{2}x^2 + \frac{1}{24}x^4 - x^2 + \frac{1}{2}x^4 + \frac{1}{2}x^4 + \cdots = 1 - \frac{3}{2}x^2 + \frac{25}{24}x^4 + \cdots$.

© 2012 Cengage Learning. All Rights Reserved. May not be scanned, copied, or duplicated, or posted to a publicly accessible website, in whole or in part.

61. $\dfrac{x}{\sin x} \overset{(15)}{=} \dfrac{x}{x - \frac{1}{6}x^3 + \frac{1}{120}x^5 - \cdots}.$

$$
\begin{array}{r}
1 + \frac{1}{6}x^2 + \frac{7}{360}x^4 + \cdots \\[4pt]
x - \frac{1}{6}x^3 + \frac{1}{120}x^5 - \cdots \enclose{longdiv}{\;x} \\[4pt]
x - \frac{1}{6}x^3 + \frac{1}{120}x^5 - \cdots \\[2pt]
\hline
\frac{1}{6}x^3 - \frac{1}{120}x^5 + \cdots \\[2pt]
\frac{1}{6}x^3 - \frac{1}{36}x^5 + \cdots \\[2pt]
\hline
\frac{7}{360}x^5 + \cdots \\[2pt]
\frac{7}{360}x^5 + \cdots \\[2pt]
\hline
\cdots
\end{array}
$$

From the long division above, $\dfrac{x}{\sin x} = 1 + \frac{1}{6}x^2 + \frac{7}{360}x^4 + \cdots.$

63. $\displaystyle\sum_{n=0}^{\infty} (-1)^n \frac{x^{4n}}{n!} = \sum_{n=0}^{\infty} \frac{\left(-x^4\right)^n}{n!} = e^{-x^4}$, by (11).

65. $\displaystyle\sum_{n=1}^{\infty} (-1)^{n-1} \frac{3^n}{n5^n} = \sum_{n=1}^{\infty} (-1)^{n-1} \frac{(3/5)^n}{n} = \ln\left(1 + \frac{3}{5}\right)$ [from Table 1] $= \ln\dfrac{8}{5}$

67. $\displaystyle\sum_{n=0}^{\infty} \frac{(-1)^n \, \pi^{2n+1}}{4^{2n+1}(2n+1)!} = \sum_{n=0}^{\infty} \frac{(-1)^n \left(\frac{\pi}{4}\right)^{2n+1}}{(2n+1)!} = \sin\frac{\pi}{4} = \frac{1}{\sqrt{2}}$, by (15).

69. $3 + \dfrac{9}{2!} + \dfrac{27}{3!} + \dfrac{81}{4!} + \cdots = \dfrac{3^1}{1!} + \dfrac{3^2}{2!} + \dfrac{3^3}{3!} + \dfrac{3^4}{4!} + \cdots = \displaystyle\sum_{n=1}^{\infty} \frac{3^n}{n!} = \sum_{n=0}^{\infty} \frac{3^n}{n!} - 1 = e^3 - 1$, by (11).

71. If p is an nth-degree polynomial, then $p^{(i)}(x) = 0$ for $i > n$, so its Taylor series at a is $p(x) = \displaystyle\sum_{i=0}^{n} \frac{p^{(i)}(a)}{i!}(x-a)^i$.

Put $x - a = 1$, so that $x = a + 1$. Then $p(a+1) = \displaystyle\sum_{i=0}^{n} \frac{p^{(i)}(a)}{i!}$.

This is true for any a, so replace a by x: $p(x+1) = \displaystyle\sum_{i=0}^{n} \frac{p^{(i)}(x)}{i!}$

73. Assume that $|f'''(x)| \le M$, so $f'''(x) \le M$ for $a \le x \le a + d$. Now $\int_a^x f'''(t)\,dt \le \int_a^x M\,dt \;\Rightarrow$

$f''(x) - f''(a) \le M(x-a) \;\Rightarrow\; f''(x) \le f''(a) + M(x-a)$. Thus, $\int_a^x f''(t)\,dt \le \int_a^x [f''(a) + M(t-a)]\,dt \;\Rightarrow$

$f'(x) - f'(a) \le f''(a)(x-a) + \frac{1}{2}M(x-a)^2 \;\Rightarrow\; f'(x) \le f'(a) + f''(a)(x-a) + \frac{1}{2}M(x-a)^2 \;\Rightarrow$

$\int_a^x f'(t)\,dt \le \int_a^x \left[f'(a) + f''(a)(t-a) + \frac{1}{2}M(t-a)^2\right] dt \;\Rightarrow$

$f(x) - f(a) \le f'(a)(x-a) + \frac{1}{2}f''(a)(x-a)^2 + \frac{1}{6}M(x-a)^3$. So

$f(x) - f(a) - f'(a)(x-a) - \frac{1}{2}f''(a)(x-a)^2 \le \frac{1}{6}M(x-a)^3$. But

$R_2(x) = f(x) - T_2(x) = f(x) - f(a) - f'(a)(x-a) - \frac{1}{2}f''(a)(x-a)^2$, so $R_2(x) \le \frac{1}{6}M(x-a)^3$.

A similar argument using $f'''(x) \ge -M$ shows that $R_2(x) \ge -\frac{1}{6}M(x-a)^3$. So $|R_2(x_2)| \le \frac{1}{6}M\,|x-a|^3$.

Although we have assumed that $x > a$, a similar calculation shows that this inequality is also true if $x < a$.

© 2012 Cengage Learning. All Rights Reserved. May not be scanned, copied, or duplicated, or posted to a publicly accessible website, in whole or in part.

75. (a) $g(x) = \sum_{n=0}^{\infty} \binom{k}{n} x^n \;\Rightarrow\; g'(x) = \sum_{n=1}^{\infty} \binom{k}{n} n x^{n-1}$, so

$$(1+x)g'(x) = (1+x) \sum_{n=1}^{\infty} \binom{k}{n} n x^{n-1} = \sum_{n=1}^{\infty} \binom{k}{n} n x^{n-1} + \sum_{n=1}^{\infty} \binom{k}{n} n x^n$$

$$= \sum_{n=0}^{\infty} \binom{k}{n+1}(n+1)x^n + \sum_{n=0}^{\infty} \binom{k}{n} n x^n \qquad \begin{bmatrix} \text{Replace } n \text{ with } n+1 \\ \text{in the first series} \end{bmatrix}$$

$$= \sum_{n=0}^{\infty} (n+1)\frac{k(k-1)(k-2)\cdots(k-n+1)(k-n)}{(n+1)!} x^n + \sum_{n=0}^{\infty} \left[(n)\frac{k(k-1)(k-2)\cdots(k-n+1)}{n!}\right] x^n$$

$$= \sum_{n=0}^{\infty} \frac{(n+1)k(k-1)(k-2)\cdots(k-n+1)}{(n+1)!} [(k-n)+n]\, x^n$$

$$= k \sum_{n=0}^{\infty} \frac{k(k-1)(k-2)\cdots(k-n+1)}{n!} x^n = k \sum_{n=0}^{\infty} \binom{k}{n} x^n = kg(x)$$

Thus, $g'(x) = \dfrac{kg(x)}{1+x}$.

(b) $h(x) = (1+x)^{-k} g(x) \;\Rightarrow$

$$h'(x) = -k(1+x)^{-k-1}g(x) + (1+x)^{-k}g'(x) \qquad \text{[Product Rule]}$$

$$= -k(1+x)^{-k-1}g(x) + (1+x)^{-k}\frac{kg(x)}{1+x} \qquad \text{[from part (a)]}$$

$$= -k(1+x)^{-k-1}g(x) + k(1+x)^{-k-1}g(x) = 0$$

(c) From part (b) we see that $h(x)$ must be constant for $x \in (-1,1)$, so $h(x) = h(0) = 1$ for $x \in (-1,1)$.

Thus, $h(x) = 1 = (1+x)^{-k} g(x) \;\Leftrightarrow\; g(x) = (1+x)^k$ for $x \in (-1,1)$.

11.11 Applications of Taylor Polynomials

1. (a)

n	$f^{(n)}(x)$	$f^{(n)}(0)$	$T_n(x)$
0	$\cos x$	1	1
1	$-\sin x$	0	1
2	$-\cos x$	-1	$1 - \frac{1}{2}x^2$
3	$\sin x$	0	$1 - \frac{1}{2}x^2$
4	$\cos x$	1	$1 - \frac{1}{2}x^2 + \frac{1}{24}x^4$
5	$-\sin x$	0	$1 - \frac{1}{2}x^2 + \frac{1}{24}x^4$
6	$-\cos x$	-1	$1 - \frac{1}{2}x^2 + \frac{1}{24}x^4 - \frac{1}{720}x^6$

© 2012 Cengage Learning. All Rights Reserved. May not be scanned, copied, or duplicated, or posted to a publicly accessible website, in whole or in part.

(b)

x	f	$T_0 = T_1$	$T_2 = T_3$	$T_4 = T_5$	T_6
$\frac{\pi}{4}$	0.7071	1	0.6916	0.7074	0.7071
$\frac{\pi}{2}$	0	1	−0.2337	0.0200	−0.0009
π	−1	1	−3.9348	0.1239	−1.2114

(c) As n increases, $T_n(x)$ is a good approximation to $f(x)$ on a larger and larger interval.

3.

n	$f^{(n)}(x)$	$f^{(n)}(2)$
0	$1/x$	$\frac{1}{2}$
1	$-1/x^2$	$-\frac{1}{4}$
2	$2/x^3$	$\frac{1}{4}$
3	$-6/x^4$	$-\frac{3}{8}$

$$T_3(x) = \sum_{n=0}^{3} \frac{f^{(n)}(2)}{n!}(x-2)^n$$

$$= \frac{\frac{1}{2}}{0!} - \frac{\frac{1}{4}}{1!}(x-2) + \frac{\frac{1}{4}}{2!}(x-2)^2 - \frac{\frac{3}{8}}{3!}(x-2)^3$$

$$= \frac{1}{2} - \frac{1}{4}(x-2) + \frac{1}{8}(x-2)^2 - \frac{1}{16}(x-2)^3$$

5.

n	$f^{(n)}(x)$	$f^{(n)}(\pi/2)$
0	$\cos x$	0
1	$-\sin x$	−1
2	$-\cos x$	0
3	$\sin x$	1

$$T_3(x) = \sum_{n=0}^{3} \frac{f^{(n)}(\pi/2)}{n!}\left(x - \frac{\pi}{2}\right)^n$$

$$= -\left(x - \frac{\pi}{2}\right) + \frac{1}{6}\left(x - \frac{\pi}{2}\right)^3$$

7.

n	$f^{(n)}(x)$	$f^{(n)}(1)$
0	$\ln x$	0
1	$1/x$	1
2	$-1/x^2$	−1
3	$2/x^3$	2

$$T_3(x) = \sum_{n=0}^{3} \frac{f^{(n)}(1)}{n!}(x-1)^n$$

$$= 0 + \frac{1}{1!}(x-1) + \frac{-1}{2!}(x-1)^2 + \frac{2}{3!}(x-1)^3$$

$$= (x-1) - \frac{1}{2}(x-1)^2 + \frac{1}{3}(x-1)^3$$

© 2012 Cengage Learning. All Rights Reserved. May not be scanned, copied, or duplicated, or posted to a publicly accessible website, in whole or in part.

9.

n	$f^{(n)}(x)$	$f^{(n)}(0)$
0	xe^{-2x}	0
1	$(1-2x)e^{-2x}$	1
2	$4(x-1)e^{-2x}$	-4
3	$4(3-2x)e^{-2x}$	12

$$T_3(x) = \sum_{n=0}^{3} \frac{f^{(n)}(0)}{n!}x^n = \frac{0}{1}\cdot 1 + \frac{1}{1}x^1 + \frac{-4}{2}x^2 + \frac{12}{6}x^3 = x - 2x^2 + 2x^3$$

11. You may be able to simply find the Taylor polynomials for

$f(x) = \cot x$ using your CAS. We will list the values of $f^{(n)}(\pi/4)$

for $n = 0$ to $n = 5$.

n	0	1	2	3	4	5
$f^{(n)}(\pi/4)$	1	-2	4	-16	80	-512

$$T_5(x) = \sum_{n=0}^{5} \frac{f^{(n)}(\pi/4)}{n!}\left(x - \tfrac{\pi}{4}\right)^n$$

$$= 1 - 2\left(x - \tfrac{\pi}{4}\right) + 2\left(x - \tfrac{\pi}{4}\right)^2 - \tfrac{8}{3}\left(x - \tfrac{\pi}{4}\right)^3 + \tfrac{10}{3}\left(x - \tfrac{\pi}{4}\right)^4 - \tfrac{64}{15}\left(x - \tfrac{\pi}{4}\right)^5$$

For $n = 2$ to $n = 5$, $T_n(x)$ is the polynomial consisting of all the terms up to and including the $\left(x - \tfrac{\pi}{4}\right)^n$ term.

13.

n	$f^{(n)}(x)$	$f^{(n)}(4)$
0	$\sqrt{x}$	2
1	$\frac{1}{2}x^{-1/2}$	$\frac{1}{4}$
2	$-\frac{1}{4}x^{-3/2}$	$-\frac{1}{32}$
3	$\frac{3}{8}x^{-5/2}$	

(a) $f(x) = \sqrt{x} \approx T_2(x) = 2 + \dfrac{1}{4}(x-4) - \dfrac{1/32}{2!}(x-4)^2$

$$= 2 + \tfrac{1}{4}(x-4) - \tfrac{1}{64}(x-4)^2$$

(b) $|R_2(x)| \leq \dfrac{M}{3!}|x-4|^3$, where $|f'''(x)| \leq M$. Now $4 \leq x \leq 4.2 \Rightarrow$

$|x-4| \leq 0.2 \Rightarrow |x-4|^3 \leq 0.008$. Since $f'''(x)$ is decreasing

on $[4, 4.2]$, we can take $M = |f'''(4)| = \tfrac{3}{8}4^{-5/2} = \tfrac{3}{256}$, so

$$|R_2(x)| \leq \dfrac{3/256}{6}(0.008) = \dfrac{0.008}{512} = 0.000\,015\,625.$$

(c)

From the graph of $|R_2(x)| = |\sqrt{x} - T_2(x)|$, it seems that the

error is less than 1.52×10^{-5} on $[4, 4.2]$.

© 2012 Cengage Learning. All Rights Reserved. May not be scanned, copied, or duplicated, or posted to a publicly accessible website, in whole or in part.

15.

n	$f^{(n)}(x)$	$f^{(n)}(1)$
0	$x^{2/3}$	1
1	$\frac{2}{3}x^{-1/3}$	$\frac{2}{3}$
2	$-\frac{2}{9}x^{-4/3}$	$-\frac{2}{9}$
3	$\frac{8}{27}x^{-7/3}$	$\frac{8}{27}$
4	$-\frac{56}{81}x^{-10/3}$	

(a) $f(x) = x^{2/3} \approx T_3(x) = 1 + \frac{2}{3}(x-1) - \frac{2/9}{2!}(x-1)^2 + \frac{8/27}{3!}(x-1)^3$

$\qquad = 1 + \frac{2}{3}(x-1) - \frac{1}{9}(x-1)^2 + \frac{4}{81}(x-1)^3$

(b) $|R_3(x)| \le \frac{M}{4!}|x-1|^4$, where $\left|f^{(4)}(x)\right| \le M$. Now $0.8 \le x \le 1.2 \Rightarrow$

$|x-1| \le 0.2 \Rightarrow |x-1|^4 \le 0.0016$. Since $\left|f^{(4)}(x)\right|$ is decreasing

on $[0.8, 1.2]$, we can take $M = \left|f^{(4)}(0.8)\right| = \frac{56}{81}(0.8)^{-10/3}$, so

$|R_3(x)| \le \frac{\frac{56}{81}(0.8)^{-10/3}}{24}(0.0016) \approx 0.000\,096\,97$.

From the graph of $|R_3(x)| = \left|x^{2/3} - T_3(x)\right|$, it seems that the

error is less than $0.000\,053\,3$ on $[0.8, 1.2]$.

(c)

0.00006

$y = |R_3(x)|$

0.8 ——— 0 ——— 1 ——— 1.2

17.

n	$f^{(n)}(x)$	$f^{(n)}(0)$
0	$\sec x$	1
1	$\sec x \tan x$	0
2	$\sec x\,(2\sec^2 x - 1)$	1
3	$\sec x \tan x\,(6\sec^2 x - 1)$	

(a) $f(x) = \sec x \approx T_2(x) = 1 + \frac{1}{2}x^2$

(b) $|R_2(x)| \le \frac{M}{3!}|x|^3$, where $\left|f^{(3)}(x)\right| \le M$. Now $-0.2 \le x \le 0.2 \Rightarrow |x| \le 0.2 \Rightarrow |x|^3 \le (0.2)^3$.

$f^{(3)}(x)$ is an odd function and it is increasing on $[0, 0.2]$ since $\sec x$ and $\tan x$ are increasing on $[0, 0.2]$,

so $\left|f^{(3)}(x)\right| \le f^{(3)}(0.2) \approx 1.085\,158\,892$. Thus, $|R_2(x)| \le \frac{f^{(3)}(0.2)}{3!}(0.2)^3 \approx 0.001\,447$.

(c)

0.0004

$y = |R_2(x)|$

-0.2 ——— 0 ——— 0.2

From the graph of $|R_2(x)| = |\sec x - T_2(x)|$, it seems that the

error is less than $0.000\,339$ on $[-0.2, 0.2]$.

19.

n	$f^{(n)}(x)$	$f^{(n)}(0)$
0	e^{x^2}	1
1	$e^{x^2}(2x)$	0
2	$e^{x^2}(2+4x^2)$	2
3	$e^{x^2}(12x+8x^3)$	0
4	$e^{x^2}(12+48x^2+16x^4)$	

(a) $f(x) = e^{x^2} \approx T_3(x) = 1 + \frac{2}{2!}x^2 = 1 + x^2$

(b) $|R_3(x)| \le \frac{M}{4!}|x|^4$, where $\left|f^{(4)}(x)\right| \le M$. Now $0 \le x \le 0.1 \Rightarrow$

$x^4 \le (0.1)^4$, and letting $x = 0.1$ gives

$|R_3(x)| \le \frac{e^{0.01}(12 + 0.48 + 0.0016)}{24}(0.1)^4 \approx 0.00006$.

© 2012 Cengage Learning. All Rights Reserved. May not be scanned, copied, or duplicated, or posted to a publicly accessible website, in whole or in part.

(c)

From the graph of $|R_3(x)| = \left|e^{x^2} - T_3(x)\right|$, it appears that the error is less than $0.000\,051$ on $[0, 0.1]$.

21.

n	$f^{(n)}(x)$	$f^{(n)}(0)$
0	$x \sin x$	0
1	$\sin x + x \cos x$	0
2	$2 \cos x - x \sin x$	2
3	$-3 \sin x - x \cos x$	0
4	$-4 \cos x + x \sin x$	-4
5	$5 \sin x + x \cos x$	

(a) $f(x) = x \sin x \approx T_4(x) = \dfrac{2}{2!}(x - 0)^2 + \dfrac{-4}{4!}(x - 0)^4 = x^2 - \dfrac{1}{6}x^4$

(b) $|R_4(x)| \leq \dfrac{M}{5!}|x|^5$, where $\left|f^{(5)}(x)\right| \leq M$. Now $-1 \leq x \leq 1 \;\Rightarrow$

$|x| \leq 1$, and a graph of $f^{(5)}(x)$ shows that $\left|f^{(5)}(x)\right| \leq 5$ for $-1 \leq x \leq 1$.

Thus, we can take $M = 5$ and get $|R_4(x)| \leq \dfrac{5}{5!} \cdot 1^5 = \dfrac{1}{24} = 0.041\overline{6}$.

(c)

From the graph of $|R_4(x)| = |x \sin x - T_4(x)|$, it seems that the error is less than 0.0082 on $[-1, 1]$.

23. From Exercise 5, $\cos x = -\left(x - \frac{\pi}{2}\right) + \frac{1}{6}\left(x - \frac{\pi}{2}\right)^3 + R_3(x)$, where $|R_3(x)| \leq \dfrac{M}{4!}\left|x - \frac{\pi}{2}\right|^4$ with

$\left|f^{(4)}(x)\right| = |\cos x| \leq M = 1$. Now $x = 80° = (90° - 10°) = \left(\frac{\pi}{2} - \frac{\pi}{18}\right) = \frac{4\pi}{9}$ radians, so the error is

$\left|R_3\left(\frac{4\pi}{9}\right)\right| \leq \frac{1}{24}\left(\frac{\pi}{18}\right)^4 \approx 0.000\,039$, which means our estimate would *not* be accurate to five decimal places. However,

$T_3 = T_4$, so we can use $\left|R_4\left(\frac{4\pi}{9}\right)\right| \leq \frac{1}{120}\left(\frac{\pi}{18}\right)^5 \approx 0.000\,001$. Therefore, to five decimal places,

$\cos 80° \approx -\left(-\frac{\pi}{18}\right) + \frac{1}{6}\left(-\frac{\pi}{18}\right)^3 \approx 0.17365$.

25. All derivatives of e^x are e^x, so $|R_n(x)| \leq \dfrac{e^x}{(n+1)!}|x|^{n+1}$, where $0 < x < 0.1$. Letting $x = 0.1$,

$R_n(0.1) \leq \dfrac{e^{0.1}}{(n+1)!}(0.1)^{n+1} < 0.00001$, and by trial and error we find that $n = 3$ satisfies this inequality since

$R_3(0.1) < 0.0000046$. Thus, by adding the four terms of the Maclaurin series for e^x corresponding to $n = 0, 1, 2,$ and 3,

we can estimate $e^{0.1}$ to within 0.00001. (In fact, this sum is $1.1051\overline{6}$ and $e^{0.1} \approx 1.10517$.)

© 2012 Cengage Learning. All Rights Reserved. May not be scanned, copied, or duplicated, or posted to a publicly accessible website, in whole or in part.

27. $\sin x = x - \dfrac{1}{3!}x^3 + \dfrac{1}{5!}x^5 - \cdots$. By the Alternating Series

Estimation Theorem, the error in the approximation

$\sin x = x - \dfrac{1}{3!}x^3$ is less than $\left|\dfrac{1}{5!}x^5\right| < 0.01 \quad \Leftrightarrow$

$\left|x^5\right| < 120(0.01) \quad \Leftrightarrow \quad |x| < (1.2)^{1/5} \approx 1.037$. The curves

$y = x - \frac{1}{6}x^3$ and $y = \sin x - 0.01$ intersect at $x \approx 1.043$, so

the graph confirms our estimate. Since both the sine function

and the given approximation are odd functions, we need to check the estimate only for $x > 0$. Thus, the desired range of

values for x is $-1.037 < x < 1.037$.

29. $\arctan x = x - \dfrac{x^3}{3} + \dfrac{x^5}{5} - \dfrac{x^7}{7} + \cdots$. By the Alternating Series

Estimation Theorem, the error is less than $\left|-\frac{1}{7}x^7\right| < 0.05 \quad \Leftrightarrow$

$\left|x^7\right| < 0.35 \quad \Leftrightarrow \quad |x| < (0.35)^{1/7} \approx 0.8607$. The curves

$y = x - \frac{1}{3}x^3 + \frac{1}{5}x^5$ and $y = \arctan x + 0.05$ intersect at

$x \approx 0.9245$, so the graph confirms our estimate. Since both the

arctangent function and the given approximation are odd functions,

we need to check the estimate only for $x > 0$. Thus, the desired

range of values for x is $-0.86 < x < 0.86$.

31. Let $s(t)$ be the position function of the car, and for convenience set $s(0) = 0$. The velocity of the car is $v(t) = s'(t)$ and the

acceleration is $a(t) = s''(t)$, so the second degree Taylor polynomial is $T_2(t) = s(0) + v(0)t + \dfrac{a(0)}{2}t^2 = 20t + t^2$. We

estimate the distance traveled during the next second to be $s(1) \approx T_2(1) = 20 + 1 = 21$ m. The function $T_2(t)$ would not be

accurate over a full minute, since the car could not possibly maintain an acceleration of 2 m/s^2 for that long (if it did, its final

speed would be 140 m/s ≈ 313 mi/h!).

33. $E = \dfrac{q}{D^2} - \dfrac{q}{(D+d)^2} = \dfrac{q}{D^2} - \dfrac{q}{D^2(1+d/D)^2} = \dfrac{q}{D^2}\left[1 - \left(1 + \dfrac{d}{D}\right)^{-2}\right]$.

We use the Binomial Series to expand $(1 + d/D)^{-2}$:

$$E = \dfrac{q}{D^2}\left[1 - \left(1 - 2\left(\dfrac{d}{D}\right) + \dfrac{2\cdot 3}{2!}\left(\dfrac{d}{D}\right)^2 - \dfrac{2\cdot 3\cdot 4}{3!}\left(\dfrac{d}{D}\right)^3 + \cdots\right)\right] = \dfrac{q}{D^2}\left[2\left(\dfrac{d}{D}\right) - 3\left(\dfrac{d}{D}\right)^2 + 4\left(\dfrac{d}{D}\right)^3 - \cdots\right]$$

$$\approx \dfrac{q}{D^2}\cdot 2\left(\dfrac{d}{D}\right) = 2qd\cdot\dfrac{1}{D^3}$$

when D is much larger than d; that is, when P is far away from the dipole.

35. (a) If the water is deep, then $2\pi d/L$ is large, and we know that $\tanh x \to 1$ as $x \to \infty$. So we can approximate

$\tanh(2\pi d/L) \approx 1$, and so $v^2 \approx gL/(2\pi) \quad \Leftrightarrow \quad v \approx \sqrt{gL/(2\pi)}$.

© 2012 Cengage Learning. All Rights Reserved. May not be scanned, copied, or duplicated, or posted to a publicly accessible website, in whole or in part.

(b) From the table, the first term in the Maclaurin series of $\tanh x$ is x, so if the water is shallow, we can approximate

$$\tanh \frac{2\pi d}{L} \approx \frac{2\pi d}{L}, \text{ and so } v^2 \approx \frac{gL}{2\pi} \cdot \frac{2\pi d}{L} \quad \Leftrightarrow \quad v \approx \sqrt{gd}.$$

n	$f^{(n)}(x)$	$f^{(n)}(0)$
0	$\tanh x$	0
1	$\operatorname{sech}^2 x$	1
2	$-2\operatorname{sech}^2 x \tanh x$	0
3	$2\operatorname{sech}^2 x (3\tanh^2 x - 1)$	-2

(c) Since $\tanh x$ is an odd function, its Maclaurin series is alternating, so the error in the approximation

$$\tanh \frac{2\pi d}{L} \approx \frac{2\pi d}{L} \text{ is less than the first neglected term, which is } \frac{|f'''(0)|}{3!}\left(\frac{2\pi d}{L}\right)^3 = \frac{1}{3}\left(\frac{2\pi d}{L}\right)^3.$$

If $L > 10d$, then $\dfrac{1}{3}\left(\dfrac{2\pi d}{L}\right)^3 < \dfrac{1}{3}\left(2\pi \cdot \dfrac{1}{10}\right)^3 = \dfrac{\pi^3}{375}$, so the error in the approximation $v^2 = gd$ is less

than $\dfrac{gL}{2\pi} \cdot \dfrac{\pi^3}{375} \approx 0.0132gL.$

37. (a) L is the length of the arc subtended by the angle θ, so $L = R\theta \quad \Rightarrow$

$\theta = L/R$. Now $\sec\theta = (R+C)/R \quad \Rightarrow \quad R\sec\theta = R + C \quad \Rightarrow$

$C = R\sec\theta - R = R\sec(L/R) - R.$

(b) First we'll find a Taylor polynomial $T_4(x)$ for $f(x) = \sec x$ at $x = 0$.

n	$f^{(n)}(x)$	$f^{(n)}(0)$
0	$\sec x$	1
1	$\sec x \tan x$	0
2	$\sec x (2\tan^2 x + 1)$	1
3	$\sec x \tan x (6\tan^2 x + 5)$	0
4	$\sec x (24\tan^4 x + 28\tan^2 x + 5)$	5

Thus, $f(x) = \sec x \approx T_4(x) = 1 + \frac{1}{2!}(x-0)^2 + \frac{5}{4!}(x-0)^4 = 1 + \frac{1}{2}x^2 + \frac{5}{24}x^4.$ By part (a),

$$C \approx R\left[1 + \frac{1}{2}\left(\frac{L}{R}\right)^2 + \frac{5}{24}\left(\frac{L}{R}\right)^4\right] - R = R + \frac{1}{2}R \cdot \frac{L^2}{R^2} + \frac{5}{24}R \cdot \frac{L^4}{R^4} - R = \frac{L^2}{2R} + \frac{5L^4}{24R^3}.$$

(c) Taking $L = 100$ km and $R = 6370$ km, the formula in part (a) says that

$C = R\sec(L/R) - R = 6370\sec(100/6370) - 6370 \approx 0.785\,009\,965\,44$ km.

The formula in part (b) says that $C \approx \dfrac{L^2}{2R} + \dfrac{5L^4}{24R^3} = \dfrac{100^2}{2 \cdot 6370} + \dfrac{5 \cdot 100^4}{24 \cdot 6370^3} \approx 0.785\,009\,957\,36$ km.

The difference between these two results is only $0.000\,000\,008\,08$ km, or $0.000\,008\,08$ m!

39. Using $f(x) = T_n(x) + R_n(x)$ with $n = 1$ and $x = r$, we have $f(r) = T_1(r) + R_1(r)$, where T_1 is the first-degree Taylor

polynomial of f at a. Because $a = x_n$, $f(r) = f(x_n) + f'(x_n)(r - x_n) + R_1(r)$. But r is a root of f, so $f(r) = 0$

and we have $0 = f(x_n) + f'(x_n)(r - x_n) + R_1(r)$. Taking the first two terms to the left side gives us

© 2012 Cengage Learning. All Rights Reserved. May not be scanned, copied, or duplicated, or posted to a publicly accessible website, in whole or in part.

$f'(x_n)(x_n - r) - f(x_n) = R_1(r)$. Dividing by $f'(x_n)$, we get $x_n - r - \dfrac{f(x_n)}{f'(x_n)} = \dfrac{R_1(r)}{f'(x_n)}$. By the formula for Newton's

method, the left side of the preceding equation is $x_{n+1} - r$, so $|x_{n+1} - r| = \left| \dfrac{R_1(r)}{f'(x_n)} \right|$. Taylor's Inequality gives us

$|R_1(r)| \leq \dfrac{|f''(r)|}{2!} |r - x_n|^2$. Combining this inequality with the facts $|f''(x)| \leq M$ and $|f'(x)| \geq K$ gives us

$|x_{n+1} - r| \leq \dfrac{M}{2K} |x_n - r|^2$.

11   Review

CONCEPT CHECK

1. (a) See Definition 11.1.1.

    (b) See Definition 11.2.2.

    (c) The terms of the sequence $\{a_n\}$ approach 3 as n becomes large.

    (d) By adding sufficiently many terms of the series, we can make the partial sums as close to 3 as we like.

2. (a) See the definition on page 721 [ET page 697].

    (b) A sequence is monotonic if it is either increasing or decreasing.

    (c) By Theorem 11.1.12, every bounded, monotonic sequence is convergent.

3. (a) See (4) in Section 11.2.

    (b) The p-series $\displaystyle\sum_{n=1}^{\infty} \dfrac{1}{n^p}$ is convergent if $p > 1$.

4. If $\sum a_n = 3$, then $\displaystyle\lim_{n \to \infty} a_n = 0$ and $\displaystyle\lim_{n \to \infty} s_n = 3$.

5. (a) *Test for Divergence:* If $\displaystyle\lim_{n \to \infty} a_n$ does not exist or if $\displaystyle\lim_{n \to \infty} a_n \neq 0$, then the series $\sum_{n=1}^{\infty} a_n$ is divergent.

    (b) *Integral Test:* Suppose f is a continuous, positive, decreasing function on $[1, \infty)$ and let $a_n = f(n)$. Then the series

    $\sum_{n=1}^{\infty} a_n$ is convergent if and only if the improper integral $\int_1^{\infty} f(x)\, dx$ is convergent. In other words:

      (i) If $\int_1^{\infty} f(x)\, dx$ is convergent, then $\sum_{n=1}^{\infty} a_n$ is convergent.

      (ii) If $\int_1^{\infty} f(x)\, dx$ is divergent, then $\sum_{n=1}^{\infty} a_n$ is divergent.

    (c) *Comparison Test:* Suppose that $\sum a_n$ and $\sum b_n$ are series with positive terms.

      (i) If $\sum b_n$ is convergent and $a_n \leq b_n$ for all n, then $\sum a_n$ is also convergent.

      (ii) If $\sum b_n$ is divergent and $a_n \geq b_n$ for all n, then $\sum a_n$ is also divergent.

    (d) *Limit Comparison Test:* Suppose that $\sum a_n$ and $\sum b_n$ are series with positive terms. If $\displaystyle\lim_{n \to \infty} (a_n/b_n) = c$, where c is a

    finite number and $c > 0$, then either both series converge or both diverge.

    (e) *Alternating Series Test:* If the alternating series $\sum_{n=1}^{\infty} (-1)^{n-1} b_n = b_1 - b_2 + b_3 - b_4 + b_5 - b_6 + \cdots \; [b_n > 0]$

    satisfies (i) $b_{n+1} \leq b_n$ for all n and (ii) $\displaystyle\lim_{n \to \infty} b_n = 0$, then the series is convergent.

© 2012 Cengage Learning. All Rights Reserved. May not be scanned, copied, or duplicated, or posted to a publicly accessible website, in whole or in part.

(f) *Ratio Test:*

(i) If $\lim\limits_{n\to\infty}\left|\dfrac{a_{n+1}}{a_n}\right| = L < 1$, then the series $\sum\limits_{n=1}^{\infty} a_n$ is absolutely convergent (and therefore convergent).

(ii) If $\lim\limits_{n\to\infty}\left|\dfrac{a_{n+1}}{a_n}\right| = L > 1$ or $\lim\limits_{n\to\infty}\left|\dfrac{a_{n+1}}{a_n}\right| = \infty$, then the series $\sum\limits_{n=1}^{\infty} a_n$ is divergent.

(iii) If $\lim\limits_{n\to\infty}\left|\dfrac{a_{n+1}}{a_n}\right| = 1$, the Ratio Test is inconclusive; that is, no conclusion can be drawn about the convergence or

divergence of $\sum a_n$.

(g) *Root Test:*

(i) If $\lim\limits_{n\to\infty} \sqrt[n]{|a_n|} = L < 1$, then the series $\sum_{n=1}^{\infty} a_n$ is absolutely convergent (and therefore convergent).

(ii) If $\lim\limits_{n\to\infty} \sqrt[n]{|a_n|} = L > 1$ or $\lim\limits_{n\to\infty} \sqrt[n]{|a_n|} = \infty$, then the series $\sum_{n=1}^{\infty} a_n$ is divergent.

(iii) If $\lim\limits_{n\to\infty} \sqrt[n]{|a_n|} = 1$, the Root Test is inconclusive.

6. (a) A series $\sum a_n$ is called *absolutely convergent* if the series of absolute values $\sum |a_n|$ is convergent.

(b) If a series $\sum a_n$ is absolutely convergent, then it is convergent.

(c) A series $\sum a_n$ is called *conditionally convergent* if it is convergent but not absolutely convergent.

7. (a) Use (3) in Section 11.3.

(b) See Example 5 in Section 11.4.

(c) By adding terms until you reach the desired accuracy given by the Alternating Series Estimation Theorem.

8. (a) $\sum\limits_{n=0}^{\infty} c_n(x-a)^n$

(b) Given the power series $\sum\limits_{n=0}^{\infty} c_n(x-a)^n$, the radius of convergence is:

(i) 0 if the series converges only when $x = a$

(ii) ∞ if the series converges for all x, or

(iii) a positive number R such that the series converges if $|x - a| < R$ and diverges if $|x - a| > R$.

(c) The interval of convergence of a power series is the interval that consists of all values of x for which the series converges. Corresponding to the cases in part (b), the interval of convergence is: (i) the single point $\{a\}$, (ii) all real numbers, that is, the real number line $(-\infty, \infty)$, or (iii) an interval with endpoints $a - R$ and $a + R$ which can contain neither, either, or both of the endpoints. In this case, we must test the series for convergence at each endpoint to determine the interval of convergence.

9. (a), (b) See Theorem 11.9.2.

10. (a) $T_n(x) = \sum\limits_{i=0}^{n} \dfrac{f^{(i)}(a)}{i!}(x-a)^i$

(b) $\sum\limits_{n=0}^{\infty} \dfrac{f^{(n)}(a)}{n!}(x-a)^n$

© 2012 Cengage Learning. All Rights Reserved. May not be scanned, copied, or duplicated, or posted to a publicly accessible website, in whole or in part.

(c) $\sum_{n=0}^{\infty} \dfrac{f^{(n)}(0)}{n!}\, x^n$ [$a = 0$ in part (b)]

(d) See Theorem 11.10.8.

(e) See Taylor's Inequality (11.10.9).

11. (a)–(f) See Table 1 on page 786 [ET 762].

12. See the binomial series (11.10.17) for the expansion. The radius of convergence for the binomial series is 1.

TRUE-FALSE QUIZ

1. False.   See Note 2 after Theorem 11.2.6.

3. True.   If $\lim\limits_{n \to \infty} a_n = L$, then as $n \to \infty$, $2n + 1 \to \infty$, so $a_{2n+1} \to L$.

5. False.   For example, take $c_n = (-1)^n/(n6^n)$.

7. False, since $\lim\limits_{n \to \infty} \left| \dfrac{a_{n+1}}{a_n} \right| = \lim\limits_{n \to \infty} \left| \dfrac{1}{(n+1)^3} \cdot \dfrac{n^3}{1} \right| = \lim\limits_{n \to \infty} \left| \dfrac{n^3}{(n+1)^3} \cdot \dfrac{1/n^3}{1/n^3} \right| = \lim\limits_{n \to \infty} \dfrac{1}{(1+1/n)^3} = 1.$

9. False.   See the note after Example 2 in Section 11.4.

11. True.   See (9) in Section 11.1.

13. True.   By Theorem 11.10.5 the coefficient of x^3 is $\dfrac{f'''(0)}{3!} = \dfrac{1}{3} \;\Rightarrow\; f'''(0) = 2.$

     Or: Use Theorem 11.9.2 to differentiate f three times.

15. False.   For example, let $a_n = b_n = (-1)^n$. Then $\{a_n\}$ and $\{b_n\}$ are divergent, but $a_n b_n = 1$, so $\{a_n b_n\}$ is convergent.

17. True by Theorem 11.6.3. $\left[\sum (-1)^n a_n \text{ is absolutely convergent and hence convergent.} \right]$

19. True. $0.99999\ldots = 0.9 + 0.9(0.1)^1 + 0.9(0.1)^2 + 0.9(0.1)^3 + \cdots = \sum\limits_{n=1}^{\infty} (0.9)(0.1)^{n-1} = \dfrac{0.9}{1 - 0.1} = 1$ by the formula

     for the sum of a geometric series $[S = a_1/(1 - r)]$ with ratio r satisfying $|r| < 1$.

21. True.   A finite number of terms doesn't affect convergence or divergence of a series.

EXERCISES

1. $\left\{ \dfrac{2 + n^3}{1 + 2n^3} \right\}$ converges since $\lim\limits_{n \to \infty} \dfrac{2 + n^3}{1 + 2n^3} = \lim\limits_{n \to \infty} \dfrac{2/n^3 + 1}{1/n^3 + 2} = \dfrac{1}{2}.$

3. $\lim\limits_{n \to \infty} a_n = \lim\limits_{n \to \infty} \dfrac{n^3}{1 + n^2} = \lim\limits_{n \to \infty} \dfrac{n}{1/n^2 + 1} = \infty$, so the sequence diverges.

© 2012 Cengage Learning. All Rights Reserved. May not be scanned, copied, or duplicated, or posted to a publicly accessible website, in whole or in part.

5. $|a_n| = \left| \dfrac{n \sin n}{n^2 + 1} \right| \le \dfrac{n}{n^2 + 1} < \dfrac{1}{n}$, so $|a_n| \to 0$ as $n \to \infty$. Thus, $\lim\limits_{n \to \infty} a_n = 0$. The sequence $\{a_n\}$ is convergent.

7. $\left\{ \left(1 + \dfrac{3}{n}\right)^{4n} \right\}$ is convergent. Let $y = \left(1 + \dfrac{3}{x}\right)^{4x}$. Then

$$\lim_{x \to \infty} \ln y = \lim_{x \to \infty} 4x \ln(1 + 3/x) = \lim_{x \to \infty} \dfrac{\ln(1 + 3/x)}{1/(4x)} \overset{\text{H}}{=} \lim_{x \to \infty} \dfrac{\dfrac{1}{1 + 3/x}\left(-\dfrac{3}{x^2}\right)}{-1/(4x^2)} = \lim_{x \to \infty} \dfrac{12}{1 + 3/x} = 12, \text{ so}$$

$$\lim_{x \to \infty} y = \lim_{n \to \infty} \left(1 + \dfrac{3}{n}\right)^{4n} = e^{12}.$$

9. We use induction, hypothesizing that $a_{n-1} < a_n < 2$. Note first that $1 < a_2 = \frac{1}{3}(1 + 4) = \frac{5}{3} < 2$, so the hypothesis holds for $n = 2$. Now assume that $a_{k-1} < a_k < 2$. Then $a_k = \frac{1}{3}(a_{k-1} + 4) < \frac{1}{3}(a_k + 4) < \frac{1}{3}(2 + 4) = 2$. So $a_k < a_{k+1} < 2$, and the induction is complete. To find the limit of the sequence, we note that $L = \lim\limits_{n \to \infty} a_n = \lim\limits_{n \to \infty} a_{n+1} \Rightarrow$

$L = \frac{1}{3}(L + 4) \Rightarrow L = 2$.

11. $\dfrac{n}{n^3 + 1} < \dfrac{n}{n^3} = \dfrac{1}{n^2}$, so $\sum\limits_{n=1}^{\infty} \dfrac{n}{n^3 + 1}$ converges by the Comparison Test with the convergent p-series $\sum\limits_{n=1}^{\infty} \dfrac{1}{n^2}$ $[p = 2 > 1]$.

13. $\lim\limits_{n \to \infty} \left| \dfrac{a_{n+1}}{a_n} \right| = \lim\limits_{n \to \infty} \left[\dfrac{(n+1)^3}{5^{n+1}} \cdot \dfrac{5^n}{n^3} \right] = \lim\limits_{n \to \infty} \left(1 + \dfrac{1}{n}\right)^3 \cdot \dfrac{1}{5} = \dfrac{1}{5} < 1$, so $\sum\limits_{n=1}^{\infty} \dfrac{n^3}{5^n}$ converges by the Ratio Test.

15. Let $f(x) = \dfrac{1}{x \sqrt{\ln x}}$. Then f is continuous, positive, and decreasing on $[2, \infty)$, so the Integral Test applies.

$$\int_2^{\infty} f(x)\, dx = \lim_{t \to \infty} \int_2^t \dfrac{1}{x \sqrt{\ln x}}\, dx \quad \left[u = \ln x, du = \dfrac{1}{x}\, dx\right] = \lim_{t \to \infty} \int_{\ln 2}^{\ln t} u^{-1/2}\, du = \lim_{t \to \infty} \left[2 \sqrt{u}\right]_{\ln 2}^{\ln t}$$

$$= \lim_{t \to \infty} \left(2 \sqrt{\ln t} - 2 \sqrt{\ln 2}\right) = \infty,$$

so the series $\sum\limits_{n=2}^{\infty} \dfrac{1}{n \sqrt{\ln n}}$ diverges.

17. $|a_n| = \left| \dfrac{\cos 3n}{1 + (1.2)^n} \right| \le \dfrac{1}{1 + (1.2)^n} < \dfrac{1}{(1.2)^n} = \left(\dfrac{5}{6}\right)^n$, so $\sum\limits_{n=1}^{\infty} |a_n|$ converges by comparison with the convergent geometric

series $\sum\limits_{n=1}^{\infty} \left(\dfrac{5}{6}\right)^n$ $\left[r = \dfrac{5}{6} < 1\right]$. It follows that $\sum\limits_{n=1}^{\infty} a_n$ converges (by Theorem 3 in Section 11.6).

19. $\lim\limits_{n \to \infty} \left| \dfrac{a_{n+1}}{a_n} \right| = \lim\limits_{n \to \infty} \dfrac{1 \cdot 3 \cdot 5 \cdot \cdots \cdot (2n - 1)(2n + 1)}{5^{n+1}(n+1)!} \cdot \dfrac{5^n\, n!}{1 \cdot 3 \cdot 5 \cdot \cdots \cdot (2n - 1)} = \lim\limits_{n \to \infty} \dfrac{2n + 1}{5(n + 1)} = \dfrac{2}{5} < 1$, so the series

converges by the Ratio Test.

21. $b_n = \dfrac{\sqrt{n}}{n + 1} > 0$, $\{b_n\}$ is decreasing, and $\lim\limits_{n \to \infty} b_n = 0$, so the series $\sum\limits_{n=1}^{\infty} (-1)^{n-1} \dfrac{\sqrt{n}}{n + 1}$ converges by the Alternating

Series Test.

© 2012 Cengage Learning. All Rights Reserved. May not be scanned, copied, or duplicated, or posted to a publicly accessible website, in whole or in part.

23. Consider the series of absolute values: $\sum\limits_{n=1}^{\infty} n^{-1/3}$ is a p-series with $p = \frac{1}{3} \le 1$ and is therefore divergent. But if we apply the

Alternating Series Test, we see that $b_n = \dfrac{1}{\sqrt[3]{n}} > 0$, $\{b_n\}$ is decreasing, and $\lim\limits_{n \to \infty} b_n = 0$, so the series $\sum\limits_{n=1}^{\infty} (-1)^{n-1} n^{-1/3}$

converges. Thus, $\sum\limits_{n=1}^{\infty} (-1)^{n-1} n^{-1/3}$ is conditionally convergent.

25. $\left| \dfrac{a_{n+1}}{a_n} \right| = \left| \dfrac{(-1)^{n+1}(n+2)3^{n+1}}{2^{2n+3}} \cdot \dfrac{2^{2n+1}}{(-1)^n (n+1)3^n} \right| = \dfrac{n+2}{n+1} \cdot \dfrac{3}{4} = \dfrac{1 + (2/n)}{1 + (1/n)} \cdot \dfrac{3}{4} \to \dfrac{3}{4} < 1$ as $n \to \infty$, so by the Ratio

Test, $\sum\limits_{n=1}^{\infty} \dfrac{(-1)^n (n+1)3^n}{2^{2n+1}}$ is absolutely convergent.

27. $\sum\limits_{n=1}^{\infty} \dfrac{(-3)^{n-1}}{2^{3n}} = \sum\limits_{n=1}^{\infty} \dfrac{(-3)^{n-1}}{(2^3)^n} = \sum\limits_{n=1}^{\infty} \dfrac{(-3)^{n-1}}{8^n} = \dfrac{1}{8} \sum\limits_{n=1}^{\infty} \dfrac{(-3)^{n-1}}{8^{n-1}} = \dfrac{1}{8} \sum\limits_{n=1}^{\infty} \left(-\dfrac{3}{8} \right)^{n-1} = \dfrac{1}{8} \left(\dfrac{1}{1 - (-3/8)} \right)$

$= \dfrac{1}{8} \cdot \dfrac{8}{11} = \dfrac{1}{11}$

29. $\sum\limits_{n=1}^{\infty} [\tan^{-1}(n+1) - \tan^{-1} n] = \lim\limits_{n \to \infty} s_n$

$= \lim\limits_{n \to \infty} [(\tan^{-1} 2 - \tan^{-1} 1) + (\tan^{-1} 3 - \tan^{-1} 2) + \cdots + (\tan^{-1}(n+1) - \tan^{-1} n)]$

$= \lim\limits_{n \to \infty} [\tan^{-1}(n+1) - \tan^{-1} 1] = \frac{\pi}{2} - \frac{\pi}{4} = \frac{\pi}{4}$

31. $1 - e + \dfrac{e^2}{2!} - \dfrac{e^3}{3!} + \dfrac{e^4}{4!} - \cdots = \sum\limits_{n=0}^{\infty} (-1)^n \dfrac{e^n}{n!} = \sum\limits_{n=0}^{\infty} \dfrac{(-e)^n}{n!} = e^{-e}$ since $e^x = \sum\limits_{n=0}^{\infty} \dfrac{x^n}{n!}$ for all x.

33. $\cosh x = \dfrac{1}{2}(e^x + e^{-x}) = \dfrac{1}{2} \left(\sum\limits_{n=0}^{\infty} \dfrac{x^n}{n!} + \sum\limits_{n=0}^{\infty} \dfrac{(-x)^n}{n!} \right)$

$= \dfrac{1}{2} \left[\left(1 + x + \dfrac{x^2}{2!} + \dfrac{x^3}{3!} + \dfrac{x^4}{4!} + \cdots \right) + \left(1 - x + \dfrac{x^2}{2!} - \dfrac{x^3}{3!} + \dfrac{x^4}{4!} - \cdots \right) \right]$

$= \dfrac{1}{2} \left(2 + 2 \cdot \dfrac{x^2}{2!} + 2 \cdot \dfrac{x^4}{4!} + \cdots \right) = 1 + \dfrac{1}{2}x^2 + \sum\limits_{n=2}^{\infty} \dfrac{x^{2n}}{(2n)!} \ge 1 + \dfrac{1}{2}x^2$ for all x

35. $\sum\limits_{n=1}^{\infty} \dfrac{(-1)^{n+1}}{n^5} = 1 - \dfrac{1}{32} + \dfrac{1}{243} - \dfrac{1}{1024} + \dfrac{1}{3125} - \dfrac{1}{7776} + \dfrac{1}{16,807} - \dfrac{1}{32,768} + \cdots$.

Since $b_8 = \dfrac{1}{8^5} = \dfrac{1}{32,768} < 0.000031$, $\sum\limits_{n=1}^{\infty} \dfrac{(-1)^{n+1}}{n^5} \approx \sum\limits_{n=1}^{7} \dfrac{(-1)^{n+1}}{n^5} \approx 0.9721$.

37. $\sum\limits_{n=1}^{\infty} \dfrac{1}{2 + 5^n} \approx \sum\limits_{n=1}^{8} \dfrac{1}{2 + 5^n} \approx 0.18976224$. To estimate the error, note that $\dfrac{1}{2 + 5^n} < \dfrac{1}{5^n}$, so the remainder term is

$R_8 = \sum\limits_{n=9}^{\infty} \dfrac{1}{2 + 5^n} < \sum\limits_{n=9}^{\infty} \dfrac{1}{5^n} = \dfrac{1/5^9}{1 - 1/5} = 6.4 \times 10^{-7}$ [geometric series with $a = \frac{1}{5^9}$ and $r = \frac{1}{5}$].

© 2012 Cengage Learning. All Rights Reserved. May not be scanned, copied, or duplicated, or posted to a publicly accessible website, in whole or in part.

39. Use the Limit Comparison Test. $\displaystyle\lim_{n\to\infty}\left|\frac{\left(\frac{n+1}{n}\right)a_n}{a_n}\right| = \lim_{n\to\infty}\frac{n+1}{n} = \lim_{n\to\infty}\left(1+\frac{1}{n}\right) = 1 > 0$.

Since $\sum|a_n|$ is convergent, so is $\sum\left|\left(\frac{n+1}{n}\right)a_n\right|$, by the Limit Comparison Test.

41. $\displaystyle\lim_{n\to\infty}\left|\frac{a_{n+1}}{a_n}\right| = \lim_{n\to\infty}\left[\frac{|x+2|^{n+1}}{(n+1)\,4^{n+1}}\cdot\frac{n\,4^n}{|x+2|^n}\right] = \lim_{n\to\infty}\left[\frac{n}{n+1}\frac{|x+2|}{4}\right] = \frac{|x+2|}{4} < 1 \ \Leftrightarrow \ |x+2| < 4$, so $R = 4$.

$|x+2| < 4 \ \Leftrightarrow \ -4 < x+2 < 4 \ \Leftrightarrow \ -6 < x < 2$. If $x = -6$, then the series $\displaystyle\sum_{n=1}^{\infty}\frac{(x+2)^n}{n\,4^n}$ becomes

$\displaystyle\sum_{n=1}^{\infty}\frac{(-4)^n}{n4^n} = \sum_{n=1}^{\infty}\frac{(-1)^n}{n}$, the alternating harmonic series, which converges by the Alternating Series Test. When $x = 2$, the

series becomes the harmonic series $\displaystyle\sum_{n=1}^{\infty}\frac{1}{n}$, which diverges. Thus, $I = [-6, 2)$.

43. $\displaystyle\lim_{n\to\infty}\left|\frac{a_{n+1}}{a_n}\right| = \lim_{n\to\infty}\left|\frac{2^{n+1}(x-3)^{n+1}}{\sqrt{n+4}}\cdot\frac{\sqrt{n+3}}{2^n(x-3)^n}\right| = 2\,|x-3|\lim_{n\to\infty}\sqrt{\frac{n+3}{n+4}} = 2\,|x-3| < 1 \ \Leftrightarrow \ |x-3| < \frac{1}{2}$,

so $R = \frac{1}{2}$. $|x-3| < \frac{1}{2} \ \Leftrightarrow \ -\frac{1}{2} < x-3 < \frac{1}{2} \ \Leftrightarrow \ \frac{5}{2} < x < \frac{7}{2}$. For $x = \frac{7}{2}$, the series $\displaystyle\sum_{n=1}^{\infty}\frac{2^n(x-3)^n}{\sqrt{n+3}}$ becomes

$\displaystyle\sum_{n=0}^{\infty}\frac{1}{\sqrt{n+3}} = \sum_{n=3}^{\infty}\frac{1}{n^{1/2}}$, which diverges $\left[p = \frac{1}{2} \le 1\right]$, but for $x = \frac{5}{2}$, we get $\displaystyle\sum_{n=0}^{\infty}\frac{(-1)^n}{\sqrt{n+3}}$, which is a convergent

alternating series, so $I = \left[\frac{5}{2}, \frac{7}{2}\right)$.

45.

n	$f^{(n)}(x)$	$f^{(n)}\left(\frac{\pi}{6}\right)$
0	$\sin x$	$\frac{1}{2}$
1	$\cos x$	$\frac{\sqrt{3}}{2}$
2	$-\sin x$	$-\frac{1}{2}$
3	$-\cos x$	$-\frac{\sqrt{3}}{2}$
4	$\sin x$	$\frac{1}{2}$
$\vdots$	$\vdots$	$\vdots$

$\sin x = f\left(\frac{\pi}{6}\right) + f'\left(\frac{\pi}{6}\right)\left(x-\frac{\pi}{6}\right) + \frac{f''\left(\frac{\pi}{6}\right)}{2!}\left(x-\frac{\pi}{6}\right)^2 + \frac{f^{(3)}\left(\frac{\pi}{6}\right)}{3!}\left(x-\frac{\pi}{6}\right)^3 + \frac{f^{(4)}\left(\frac{\pi}{6}\right)}{4!}\left(x-\frac{\pi}{6}\right)^4 + \cdots$

$= \frac{1}{2}\left[1 - \frac{1}{2!}\left(x-\frac{\pi}{6}\right)^2 + \frac{1}{4!}\left(x-\frac{\pi}{6}\right)^4 - \cdots\right] + \frac{\sqrt{3}}{2}\left[\left(x-\frac{\pi}{6}\right) - \frac{1}{3!}\left(x-\frac{\pi}{6}\right)^3 + \cdots\right]$

$= \frac{1}{2}\sum_{n=0}^{\infty}(-1)^n\frac{1}{(2n)!}\left(x-\frac{\pi}{6}\right)^{2n} + \frac{\sqrt{3}}{2}\sum_{n=0}^{\infty}(-1)^n\frac{1}{(2n+1)!}\left(x-\frac{\pi}{6}\right)^{2n+1}$

47. $\dfrac{1}{1+x} = \dfrac{1}{1-(-x)} = \displaystyle\sum_{n=0}^{\infty}(-x)^n = \sum_{n=0}^{\infty}(-1)^n\,x^n$ for $|x| < 1 \ \Rightarrow \ \dfrac{x^2}{1+x} = \displaystyle\sum_{n=0}^{\infty}(-1)^n\,x^{n+2}$ with $R = 1$.

© 2012 Cengage Learning. All Rights Reserved. May not be scanned, copied, or duplicated, or posted to a publicly accessible website, in whole or in part.

49. $\int \dfrac{1}{4-x}\,dx = -\ln(4-x) + C$ and

$$\int \frac{1}{4-x}\,dx = \frac{1}{4}\int \frac{1}{1-x/4}\,dx = \frac{1}{4}\int \sum_{n=0}^{\infty}\left(\frac{x}{4}\right)^n dx = \frac{1}{4}\int \sum_{n=0}^{\infty}\frac{x^n}{4^n}\,dx = \frac{1}{4}\sum_{n=0}^{\infty}\frac{x^{n+1}}{4^n(n+1)} + C.\ \text{So}$$

$$\ln(4-x) = -\frac{1}{4}\sum_{n=0}^{\infty}\frac{x^{n+1}}{4^n(n+1)} + C = -\sum_{n=0}^{\infty}\frac{x^{n+1}}{4^{n+1}(n+1)} + C = -\sum_{n=1}^{\infty}\frac{x^n}{n4^n} + C.\ \text{Putting } x = 0,\ \text{we get } C = \ln 4.$$

Thus, $f(x) = \ln(4-x) = \ln 4 - \displaystyle\sum_{n=1}^{\infty}\frac{x^n}{n4^n}$. The series converges for $|x/4| < 1 \ \Leftrightarrow \ |x| < 4$, so $R = 4$.

Another solution:

$$\ln(4-x) = \ln[4(1-x/4)] = \ln 4 + \ln(1-x/4) = \ln 4 + \ln[1 + (-x/4)]$$

$$= \ln 4 + \sum_{n=1}^{\infty}(-1)^{n+1}\frac{(-x/4)^n}{n} \ \text{[from Table 1]} \ = \ln 4 + \sum_{n=1}^{\infty}(-1)^{2n+1}\frac{x^n}{n4^n} = \ln 4 - \sum_{n=1}^{\infty}\frac{x^n}{n4^n}.$$

51. $\sin x = \displaystyle\sum_{n=0}^{\infty}\frac{(-1)^n\,x^{2n+1}}{(2n+1)!} \ \Rightarrow \ \sin(x^4) = \sum_{n=0}^{\infty}\frac{(-1)^n\,(x^4)^{2n+1}}{(2n+1)!} = \sum_{n=0}^{\infty}\frac{(-1)^n\,x^{8n+4}}{(2n+1)!}$ for all x, so the radius of

convergence is ∞.

53. $f(x) = \dfrac{1}{\sqrt[4]{16-x}} = \dfrac{1}{\sqrt[4]{16(1-x/16)}} = \dfrac{1}{\sqrt[4]{16}\,(1-\frac{1}{16}x)^{1/4}} = \dfrac{1}{2}\left(1 - \frac{1}{16}x\right)^{-1/4}$

$$= \frac{1}{2}\left[1 + \left(-\frac{1}{4}\right)\left(-\frac{x}{16}\right) + \frac{\left(-\frac{1}{4}\right)\left(-\frac{5}{4}\right)}{2!}\left(-\frac{x}{16}\right)^2 + \frac{\left(-\frac{1}{4}\right)\left(-\frac{5}{4}\right)\left(-\frac{9}{4}\right)}{3!}\left(-\frac{x}{16}\right)^3 + \cdots\right]$$

$$= \frac{1}{2} + \sum_{n=1}^{\infty}\frac{1\cdot 5\cdot 9\cdot\cdots\cdot(4n-3)}{2\cdot 4^n\cdot n!\cdot 16^n}\,x^n = \frac{1}{2} + \sum_{n=1}^{\infty}\frac{1\cdot 5\cdot 9\cdot\cdots\cdot(4n-3)}{2^{6n+1}\,n!}\,x^n$$

for $\left|-\dfrac{x}{16}\right| < 1 \ \Leftrightarrow \ |x| < 16$, so $R = 16$.

55. $e^x = \displaystyle\sum_{n=0}^{\infty}\frac{x^n}{n!}$, so $\dfrac{e^x}{x} = \dfrac{1}{x}\displaystyle\sum_{n=0}^{\infty}\frac{x^n}{n!} = \sum_{n=0}^{\infty}\frac{x^{n-1}}{n!} = x^{-1} + \sum_{n=1}^{\infty}\frac{x^{n-1}}{n!} = \frac{1}{x} + \sum_{n=1}^{\infty}\frac{x^{n-1}}{n!}$ and

$$\int \frac{e^x}{x}\,dx = C + \ln|x| + \sum_{n=1}^{\infty}\frac{x^n}{n\cdot n!}.$$

57. (a)

n	$f^{(n)}(x)$	$f^{(n)}(1)$
0	$x^{1/2}$	1
1	$\frac{1}{2}x^{-1/2}$	$\frac{1}{2}$
2	$-\frac{1}{4}x^{-3/2}$	$-\frac{1}{4}$
3	$\frac{3}{8}x^{-5/2}$	$\frac{3}{8}$
4	$-\frac{15}{16}x^{-7/2}$	$-\frac{15}{16}$
⋮	⋮	⋮

$$\sqrt{x} \approx T_3(x) = 1 + \frac{1/2}{1!}(x-1) - \frac{1/4}{2!}(x-1)^2 + \frac{3/8}{3!}(x-1)^3$$

$$= 1 + \tfrac{1}{2}(x-1) - \tfrac{1}{8}(x-1)^2 + \tfrac{1}{16}(x-1)^3$$

© 2012 Cengage Learning. All Rights Reserved. May not be scanned, copied, or duplicated, or posted to a publicly accessible website, in whole or in part.

(b)

(c) $|R_3(x)| \leq \dfrac{M}{4!}|x-1|^4$, where $\left|f^{(4)}(x)\right| \leq M$ with

$f^{(4)}(x) = -\frac{15}{16}x^{-7/2}$. Now $0.9 \leq x \leq 1.1$ $\Rightarrow$

$-0.1 \leq x - 1 \leq 0.1$ $\Rightarrow$ $(x-1)^4 \leq (0.1)^4$,

and letting $x = 0.9$ gives $M = \dfrac{15}{16(0.9)^{7/2}}$, so

$|R_3(x)| \leq \dfrac{15}{16(0.9)^{7/2}\, 4!}(0.1)^4 \approx 0.000\,005\,648$

$\approx 0.000\,006 = 6 \times 10^{-6}$

(d)

From the graph of $|R_3(x)| = |\sqrt{x} - T_3(x)|$, it appears that

the error is less than 5×10^{-6} on $[0.9, 1.1]$.

59. $\sin x = \displaystyle\sum_{n=0}^{\infty}(-1)^n\frac{x^{2n+1}}{(2n+1)!} = x - \frac{x^3}{3!} + \frac{x^5}{5!} - \frac{x^7}{7!} + \cdots$, so $\sin x - x = -\dfrac{x^3}{3!} + \dfrac{x^5}{5!} - \dfrac{x^7}{7!} + \cdots$ and

$\dfrac{\sin x - x}{x^3} = -\dfrac{1}{3!} + \dfrac{x^2}{5!} - \dfrac{x^4}{7!} + \cdots$. Thus, $\displaystyle\lim_{x\to 0}\frac{\sin x - x}{x^3} = \lim_{x\to 0}\left(-\frac{1}{6} + \frac{x^2}{120} - \frac{x^4}{5040} + \cdots\right) = -\frac{1}{6}$.

61. $f(x) = \displaystyle\sum_{n=0}^{\infty} c_n\, x^n$ $\Rightarrow$ $f(-x) = \displaystyle\sum_{n=0}^{\infty} c_n(-x)^n = \sum_{n=0}^{\infty}(-1)^n c_n\, x^n$

(a) If f is an odd function, then $f(-x) = -f(x)$ $\Rightarrow$ $\displaystyle\sum_{n=0}^{\infty}(-1)^n c_n x^n = \sum_{n=0}^{\infty} -c_n x^n$. The coefficients of any power series

are uniquely determined (by Theorem 11.10.5), so $(-1)^n c_n = -c_n$.

If n is even, then $(-1)^n = 1$, so $c_n = -c_n$ $\Rightarrow$ $2c_n = 0$ $\Rightarrow$ $c_n = 0$. Thus, all even coefficients are 0, that is,

$c_0 = c_2 = c_4 = \cdots = 0$.

(b) If f is even, then $f(-x) = f(x)$ $\Rightarrow$ $\displaystyle\sum_{n=0}^{\infty}(-1)^n c_n x^n = \sum_{n=0}^{\infty} c_n x^n$ $\Rightarrow$ $(-1)^n c_n = c_n$.

If n is odd, then $(-1)^n = -1$, so $-c_n = c_n$ $\Rightarrow$ $2c_n = 0$ $\Rightarrow$ $c_n = 0$. Thus, all odd coefficients are 0,

that is, $c_1 = c_3 = c_5 = \cdots = 0$.

© 2012 Cengage Learning. All Rights Reserved. May not be scanned, copied, or duplicated, or posted to a publicly accessible website, in whole or in part.

◻ PROBLEMS PLUS

1. It would be far too much work to compute 15 derivatives of f. The key idea is to remember that $f^{(n)}(0)$ occurs in the

coefficient of x^n in the Maclaurin series of f. We start with the Maclaurin series for sin: $\sin x = x - \dfrac{x^3}{3!} + \dfrac{x^5}{5!} - \cdots$.

Then $\sin(x^3) = x^3 - \dfrac{x^9}{3!} + \dfrac{x^{15}}{5!} - \cdots$, and so the coefficient of x^{15} is $\dfrac{f^{(15)}(0)}{15!} = \dfrac{1}{5!}$. Therefore,

$f^{(15)}(0) = \dfrac{15!}{5!} = 6 \cdot 7 \cdot 8 \cdot 9 \cdot 10 \cdot 11 \cdot 12 \cdot 13 \cdot 14 \cdot 15 = 10{,}897{,}286{,}400$.

3. (a) From Formula 14a in Appendix D, with $x = y = \theta$, we get $\tan 2\theta = \dfrac{2\tan\theta}{1 - \tan^2\theta}$, so $\cot 2\theta = \dfrac{1 - \tan^2\theta}{2\tan\theta}$ $\Rightarrow$

$2\cot 2\theta = \dfrac{1 - \tan^2\theta}{\tan\theta} = \cot\theta - \tan\theta$. Replacing θ by $\tfrac{1}{2}x$, we get $2\cot x = \cot\tfrac{1}{2}x - \tan\tfrac{1}{2}x$, or

$\tan\tfrac{1}{2}x = \cot\tfrac{1}{2}x - 2\cot x$.

(b) From part (a) with $\dfrac{x}{2^{n-1}}$ in place of x, $\tan\dfrac{x}{2^n} = \cot\dfrac{x}{2^n} - 2\cot\dfrac{x}{2^{n-1}}$, so the nth partial sum of $\displaystyle\sum_{n=1}^{\infty}\dfrac{1}{2^n}\tan\dfrac{x}{2^n}$ is

$$s_n = \frac{\tan(x/2)}{2} + \frac{\tan(x/4)}{4} + \frac{\tan(x/8)}{8} + \cdots + \frac{\tan(x/2^n)}{2^n}$$

$$= \left[\frac{\cot(x/2)}{2} - \cot x\right] + \left[\frac{\cot(x/4)}{4} - \frac{\cot(x/2)}{2}\right] + \left[\frac{\cot(x/8)}{8} - \frac{\cot(x/4)}{4}\right] + \cdots$$

$$+ \left[\frac{\cot(x/2^n)}{2^n} - \frac{\cot(x/2^{n-1})}{2^{n-1}}\right] = -\cot x + \frac{\cot(x/2^n)}{2^n} \quad \text{[telescoping sum]}$$

Now $\dfrac{\cot(x/2^n)}{2^n} = \dfrac{\cos(x/2^n)}{2^n\sin(x/2^n)} = \dfrac{\cos(x/2^n)}{x} \cdot \dfrac{x/2^n}{\sin(x/2^n)} \to \dfrac{1}{x} \cdot 1 = \dfrac{1}{x}$ as $n \to \infty$ since $x/2^n \to 0$

for $x \neq 0$. Therefore, if $x \neq 0$ and $x \neq k\pi$ where k is any integer, then

$$\sum_{n=1}^{\infty}\frac{1}{2^n}\tan\frac{x}{2^n} = \lim_{n\to\infty} s_n = \lim_{n\to\infty}\left(-\cot x + \frac{1}{2^n}\cot\frac{x}{2^n}\right) = -\cot x + \frac{1}{x}$$

If $x = 0$, then all terms in the series are 0, so the sum is 0.

5. (a) At each stage, each side is replaced by four shorter sides, each of length

$\tfrac{1}{3}$ of the side length at the preceding stage. Writing s_0 and ℓ_0 for the

number of sides and the length of the side of the initial triangle, we

generate the table at right. In general, we have $s_n = 3 \cdot 4^n$ and

$s_0 = 3$	$\ell_0 = 1$
$s_1 = 3 \cdot 4$	$\ell_1 = 1/3$
$s_2 = 3 \cdot 4^2$	$\ell_2 = 1/3^2$
$s_3 = 3 \cdot 4^3$	$\ell_3 = 1/3^3$
$\vdots$	$\vdots$

$\ell_n = \left(\tfrac{1}{3}\right)^n$, so the length of the perimeter at the nth stage of construction

is $p_n = s_n\ell_n = 3 \cdot 4^n \cdot \left(\tfrac{1}{3}\right)^n = 3 \cdot \left(\tfrac{4}{3}\right)^n$.

(b) $p_n = \dfrac{4^n}{3^{n-1}} = 4\left(\dfrac{4}{3}\right)^{n-1}$. Since $\tfrac{4}{3} > 1$, $p_n \to \infty$ as $n \to \infty$.

(c) The area of each of the small triangles added at a given stage is one-ninth of the area of the triangle added at the preceding

stage. Let a be the area of the original triangle. Then the area a_n of each of the small triangles added at stage n is

© 2012 Cengage Learning. All Rights Reserved. May not be scanned, copied, or duplicated, or posted to a publicly accessible website, in whole or in part.

$a_n = a \cdot \dfrac{1}{9^n} = \dfrac{a}{9^n}$. Since a small triangle is added to each side at every stage, it follows that the total area A_n added to the

figure at the nth stage is $A_n = s_{n-1} \cdot a_n = 3 \cdot 4^{n-1} \cdot \dfrac{a}{9^n} = a \cdot \dfrac{4^{n-1}}{3^{2n-1}}$. Then the total area enclosed by the snowflake

curve is $A = a + A_1 + A_2 + A_3 + \cdots = a + a \cdot \dfrac{1}{3} + a \cdot \dfrac{4}{3^3} + a \cdot \dfrac{4^2}{3^5} + a \cdot \dfrac{4^3}{3^7} + \cdots$. After the first term, this is a

geometric series with common ratio $\dfrac{4}{9}$, so $A = a + \dfrac{a/3}{1 - \frac{4}{9}} = a + \dfrac{a}{3} \cdot \dfrac{9}{5} = \dfrac{8a}{5}$. But the area of the original equilateral

triangle with side 1 is $a = \dfrac{1}{2} \cdot 1 \cdot \sin \dfrac{\pi}{3} = \dfrac{\sqrt{3}}{4}$. So the area enclosed by the snowflake curve is $\dfrac{8}{5} \cdot \dfrac{\sqrt{3}}{4} = \dfrac{2\sqrt{3}}{5}$.

7. (a) Let $a = \arctan x$ and $b = \arctan y$. Then, from Formula 14b in Appendix D,

$$\tan(a - b) = \dfrac{\tan a - \tan b}{1 + \tan a \tan b} = \dfrac{\tan(\arctan x) - \tan(\arctan y)}{1 + \tan(\arctan x) \tan(\arctan y)} = \dfrac{x - y}{1 + xy}$$

Now $\arctan x - \arctan y = a - b = \arctan(\tan(a - b)) = \arctan \dfrac{x - y}{1 + xy}$ since $-\dfrac{\pi}{2} < a - b < \dfrac{\pi}{2}$.

(b) From part (a) we have

$$\arctan \tfrac{120}{119} - \arctan \tfrac{1}{239} = \arctan \dfrac{\frac{120}{119} - \frac{1}{239}}{1 + \frac{120}{119} \cdot \frac{1}{239}} = \arctan \dfrac{\frac{28,561}{28,441}}{\frac{28,561}{28,441}} = \arctan 1 = \tfrac{\pi}{4}$$

(c) Replacing y by $-y$ in the formula of part (a), we get $\arctan x + \arctan y = \arctan \dfrac{x + y}{1 - xy}$. So

$$4 \arctan \tfrac{1}{5} = 2\left(\arctan \tfrac{1}{5} + \arctan \tfrac{1}{5}\right) = 2 \arctan \dfrac{\frac{1}{5} + \frac{1}{5}}{1 - \frac{1}{5} \cdot \frac{1}{5}} = 2 \arctan \tfrac{5}{12} = \arctan \tfrac{5}{12} + \arctan \tfrac{5}{12}$$

$$= \arctan \dfrac{\frac{5}{12} + \frac{5}{12}}{1 - \frac{5}{12} \cdot \frac{5}{12}} = \arctan \tfrac{120}{119}$$

Thus, from part (b), we have $4 \arctan \tfrac{1}{5} - \arctan \tfrac{1}{239} = \arctan \tfrac{120}{119} - \arctan \tfrac{1}{239} = \tfrac{\pi}{4}$.

(d) From Example 7 in Section 11.9 we have $\arctan x = x - \dfrac{x^3}{3} + \dfrac{x^5}{5} - \dfrac{x^7}{7} + \dfrac{x^9}{9} - \dfrac{x^{11}}{11} + \cdots$, so

$$\arctan \tfrac{1}{5} = \dfrac{1}{5} - \dfrac{1}{3 \cdot 5^3} + \dfrac{1}{5 \cdot 5^5} - \dfrac{1}{7 \cdot 5^7} + \dfrac{1}{9 \cdot 5^9} - \dfrac{1}{11 \cdot 5^{11}} + \cdots$$

This is an alternating series and the size of the terms decreases to 0, so by the Alternating Series Estimation Theorem, the sum lies between s_5 and s_6, that is, $0.197395560 < \arctan \tfrac{1}{5} < 0.197395562$.

(e) From the series in part (d) we get $\arctan \dfrac{1}{239} = \dfrac{1}{239} - \dfrac{1}{3 \cdot 239^3} + \dfrac{1}{5 \cdot 239^5} - \cdots$. The third term is less than

2.6×10^{-13}, so by the Alternating Series Estimation Theorem, we have, to nine decimal places,

$\arctan \tfrac{1}{239} \approx s_2 \approx 0.004184076$. Thus, $0.004184075 < \arctan \tfrac{1}{239} < 0.004184077$.

(f) From part (c) we have $\pi = 16 \arctan \tfrac{1}{5} - 4 \arctan \tfrac{1}{239}$, so from parts (d) and (e) we have

$16(0.197395560) - 4(0.004184077) < \pi < 16(0.197395562) - 4(0.004184075) \quad \Rightarrow$

$3.141592652 < \pi < 3.141592692$. So, to 7 decimal places, $\pi \approx 3.1415927$.

© 2012 Cengage Learning. All Rights Reserved. May not be scanned, copied, or duplicated, or posted to a publicly accessible website, in whole or in part.

9. We start with the geometric series $\displaystyle\sum_{n=0}^{\infty} x^n = \frac{1}{1-x}$, $|x| < 1$, and differentiate:

$$\sum_{n=1}^{\infty} nx^{n-1} = \frac{d}{dx}\left(\sum_{n=0}^{\infty} x^n\right) = \frac{d}{dx}\left(\frac{1}{1-x}\right) = \frac{1}{(1-x)^2} \text{ for } |x| < 1 \ \Rightarrow \ \sum_{n=1}^{\infty} nx^{n} = x\sum_{n=1}^{\infty} nx^{n-1} = \frac{x}{(1-x)^2}$$

for $|x| < 1$. Differentiate again:

$$\sum_{n=1}^{\infty} n^2 x^{n-1} = \frac{d}{dx}\frac{x}{(1-x)^2} = \frac{(1-x)^2 - x\cdot 2(1-x)(-1)}{(1-x)^4} = \frac{x+1}{(1-x)^3} \ \Rightarrow \ \sum_{n=1}^{\infty} n^2 x^{n} = \frac{x^2+x}{(1-x)^3} \ \Rightarrow$$

$$\sum_{n=1}^{\infty} n^3 x^{n-1} = \frac{d}{dx}\frac{x^2+x}{(1-x)^3} = \frac{(1-x)^3(2x+1) - (x^2+x)3(1-x)^2(-1)}{(1-x)^6} = \frac{x^2+4x+1}{(1-x)^4} \ \Rightarrow$$

$$\sum_{n=1}^{\infty} n^3 x^{n} = \frac{x^3+4x^2+x}{(1-x)^4}, \ |x| < 1.$$ The radius of convergence is 1 because that is the radius of convergence for the

geometric series we started with. If $x = \pm 1$, the series is $\sum n^3(\pm 1)^n$, which diverges by the Test For Divergence, so the

interval of convergence is $(-1, 1)$.

11. $\displaystyle\ln\left(1 - \frac{1}{n^2}\right) = \ln\left(\frac{n^2-1}{n^2}\right) = \ln\frac{(n+1)(n-1)}{n^2} = \ln[(n+1)(n-1)] - \ln n^2$

$$= \ln(n+1) + \ln(n-1) - 2\ln n = \ln(n-1) - \ln n - \ln n + \ln(n+1)$$

$$= \ln\frac{n-1}{n} - [\ln n - \ln(n+1)] = \ln\frac{n-1}{n} - \ln\frac{n}{n+1}.$$

Let $\displaystyle s_k = \sum_{n=2}^{k} \ln\left(1 - \frac{1}{n^2}\right) = \sum_{n=2}^{k}\left(\ln\frac{n-1}{n} - \ln\frac{n}{n+1}\right)$ for $k \ge 2$. Then

$$s_k = \left(\ln\frac{1}{2} - \ln\frac{2}{3}\right) + \left(\ln\frac{2}{3} - \ln\frac{3}{4}\right) + \cdots + \left(\ln\frac{k-1}{k} - \ln\frac{k}{k+1}\right) = \ln\frac{1}{2} - \ln\frac{k}{k+1}, \text{ so}$$

$$\sum_{n=2}^{\infty} \ln\left(1 - \frac{1}{n^2}\right) = \lim_{k\to\infty} s_k = \lim_{k\to\infty}\left(\ln\frac{1}{2} - \ln\frac{k}{k+1}\right) = \ln\frac{1}{2} - \ln 1 = \ln 1 - \ln 2 - \ln 1 = -\ln 2.$$

13. (a)

The x-intercepts of the curve occur where $\sin x = 0 \ \Leftrightarrow \ x = n\pi$,

n an integer. So using the formula for disks (and either a CAS or

$\sin^2 x = \frac{1}{2}(1 - \cos 2x)$ and Formula 99 to evaluate the integral),

the volume of the nth bead is

$$V_n = \pi\int_{(n-1)\pi}^{n\pi}(e^{-x/10}\sin x)^2\,dx = \pi\int_{(n-1)\pi}^{n\pi} e^{-x/5}\sin^2 x\,dx$$

$$= \frac{250\pi}{101}\left(e^{-(n-1)\pi/5} - e^{-n\pi/5}\right)$$

(b) The total volume is

$$\pi\int_0^{\infty} e^{-x/5}\sin^2 x\,dx = \sum_{n=1}^{\infty} V_n = \frac{250\pi}{101}\sum_{n=1}^{\infty}\left[e^{-(n-1)\pi/5} - e^{-n\pi/5}\right] = \frac{250\pi}{101} \quad \text{[telescoping sum]}.$$

Another method: If the volume in part (a) has been written as $V_n = \frac{250\pi}{101}e^{-n\pi/5}(e^{\pi/5} - 1)$, then we recognize $\displaystyle\sum_{n=1}^{\infty} V_n$

as a geometric series with $a = \frac{250\pi}{101}(1 - e^{-\pi/5})$ and $r = e^{-\pi/5}$.

© 2012 Cengage Learning. All Rights Reserved. May not be scanned, copied, or duplicated, or posted to a publicly accessible website, in whole or in part.

15. If L is the length of a side of the equilateral triangle, then the area is $A = \frac{1}{2}L \cdot \frac{\sqrt{3}}{2}L = \frac{\sqrt{3}}{4}L^2$ and so $L^2 = \frac{4}{\sqrt{3}}A$.

Let r be the radius of one of the circles. When there are n rows of circles, the figure shows that

$$L = \sqrt{3}\,r + r + (n-2)(2r) + r + \sqrt{3}\,r = r\big(2n - 2 + 2\sqrt{3}\big), \text{ so } r = \frac{L}{2\big(n + \sqrt{3} - 1\big)}.$$

The number of circles is $1 + 2 + \cdots + n = \dfrac{n(n+1)}{2}$, and so the total area of the circles is

$$
\begin{aligned}
A_n &= \frac{n(n+1)}{2}\pi r^2 = \frac{n(n+1)}{2}\,\pi\,\frac{L^2}{4\big(n + \sqrt{3} - 1\big)^2}\\[2mm]
&= \frac{n(n+1)}{2}\,\pi\,\frac{4A/\sqrt{3}}{4\big(n + \sqrt{3} - 1\big)^2} = \frac{n(n+1)}{\big(n + \sqrt{3} - 1\big)^2}\,\frac{\pi A}{2\sqrt{3}} \quad\Rightarrow
\end{aligned}
$$

$$
\begin{aligned}
\frac{A_n}{A} &= \frac{n(n+1)}{\big(n + \sqrt{3} - 1\big)^2}\,\frac{\pi}{2\sqrt{3}}\\[2mm]
&= \frac{1 + 1/n}{\big[1 + (\sqrt{3}-1)/n\big]^2}\,\frac{\pi}{2\sqrt{3}} \to \frac{\pi}{2\sqrt{3}} \quad \text{as } n \to \infty
\end{aligned}
$$

17. As in Section 11.9 we have to integrate the function x^x by integrating series. Writing $x^x = (e^{\ln x})^x = e^{x\ln x}$ and using the

Maclaurin series for e^x, we have $x^x = (e^{\ln x})^x = e^{x\ln x} = \displaystyle\sum_{n=0}^{\infty} \frac{(x\ln x)^n}{n!} = \sum_{n=0}^{\infty} \frac{x^n(\ln x)^n}{n!}$. As with power series, we can

integrate this series term-by-term: $\displaystyle\int_0^1 x^x\,dx = \sum_{n=0}^{\infty} \int_0^1 \frac{x^n(\ln x)^n}{n!}\,dx = \sum_{n=0}^{\infty} \frac{1}{n!}\int_0^1 x^n(\ln x)^n\,dx$. We integrate by parts

with $u = (\ln x)^n$, $dv = x^n\,dx$, so $du = \dfrac{n(\ln x)^{n-1}}{x}\,dx$ and $v = \dfrac{x^{n+1}}{n+1}$:

$$
\begin{aligned}
\int_0^1 x^n(\ln x)^n\,dx &= \lim_{t\to0^+}\int_t^1 x^n(\ln x)^n\,dx = \lim_{t\to0^+}\left[\frac{x^{n+1}}{n+1}(\ln x)^n\right]_t^1 - \lim_{t\to0^+}\int_t^1 \frac{n}{n+1}x^n(\ln x)^{n-1}\,dx\\[2mm]
&= 0 - \frac{n}{n+1}\int_0^1 x^n(\ln x)^{n-1}\,dx
\end{aligned}
$$

(where l'Hospital's Rule was used to help evaluate the first limit). Further integration by parts gives

$$\int_0^1 x^n(\ln x)^k\,dx = -\frac{k}{n+1}\int_0^1 x^n(\ln x)^{k-1}\,dx \text{ and, combining these steps, we get}$$

$$\int_0^1 x^n(\ln x)^n\,dx = \frac{(-1)^n\,n!}{(n+1)^n}\int_0^1 x^n\,dx = \frac{(-1)^n\,n!}{(n+1)^{n+1}} \quad\Rightarrow$$

$$\int_0^1 x^x\,dx = \sum_{n=0}^{\infty}\frac{1}{n!}\int_0^1 x^n(\ln x)^n\,dx = \sum_{n=0}^{\infty}\frac{1}{n!}\frac{(-1)^n\,n!}{(n+1)^{n+1}} = \sum_{n=0}^{\infty}\frac{(-1)^n}{(n+1)^{n+1}} = \sum_{n=1}^{\infty}\frac{(-1)^{n-1}}{n^n}.$$

19. By Table 1 in Section 11.10, $\tan^{-1} x = \displaystyle\sum_{n=0}^{\infty}(-1)^n\frac{x^{2n+1}}{2n+1}$ for $|x| < 1$. In particular, for $x = \dfrac{1}{\sqrt{3}}$, we

have $\dfrac{\pi}{6} = \tan^{-1}\left(\dfrac{1}{\sqrt{3}}\right) = \displaystyle\sum_{n=0}^{\infty}(-1)^n\frac{(1/\sqrt{3})^{2n+1}}{2n+1} = \sum_{n=0}^{\infty}(-1)^n\left(\frac{1}{3}\right)^n\frac{1}{\sqrt{3}}\frac{1}{2n+1}$, so

$$\pi = \frac{6}{\sqrt{3}}\sum_{n=0}^{\infty}\frac{(-1)^n}{(2n+1)3^n} = 2\sqrt{3}\sum_{n=0}^{\infty}\frac{(-1)^n}{(2n+1)3^n} = 2\sqrt{3}\left(1 + \sum_{n=1}^{\infty}\frac{(-1)^n}{(2n+1)3^n}\right) \quad\Rightarrow\quad \sum_{n=1}^{\infty}\frac{(-1)^n}{(2n+1)3^n} = \frac{\pi}{2\sqrt{3}} - 1.$$

© 2012 Cengage Learning. All Rights Reserved. May not be scanned, copied, or duplicated, or posted to a publicly accessible website, in whole or in part.

21. Let $f(x)$ denote the left-hand side of the equation $1 + \dfrac{x}{2!} + \dfrac{x^2}{4!} + \dfrac{x^3}{6!} + \dfrac{x^4}{8!} + \cdots = 0$. If $x \geq 0$, then $f(x) \geq 1$ and there are

no solutions of the equation. Note that $f(-x^2) = 1 - \dfrac{x^2}{2!} + \dfrac{x^4}{4!} - \dfrac{x^6}{6!} + \dfrac{x^8}{8!} - \cdots = \cos x$. The solutions of $\cos x = 0$ for

$x < 0$ are given by $x = \dfrac{\pi}{2} - \pi k$, where k is a positive integer. Thus, the solutions of $f(x) = 0$ are $x = -\left(\dfrac{\pi}{2} - \pi k\right)^2$, where

k is a positive integer.

23. Call the series S. We group the terms according to the number of digits in their denominators:

$$S = \underbrace{\left(\tfrac{1}{1} + \tfrac{1}{2} + \cdots + \tfrac{1}{8} + \tfrac{1}{9}\right)}_{g_1} + \underbrace{\left(\tfrac{1}{11} + \cdots + \tfrac{1}{99}\right)}_{g_2} + \underbrace{\left(\tfrac{1}{111} + \cdots + \tfrac{1}{999}\right)}_{g_3} + \cdots$$

Now in the group g_n, since we have 9 choices for each of the n digits in the denominator, there are 9^n terms.

Furthermore, each term in g_n is less than $\dfrac{1}{10^{n-1}}$ [except for the first term in g_1]. So $g_n < 9^n \cdot \dfrac{1}{10^{n-1}} = 9\left(\dfrac{9}{10}\right)^{n-1}$.

Now $\displaystyle\sum_{n=1}^{\infty} 9\left(\dfrac{9}{10}\right)^{n-1}$ is a geometric series with $a = 9$ and $r = \dfrac{9}{10} < 1$. Therefore, by the Comparison Test,

$$S = \sum_{n=1}^{\infty} g_n < \sum_{n=1}^{\infty} 9\left(\tfrac{9}{10}\right)^{n-1} = \dfrac{9}{1 - 9/10} = 90.$$

25. $u = 1 + \dfrac{x^3}{3!} + \dfrac{x^6}{6!} + \dfrac{x^9}{9!} + \cdots, \ v = x + \dfrac{x^4}{4!} + \dfrac{x^7}{7!} + \dfrac{x^{10}}{10!} + \cdots, \ w = \dfrac{x^2}{2!} + \dfrac{x^5}{5!} + \dfrac{x^8}{8!} + \cdots.$

Use the Ratio Test to show that the series for u, v, and w have positive radii of convergence (∞ in each case), so

Theorem 11.9.2 applies, and hence, we may differentiate each of these series:

$$\dfrac{du}{dx} = \dfrac{3x^2}{3!} + \dfrac{6x^5}{6!} + \dfrac{9x^8}{9!} + \cdots = \dfrac{x^2}{2!} + \dfrac{x^5}{5!} + \dfrac{x^8}{8!} + \cdots = w$$

Similarly, $\dfrac{dv}{dx} = 1 + \dfrac{x^3}{3!} + \dfrac{x^6}{6!} + \dfrac{x^9}{9!} + \cdots = u$, and $\dfrac{dw}{dx} = x + \dfrac{x^4}{4!} + \dfrac{x^7}{7!} + \dfrac{x^{10}}{10!} + \cdots = v$.

So $u' = w$, $v' = u$, and $w' = v$. Now differentiate the left-hand side of the desired equation:

$$\dfrac{d}{dx}(u^3 + v^3 + w^3 - 3uvw) = 3u^2u' + 3v^2v' + 3w^2w' - 3(u'vw + uv'w + uvw')$$

$$= 3u^2w + 3v^2u + 3w^2v - 3(vw^2 + u^2w + uv^2) = 0 \quad \Rightarrow$$

$u^3 + v^3 + w^3 - 3uvw = C$. To find the value of the constant C, we put $x = 0$ in the last equation and get

$1^3 + 0^3 + 0^3 - 3(1 \cdot 0 \cdot 0) = C \quad \Rightarrow \quad C = 1$, so $u^3 + v^3 + w^3 - 3uvw = 1$.

© 2012 Cengage Learning. All Rights Reserved. May not be scanned, copied, or duplicated, or posted to a publicly accessible website, in whole or in part.

12 □ VECTORS AND THE GEOMETRY OF SPACE

12.1 Three-Dimensional Coordinate Systems

1. We start at the origin, which has coordinates $(0, 0, 0)$. First we move 4 units along the positive x-axis, affecting only the x-coordinate, bringing us to the point $(4, 0, 0)$. We then move 3 units straight downward, in the negative z-direction. Thus only the z-coordinate is affected, and we arrive at $(4, 0, -3)$.

3. The distance from a point to the yz-plane is the absolute value of the x-coordinate of the point. $C(2, 4, 6)$ has the x-coordinate with the smallest absolute value, so C is the point closest to the yz-plane. $A(-4, 0, -1)$ must lie in the xz-plane since the distance from A to the xz-plane, given by the y-coordinate of A, is 0.

5. The equation $x + y = 2$ represents the set of all points in $\mathbb{R}^3$ whose x- and y-coordinates have a sum of 2, or equivalently where $y = 2 - x$. This is the set $\{(x, 2 - x, z) \mid x \in \mathbb{R}, z \in \mathbb{R}\}$ which is a vertical plane that intersects the xy-plane in the line $y = 2 - x$, $z = 0$.

7. We can find the lengths of the sides of the triangle by using the distance formula between pairs of vertices:

$$|PQ| = \sqrt{(7 - 3)^2 + [0 - (-2)]^2 + [1 - (-3)]^2} = \sqrt{16 + 4 + 16} = 6$$

$$|QR| = \sqrt{(1 - 7)^2 + (2 - 0)^2 + (1 - 1)^2} = \sqrt{36 + 4 + 0} = \sqrt{40} = 2\sqrt{10}$$

$$|RP| = \sqrt{(3 - 1)^2 + (-2 - 2)^2 + (-3 - 1)^2} = \sqrt{4 + 16 + 16} = 6$$

The longest side is QR, but the Pythagorean Theorem is not satisfied: $|PQ|^2 + |RP|^2 \neq |QR|^2$. Thus PQR is not a right triangle. PQR is isosceles, as two sides have the same length.

9. (a) First we find the distances between points:

$$|AB| = \sqrt{(3 - 2)^2 + (7 - 4)^2 + (-2 - 2)^2} = \sqrt{26}$$

$$|BC| = \sqrt{(1 - 3)^2 + (3 - 7)^2 + [3 - (-2)]^2} = \sqrt{45} = 3\sqrt{5}$$

$$|AC| = \sqrt{(1 - 2)^2 + (3 - 4)^2 + (3 - 2)^2} = \sqrt{3}$$

In order for the points to lie on a straight line, the sum of the two shortest distances must be equal to the longest distance. Since $\sqrt{26} + \sqrt{3} \neq 3\sqrt{5}$, the three points do not lie on a straight line.

© 2012 Cengage Learning. All Rights Reserved. May not be scanned, copied, or duplicated, or posted to a publicly accessible website, in whole or in part.

(b) First we find the distances between points:

$$|DE| = \sqrt{(1-0)^2 + [-2-(-5)]^2 + (4-5)^2} = \sqrt{11}$$

$$|EF| = \sqrt{(3-1)^2 + [4-(-2)]^2 + (2-4)^2} = \sqrt{44} = 2\sqrt{11}$$

$$|DF| = \sqrt{(3-0)^2 + [4-(-5)]^2 + (2-5)^2} = \sqrt{99} = 3\sqrt{11}$$

Since $|DE| + |EF| = |DF|$, the three points lie on a straight line.

11. An equation of the sphere with center $(-3, 2, 5)$ and radius 4 is $[x-(-3)]^2 + (y-2)^2 + (z-5)^2 = 4^2$ or

$(x+3)^2 + (y-2)^2 + (z-5)^2 = 16$. The intersection of this sphere with the yz-plane is the set of points on the sphere

whose x-coordinate is 0. Putting $x = 0$ into the equation, we have $9 + (y-2)^2 + (z-5)^2 = 16$, $x = 0$ or

$(y-2)^2 + (z-5)^2 = 7$, $x = 0$, which represents a circle in the yz-plane with center $(0, 2, 5)$ and radius $\sqrt{7}$.

13. The radius of the sphere is the distance between $(4, 3, -1)$ and $(3, 8, 1)$: $r = \sqrt{(3-4)^2 + (8-3)^2 + [1-(-1)]^2} = \sqrt{30}$.

Thus, an equation of the sphere is $(x-3)^2 + (y-8)^2 + (z-1)^2 = 30$.

15. Completing squares in the equation $x^2 + y^2 + z^2 - 2x - 4y + 8z = 15$ gives

$(x^2 - 2x + 1) + (y^2 - 4y + 4) + (z^2 + 8z + 16) = 15 + 1 + 4 + 16$ ⇒ $(x-1)^2 + (y-2)^2 + (z+4)^2 = 36$, which we

recognize as an equation of a sphere with center $(1, 2, -4)$ and radius 6.

17. Completing squares in the equation $2x^2 - 8x + 2y^2 + 2z^2 + 24z = 1$ gives

$2(x^2 - 4x + 4) + 2y^2 + 2(z^2 + 12z + 36) = 1 + 8 + 72$ ⇒ $2(x-2)^2 + 2y^2 + 2(z+6)^2 = 81$ ⇒

$(x-2)^2 + y^2 + (z+6)^2 = \frac{81}{2}$, which we recognize as an equation of a sphere with center $(2, 0, -6)$ and

radius $\sqrt{\frac{81}{2}} = 9/\sqrt{2}$.

19. (a) If the midpoint of the line segment from $P_1(x_1, y_1, z_1)$ to $P_2(x_2, y_2, z_2)$ is $Q = \left(\dfrac{x_1+x_2}{2}, \dfrac{y_1+y_2}{2}, \dfrac{z_1+z_2}{2}\right)$,

then the distances $|P_1Q|$ and $|QP_2|$ are equal, and each is half of $|P_1P_2|$. We verify that this is the case:

$$|P_1P_2| = \sqrt{(x_2-x_1)^2 + (y_2-y_1)^2 + (z_2-z_1)^2}$$

$$|P_1Q| = \sqrt{\left[\tfrac{1}{2}(x_1+x_2)-x_1\right]^2 + \left[\tfrac{1}{2}(y_1+y_2)-y_1\right]^2 + \left[\tfrac{1}{2}(z_1+z_2)-z_1\right]^2}$$

$$= \sqrt{\left(\tfrac{1}{2}x_2 - \tfrac{1}{2}x_1\right)^2 + \left(\tfrac{1}{2}y_2 - \tfrac{1}{2}y_1\right)^2 + \left(\tfrac{1}{2}z_2 - \tfrac{1}{2}z_1\right)^2}$$

$$= \sqrt{\left(\tfrac{1}{2}\right)^2\left[(x_2-x_1)^2 + (y_2-y_1)^2 + (z_2-z_1)^2\right]} = \tfrac{1}{2}\sqrt{(x_2-x_1)^2 + (y_2-y_1)^2 + (z_2-z_1)^2}$$

$$= \tfrac{1}{2}|P_1P_2|$$

© 2012 Cengage Learning. All Rights Reserved. May not be scanned, copied, or duplicated, or posted to a publicly accessible website, in whole or in part.

$$|QP_2| = \sqrt{\left[x_2 - \tfrac{1}{2}(x_1 + x_2)\right]^2 + \left[y_2 - \tfrac{1}{2}(y_1 + y_2)\right]^2 + \left[z_2 - \tfrac{1}{2}(z_1 + z_2)\right]^2}$$

$$= \sqrt{\left(\tfrac{1}{2}x_2 - \tfrac{1}{2}x_1\right)^2 + \left(\tfrac{1}{2}y_2 - \tfrac{1}{2}y_1\right)^2 + \left(\tfrac{1}{2}z_2 - \tfrac{1}{2}z_1\right)^2} = \sqrt{\left(\tfrac{1}{2}\right)^2 \left[(x_2 - x_1)^2 + (y_2 - y_1)^2 + (z_2 - z_1)^2\right]}$$

$$= \tfrac{1}{2}\sqrt{(x_2 - x_1)^2 + (y_2 - y_1)^2 + (z_2 - z_1)^2} = \tfrac{1}{2}|P_1 P_2|$$

So Q is indeed the midpoint of $P_1 P_2$.

(b) By part (a), the midpoints of sides AB, BC and CA are $P_1\left(-\tfrac{1}{2}, 1, 4\right)$, $P_2\left(1, \tfrac{1}{2}, 5\right)$ and $P_3\left(\tfrac{5}{2}, \tfrac{3}{2}, 4\right)$. (Recall that a median of a triangle is a line segment from a vertex to the midpoint of the opposite side.) Then the lengths of the medians are:

$$|AP_2| = \sqrt{0^2 + \left(\tfrac{1}{2} - 2\right)^2 + (5 - 3)^2} = \sqrt{\tfrac{9}{4} + 4} = \sqrt{\tfrac{25}{4}} = \tfrac{5}{2}$$

$$|BP_3| = \sqrt{\left(\tfrac{5}{2} + 2\right)^2 + \left(\tfrac{3}{2}\right)^2 + (4 - 5)^2} = \sqrt{\tfrac{81}{4} + \tfrac{9}{4} + 1} = \sqrt{\tfrac{94}{4}} = \tfrac{1}{2}\sqrt{94}$$

$$|CP_1| = \sqrt{\left(-\tfrac{1}{2} - 4\right)^2 + (1 - 1)^2 + (4 - 5)^2} = \sqrt{\tfrac{81}{4} + 1} = \tfrac{1}{2}\sqrt{85}$$

21. (a) Since the sphere touches the xy-plane, its radius is the distance from its center, $(2, -3, 6)$, to the xy-plane, namely 6. Therefore $r = 6$ and an equation of the sphere is $(x - 2)^2 + (y + 3)^2 + (z - 6)^2 = 6^2 = 36$.

(b) The radius of this sphere is the distance from its center $(2, -3, 6)$ to the yz-plane, which is 2. Therefore, an equation is $(x - 2)^2 + (y + 3)^2 + (z - 6)^2 = 4$.

(c) Here the radius is the distance from the center $(2, -3, 6)$ to the xz-plane, which is 3. Therefore, an equation is $(x - 2)^2 + (y + 3)^2 + (z - 6)^2 = 9$.

23. The equation $x = 5$ represents a plane parallel to the yz-plane and 5 units in front of it.

25. The inequality $y < 8$ represents a half-space consisting of all points to the left of the plane $y = 8$.

27. The inequality $0 \le z \le 6$ represents all points on or between the horizontal planes $z = 0$ (the xy-plane) and $z = 6$.

29. Because $z = -1$, all points in the region must lie in the horizontal plane $z = -1$. In addition, $x^2 + y^2 = 4$, so the region consists of all points that lie on a circle with radius 2 and center on the z-axis that is contained in the plane $z = -1$.

31. The inequality $x^2 + y^2 + z^2 \le 3$ is equivalent to $\sqrt{x^2 + y^2 + z^2} \le \sqrt{3}$, so the region consists of those points whose distance from the origin is at most $\sqrt{3}$. This is the set of all points on or inside the sphere with radius $\sqrt{3}$ and center $(0, 0, 0)$.

33. Here $x^2 + z^2 \le 9$ or equivalently $\sqrt{x^2 + z^2} \le 3$ which describes the set of all points in $\mathbb{R}^3$ whose distance from the y-axis is at most 3. Thus, the inequality represents the region consisting of all points on or inside a circular cylinder of radius 3 with axis the y-axis.

35. This describes all points whose x-coordinate is between 0 and 5, that is, $0 < x < 5$.

37. This describes a region all of whose points have a distance to the origin which is greater than r, but smaller than R. So inequalities describing the region are $r < \sqrt{x^2 + y^2 + z^2} < R$, or $r^2 < x^2 + y^2 + z^2 < R^2$.

© 2012 Cengage Learning. All Rights Reserved. May not be scanned, copied, or duplicated, or posted to a publicly accessible website, in whole or in part.

39. (a) To find the x- and y-coordinates of the point P, we project it onto L_2
and project the resulting point Q onto the x- and y-axes. To find the
z-coordinate, we project P onto either the xz-plane or the yz-plane
(using our knowledge of its x- or y-coordinate) and then project the
resulting point onto the z-axis. (Or, we could draw a line parallel to
QO from P to the z-axis.) The coordinates of P are $(2, 1, 4)$.

(b) A is the intersection of L_1 and L_2, B is directly below the
y-intercept of L_2, and C is directly above the x-intercept of L_2.

41. We need to find a set of points $\left\{ P(x, y, z) \mid |AP| = |BP| \right\}$.

$$\sqrt{(x+1)^2 + (y-5)^2 + (z-3)^2} = \sqrt{(x-6)^2 + (y-2)^2 + (z+2)^2} \quad \Rightarrow$$

$$(x+1)^2 + (y-5) + (z-3)^2 = (x-6)^2 + (y-2)^2 + (z+2)^2 \quad \Rightarrow$$

$$x^2 + 2x + 1 + y^2 - 10y + 25 + z^2 - 6z + 9 = x^2 - 12x + 36 + y^2 - 4y + 4 + z^2 + 4z + 4 \quad \Rightarrow \quad 14x - 6y - 10z = 9.$$

Thus the set of points is a plane perpendicular to the line segment joining A and B (since this plane must contain the
perpendicular bisector of the line segment AB).

43. The sphere $x^2 + y^2 + z^2 = 4$ has center $(0, 0, 0)$ and radius 2. Completing squares in $x^2 - 4x + y^2 - 4y + z^2 - 4z = -11$
gives $(x^2 - 4x + 4) + (y^2 - 4y + 4) + (z^2 - 4z + 4) = -11 + 4 + 4 + 4 \quad \Rightarrow \quad (x-2)^2 + (y-2)^2 + (z-2)^2 = 1$,
so this is the sphere with center $(2, 2, 2)$ and radius 1. The (shortest) distance between the spheres is measured along
the line segment connecting their centers. The distance between $(0, 0, 0)$ and $(2, 2, 2)$ is
$\sqrt{(2-0)^2 + (2-0)^2 + (2-0)^2} = \sqrt{12} = 2\sqrt{3}$, and subtracting the radius of each circle, the distance between the
spheres is $2\sqrt{3} - 2 - 1 = 2\sqrt{3} - 3$.

12.2 Vectors

1. (a) The cost of a theater ticket is a scalar, because it has only magnitude.

(b) The current in a river is a vector, because it has both magnitude (the speed of the current) and direction at any given
location.

(c) If we assume that the initial path is linear, the initial flight path from Houston to Dallas is a vector, because it has both
magnitude (distance) and direction.

(d) The population of the world is a scalar, because it has only magnitude.

3. Vectors are equal when they share the same length and direction (but not necessarily location). Using the symmetry of the
parallelogram as a guide, we see that $\overrightarrow{AB} = \overrightarrow{DC}$, $\overrightarrow{DA} = \overrightarrow{CB}$, $\overrightarrow{DE} = \overrightarrow{EB}$, and $\overrightarrow{EA} = \overrightarrow{CE}$.

© 2012 Cengage Learning. All Rights Reserved. May not be scanned, copied, or duplicated, or posted to a publicly accessible website, in whole or in part.

5. (a)

(b)

(c)

(d)

(e)

(f)

7. Because the tail of **d** is the midpoint of QR we have $\overrightarrow{QR} = 2\mathbf{d}$, and by the Triangle Law,

$\mathbf{a} + 2\mathbf{d} = \mathbf{b} \quad \Rightarrow \quad 2\mathbf{d} = \mathbf{b} - \mathbf{a} \quad \Rightarrow \quad \mathbf{d} = \frac{1}{2}(\mathbf{b} - \mathbf{a}) = \frac{1}{2}\mathbf{b} - \frac{1}{2}\mathbf{a}$. Again by the Triangle Law we have $\mathbf{c} + \mathbf{d} = \mathbf{b}$ so

$\mathbf{c} = \mathbf{b} - \mathbf{d} = \mathbf{b} - \left(\frac{1}{2}\mathbf{b} - \frac{1}{2}\mathbf{a}\right) = \frac{1}{2}\mathbf{a} + \frac{1}{2}\mathbf{b}$.

9. $\mathbf{a} = \langle 3 - (-1), 2 - 1 \rangle = \langle 4, 1 \rangle$

11. $\mathbf{a} = \langle 2 - (-1), 2 - 3 \rangle = \langle 3, -1 \rangle$

13. $\mathbf{a} = \langle 2 - 0, 3 - 3, -1 - 1 \rangle = \langle 2, 0, -2 \rangle$

15. $\langle -1, 4 \rangle + \langle 6, -2 \rangle = \langle -1 + 6, 4 + (-2) \rangle = \langle 5, 2 \rangle$

17. $\langle 3, 0, 1 \rangle + \langle 0, 8, 0 \rangle = \langle 3 + 0, 0 + 8, 1 + 0 \rangle$
$= \langle 3, 8, 1 \rangle$

© 2012 Cengage Learning. All Rights Reserved. May not be scanned, copied, or duplicated, or posted to a publicly accessible website, in whole or in part.

19. $\mathbf{a} + \mathbf{b} = \langle 5 + (-3), -12 + (-6) \rangle = \langle 2, -18 \rangle$

$2\mathbf{a} + 3\mathbf{b} = \langle 10, -24 \rangle + \langle -9, -18 \rangle = \langle 1, -42 \rangle$

$|\mathbf{a}| = \sqrt{5^2 + (-12)^2} = \sqrt{169} = 13$

$|\mathbf{a} - \mathbf{b}| = |\langle 5 - (-3), -12 - (-6) \rangle| = |\langle 8, -6 \rangle| = \sqrt{8^2 + (-6)^2} = \sqrt{100} = 10$

21. $\mathbf{a} + \mathbf{b} = (\mathbf{i} + 2\mathbf{j} - 3\mathbf{k}) + (-2\mathbf{i} - \mathbf{j} + 5\mathbf{k}) = -\mathbf{i} + \mathbf{j} + 2\mathbf{k}$

$2\mathbf{a} + 3\mathbf{b} = 2(\mathbf{i} + 2\mathbf{j} - 3\mathbf{k}) + 3(-2\mathbf{i} - \mathbf{j} + 5\mathbf{k}) = 2\mathbf{i} + 4\mathbf{j} - 6\mathbf{k} - 6\mathbf{i} - 3\mathbf{j} + 15\mathbf{k} = -4\mathbf{i} + \mathbf{j} + 9\mathbf{k}$

$|\mathbf{a}| = \sqrt{1^2 + 2^2 + (-3)^2} = \sqrt{14}$

$|\mathbf{a} - \mathbf{b}| = |(\mathbf{i} + 2\mathbf{j} - 3\mathbf{k}) - (-2\mathbf{i} - \mathbf{j} + 5\mathbf{k})| = |3\mathbf{i} + 3\mathbf{j} - 8\mathbf{k}| = \sqrt{3^2 + 3^2 + (-8)^2} = \sqrt{82}$

23. The vector $-3\mathbf{i} + 7\mathbf{j}$ has length $|-3\mathbf{i} + 7\mathbf{j}| = \sqrt{(-3)^2 + 7^2} = \sqrt{58}$, so by Equation 4 the unit vector with the same

direction is $\dfrac{1}{\sqrt{58}}(-3\mathbf{i} + 7\mathbf{j}) = -\dfrac{3}{\sqrt{58}}\mathbf{i} + \dfrac{7}{\sqrt{58}}\mathbf{j}$.

25. The vector $8\mathbf{i} - \mathbf{j} + 4\mathbf{k}$ has length $|8\mathbf{i} - \mathbf{j} + 4\mathbf{k}| = \sqrt{8^2 + (-1)^2 + 4^2} = \sqrt{81} = 9$, so by Equation 4 the unit vector with

the same direction is $\frac{1}{9}(8\mathbf{i} - \mathbf{j} + 4\mathbf{k}) = \frac{8}{9}\mathbf{i} - \frac{1}{9}\mathbf{j} + \frac{4}{9}\mathbf{k}$.

27.

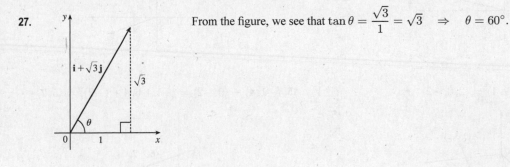

From the figure, we see that $\tan\theta = \dfrac{\sqrt{3}}{1} = \sqrt{3} \quad \Rightarrow \quad \theta = 60°$.

29. From the figure, we see that the x-component of $\mathbf{v}$ is

$v_1 = |\mathbf{v}| \cos(\pi/3) = 4 \cdot \frac{1}{2} = 2$ and the y-component is

$v_2 = |\mathbf{v}| \sin(\pi/3) = 4 \cdot \frac{\sqrt{3}}{2} = 2\sqrt{3}$. Thus

$\mathbf{v} = \langle v_1, v_2 \rangle = \langle 2, 2\sqrt{3} \rangle$.

31. The velocity vector $\mathbf{v}$ makes an angle of $40°$ with the horizontal and

has magnitude equal to the speed at which the football was thrown.

From the figure, we see that the horizontal component of $\mathbf{v}$ is

$|\mathbf{v}| \cos 40° = 60 \cos 40° \approx 45.96$ ft/s and the vertical component

is $|\mathbf{v}| \sin 40° = 60 \sin 40° \approx 38.57$ ft/s.

© 2012 Cengage Learning. All Rights Reserved. May not be scanned, copied, or duplicated, or posted to a publicly accessible website, in whole or in part.

33. The given force vectors can be expressed in terms of their horizontal and vertical components as $-300\,\mathbf{i}$ and

$200\cos 60°\,\mathbf{i} + 200\sin 60°\,\mathbf{j} = 200\left(\frac{1}{2}\right)\mathbf{i} + 200\left(\frac{\sqrt{3}}{2}\right)\mathbf{j} = 100\,\mathbf{i} + 100\sqrt{3}\,\mathbf{j}$. The resultant force $\mathbf{F}$ is the sum of

these two vectors: $\mathbf{F} = (-300 + 100)\,\mathbf{i} + (0 + 100\sqrt{3})\,\mathbf{j} = -200\,\mathbf{i} + 100\sqrt{3}\,\mathbf{j}$. Then we have

$|\mathbf{F}| \approx \sqrt{(-200)^2 + (100\sqrt{3})^2} = \sqrt{70{,}000} = 100\sqrt{7} \approx 264.6$ N. Let θ be the angle $\mathbf{F}$ makes with the

positive x-axis. Then $\tan\theta = \dfrac{100\sqrt{3}}{-200} = -\dfrac{\sqrt{3}}{2}$ and the terminal point of $\mathbf{F}$ lies in the second quadrant, so

$\theta = \tan^{-1}\left(-\dfrac{\sqrt{3}}{2}\right) + 180° \approx -40.9° + 180° = 139.1°$.

35. With respect to the water's surface, the woman's velocity is the vector sum of the velocity of the ship with respect

to the water, and the woman's velocity with respect to the ship. If we let north be the positive y-direction, then

$\mathbf{v} = \langle 0, 22\rangle + \langle -3, 0\rangle = \langle -3, 22\rangle$. The woman's speed is $|\mathbf{v}| = \sqrt{9 + 484} \approx 22.2$ mi/h. The vector $\mathbf{v}$ makes an angle θ

with the east, where $\theta = \tan^{-1}\left(\dfrac{22}{-3}\right) \approx 98°$. Therefore, the woman's direction is about $\mathrm{N}(98-90)°\mathrm{W} = \mathrm{N}8°\mathrm{W}$.

37. Let $\mathbf{T}_1$ and $\mathbf{T}_2$ represent the tension vectors in each side of the

clothesline as shown in the figure. $\mathbf{T}_1$ and $\mathbf{T}_2$ have equal vertical

components and opposite horizontal components, so we can write

$\mathbf{T}_1 = -a\,\mathbf{i} + b\,\mathbf{j}$ and $\mathbf{T}_2 = a\,\mathbf{i} + b\,\mathbf{j}$ $[a, b > 0]$. By similar triangles, $\dfrac{b}{a} = \dfrac{0.08}{4} \;\Rightarrow\; a = 50b$. The force due to gravity

acting on the shirt has magnitude $0.8g \approx (0.8)(9.8) = 7.84$ N, hence we have $\mathbf{w} = -7.84\,\mathbf{j}$. The resultant $\mathbf{T}_1 + \mathbf{T}_2$

of the tensile forces counterbalances $\mathbf{w}$, so $\mathbf{T}_1 + \mathbf{T}_2 = -\mathbf{w} \;\Rightarrow\; (-a\,\mathbf{i} + b\,\mathbf{j}) + (a\,\mathbf{i} + b\,\mathbf{j}) = 7.84\,\mathbf{j} \;\Rightarrow\;$

$(-50b\,\mathbf{i} + b\,\mathbf{j}) + (50b\,\mathbf{i} + b\,\mathbf{j}) = 2b\,\mathbf{j} = 7.84\,\mathbf{j} \;\Rightarrow\; b = \frac{7.84}{2} = 3.92$ and $a = 50b = 196$. Thus the tensions are

$\mathbf{T}_1 = -a\,\mathbf{i} + b\,\mathbf{j} = -196\,\mathbf{i} + 3.92\,\mathbf{j}$ and $\mathbf{T}_2 = a\,\mathbf{i} + b\,\mathbf{j} = 196\,\mathbf{i} + 3.92\,\mathbf{j}$.

Alternatively, we can find the value of θ and proceed as in Example 7.

39. (a) Set up coordinate axes so that the boatman is at the origin, the canal is

bordered by the y-axis and the line $x = 3$, and the current flows in the

negative y-direction. The boatman wants to reach the point $(3, 2)$. Let θ be

the angle, measured from the positive y-axis, in the direction he should

steer. (See the figure.)

In still water, the boat has velocity $\mathbf{v}_b = \langle 13\sin\theta, 13\cos\theta\rangle$ and the velocity of the current is $\mathbf{v}_c\,\langle 0, -3.5\rangle$, so the true path

of the boat is determined by the velocity vector $\mathbf{v} = \mathbf{v}_b + \mathbf{v}_c = \langle 13\sin\theta, 13\cos\theta - 3.5\rangle$. Let t be the time (in hours)

after the boat departs; then the position of the boat at time t is given by $t\mathbf{v}$ and the boat crosses the canal when

$t\mathbf{v} = \langle 13\sin\theta, 13\cos\theta - 3.5\rangle\,t = \langle 3, 2\rangle$. Thus $13(\sin\theta)t = 3 \;\Rightarrow\; t = \dfrac{3}{13\sin\theta}$ and $(13\cos\theta - 3.5)\,t = 2$.

© 2012 Cengage Learning. All Rights Reserved. May not be scanned, copied, or duplicated, or posted to a publicly accessible website, in whole or in part.

Substituting gives $(13\cos\theta - 3.5)\dfrac{3}{13\sin\theta} = 2 \quad \Rightarrow \quad 39\cos\theta - 10.5 = 26\sin\theta$ **(1)**. Squaring both sides, we have

$$1521\cos^2\theta - 819\cos\theta + 110.25 = 676\sin^2\theta = 676\left(1 - \cos^2\theta\right)$$

$$2197\cos^2\theta - 819\cos\theta - 565.75 = 0$$

The quadratic formula gives

$$\cos\theta = \frac{819 \pm \sqrt{(-819)^2 - 4(2197)(-565.75)}}{2(2197)}$$

$$= \frac{819 \pm \sqrt{5,642,572}}{4394} \approx 0.72699 \text{ or } -0.35421$$

The acute value for θ is approximately $\cos^{-1}(0.72699) \approx 43.4°$. Thus the boatman should steer in the direction that is

43.4° from the bank, toward upstream.

Alternate solution: We could solve **(1)** graphically by plotting $y = 39\cos\theta - 10.5$ and $y = 26\sin\theta$ on a graphing device

and finding the appoximate intersection point $(0.757, 17.85)$. Thus $\theta \approx 0.757$ radians or equivalently 43.4°.

(b) From part (a) we know the trip is completed when $t = \dfrac{3}{13\sin\theta}$. But $\theta \approx 43.4°$, so the time required is approximately

$\dfrac{3}{13\sin 43.4°} \approx 0.336$ hours or 20.2 minutes.

41. The slope of the tangent line to the graph of $y = x^2$ at the point $(2, 4)$ is

$$\left.\frac{dy}{dx}\right|_{x=2} = 2x\Big|_{x=2} = 4$$

and a parallel vector is $\mathbf{i} + 4\mathbf{j}$ which has length $|\mathbf{i} + 4\mathbf{j}| = \sqrt{1^2 + 4^2} = \sqrt{17}$, so unit vectors parallel to the tangent line

are $\pm\frac{1}{\sqrt{17}}(\mathbf{i} + 4\mathbf{j})$.

43. By the Triangle Law, $\overrightarrow{AB} + \overrightarrow{BC} = \overrightarrow{AC}$. Then $\overrightarrow{AB} + \overrightarrow{BC} + \overrightarrow{CA} = \overrightarrow{AC} + \overrightarrow{CA}$, but $\overrightarrow{AC} + \overrightarrow{CA} = \overrightarrow{AC} + \left(-\overrightarrow{AC}\right) = \mathbf{0}$.

So $\overrightarrow{AB} + \overrightarrow{BC} + \overrightarrow{CA} = \mathbf{0}$.

45. (a), (b)

(c) From the sketch, we estimate that $s \approx 1.3$ and $t \approx 1.6$.

(d) $\mathbf{c} = s\mathbf{a} + t\mathbf{b} \quad \Leftrightarrow \quad 7 = 3s + 2t$ and $1 = 2s - t$.

Solving these equations gives $s = \frac{9}{7}$ and $t = \frac{11}{7}$.

47. $|\mathbf{r} - \mathbf{r}_0|$ is the distance between the points (x, y, z) and (x_0, y_0, z_0), so the set of points is a sphere with radius 1 and

center (x_0, y_0, z_0).

Alternate method: $|\mathbf{r} - \mathbf{r}_0| = 1 \quad \Leftrightarrow \quad \sqrt{(x - x_0)^2 + (y - y_0)^2 + (z - z_0)^2} = 1 \quad \Leftrightarrow$

$(x - x_0)^2 + (y - y_0)^2 + (z - z_0)^2 = 1$, which is the equation of a sphere with radius 1 and center (x_0, y_0, z_0).

© 2012 Cengage Learning. All Rights Reserved. May not be scanned, copied, or duplicated, or posted to a publicly accessible website, in whole or in part.

49. $\mathbf{a} + (\mathbf{b} + \mathbf{c}) = \langle a_1, a_2 \rangle + (\langle b_1, b_2 \rangle + \langle c_1, c_2 \rangle) = \langle a_1, a_2 \rangle + \langle b_1 + c_1, b_2 + c_2 \rangle$

$\qquad = \langle a_1 + b_1 + c_1, a_2 + b_2 + c_2 \rangle = \langle (a_1 + b_1) + c_1, (a_2 + b_2) + c_2 \rangle$

$\qquad = \langle a_1 + b_1, a_2 + b_2 \rangle + \langle c_1, c_2 \rangle = (\langle a_1, a_2 \rangle + \langle b_1, b_2 \rangle) + \langle c_1, c_2 \rangle$

$\qquad = (\mathbf{a} + \mathbf{b}) + \mathbf{c}$

51. Consider triangle ABC, where D and E are the midpoints of AB and BC. We know that $\overrightarrow{AB} + \overrightarrow{BC} = \overrightarrow{AC}$ **(1)**   and

$\overrightarrow{DB} + \overrightarrow{BE} = \overrightarrow{DE}$ **(2)**.   However, $\overrightarrow{DB} = \frac{1}{2}\overrightarrow{AB}$, and $\overrightarrow{BE} = \frac{1}{2}\overrightarrow{BC}$. Substituting these expressions for $\overrightarrow{DB}$ and $\overrightarrow{BE}$ into

(2) gives $\frac{1}{2}\overrightarrow{AB} + \frac{1}{2}\overrightarrow{BC} = \overrightarrow{DE}$. Comparing this with **(1)** gives $\overrightarrow{DE} = \frac{1}{2}\overrightarrow{AC}$. Therefore $\overrightarrow{AC}$ and $\overrightarrow{DE}$ are parallel and

$\left| \overrightarrow{DE} \right| = \frac{1}{2}\left| \overrightarrow{AC} \right|$.

12.3   The Dot Product

1. (a) $\mathbf{a} \cdot \mathbf{b}$ is a scalar, and the dot product is defined only for vectors, so $(\mathbf{a} \cdot \mathbf{b}) \cdot \mathbf{c}$ has no meaning.

(b) $(\mathbf{a} \cdot \mathbf{b})\mathbf{c}$ is a scalar multiple of a vector, so it does have meaning.

(c) Both $|\mathbf{a}|$ and $\mathbf{b} \cdot \mathbf{c}$ are scalars, so $|\mathbf{a}|(\mathbf{b} \cdot \mathbf{c})$ is an ordinary product of real numbers, and has meaning.

(d) Both $\mathbf{a}$ and $\mathbf{b} + \mathbf{c}$ are vectors, so the dot product $\mathbf{a} \cdot (\mathbf{b} + \mathbf{c})$ has meaning.

(e) $\mathbf{a} \cdot \mathbf{b}$ is a scalar, but $\mathbf{c}$ is a vector, and so the two quantities cannot be added and $\mathbf{a} \cdot \mathbf{b} + \mathbf{c}$ has no meaning.

(f) $|\mathbf{a}|$ is a scalar, and the dot product is defined only for vectors, so $|\mathbf{a}| \cdot (\mathbf{b} + \mathbf{c})$ has no meaning.

3. $\mathbf{a} \cdot \mathbf{b} = \left\langle -2, \frac{1}{3} \right\rangle \cdot \langle -5, 12 \rangle = (-2)(-5) + \left(\frac{1}{3}\right)(12) = 10 + 4 = 14$

5. $\mathbf{a} \cdot \mathbf{b} = \left\langle 4, 1, \frac{1}{4} \right\rangle \cdot \langle 6, -3, -8 \rangle = (4)(6) + (1)(-3) + \left(\frac{1}{4}\right)(-8) = 19$

7. $\mathbf{a} \cdot \mathbf{b} = (2\mathbf{i} + \mathbf{j}) \cdot (\mathbf{i} - \mathbf{j} + \mathbf{k}) = (2)(1) + (1)(-1) + (0)(1) = 1$

9. By Theorem 3, $\mathbf{a} \cdot \mathbf{b} = |\mathbf{a}|\,|\mathbf{b}| \cos \theta = (6)(5) \cos \frac{2\pi}{3} = 30 \left(-\frac{1}{2}\right) = -15$.

11. $\mathbf{u}$, $\mathbf{v}$, and $\mathbf{w}$ are all unit vectors, so the triangle is an equilateral triangle. Thus the angle between $\mathbf{u}$ and $\mathbf{v}$ is $60°$ and

$\mathbf{u} \cdot \mathbf{v} = |\mathbf{u}|\,|\mathbf{v}| \cos 60° = (1)(1)\left(\frac{1}{2}\right) = \frac{1}{2}$. If $\mathbf{w}$ is moved so it has the same initial point as $\mathbf{u}$, we can see that the angle

between them is $120°$ and we have $\mathbf{u} \cdot \mathbf{w} = |\mathbf{u}|\,|\mathbf{w}| \cos 120° = (1)(1)\left(-\frac{1}{2}\right) = -\frac{1}{2}$.

13. (a) $\mathbf{i} \cdot \mathbf{j} = \langle 1, 0, 0 \rangle \cdot \langle 0, 1, 0 \rangle = (1)(0) + (0)(1) + (0)(0) = 0$. Similarly, $\mathbf{j} \cdot \mathbf{k} = (0)(0) + (1)(0) + (0)(1) = 0$ and

$\mathbf{k} \cdot \mathbf{i} = (0)(1) + (0)(0) + (1)(0) = 0$.

Another method: Because $\mathbf{i}$, $\mathbf{j}$, and $\mathbf{k}$ are mutually perpendicular, the cosine factor in each dot product (see Theorem 3)

is $\cos \frac{\pi}{2} = 0$.

(b) By Property 1 of the dot product, $\mathbf{i} \cdot \mathbf{i} = |\mathbf{i}|^2 = 1^2 = 1$ since $\mathbf{i}$ is a unit vector. Similarly, $\mathbf{j} \cdot \mathbf{j} = |\mathbf{j}|^2 = 1$ and

$\mathbf{k} \cdot \mathbf{k} = |\mathbf{k}|^2 = 1$.

© 2012 Cengage Learning. All Rights Reserved. May not be scanned, copied, or duplicated, or posted to a publicly accessible website, in whole or in part.

15. $|\mathbf{a}| = \sqrt{4^2 + 3^2} = 5$, $|\mathbf{b}| = \sqrt{2^2 + (-1)^2} = \sqrt{5}$, and $\mathbf{a} \cdot \mathbf{b} = (4)(2) + (3)(-1) = 5$. From Corollary 6, we have

$$\cos\theta = \frac{\mathbf{a} \cdot \mathbf{b}}{|\mathbf{a}|\,|\mathbf{b}|} = \frac{5}{5 \cdot \sqrt{5}} = \frac{1}{\sqrt{5}}. \text{ So the angle between } \mathbf{a} \text{ and } \mathbf{b} \text{ is } \theta = \cos^{-1}\left(\frac{1}{\sqrt{5}}\right) \approx 63°.$$

17. $|\mathbf{a}| = \sqrt{3^2 + (-1)^2 + 5^2} = \sqrt{35}$, $|\mathbf{b}| = \sqrt{(-2)^2 + 4^2 + 3^2} = \sqrt{29}$, and $\mathbf{a} \cdot \mathbf{b} = (3)(-2) + (-1)(4) + (5)(3) = 5$. Then

$$\cos\theta = \frac{\mathbf{a} \cdot \mathbf{b}}{|\mathbf{a}|\,|\mathbf{b}|} = \frac{5}{\sqrt{35} \cdot \sqrt{29}} = \frac{5}{\sqrt{1015}} \text{ and the angle between } \mathbf{a} \text{ and } \mathbf{b} \text{ is } \theta = \cos^{-1}\left(\frac{5}{\sqrt{1015}}\right) \approx 81°.$$

19. $|\mathbf{a}| = \sqrt{4^2 + (-3)^2 + 1^2} = \sqrt{26}$, $|\mathbf{b}| = \sqrt{2^2 + 0^2 + (-1)^2} = \sqrt{5}$, and $\mathbf{a} \cdot \mathbf{b} = (4)(2) + (-3)(0) + (1)(-1) = 7$.

Then $\cos\theta = \dfrac{\mathbf{a} \cdot \mathbf{b}}{|\mathbf{a}|\,|\mathbf{b}|} = \dfrac{7}{\sqrt{26} \cdot \sqrt{5}} = \dfrac{7}{\sqrt{130}}$ and $\theta = \cos^{-1}\left(\dfrac{7}{\sqrt{130}}\right) \approx 52°$.

21. Let p, q, and r be the angles at vertices P, Q, and R respectively.

Then p is the angle between vectors $\overrightarrow{PQ}$ and $\overrightarrow{PR}$, q is the angle

between vectors $\overrightarrow{QP}$ and $\overrightarrow{QR}$, and r is the angle between vectors

$\overrightarrow{RP}$ and $\overrightarrow{RQ}$.

Thus $\cos p = \dfrac{\overrightarrow{PQ} \cdot \overrightarrow{PR}}{\left|\overrightarrow{PQ}\right|\left|\overrightarrow{PR}\right|} = \dfrac{\langle -2, 3\rangle \cdot \langle 1, 4\rangle}{\sqrt{(-2)^2 + 3^2}\,\sqrt{1^2 + 4^2}} = \dfrac{-2 + 12}{\sqrt{13}\,\sqrt{17}} = \dfrac{10}{\sqrt{221}}$ and $p = \cos^{-1}\left(\dfrac{10}{\sqrt{221}}\right) \approx 48°$. Similarly,

$$\cos q = \frac{\overrightarrow{QP} \cdot \overrightarrow{QR}}{\left|\overrightarrow{QP}\right|\left|\overrightarrow{QR}\right|} = \frac{\langle 2, -3\rangle \cdot \langle 3, 1\rangle}{\sqrt{4+9}\,\sqrt{9+1}} = \frac{6-3}{\sqrt{13}\,\sqrt{10}} = \frac{3}{\sqrt{130}} \text{ so } q = \cos^{-1}\left(\frac{3}{\sqrt{130}}\right) \approx 75° \text{ and}$$

$r \approx 180° - (48° + 75°) = 57°$.

Alternate solution: Apply the Law of Cosines three times as follows: $\cos p = \dfrac{\left|\overrightarrow{QR}\right|^2 - \left|\overrightarrow{PQ}\right|^2 - \left|\overrightarrow{PR}\right|^2}{2\left|\overrightarrow{PQ}\right|\left|\overrightarrow{PR}\right|}$,

$$\cos q = \frac{\left|\overrightarrow{PR}\right|^2 - \left|\overrightarrow{PQ}\right|^2 - \left|\overrightarrow{QR}\right|^2}{2\left|\overrightarrow{PQ}\right|\left|\overrightarrow{QR}\right|}, \text{ and } \cos r = \frac{\left|\overrightarrow{PQ}\right|^2 - \left|\overrightarrow{PR}\right|^2 - \left|\overrightarrow{QR}\right|^2}{2\left|\overrightarrow{PR}\right|\left|\overrightarrow{QR}\right|}.$$

23. (a) $\mathbf{a} \cdot \mathbf{b} = (-5)(6) + (3)(-8) + (7)(2) = -40 \neq 0$, so $\mathbf{a}$ and $\mathbf{b}$ are not orthogonal. Also, since $\mathbf{a}$ is not a scalar multiple

of $\mathbf{b}$, $\mathbf{a}$ and $\mathbf{b}$ are not parallel.

(b) $\mathbf{a} \cdot \mathbf{b} = (4)(-3) + (6)(2) = 0$, so $\mathbf{a}$ and $\mathbf{b}$ are orthogonal (and not parallel).

(c) $\mathbf{a} \cdot \mathbf{b} = (-1)(3) + (2)(4) + (5)(-1) = 0$, so $\mathbf{a}$ and $\mathbf{b}$ are orthogonal (and not parallel).

(d) Because $\mathbf{a} = -\frac{2}{3}\mathbf{b}$, $\mathbf{a}$ and $\mathbf{b}$ are parallel.

25. $\overrightarrow{QP} = \langle -1, -3, 2\rangle$, $\overrightarrow{QR} = \langle 4, -2, -1\rangle$, and $\overrightarrow{QP} \cdot \overrightarrow{QR} = -4 + 6 - 2 = 0$. Thus $\overrightarrow{QP}$ and $\overrightarrow{QR}$ are orthogonal, so the angle of

the triangle at vertex Q is a right angle.

© 2012 Cengage Learning. All Rights Reserved. May not be scanned, copied, or duplicated, or posted to a publicly accessible website, in whole or in part.

27. Let $\mathbf{a} = a_1\,\mathbf{i} + a_2\,\mathbf{j} + a_3\,\mathbf{k}$ be a vector orthogonal to both $\mathbf{i} + \mathbf{j}$ and $\mathbf{i} + \mathbf{k}$. Then $\mathbf{a} \cdot (\mathbf{i} + \mathbf{j}) = 0$ $\Leftrightarrow$ $a_1 + a_2 = 0$ and

$\mathbf{a} \cdot (\mathbf{i} + \mathbf{k}) = 0$ $\Leftrightarrow$ $a_1 + a_3 = 0$, so $a_1 = -a_2 = -a_3$. Furthermore $\mathbf{a}$ is to be a unit vector, so $1 = a_1^2 + a_2^2 + a_3^2 = 3a_1^2$

implies $a_1 = \pm\frac{1}{\sqrt{3}}$. Thus $\mathbf{a} = \frac{1}{\sqrt{3}}\,\mathbf{i} - \frac{1}{\sqrt{3}}\,\mathbf{j} - \frac{1}{\sqrt{3}}\,\mathbf{k}$ and $\mathbf{a} = -\frac{1}{\sqrt{3}}\,\mathbf{i} + \frac{1}{\sqrt{3}}\,\mathbf{j} + \frac{1}{\sqrt{3}}\,\mathbf{k}$ are two such unit vectors.

29. The line $2x - y = 3$ $\Leftrightarrow$ $y = 2x - 3$ has slope 2, so a vector parallel to the line is $\mathbf{a} = \langle 1, 2 \rangle$. The line $3x + y = 7$ $\Leftrightarrow$

$y = -3x + 7$ has slope -3, so a vector parallel to the line is $\mathbf{b} = \langle 1, -3 \rangle$. The angle between the lines is the same as the

angle θ between the vectors. Here we have $\mathbf{a} \cdot \mathbf{b} = (1)(1) + (2)(-3) = -5$, $|\mathbf{a}| = \sqrt{1^2 + 2^2} = \sqrt{5}$, and

$|\mathbf{b}| = \sqrt{1^2 + (-3)^2} = \sqrt{10}$, so $\cos\theta = \dfrac{\mathbf{a} \cdot \mathbf{b}}{|\mathbf{a}|\,|\mathbf{b}|} = \dfrac{-5}{\sqrt{5} \cdot \sqrt{10}} = \dfrac{-5}{5\sqrt{2}} = -\dfrac{1}{\sqrt{2}}$ or $-\dfrac{\sqrt{2}}{2}$. Thus $\theta = 135°$, and the

acute angle between the lines is $180° - 135° = 45°$.

31. The curves $y = x^2$ and $y = x^3$ meet when $x^2 = x^3$ $\Leftrightarrow$ $x^3 - x^2 = 0$ $\Leftrightarrow$ $x^2(x - 1) = 0$ $\Leftrightarrow$ $x = 0, x = 1$. We have

$\dfrac{d}{dx}x^2 = 2x$ and $\dfrac{d}{dx}x^3 = 3x^2$, so the tangent lines of both curves have slope 0 at $x = 0$. Thus the angle between the curves is

$0°$ at the point $(0, 0)$. For $x = 1$, $\dfrac{d}{dx}\,x^2\Big|_{x=1} = 2$ and $\dfrac{d}{dx}\,x^3\Big|_{x=1} = 3$ so the tangent lines at the point $(1, 1)$ have slopes 2 and

3. Vectors parallel to the tangent lines are $\langle 1, 2 \rangle$ and $\langle 1, 3 \rangle$, and the angle θ between them is given by

$$\cos\theta = \frac{\langle 1, 2 \rangle \cdot \langle 1, 3 \rangle}{|\langle 1, 2 \rangle|\,|\langle 1, 3 \rangle|} = \frac{1 + 6}{\sqrt{5}\,\sqrt{10}} - \frac{7}{5\sqrt{2}}$$

Thus $\theta = \cos^{-1}\left(\dfrac{7}{5\sqrt{2}}\right) \approx 8.1°$.

33. Since $|\langle 2, 1, 2 \rangle| = \sqrt{4 + 1 + 4} = \sqrt{9} = 3$, using Equations 8 and 9 we have $\cos\alpha = \frac{2}{3}$, $\cos\beta = \frac{1}{3}$, and $\cos\gamma = \frac{2}{3}$. The

direction angles are given by $\alpha = \cos^{-1}\left(\frac{2}{3}\right) \approx 48°$, $\beta = \cos^{-1}\left(\frac{1}{3}\right) \approx 71°$, and $\gamma = \cos^{-1}\left(\frac{2}{3}\right) = 48°$.

35. Since $|\mathbf{i} - 2\,\mathbf{j} - 3\,\mathbf{k}| = \sqrt{1 + 4 + 9} = \sqrt{14}$, Equations 8 and 9 give $\cos\alpha = \frac{1}{\sqrt{14}}$, $\cos\beta = \frac{-2}{\sqrt{14}}$, and $\cos\gamma = \frac{-3}{\sqrt{14}}$, while

$\alpha = \cos^{-1}\left(\frac{1}{\sqrt{14}}\right) \approx 74°$, $\beta = \cos^{-1}\left(-\frac{2}{\sqrt{14}}\right) \approx 122°$, and $\gamma = \cos^{-1}\left(-\frac{3}{\sqrt{14}}\right) \approx 143°$.

37. $|\langle c, c, c \rangle| = \sqrt{c^2 + c^2 + c^2} = \sqrt{3}\,c$ [since $c > 0$], so $\cos\alpha = \cos\beta = \cos\gamma = \dfrac{c}{\sqrt{3}\,c} = \dfrac{1}{\sqrt{3}}$ and

$\alpha = \beta = \gamma = \cos^{-1}\left(\dfrac{1}{\sqrt{3}}\right) \approx 55°$.

39. $|\mathbf{a}| = \sqrt{(-5)^2 + 12^2} = \sqrt{169} = 13$. The scalar projection of $\mathbf{b}$ onto $\mathbf{a}$ is $\text{comp}_\mathbf{a}\,\mathbf{b} = \dfrac{\mathbf{a} \cdot \mathbf{b}}{|\mathbf{a}|} = \dfrac{-5 \cdot 4 + 12 \cdot 6}{13} = 4$ and the

vector projection of $\mathbf{b}$ onto $\mathbf{a}$ is $\text{proj}_\mathbf{a}\,\mathbf{b} = \left(\dfrac{\mathbf{a} \cdot \mathbf{b}}{|\mathbf{a}|}\right)\dfrac{\mathbf{a}}{|\mathbf{a}|} = 4 \cdot \frac{1}{13}\langle -5, 12 \rangle = \left\langle -\frac{20}{13}, -\frac{48}{13} \right\rangle$.

41. $|\mathbf{a}| = \sqrt{9 + 36 + 4} = 7$ so the scalar projection of $\mathbf{b}$ onto $\mathbf{a}$ is $\text{comp}_\mathbf{a}\mathbf{b} = \dfrac{\mathbf{a} \cdot \mathbf{b}}{|\mathbf{a}|} = \frac{1}{7}(3 + 12 - 6) = \frac{9}{7}$. The vector

projection of $\mathbf{b}$ onto $\mathbf{a}$ is $\text{proj}_\mathbf{a}\mathbf{b} = \dfrac{9}{7}\dfrac{\mathbf{a}}{|\mathbf{a}|} = \frac{9}{7} \cdot \frac{1}{7}\langle 3, 6, -2 \rangle = \frac{9}{49}\langle 3, 6, -2 \rangle = \left\langle \frac{27}{49}, \frac{54}{49}, -\frac{18}{49} \right\rangle$.

© 2012 Cengage Learning. All Rights Reserved. May not be scanned, copied, or duplicated, or posted to a publicly accessible website, in whole or in part.

43. $|\mathbf{a}| = \sqrt{4 + 1 + 16} = \sqrt{21}$ so the scalar projection of $\mathbf{b}$ onto $\mathbf{a}$ is $\text{comp}_\mathbf{a}\,\mathbf{b} = \dfrac{\mathbf{a} \cdot \mathbf{b}}{|\mathbf{a}|} = \dfrac{0 - 1 + 2}{\sqrt{21}} = \dfrac{1}{\sqrt{21}}$ while the vector

projection of $\mathbf{b}$ onto $\mathbf{a}$ is $\text{proj}_\mathbf{a}\,\mathbf{b} = \dfrac{1}{\sqrt{21}}\dfrac{\mathbf{a}}{|\mathbf{a}|} = \dfrac{1}{\sqrt{21}} \cdot \dfrac{2\,\mathbf{i} - \mathbf{j} + 4\,\mathbf{k}}{\sqrt{21}} = \tfrac{1}{21}(2\,\mathbf{i} - \mathbf{j} + 4\,\mathbf{k}) = \tfrac{2}{21}\,\mathbf{i} - \tfrac{1}{21}\,\mathbf{j} + \tfrac{4}{21}\,\mathbf{k}$.

45. $(\text{orth}_\mathbf{a}\,\mathbf{b}) \cdot \mathbf{a} = (\mathbf{b} - \text{proj}_\mathbf{a}\,\mathbf{b}) \cdot \mathbf{a} = \mathbf{b} \cdot \mathbf{a} - (\text{proj}_\mathbf{a}\,\mathbf{b}) \cdot \mathbf{a} = \mathbf{b} \cdot \mathbf{a} - \dfrac{\mathbf{a} \cdot \mathbf{b}}{|\mathbf{a}|^2}\,\mathbf{a} \cdot \mathbf{a} = \mathbf{b} \cdot \mathbf{a} - \dfrac{\mathbf{a} \cdot \mathbf{b}}{|\mathbf{a}|^2}\,|\mathbf{a}|^2 = \mathbf{b} \cdot \mathbf{a} - \mathbf{a} \cdot \mathbf{b} = 0$.

So they are orthogonal by (7).

47. $\text{comp}_\mathbf{a}\,\mathbf{b} = \dfrac{\mathbf{a} \cdot \mathbf{b}}{|\mathbf{a}|} = 2 \quad \Leftrightarrow \quad \mathbf{a} \cdot \mathbf{b} = 2\,|\mathbf{a}| = 2\sqrt{10}$. If $\mathbf{b} = \langle b_1, b_2, b_3 \rangle$, then we need $3b_1 + 0b_2 - 1b_3 = 2\sqrt{10}$.

One possible solution is obtained by taking $b_1 = 0$, $b_2 = 0$, $b_3 = -2\sqrt{10}$. In general, $\mathbf{b} = \langle s, t, 3s - 2\sqrt{10} \rangle$, $s, t \in \mathbb{R}$.

49. The displacement vector is $\mathbf{D} = (6 - 0)\,\mathbf{i} + (12 - 10)\,\mathbf{j} + (20 - 8)\,\mathbf{k} = 6\,\mathbf{i} + 2\,\mathbf{j} + 12\,\mathbf{k}$ so, by Equation 12, the work done is

$W = \mathbf{F} \cdot \mathbf{D} = (8\,\mathbf{i} - 6\,\mathbf{j} + 9\,\mathbf{k}) \cdot (6\,\mathbf{i} + 2\,\mathbf{j} + 12\,\mathbf{k}) = 48 - 12 + 108 = 144$ joules.

51. Here $|\mathbf{D}| = 80$ ft, $|\mathbf{F}| = 30$ lb, and $\theta = 40°$. Thus

$W = \mathbf{F} \cdot \mathbf{D} = |\mathbf{F}|\,|\mathbf{D}| \cos\theta = (30)(80)\cos 40° = 2400 \cos 40° \approx 1839$ ft-lb.

53. First note that $\mathbf{n} = \langle a, b \rangle$ is perpendicular to the line, because if $Q_1 = (a_1, b_1)$ and $Q_2 = (a_2, b_2)$ lie on the line, then

$\mathbf{n} \cdot \overrightarrow{Q_1 Q_2} = aa_2 - aa_1 + bb_2 - bb_1 = 0$, since $aa_2 + bb_2 = -c = aa_1 + bb_1$ from the equation of the line.

Let $P_2 = (x_2, y_2)$ lie on the line. Then the distance from P_1 to the line is the absolute value of the scalar projection

of $\overrightarrow{P_1 P_2}$ onto $\mathbf{n}$. $\text{comp}_\mathbf{n}\left(\overrightarrow{P_1 P_2}\right) = \dfrac{|\mathbf{n} \cdot \langle x_2 - x_1, y_2 - y_1 \rangle|}{|\mathbf{n}|} = \dfrac{|ax_2 - ax_1 + by_2 - by_1|}{\sqrt{a^2 + b^2}} = \dfrac{|ax_1 + by_1 + c|}{\sqrt{a^2 + b^2}}$

since $ax_2 + by_2 = -c$. The required distance is $\dfrac{|(3)(-2) + (-4)(3) + 5|}{\sqrt{3^2 + (-4)^2}} = \dfrac{13}{5}$.

55. For convenience, consider the unit cube positioned so that its back left corner is at the origin, and its edges lie along the

coordinate axes. The diagonal of the cube that begins at the origin and ends at $(1, 1, 1)$ has vector representation $\langle 1, 1, 1 \rangle$.

The angle θ between this vector and the vector of the edge which also begins at the origin and runs along the x-axis [that is,

$\langle 1, 0, 0 \rangle$] is given by $\cos\theta = \dfrac{\langle 1, 1, 1 \rangle \cdot \langle 1, 0, 0 \rangle}{|\langle 1, 1, 1 \rangle|\,|\langle 1, 0, 0 \rangle|} = \dfrac{1}{\sqrt{3}} \quad \Rightarrow \quad \theta = \cos^{-1}\left(\tfrac{1}{\sqrt{3}}\right) \approx 55°$.

57. Consider the H—C—H combination consisting of the sole carbon atom and the two hydrogen atoms that are at $(1, 0, 0)$ and

$(0, 1, 0)$ (or any H—C—H combination, for that matter). Vector representations of the line segments emanating from the

carbon atom and extending to these two hydrogen atoms are $\left\langle 1 - \tfrac{1}{2}, 0 - \tfrac{1}{2}, 0 - \tfrac{1}{2} \right\rangle = \left\langle \tfrac{1}{2}, -\tfrac{1}{2}, -\tfrac{1}{2} \right\rangle$ and

$\left\langle 0 - \tfrac{1}{2}, 1 - \tfrac{1}{2}, 0 - \tfrac{1}{2} \right\rangle = \left\langle -\tfrac{1}{2}, \tfrac{1}{2}, -\tfrac{1}{2} \right\rangle$. The bond angle, θ, is therefore given by

$\cos\theta = \dfrac{\left\langle \tfrac{1}{2}, -\tfrac{1}{2}, -\tfrac{1}{2} \right\rangle \cdot \left\langle -\tfrac{1}{2}, \tfrac{1}{2}, -\tfrac{1}{2} \right\rangle}{\left|\left\langle \tfrac{1}{2}, -\tfrac{1}{2}, -\tfrac{1}{2} \right\rangle\right|\left|\left\langle -\tfrac{1}{2}, \tfrac{1}{2}, -\tfrac{1}{2} \right\rangle\right|} = \dfrac{-\tfrac{1}{4} - \tfrac{1}{4} + \tfrac{1}{4}}{\sqrt{\tfrac{3}{4}}\sqrt{\tfrac{3}{4}}} = -\dfrac{1}{3} \quad \Rightarrow \quad \theta = \cos^{-1}\left(-\tfrac{1}{3}\right) \approx 109.5°$.

© 2012 Cengage Learning. All Rights Reserved. May not be scanned, copied, or duplicated, or posted to a publicly accessible website, in whole or in part.

59. Let $\mathbf{a} = \langle a_1, a_2, a_3 \rangle$ and $= \langle b_1, b_2, b_3 \rangle$.

Property 2: $\mathbf{a} \cdot \mathbf{b} = \langle a_1, a_2, a_3 \rangle \cdot \langle b_1, b_2, b_3 \rangle = a_1 b_1 + a_2 b_2 + a_3 b_3$

$$= b_1 a_1 + b_2 a_2 + b_3 a_3 = \langle b_1, b_2, b_3 \rangle \cdot \langle a_1, a_2, a_3 \rangle = \mathbf{b} \cdot \mathbf{a}$$

Property 4: $(c\,\mathbf{a}) \cdot \mathbf{b} = \langle ca_1, ca_2, ca_3 \rangle \cdot \langle b_1, b_2, b_3 \rangle = (ca_1)b_1 + (ca_2)b_2 + (ca_3)b_3$

$$= c\,(a_1 b_1 + a_2 b_2 + a_3 b_3) = c\,(\mathbf{a} \cdot \mathbf{b}) = a_1(cb_1) + a_2(cb_2) + a_3(cb_3)$$

$$= \langle a_1, a_2, a_3 \rangle \cdot \langle cb_1, cb_2, cb_3 \rangle = \mathbf{a} \cdot (c\,\mathbf{b})$$

Property 5: $\mathbf{0} \cdot \mathbf{a} = \langle 0, 0, 0 \rangle \cdot \langle a_1, a_2, a_3 \rangle = (0)(a_1) + (0)(a_2) + (0)(a_3) = 0$

61. $|\mathbf{a} \cdot \mathbf{b}| = \big| \, |\mathbf{a}|\,|\mathbf{b}| \cos\theta \, \big| = |\mathbf{a}|\,|\mathbf{b}|\,|\cos\theta|$. Since $|\cos\theta| \le 1$, $|\mathbf{a} \cdot \mathbf{b}| = |\mathbf{a}|\,|\mathbf{b}|\,|\cos\theta| \le |\mathbf{a}|\,|\mathbf{b}|$.

Note: We have equality in the case of $\cos\theta = \pm 1$, so $\theta = 0$ or $\theta = \pi$, thus equality when $\mathbf{a}$ and $\mathbf{b}$ are parallel.

63. (a)

The Parallelogram Law states that the sum of the squares of the lengths of the diagonals of a parallelogram equals the sum of the squares of its (four) sides.

(b) $|\mathbf{a} + \mathbf{b}|^2 = (\mathbf{a} + \mathbf{b}) \cdot (\mathbf{a} + \mathbf{b}) = |\mathbf{a}|^2 + 2(\mathbf{a} \cdot \mathbf{b}) + |\mathbf{b}|^2$ and $|\mathbf{a} - \mathbf{b}|^2 = (\mathbf{a} - \mathbf{b}) \cdot (\mathbf{a} - \mathbf{b}) = |\mathbf{a}|^2 - 2(\mathbf{a} \cdot \mathbf{b}) + |\mathbf{b}|^2$.

Adding these two equations gives $|\mathbf{a} + \mathbf{b}|^2 + |\mathbf{a} - \mathbf{b}|^2 = 2\,|\mathbf{a}|^2 + 2\,|\mathbf{b}|^2$.

12.4 The Cross Product

1. $\mathbf{a} \times \mathbf{b} = \begin{vmatrix} \mathbf{i} & \mathbf{j} & \mathbf{k} \\ 6 & 0 & -2 \\ 0 & 8 & 0 \end{vmatrix} = \begin{vmatrix} 0 & -2 \\ 8 & 0 \end{vmatrix} \mathbf{i} - \begin{vmatrix} 6 & -2 \\ 0 & 0 \end{vmatrix} \mathbf{j} + \begin{vmatrix} 6 & 0 \\ 0 & 8 \end{vmatrix} \mathbf{k}$

$$= [0 - (-16)]\,\mathbf{i} - (0 - 0)\,\mathbf{j} + (48 - 0)\,\mathbf{k} = 16\,\mathbf{i} + 48\,\mathbf{k}$$

Now $(\mathbf{a} \times \mathbf{b}) \cdot \mathbf{a} = \langle 16, 0, 48 \rangle \cdot \langle 6, 0, -2 \rangle = 96 + 0 - 96 = 0$ and $(\mathbf{a} \times \mathbf{b}) \cdot \mathbf{b} = \langle 16, 0, 48 \rangle \cdot \langle 0, 8, 0 \rangle = 0 + 0 + 0 = 0$, so $\mathbf{a} \times \mathbf{b}$ is orthogonal to both $\mathbf{a}$ and $\mathbf{b}$.

3. $\mathbf{a} \times \mathbf{b} = \begin{vmatrix} \mathbf{i} & \mathbf{j} & \mathbf{k} \\ 1 & 3 & -2 \\ -1 & 0 & 5 \end{vmatrix} = \begin{vmatrix} 3 & -2 \\ 0 & 5 \end{vmatrix} \mathbf{i} - \begin{vmatrix} 1 & -2 \\ -1 & 5 \end{vmatrix} \mathbf{j} + \begin{vmatrix} 1 & 3 \\ -1 & 0 \end{vmatrix} \mathbf{k}$

$$= (15 - 0)\,\mathbf{i} - (5 - 2)\,\mathbf{j} + [0 - (-3)]\,\mathbf{k} = 15\,\mathbf{i} - 3\,\mathbf{j} + 3\,\mathbf{k}$$

Since $(\mathbf{a} \times \mathbf{b}) \cdot \mathbf{a} = (15\,\mathbf{i} - 3\,\mathbf{j} + 3\,\mathbf{k}) \cdot (\mathbf{i} + 3\,\mathbf{j} - 2\,\mathbf{k}) = 15 - 9 - 6 = 0$, $\mathbf{a} \times \mathbf{b}$ is orthogonal to $\mathbf{a}$.

Since $(\mathbf{a} \times \mathbf{b}) \cdot \mathbf{b} = (15\,\mathbf{i} - 3\,\mathbf{j} + 3\,\mathbf{k}) \cdot (-\mathbf{i} + 5\,\mathbf{k}) = -15 + 0 + 15 = 0$, $\mathbf{a} \times \mathbf{b}$ is orthogonal to $\mathbf{b}$.

© 2012 Cengage Learning. All Rights Reserved. May not be scanned, copied, or duplicated, or posted to a publicly accessible website, in whole or in part.

5. $\mathbf{a} \times \mathbf{b} = \begin{vmatrix} \mathbf{i} & \mathbf{j} & \mathbf{k} \\ 1 & -1 & -1 \\ \frac{1}{2} & 1 & \frac{1}{2} \end{vmatrix} = \begin{vmatrix} -1 & -1 \\ 1 & \frac{1}{2} \end{vmatrix} \mathbf{i} - \begin{vmatrix} 1 & -1 \\ \frac{1}{2} & \frac{1}{2} \end{vmatrix} \mathbf{j} + \begin{vmatrix} 1 & -1 \\ \frac{1}{2} & 1 \end{vmatrix} \mathbf{k}$

$= \left[-\frac{1}{2} - (-1) \right] \mathbf{i} - \left[\frac{1}{2} - \left(-\frac{1}{2} \right) \right] \mathbf{j} + \left[1 - \left(-\frac{1}{2} \right) \right] \mathbf{k} = \frac{1}{2} \mathbf{i} - \mathbf{j} + \frac{3}{2} \mathbf{k}$

Now $(\mathbf{a} \times \mathbf{b}) \cdot \mathbf{a} = \left(\frac{1}{2} \mathbf{i} - \mathbf{j} + \frac{3}{2} \mathbf{k} \right) \cdot (\mathbf{i} - \mathbf{j} - \mathbf{k}) = \frac{1}{2} + 1 - \frac{3}{2} = 0$ and

$(\mathbf{a} \times \mathbf{b}) \cdot \mathbf{b} = \left(\frac{1}{2} \mathbf{i} - \mathbf{j} + \frac{3}{2} \mathbf{k} \right) \cdot \left(\frac{1}{2} \mathbf{i} + \mathbf{j} + \frac{1}{2} \mathbf{k} \right) = \frac{1}{4} - 1 + \frac{3}{4} = 0$, so $\mathbf{a} \times \mathbf{b}$ is orthogonal to both $\mathbf{a}$ and $\mathbf{b}$.

7. $\mathbf{a} \times \mathbf{b} = \begin{vmatrix} \mathbf{i} & \mathbf{j} & \mathbf{k} \\ t & 1 & 1/t \\ t^2 & t^2 & 1 \end{vmatrix} = \begin{vmatrix} 1 & 1/t \\ t^2 & 1 \end{vmatrix} \mathbf{i} - \begin{vmatrix} t & 1/t \\ t^2 & 1 \end{vmatrix} \mathbf{j} + \begin{vmatrix} t & 1 \\ t^2 & t^2 \end{vmatrix} \mathbf{k}$

$= (1 - t) \mathbf{i} - (t - t) \mathbf{j} + (t^3 - t^2) \mathbf{k} = (1 - t) \mathbf{i} + (t^3 - t^2) \mathbf{k}$

Since $(\mathbf{a} \times \mathbf{b}) \cdot \mathbf{a} = \langle 1 - t, 0, t^3 - t^2 \rangle \cdot \langle t, 1, 1/t \rangle = t - t^2 + 0 + t^2 - t = 0$, $\mathbf{a} \times \mathbf{b}$ is orthogonal to $\mathbf{a}$.

Since $(\mathbf{a} \times \mathbf{b}) \cdot \mathbf{b} = \langle 1 - t, 0, t^3 - t^2 \rangle \cdot \langle t^2, t^2, 1 \rangle = t^2 - t^3 + 0 + t^3 - t^2 = 0$, $\mathbf{a} \times \mathbf{b}$ is orthogonal to $\mathbf{b}$.

9. According to the discussion preceding Theorem 11, $\mathbf{i} \times \mathbf{j} = \mathbf{k}$, so $(\mathbf{i} \times \mathbf{j}) \times \mathbf{k} = \mathbf{k} \times \mathbf{k} = \mathbf{0}$ [by Example 2].

11. $(\mathbf{j} - \mathbf{k}) \times (\mathbf{k} - \mathbf{i}) = (\mathbf{j} - \mathbf{k}) \times \mathbf{k} + (\mathbf{j} - \mathbf{k}) \times (-\mathbf{i})$ by Property 3 of Theorem 11

$= \mathbf{j} \times \mathbf{k} + (-\mathbf{k}) \times \mathbf{k} + \mathbf{j} \times (-\mathbf{i}) + (-\mathbf{k}) \times (-\mathbf{i})$ by Property 4 of Theorem 11

$= (\mathbf{j} \times \mathbf{k}) + (-1)(\mathbf{k} \times \mathbf{k}) + (-1)(\mathbf{j} \times \mathbf{i}) + (-1)^2 (\mathbf{k} \times \mathbf{i})$ by Property 2 of Theorem 11

$= \mathbf{i} + (-1)\mathbf{0} + (-1)(-\mathbf{k}) + \mathbf{j} = \mathbf{i} + \mathbf{j} + \mathbf{k}$ by Example 2 and
the discussion preceeding Theorem 11

13. (a) Since $\mathbf{b} \times \mathbf{c}$ is a vector, the dot product $\mathbf{a} \cdot (\mathbf{b} \times \mathbf{c})$ is meaningful and is a scalar.

(b) $\mathbf{b} \cdot \mathbf{c}$ is a scalar, so $\mathbf{a} \times (\mathbf{b} \cdot \mathbf{c})$ is meaningless, as the cross product is defined only for two *vectors*.

(c) Since $\mathbf{b} \times \mathbf{c}$ is a vector, the cross product $\mathbf{a} \times (\mathbf{b} \times \mathbf{c})$ is meaningful and results in another vector.

(d) $\mathbf{b} \cdot \mathbf{c}$ is a scalar, so the dot product $\mathbf{a} \cdot (\mathbf{b} \cdot \mathbf{c})$ is meaningless, as the dot product is defined only for two vectors.

(e) Since $(\mathbf{a} \cdot \mathbf{b})$ and $(\mathbf{c} \cdot \mathbf{d})$ are both scalars, the cross product $(\mathbf{a} \cdot \mathbf{b}) \times (\mathbf{c} \cdot \mathbf{d})$ is meaningless.

(f) $\mathbf{a} \times \mathbf{b}$ and $\mathbf{c} \times \mathbf{d}$ are both vectors, so the dot product $(\mathbf{a} \times \mathbf{b}) \cdot (\mathbf{c} \times \mathbf{d})$ is meaningful and is a scalar.

15. If we sketch $\mathbf{u}$ and $\mathbf{v}$ starting from the same initial point, we see that the

angle between them is $60°$. Using Theorem 9, we have

$|\mathbf{u} \times \mathbf{v}| = |\mathbf{u}| \, |\mathbf{v}| \sin \theta = (12)(16) \sin 60° = 192 \cdot \dfrac{\sqrt{3}}{2} = 96 \sqrt{3}.$

By the right-hand rule, $\mathbf{u} \times \mathbf{v}$ is directed into the page.

© 2012 Cengage Learning. All Rights Reserved. May not be scanned, copied, or duplicated, or posted to a publicly accessible website, in whole or in part.

17. $\mathbf{a} \times \mathbf{b} = \begin{vmatrix} \mathbf{i} & \mathbf{j} & \mathbf{k} \\ 2 & -1 & 3 \\ 4 & 2 & 1 \end{vmatrix} = \begin{vmatrix} -1 & 3 \\ 2 & 1 \end{vmatrix} \mathbf{i} - \begin{vmatrix} 2 & 3 \\ 4 & 1 \end{vmatrix} \mathbf{j} + \begin{vmatrix} 2 & -1 \\ 4 & 2 \end{vmatrix} \mathbf{k} = (-1-6)\mathbf{i} - (2-12)\mathbf{j} + [4-(-4)]\mathbf{k} = -7\mathbf{i} + 10\mathbf{j} + 8\mathbf{k}$

$\mathbf{b} \times \mathbf{a} = \begin{vmatrix} \mathbf{i} & \mathbf{j} & \mathbf{k} \\ 4 & 2 & 1 \\ 2 & -1 & 3 \end{vmatrix} = \begin{vmatrix} 2 & 1 \\ -1 & 3 \end{vmatrix} \mathbf{i} - \begin{vmatrix} 4 & 1 \\ 2 & 3 \end{vmatrix} \mathbf{j} + \begin{vmatrix} 4 & 2 \\ 2 & -1 \end{vmatrix} \mathbf{k} = [6-(-1)]\mathbf{i} - (12-2)\mathbf{j} + (-4-4)\mathbf{k} = 7\mathbf{i} - 10\mathbf{j} - 8\mathbf{k}$

Notice $\mathbf{a} \times \mathbf{b} = -\mathbf{b} \times \mathbf{a}$ here, as we know is always true by Property 1 of Theorem 11.

19. By Theorem 8, the cross product of two vectors is orthogonal to both vectors. So we calculate

$\langle 3, 2, 1 \rangle \times \langle -1, 1, 0 \rangle = \begin{vmatrix} \mathbf{i} & \mathbf{j} & \mathbf{k} \\ 3 & 2 & 1 \\ -1 & 1 & 0 \end{vmatrix} = \begin{vmatrix} 2 & 1 \\ 1 & 0 \end{vmatrix} \mathbf{i} - \begin{vmatrix} 3 & 1 \\ -1 & 0 \end{vmatrix} \mathbf{j} + \begin{vmatrix} 3 & 2 \\ -1 & 1 \end{vmatrix} \mathbf{k} = -\mathbf{i} - \mathbf{j} + 5\mathbf{k}.$

So two unit vectors orthogonal to both are $\pm \dfrac{\langle -1, -1, 5 \rangle}{\sqrt{1+1+25}} = \pm \dfrac{\langle -1, -1, 5 \rangle}{3\sqrt{3}}$, that is, $\left\langle -\dfrac{1}{3\sqrt{3}}, -\dfrac{1}{3\sqrt{3}}, \dfrac{5}{3\sqrt{3}} \right\rangle$

and $\left\langle \dfrac{1}{3\sqrt{3}}, \dfrac{1}{3\sqrt{3}}, -\dfrac{5}{3\sqrt{3}} \right\rangle$.

21. Let $\mathbf{a} = \langle a_1, a_2, a_3 \rangle$. Then

$\mathbf{0} \times \mathbf{a} = \begin{vmatrix} \mathbf{i} & \mathbf{j} & \mathbf{k} \\ 0 & 0 & 0 \\ a_1 & a_2 & a_3 \end{vmatrix} = \begin{vmatrix} 0 & 0 \\ a_2 & a_3 \end{vmatrix} \mathbf{i} - \begin{vmatrix} 0 & 0 \\ a_1 & a_3 \end{vmatrix} \mathbf{j} + \begin{vmatrix} 0 & 0 \\ a_1 & a_2 \end{vmatrix} \mathbf{k} = \mathbf{0},$

$\mathbf{a} \times \mathbf{0} = \begin{vmatrix} \mathbf{i} & \mathbf{j} & \mathbf{k} \\ a_1 & a_2 & a_3 \\ 0 & 0 & 0 \end{vmatrix} = \begin{vmatrix} a_2 & a_3 \\ 0 & 0 \end{vmatrix} \mathbf{i} - \begin{vmatrix} a_1 & a_3 \\ 0 & 0 \end{vmatrix} \mathbf{j} + \begin{vmatrix} a_1 & a_2 \\ 0 & 0 \end{vmatrix} \mathbf{k} = \mathbf{0}.$

23. $\mathbf{a} \times \mathbf{b} = \langle a_2 b_3 - a_3 b_2, a_3 b_1 - a_1 b_3, a_1 b_2 - a_2 b_1 \rangle$

$= \langle (-1)(b_2 a_3 - b_3 a_2), (-1)(b_3 a_1 - b_1 a_3), (-1)(b_1 a_2 - b_2 a_1) \rangle$

$= -\langle b_2 a_3 - b_3 a_2, b_3 a_1 - b_1 a_3, b_1 a_2 - b_2 a_1 \rangle = -\mathbf{b} \times \mathbf{a}$

25. $\mathbf{a} \times (\mathbf{b} + \mathbf{c}) = \mathbf{a} \times \langle b_1 + c_1, b_2 + c_2, b_3 + c_3 \rangle$

$= \langle a_2(b_3 + c_3) - a_3(b_2 + c_2), a_3(b_1 + c_1) - a_1(b_3 + c_3), a_1(b_2 + c_2) - a_2(b_1 + c_1) \rangle$

$= \langle a_2 b_3 + a_2 c_3 - a_3 b_2 - a_3 c_2, a_3 b_1 + a_3 c_1 - a_1 b_3 - a_1 c_3, a_1 b_2 + a_1 c_2 - a_2 b_1 - a_2 c_1 \rangle$

$= \langle (a_2 b_3 - a_3 b_2) + (a_2 c_3 - a_3 c_2), (a_3 b_1 - a_1 b_3) + (a_3 c_1 - a_1 c_3), (a_1 b_2 - a_2 b_1) + (a_1 c_2 - a_2 c_1) \rangle$

$= \langle a_2 b_3 - a_3 b_2, a_3 b_1 - a_1 b_3, a_1 b_2 - a_2 b_1 \rangle + \langle a_2 c_3 - a_3 c_2, a_3 c_1 - a_1 c_3, a_1 c_2 - a_2 c_1 \rangle$

$= (\mathbf{a} \times \mathbf{b}) + (\mathbf{a} \times \mathbf{c})$

© 2012 Cengage Learning. All Rights Reserved. May not be scanned, copied, or duplicated, or posted to a publicly accessible website, in whole or in part.

27. By plotting the vertices, we can see that the parallelogram is determined by the

vectors $\overrightarrow{AB} = \langle 2, 3 \rangle$ and $\overrightarrow{AD} = \langle 4, -2 \rangle$. We know that the area of the parallelogram

determined by two vectors is equal to the length of the cross product of these vectors.

In order to compute the cross product, we consider the vector $\overrightarrow{AB}$ as the three-

dimensional vector $\langle 2, 3, 0 \rangle$ (and similarly for $\overrightarrow{AD}$), and then the area of

parallelogram $ABCD$ is

$$\left| \overrightarrow{AB} \times \overrightarrow{AD} \right| = \begin{vmatrix} \mathbf{i} & \mathbf{j} & \mathbf{k} \\ 2 & 3 & 0 \\ 4 & -2 & 0 \end{vmatrix} = |(0)\,\mathbf{i} - (0)\,\mathbf{j} + (-4 - 12)\,\mathbf{k}| = |-16\,\mathbf{k}| = 16$$

29. (a) Because the plane through P, Q, and R contains the vectors $\overrightarrow{PQ}$ and $\overrightarrow{PR}$, a vector orthogonal to both of these vectors

(such as their cross product) is also orthogonal to the plane. Here $\overrightarrow{PQ} = \langle -3, 1, 2 \rangle$ and $\overrightarrow{PR} = \langle 3, 2, 4 \rangle$, so

$$\overrightarrow{PQ} \times \overrightarrow{PR} = \langle (1)(4) - (2)(2), (2)(3) - (-3)(4), (-3)(2) - (1)(3) \rangle = \langle 0, 18, -9 \rangle$$

Therefore, $\langle 0, 18, -9 \rangle$ (or any nonzero scalar multiple thereof, such as $\langle 0, 2, -1 \rangle$) is orthogonal to the plane through P, Q,

and R.

(b) Note that the area of the triangle determined by P, Q, and R is equal to half of the area of the

parallelogram determined by the three points. From part (a), the area of the parallelogram is

$\left| \overrightarrow{PQ} \times \overrightarrow{PR} \right| = |\langle 0, 18, -9 \rangle| = \sqrt{0 + 324 + 81} = \sqrt{405} = 9\sqrt{5}$, so the area of the triangle is $\frac{1}{2} \cdot 9\sqrt{5} = \frac{9}{2}\sqrt{5}$.

31. (a) $\overrightarrow{PQ} = \langle 4, 3, -2 \rangle$ and $\overrightarrow{PR} = \langle 5, 5, 1 \rangle$, so a vector orthogonal to the plane through P, Q, and R is

$$\overrightarrow{PQ} \times \overrightarrow{PR} = \langle (3)(1) - (-2)(5), (-2)(5) - (4)(1), (4)(5) - (3)(5) \rangle = \langle 13, -14, 5 \rangle \text{ [or any scalar mutiple thereof]}.$$

(b) The area of the parallelogram determined by $\overrightarrow{PQ}$ and $\overrightarrow{PR}$ is

$\left| \overrightarrow{PQ} \times \overrightarrow{PR} \right| = |\langle 13, -14, 5 \rangle| = \sqrt{13^2 + (-14)^2 + 5^2} = \sqrt{390}$, so the area of triangle PQR is $\frac{1}{2}\sqrt{390}$.

33. By Equation 14, the volume of the parallelepiped determined by $\mathbf{a}$, $\mathbf{b}$, and $\mathbf{c}$ is the magnitude of their scalar triple product,

which is $\mathbf{a} \cdot (\mathbf{b} \times \mathbf{c}) = \begin{vmatrix} 1 & 2 & 3 \\ -1 & 1 & 2 \\ 2 & 1 & 4 \end{vmatrix} = 1 \begin{vmatrix} 1 & 2 \\ 1 & 4 \end{vmatrix} - 2 \begin{vmatrix} -1 & 2 \\ 2 & 4 \end{vmatrix} + 3 \begin{vmatrix} -1 & 1 \\ 2 & 1 \end{vmatrix} = 1(4 - 2) - 2(-4 - 4) + 3(-1 - 2) = 9.$

Thus the volume of the parallelepiped is 9 cubic units.

35. $\mathbf{a} = \overrightarrow{PQ} = \langle 4, 2, 2 \rangle$, $\mathbf{b} = \overrightarrow{PR} = \langle 3, 3, -1 \rangle$, and $\mathbf{c} = \overrightarrow{PS} = \langle 5, 5, 1 \rangle$.

$$\mathbf{a} \cdot (\mathbf{b} \times \mathbf{c}) = \begin{vmatrix} 4 & 2 & 2 \\ 3 & 3 & -1 \\ 5 & 5 & 1 \end{vmatrix} = 4 \begin{vmatrix} 3 & -1 \\ 5 & 1 \end{vmatrix} - 2 \begin{vmatrix} 3 & -1 \\ 5 & 1 \end{vmatrix} + 2 \begin{vmatrix} 3 & 3 \\ 5 & 5 \end{vmatrix} = 32 - 16 + 0 = 16,$$

so the volume of the parallelepiped is 16 cubic units.

© 2012 Cengage Learning. All Rights Reserved. May not be scanned, copied, or duplicated, or posted to a publicly accessible website, in whole or in part.

37. $\mathbf{u} \cdot (\mathbf{v} \times \mathbf{w}) = \begin{vmatrix} 1 & 5 & -2 \\ 3 & -1 & 0 \\ 5 & 9 & -4 \end{vmatrix} = 1 \begin{vmatrix} -1 & 0 \\ 9 & -4 \end{vmatrix} - 5 \begin{vmatrix} 3 & 0 \\ 5 & -4 \end{vmatrix} + (-2) \begin{vmatrix} 3 & -1 \\ 5 & 9 \end{vmatrix} = 4 + 60 - 64 = 0$, which says that the volume

of the parallelepiped determined by $\mathbf{u}$, $\mathbf{v}$ and $\mathbf{w}$ is 0, and thus these three vectors are coplanar.

39. The magnitude of the torque is $|\boldsymbol{\tau}| = |\mathbf{r} \times \mathbf{F}| = |\mathbf{r}| \, |\mathbf{F}| \sin \theta = (0.18 \text{ m})(60 \text{ N}) \sin(70 + 10)° = 10.8 \sin 80° \approx 10.6 \text{ N·m}$.

41. Using the notation of the text, $\mathbf{r} = \langle 0, 0.3, 0 \rangle$ and $\mathbf{F}$ has direction $\langle 0, 3, -4 \rangle$. The angle θ between them can be determined by

$\cos \theta = \dfrac{\langle 0, 0.3, 0 \rangle \cdot \langle 0, 3, -4 \rangle}{|\langle 0, 0.3, 0 \rangle| \, |\langle 0, 3, -4 \rangle|} \;\Rightarrow\; \cos \theta = \dfrac{0.9}{(0.3)(5)} \;\Rightarrow\; \cos \theta = 0.6 \;\Rightarrow\; \theta \approx 53.1°$. Then $|\boldsymbol{\tau}| = |\mathbf{r}| \, |\mathbf{F}| \sin \theta \;\Rightarrow\;$

$100 = 0.3 \, |\mathbf{F}| \sin 53.1° \;\Rightarrow\; |\mathbf{F}| \approx 417 \text{ N}$.

43. From Theorem 9 we have $|\mathbf{a} \times \mathbf{b}| = |\mathbf{a}| \, |\mathbf{b}| \sin \theta$, where θ is the angle between $\mathbf{a}$ and $\mathbf{b}$, and from Theorem 12.3.3 we have

$\mathbf{a} \cdot \mathbf{b} = |\mathbf{a}| \, |\mathbf{b}| \cos \theta \;\Rightarrow\; |\mathbf{a}| \, |\mathbf{b}| = \dfrac{\mathbf{a} \cdot \mathbf{b}}{\cos \theta}$. Substituting the second equation into the first gives $|\mathbf{a} \times \mathbf{b}| = \dfrac{\mathbf{a} \cdot \mathbf{b}}{\cos \theta} \sin \theta$, so

$\dfrac{|\mathbf{a} \times \mathbf{b}|}{\mathbf{a} \cdot \mathbf{b}} = \tan \theta$. Here $|\mathbf{a} \times \mathbf{b}| = |\langle 1, 2, 2 \rangle| = \sqrt{1 + 4 + 4} = 3$, so $\tan \theta = \dfrac{|\mathbf{a} \times \mathbf{b}|}{\mathbf{a} \cdot \mathbf{b}} = \dfrac{3}{\sqrt{3}} = \sqrt{3} \;\Rightarrow\; \theta = 60°$.

45. (a)

The distance between a point and a line is the length of the perpendicular

from the point to the line, here $\left| \overrightarrow{PS} \right| = d$. But referring to triangle PQS,

$d = \left| \overrightarrow{PS} \right| = \left| \overrightarrow{QP} \right| \sin \theta = |\mathbf{b}| \sin \theta$. But θ is the angle between $\overrightarrow{QP} = \mathbf{b}$

and $\overrightarrow{QR} = \mathbf{a}$. Thus by Theorem 9, $\sin \theta = \dfrac{|\mathbf{a} \times \mathbf{b}|}{|\mathbf{a}| \, |\mathbf{b}|}$

and so $d = |\mathbf{b}| \sin \theta = \dfrac{|\mathbf{b}| \, |\mathbf{a} \times \mathbf{b}|}{|\mathbf{a}| \, |\mathbf{b}|} = \dfrac{|\mathbf{a} \times \mathbf{b}|}{|\mathbf{a}|}$.

(b) $\mathbf{a} = \overrightarrow{QR} = \langle -1, -2, -1 \rangle$ and $\mathbf{b} = \overrightarrow{QP} = \langle 1, -5, -7 \rangle$. Then

$\mathbf{a} \times \mathbf{b} = \langle (-2)(-7) - (-1)(-5), \, (-1)(1) - (-1)(-7), \, (-1)(-5) - (-2)(1) \rangle = \langle 9, -8, 7 \rangle$.

Thus the distance is $d = \dfrac{|\mathbf{a} \times \mathbf{b}|}{|\mathbf{a}|} = \dfrac{1}{\sqrt{6}} \sqrt{81 + 64 + 49} = \sqrt{\dfrac{194}{6}} = \sqrt{\dfrac{97}{3}}$.

47. From Theorem 9 we have $|\mathbf{a} \times \mathbf{b}| = |\mathbf{a}| \, |\mathbf{b}| \sin \theta$ so

$$|\mathbf{a} \times \mathbf{b}|^2 = |\mathbf{a}|^2 \, |\mathbf{b}|^2 \sin^2 \theta = |\mathbf{a}|^2 \, |\mathbf{b}|^2 \left(1 - \cos^2 \theta\right)$$
$$= |\mathbf{a}|^2 \, |\mathbf{b}|^2 - \left(|\mathbf{a}| \, |\mathbf{b}| \cos \theta\right)^2 = |\mathbf{a}|^2 \, |\mathbf{b}|^2 - (\mathbf{a} \cdot \mathbf{b})^2$$

by Theorem 12.3.3.

© 2012 Cengage Learning. All Rights Reserved. May not be scanned, copied, or duplicated, or posted to a publicly accessible website, in whole or in part.

49. $(\mathbf{a} - \mathbf{b}) \times (\mathbf{a} + \mathbf{b}) = (\mathbf{a} - \mathbf{b}) \times \mathbf{a} + (\mathbf{a} - \mathbf{b}) \times \mathbf{b}$ by Property 3 of Theorem 11

$\qquad\qquad\qquad\qquad = \mathbf{a} \times \mathbf{a} + (-\mathbf{b}) \times \mathbf{a} + \mathbf{a} \times \mathbf{b} + (-\mathbf{b}) \times \mathbf{b}$ by Property 4 of Theorem 11

$\qquad\qquad\qquad\qquad = (\mathbf{a} \times \mathbf{a}) - (\mathbf{b} \times \mathbf{a}) + (\mathbf{a} \times \mathbf{b}) - (\mathbf{b} \times \mathbf{b})$ by Property 2 of Theorem 11 (with $c = -1$)

$\qquad\qquad\qquad\qquad = \mathbf{0} - (\mathbf{b} \times \mathbf{a}) + (\mathbf{a} \times \mathbf{b}) - \mathbf{0}$ by Example 2

$\qquad\qquad\qquad\qquad = (\mathbf{a} \times \mathbf{b}) + (\mathbf{a} \times \mathbf{b})$ by Property 1 of Theorem 11

$\qquad\qquad\qquad\qquad = 2(\mathbf{a} \times \mathbf{b})$

51. $\mathbf{a} \times (\mathbf{b} \times \mathbf{c}) + \mathbf{b} \times (\mathbf{c} \times \mathbf{a}) + \mathbf{c} \times (\mathbf{a} \times \mathbf{b})$

$\qquad\qquad = [(\mathbf{a} \cdot \mathbf{c})\mathbf{b} - (\mathbf{a} \cdot \mathbf{b})\mathbf{c}] + [(\mathbf{b} \cdot \mathbf{a})\mathbf{c} - (\mathbf{b} \cdot \mathbf{c})\mathbf{a}] + [(\mathbf{c} \cdot \mathbf{b})\mathbf{a} - (\mathbf{c} \cdot \mathbf{a})\mathbf{b}]$ by Exercise 50

$\qquad\qquad = (\mathbf{a} \cdot \mathbf{c})\mathbf{b} - (\mathbf{a} \cdot \mathbf{b})\mathbf{c} + (\mathbf{a} \cdot \mathbf{b})\mathbf{c} - (\mathbf{b} \cdot \mathbf{c})\mathbf{a} + (\mathbf{b} \cdot \mathbf{c})\mathbf{a} - (\mathbf{a} \cdot \mathbf{c})\mathbf{b} = \mathbf{0}$

53. (a) No. If $\mathbf{a} \cdot \mathbf{b} = \mathbf{a} \cdot \mathbf{c}$, then $\mathbf{a} \cdot (\mathbf{b} - \mathbf{c}) = 0$, so $\mathbf{a}$ is perpendicular to $\mathbf{b} - \mathbf{c}$, which can happen if $\mathbf{b} \neq \mathbf{c}$. For example,

let $\mathbf{a} = \langle 1, 1, 1 \rangle$, $\mathbf{b} = \langle 1, 0, 0 \rangle$ and $\mathbf{c} = \langle 0, 1, 0 \rangle$.

(b) No. If $\mathbf{a} \times \mathbf{b} = \mathbf{a} \times \mathbf{c}$ then $\mathbf{a} \times (\mathbf{b} - \mathbf{c}) = \mathbf{0}$, which implies that $\mathbf{a}$ is parallel to $\mathbf{b} - \mathbf{c}$, which of course can happen

if $\mathbf{b} \neq \mathbf{c}$.

(c) Yes. Since $\mathbf{a} \cdot \mathbf{c} = \mathbf{a} \cdot \mathbf{b}$, $\mathbf{a}$ is perpendicular to $\mathbf{b} - \mathbf{c}$, by part (a). From part (b), $\mathbf{a}$ is also parallel to $\mathbf{b} - \mathbf{c}$. Thus since

$\mathbf{a} \neq \mathbf{0}$ but is both parallel and perpendicular to $\mathbf{b} - \mathbf{c}$, we have $\mathbf{b} - \mathbf{c} = \mathbf{0}$, so $\mathbf{b} = \mathbf{c}$.

12.5   Equations of Lines and Planes

1. (a) True; each of the first two lines has a direction vector parallel to the direction vector of the third line, so these vectors are

each scalar multiples of the third direction vector. Then the first two direction vectors are also scalar multiples of each

other, so these vectors, and hence the two lines, are parallel.

(b) False; for example, the x- and y-axes are both perpendicular to the z-axis, yet the x- and y-axes are not parallel.

(c) True; each of the first two planes has a normal vector parallel to the normal vector of the third plane, so these two normal

vectors are parallel to each other and the planes are parallel.

(d) False; for example, the xy- and yz-planes are not parallel, yet they are both perpendicular to the xz-plane.

(e) False; the x- and y-axes are not parallel, yet they are both parallel to the plane $z = 1$.

(f) True; if each line is perpendicular to a plane, then the lines' direction vectors are both parallel to a normal vector for the

plane. Thus, the direction vectors are parallel to each other and the lines are parallel.

(g) False; the planes $y = 1$ and $z = 1$ are not parallel, yet they are both parallel to the x-axis.

(h) True; if each plane is perpendicular to a line, then any normal vector for each plane is parallel to a direction vector for the

line. Thus, the normal vectors are parallel to each other and the planes are parallel.

(i) True; see Figure 9 and the accompanying discussion.

© 2012 Cengage Learning. All Rights Reserved. May not be scanned, copied, or duplicated, or posted to a publicly accessible website, in whole or in part.

(j) False; they can be skew, as in Example 3.

(k) True. Consider any normal vector for the plane and any direction vector for the line. If the normal vector is perpendicular to the direction vector, the line and plane are parallel. Otherwise, the vectors meet at an angle θ, $0° \le \theta < 90°$, and the line will intersect the plane at an angle $90° - \theta$.

3. For this line, we have $\mathbf{r}_0 = 2\,\mathbf{i} + 2.4\,\mathbf{j} + 3.5\,\mathbf{k}$ and $\mathbf{v} = 3\,\mathbf{i} + 2\,\mathbf{j} - \mathbf{k}$, so a vector equation is

$\mathbf{r} = \mathbf{r}_0 + t\,\mathbf{v} = (2\,\mathbf{i} + 2.4\,\mathbf{j} + 3.5\,\mathbf{k}) + t(3\,\mathbf{i} + 2\,\mathbf{j} - \mathbf{k}) = (2 + 3t)\,\mathbf{i} + (2.4 + 2t)\,\mathbf{j} + (3.5 - t)\,\mathbf{k}$ and parametric equations are $x = 2 + 3t$, $y = 2.4 + 2t$, $z = 3.5 - t$.

5. A line perpendicular to the given plane has the same direction as a normal vector to the plane, such as $\mathbf{n} = \langle 1, 3, 1 \rangle$. So $\mathbf{r}_0 = \mathbf{i} + 6\,\mathbf{k}$, and we can take $\mathbf{v} = \mathbf{i} + 3\,\mathbf{j} + \mathbf{k}$. Then a vector equation is

$\mathbf{r} = (\mathbf{i} + 6\,\mathbf{k}) + t(\mathbf{i} + 3\,\mathbf{j} + \mathbf{k}) = (1 + t)\,\mathbf{i} + 3t\,\mathbf{j} + (6 + t)\,\mathbf{k}$, and parametric equations are $x = 1 + t$, $y = 3t$, $z = 6 + t$.

7. The vector $\mathbf{v} = \langle 2 - 0, 1 - \frac{1}{2}, -3 - 1 \rangle = \langle 2, \frac{1}{2}, -4 \rangle$ is parallel to the line. Letting $P_0 = (2, 1, -3)$, parametric equations are $x = 2 + 2t$, $y = 1 + \frac{1}{2}t$, $z = -3 - 4t$, while symmetric equations are $\dfrac{x - 2}{2} = \dfrac{y - 1}{1/2} = \dfrac{z + 3}{-4}$ or

$\dfrac{x - 2}{2} = 2y - 2 = \dfrac{z + 3}{-4}$.

9. $\mathbf{v} = \langle 3 - (-8), -2 - 1, 4 - 4 \rangle = \langle 11, -3, 0 \rangle$, and letting $P_0 = (-8, 1, 4)$, parametric equations are $x = -8 + 11t$, $y = 1 - 3t$, $z = 4 + 0t = 4$, while symmetric equations are $\dfrac{x + 8}{11} = \dfrac{y - 1}{-3}$, $z = 4$. Notice here that the direction number $c = 0$, so rather than writing $\dfrac{z - 4}{0}$ in the symmetric equation we must write the equation $z = 4$ separately.

11. The line has direction $\mathbf{v} = \langle 1, 2, 1 \rangle$. Letting $P_0 = (1, \;1, 1)$, parametric equations are $x = 1 + t$, $y = -1 + 2t$, $z = 1 + t$ and symmetric equations are $x - 1 = \dfrac{y + 1}{2} = z - 1$.

13. Direction vectors of the lines are $\mathbf{v}_1 = \langle -2 - (-4), 0 - (-6), -3 - 1 \rangle = \langle 2, 6, -4 \rangle$ and $\mathbf{v}_2 = \langle 5 - 10, 3 - 18, 14 - 4 \rangle = \langle -5, -15, 10 \rangle$, and since $\mathbf{v}_2 = -\frac{5}{2}\mathbf{v}_1$, the direction vectors and thus the lines are parallel.

15. (a) The line passes through the point $(1, -5, 6)$ and a direction vector for the line is $\langle -1, 2, -3 \rangle$, so symmetric equations for the line are $\dfrac{x - 1}{-1} = \dfrac{y + 5}{2} = \dfrac{z - 6}{-3}$.

(b) The line intersects the xy-plane when $z = 0$, so we need $\dfrac{x - 1}{-1} = \dfrac{y + 5}{2} = \dfrac{0 - 6}{-3}$ or $\dfrac{x - 1}{-1} = 2 \;\Rightarrow\; x = -1$, $\dfrac{y + 5}{2} = 2 \;\Rightarrow\; y = -1$. Thus the point of intersection with the xy-plane is $(-1, -1, 0)$. Similarly for the yz-plane, we need $x = 0 \;\Rightarrow\; 1 = \dfrac{y + 5}{2} = \dfrac{z - 6}{-3} \;\Rightarrow\; y = -3$, $z = 3$. Thus the line intersects the yz-plane at $(0, -3, 3)$. For the xz-plane, we need $y = 0 \;\Rightarrow\; \dfrac{x - 1}{-1} = \dfrac{5}{2} = \dfrac{z - 6}{-3} \;\Rightarrow\; x = -\frac{3}{2}$, $z = -\frac{3}{2}$. So the line intersects the xz-plane at $\left(-\frac{3}{2}, 0, -\frac{3}{2}\right)$.

© 2012 Cengage Learning. All Rights Reserved. May not be scanned, copied, or duplicated, or posted to a publicly accessible website, in whole or in part.

17. From Equation 4, the line segment from $\mathbf{r}_0 = 2\mathbf{i} - \mathbf{j} + 4\mathbf{k}$ to $\mathbf{r}_1 = 4\mathbf{i} + 6\mathbf{j} + \mathbf{k}$ is

$$\mathbf{r}(t) = (1 - t)\,\mathbf{r}_0 + t\,\mathbf{r}_1 = (1 - t)(2\mathbf{i} - \mathbf{j} + 4\mathbf{k}) + t(4\mathbf{i} + 6\mathbf{j} + \mathbf{k}) = (2\mathbf{i} - \mathbf{j} + 4\mathbf{k}) + t(2\mathbf{i} + 7\mathbf{j} - 3\mathbf{k}), 0 \le t \le 1.$$

19. Since the direction vectors $\langle 2, -1, 3 \rangle$ and $\langle 4, -2, 5 \rangle$ are not scalar multiples of each other, the lines aren't parallel. For the

lines to intersect, we must be able to find one value of t and one value of s that produce the same point from the respective

parametric equations. Thus we need to satisfy the following three equations: $3 + 2t = 1 + 4s$, $4 - t = 3 - 2s$,

$1 + 3t = 4 + 5s$. Solving the last two equations we get $t = 1$, $s = 0$ and checking, we see that these values don't satisfy the

first equation. Thus the lines aren't parallel and don't intersect, so they must be skew lines.

21. Since the direction vectors $\langle 1, -2, -3 \rangle$ and $\langle 1, 3, -7 \rangle$ aren't scalar multiples of each other, the lines aren't parallel. Parametric

equations of the lines are $L_1: x = 2 + t$, $y = 3 - 2t$, $z = 1 - 3t$ and $L_2: x = 3 + s$, $y = -4 + 3s$, $z = 2 - 7s$. Thus, for the

lines to intersect, the three equations $2 + t = 3 + s$, $3 - 2t = -4 + 3s$, and $1 - 3t = 2 - 7s$ must be satisfied simultaneously.

Solving the first two equations gives $t = 2$, $s = 1$ and checking, we see that these values do satisfy the third equation, so the

lines intersect when $t = 2$ and $s = 1$, that is, at the point $(4, -1, -5)$.

23. Since the plane is perpendicular to the vector $\langle 1, -2, 5 \rangle$, we can take $\langle 1, -2, 5 \rangle$ as a normal vector to the plane.

$(0, 0, 0)$ is a point on the plane, so setting $a = 1$, $b = -2$, $c = 5$ and $x_0 = 0$, $y_0 = 0$, $z_0 = 0$ in Equation 7 gives

$1(x - 0) + (-2)(y - 0) + 5(z - 0) = 0$ or $x - 2y + 5z = 0$ as an equation of the plane.

25. $\mathbf{i} + 4\mathbf{j} + \mathbf{k} = \langle 1, 4, 1 \rangle$ is a normal vector to the plane and $\left(-1, \frac{1}{2}, 3\right)$ is a point on the plane, so setting $a = 1$, $b = 4$, $c = 1$,

$x_0 = -1$, $y_0 = \frac{1}{2}$, $z_0 = 3$ in Equation 7 gives $1[x - (-1)] + 4\left(y - \frac{1}{2}\right) + 1(z - 3) = 0$ or $x + 4y + z = 4$ as an equation of

the plane.

27. Since the two planes are parallel, they will have the same normal vectors. So we can take $\mathbf{n} = \langle 5, -1, -1 \rangle$, and an equation of

the plane is $5(x - 1) - 1[y - (-1)] - 1[z - (-1)] = 0$ or $5x - y - z = 7$.

29. Since the two planes are parallel, they will have the same normal vectors. So we can take $\mathbf{n} = \langle 1, 1, 1 \rangle$, and an equation of the

plane is $1(x - 1) + 1\left(y - \frac{1}{2}\right) + 1\left(z - \frac{1}{3}\right) = 0$ or $x + y + z = \frac{11}{6}$ or $6x + 6y + 6z = 11$.

31. Here the vectors $\mathbf{a} = \langle 1 - 0, 0 - 1, 1 - 1 \rangle = \langle 1, -1, 0 \rangle$ and $\mathbf{b} = \langle 1 - 0, 1 - 1, 0 - 1 \rangle = \langle 1, 0, -1 \rangle$ lie in the plane, so

$\mathbf{a} \times \mathbf{b}$ is a normal vector to the plane. Thus, we can take $\mathbf{n} = \mathbf{a} \times \mathbf{b} = \langle 1 - 0, 0 + 1, 0 + 1 \rangle = \langle 1, 1, 1 \rangle$. If P_0 is the point

$(0, 1, 1)$, an equation of the plane is $1(x - 0) + 1(y - 1) + 1(z - 1) = 0$ or $x + y + z = 2$.

33. Here the vectors $\mathbf{a} = \langle 8 - 3, 2 - (-1), 4 - 2 \rangle = \langle 5, 3, 2 \rangle$ and $\mathbf{b} = \langle -1 - 3, -2 - (-1), -3 - 2 \rangle = \langle -4, -1, -5 \rangle$ lie in

the plane, so a normal vector to the plane is $\mathbf{n} = \mathbf{a} \times \mathbf{b} = \langle -15 + 2, -8 + 25, -5 + 12 \rangle = \langle -13, 17, 7 \rangle$ and an equation of

the plane is $-13(x - 3) + 17[y - (-1)] + 7(z - 2) = 0$ or $-13x + 17y + 7z = -42$.

35. If we first find two nonparallel vectors in the plane, their cross product will be a normal vector to the plane. Since the given

line lies in the plane, its direction vector $\mathbf{a} = \langle -2, 5, 4 \rangle$ is one vector in the plane. We can verify that the given point $(6, 0, -2)$

© 2012 Cengage Learning. All Rights Reserved. May not be scanned, copied, or duplicated, or posted to a publicly accessible website, in whole or in part.

does not lie on this line, so to find another nonparallel vector **b** which lies in the plane, we can pick any point on the line and find a vector connecting the points. If we put $t = 0$, we see that $(4, 3, 7)$ is on the line, so

$\mathbf{b} = \langle 6 - 4, 0 - 3, -2 - 7 \rangle = \langle 2, -3, -9 \rangle$ and $\mathbf{n} = \mathbf{a} \times \mathbf{b} = \langle -45 + 12, 8 - 18, 6 - 10 \rangle = \langle -33, -10, -4 \rangle$. Thus, an equation of the plane is $-33(x - 6) - 10(y - 0) - 4[z - (-2)] = 0$ or $33x + 10y + 4z = 190$.

37. A direction vector for the line of intersection is $\mathbf{a} = \mathbf{n_1} \times \mathbf{n_2} = \langle 1, 1, -1 \rangle \times \langle 2, -1, 3 \rangle = \langle 2, -5, -3 \rangle$, and $\mathbf{a}$ is parallel to the desired plane. Another vector parallel to the plane is the vector connecting any point on the line of intersection to the given point $(-1, 2, 1)$ in the plane. Setting $x = 0$, the equations of the planes reduce to $y - z = 2$ and $-y + 3z = 1$ with simultaneous solution $y = \frac{7}{2}$ and $z = \frac{3}{2}$. So a point on the line is $\left(0, \frac{7}{2}, \frac{3}{2}\right)$ and another vector parallel to the plane is $\left\langle -1, -\frac{3}{2}, -\frac{1}{2} \right\rangle$. Then a normal vector to the plane is $\mathbf{n} = \langle 2, -5, -3 \rangle \times \left\langle -1, -\frac{3}{2}, -\frac{1}{2} \right\rangle = \langle -2, 4, -8 \rangle$ and an equation of the plane is $-2(x + 1) + 4(y - 2) - 8(z - 1) = 0$ or $x - 2y + 4z = -1$.

39. If a plane is perpendicular to two other planes, its normal vector is perpendicular to the normal vectors of the other two planes. Thus $\langle 2, 1, -2 \rangle \times \langle 1, 0, 3 \rangle = \langle 3 - 0, -2 - 6, 0 - 1 \rangle = \langle 3, -8, -1 \rangle$ is a normal vector to the desired plane. The point $(1, 5, 1)$ lies on the plane, so an equation is $3(x - 1) - 8(y - 5) - (z - 1) = 0$ or $3x - 8y - z = -38$.

41. To find the x-intercept we set $y = z = 0$ in the equation $2x + 5y + z = 10$ and obtain $2x = 10 \Rightarrow x = 5$ so the x-intercept is $(5, 0, 0)$. When $x = z = 0$ we get $5y = 10 \Rightarrow y = 2$, so the y-intercept is $(0, 2, 0)$. Setting $x = y = 0$ gives $z = 10$, so the z-intercept is $(0, 0, 10)$ and we graph the portion of the plane that lies in the first octant.

43. Setting $y = z = 0$ in the equation $6x - 3y + 4z = 6$ gives $6x = 6 \Rightarrow x = 1$, when $x = z = 0$ we have $-3y = 6 \Rightarrow y = -2$, and $x = y = 0$ implies $4z = 6 \Rightarrow z = \frac{3}{2}$, so the intercepts are $(1, 0, 0)$, $(0, -2, 0)$, and $\left(0, 0, \frac{3}{2}\right)$. The figure shows the portion of the plane cut off by the coordinate planes.

45. Substitute the parametric equations of the line into the equation of the plane: $(3 - t) - (2 + t) + 2(5t) = 9 \Rightarrow 8t = 8 \Rightarrow t = 1$. Therefore, the point of intersection of the line and the plane is given by $x = 3 - 1 = 2$, $y = 2 + 1 = 3$, and $z = 5(1) = 5$, that is, the point $(2, 3, 5)$.

47. Parametric equations for the line are $x = t$, $y = 1 + t$, $z = \frac{1}{2}t$ and substituting into the equation of the plane gives $4(t) - (1 + t) + 3\left(\frac{1}{2}t\right) = 8 \Rightarrow \frac{9}{2}t = 9 \Rightarrow t = 2$. Thus $x = 2$, $y = 1 + 2 = 3$, $z = \frac{1}{2}(2) = 1$ and the point of intersection is $(2, 3, 1)$.

© 2012 Cengage Learning. All Rights Reserved. May not be scanned, copied, or duplicated, or posted to a publicly accessible website, in whole or in part.

49. Setting $x = 0$, we see that $(0, 1, 0)$ satisfies the equations of both planes, so that they do in fact have a line of intersection.

$\mathbf{v} = \mathbf{n}_1 \times \mathbf{n}_2 = \langle 1, 1, 1 \rangle \times \langle 1, 0, 1 \rangle = \langle 1, 0, -1 \rangle$ is the direction of this line. Therefore, direction numbers of the intersecting line are $1, 0, -1$.

51. Normal vectors for the planes are $\mathbf{n}_1 = \langle 1, 4, -3 \rangle$ and $\mathbf{n}_2 = \langle -3, 6, 7 \rangle$, so the normals (and thus the planes) aren't parallel.

But $\mathbf{n}_1 \cdot \mathbf{n}_2 = -3 + 24 - 21 = 0$, so the normals (and thus the planes) are perpendicular.

53. Normal vectors for the planes are $\mathbf{n}_1 = \langle 1, 1, 1 \rangle$ and $\mathbf{n}_2 = \langle 1, -1, 1 \rangle$. The normals are not parallel, so neither are the planes.

Furthermore, $\mathbf{n}_1 \cdot \mathbf{n}_2 = 1 - 1 + 1 = 1 \neq 0$, so the planes aren't perpendicular. The angle between them is given by

$\cos \theta = \dfrac{\mathbf{n}_1 \cdot \mathbf{n}_2}{|\mathbf{n}_1| \, |\mathbf{n}_2|} = \dfrac{1}{\sqrt{3}\,\sqrt{3}} = \dfrac{1}{3} \quad \Rightarrow \quad \theta = \cos^{-1}\left(\tfrac{1}{3}\right) \approx 70.5°$.

55. The normals are $\mathbf{n}_1 = \langle 1, -4, 2 \rangle$ and $\mathbf{n}_2 = \langle 2, -8, 4 \rangle$. Since $\mathbf{n}_2 = 2\mathbf{n}_1$, the normals (and thus the planes) are parallel.

57. (a) To find a point on the line of intersection, set one of the variables equal to a constant, say $z = 0$. (This will fail if the line of intersection does not cross the xy-plane; in that case, try setting x or y equal to 0.) The equations of the two planes reduce to $x + y = 1$ and $x + 2y = 1$. Solving these two equations gives $x = 1$, $y = 0$. Thus a point on the line is $(1, 0, 0)$.

A vector $\mathbf{v}$ in the direction of this intersecting line is perpendicular to the normal vectors of both planes, so we can take $\mathbf{v} = \mathbf{n}_1 \times \mathbf{n}_2 = \langle 1, 1, 1 \rangle \times \langle 1, 2, 2 \rangle = \langle 2 - 2, 1 - 2, 2 - 1 \rangle = \langle 0, -1, 1 \rangle$. By Equations 2, parametric equations for the line are $x = 1$, $y = -t$, $z = t$.

(b) The angle between the planes satisfies $\cos \theta = \dfrac{\mathbf{n}_1 \cdot \mathbf{n}_2}{|\mathbf{n}_1| \, |\mathbf{n}_2|} = \dfrac{1 + 2 + 2}{\sqrt{3}\,\sqrt{9}} = \dfrac{5}{3\sqrt{3}}$. Therefore $\theta = \cos^{-1}\left(\dfrac{5}{3\sqrt{3}}\right) \approx 15.8°$.

59. Setting $z = 0$, the equations of the two planes become $5x - 2y = 1$ and $4x + y = 6$. Solving these two equations gives $x = 1$, $y = 2$ so a point on the line of intersection is $(1, 2, 0)$. A vector $\mathbf{v}$ in the direction of this intersecting line is perpendicular to the normal vectors of both planes. So we can use $\mathbf{v} = \mathbf{n}_1 \times \mathbf{n}_2 = \langle 5, -2, -2 \rangle \times \langle 4, 1, 1 \rangle = \langle 0, -13, 13 \rangle$ or equivalently we can take $\mathbf{v} = \langle 0, -1, 1 \rangle$, and symmetric equations for the line are $x = 1$, $\dfrac{y - 2}{-1} = \dfrac{z}{1}$ or $x = 1$, $y - 2 = -z$.

61. The distance from a point (x, y, z) to $(1, 0, -2)$ is $d_1 = \sqrt{(x - 1)^2 + y^2 + (z + 2)^2}$ and the distance from (x, y, z) to $(3, 4, 0)$ is $d_2 = \sqrt{(x - 3)^2 + (y - 4)^2 + z^2}$. The plane consists of all points (x, y, z) where $d_1 = d_2 \;\Rightarrow\; d_1^2 = d_2^2 \;\Leftrightarrow$

$(x - 1)^2 + y^2 + (z + 2)^2 = (x - 3)^2 + (y - 4)^2 + z^2 \;\Leftrightarrow$

$x^2 - 2x + y^2 + z^2 + 4z + 5 = x^2 - 6x + y^2 - 8y + z^2 + 25 \;\Leftrightarrow\; 4x + 8y + 4z = 20$ so an equation for the plane is $4x + 8y + 4z = 20$ or equivalently $x + 2y + z = 5$.

Alternatively, you can argue that the segment joining points $(1, 0, -2)$ and $(3, 4, 0)$ is perpendicular to the plane and the plane includes the midpoint of the segment.

63. The plane contains the points $(a, 0, 0)$, $(0, b, 0)$ and $(0, 0, c)$. Thus the vectors $\mathbf{a} = \langle -a, b, 0 \rangle$ and $\mathbf{b} = \langle -a, 0, c \rangle$ lie in the plane, and $\mathbf{n} = \mathbf{a} \times \mathbf{b} = \langle bc - 0, 0 + ac, 0 + ab \rangle = \langle bc, ac, ab \rangle$ is a normal vector to the plane. The equation of the plane is

© 2012 Cengage Learning. All Rights Reserved. May not be scanned, copied, or duplicated, or posted to a publicly accessible website, in whole or in part.

therefore $bcx + acy + abz = abc + 0 + 0$ or $bcx + acy + abz = abc$. Notice that if $a \neq 0$, $b \neq 0$ and $c \neq 0$ then we can

rewrite the equation as $\dfrac{x}{a} + \dfrac{y}{b} + \dfrac{z}{c} = 1$. This is a good equation to remember!

65. Two vectors which are perpendicular to the required line are the normal of the given plane, $\langle 1, 1, 1 \rangle$, and a direction vector for

the given line, $\langle 1, -1, 2 \rangle$. So a direction vector for the required line is $\langle 1, 1, 1 \rangle \times \langle 1, -1, 2 \rangle = \langle 3, -1, -2 \rangle$. Thus L is given

by $\langle x, y, z \rangle = \langle 0, 1, 2 \rangle + t\langle 3, -1, -2 \rangle$, or in parametric form, $x = 3t$, $y = 1 - t$, $z = 2 - 2t$.

67. Let P_i have normal vector $\mathbf{n}_i$. Then $\mathbf{n}_1 = \langle 3, 6, -3 \rangle$, $\mathbf{n}_2 = \langle 4, -12, 8 \rangle$, $\mathbf{n}_3 = \langle 3, -9, 6 \rangle$, $\mathbf{n}_4 = \langle 1, 2, -1 \rangle$. Now $\mathbf{n}_1 = 3\mathbf{n}_4$,

so $\mathbf{n}_1$ and $\mathbf{n}_4$ are parallel, and hence P_1 and P_4 are parallel; similarly P_2 and P_3 are parallel because $\mathbf{n}_2 = \frac{4}{3}\mathbf{n}_3$. However, $\mathbf{n}_1$

and $\mathbf{n}_2$ are not parallel (so not all four planes are parallel). Notice that the point $(2, 0, 0)$ lies on both P_1 and P_4, so these two

planes are identical. The point $\left(\frac{5}{4}, 0, 0\right)$ lies on P_2 but not on P_3, so these are different planes.

69. Let $Q = (1, 3, 4)$ and $R = (2, 1, 1)$, points on the line corresponding to $t = 0$ and $t = 1$. Let

$P = (4, 1, -2)$. Then $\mathbf{a} = \overrightarrow{QR} = \langle 1, -2, -3 \rangle$, $\mathbf{b} = \overrightarrow{QP} = \langle 3, -2, -6 \rangle$. The distance is

$$d = \frac{|\mathbf{a} \times \mathbf{b}|}{|\mathbf{a}|} = \frac{|\langle 1, -2, -3 \rangle \times \langle 3, -2, -6 \rangle|}{|\langle 1, -2, -3 \rangle|} = \frac{|\langle 6, -3, 4 \rangle|}{|\langle 1, -2, -3 \rangle|} = \frac{\sqrt{6^2 + (-3)^2 + 4^2}}{\sqrt{1^2 + (-2)^2 + (-3)^2}} = \frac{\sqrt{61}}{\sqrt{14}} = \sqrt{\frac{61}{14}}.$$

71. By Equation 9, the distance is $D = \dfrac{|ax_1 + by_1 + cz_1 + d|}{\sqrt{a^2 + b^2 + c^2}} = \dfrac{|3(1) + 2(-2) + 6(4) - 5|}{\sqrt{3^2 + 2^2 + 6^2}} = \dfrac{|18|}{\sqrt{49}} = \dfrac{18}{7}$.

73. Put $y = z = 0$ in the equation of the first plane to get the point $(2, 0, 0)$ on the plane. Because the planes are parallel, the

distance D between them is the distance from $(2, 0, 0)$ to the second plane. By Equation 9,

$$D = \frac{|4(2) - 6(0) + 2(0) - 3|}{\sqrt{4^2 + (-6)^2 + (2)^2}} = \frac{5}{\sqrt{56}} = \frac{5}{2\sqrt{14}} \text{ or } \frac{5\sqrt{14}}{28}.$$

75. The distance between two parallel planes is the same as the distance between a point on one of the planes and the other plane.

Let $P_0 = (x_0, y_0, z_0)$ be a point on the plane given by $ax + by + cz + d_1 = 0$. Then $ax_0 + by_0 + cz_0 + d_1 = 0$ and the

distance between P_0 and the plane given by $ax + by + cz + d_2 = 0$ is, from Equation 9,

$$D = \frac{|ax_0 + by_0 + cz_0 + d_2|}{\sqrt{a^2 + b^2 + c^2}} = \frac{|-d_1 + d_2|}{\sqrt{a^2 + b^2 + c^2}} = \frac{|d_1 - d_2|}{\sqrt{a^2 + b^2 + c^2}}.$$

77. L_1: $x = y = z$ $\Rightarrow$ $x = y$ **(1)**. L_2: $x + 1 = y/2 = z/3$ $\Rightarrow$ $x + 1 = y/2$ **(2)**. The solution of **(1)** and **(2)** is

$x = y = -2$. However, when $x = -2$, $x = z$ $\Rightarrow$ $z = -2$, but $x + 1 = z/3$ $\Rightarrow$ $z = -3$, a contradiction. Hence the

lines do not intersect. For L_1, $\mathbf{v}_1 = \langle 1, 1, 1 \rangle$, and for L_2, $\mathbf{v}_2 = \langle 1, 2, 3 \rangle$, so the lines are not parallel. Thus the lines are skew

lines. If two lines are skew, they can be viewed as lying in two parallel planes and so the distance between the skew lines

would be the same as the distance between these parallel planes. The common normal vector to the planes must be

perpendicular to both $\langle 1, 1, 1 \rangle$ and $\langle 1, 2, 3 \rangle$, the direction vectors of the two lines. So set

© 2012 Cengage Learning. All Rights Reserved. May not be scanned, copied, or duplicated, or posted to a publicly accessible website, in whole or in part.

$\mathbf{n} = \langle 1, 1, 1 \rangle \times \langle 1, 2, 3 \rangle = \langle 3 - 2, -3 + 1, 2 - 1 \rangle = \langle 1, -2, 1 \rangle$. From above, we know that $(-2, -2, -2)$ and $(-2, -2, -3)$ are points of L_1 and L_2 respectively. So in the notation of Equation 8, $1(-2) - 2(-2) + 1(-2) + d_1 = 0 \Rightarrow d_1 = 0$ and $1(-2) - 2(-2) + 1(-3) + d_2 = 0 \Rightarrow d_2 = 1$.

By Exercise 75, the distance between these two skew lines is $D = \dfrac{|0 - 1|}{\sqrt{1 + 4 + 1}} = \dfrac{1}{\sqrt{6}}$.

Alternate solution (without reference to planes): A vector which is perpendicular to both of the lines is

$\mathbf{n} = \langle 1, 1, 1 \rangle \times \langle 1, 2, 3 \rangle = \langle 1, -2, 1 \rangle$. Pick any point on each of the lines, say $(-2, -2, -2)$ and $(-2, -2, -3)$, and form the

vector $\mathbf{b} = \langle 0, 0, 1 \rangle$ connecting the two points. The distance between the two skew lines is the absolute value of the scalar

projection of $\mathbf{b}$ along $\mathbf{n}$, that is, $D = \dfrac{|\mathbf{n} \cdot \mathbf{b}|}{|\mathbf{n}|} = \dfrac{|1 \cdot 0 - 2 \cdot 0 + 1 \cdot 1|}{\sqrt{1 + 4 + 1}} = \dfrac{1}{\sqrt{6}}$.

79. A direction vector for L_1 is $\mathbf{v}_1 = \langle 2, 0, -1 \rangle$ and a direction vector for L_2 is $\mathbf{v}_2 = \langle 3, 2, 2 \rangle$. These vectors are not parallel so neither are the lines. Parametric equations for the lines are L_1: $x = 2t$, $y = 0$, $z = -t$, and L_2: $x = 1 + 3s$, $y = -1 + 2s$, $z = 1 + 2s$. No values of t and s satisfy these equations simultaneously, so the lines don't intersect and hence are skew. We can view the lines as lying in two parallel planes; a common normal vector to the planes is $\mathbf{n} = \mathbf{v}_1 \times \mathbf{v}_2 = \langle 2, -7, 4 \rangle$. Line L_1 passes through the origin, so $(0, 0, 0)$ lies on one of the planes, and $(1, -1, 1)$ is a point on L_2 and therefore on the other plane. Equations of the planes then are $2x - 7y + 4z = 0$ and $2x - 7y + 4z - 13 = 0$, and by Exercise 75, the distance between the two skew lines is $D = \dfrac{|0 - (-13)|}{\sqrt{4 + 49 + 16}} = \dfrac{13}{\sqrt{69}}$.

Alternate solution (without reference to planes): Direction vectors of the two lines are $\mathbf{v}_1 = \langle 2, 0, -1 \rangle$ and $\mathbf{v}_2 = \langle 3, 2, 2 \rangle$. Then $\mathbf{n} = \mathbf{v}_1 \times \mathbf{v}_2 = \langle 2, -7, 4 \rangle$ is perpendicular to both lines. Pick any point on each of the lines, say $(0, 0, 0)$ and $(1, -1, 1)$, and form the vector $\mathbf{b} = \langle 1, -1, 1 \rangle$ connecting the two points. Then the distance between the two skew lines is the absolute value of the scalar projection of $\mathbf{b}$ along $\mathbf{n}$, that is, $D = \dfrac{|\mathbf{n} \cdot \mathbf{b}|}{|\mathbf{n}|} = \dfrac{|2 + 7 + 4|}{\sqrt{4 + 49 + 16}} = \dfrac{13}{\sqrt{69}}$.

81. If $a \neq 0$, then $ax + by + cz + d = 0 \Rightarrow a(x + d/a) + b(y - 0) + c(z - 0) = 0$ which by (7) is the scalar equation of the plane through the point $(-d/a, 0, 0)$ with normal vector $\langle a, b, c \rangle$. Similarly, if $b \neq 0$ (or if $c \neq 0$) the equation of the plane can be rewritten as $a(x - 0) + b(y + d/b) + c(z - 0) = 0$ [or as $a(x - 0) + b(y - 0) + c(z + d/c) = 0$] which by (7) is the scalar equation of a plane through the point $(0, -d/b, 0)$ [or the point $(0, 0, -d/c)$] with normal vector $\langle a, b, c \rangle$.

© 2012 Cengage Learning. All Rights Reserved. May not be scanned, copied, or duplicated, or posted to a publicly accessible website, in whole or in part.

12.6 Cylinders and Quadric Surfaces

1. (a) In $\mathbb{R}^2$, the equation $y = x^2$ represents a parabola.

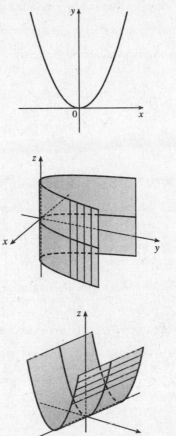

(b) In $\mathbb{R}^3$, the equation $y = x^2$ doesn't involve z, so any horizontal plane with equation $z = k$ intersects the graph in a curve with equation $y = x^2$. Thus, the surface is a parabolic cylinder, made up of infinitely many shifted copies of the same parabola. The rulings are parallel to the z-axis.

(c) In $\mathbb{R}^3$, the equation $z = y^2$ also represents a parabolic cylinder. Since x doesn't appear, the graph is formed by moving the parabola $z = y^2$ in the direction of the x-axis. Thus, the rulings of the cylinder are parallel to the x-axis.

3. Since y is missing from the equation, the vertical traces $x^2 + z^2 = 1$, $y = k$, are copies of the same circle in the plane $y = k$. Thus the surface $x^2 + z^2 = 1$ is a circular cylinder with rulings parallel to the y-axis.

5. Since x is missing, each vertical trace $z = 1 - y^2$, $x = k$, is a copy of the same parabola in the plane $x = k$. Thus the surface $z = 1 - y^2$ is a parabolic cylinder with rulings parallel to the x-axis.

© 2012 Cengage Learning. All Rights Reserved. May not be scanned, copied, or duplicated, or posted to a publicly accessible website, in whole or in part.

7. Since z is missing, each horizontal trace $xy = 1$,

$z = k$, is a copy of the same hyperbola in the plane

$z = k$. Thus the surface $xy = 1$ is a hyperbolic

cylinder with rulings parallel to the z-axis.

9. (a) The traces of $x^2 + y^2 - z^2 = 1$ in $x = k$ are $y^2 - z^2 = 1 - k^2$, a family of hyperbolas. (Note that the hyperbolas are

oriented differently for $-1 < k < 1$ than for $k < -1$ or $k > 1$.) The traces in $y = k$ are $x^2 - z^2 = 1 - k^2$, a similar

family of hyperbolas. The traces in $z = k$ are $x^2 + y^2 = 1 + k^2$, a family of circles. For $k = 0$, the trace in the

xy-plane, the circle is of radius 1. As $|k|$ increases, so does the radius of the circle. This behavior, combined with the

hyperbolic vertical traces, gives the graph of the hyperboloid of one sheet in Table 1.

(b) The shape of the surface is unchanged, but the hyperboloid is

rotated so that its axis is the y-axis. Traces in $y = k$ are circles,

while traces in $x = k$ and $z = k$ are hyperbolas.

(c) Completing the square in y gives $x^2 + (y + 1)^2 - z^2 = 1$. The

surface is a hyperboloid identical to the one in part (a) but shifted

one unit in the negative y-direction.

11. For $x = y^2 + 4z^2$, the traces in $x = k$ are $y^2 + 4z^2 = k$. When $k > 0$ we

have a family of ellipses. When $k = 0$ we have just a point at the origin, and

the trace is empty for $k < 0$. The traces in $y = k$ are $x = 4z^2 + k^2$, a

family of parabolas opening in the positive x-direction. Similarly, the traces

in $z = k$ are $x = y^2 + 4k^2$, a family of parabolas opening in the positive

x-direction. We recognize the graph as an elliptic paraboloid with axis the

x-axis and vertex the origin.

ⓒ 2012 Cengage Learning. All Rights Reserved. May not be scanned, copied, or duplicated, or posted to a publicly accessible website, in whole or in part.

13. $x^2 = y^2 + 4z^2$. The traces in $x = k$ are the ellipses $y^2 + 4z^2 = k^2$. The

traces in $y = k$ are $x^2 - 4z^2 = k^2$, hyperbolas for $k \neq 0$ and two

intersecting lines if $k = 0$. Similarly, the traces in $z = k$ are

$x^2 - y^2 = 4k^2$, hyperbolas for $k \neq 0$ and two intersecting lines if $k = 0$.

We recognize the graph as an elliptic cone with axis the x-axis and vertex

the origin.

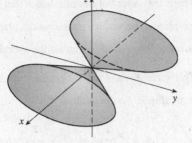

15. $-x^2 + 4y^2 - z^2 = 4$. The traces in $x = k$ are the hyperbolas

$4y^2 - z^2 = 4 + k^2$. The traces in $y = k$ are $x^2 + z^2 = 4k^2 - 4$, a family of

circles for $|k| > 1$, and the traces in $z = k$ are $4y^2 - x^2 = 4 + k^2$, a family

of hyperbolas. Thus the surface is a hyperboloid of two sheets with

axis the y-axis.

17. $36x^2 + y^2 + 36z^2 = 36$. The traces in $x = k$ are $y^2 + 36z^2 = 36(1 - k^2)$,

a family of ellipses for $|k| < 1$. (The traces are a single point for $|k| = 1$

and are empty for $|k| > 1$.) The traces in $y = k$ are the circles

$36x^2 + 36z^2 = 36 - k^2 \quad \Leftrightarrow \quad x^2 + z^2 = 1 - \frac{1}{36}k^2, |k| < 6$, and the

traces in $z = k$ are the ellipses $36x^2 + y^2 = 36(1 - k^2), |k| < 1$. The

graph is an ellipsoid centered at the origin with intercepts $x = \pm 1$, $y = \pm 6$,

$z = \pm 1$.

19. $y = z^2 - x^2$. The traces in $x = k$ are the parabolas $y = z^2 - k^2$;

the traces in $y = k$ are $k = z^2 - x^2$, which are hyperbolas (note the hyperbolas

are oriented differently for $k > 0$ than for $k < 0$); and the traces in $z = k$ are

the parabolas $y = k^2 - x^2$. Thus, $\dfrac{y}{1} = \dfrac{z^2}{1^2} - \dfrac{x^2}{1^2}$ is a hyperbolic paraboloid.

21. This is the equation of an ellipsoid: $x^2 + 4y^2 + 9z^2 = x^2 + \dfrac{y^2}{(1/2)^2} + \dfrac{z^2}{(1/3)^2} = 1$, with x-intercepts ± 1, y-intercepts $\pm \frac{1}{2}$

and z-intercepts $\pm \frac{1}{3}$. So the major axis is the x-axis and the only possible graph is VII.

23. This is the equation of a hyperboloid of one sheet, with $a = b = c = 1$. Since the coefficient of y^2 is negative, the axis of the

hyperboloid is the y-axis, hence the correct graph is II.

© 2012 Cengage Learning. All Rights Reserved. May not be scanned, copied, or duplicated, or posted to a publicly accessible website, in whole or in part.

25. There are no real values of x and z that satisfy this equation for $y < 0$, so this surface does not extend to the left of the xz-plane. The surface intersects the plane $y = k > 0$ in an ellipse. Notice that y occurs to the first power whereas x and z occur to the second power. So the surface is an elliptic paraboloid with axis the y-axis. Its graph is VI.

27. This surface is a cylinder because the variable y is missing from the equation. The intersection of the surface and the xz-plane is an ellipse. So the graph is VIII.

29. $y^2 = x^2 + \frac{1}{9}z^2$ or $y^2 = x^2 + \frac{z^2}{9}$ represents an elliptic cone with vertex $(0, 0, 0)$ and axis the y-axis.

31. $x^2 + 2y - 2z^2 = 0$ or $2y = 2z^2 - x^2$ or $y = z^2 - \frac{x^2}{2}$ represents a hyperbolic paraboloid with center $(0, 0, 0)$.

33. Completing squares in y and z gives
$4x^2 + (y - 2)^2 + 4(z - 3)^2 = 4$ or
$x^2 + \frac{(y - 2)^2}{4} + (z - 3)^2 = 1$, an ellipsoid with center $(0, 2, 3)$.

35. Completing squares in all three variables gives
$(x - 2)^2 - (y + 1)^2 + (z - 1)^2 = 0$ or
$(y + 1)^2 = (x - 2)^2 + (z - 1)^2$, a circular cone with center $(2, -1, 1)$ and axis the horizontal line $x = 2$, $z = 1$.

37. Solving the equation for z we get $z = \pm\sqrt{1 + 4x^2 + y^2}$, so we plot separately $z = \sqrt{1 + 4x^2 + y^2}$ and $z = -\sqrt{1 + 4x^2 + y^2}$.

© 2012 Cengage Learning. All Rights Reserved. May not be scanned, copied, or duplicated, or posted to a publicly accessible website, in whole or in part.

To restrict the z-range as in the second graph, we can use the option `view=-4..4` in Maple's `plot3d` command, or `PlotRange -> {-4,4}` in Mathematica's `Plot3D` command.

39. Solving the equation for z we get $z = \pm\sqrt{4x^2 + y^2}$, so we plot separately $z = \sqrt{4x^2 + y^2}$ and $z = -\sqrt{4x^2 + y^2}$.

41.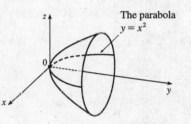

43. The surface is a paraboloid of revolution (circular paraboloid) with vertex at the origin, axis the y-axis and opens to the right. Thus the trace in the yz-plane is also a parabola: $y = z^2$, $x = 0$. The equation is $y = x^2 + z^2$.

45. Let $P = (x, y, z)$ be an arbitrary point equidistant from $(-1, 0, 0)$ and the plane $x = 1$. Then the distance from P to $(-1, 0, 0)$ is $\sqrt{(x+1)^2 + y^2 + z^2}$ and the distance from P to the plane $x = 1$ is $|x - 1|\,/\sqrt{1^2} = |x - 1|$ (by Equation 12.5.9). So $|x - 1| = \sqrt{(x+1)^2 + y^2 + z^2}$ $\Leftrightarrow$ $(x-1)^2 = (x+1)^2 + y^2 + z^2$ $\Leftrightarrow$ $x^2 - 2x + 1 = x^2 + 2x + 1 + y^2 + z^2$ $\Leftrightarrow$ $-4x = y^2 + z^2$. Thus the collection of all such points P is a circular paraboloid with vertex at the origin, axis the x-axis, which opens in the negative direction.

47. (a) An equation for an ellipsoid centered at the origin with intercepts $x = \pm a$, $y = \pm b$, and $z = \pm c$ is $\dfrac{x^2}{a^2} + \dfrac{y^2}{b^2} + \dfrac{z^2}{c^2} = 1$.

Here the poles of the model intersect the z-axis at $z = \pm 6356.523$ and the equator intersects the x- and y-axes at $x = \pm 6378.137$, $y = \pm 6378.137$, so an equation is

$$\frac{x^2}{(6378.137)^2} + \frac{y^2}{(6378.137)^2} + \frac{z^2}{(6356.523)^2} = 1$$

(b) Traces in $z = k$ are the circles $\dfrac{x^2}{(6378.137)^2} + \dfrac{y^2}{(6378.137)^2} = 1 - \dfrac{k^2}{(6356.523)^2}$ $\Leftrightarrow$

$$x^2 + y^2 = (6378.137)^2 - \left(\frac{6378.137}{6356.523}\right)^2 k^2.$$

© 2012 Cengage Learning. All Rights Reserved. May not be scanned, copied, or duplicated, or posted to a publicly accessible website, in whole or in part.

(c) To identify the traces in $y = mx$ we substitute $y = mx$ into the equation of the ellipsoid:

$$\frac{x^2}{(6378.137)^2} + \frac{(mx)^2}{(6378.137)^2} + \frac{z^2}{(6356.523)^2} = 1$$

$$\frac{(1+m^2)x^2}{(6378.137)^2} + \frac{z^2}{(6356.523)^2} = 1$$

$$\frac{x^2}{(6378.137)^2/(1+m^2)} + \frac{z^2}{(6356.523)^2} = 1$$

As expected, this is a family of ellipses.

49. If (a, b, c) satisfies $z = y^2 - x^2$, then $c = b^2 - a^2$. L_1: $x = a + t$, $y = b + t$, $z = c + 2(b-a)t$,

L_2: $x = a + t$, $y = b - t$, $z = c - 2(b+a)t$. Substitute the parametric equations of L_1 into the equation

of the hyperbolic paraboloid in order to find the points of intersection: $z = y^2 - x^2$ $\Rightarrow$

$c + 2(b-a)t = (b+t)^2 - (a+t)^2 = b^2 - a^2 + 2(b-a)t$ $\Rightarrow$ $c = b^2 - a^2$. As this is true for all values of t,

L_1 lies on $z = y^2 - x^2$. Performing similar operations with L_2 gives: $z = y^2 - x^2$ $\Rightarrow$

$c - 2(b+a)t = (b-t)^2 - (a+t)^2 = b^2 - a^2 - 2(b+a)t$ $\Rightarrow$ $c = b^2 - a^2$. This tells us that all of L_2 also lies on

$z = y^2 - x^2$.

51.

The curve of intersection looks like a bent ellipse. The projection of this curve onto the xy-plane is the set of points $(x, y, 0)$ which satisfy $x^2 + y^2 = 1 - y^2$ $\Leftrightarrow$ $x^2 + 2y^2 = 1$ $\Leftrightarrow$

$x^2 + \dfrac{y^2}{\left(1/\sqrt{2}\right)^2} = 1$. This is an equation of an ellipse.

12 Review

CONCEPT CHECK

1. A scalar is a real number, while a vector is a quantity that has both a real-valued magnitude and a direction.

2. To add two vectors geometrically, we can use either the Triangle Law or the Parallelogram Law, as illustrated in Figures 3 and 4 in Section 12.2. Algebraically, we add the corresponding components of the vectors.

3. For $c > 0$, $c\mathbf{a}$ is a vector with the same direction as $\mathbf{a}$ and length c times the length of $\mathbf{a}$. If $c < 0$, $c\mathbf{a}$ points in the opposite direction as $\mathbf{a}$ and has length $|c|$ times the length of $\mathbf{a}$. (See Figures 7 and 15 in Section 12.2.) Algebraically, to find $c\mathbf{a}$ we multiply each component of $\mathbf{a}$ by c.

4. See (1) in Section 12.2.

5. See Theorem 12.3.3 and Definition 12.3.1.

© 2012 Cengage Learning. All Rights Reserved. May not be scanned, copied, or duplicated, or posted to a publicly accessible website, in whole or in part.

6. The dot product can be used to find the angle between two vectors and the scalar projection of one vector onto another. In particular, the dot product can determine if two vectors are orthogonal. Also, the dot product can be used to determine the work done moving an object given the force and displacement vectors.

7. See the boxed equations as well as Figures 4 and 5 and the accompanying discussion on page 828 [ET 804].

8. See Theorem 12.4.9 and the preceding discussion; use either (4) or (7) in Section 12.4.

9. The cross product can be used to create a vector orthogonal to two given vectors as well as to determine if two vectors are parallel. The cross product can also be used to find the area of a parallelogram determined by two vectors. In addition, the cross product can be used to determine torque if the force and position vectors are known.

10. (a) The area of the parallelogram determined by $\mathbf{a}$ and $\mathbf{b}$ is the length of the cross product: $|\mathbf{a} \times \mathbf{b}|$.

    (b) The volume of the parallelepiped determined by $\mathbf{a}$, $\mathbf{b}$, and $\mathbf{c}$ is the magnitude of their scalar triple product: $|\mathbf{a} \cdot (\mathbf{b} \times \mathbf{c})|$.

11. If an equation of the plane is known, it can be written as $ax + by + cz + d = 0$. A normal vector, which is perpendicular to the plane, is $\langle a, b, c \rangle$ (or any scalar multiple of $\langle a, b, c \rangle$). If an equation is not known, we can use points on the plane to find two non-parallel vectors which lie in the plane. The cross product of these vectors is a vector perpendicular to the plane.

12. The angle between two intersecting planes is defined as the acute angle between their normal vectors. We can find this angle using Corollary 12.3.6.

13. See (1), (2), and (3) in Section 12.5.

14. See (5), (6), and (7) in Section 12.5.

15. (a) Two (nonzero) vectors are parallel if and only if one is a scalar multiple of the other. In addition, two nonzero vectors are parallel if and only if their cross product is $\mathbf{0}$.

    (b) Two vectors are perpendicular if and only if their dot product is 0.

    (c) Two planes are parallel if and only if their normal vectors are parallel.

16. (a) Determine the vectors $\overrightarrow{PQ} = \langle a_1, a_2, a_3 \rangle$ and $\overrightarrow{PR} = \langle b_1, b_2, b_3 \rangle$. If there is a scalar t such that $\langle a_1, a_2, a_3 \rangle = t \langle b_1, b_2, b_3 \rangle$, then the vectors are parallel and the points must all lie on the same line.

    Alternatively, if $\overrightarrow{PQ} \times \overrightarrow{PR} = \mathbf{0}$, then $\overrightarrow{PQ}$ and $\overrightarrow{PR}$ are parallel, so P, Q, and R are collinear.

    Thirdly, an algebraic method is to determine an equation of the line joining two of the points, and then check whether or not the third point satisfies this equation.

    (b) Find the vectors $\overrightarrow{PQ} = \mathbf{a}$, $\overrightarrow{PR} = \mathbf{b}$, $\overrightarrow{PS} = \mathbf{c}$. $\mathbf{a} \times \mathbf{b}$ is normal to the plane formed by P, Q and R, and so S lies on this plane if $\mathbf{a} \times \mathbf{b}$ and $\mathbf{c}$ are orthogonal, that is, if $(\mathbf{a} \times \mathbf{b}) \cdot \mathbf{c} = 0$. (Or use the reasoning in Example 5 in Section 12.4.) Alternatively, find an equation for the plane determined by three of the points and check whether or not the fourth point satisfies this equation.

© 2012 Cengage Learning. All Rights Reserved. May not be scanned, copied, or duplicated, or posted to a publicly accessible website, in whole or in part.

17. (a) See Exercise 12.4.45.

(b) See Example 8 in Section 12.5.

(c) See Example 10 in Section 12.5.

18. The traces of a surface are the curves of intersection of the surface with planes parallel to the coordinate planes. We can find the trace in the plane $x = k$ (parallel to the yz-plane) by setting $x = k$ and determining the curve represented by the resulting equation. Traces in the planes $y = k$ (parallel to the xz-plane) and $z = k$ (parallel to the xy-plane) are found similarly.

19. See Table 1 in Section 12.6.

TRUE-FALSE QUIZ

1. This is false, as the dot product of two vectors is a scalar, not a vector.

3. False. For example, if $\mathbf{u} = \mathbf{i}$ and $\mathbf{v} = \mathbf{j}$ then $|\mathbf{u} \cdot \mathbf{v}| = |0| = 0$ but $|\mathbf{u}|\,|\mathbf{v}| = 1 \cdot 1 = 1$. In fact, by Theorem 12.3.3,

$|\mathbf{u} \cdot \mathbf{v}| = |\,|\mathbf{u}|\,|\mathbf{v}| \cos \theta\,|$.

5. True, by Theorem 12.3.2, property 2.

7. True. If θ is the angle between $\mathbf{u}$ and $\mathbf{v}$, then by Theorem 12.4.9, $|\mathbf{u} \times \mathbf{v}| = |\mathbf{u}|\,|\mathbf{v}| \sin \theta = |\mathbf{v}|\,|\mathbf{u}| \sin \theta = |\mathbf{v} \times \mathbf{u}|$.

(Or, by Theorem 12.4.11, $|\mathbf{u} \times \mathbf{v}| = |-\mathbf{v} \times \mathbf{u}| = |-1|\,|\mathbf{v} \times \mathbf{u}| = |\mathbf{v} \times \mathbf{u}|$.)

9. Theorem 12.4.11, property 2 tells us that this is true.

11. This is true by Theorem 12.4.11, property 5.

13. This is true because $\mathbf{u} \times \mathbf{v}$ is orthogonal to $\mathbf{u}$ (see Theorem 12.4.8), and the dot product of two orthogonal vectors is 0.

15. This is false. A normal vector to the plane is $\mathbf{n} = \langle 6, -2, 4 \rangle$. Because $\langle 3, -1, 2 \rangle = \frac{1}{2}\mathbf{n}$, the vector is parallel to $\mathbf{n}$ and hence perpendicular to the plane.

17. This is false. In $\mathbb{R}^2$, $x^2 + y^2 = 1$ represents a circle, but $\{(x, y, z) \mid x^2 + y^2 = 1\}$ represents a *three-dimensional surface*, namely, a circular cylinder with axis the z-axis.

19. False. For example, $\mathbf{i} \cdot \mathbf{j} = 0$ but $\mathbf{i} \neq \mathbf{0}$ and $\mathbf{j} \neq \mathbf{0}$.

21. This is true. If $\mathbf{u}$ and $\mathbf{v}$ are both nonzero, then by (7) in Section 12.3, $\mathbf{u} \cdot \mathbf{v} = 0$ implies that $\mathbf{u}$ and $\mathbf{v}$ are orthogonal. But $\mathbf{u} \times \mathbf{v} = \mathbf{0}$ implies that $\mathbf{u}$ and $\mathbf{v}$ are parallel (see Corollary 12.4.10). Two nonzero vectors can't be both parallel and orthogonal, so at least one of $\mathbf{u}$, $\mathbf{v}$ must be $\mathbf{0}$.

© 2012 Cengage Learning. All Rights Reserved. May not be scanned, copied, or duplicated, or posted to a publicly accessible website, in whole or in part.

EXERCISES

1. (a) The radius of the sphere is the distance between the points $(-1, 2, 1)$ and $(6, -2, 3)$, namely,

$$\sqrt{[6 - (-1)]^2 + (-2 - 2)^2 + (3 - 1)^2} = \sqrt{69}.$$ By the formula for an equation of a sphere (see page 813 [ET 789]),

an equation of the sphere with center $(-1, 2, 1)$ and radius $\sqrt{69}$ is $(x + 1)^2 + (y - 2)^2 + (z - 1)^2 = 69$.

(b) The intersection of this sphere with the yz-plane is the set of points on the sphere whose x-coordinate is 0. Putting $x = 0$

into the equation, we have $(y - 2)^2 + (z - 1)^2 = 68, x = 0$ which represents a circle in the yz-plane with center $(0, 2, 1)$

and radius $\sqrt{68}$.

(c) Completing squares gives $(x - 4)^2 + (y + 1)^2 + (z + 3)^2 = -1 + 16 + 1 + 9 = 25$. Thus the sphere is centered at

$(4, -1, -3)$ and has radius 5.

3. $\mathbf{u} \cdot \mathbf{v} = |\mathbf{u}|\,|\mathbf{v}| \cos 45° = (2)(3)\frac{\sqrt{2}}{2} = 3\sqrt{2}.$ $|\mathbf{u} \times \mathbf{v}| = |\mathbf{u}|\,|\mathbf{v}| \sin 45° = (2)(3)\frac{\sqrt{2}}{2} = 3\sqrt{2}.$

By the right-hand rule, $\mathbf{u} \times \mathbf{v}$ is directed out of the page.

5. For the two vectors to be orthogonal, we need $\langle 3, 2, x \rangle \cdot \langle 2x, 4, x \rangle = 0$ $\Leftrightarrow$ $(3)(2x) + (2)(4) + (x)(x) = 0$ $\Leftrightarrow$

$x^2 + 6x + 8 = 0$ $\Leftrightarrow$ $(x + 2)(x + 4) = 0$ $\Leftrightarrow$ $x = -2$ or $x = -4$.

7. (a) $(\mathbf{u} \times \mathbf{v}) \cdot \mathbf{w} = \mathbf{u} \cdot (\mathbf{v} \times \mathbf{w}) = 2$

(b) $\mathbf{u} \cdot (\mathbf{w} \times \mathbf{v}) = \mathbf{u} \cdot [-(\mathbf{v} \times \mathbf{w})] = -\mathbf{u} \cdot (\mathbf{v} \times \mathbf{w}) = -2$

(c) $\mathbf{v} \cdot (\mathbf{u} \times \mathbf{w}) = (\mathbf{v} \times \mathbf{u}) \cdot \mathbf{w} = -(\mathbf{u} \times \mathbf{v}) \cdot \mathbf{w} = -2$

(d) $(\mathbf{u} \times \mathbf{v}) \cdot \mathbf{v} = \mathbf{u} \cdot (\mathbf{v} \times \mathbf{v}) = \mathbf{u} \cdot \mathbf{0} = 0$

9. For simplicity, consider a unit cube positioned with its back left corner at the origin. Vector representations of the diagonals

joining the points $(0, 0, 0)$ to $(1, 1, 1)$ and $(1, 0, 0)$ to $(0, 1, 1)$ are $\langle 1, 1, 1 \rangle$ and $\langle -1, 1, 1 \rangle$. Let θ be the angle between these

two vectors. $\langle 1, 1, 1 \rangle \cdot \langle -1, 1, 1 \rangle = -1 + 1 + 1 = 1 = |\langle 1, 1, 1 \rangle|\,|\langle -1, 1, 1 \rangle| \cos \theta = 3 \cos \theta$ $\Rightarrow$ $\cos \theta = \frac{1}{3}$ $\Rightarrow$

$\theta = \cos^{-1}\left(\frac{1}{3}\right) \approx 71°$.

11. $\overrightarrow{AB} = \langle 1, 0, -1 \rangle$, $\overrightarrow{AC} = \langle 0, 4, 3 \rangle$, so

(a) a vector perpendicular to the plane is $\overrightarrow{AB} \times \overrightarrow{AC} = \langle 0 + 4, -(3 + 0), 4 - 0 \rangle = \langle 4, -3, 4 \rangle$.

(b) $\frac{1}{2}\left|\overrightarrow{AB} \times \overrightarrow{AC}\right| = \frac{1}{2}\sqrt{16 + 9 + 16} = \frac{\sqrt{41}}{2}$.

13. Let F_1 be the magnitude of the force directed $20°$ away from the direction of shore, and let F_2 be the magnitude of the other

force. Separating these forces into components parallel to the direction of the resultant force and perpendicular to it gives

$F_1 \cos 20° + F_2 \cos 30° = 255$ **(1)**, and $F_1 \sin 20° - F_2 \sin 30° = 0$ $\Rightarrow$ $F_1 = F_2 \dfrac{\sin 30°}{\sin 20°}$ **(2)**. Substituting **(2)**

into **(1)** gives $F_2(\sin 30° \cot 20° + \cos 30°) = 255$ $\Rightarrow$ $F_2 \approx 114$ N. Substituting this into **(2)** gives $F_1 \approx 166$ N.

© 2012 Cengage Learning. All Rights Reserved. May not be scanned, copied, or duplicated, or posted to a publicly accessible website, in whole or in part.

15. The line has direction $\mathbf{v} = \langle -3, 2, 3 \rangle$. Letting $P_0 = (4, -1, 2)$, parametric equations are

$x = 4 - 3t, \ y = -1 + 2t, \ z = 2 + 3t$.

17. A direction vector for the line is a normal vector for the plane, $\mathbf{n} = \langle 2, -1, 5 \rangle$, and parametric equations for the line are

$x = -2 + 2t, \ y = 2 - t, \ z = 4 + 5t$.

19. Here the vectors $\mathbf{a} = \langle 4 - 3, 0 - (-1), 2 - 1 \rangle = \langle 1, 1, 1 \rangle$ and $\mathbf{b} = \langle 6 - 3, 3 - (-1), 1 - 1 \rangle = \langle 3, 4, 0 \rangle$ lie in the plane,

so $\mathbf{n} = \mathbf{a} \times \mathbf{b} = \langle -4, 3, 1 \rangle$ is a normal vector to the plane and an equation of the plane is

$-4(x - 3) + 3(y - (-1)) + 1(z - 1) = 0$ or $-4x + 3y + z = -14$.

21. Substitution of the parametric equations into the equation of the plane gives $2x - y + z = 2(2 - t) - (1 + 3t) + 4t = 2 \ \Rightarrow$

$-t + 3 = 2 \ \Rightarrow \ t = 1$. When $t = 1$, the parametric equations give $x = 2 - 1 = 1, y = 1 + 3 = 4$ and $z = 4$. Therefore,

the point of intersection is $(1, 4, 4)$.

23. Since the direction vectors $\langle 2, 3, 4 \rangle$ and $\langle 6, -1, 2 \rangle$ aren't parallel, neither are the lines. For the lines to intersect, the three

equations $1 + 2t = -1 + 6s, 2 + 3t = 3 - s, 3 + 4t = -5 + 2s$ must be satisfied simultaneously. Solving the first two

equations gives $t = \frac{1}{5}, s = \frac{2}{5}$ and checking we see these values don't satisfy the third equation. Thus the lines aren't parallel

and they don't intersect, so they must be skew.

25. $\mathbf{n}_1 = \langle 1, 0, -1 \rangle$ and $\mathbf{n}_2 = \langle 0, 1, 2 \rangle$. Setting $z = 0$, it is easy to see that $(1, 3, 0)$ is a point on the line of intersection of

$x - z = 1$ and $y + 2z = 3$. The direction of this line is $\mathbf{v}_1 = \mathbf{n}_1 \times \mathbf{n}_2 = \langle 1, -2, 1 \rangle$. A second vector parallel to the desired

plane is $\mathbf{v}_2 = \langle 1, 1, -2 \rangle$, since it is perpendicular to $x + y - 2z = 1$. Therefore, the normal of the plane in question is

$\mathbf{n} = \mathbf{v}_1 \times \mathbf{v}_2 = \langle 4 - 1, 1 + 2, 1 + 2 \rangle = 3 \langle 1, 1, 1 \rangle$. Taking $(x_0, y_0, z_0) = (1, 3, 0)$, the equation we are looking for is

$(x - 1) + (y - 3) + z = 0 \ \Leftrightarrow \ x + y + z = 4$.

27. By Exercise 12.5.75, $D = \dfrac{|-2 - (-24)|}{\sqrt{3^2 + 1^2 + (-4)^2}} = \dfrac{22}{\sqrt{26}}$.

29. The equation $x = z$ represents a plane perpendicular to

the xz-plane and intersecting the xz-plane in the line

$x = z, y = 0$.

31. The equation $x^2 = y^2 + 4z^2$ represents a (right elliptical)

cone with vertex at the origin and axis the x-axis.

© 2012 Cengage Learning. All Rights Reserved. May not be scanned, copied, or duplicated, or posted to a publicly accessible website, in whole or in part.

33. An equivalent equation is $-x^2 + \dfrac{y^2}{4} - z^2 = 1$, a hyperboloid of two sheets with axis the y-axis. For $|y| > 2$, traces parallel to the xz-plane are circles.

35. Completing the square in y gives

$$4x^2 + 4(y-1)^2 + z^2 = 4 \text{ or } x^2 + (y-1)^2 + \frac{z^2}{4} = 1,$$

an ellipsoid centered at $(0, 1, 0)$.

37. $4x^2 + y^2 = 16 \;\Leftrightarrow\; \dfrac{x^2}{4} + \dfrac{y^2}{16} = 1$. The equation of the ellipsoid is $\dfrac{x^2}{4} + \dfrac{y^2}{16} + \dfrac{z^2}{c^2} = 1$, since the horizontal trace in the

plane $z = 0$ must be the original ellipse. The traces of the ellipsoid in the yz-plane must be circles since the surface is obtained

by rotation about the x-axis. Therefore, $c^2 = 16$ and the equation of the ellipsoid is $\dfrac{x^2}{4} + \dfrac{y^2}{16} + \dfrac{z^2}{16} = 1 \;\Leftrightarrow$

$4x^2 + y^2 + z^2 = 16$.

© 2012 Cengage Learning. All Rights Reserved. May not be scanned, copied, or duplicated, or posted to a publicly accessible website, in whole or in part.

□ PROBLEMS PLUS

1. Since three-dimensional situations are often difficult to visualize and work with, let us first try to find an analogous problem in two dimensions. The analogue of a cube is a square and the analogue of a sphere is a circle. Thus a similar problem in two dimensions is the following: if five circles with the same radius r are contained in a square of side 1 m so that the circles touch each other and four of the circles touch two sides of the square, find r.

The diagonal of the square is $\sqrt{2}$. The diagonal is also $4r + 2x$. But x is the diagonal of a smaller square of side r. Therefore

$$x = \sqrt{2}r \quad \Rightarrow \quad \sqrt{2} = 4r + 2x = 4r + 2\sqrt{2}r = \left(4 + 2\sqrt{2}\right)r \quad \Rightarrow \quad r = \tfrac{\sqrt{2}}{4 + 2\sqrt{2}}.$$

Let's use these ideas to solve the original three-dimensional problem. The diagonal of the cube is $\sqrt{1^2 + 1^2 + 1^2} = \sqrt{3}$. The diagonal of the cube is also $4r + 2x$ where x is the diagonal of a smaller cube with edge r. Therefore

$$x = \sqrt{r^2 + r^2 + r^2} = \sqrt{3}r \quad \Rightarrow \quad \sqrt{3} = 4r + 2x = 4r + 2\sqrt{3}r = \left(4 + 2\sqrt{3}\right)r. \text{ Thus } r = \frac{\sqrt{3}}{4 + 2\sqrt{3}} = \frac{2\sqrt{3} - 3}{2}.$$

The radius of each ball is $\left(\sqrt{3} - \tfrac{3}{2}\right)$ m.

3. (a) We find the line of intersection L as in Example 12.5.7(b). Observe that the point $(-1, c, c)$ lies on both planes. Now since L lies in both planes, it is perpendicular to both of the normal vectors $\mathbf{n}_1$ and $\mathbf{n}_2$, and thus parallel to their cross product

$$\mathbf{n}_1 \times \mathbf{n}_2 = \begin{vmatrix} \mathbf{i} & \mathbf{j} & \mathbf{k} \\ c & 1 & 1 \\ 1 & -c & c \end{vmatrix} = \langle 2c, -c^2 + 1, -c^2 - 1 \rangle. \text{ So symmetric equations of } L \text{ can be written as}$$

$$\frac{x + 1}{-2c} = \frac{y - c}{c^2 - 1} - \frac{z - c}{c^2 + 1}, \text{ provided that } c \neq 0, \pm 1.$$

If $c = 0$, then the two planes are given by $y + z = 0$ and $x = -1$, so symmetric equations of L are $x = -1$, $y = -z$. If $c = -1$, then the two planes are given by $-x + y + z = -1$ and $x + y + z = -1$, and they intersect in the line $x = 0$, $y = -z - 1$. If $c = 1$, then the two planes are given by $x + y + z = 1$ and $x - y + z = 1$, and they intersect in the line $y = 0$, $x = 1 - z$.

(b) If we set $z = t$ in the symmetric equations and solve for x and y separately, we get $x + 1 = \dfrac{(t - c)(-2c)}{c^2 + 1}$,

$y - c = \dfrac{(t - c)(c^2 - 1)}{c^2 + 1} \quad \Rightarrow \quad x = \dfrac{-2ct + (c^2 - 1)}{c^2 + 1}, \; y = \dfrac{(c^2 - 1)t + 2c}{c^2 + 1}.$ Eliminating c from these equations, we have $x^2 + y^2 = t^2 + 1$. So the curve traced out by L in the plane $z = t$ is a circle with center at $(0, 0, t)$ and radius $\sqrt{t^2 + 1}$.

© 2012 Cengage Learning. All Rights Reserved. May not be scanned, copied, or duplicated, or posted to a publicly accessible website, in whole or in part.

(c) The area of a horizontal cross-section of the solid is $A(z) = \pi(z^2 + 1)$, so $V = \int_0^1 A(z)\,dz = \pi\left[\frac{1}{3}z^3 + z\right]_0^1 = \frac{4\pi}{3}$.

5. $\mathbf{v}_3 = \text{proj}_{\mathbf{v}_1}\mathbf{v}_2 = \dfrac{\mathbf{v}_1 \cdot \mathbf{v}_2}{|\mathbf{v}_1|^2}\,\mathbf{v}_1 = \dfrac{5}{2^2}\,\mathbf{v}_1$ so $|\mathbf{v}_3| = \dfrac{5}{2^2}\,|\mathbf{v}_1| = \dfrac{5}{2}$,

$\mathbf{v}_4 = \text{proj}_{\mathbf{v}_2}\mathbf{v}_3 = \dfrac{\mathbf{v}_2 \cdot \mathbf{v}_3}{|\mathbf{v}_2|^2}\,\mathbf{v}_2 = \dfrac{\mathbf{v}_2 \cdot \frac{5}{2^2}\mathbf{v}_1}{|\mathbf{v}_2|^2}\,\mathbf{v}_2 = \dfrac{5}{2^2 \cdot 3^2}\,(\mathbf{v}_1 \cdot \mathbf{v}_2)\,\mathbf{v}_2 = \dfrac{5^2}{2^2 \cdot 3^2}\,\mathbf{v}_2 \;\Rightarrow\; |\mathbf{v}_4| = \dfrac{5^2}{2^2 \cdot 3^2}\,|\mathbf{v}_2| = \dfrac{5^2}{2^2 \cdot 3}$,

$\mathbf{v}_5 = \text{proj}_{\mathbf{v}_3}\mathbf{v}_4 = \dfrac{\mathbf{v}_3 \cdot \mathbf{v}_4}{|\mathbf{v}_3|^2}\,\mathbf{v}_3 = \dfrac{\frac{5}{2^2}\mathbf{v}_1 \cdot \frac{5^2}{2^2 3^2}\mathbf{v}_2}{\left(\frac{5}{2}\right)^2}\,\left(\dfrac{5}{2^2}\,\mathbf{v}_1\right) = \dfrac{5^2}{2^4 \cdot 3^2}\,(\mathbf{v}_1 \cdot \mathbf{v}_2)\,\mathbf{v}_1 = \dfrac{5^3}{2^4 \cdot 3^2}\,\mathbf{v}_1 \;\Rightarrow$

$|\mathbf{v}_5| = \dfrac{5^3}{2^4 \cdot 3^2}\,|\mathbf{v}_1| = \dfrac{5^3}{2^3 \cdot 3^2}$. Similarly, $|\mathbf{v}_6| = \dfrac{5^4}{2^4 \cdot 3^3}$, $|\mathbf{v}_7| = \dfrac{5^5}{2^5 \cdot 3^4}$, and in general, $|\mathbf{v}_n| = \dfrac{5^{n-2}}{2^{n-2} \cdot 3^{n-3}} = 3\left(\frac{5}{6}\right)^{n-2}$.

Thus

$$\sum_{n=1}^{\infty} |\mathbf{v}_n| = |\mathbf{v}_1| + |\mathbf{v}_2| + \sum_{n=3}^{\infty} 3\left(\tfrac{5}{6}\right)^{n-2} = 2 + 3 + \sum_{n=1}^{\infty} 3\left(\tfrac{5}{6}\right)^n$$

$$= 5 + \sum_{n=1}^{\infty} \tfrac{5}{2}\left(\tfrac{5}{6}\right)^{n-1} = 5 + \frac{\frac{5}{2}}{1 - \frac{5}{6}} \quad \text{[sum of a geometric series]} \quad = 5 + 15 = 20$$

7. (a) When $\theta = \theta_s$, the block is not moving, so the sum of the forces on the block

must be $\mathbf{0}$, thus $\mathbf{N} + \mathbf{F} + \mathbf{W} = \mathbf{0}$. This relationship is illustrated

geometrically in the figure. Since the vectors form a right triangle, we have

$$\tan(\theta_s) = \frac{|\mathbf{F}|}{|\mathbf{N}|} = \frac{\mu_s n}{n} = \mu_s.$$

(b) We place the block at the origin and sketch the force vectors acting on the block, including the additional horizontal force

$\mathbf{H}$, with initial points at the origin. We then rotate this system so that $\mathbf{F}$ lies along the positive x-axis and the inclined plane

is parallel to the x-axis. (See the following figure.)

$|\mathbf{F}|$ is maximal, so $|\mathbf{F}| = \mu_s n$ for $\theta > \theta_s$. Then the vectors, in terms of components parallel and perpendicular to the

inclined plane, are

$$\mathbf{N} = n\,\mathbf{j} \qquad\qquad \mathbf{F} = (\mu_s n)\,\mathbf{i}$$

$$\mathbf{W} = (-mg\sin\theta)\,\mathbf{i} + (-mg\cos\theta)\,\mathbf{j} \qquad\qquad \mathbf{H} = (h_{\min}\cos\theta)\,\mathbf{i} + (-h_{\min}\sin\theta)\,\mathbf{j}$$

© 2012 Cengage Learning. All Rights Reserved. May not be scanned, copied, or duplicated, or posted to a publicly accessible website, in whole or in part.

Equating components, we have

$$\mu_s n - mg\sin\theta + h_{\min}\cos\theta = 0 \quad\Rightarrow\quad h_{\min}\cos\theta + \mu_s n = mg\sin\theta \tag{1}$$

$$n - mg\cos\theta - h_{\min}\sin\theta = 0 \quad\Rightarrow\quad h_{\min}\sin\theta + mg\cos\theta = n \tag{2}$$

(c) Since **(2)** is solved for n, we substitute into **(1)**:

$$h_{\min}\cos\theta + \mu_s(h_{\min}\sin\theta + mg\cos\theta) = mg\sin\theta \quad\Rightarrow$$

$$h_{\min}\cos\theta + h_{\min}\mu_s\sin\theta = mg\sin\theta - mg\mu_s\cos\theta \quad\Rightarrow$$

$$h_{\min} = mg\left(\frac{\sin\theta - \mu_s\cos\theta}{\cos\theta + \mu_s\sin\theta}\right) = mg\left(\frac{\tan\theta - \mu_s}{1 + \mu_s\tan\theta}\right)$$

From part (a) we know $\mu_s = \tan\theta_s$, so this becomes $h_{\min} = mg\left(\dfrac{\tan\theta - \tan\theta_s}{1 + \tan\theta_s\tan\theta}\right)$ and using a trigonometric identity,

this is $mg\tan(\theta - \theta_s)$ as desired.

Note for $\theta = \theta_s$, $h_{\min} = mg\tan 0 = 0$, which makes sense since the block is at rest for θ_s, thus no additional force **H** is necessary to prevent it from moving. As θ increases, the factor $\tan(\theta - \theta_s)$, and hence the value of $h_{\min}$, increases slowly for small values of $\theta - \theta_s$ but much more rapidly as $\theta - \theta_s$ becomes significant. This seems reasonable, as the steeper the inclined plane, the less the horizontal components of the various forces affect the movement of the block, so we would need a much larger magnitude of horizontal force to keep the block motionless. If we allow $\theta \to 90°$, corresponding to the inclined plane being placed vertically, the value of $h_{\min}$ is quite large; this is to be expected, as it takes a great amount of horizontal force to keep an object from moving vertically. In fact, without friction (so $\theta_s = 0$), we would have $\theta \to 90° \Rightarrow h_{\min} \to \infty$, and it would be impossible to keep the block from slipping.

(d) Since $h_{\max}$ is the largest value of h that keeps the block from slipping, the force of friction is keeping the block from moving *up* the inclined plane; thus, **F** is directed *down* the plane. Our system of forces is similar to that in part (b), then, except that we have $\mathbf{F} = -(\mu_s n)\,\mathbf{i}$. (Note that $|\mathbf{F}|$ is again maximal.) Following our procedure in parts (b) and (c), we equate components:

$$-\mu_s n - mg\sin\theta + h_{\max}\cos\theta = 0 \quad\Rightarrow\quad h_{\max}\cos\theta - \mu_s n = mg\sin\theta$$

$$n - mg\cos\theta - h_{\max}\sin\theta = 0 \quad\Rightarrow\quad h_{\max}\sin\theta + mg\cos\theta = n$$

Then substituting,

$$h_{\max}\cos\theta - \mu_s(h_{\max}\sin\theta + mg\cos\theta) = mg\sin\theta \quad\Rightarrow$$

$$h_{\max}\cos\theta - h_{\max}\mu_s\sin\theta = mg\sin\theta + mg\mu_s\cos\theta \quad\Rightarrow$$

© 2012 Cengage Learning. All Rights Reserved. May not be scanned, copied, or duplicated, or posted to a publicly accessible website, in whole or in part.

$$h_{max} = mg\left(\frac{\sin\theta + \mu_s\cos\theta}{\cos\theta - \mu_s\sin\theta}\right) = mg\left(\frac{\tan\theta + \mu_s}{1 - \mu_s\tan\theta}\right)$$

$$= mg\left(\frac{\tan\theta + \tan\theta_s}{1 - \tan\theta_s\tan\theta}\right) = mg\tan(\theta + \theta_s)$$

We would expect h_{max} to increase as θ increases, with similar behavior as we established for h_{min}, but with h_{max} values always larger than h_{min}. We can see that this is the case if we graph h_{max} as a function of θ, as the curve is the graph of h_{min} translated $2\theta_s$ to the left, so the equation does seem reasonable. Notice that the equation predicts $h_{max} \to \infty$ as $\theta \to (90° - \theta_s)$. In fact, as h_{max} increases, the normal force increases as well. When $(90° - \theta_s) \leq \theta \leq 90°$, the horizontal force is completely counteracted by the sum of the normal and frictional forces, so no part of the horizontal force contributes to moving the block up the plane no matter how large its magnitude.

© 2012 Cengage Learning. All Rights Reserved. May not be scanned, copied, or duplicated, or posted to a publicly accessible website, in whole or in part.

13 ☐ VECTOR FUNCTIONS

13.1 Vector Functions and Space Curves

1. The component functions $\sqrt{4-t^2}$, e^{-3t}, and $\ln(t+1)$ are all defined when $4-t^2 \geq 0 \Rightarrow -2 \leq t \leq 2$ and

$t+1 > 0 \Rightarrow t > -1$, so the domain of $\mathbf{r}$ is $(-1, 2]$.

3. $\lim\limits_{t \to 0} e^{-3t} = e^0 = 1$, $\lim\limits_{t \to 0} \dfrac{t^2}{\sin^2 t} = \lim\limits_{t \to 0} \dfrac{1}{\frac{\sin^2 t}{t^2}} = \dfrac{1}{\lim\limits_{t \to 0} \dfrac{\sin^2 t}{t^2}} = \dfrac{1}{\left(\lim\limits_{t \to 0} \dfrac{\sin t}{t}\right)^2} = \dfrac{1}{1^2} = 1$,

and $\lim\limits_{t \to 0} \cos 2t = \cos 0 = 1$. Thus

$$\lim_{t \to 0}\left(e^{-3t}\,\mathbf{i} + \frac{t^2}{\sin^2 t}\,\mathbf{j} + \cos 2t\,\mathbf{k}\right) = \left[\lim_{t \to 0} e^{-3t}\right]\mathbf{i} + \left[\lim_{t \to 0} \frac{t^2}{\sin^2 t}\right]\mathbf{j} + \left[\lim_{t \to 0} \cos 2t\right]\mathbf{k} = \mathbf{i} + \mathbf{j} + \mathbf{k}.$$

5. $\lim\limits_{t \to \infty} \dfrac{1+t^2}{1-t^2} = \lim\limits_{t \to \infty} \dfrac{(1/t^2)+1}{(1/t^2)-1} = \dfrac{0+1}{0-1} = -1$, $\lim\limits_{t \to \infty} \tan^{-1} t = \dfrac{\pi}{2}$, $\lim\limits_{t \to \infty} \dfrac{1-e^{-2t}}{t} = \lim\limits_{t \to \infty} \dfrac{1}{t} - \dfrac{1}{te^{2t}} = 0 - 0 = 0$. Thus

$$\lim_{t \to \infty}\left\langle \frac{1+t^2}{1-t^2}, \tan^{-1} t, \frac{1-e^{-2t}}{t}\right\rangle = \left\langle -1, \tfrac{\pi}{2}, 0\right\rangle.$$

7. The corresponding parametric equations for this curve are $x = \sin t$, $y = t$.

We can make a table of values, or we can eliminate the parameter: $t = y \Rightarrow$

$x = \sin y$, with $y \in \mathbb{R}$. By comparing different values of t, we find the direction in

which t increases as indicated in the graph.

9. The corresponding parametric equations are $x = t$, $y = 2-t$, $z = 2t$, which are

parametric equations of a line through the point $(0, 2, 0)$ and with direction vector

$\langle 1, -1, 2 \rangle$.

© 2012 Cengage Learning. All Rights Reserved. May not be scanned, copied, or duplicated, or posted to a publicly accessible website, in whole or in part.

11. The corresponding parametric equations are $x = 1$, $y = \cos t$, $z = 2\sin t$.

Eliminating the parameter in y and z gives $y^2 + (z/2)^2 = \cos^2 t + \sin^2 t = 1$

or $y^2 + z^2/4 = 1$. Since $x = 1$, the curve is an ellipse centered at $(1, 0, 0)$ in

the plane $x = 1$.

13. The parametric equations are $x = t^2$, $y = t^4$, $z = t^6$. These are positive

for $t \neq 0$ and 0 when $t = 0$. So the curve lies entirely in the first octant.

The projection of the graph onto the xy-plane is $y = x^2$, $y > 0$, a half parabola.

Onto the xz-plane $z = x^3$, $z > 0$, a half cubic, and the yz-plane, $y^3 = z^2$.

15. The projection of the curve onto the xy-plane is given by $\mathbf{r}(t) = \langle t, \sin t, 0 \rangle$ [we use 0 for the z-component] whose graph

is the curve $y = \sin x$, $z = 0$. Similarly, the projection onto the xz-plane is $\mathbf{r}(t) = \langle t, 0, 2\cos t \rangle$, whose graph is the cosine

wave $z = 2\cos x$, $y = 0$, and the projection onto the yz-plane is $\mathbf{r}(t) = \langle 0, \sin t, 2\cos t \rangle$ whose graph is the ellipse

$y^2 + \frac{1}{4}z^2 = 1$, $x = 0$.

| xy-plane | xz-plane | yz-plane |

From the projection onto the yz-plane we see that the curve lies on an elliptical

cylinder with axis the x-axis. The other two projections show that the curve

oscillates both vertically and horizontally as we move in the x-direction,

suggesting that the curve is an elliptical helix that spirals along the cylinder.

17. Taking $\mathbf{r}_0 = \langle 2, 0, 0 \rangle$ and $\mathbf{r}_1 = \langle 6, 2, -2 \rangle$, we have from Equation 12.5.4

$\mathbf{r}(t) = (1-t)\,\mathbf{r}_0 + t\,\mathbf{r}_1 = (1-t)\,\langle 2, 0, 0 \rangle + t\,\langle 6, 2, -2 \rangle$, $0 \le t \le 1$ or $\mathbf{r}(t) = \langle 2 + 4t, 2t, -2t \rangle$, $0 \le t \le 1$.

Parametric equations are $x = 2 + 4t$, $y = 2t$, $z = -2t$, $0 \le t \le 1$.

© 2012 Cengage Learning. All Rights Reserved. May not be scanned, copied, or duplicated, or posted to a publicly accessible website, in whole or in part.

19. Taking $\mathbf{r}_0 = \langle 0, -1, 1 \rangle$ and $\mathbf{r}_1 = \langle \frac{1}{2}, \frac{1}{3}, \frac{1}{4} \rangle$, we have

$\mathbf{r}(t) = (1-t)\,\mathbf{r}_0 + t\,\mathbf{r}_1 = (1-t)\,\langle 0, -1, 1 \rangle + t\,\langle \frac{1}{2}, \frac{1}{3}, \frac{1}{4} \rangle,\ 0 \le t \le 1$ or $\mathbf{r}(t) = \langle \frac{1}{2}t, -1 + \frac{4}{3}t, 1 - \frac{3}{4}t \rangle,\ 0 \le t \le 1$.

Parametric equations are $x = \frac{1}{2}t,\ y = -1 + \frac{4}{3}t,\ z = 1 - \frac{3}{4}t,\ 0 \le t \le 1$.

21. $x = t\cos t,\ y = t,\ z = t\sin t,\ t \ge 0$. At any point (x, y, z) on the curve, $x^2 + z^2 = t^2\cos^2 t + t^2\sin^2 t = t^2 = y^2$ so the

curve lies on the circular cone $x^2 + z^2 = y^2$ with axis the y-axis. Also notice that $y \ge 0$; the graph is II.

23. $x = t,\ y = 1/(1+t^2),\ z = t^2$. At any point on the curve we have $z = x^2$, so the curve lies on a parabolic cylinder parallel

to the y-axis. Notice that $0 < y \le 1$ and $z \ge 0$. Also the curve passes through $(0, 1, 0)$ when $t = 0$ and $y \to 0,\ z \to \infty$ as

$t \to \pm\infty$, so the graph must be V.

25. $x = \cos 8t,\ y = \sin 8t,\ z = e^{0.8t},\ t \ge 0$. $x^2 + y^2 = \cos^2 8t + \sin^2 8t = 1$, so the curve lies on a circular cylinder with

axis the z-axis. A point (x, y, z) on the curve lies directly above the point $(x, y, 0)$, which moves counterclockwise around the

unit circle in the xy-plane as t increases. The curve starts at $(1, 0, 1)$, when $t = 0$, and $z \to \infty$ (at an increasing rate) as

$t \to \infty$, so the graph is IV.

27. If $x = t\cos t,\ y = t\sin t,\ z = t$, then $x^2 + y^2 = t^2\cos^2 t + t^2\sin^2 t = t^2 = z^2$,

so the curve lies on the cone $z^2 = x^2 + y^2$. Since $z = t$, the curve is a spiral on

this cone.

29. Parametric equations for the curve are $x = t,\ y = 0,\ z = 2t - t^2$. Substituting into the equation of the paraboloid

gives $2t - t^2 = t^2 \ \Rightarrow\ 2t = 2t^2 \ \to\ t = 0, 1$. Since $\mathbf{r}(0) = \mathbf{0}$ and $\mathbf{r}(1) = \mathbf{i} + \mathbf{k}$, the points of intersection

are $(0, 0, 0)$ and $(1, 0, 1)$.

31. $\mathbf{r}(t) = \langle \cos t \sin 2t, \sin t \sin 2t, \cos 2t \rangle$.

We include both a regular plot and a plot

showing a tube of radius 0.08 around the

curve.

© 2012 Cengage Learning. All Rights Reserved. May not be scanned, copied, or duplicated, or posted to a publicly accessible website, in whole or in part.

33. $\mathbf{r}(t) = \langle t, t\sin t, t\cos t \rangle$

35. $\mathbf{r}(t) = \langle \cos 2t, \cos 3t, \cos 4t \rangle$

37.

$x = (1 + \cos 16t)\cos t$, $y = (1 + \cos 16t)\sin t$, $z = 1 + \cos 16t$. At any point on the graph,

$$x^2 + y^2 = (1 + \cos 16t)^2 \cos^2 t + (1 + \cos 16t)^2 \sin^2 t$$
$$= (1 + \cos 16t)^2 = z^2, \text{ so the graph lies on the cone } x^2 + y^2 = z^2.$$

From the graph at left, we see that this curve looks like the projection of a leaved two-dimensional curve onto a cone.

39. If $t = -1$, then $x = 1$, $y = 4$, $z = 0$, so the curve passes through the point $(1, 4, 0)$. If $t = 3$, then $x = 9$, $y = -8$, $z = 28$, so the curve passes through the point $(9, -8, 28)$. For the point $(4, 7, -6)$ to be on the curve, we require $y = 1 - 3t = 7 \Rightarrow$ $t = -2$. But then $z = 1 + (-2)^3 = -7 \neq -6$, so $(4, 7, -6)$ is not on the curve.

41. Both equations are solved for z, so we can substitute to eliminate z: $\sqrt{x^2 + y^2} = 1 + y \Rightarrow x^2 + y^2 = 1 + 2y + y^2 \Rightarrow$ $x^2 = 1 + 2y \Rightarrow y = \frac{1}{2}(x^2 - 1)$. We can form parametric equations for the curve C of intersection by choosing a parameter $x = t$, then $y = \frac{1}{2}(t^2 - 1)$ and $z = 1 + y = 1 + \frac{1}{2}(t^2 - 1) = \frac{1}{2}(t^2 + 1)$. Thus a vector function representing C is $\mathbf{r}(t) = t\,\mathbf{i} + \frac{1}{2}(t^2 - 1)\,\mathbf{j} + \frac{1}{2}(t^2 + 1)\,\mathbf{k}$.

43. The projection of the curve C of intersection onto the xy-plane is the circle $x^2 + y^2 = 1$, $z = 0$, so we can write $x = \cos t$, $y = \sin t$, $0 \leq t \leq 2\pi$. Since C also lies on the surface $z = x^2 - y^2$, we have $z = x^2 - y^2 = \cos^2 t - \sin^2 t$ or $\cos 2t$. Thus parametric equations for C are $x = \cos t$, $y = \sin t$, $z = \cos 2t$, $0 \leq t \leq 2\pi$, and the corresponding vector function is $\mathbf{r}(t) = \cos t\,\mathbf{i} + \sin t\,\mathbf{j} + \cos 2t\,\mathbf{k}$, $0 \leq t \leq 2\pi$.

45.

The projection of the curve C of intersection onto the xy-plane is the circle $x^2 + y^2 = 4$, $z = 0$. Then we can write $x = 2\cos t$, $y = 2\sin t$, $0 \leq t \leq 2\pi$. Since C also lies on the surface $z = x^2$, we have $z = x^2 = (2\cos t)^2 = 4\cos^2 t$. Then parametric equations for C are $x = 2\cos t$, $y = 2\sin t$, $z = 4\cos^2 t$, $0 \leq t \leq 2\pi$.

© 2012 Cengage Learning. All Rights Reserved. May not be scanned, copied, or duplicated, or posted to a publicly accessible website, in whole or in part.

47. For the particles to collide, we require $\mathbf{r}_1(t) = \mathbf{r}_2(t)$ $\Leftrightarrow$ $\langle t^2, 7t - 12, t^2 \rangle = \langle 4t - 3, t^2, 5t - 6 \rangle$. Equating components

gives $t^2 = 4t - 3$, $7t - 12 = t^2$, and $t^2 = 5t - 6$. From the first equation, $t^2 - 4t + 3 = 0$ $\Leftrightarrow$ $(t - 3)(t - 1) = 0$ so $t = 1$

or $t = 3$. $t = 1$ does not satisfy the other two equations, but $t = 3$ does. The particles collide when $t = 3$, at the

point $(9, 9, 9)$.

49. Let $\mathbf{u}(t) = \langle u_1(t), u_2(t), u_3(t) \rangle$ and $\mathbf{v}(t) = \langle v_1(t), v_2(t), v_3(t) \rangle$. In each part of this problem the basic procedure is to use

Equation 1 and then analyze the individual component functions using the limit properties we have already developed for

real-valued functions.

(a) $\lim\limits_{t \to a} \mathbf{u}(t) + \lim\limits_{t \to a} \mathbf{v}(t) = \left\langle \lim\limits_{t \to a} u_1(t), \lim\limits_{t \to a} u_2(t), \lim\limits_{t \to a} u_3(t) \right\rangle + \left\langle \lim\limits_{t \to a} v_1(t), \lim\limits_{t \to a} v_2(t), \lim\limits_{t \to a} v_3(t) \right\rangle$ and the limits of these

component functions must each exist since the vector functions both possess limits as $t \to a$. Then adding the two vectors

and using the addition property of limits for real-valued functions, we have that

$$\lim\limits_{t \to a} \mathbf{u}(t) + \lim\limits_{t \to a} \mathbf{v}(t) = \left\langle \lim\limits_{t \to a} u_1(t) + \lim\limits_{t \to a} v_1(t), \lim\limits_{t \to a} u_2(t) + \lim\limits_{t \to a} v_2(t), \lim\limits_{t \to a} u_3(t) + \lim\limits_{t \to a} v_3(t) \right\rangle$$

$$= \left\langle \lim\limits_{t \to a} [u_1(t) + v_1(t)], \lim\limits_{t \to a} [u_2(t) + v_2(t)], \lim\limits_{t \to a} [u_3(t) + v_3(t)] \right\rangle$$

$$= \lim\limits_{t \to a} \langle u_1(t) + v_1(t), u_2(t) + v_2(t), u_3(t) + v_3(t) \rangle \qquad \text{[using (1) backward]}$$

$$= \lim\limits_{t \to a} [\mathbf{u}(t) + \mathbf{v}(t)]$$

(b) $\lim\limits_{t \to a} c\mathbf{u}(t) = \lim\limits_{t \to a} \langle cu_1(t), cu_2(t), cu_3(t) \rangle = \left\langle \lim\limits_{t \to a} cu_1(t), \lim\limits_{t \to a} cu_2(t), \lim\limits_{t \to a} cu_3(t) \right\rangle$

$$= \left\langle c \lim\limits_{t \to a} u_1(t), c \lim\limits_{t \to a} u_2(t), c \lim\limits_{t \to u} u_3(t) \right\rangle = c \left\langle \lim\limits_{t \to a} u_1(t), \lim\limits_{t \to a} u_2(t), \lim\limits_{t \to a} u_3(t) \right\rangle$$

$$= c \lim\limits_{t \to a} \langle u_1(t), u_2(t), u_3(t) \rangle = c \lim\limits_{t \to a} \mathbf{u}(t)$$

(c) $\lim\limits_{t \to a} \mathbf{u}(t) \cdot \lim\limits_{t \to a} \mathbf{v}(t) = \left\langle \lim\limits_{t \to a} u_1(t), \lim\limits_{t \to a} u_2(t), \lim\limits_{t \to a} u_3(t) \right\rangle \cdot \left\langle \lim\limits_{t \to a} v_1(t), \lim\limits_{t \to a} v_2(t), \lim\limits_{t \to a} v_3(t) \right\rangle$

$$= \left[\lim\limits_{t \to a} u_1(t) \right] \left[\lim\limits_{t \to a} v_1(t) \right] + \left[\lim\limits_{t \to a} u_2(t) \right] \left[\lim\limits_{t \to a} v_2(t) \right] + \left[\lim\limits_{t \to a} u_3(t) \right] \left[\lim\limits_{t \to a} v_3(t) \right]$$

$$= \lim\limits_{t \to a} u_1(t)v_1(t) + \lim\limits_{t \to a} u_2(t)v_2(t) + \lim\limits_{t \to a} u_3(t)v_3(t)$$

$$= \lim\limits_{t \to a} [u_1(t)v_1(t) + u_2(t)v_2(t) + u_3(t)v_3(t)] = \lim\limits_{t \to a} [\mathbf{u}(t) \cdot \mathbf{v}(t)]$$

© 2012 Cengage Learning. All Rights Reserved. May not be scanned, copied, or duplicated, or posted to a publicly accessible website, in whole or in part.

(d) $\displaystyle\lim_{t\to a}\mathbf{u}(t)\times\lim_{t\to a}\mathbf{v}(t)=\left\langle\lim_{t\to a}u_1(t),\lim_{t\to a}u_2(t),\lim_{t\to a}u_3(t)\right\rangle\times\left\langle\lim_{t\to a}v_1(t),\lim_{t\to a}v_2(t),\lim_{t\to a}v_3(t)\right\rangle$

$$=\left\langle\left[\lim_{t\to a}u_2(t)\right]\left[\lim_{t\to a}v_3(t)\right]-\left[\lim_{t\to a}u_3(t)\right]\left[\lim_{t\to a}v_2(t)\right],\right.$$

$$\left[\lim_{t\to a}u_3(t)\right]\left[\lim_{t\to a}v_1(t)\right]-\left[\lim_{t\to a}u_1(t)\right]\left[\lim_{t\to a}v_3(t)\right],$$

$$\left.\left[\lim_{t\to a}u_1(t)\right]\left[\lim_{t\to a}v_2(t)\right]-\left[\lim_{t\to a}u_2(t)\right]\left[\lim_{t\to a}v_1(t)\right]\right\rangle$$

$$=\left\langle\lim_{t\to a}[u_2(t)v_3(t)-u_3(t)v_2(t)],\lim_{t\to a}[u_3(t)v_1(t)-u_1(t)v_3(t)],\right.$$

$$\left.\lim_{t\to a}[u_1(t)v_2(t)-u_2(t)v_1(t)]\right\rangle$$

$$=\lim_{t\to a}\langle u_2(t)v_3(t)-u_3(t)v_2(t),u_3(t)v_1(t)-u_1(t)v_3(t),u_1(t)v_2(t)-u_2(t)v_1(t)\rangle$$

$$=\lim_{t\to a}[\mathbf{u}(t)\times\mathbf{v}(t)]$$

51. Let $\mathbf{r}(t)=\langle f(t),g(t),h(t)\rangle$ and $\mathbf{b}=\langle b_1,b_2,b_3\rangle$. If $\displaystyle\lim_{t\to a}\mathbf{r}(t)=\mathbf{b}$, then $\displaystyle\lim_{t\to a}\mathbf{r}(t)$ exists, so by (1),

$\mathbf{b}=\displaystyle\lim_{t\to a}\mathbf{r}(t)=\left\langle\lim_{t\to a}f(t),\lim_{t\to a}g(t),\lim_{t\to a}h(t)\right\rangle$. By the definition of equal vectors we have $\displaystyle\lim_{t\to a}f(t)=b_1,\lim_{t\to a}g(t)=b_2$

and $\displaystyle\lim_{t\to a}h(t)=b_3$. But these are limits of real-valued functions, so by the definition of limits, for every $\varepsilon>0$ there exists

$\delta_1>0,\delta_2>0,\delta_3>0$ so that if $0<|t-a|<\delta_1$ then $|f(t)-b_1|<\varepsilon/3$, if $0<|t-a|<\delta_2$ then $|g(t)-b_2|<\varepsilon/3$, and

if $0<|t-a|<\delta_3$ then $|h(t)-b_3|<\varepsilon/3$. Letting $\delta=$ minimum of $\{\delta_1,\delta_2,\delta_3\}$, then if $0<|t-a|<\delta$ we have

$|f(t)-b_1|+|g(t)-b_2|+|h(t)-b_3|<\varepsilon/3+\varepsilon/3+\varepsilon/3=\varepsilon$. But

$$|\mathbf{r}(t)-\mathbf{b}|=|\langle f(t)-b_1,g(t)-b_2,h(t)-b_3\rangle|=\sqrt{(f(t)-b_1)^2+(g(t)-b_2)^2+(h(t)-b_3)^2}$$

$$\le\sqrt{[f(t)-b_1]^2}+\sqrt{[g(t)-b_2]^2}+\sqrt{[h(t)-b_3]^2}=|f(t)-b_1|+|g(t)-b_2|+|h(t)-b_3|$$

Thus for every $\varepsilon>0$ there exists $\delta>0$ such that if $0<|t-a|<\delta$ then

$|\mathbf{r}(t)-\mathbf{b}|\le|f(t)-b_1|+|g(t)-b_2|+|h(t)-b_3|<\varepsilon$. Conversely, suppose for every $\varepsilon>0$, there exists $\delta>0$ such

that if $0<|t-a|<\delta$ then $|\mathbf{r}(t)-\mathbf{b}|<\varepsilon\ \Leftrightarrow\ |\langle f(t)-b_1,g(t)-b_2,h(t)-b_3\rangle|<\varepsilon\ \Leftrightarrow$

$\sqrt{[f(t)-b_1]^2+[g(t)-b_2]^2+[h(t)-b_3]^2}<\varepsilon\ \Leftrightarrow\ [f(t)-b_1]^2+[g(t)-b_2]^2+[h(t)-b_3]^2<\varepsilon^2$. But each term

on the left side of the last inequality is positive, so if $0<|t-a|<\delta$, then $[f(t)-b_1]^2<\varepsilon^2,[g(t)-b_2]^2<\varepsilon^2$ and

$[h(t)-b_3]^2<\varepsilon^2$ or, taking the square root of both sides in each of the above, $|f(t)-b_1|<\varepsilon,|g(t)-b_2|<\varepsilon$ and

$|h(t)-b_3|<\varepsilon$. And by definition of limits of real-valued functions we have $\displaystyle\lim_{t\to a}f(t)=b_1,\lim_{t\to a}g(t)=b_2$ and

$\displaystyle\lim_{t\to a}h(t)=b_3$. But by (1), $\displaystyle\lim_{t\to a}\mathbf{r}(t)=\left\langle\lim_{t\to a}f(t),\lim_{t\to a}g(t),\lim_{t\to a}h(t)\right\rangle$, so $\displaystyle\lim_{t\to a}\mathbf{r}(t)=\langle b_1,b_2,b_3\rangle=\mathbf{b}$.

© 2012 Cengage Learning. All Rights Reserved. May not be scanned, copied, or duplicated, or posted to a publicly accessible website, in whole or in part.

13.2 Derivatives and Integrals of Vector Functions

1. (a)

(b) $\dfrac{\mathbf{r}(4.5) - \mathbf{r}(4)}{0.5} = 2[\mathbf{r}(4.5) - \mathbf{r}(4)]$, so we draw a vector in the same

direction but with twice the length of the vector $\mathbf{r}(4.5) - \mathbf{r}(4)$.

$\dfrac{\mathbf{r}(4.2) - \mathbf{r}(4)}{0.2} = 5[\mathbf{r}(4.2) - \mathbf{r}(4)]$, so we draw a vector in the same

direction but with 5 times the length of the vector $\mathbf{r}(4.2) - \mathbf{r}(4)$.

(c) By Definition 1, $\mathbf{r}'(4) = \lim\limits_{h \to 0} \dfrac{\mathbf{r}(4 + h) - \mathbf{r}(4)}{h}$. $\quad \mathbf{T}(4) = \dfrac{\mathbf{r}'(4)}{|\mathbf{r}'(4)|}$.

(d) $\mathbf{T}(4)$ is a unit vector in the same direction as $\mathbf{r}'(4)$, that is, parallel to the

tangent line to the curve at $\mathbf{r}(4)$ with length 1.

3. Since $(x + 2)^2 = t^2 = y - 1 \;\Rightarrow$

$y = (x + 2)^2 + 1$, the curve is a

parabola.

(a), (c)

(b) $\mathbf{r}'(t) = \langle 1, 2t \rangle$,

$\quad\mathbf{r}'(-1) = \langle 1, -2 \rangle$

5. $x = \sin t, \; y = 2 \cos t$ so

$x^2 + (y/2)^2 = 1$ and the curve is

an ellipse.

(a), (c)

(b) $\mathbf{r}'(t) = \cos t\,\mathbf{i} - 2 \sin t\,\mathbf{j}$,

$\quad\mathbf{r}'\!\left(\dfrac{\pi}{4}\right) = \dfrac{\sqrt{2}}{2}\mathbf{i} - \sqrt{2}\,\mathbf{j}$

© 2012 Cengage Learning. All Rights Reserved. May not be scanned, copied, or duplicated, or posted to a publicly accessible website, in whole or in part.

7. Since $x = e^{2t} = (e^t)^2 = y^2$, the **(a), (c)** **(b)** $\mathbf{r}'(t) = 2e^{2t}\,\mathbf{i} + e^t\,\mathbf{j}$,

curve is part of a parabola. Note $\mathbf{r}'(0) = 2\,\mathbf{i} + \mathbf{j}$

that here $x > 0$, $y > 0$.

9. $\mathbf{r}'(t) = \left\langle \dfrac{d}{dt}\left[t\sin t\right], \dfrac{d}{dt}\left[t^2\right], \dfrac{d}{dt}\left[t\cos 2t\right] \right\rangle = \langle t\cos t + \sin t, 2t, t(-\sin 2t)\cdot 2 + \cos 2t \rangle$

$\qquad = \langle t\cos t + \sin t, 2t, \cos 2t - 2t\sin 2t \rangle$

11. $\mathbf{r}(t) = t\,\mathbf{i} + \mathbf{j} + 2\sqrt{t}\,\mathbf{k} \ \Rightarrow \ \mathbf{r}'(t) = 1\,\mathbf{i} + 0\,\mathbf{j} + 2\left(\frac{1}{2}t^{-1/2}\right)\mathbf{k} = \mathbf{i} + \dfrac{1}{\sqrt{t}}\,\mathbf{k}$

13. $\mathbf{r}(t) = e^{t^2}\,\mathbf{i} - \mathbf{j} + \ln(1+3t)\,\mathbf{k} \ \Rightarrow \ \mathbf{r}'(t) = 2te^{t^2}\,\mathbf{i} + \dfrac{3}{1+3t}\,\mathbf{k}$

15. $\mathbf{r}'(t) = \mathbf{0} + \mathbf{b} + 2t\,\mathbf{c} = \mathbf{b} + 2t\,\mathbf{c}$ by Formulas 1 and 3 of Theorem 3.

17. $\mathbf{r}'(t) = \langle -te^{-t} + e^{-t}, 2/(1+t^2), 2e^t \rangle \ \Rightarrow \ \mathbf{r}'(0) = \langle 1, 2, 2 \rangle$. So $|\mathbf{r}'(0)| = \sqrt{1^2 + 2^2 + 2^2} = \sqrt{9} = 3$ and

$\qquad \mathbf{T}(0) = \dfrac{\mathbf{r}'(0)}{|\mathbf{r}'(0)|} = \frac{1}{3}\langle 1, 2, 2 \rangle = \langle \frac{1}{3}, \frac{2}{3}, \frac{2}{3} \rangle.$

19. $\mathbf{r}'(t) = -\sin t\,\mathbf{i} + 3\,\mathbf{j} + 4\cos 2t\,\mathbf{k} \ \Rightarrow \ \mathbf{r}'(0) = 3\,\mathbf{j} + 4\,\mathbf{k}$. Thus

$\qquad \mathbf{T}(0) = \dfrac{\mathbf{r}'(0)}{|\mathbf{r}'(0)|} = \dfrac{1}{\sqrt{0^2 + 3^2 + 4^2}}\,(3\,\mathbf{j} + 4\,\mathbf{k}) = \frac{1}{5}(3\,\mathbf{j} + 4\,\mathbf{k}) = \frac{3}{5}\,\mathbf{j} + \frac{4}{5}\,\mathbf{k}.$

21. $\mathbf{r}(t) = \langle t, t^2, t^3 \rangle \ \Rightarrow \ \mathbf{r}'(t) = \langle 1, 2t, 3t^2 \rangle$. Then $\mathbf{r}'(1) = \langle 1, 2, 3 \rangle$ and $|\mathbf{r}'(1)| = \sqrt{1^2 + 2^2 + 3^2} = \sqrt{14}$, so

$\qquad \mathbf{T}(1) = \dfrac{\mathbf{r}'(1)}{|\mathbf{r}'(1)|} = \frac{1}{\sqrt{14}}\langle 1, 2, 3 \rangle = \left\langle \frac{1}{\sqrt{14}}, \frac{2}{\sqrt{14}}, \frac{3}{\sqrt{14}} \right\rangle.$ $\mathbf{r}''(t) = \langle 0, 2, 6t \rangle$, so

$$\mathbf{r}'(t) \times \mathbf{r}''(t) = \begin{vmatrix} \mathbf{i} & \mathbf{j} & \mathbf{k} \\ 1 & 2t & 3t^2 \\ 0 & 2 & 6t \end{vmatrix} = \begin{vmatrix} 2t & 3t^2 \\ 2 & 6t \end{vmatrix}\mathbf{i} - \begin{vmatrix} 1 & 3t^2 \\ 0 & 6t \end{vmatrix}\mathbf{j} + \begin{vmatrix} 1 & 2t \\ 0 & 2 \end{vmatrix}\mathbf{k}$$

$$= (12t^2 - 6t^2)\,\mathbf{i} - (6t - 0)\,\mathbf{j} + (2 - 0)\,\mathbf{k} = \langle 6t^2, -6t, 2 \rangle$$

23. The vector equation for the curve is $\mathbf{r}(t) = \langle 1 + 2\sqrt{t}, t^3 - t, t^3 + t \rangle$, so $\mathbf{r}'(t) = \langle 1/\sqrt{t}, 3t^2 - 1, 3t^2 + 1 \rangle$. The point

$(3, 0, 2)$ corresponds to $t = 1$, so the tangent vector there is $\mathbf{r}'(1) = \langle 1, 2, 4 \rangle$. Thus, the tangent line goes through the point

$(3, 0, 2)$ and is parallel to the vector $\langle 1, 2, 4 \rangle$. Parametric equations are $x = 3 + t$, $y = 2t$, $z = 2 + 4t$.

25. The vector equation for the curve is $\mathbf{r}(t) = \langle e^{-t}\cos t, e^{-t}\sin t, e^{-t} \rangle$, so

$$\mathbf{r}'(t) = \langle e^{-t}(-\sin t) + (\cos t)(-e^{-t}), e^{-t}\cos t + (\sin t)(-e^{-t}), (-e^{-t}) \rangle$$

$$= \langle -e^{-t}(\cos t + \sin t), e^{-t}(\cos t - \sin t), -e^{-t} \rangle$$

The point $(1, 0, 1)$ corresponds to $t = 0$, so the tangent vector there is

$\mathbf{r}'(0) = \langle -e^0(\cos 0 + \sin 0), e^0(\cos 0 - \sin 0), -e^0 \rangle = \langle -1, 1, -1 \rangle$. Thus, the tangent line is parallel to the vector

$\langle -1, 1, -1 \rangle$ and parametric equations are $x = 1 + (-1)t = 1 - t$, $y = 0 + 1\cdot t = t$, $z = 1 + (-1)t = 1 - t$.

© 2012 Cengage Learning. All Rights Reserved. May not be scanned, copied, or duplicated, or posted to a publicly accessible website, in whole or in part.

27. First we parametrize the curve C of intersection. The projection of C onto the xy-plane is contained in the circle

$x^2 + y^2 = 25$, $z = 0$, so we can write $x = 5\cos t$, $y = 5\sin t$. C also lies on the cylinder $y^2 + z^2 = 20$, and $z \ge 0$

near the point $(3, 4, 2)$, so we can write $z = \sqrt{20 - y^2} = \sqrt{20 - 25\sin^2 t}$. A vector equation then for C is

$\mathbf{r}(t) = \left\langle 5\cos t, 5\sin t, \sqrt{20 - 25\sin^2 t} \right\rangle \Rightarrow \mathbf{r}'(t) = \left\langle -5\sin t, 5\cos t, \frac{1}{2}(20 - 25\sin^2 t)^{-1/2}(-50\sin t \cos t) \right\rangle$.

The point $(3, 4, 2)$ corresponds to $t = \cos^{-1}\left(\frac{3}{5}\right)$, so the tangent vector there is

$\mathbf{r}'\left(\cos^{-1}\left(\frac{3}{5}\right)\right) = \left\langle -5\left(\frac{4}{5}\right), 5\left(\frac{3}{5}\right), \frac{1}{2}\left(20 - 25\left(\frac{4}{5}\right)^2\right)^{-1/2}\left(-50\left(\frac{4}{5}\right)\left(\frac{3}{5}\right)\right) \right\rangle = \langle -4, 3, -6 \rangle$.

The tangent line is parallel to this vector and passes through $(3, 4, 2)$, so a vector equation for the line

is $\mathbf{r}(t) = (3 - 4t)\mathbf{i} + (4 + 3t)\mathbf{j} + (2 - 6t)\mathbf{k}$.

29. $\mathbf{r}(t) = \langle t, e^{-t}, 2t - t^2 \rangle \Rightarrow \mathbf{r}'(t) = \langle 1, -e^{-t}, 2 - 2t \rangle$. At $(0, 1, 0)$,

$t = 0$ and $\mathbf{r}'(0) = \langle 1, -1, 2 \rangle$. Thus, parametric equations of the tangent

line are $x = t$, $y = 1 - t$, $z = 2t$.

31. $\mathbf{r}(t) = \langle t\cos t, t, t\sin t \rangle \Rightarrow \mathbf{r}'(t) = \langle \cos t - t\sin t, 1, t\cos t + \sin t \rangle$.

At $(-\pi, \pi, 0)$, $t = \pi$ and $\mathbf{r}'(\pi) = \langle -1, 1, -\pi \rangle$. Thus, parametric equations

of the tangent line are $x = -\pi - t$, $y = \pi + t$, $z = -\pi t$.

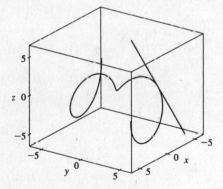

33. The angle of intersection of the two curves is the angle between the two tangent vectors to the curves at the point of

intersection. Since $\mathbf{r}_1'(t) = \langle 1, 2t, 3t^2 \rangle$ and $t = 0$ at $(0, 0, 0)$, $\mathbf{r}_1'(0) = \langle 1, 0, 0 \rangle$ is a tangent vector to $\mathbf{r}_1$ at $(0, 0, 0)$. Similarly,

$\mathbf{r}_2'(t) = \langle \cos t, 2\cos 2t, 1 \rangle$ and since $\mathbf{r}_2(0) = \langle 0, 0, 0 \rangle$, $\mathbf{r}_2'(0) = \langle 1, 2, 1 \rangle$ is a tangent vector to $\mathbf{r}_2$ at $(0, 0, 0)$. If θ is the angle

between these two tangent vectors, then $\cos\theta = \frac{1}{\sqrt{1}\sqrt{6}}\langle 1, 0, 0 \rangle \cdot \langle 1, 2, 1 \rangle = \frac{1}{\sqrt{6}}$ and $\theta = \cos^{-1}\left(\frac{1}{\sqrt{6}}\right) \approx 66°$.

35. $\int_0^2 (t\mathbf{i} - t^3\mathbf{j} + 3t^5\mathbf{k})\, dt = \left(\int_0^2 t\, dt\right)\mathbf{i} - \left(\int_0^2 t^3\, dt\right)\mathbf{j} + \left(\int_0^2 3t^5\, dt\right)\mathbf{k}$

$\qquad = \left[\frac{1}{2}t^2\right]_0^2 \mathbf{i} - \left[\frac{1}{4}t^4\right]_0^2 \mathbf{j} + \left[\frac{1}{2}t^6\right]_0^2 \mathbf{k}$

$\qquad = \frac{1}{2}(4 - 0)\mathbf{i} - \frac{1}{4}(16 - 0)\mathbf{j} + \frac{1}{2}(64 - 0)\mathbf{k} = 2\mathbf{i} - 4\mathbf{j} + 32\mathbf{k}$

© 2012 Cengage Learning. All Rights Reserved. May not be scanned, copied, or duplicated, or posted to a publicly accessible website, in whole or in part.

37. $\int_0^{\pi/2}(3\sin^2 t\,\cos t\,\mathbf{i}+3\sin t\,\cos^2 t\,\mathbf{j}+2\sin t\,\cos t\,\mathbf{k})\,dt$

$$=\left(\int_0^{\pi/2}3\sin^2 t\,\cos t\,dt\right)\mathbf{i}+\left(\int_0^{\pi/2}3\sin t\,\cos^2 t\,dt\right)\mathbf{j}+\left(\int_0^{\pi/2}2\sin t\,\cos t\,dt\right)\mathbf{k}$$

$$=\left[\sin^3 t\right]_0^{\pi/2}\mathbf{i}+\left[-\cos^3 t\right]_0^{\pi/2}\mathbf{j}+\left[\sin^2 t\right]_0^{\pi/2}\mathbf{k}=(1-0)\mathbf{i}+(0+1)\mathbf{j}+(1-0)\mathbf{k}=\mathbf{i}+\mathbf{j}+\mathbf{k}$$

39. $\int(\sec^2 t\,\mathbf{i}+t(t^2+1)^3\,\mathbf{j}+t^2\ln t\,\mathbf{k})\,dt=\left(\int\sec^2 t\,dt\right)\mathbf{i}+\left(\int t(t^2+1)^3\,dt\right)\mathbf{j}+\left(\int t^2\ln t\,dt\right)\mathbf{k}$

$$=\tan t\,\mathbf{i}+\tfrac{1}{8}(t^2+1)^4\,\mathbf{j}+\left(\tfrac{1}{3}t^3\ln t-\tfrac{1}{9}t^3\right)\mathbf{k}+\mathbf{C},$$

where $\mathbf{C}$ is a vector constant of integration. [For the z-component, integrate by parts with $u=\ln t$, $dv=t^2\,dt$.]

41. $\mathbf{r}'(t)=2t\,\mathbf{i}+3t^2\,\mathbf{j}+\sqrt{t}\,\mathbf{k}\ \Rightarrow\ \mathbf{r}(t)=t^2\,\mathbf{i}+t^3\,\mathbf{j}+\tfrac{2}{3}t^{3/2}\,\mathbf{k}+\mathbf{C},$ where $\mathbf{C}$ is a constant vector.

But $\mathbf{i}+\mathbf{j}=\mathbf{r}(1)=\mathbf{i}+\mathbf{j}+\tfrac{2}{3}\mathbf{k}+\mathbf{C}$. Thus $\mathbf{C}=-\tfrac{2}{3}\mathbf{k}$ and $\mathbf{r}(t)=t^2\,\mathbf{i}+t^3\,\mathbf{j}+\left(\tfrac{2}{3}t^{3/2}-\tfrac{2}{3}\right)\mathbf{k}.$

For Exercises 43–46, let $\mathbf{u}(t)=\langle u_1(t),u_2(t),u_3(t)\rangle$ and $\mathbf{v}(t)=\langle v_1(t),v_2(t),v_3(t)\rangle$. In each of these exercises, the procedure is to apply Theorem 2 so that the corresponding properties of derivatives of real-valued functions can be used.

43. $\dfrac{d}{dt}[\mathbf{u}(t)+\mathbf{v}(t)]=\dfrac{d}{dt}\langle u_1(t)+v_1(t),u_2(t)+v_2(t),u_3(t)+v_3(t)\rangle$

$$=\left\langle\frac{d}{dt}[u_1(t)+v_1(t)],\frac{d}{dt}[u_2(t)+v_2(t)],\frac{d}{dt}[u_3(t)+v_3(t)]\right\rangle$$

$$=\langle u_1'(t)+v_1'(t),u_2'(t)+v_2'(t),u_3'(t)+v_3'(t)\rangle$$

$$=\langle u_1'(t),u_2'(t),u_3'(t)\rangle+\langle v_1'(t),v_2'(t),v_3'(t)\rangle=\mathbf{u}'(t)+\mathbf{v}'(t)$$

45. $\dfrac{d}{dt}[\mathbf{u}(t)\times\mathbf{v}(t)]=\dfrac{d}{dt}\langle u_2(t)v_3(t)-u_3(t)v_2(t),u_3(t)v_1(t)-u_1(t)v_3(t),u_1(t)v_2(t)-u_2(t)v_1(t)\rangle$

$$=\langle u_2'v_3(t)+u_2(t)v_3'(t)-u_3'(t)v_2(t)-u_3(t)v_2'(t),$$

$$u_3'(t)v_1(t)+u_3(t)v_1'(t)-u_1'(t)v_3(t)-u_1(t)v_3'(t),$$

$$u_1'(t)v_2(t)+u_1(t)v_2'(t)-u_2'(t)v_1(t)-u_2(t)v_1'(t)\rangle$$

$$=\langle u_2'(t)v_3(t)-u_3'(t)v_2(t),u_3'(t)v_1(t)-u_1'(t)v_3(t),u_1'(t)v_2(t)-u_2'(t)v_1(t)\rangle$$

$$+\langle u_2(t)v_3'(t)-u_3(t)v_2'(t),u_3(t)v_1'(t)-u_1(t)v_3'(t),u_1(t)v_2'(t)-u_2(t)v_1'(t)\rangle$$

$$=\mathbf{u}'(t)\times\mathbf{v}(t)+\mathbf{u}(t)\times\mathbf{v}'(t)$$

Alternate solution: Let $\mathbf{r}(t)=\mathbf{u}(t)\times\mathbf{v}(t)$. Then

$$\mathbf{r}(t+h)-\mathbf{r}(t)=[\mathbf{u}(t+h)\times\mathbf{v}(t+h)]-[\mathbf{u}(t)\times\mathbf{v}(t)]$$

$$=[\mathbf{u}(t+h)\times\mathbf{v}(t+h)]-[\mathbf{u}(t)\times\mathbf{v}(t)]+[\mathbf{u}(t+h)\times\mathbf{v}(t)]-[\mathbf{u}(t+h)\times\mathbf{v}(t)]$$

$$=\mathbf{u}(t+h)\times[\mathbf{v}(t+h)-\mathbf{v}(t)]+[\mathbf{u}(t+h)-\mathbf{u}(t)]\times\mathbf{v}(t)$$

(Be careful of the order of the cross product.) Dividing through by h and taking the limit as $h\to 0$ we have

$$\mathbf{r}'(t)=\lim_{h\to 0}\frac{\mathbf{u}(t+h)\times[\mathbf{v}(t+h)-\mathbf{v}(t)]}{h}+\lim_{h\to 0}\frac{[\mathbf{u}(t+h)-\mathbf{u}(t)]\times\mathbf{v}(t)}{h}=\mathbf{u}(t)\times\mathbf{v}'(t)+\mathbf{u}'(t)\times\mathbf{v}(t)$$

by Exercise 13.1.49(a) and Definition 1.

© 2012 Cengage Learning. All Rights Reserved. May not be scanned, copied, or duplicated, or posted to a publicly accessible website, in whole or in part.

47. $\dfrac{d}{dt}[\mathbf{u}(t) \cdot \mathbf{v}(t)] = \mathbf{u}'(t) \cdot \mathbf{v}(t) + \mathbf{u}(t) \cdot \mathbf{v}'(t)$ [by Formula 4 of Theorem 3]

$\qquad = \langle \cos t, -\sin t, 1 \rangle \cdot \langle t, \cos t, \sin t \rangle + \langle \sin t, \cos t, t \rangle \cdot \langle 1, -\sin t, \cos t \rangle$

$\qquad = t \cos t - \cos t \, \sin t + \sin t + \sin t - \cos t \, \sin t + t \cos t$

$\qquad = 2t \cos t + 2 \sin t - 2 \cos t \, \sin t$

49. By Formula 4 of Theorem 3, $f'(t) = \mathbf{u}'(t) \cdot \mathbf{v}(t) + \mathbf{u}(t) \cdot \mathbf{v}'(t)$, and $\mathbf{v}'(t) = \langle 1, 2t, 3t^2 \rangle$, so

$\quad f'(2) = \mathbf{u}'(2) \cdot \mathbf{v}(2) + \mathbf{u}(2) \cdot \mathbf{v}'(2) = \langle 3, 0, 4 \rangle \cdot \langle 2, 4, 8 \rangle + \langle 1, 2, -1 \rangle \cdot \langle 1, 4, 12 \rangle = 6 + 0 + 32 + 1 + 8 - 12 = 35.$

51. $\dfrac{d}{dt}[\mathbf{r}(t) \times \mathbf{r}'(t)] = \mathbf{r}'(t) \times \mathbf{r}'(t) + \mathbf{r}(t) \times \mathbf{r}''(t)$ by Formula 5 of Theorem 3. But $\mathbf{r}'(t) \times \mathbf{r}'(t) = \mathbf{0}$ (by Example 2 in

Section 12.4). Thus, $\dfrac{d}{dt}[\mathbf{r}(t) \times \mathbf{r}'(t)] = \mathbf{r}(t) \times \mathbf{r}''(t)$.

53. $\dfrac{d}{dt}|\mathbf{r}(t)| = \dfrac{d}{dt}[\mathbf{r}(t) \cdot \mathbf{r}(t)]^{1/2} = \tfrac{1}{2}[\mathbf{r}(t) \cdot \mathbf{r}(t)]^{-1/2}[2\mathbf{r}(t) \cdot \mathbf{r}'(t)] = \dfrac{1}{|\mathbf{r}(t)|}\,\mathbf{r}(t) \cdot \mathbf{r}'(t)$

55. Since $\mathbf{u}(t) = \mathbf{r}(t) \cdot [\mathbf{r}'(t) \times \mathbf{r}''(t)]$,

$\qquad \mathbf{u}'(t) = \mathbf{r}'(t) \cdot [\mathbf{r}'(t) \times \mathbf{r}''(t)] + \mathbf{r}(t) \cdot \dfrac{d}{dt}[\mathbf{r}'(t) \times \mathbf{r}''(t)]$

$\qquad\quad = 0 + \mathbf{r}(t) \cdot [\mathbf{r}''(t) \times \mathbf{r}''(t) + \mathbf{r}'(t) \times \mathbf{r}'''(t)]$ [since $\mathbf{r}'(t) \perp \mathbf{r}'(t) \times \mathbf{r}''(t)$]

$\qquad\quad = \mathbf{r}(t) \cdot [\mathbf{r}'(t) \times \mathbf{r}'''(t)]$ [since $\mathbf{r}''(t) \times \mathbf{r}''(t) = \mathbf{0}$]

13.3 Arc Length and Curvature

1. $\mathbf{r}(t) = \langle t, 3\cos t, 3\sin t \rangle \;\Rightarrow\; \mathbf{r}'(t) = \langle 1, -3\sin t, 3\cos t \rangle \;\Rightarrow$

$\quad |\mathbf{r}'(t)| = \sqrt{1^2 + (-3\sin t)^2 + (3\cos t)^2} = \sqrt{1 + 9(\sin^2 t + \cos^2 t)} = \sqrt{10}.$

Then using Formula 3, we have $L = \int_{-5}^{5}|\mathbf{r}'(t)|\,dt = \int_{-5}^{5}\sqrt{10}\,dt = \sqrt{10}\,t\Big]_{-5}^{5} = 10\sqrt{10}.$

3. $\mathbf{r}(t) = \sqrt{2}\,t\,\mathbf{i} + e^t\mathbf{j} + e^{-t}\mathbf{k} \;\Rightarrow\; \mathbf{r}'(t) = \sqrt{2}\,\mathbf{i} + e^t\mathbf{j} - e^{-t}\mathbf{k} \;\Rightarrow$

$\quad |\mathbf{r}'(t)| = \sqrt{\left(\sqrt{2}\right)^2 + (e^t)^2 + (-e^{-t})^2} = \sqrt{2 + e^{2t} + e^{-2t}} = \sqrt{(e^t + e^{-t})^2} = e^t + e^{-t}$ [since $e^t + e^{-t} > 0$].

Then $L = \int_0^1 |\mathbf{r}'(t)|\,dt = \int_0^1 (e^t + e^{-t})\,dt = \left[e^t - e^{-t}\right]_0^1 = e - e^{-1}.$

5. $\mathbf{r}(t) = \mathbf{i} + t^2\mathbf{j} + t^3\mathbf{k} \;\Rightarrow\; \mathbf{r}'(t) = 2t\mathbf{j} + 3t^2\mathbf{k} \;\Rightarrow\; |\mathbf{r}'(t)| = \sqrt{4t^2 + 9t^4} = t\sqrt{4 + 9t^2}$ [since $t \geq 0$].

Then $L = \int_0^1 |\mathbf{r}'(t)|\,dt = \int_0^1 t\sqrt{4 + 9t^2}\,dt = \tfrac{1}{18} \cdot \tfrac{2}{3}(4 + 9t^2)^{3/2}\Big]_0^1 = \tfrac{1}{27}(13^{3/2} - 4^{3/2}) = \tfrac{1}{27}(13^{3/2} - 8).$

7. $\mathbf{r}(t) = \langle t^2, t^3, t^4 \rangle \;\Rightarrow\; \mathbf{r}'(t) = \langle 2t, 3t^2, 4t^3 \rangle \;\Rightarrow\; |\mathbf{r}'(t)| = \sqrt{(2t)^2 + (3t^2)^2 + (4t^3)^2} = \sqrt{4t^2 + 9t^4 + 16t^6}$, so

$\quad L = \int_0^2 |\mathbf{r}'(t)|\,dt = \int_0^2 \sqrt{4t^2 + 9t^4 + 16t^6}\,dt \approx 18.6833.$

© 2012 Cengage Learning. All Rights Reserved. May not be scanned, copied, or duplicated, or posted to a publicly accessible website, in whole or in part.

9. $\mathbf{r}(t) = \langle \sin t, \cos t, \tan t \rangle \;\Rightarrow\; \mathbf{r}'(t) = \langle \cos t, -\sin t, \sec^2 t \rangle \;\Rightarrow$

$|\mathbf{r}'(t)| = \sqrt{\cos^2 t + (-\sin t)^2 + (\sec^2 t)^2} = \sqrt{1 + \sec^4 t}$ and $L = \int_0^{\pi/4} |\mathbf{r}'(t)|\, dt = \int_0^{\pi/4} \sqrt{1 + \sec^4 t}\, dt \approx 1.2780$.

11. The projection of the curve C onto the xy-plane is the curve $x^2 = 2y$ or $y = \frac{1}{2}x^2$, $z = 0$. Then we can choose the parameter

$x = t \;\Rightarrow\; y = \frac{1}{2}t^2$. Since C also lies on the surface $3z = xy$, we have $z = \frac{1}{3}xy = \frac{1}{3}(t)(\frac{1}{2}t^2) = \frac{1}{6}t^3$. Then parametric

equations for C are $x = t$, $y = \frac{1}{2}t^2$, $z = \frac{1}{6}t^3$ and the corresponding vector equation is $\mathbf{r}(t) = \langle t, \frac{1}{2}t^2, \frac{1}{6}t^3 \rangle$. The origin

corresponds to $t = 0$ and the point $(6, 18, 36)$ corresponds to $t = 6$, so

$$L = \int_0^6 |\mathbf{r}'(t)|\, dt = \int_0^6 \left| \langle 1, t, \tfrac{1}{2}t^2 \rangle \right| dt = \int_0^6 \sqrt{1^2 + t^2 + \left(\tfrac{1}{2}t^2\right)^2}\, dt = \int_0^6 \sqrt{1 + t^2 + \tfrac{1}{4}t^4}\, dt$$

$$= \int_0^6 \sqrt{(1 + \tfrac{1}{2}t^2)^2}\, dt = \int_0^6 (1 + \tfrac{1}{2}t^2)\, dt = \left[t + \tfrac{1}{6}t^3 \right]_0^6 = 6 + 36 = 42$$

13. $\mathbf{r}(t) = 2t\,\mathbf{i} + (1 - 3t)\,\mathbf{j} + (5 + 4t)\,\mathbf{k} \;\Rightarrow\; \mathbf{r}'(t) = 2\,\mathbf{i} - 3\,\mathbf{j} + 4\,\mathbf{k}$ and $\frac{ds}{dt} = |\mathbf{r}'(t)| = \sqrt{4 + 9 + 16} = \sqrt{29}$. Then

$s = s(t) = \int_0^t |\mathbf{r}'(u)|\, du = \int_0^t \sqrt{29}\, du = \sqrt{29}\, t$. Therefore, $t = \frac{1}{\sqrt{29}}s$, and substituting for t in the original equation, we

have $\mathbf{r}(t(s)) = \frac{2}{\sqrt{29}}s\,\mathbf{i} + \left(1 - \frac{3}{\sqrt{29}}s\right)\mathbf{j} + \left(5 + \frac{4}{\sqrt{29}}s\right)\mathbf{k}$.

15. Here $\mathbf{r}(t) = \langle 3\sin t, 4t, 3\cos t \rangle$, so $\mathbf{r}'(t) = \langle 3\cos t, 4, -3\sin t \rangle$ and $|\mathbf{r}'(t)| = \sqrt{9\cos^2 t + 16 + 9\sin^2 t} = \sqrt{25} = 5$.

The point $(0, 0, 3)$ corresponds to $t = 0$, so the arc length function beginning at $(0, 0, 3)$ and measuring in the positive

direction is given by $s(t) = \int_0^t |\mathbf{r}'(u)|\, du = \int_0^t 5\, du = 5t$. $s(t) = 5 \;\Rightarrow\; 5t = 5 \;\Rightarrow\; t = 1$, thus your location after

moving 5 units along the curve is $(3\sin 1, 4, 3\cos 1)$.

17. (a) $\mathbf{r}(t) = \langle t, 3\cos t, 3\sin t \rangle \;\Rightarrow\; \mathbf{r}'(t) = \langle 1, -3\sin t, 3\cos t \rangle \;\Rightarrow\; |\mathbf{r}'(t)| = \sqrt{1 + 9\sin^2 t + 9\cos^2 t} = \sqrt{10}$.

Then $\mathbf{T}(t) = \dfrac{\mathbf{r}'(t)}{|\mathbf{r}'(t)|} = \frac{1}{\sqrt{10}} \langle 1, -3\sin t, 3\cos t \rangle$ or $\left\langle \frac{1}{\sqrt{10}}, -\frac{3}{\sqrt{10}}\sin t, \frac{3}{\sqrt{10}}\cos t \right\rangle$.

$\mathbf{T}'(t) = \frac{1}{\sqrt{10}} \langle 0, -3\cos t, -3\sin t \rangle \;\Rightarrow\; |\mathbf{T}'(t)| = \frac{1}{\sqrt{10}} \sqrt{0 + 9\cos^2 t + 9\sin^2 t} = \frac{3}{\sqrt{10}}$. Thus

$\mathbf{N}(t) = \dfrac{\mathbf{T}'(t)}{|\mathbf{T}'(t)|} = \dfrac{1/\sqrt{10}}{3/\sqrt{10}} \langle 0, -3\cos t, -3\sin t \rangle = \langle 0, -\cos t, -\sin t \rangle$.

(b) $\kappa(t) = \dfrac{|\mathbf{T}'(t)|}{|\mathbf{r}'(t)|} = \dfrac{3/\sqrt{10}}{\sqrt{10}} = \dfrac{3}{10}$

19. (a) $\mathbf{r}(t) = \langle \sqrt{2}\,t, e^t, e^{-t} \rangle \;\Rightarrow\; \mathbf{r}'(t) = \langle \sqrt{2}, e^t, -e^{-t} \rangle \;\Rightarrow\; |\mathbf{r}'(t)| = \sqrt{2 + e^{2t} + e^{-2t}} = \sqrt{(e^t + e^{-t})^2} = e^t + e^{-t}$.
Then

$\mathbf{T}(t) = \dfrac{\mathbf{r}'(t)}{|\mathbf{r}'(t)|} = \dfrac{1}{e^t + e^{-t}} \langle \sqrt{2}, e^t, -e^{-t} \rangle = \dfrac{1}{e^{2t} + 1} \langle \sqrt{2}\,e^t, e^{2t}, -1 \rangle \quad \left[\text{after multiplying by } \dfrac{e^t}{e^t}\right]$ and

$\mathbf{T}'(t) = \dfrac{1}{e^{2t} + 1} \langle \sqrt{2}\,e^t, 2e^{2t}, 0 \rangle - \dfrac{2e^{2t}}{(e^{2t} + 1)^2} \langle \sqrt{2}\,e^t, e^{2t}, -1 \rangle$

$= \dfrac{1}{(e^{2t} + 1)^2} \left[(e^{2t} + 1) \langle \sqrt{2}\,e^t, 2e^{2t}, 0 \rangle - 2e^{2t} \langle \sqrt{2}\,e^t, e^{2t}, -1 \rangle \right] = \dfrac{1}{(e^{2t} + 1)^2} \langle \sqrt{2}\,e^t (1 - e^{2t}), 2e^{2t}, 2e^{2t} \rangle$

© 2012 Cengage Learning. All Rights Reserved. May not be scanned, copied, or duplicated, or posted to a publicly accessible website, in whole or in part.

Then

$$|\mathbf{T}'(t)| = \frac{1}{(e^{2t}+1)^2}\sqrt{2e^{2t}(1-2e^{2t}+e^{4t})+4e^{4t}+4e^{4t}} = \frac{1}{(e^{2t}+1)^2}\sqrt{2e^{2t}(1+2e^{2t}+e^{4t})}$$

$$= \frac{1}{(e^{2t}+1)^2}\sqrt{2e^{2t}(1+e^{2t})^2} = \frac{\sqrt{2}\,e^t(1+e^{2t})}{(e^{2t}+1)^2} = \frac{\sqrt{2}\,e^t}{e^{2t}+1}$$

Therefore

$$\mathbf{N}(t) = \frac{\mathbf{T}'(t)}{|\mathbf{T}'(t)|} = \frac{e^{2t}+1}{\sqrt{2}\,e^t}\frac{1}{(e^{2t}+1)^2}\left\langle \sqrt{2}\,e^t(1-e^{2t}), 2e^{2t}, 2e^{2t}\right\rangle$$

$$= \frac{1}{\sqrt{2}\,e^t(e^{2t}+1)}\left\langle \sqrt{2}\,e^t(1-e^{2t}), 2e^{2t}, 2e^{2t}\right\rangle = \frac{1}{e^{2t}+1}\left\langle 1-e^{2t}, \sqrt{2}\,e^t, \sqrt{2}\,e^t\right\rangle$$

(b) $\kappa(t) = \dfrac{|\mathbf{T}'(t)|}{|\mathbf{r}'(t)|} = \dfrac{\sqrt{2}\,e^t}{e^{2t}+1}\cdot\dfrac{1}{e^t+e^{-t}} = \dfrac{\sqrt{2}\,e^t}{e^{3t}+2e^t+e^{-t}} = \dfrac{\sqrt{2}\,e^{2t}}{e^{4t}+2e^{2t}+1} = \dfrac{\sqrt{2}\,e^{2t}}{(e^{2t}+1)^2}$

21. $\mathbf{r}(t) = t^3\,\mathbf{j} + t^2\,\mathbf{k} \;\Rightarrow\; \mathbf{r}'(t) = 3t^2\,\mathbf{j} + 2t\,\mathbf{k},\;\; \mathbf{r}''(t) = 6t\,\mathbf{j} + 2\,\mathbf{k},\;\; |\mathbf{r}'(t)| = \sqrt{0^2 + (3t^2)^2 + (2t)^2} = \sqrt{9t^4 + 4t^2},$

$\mathbf{r}'(t) \times \mathbf{r}''(t) = -6t^2\,\mathbf{i},\;\; |\mathbf{r}'(t) \times \mathbf{r}''(t)| = 6t^2.$ Then $\kappa(t) = \dfrac{|\mathbf{r}'(t) \times \mathbf{r}''(t)|}{|\mathbf{r}'(t)|^3} = \dfrac{6t^2}{\left(\sqrt{9t^4+4t^2}\right)^3} = \dfrac{6t^2}{(9t^4+4t^2)^{3/2}}.$

23. $\mathbf{r}(t) = 3t\,\mathbf{i} + 4\sin t\,\mathbf{j} + 4\cos t\,\mathbf{k} \;\Rightarrow\; \mathbf{r}'(t) = 3\,\mathbf{i} + 4\cos t\,\mathbf{j} - 4\sin t\,\mathbf{k},\;\; \mathbf{r}''(t) = -4\sin t\,\mathbf{j} - 4\cos t\,\mathbf{k},$

$|\mathbf{r}'(t)| = \sqrt{9 + 16\cos^2 t + 16\sin^2 t} = \sqrt{9+16} = 5,\;\; \mathbf{r}'(t) \times \mathbf{r}''(t) = -16\,\mathbf{i} + 12\cos t\,\mathbf{j} - 12\sin t\,\mathbf{k},$

$|\mathbf{r}'(t) \times \mathbf{r}''(t)| = \sqrt{256 + 144\cos^2 t + 144\sin^2 t} = \sqrt{400} = 20.$ Then $\kappa(t) = \dfrac{|\mathbf{r}'(t) \times \mathbf{r}''(t)|}{|\mathbf{r}'(t)|^3} = \dfrac{20}{5^3} = \dfrac{4}{25}.$

25. $\mathbf{r}(t) = \langle t, t^2, t^3\rangle \;\Rightarrow\; \mathbf{r}'(t) = \langle 1, 2t, 3t^2\rangle.$ The point $(1,1,1)$ corresponds to $t = 1$, and $\mathbf{r}'(1) = \langle 1, 2, 3\rangle \;\Rightarrow$

$|\mathbf{r}'(1)| = \sqrt{1+4+9} = \sqrt{14}.\;\; \mathbf{r}''(t) = \langle 0, 2, 6t\rangle \;\Rightarrow\; \mathbf{r}''(1) = \langle 0, 2, 6\rangle.\;\; \mathbf{r}'(1) \times \mathbf{r}''(1) = \langle 6, -6, 2\rangle,$ so

$|\mathbf{r}'(1) \times \mathbf{r}''(1)| = \sqrt{36+36+4} = \sqrt{76}.$ Then $\kappa(1) = \dfrac{|\mathbf{r}'(1) \times \mathbf{r}''(1)|}{|\mathbf{r}'(1)|^3} = \dfrac{\sqrt{76}}{\sqrt{14}^3} = \dfrac{1}{7}\sqrt{\dfrac{19}{14}}.$

27. $f(x) = x^4,\;\; f'(x) = 4x^3,\;\; f''(x) = 12x^2,\;\; \kappa(x) = \dfrac{|f''(x)|}{[1 + (f'(x))^2]^{3/2}} = \dfrac{|12x^2|}{[1 + (4x^3)^2]^{3/2}} = \dfrac{12x^2}{(1 + 16x^6)^{3/2}}$

29. $f(x) = xe^x,\;\; f'(x) = xe^x + e^x,\;\; f''(x) = xe^x + 2e^x,$

$\kappa(x) = \dfrac{|f''(x)|}{[1 + (f'(x))^2]^{3/2}} = \dfrac{|xe^x + 2e^x|}{[1 + (xe^x + e^x)^2]^{3/2}} = \dfrac{|x + 2|\,e^x}{[1 + (xe^x + e^x)^2]^{3/2}}$

31. Since $y' = y'' = e^x$, the curvature is $\kappa(x) = \dfrac{|y''(x)|}{[1 + (y'(x))^2]^{3/2}} = \dfrac{e^x}{(1 + e^{2x})^{3/2}} = e^x(1 + e^{2x})^{-3/2}.$

To find the maximum curvature, we first find the critical numbers of $\kappa(x)$:

$\kappa'(x) = e^x(1 + e^{2x})^{-3/2} + e^x\left(-\tfrac{3}{2}\right)(1 + e^{2x})^{-5/2}(2e^{2x}) = e^x\dfrac{1 + e^{2x} - 3e^{2x}}{(1 + e^{2x})^{5/2}} = e^x\dfrac{1 - 2e^{2x}}{(1 + e^{2x})^{5/2}}.$

$\kappa'(x) = 0$ when $1 - 2e^{2x} = 0$, so $e^{2x} = \tfrac{1}{2}$ or $x = -\tfrac{1}{2}\ln 2$. And since $1 - 2e^{2x} > 0$ for $x < -\tfrac{1}{2}\ln 2$ and $1 - 2e^{2x} < 0$

© 2012 Cengage Learning. All Rights Reserved. May not be scanned, copied, or duplicated, or posted to a publicly accessible website, in whole or in part.

for $x > -\frac{1}{2}\ln 2$, the maximum curvature is attained at the point $\left(-\frac{1}{2}\ln 2, e^{(-\ln 2)/2}\right) = \left(-\frac{1}{2}\ln 2, \frac{1}{\sqrt{2}}\right)$.

Since $\lim\limits_{x \to \infty} e^x(1 + e^{2x})^{-3/2} = 0$, $\kappa(x)$ approaches 0 as $x \to \infty$.

33. (a) C appears to be changing direction more quickly at P than Q, so we would expect the curvature to be greater at P.

(b) First we sketch approximate osculating circles at P and Q. Using the

axes scale as a guide, we measure the radius of the osculating circle

at P to be approximately 0.8 units, thus $\rho = \dfrac{1}{\kappa}$ $\Rightarrow$

$\kappa = \dfrac{1}{\rho} \approx \dfrac{1}{0.8} \approx 1.3$. Similarly, we estimate the radius of the

osculating circle at Q to be 1.4 units, so $\kappa = \dfrac{1}{\rho} \approx \dfrac{1}{1.4} \approx 0.7$.

35. $y = x^{-2}$ $\Rightarrow$ $y' = -2x^{-3}$, $y'' = 6x^{-4}$, and

$$\kappa(x) = \frac{|y''|}{\left[1 + (y')^2\right]^{3/2}} = \frac{\left|6x^{-4}\right|}{\left[1 + (-2x^{-3})^2\right]^{3/2}} = \frac{6}{x^4\left(1 + 4x^{-6}\right)^{3/2}}.$$

The appearance of the two humps in this graph is perhaps a little surprising, but it is

explained by the fact that $y = x^{-2}$ increases asymptotically at the origin from both

directions, and so its graph has very little bend there. [Note that $\kappa(0)$ is undefined.]

37. $\mathbf{r}(t) = \langle te^t, e^{-t}, \sqrt{2}\,t \rangle$ $\Rightarrow$ $\mathbf{r}'(t) = \langle (t+1)e^t, -e^{-t}, \sqrt{2} \rangle$, $\mathbf{r}''(t) = \langle (t+2)e^t, e^{-t}, 0 \rangle$. Then

$\mathbf{r}'(t) \times \mathbf{r}''(t) = \langle -\sqrt{2}e^{-t}, \sqrt{2}(t+2)e^t, 2t+3 \rangle$, $|\mathbf{r}'(t) \times \mathbf{r}''(t)| = \sqrt{2e^{-2t} + 2(t+2)^2 e^{2t} + (2t+3)^2}$,

$|\mathbf{r}'(t)| = \sqrt{(t+1)^2 e^{2t} + e^{-2t} + 2}$, and $\kappa(t) = \dfrac{|\mathbf{r}'(t) \times \mathbf{r}''(t)|}{|\mathbf{r}'(t)|^3} = \dfrac{\sqrt{2e^{-2t} + 2(t+2)^2 e^{2t} + (2t+3)^2}}{\left[(t+1)^2 e^{2t} + e^{-2t} + 2\right]^{3/2}}$.

We plot the space curve and its curvature function for $-5 \le t \le 5$ below.

From the graph of $\kappa(t)$ we see that curvature is maximized for $t = 0$, so the curve bends most sharply at the point $(0, 1, 0)$.

The curve bends more gradually as we move away from this point, becoming almost linear. This is reflected in the curvature

graph, where $\kappa(t)$ becomes nearly 0 as $|t|$ increases.

39. Notice that the curve b has two inflection points at which the graph appears almost straight. We would expect the curvature to

be 0 or nearly 0 at these values, but the curve a isn't near 0 there. Thus, a must be the graph of $y = f(x)$ rather than the graph

of curvature, and b is the graph of $y = \kappa(x)$.

© 2012 Cengage Learning. All Rights Reserved. May not be scanned, copied, or duplicated, or posted to a publicly accessible website, in whole or in part.

41. Using a CAS, we find (after simplifying)

$$\kappa(t) = \frac{6\sqrt{4\cos^2 t - 12\cos t + 13}}{(17 - 12\cos t)^{3/2}}. \text{ (To compute cross}$$

products in Maple, use the `VectorCalculus` or

`LinearAlgebra` package and the `CrossProduct(a,b)`

command; in Mathematica, use `Cross[a,b]`.) Curvature is

largest at integer multiples of 2π.

43. $x = t^2 \;\Rightarrow\; \dot{x} = 2t \;\Rightarrow\; \ddot{x} = 2, \;\; y = t^3 \;\Rightarrow\; \dot{y} = 3t^2 \;\Rightarrow\; \ddot{y} = 6t.$

Then $\kappa(t) = \dfrac{|\dot{x}\ddot{y} - \dot{y}\ddot{x}|}{[\dot{x}^2 + \dot{y}^2]^{3/2}} = \dfrac{|(2t)(6t) - (3t^2)(2)|}{[(2t)^2 + (3t^2)^2]^{3/2}} = \dfrac{|12t^2 - 6t^2|}{(4t^2 + 9t^4)^{3/2}} = \dfrac{6t^2}{(4t^2 + 9t^4)^{3/2}}.$

45. $x = e^t\cos t \;\Rightarrow\; \dot{x} = e^t(\cos t - \sin t) \;\Rightarrow\; \ddot{x} = e^t(-\sin t - \cos t) + e^t(\cos t - \sin t) = -2e^t\sin t,$

$y = e^t\sin t \;\Rightarrow\; \dot{y} = e^t(\cos t + \sin t) \;\Rightarrow\; \ddot{y} = e^t(-\sin t + \cos t) + e^t(\cos t + \sin t) = 2e^t\cos t.$ Then

$$\kappa(t) = \frac{|\dot{x}\ddot{y} - \dot{y}\ddot{x}|}{[\dot{x}^2 + \dot{y}^2]^{3/2}} = \frac{\left|e^t(\cos t - \sin t)(2e^t\cos t) - e^t(\cos t + \sin t)(-2e^t\sin t)\right|}{\left([e^t(\cos t - \sin t)]^2 + [e^t(\cos t + \sin t)]^2\right)^{3/2}}$$

$$= \frac{\left|2e^{2t}(\cos^2 t - \sin t\cos t + \sin t\cos t + \sin^2 t)\right|}{\left[e^{2t}(\cos^2 t - 2\cos t\sin t + \sin^2 t + \cos^2 t + 2\cos t\sin t + \sin^2 t)\right]^{3/2}} - \frac{\left|2e^{2t}(1)\right|}{\left[e^{2t}(1 + 1)\right]^{3/2}} - \frac{2e^{2t}}{e^{3t}(2)^{3/2}} = \frac{1}{\sqrt{2}\,e^t}$$

47. $\left(1, \frac{2}{3}, 1\right)$ corresponds to $t = 1$. $\mathbf{T}(t) = \dfrac{\mathbf{r}'(t)}{|\mathbf{r}'(t)|} = \dfrac{\langle 2t, 2t^2, 1\rangle}{\sqrt{4t^2 + 4t^4 + 1}} = \dfrac{\langle 2t, 2t^2, 1\rangle}{2t^2 + 1},$ so $\mathbf{T}(1) = \left\langle \frac{2}{3}, \frac{2}{3}, \frac{1}{3}\right\rangle.$

$\mathbf{T}'(t) = -4t(2t^2 + 1)^{-2}\langle 2t, 2t^2, 1\rangle + (2t^2 + 1)^{-1}\langle 2, 4t, 0\rangle$ [by Formula 3 of Theorem 13.2.3]

$\qquad = (2t^2 + 1)^{-2}\langle -8t^2 + 4t^2 + 2, -8t^3 + 8t^3 + 4t, -4t\rangle = 2(2t^2 + 1)^{-2}\langle 1 - 2t^2, 2t, -2t\rangle$

$\mathbf{N}(t) = \dfrac{\mathbf{T}'(t)}{|\mathbf{T}'(t)|} = \dfrac{2(2t^2 + 1)^{-2}\langle 1 - 2t^2, 2t, -2t\rangle}{2(2t^2 + 1)^{-2}\sqrt{(1 - 2t^2)^2 + (2t)^2 + (-2t)^2}} = \dfrac{\langle 1 - 2t^2, 2t, -2t\rangle}{\sqrt{1 - 4t^2 + 4t^4 + 8t^2}} = \dfrac{\langle 1 - 2t^2, 2t, -2t\rangle}{1 + 2t^2}$

$\mathbf{N}(1) = \left\langle -\frac{1}{3}, \frac{2}{3}, -\frac{2}{3}\right\rangle$ and $\mathbf{B}(1) = \mathbf{T}(1) \times \mathbf{N}(1) = \left\langle -\frac{4}{9} - \frac{2}{9}, -\left(-\frac{4}{9} + \frac{1}{9}\right), \frac{4}{9} + \frac{2}{9}\right\rangle = \left\langle -\frac{2}{3}, \frac{1}{3}, \frac{2}{3}\right\rangle.$

49. $(0, \pi, -2)$ corresponds to $t = \pi$. $\mathbf{r}(t) = \langle 2\sin 3t, t, 2\cos 3t\rangle \;\Rightarrow$

$\mathbf{T}(t) = \dfrac{\mathbf{r}'(t)}{|\mathbf{r}'(t)|} = \dfrac{\langle 6\cos 3t, 1, -6\sin 3t\rangle}{\sqrt{36\cos^2 3t + 1 + 36\sin^2 3t}} = \dfrac{1}{\sqrt{37}}\langle 6\cos 3t, 1, -6\sin 3t\rangle.$

$\mathbf{T}(\pi) = \frac{1}{\sqrt{37}}\langle -6, 1, 0\rangle$ is a normal vector for the normal plane, and so $\langle -6, 1, 0\rangle$ is also normal. Thus an equation for the

plane is $-6(x - 0) + 1(y - \pi) + 0(z + 2) = 0$ or $y - 6x = \pi.$

$\mathbf{T}'(t) = \frac{1}{\sqrt{37}}\langle -18\sin 3t, 0, -18\cos 3t\rangle \;\Rightarrow\; |\mathbf{T}'(t)| = \dfrac{\sqrt{18^2\sin^2 3t + 18^2\cos^2 3t}}{\sqrt{37}} = \dfrac{18}{\sqrt{37}} \;\Rightarrow$

$\mathbf{N}(t) = \dfrac{\mathbf{T}'(t)}{|\mathbf{T}'(t)|} = \langle -\sin 3t, 0, -\cos 3t\rangle.$ So $\mathbf{N}(\pi) = \langle 0, 0, 1\rangle$ and $\mathbf{B}(\pi) = \frac{1}{\sqrt{37}}\langle -6, 1, 0\rangle \times \langle 0, 0, 1\rangle = \frac{1}{\sqrt{37}}\langle 1, 6, 0\rangle.$

Since $\mathbf{B}(\pi)$ is a normal to the osculating plane, so is $\langle 1, 6, 0\rangle.$

An equation for the plane is $1(x - 0) + 6(y - \pi) + 0(z + 2) = 0$ or $x + 6y = 6\pi.$

© 2012 Cengage Learning. All Rights Reserved. May not be scanned, copied, or duplicated, or posted to a publicly accessible website, in whole or in part.

51. The ellipse is given by the parametric equations $x = 2\cos t$, $y = 3\sin t$, so using the result from Exercise 42,

$$\kappa(t) = \frac{|\dot{x}\ddot{y} - \ddot{x}\dot{y}|}{[\dot{x}^2 + \dot{y}^2]^{3/2}} = \frac{|(-2\sin t)(-3\sin t) - (3\cos t)(-2\cos t)|}{(4\sin^2 t + 9\cos^2 t)^{3/2}} = \frac{6}{(4\sin^2 t + 9\cos^2 t)^{3/2}}.$$

At $(2, 0)$, $t = 0$. Now $\kappa(0) = \frac{6}{27} = \frac{2}{9}$, so the radius of the osculating circle is

$1/\kappa(0) = \frac{9}{2}$ and its center is $\left(-\frac{5}{2}, 0\right)$. Its equation is therefore $\left(x + \frac{5}{2}\right)^2 + y^2 = \frac{81}{4}$.

At $(0, 3)$, $t = \frac{\pi}{2}$, and $\kappa\left(\frac{\pi}{2}\right) = \frac{6}{8} = \frac{3}{4}$. So the radius of the osculating circle is $\frac{4}{3}$ and

its center is $\left(0, \frac{5}{3}\right)$. Hence its equation is $x^2 + \left(y - \frac{5}{3}\right)^2 = \frac{16}{9}$.

53. The tangent vector is normal to the normal plane, and the vector $\langle 6, 6, -8 \rangle$ is normal to the given plane.

But $\mathbf{T}(t) \parallel \mathbf{r}'(t)$ and $\langle 6, 6, -8 \rangle \parallel \langle 3, 3, -4 \rangle$, so we need to find t such that $\mathbf{r}'(t) \parallel \langle 3, 3, -4 \rangle$.

$\mathbf{r}(t) = \langle t^3, 3t, t^4 \rangle \Rightarrow \mathbf{r}'(t) = \langle 3t^2, 3, 4t^3 \rangle \parallel \langle 3, 3, -4 \rangle$ when $t = -1$. So the planes are parallel at the point $(-1, -3, 1)$.

55. First we parametrize the curve of intersection. We can choose $y = t$; then $x = y^2 = t^2$ and $z = x^2 = t^4$, and the curve is

given by $\mathbf{r}(t) = \langle t^2, t, t^4 \rangle$. $\mathbf{r}'(t) = \langle 2t, 1, 4t^3 \rangle$ and the point $(1, 1, 1)$ corresponds to $t = 1$, so $\mathbf{r}'(1) = \langle 2, 1, 4 \rangle$ is a normal

vector for the normal plane. Thus an equation of the normal plane is

$2(x - 1) + 1(y - 1) + 4(z - 1) = 0$ or $2x + y + 4z = 7$. $\mathbf{T}(t) = \dfrac{\mathbf{r}'(t)}{|\mathbf{r}'(t)|} = \dfrac{1}{\sqrt{4t^2 + 1 + 16t^6}} \langle 2t, 1, 4t^3 \rangle$ and

$\mathbf{T}'(t) = -\frac{1}{2}(4t^2 + 1 + 16t^6)^{-3/2}(8t + 96t^5) \langle 2t, 1, 4t^3 \rangle + (4t^2 + 1 + 16t^6)^{-1/2} \langle 2, 0, 12t^2 \rangle$. A normal vector for

the osculating plane is $\mathbf{B}(1) = \mathbf{T}(1) \times \mathbf{N}(1)$, but $\mathbf{r}'(1) = \langle 2, 1, 4 \rangle$ is parallel to $\mathbf{T}(1)$ and

$\mathbf{T}'(1) = -\frac{1}{2}(21)^{-3/2}(104)\langle 2, 1, 4 \rangle + (21)^{-1/2}\langle 2, 0, 12 \rangle = \frac{2}{21\sqrt{21}} \langle -31, -26, 22 \rangle$ is parallel to $\mathbf{N}(1)$ as is $\langle -31, -26, 22 \rangle$,

so $\langle 2, 1, 4 \rangle \times \langle -31, -26, 22 \rangle = \langle 126, -168, -21 \rangle$ is normal to the osculating plane. Thus an equation for the osculating

plane is $126(x - 1) - 168(y - 1) - 21(z - 1) = 0$ or $6x - 8y - z = -3$.

57. $\kappa = \left|\dfrac{d\mathbf{T}}{ds}\right| = \left|\dfrac{d\mathbf{T}/dt}{ds/dt}\right| = \dfrac{|d\mathbf{T}/dt|}{ds/dt}$ and $\mathbf{N} = \dfrac{d\mathbf{T}/dt}{|d\mathbf{T}/dt|}$, so $\kappa\mathbf{N} = \dfrac{\left|\dfrac{d\mathbf{T}}{dt}\right|}{\left|\dfrac{d\mathbf{T}}{dt}\right|} \dfrac{\dfrac{d\mathbf{T}}{dt}}{\dfrac{ds}{dt}} = \dfrac{d\mathbf{T}/dt}{ds/dt} = \dfrac{d\mathbf{T}}{ds}$ by the Chain Rule.

59. (a) $|\mathbf{B}| = 1 \Rightarrow \mathbf{B} \cdot \mathbf{B} = 1 \Rightarrow \dfrac{d}{ds}(\mathbf{B} \cdot \mathbf{B}) = 0 \Rightarrow 2\dfrac{d\mathbf{B}}{ds} \cdot \mathbf{B} = 0 \Rightarrow \dfrac{d\mathbf{B}}{ds} \perp \mathbf{B}$

(b) $\mathbf{B} = \mathbf{T} \times \mathbf{N} \Rightarrow$

$\dfrac{d\mathbf{B}}{ds} = \dfrac{d}{ds}(\mathbf{T} \times \mathbf{N}) = \dfrac{d}{dt}(\mathbf{T} \times \mathbf{N})\dfrac{1}{ds/dt} = \dfrac{d}{dt}(\mathbf{T} \times \mathbf{N})\dfrac{1}{|\mathbf{r}'(t)|} = [(\mathbf{T}' \times \mathbf{N}) + (\mathbf{T} \times \mathbf{N}')]\dfrac{1}{|\mathbf{r}'(t)|}$

$= \left[\left(\mathbf{T}' \times \dfrac{\mathbf{T}'}{|\mathbf{T}'|}\right) + (\mathbf{T} \times \mathbf{N}')\right]\dfrac{1}{|\mathbf{r}'(t)|} = \dfrac{\mathbf{T} \times \mathbf{N}'}{|\mathbf{r}'(t)|} \Rightarrow \dfrac{d\mathbf{B}}{ds} \perp \mathbf{T}$

© 2012 Cengage Learning. All Rights Reserved. May not be scanned, copied, or duplicated, or posted to a publicly accessible website, in whole or in part.

(c) $\mathbf{B} = \mathbf{T} \times \mathbf{N}$ $\Rightarrow$ $\mathbf{T} \perp \mathbf{N}, \mathbf{B} \perp \mathbf{T}$ and $\mathbf{B} \perp \mathbf{N}$. So $\mathbf{B}, \mathbf{T}$ and $\mathbf{N}$ form an orthogonal set of vectors in the three-

dimensional space $\mathbb{R}^3$. From parts (a) and (b), $d\mathbf{B}/ds$ is perpendicular to both $\mathbf{B}$ and $\mathbf{T}$, so $d\mathbf{B}/ds$ is parallel to $\mathbf{N}$.

Therefore, $d\mathbf{B}/ds = -\tau(s)\mathbf{N}$, where $\tau(s)$ is a scalar.

(d) Since $\mathbf{B} = \mathbf{T} \times \mathbf{N}, \mathbf{T} \perp \mathbf{N}$ and both $\mathbf{T}$ and $\mathbf{N}$ are unit vectors, $\mathbf{B}$ is a unit vector mutually perpendicular to both $\mathbf{T}$ and

$\mathbf{N}$. For a plane curve, $\mathbf{T}$ and $\mathbf{N}$ always lie in the plane of the curve, so that $\mathbf{B}$ is a constant unit vector always

perpendicular to the plane. Thus $d\mathbf{B}/ds = \mathbf{0}$, but $d\mathbf{B}/ds = -\tau(s)\mathbf{N}$ and $\mathbf{N} \neq \mathbf{0}$, so $\tau(s) = 0$.

61. (a) $\mathbf{r}' = s'\,\mathbf{T}$ $\Rightarrow$ $\mathbf{r}'' = s''\,\mathbf{T} + s'\,\mathbf{T}' = s''\,\mathbf{T} + s'\dfrac{d\mathbf{T}}{ds}s' = s''\,\mathbf{T} + \kappa(s')^2\,\mathbf{N}$ by the first Serret-Frenet formula.

(b) Using part (a), we have

$$\mathbf{r}' \times \mathbf{r}'' = (s'\,\mathbf{T}) \times [s''\,\mathbf{T} + \kappa(s')^2\,\mathbf{N}]$$
$$= [(s'\,\mathbf{T}) \times (s''\,\mathbf{T})] + [(s'\mathbf{T}) \times (\kappa(s')^2\,\mathbf{N})] \qquad \text{[by Property 3 of Theorem 12.4.11]}$$
$$= (s's'')(\mathbf{T} \times \mathbf{T}) + \kappa(s')^3(\mathbf{T} \times \mathbf{N}) = \mathbf{0} + \kappa(s')^3\,\mathbf{B} = \kappa(s')^3\,\mathbf{B}$$

(c) Using part (a), we have

$$\mathbf{r}''' = [s''\,\mathbf{T} + \kappa(s')^2\,\mathbf{N}]' = s'''\,\mathbf{T} + s''\,\mathbf{T}' + \kappa'(s')^2\,\mathbf{N} + 2\kappa s's''\,\mathbf{N} + \kappa(s')^2\,\mathbf{N}'$$
$$= s'''\,\mathbf{T} + s''\dfrac{d\mathbf{T}}{ds}s' + \kappa'(s')^2\,\mathbf{N} + 2\kappa s's''\,\mathbf{N} + \kappa(s')^2\dfrac{d\mathbf{N}}{ds}s'$$
$$= s'''\,\mathbf{T} + s''s'\kappa\,\mathbf{N} + \kappa'(s')^2\,\mathbf{N} + 2\kappa s's''\,\mathbf{N} + \kappa(s')^3(-\kappa\,\mathbf{T} + \tau\,\mathbf{B}) \qquad \text{[by the second formula]}$$
$$= [s''' - \kappa^2(s')^3]\,\mathbf{T} + [3\kappa s's'' + \kappa'(s')^2]\,\mathbf{N} + \kappa\tau(s')^3\,\mathbf{B}$$

(d) Using parts (b) and (c) and the facts that $\mathbf{B} \cdot \mathbf{T} = 0, \mathbf{B} \cdot \mathbf{N} = 0$, and $\mathbf{B} \cdot \mathbf{B} = 1$, we get

$$\dfrac{(\mathbf{r}' \times \mathbf{r}'') \cdot \mathbf{r}'''}{|\mathbf{r}' \times \mathbf{r}''|^2} = \dfrac{\kappa(s')^3\,\mathbf{B} \cdot \{[s''' - \kappa^2(s')^3]\,\mathbf{T} + [3\kappa s' s'' + \kappa'(s')^2]\,\mathbf{N} + \kappa\tau(s')^3\,\mathbf{B}\}}{|\kappa(s')^3\,\mathbf{B}|^2} = \dfrac{\kappa(s')^3\kappa\tau(s')^3}{[\kappa(s')^3]^2} = \tau.$$

63. $\mathbf{r} = \left\langle t, \tfrac{1}{2}t^2, \tfrac{1}{3}t^3 \right\rangle$ $\Rightarrow$ $\mathbf{r}' = \left\langle 1, t, t^2 \right\rangle, \mathbf{r}'' = \left\langle 0, 1, 2t \right\rangle, \mathbf{r}''' = \left\langle 0, 0, 2 \right\rangle$ $\Rightarrow$ $\mathbf{r}' \times \mathbf{r}'' = \left\langle t^2, -2t, 1 \right\rangle$ $\Rightarrow$

$$\tau = \dfrac{(\mathbf{r}' \times \mathbf{r}'') \cdot \mathbf{r}'''}{|\mathbf{r}' \times \mathbf{r}''|^2} = \dfrac{\left\langle t^2, -2t, 1 \right\rangle \cdot \left\langle 0, 0, 2 \right\rangle}{t^4 + 4t^2 + 1} = \dfrac{2}{t^4 + 4t^2 + 1}$$

65. For one helix, the vector equation is $\mathbf{r}(t) = \left\langle 10\cos t, 10\sin t, 34t/(2\pi) \right\rangle$ (measuring in angstroms), because the radius of each

helix is 10 angstroms, and z increases by 34 angstroms for each increase of 2π in t. Using the arc length formula, letting t go

from 0 to $2.9 \times 10^8 \times 2\pi$, we find the approximate length of each helix to be

$$L = \int_0^{2.9 \times 10^8 \times 2\pi} |\mathbf{r}'(t)|\, dt = \int_0^{2.9 \times 10^8 \times 2\pi} \sqrt{(-10\sin t)^2 + (10\cos t)^2 + \left(\tfrac{34}{2\pi}\right)^2}\, dt = \sqrt{100 + \left(\tfrac{34}{2\pi}\right)^2}\, t \Big]_0^{2.9 \times 10^8 \times 2\pi}$$

$$= 2.9 \times 10^8 \times 2\pi \sqrt{100 + \left(\tfrac{34}{2\pi}\right)^2} \approx 2.07 \times 10^{10} \text{ Å — more than two meters!}$$

© 2012 Cengage Learning. All Rights Reserved. May not be scanned, copied, or duplicated, or posted to a publicly accessible website, in whole or in part.

13.4 Motion in Space: Velocity and Acceleration

1. (a) If $\mathbf{r}(t) = x(t)\,\mathbf{i} + y(t)\,\mathbf{j} + z(t)\,\mathbf{k}$ is the position vector of the particle at time t, then the average velocity over the time

interval $[0, 1]$ is

$$\mathbf{v}_{ave} = \frac{\mathbf{r}(1) - \mathbf{r}(0)}{1 - 0} = \frac{(4.5\,\mathbf{i} + 6.0\,\mathbf{j} + 3.0\,\mathbf{k}) - (2.7\,\mathbf{i} + 9.8\,\mathbf{j} + 3.7\,\mathbf{k})}{1} = 1.8\,\mathbf{i} - 3.8\,\mathbf{j} - 0.7\,\mathbf{k}.$$ Similarly, over the other

intervals we have

$$[0.5, 1]: \quad \mathbf{v}_{ave} = \frac{\mathbf{r}(1) - \mathbf{r}(0.5)}{1 - 0.5} = \frac{(4.5\,\mathbf{i} + 6.0\,\mathbf{j} + 3.0\,\mathbf{k}) - (3.5\,\mathbf{i} + 7.2\,\mathbf{j} + 3.3\,\mathbf{k})}{0.5} = 2.0\,\mathbf{i} - 2.4\,\mathbf{j} - 0.6\,\mathbf{k}$$

$$[1, 2]: \quad \mathbf{v}_{ave} = \frac{\mathbf{r}(2) - \mathbf{r}(1)}{2 - 1} = \frac{(7.3\,\mathbf{i} + 7.8\,\mathbf{j} + 2.7\,\mathbf{k}) - (4.5\,\mathbf{i} + 6.0\,\mathbf{j} + 3.0\,\mathbf{k})}{1} = 2.8\,\mathbf{i} + 1.8\,\mathbf{j} - 0.3\,\mathbf{k}$$

$$[1, 1.5]: \quad \mathbf{v}_{ave} = \frac{\mathbf{r}(1.5) - \mathbf{r}(1)}{1.5 - 1} = \frac{(5.9\,\mathbf{i} + 6.4\,\mathbf{j} + 2.8\,\mathbf{k}) - (4.5\,\mathbf{i} + 6.0\,\mathbf{j} + 3.0\,\mathbf{k})}{0.5} = 2.8\,\mathbf{i} + 0.8\,\mathbf{j} - 0.4\,\mathbf{k}$$

(b) We can estimate the velocity at $t = 1$ by averaging the average velocities over the time intervals $[0.5, 1]$ and $[1, 1.5]$:

$$\mathbf{v}(1) \approx \tfrac{1}{2}[(2\,\mathbf{i} - 2.4\,\mathbf{j} - 0.6\,\mathbf{k}) + (2.8\,\mathbf{i} + 0.8\,\mathbf{j} - 0.4\,\mathbf{k})] = 2.4\,\mathbf{i} - 0.8\,\mathbf{j} - 0.5\,\mathbf{k}.$$ Then the speed is

$$|\mathbf{v}(1)| \approx \sqrt{(2.4)^2 + (-0.8)^2 + (-0.5)^2} \approx 2.58.$$

3. $\mathbf{r}(t) = \left\langle -\tfrac{1}{2}t^2, t \right\rangle \quad \Rightarrow$ At $t = 2$:

$\mathbf{v}(t) = \mathbf{r}'(t) = \langle -t, 1 \rangle$ $\mathbf{v}(2) = \langle -2, 1 \rangle$

$\mathbf{a}(t) = \mathbf{r}''(t) = \langle -1, 0 \rangle$ $\mathbf{a}(2) = \langle -1, 0 \rangle$

$|\mathbf{v}(t)| = \sqrt{t^2 + 1}$

5. $\mathbf{r}(t) = 3\cos t\,\mathbf{i} + 2\sin t\,\mathbf{j} \quad \Rightarrow$ At $t = \pi/3$:

$\mathbf{v}(t) = -3\sin t\,\mathbf{i} + 2\cos t\,\mathbf{j}$ $\mathbf{v}\!\left(\tfrac{\pi}{3}\right) = -\tfrac{3\sqrt{3}}{2}\,\mathbf{i} + \mathbf{j}$

$\mathbf{a}(t) = -3\cos t\,\mathbf{i} - 2\sin t\,\mathbf{j}$ $\mathbf{a}\!\left(\tfrac{\pi}{3}\right) = -\tfrac{3}{2}\,\mathbf{i} - \sqrt{3}\,\mathbf{j}$

$|\mathbf{v}(t)| = \sqrt{9\sin^2 t + 4\cos^2 t} = \sqrt{4 + 5\sin^2 t}$

Notice that $x^2/9 + y^2/4 = \sin^2 t + \cos^2 t = 1$, so the path is an ellipse.

7. $\mathbf{r}(t) = t\,\mathbf{i} + t^2\,\mathbf{j} + 2\,\mathbf{k} \quad \Rightarrow$ At $t = 1$:

$\mathbf{v}(t) = \mathbf{i} + 2t\,\mathbf{j}$ $\mathbf{v}(1) = \mathbf{i} + 2\,\mathbf{j}$

$\mathbf{a}(t) = 2\,\mathbf{j}$ $\mathbf{a}(1) = 2\,\mathbf{j}$

$|\mathbf{v}(t)| = \sqrt{1 + 4t^2}$

Here $x = t$, $y = t^2 \quad \Rightarrow \quad y = x^2$ and $z = 2$, so the path of the particle is a

parabola in the plane $z = 2$.

© 2012 Cengage Learning. All Rights Reserved. May not be scanned, copied, or duplicated, or posted to a publicly accessible website, in whole or in part.

9. $\mathbf{r}(t) = \langle t^2 + t, t^2 - t, t^3 \rangle \;\Rightarrow\; \mathbf{v}(t) = \mathbf{r}'(t) = \langle 2t + 1, 2t - 1, 3t^2 \rangle, \quad \mathbf{a}(t) = \mathbf{v}'(t) = \langle 2, 2, 6t \rangle,$

$|\mathbf{v}(t)| = \sqrt{(2t+1)^2 + (2t-1)^2 + (3t^2)^2} = \sqrt{9t^4 + 8t^2 + 2}.$

11. $\mathbf{r}(t) = \sqrt{2}\,t\,\mathbf{i} + e^t\,\mathbf{j} + e^{-t}\,\mathbf{k} \;\Rightarrow\; \mathbf{v}(t) = \mathbf{r}'(t) = \sqrt{2}\,\mathbf{i} + e^t\,\mathbf{j} - e^{-t}\,\mathbf{k}, \quad \mathbf{a}(t) = \mathbf{v}'(t) = e^t\,\mathbf{j} + e^{-t}\,\mathbf{k},$

$|\mathbf{v}(t)| = \sqrt{2 + e^{2t} + e^{-2t}} = \sqrt{(e^t + e^{-t})^2} = e^t + e^{-t}.$

13. $\mathbf{r}(t) = e^t\langle \cos t, \sin t, t \rangle \;\Rightarrow$

$\mathbf{v}(t) = \mathbf{r}'(t) = e^t\langle \cos t, \sin t, t \rangle + e^t\langle -\sin t, \cos t, 1 \rangle = e^t\langle \cos t - \sin t, \sin t + \cos t, t + 1 \rangle$

$\mathbf{a}(t) = \mathbf{v}'(t) = e^t\langle \cos t - \sin t - \sin t - \cos t, \sin t + \cos t + \cos t - \sin t, t + 1 + 1 \rangle$

$\qquad = e^t\langle -2\sin t, 2\cos t, t + 2 \rangle$

$|\mathbf{v}(t)| = e^t\sqrt{\cos^2 t + \sin^2 t - 2\cos t \sin t + \sin^2 t + \cos^2 t + 2\sin t \cos t + t^2 + 2t + 1}$

$\qquad = e^t\sqrt{t^2 + 2t + 3}$

15. $\mathbf{a}(t) = \mathbf{i} + 2\mathbf{j} \;\Rightarrow\; \mathbf{v}(t) = \int \mathbf{a}(t)\,dt = \int (\mathbf{i} + 2\mathbf{j})\,dt = t\,\mathbf{i} + 2t\,\mathbf{j} + \mathbf{C}$ and $\mathbf{k} = \mathbf{v}(0) = \mathbf{C}$,

so $\mathbf{C} = \mathbf{k}$ and $\mathbf{v}(t) = t\,\mathbf{i} + 2t\,\mathbf{j} + \mathbf{k}$. $\mathbf{r}(t) = \int \mathbf{v}(t)\,dt = \int (t\,\mathbf{i} + 2t\,\mathbf{j} + \mathbf{k})\,dt = \frac{1}{2}t^2\,\mathbf{i} + t^2\,\mathbf{j} + t\,\mathbf{k} + \mathbf{D}$.

But $\mathbf{i} = \mathbf{r}(0) = \mathbf{D}$, so $\mathbf{D} = \mathbf{i}$ and $\mathbf{r}(t) = \left(\frac{1}{2}t^2 + 1\right)\mathbf{i} + t^2\,\mathbf{j} + t\,\mathbf{k}$.

17. (a) $\mathbf{a}(t) = 2t\,\mathbf{i} + \sin t\,\mathbf{j} + \cos 2t\,\mathbf{k} \;\Rightarrow$ (b)

$\mathbf{v}(t) = \int (2t\,\mathbf{i} + \sin t\,\mathbf{j} + \cos 2t\,\mathbf{k})\,dt = t^2\,\mathbf{i} - \cos t\,\mathbf{j} + \frac{1}{2}\sin 2t\,\mathbf{k} + \mathbf{C}$

and $\mathbf{i} = \mathbf{v}(0) = -\mathbf{j} + \mathbf{C}$, so $\mathbf{C} = \mathbf{i} + \mathbf{j}$

and $\mathbf{v}(t) = (t^2 + 1)\,\mathbf{i} + (1 - \cos t)\,\mathbf{j} + \frac{1}{2}\sin 2t\,\mathbf{k}$.

$\mathbf{r}(t) = \int [(t^2 + 1)\,\mathbf{i} + (1 - \cos t)\,\mathbf{j} + \frac{1}{2}\sin 2t\,\mathbf{k}]\,dt$

$\qquad = \left(\frac{1}{3}t^3 + t\right)\mathbf{i} + (t - \sin t)\,\mathbf{j} - \frac{1}{4}\cos 2t\,\mathbf{k} + \mathbf{D}$

But $\mathbf{j} = \mathbf{r}(0) = -\frac{1}{4}\mathbf{k} + \mathbf{D}$, so $\mathbf{D} = \mathbf{j} + \frac{1}{4}\mathbf{k}$ and $\mathbf{r}(t) = \left(\frac{1}{3}t^3 + t\right)\mathbf{i} + (t - \sin t + 1)\,\mathbf{j} + \left(\frac{1}{4} - \frac{1}{4}\cos 2t\right)\mathbf{k}$.

19. $\mathbf{r}(t) = \langle t^2, 5t, t^2 - 16t \rangle \;\Rightarrow\; \mathbf{v}(t) = \langle 2t, 5, 2t - 16 \rangle, |\mathbf{v}(t)| = \sqrt{4t^2 + 25 + 4t^2 - 64t + 256} = \sqrt{8t^2 - 64t + 281}$

and $\dfrac{d}{dt}|\mathbf{v}(t)| = \frac{1}{2}(8t^2 - 64t + 281)^{-1/2}(16t - 64)$. This is zero if and only if the numerator is zero, that is,

$16t - 64 = 0$ or $t = 4$. Since $\dfrac{d}{dt}|\mathbf{v}(t)| < 0$ for $t < 4$ and $\dfrac{d}{dt}|\mathbf{v}(t)| > 0$ for $t > 4$, the minimum speed of $\sqrt{153}$ is attained

at $t = 4$ units of time.

21. $|\mathbf{F}(t)| = 20$ N in the direction of the positive z-axis, so $\mathbf{F}(t) = 20\,\mathbf{k}$. Also $m = 4$ kg, $\mathbf{r}(0) = \mathbf{0}$ and $\mathbf{v}(0) = \mathbf{i} - \mathbf{j}$.

Since $20\mathbf{k} = \mathbf{F}(t) = 4\,\mathbf{a}(t)$, $\mathbf{a}(t) = 5\,\mathbf{k}$. Then $\mathbf{v}(t) = 5t\,\mathbf{k} + \mathbf{c}_1$ where $\mathbf{c}_1 = \mathbf{i} - \mathbf{j}$ so $\mathbf{v}(t) = \mathbf{i} - \mathbf{j} + 5t\,\mathbf{k}$ and the

speed is $|\mathbf{v}(t)| = \sqrt{1 + 1 + 25t^2} = \sqrt{25t^2 + 2}$. Also $\mathbf{r}(t) = t\,\mathbf{i} - t\,\mathbf{j} + \frac{5}{2}t^2\,\mathbf{k} + \mathbf{c}_2$ and $\mathbf{0} = \mathbf{r}(0)$, so $\mathbf{c}_2 = \mathbf{0}$

and $\mathbf{r}(t) = t\,\mathbf{i} - t\,\mathbf{j} + \frac{5}{2}t^2\,\mathbf{k}$.

© 2012 Cengage Learning. All Rights Reserved. May not be scanned, copied, or duplicated, or posted to a publicly accessible website, in whole or in part.

23. $|\mathbf{v}(0)| = 200$ m/s and, since the angle of elevation is $60°$, a unit vector in the direction of the velocity is

$(\cos 60°)\mathbf{i} + (\sin 60°)\mathbf{j} = \frac{1}{2}\mathbf{i} + \frac{\sqrt{3}}{2}\mathbf{j}$. Thus $\mathbf{v}(0) = 200\left(\frac{1}{2}\mathbf{i} + \frac{\sqrt{3}}{2}\mathbf{j}\right) = 100\mathbf{i} + 100\sqrt{3}\mathbf{j}$ and if we set up the axes so that the

projectile starts at the origin, then $\mathbf{r}(0) = \mathbf{0}$. Ignoring air resistance, the only force is that due to gravity, so

$\mathbf{F}(t) = m\mathbf{a}(t) = -mg\mathbf{j}$ where $g \approx 9.8$ m/s^2. Thus $\mathbf{a}(t) = -9.8\mathbf{j}$ and, integrating, we have $\mathbf{v}(t) = -9.8t\mathbf{j} + \mathbf{C}$. But

$100\mathbf{i} + 100\sqrt{3}\mathbf{j} = \mathbf{v}(0) = \mathbf{C}$, so $\mathbf{v}(t) = 100\mathbf{i} + \left(100\sqrt{3} - 9.8t\right)\mathbf{j}$ and then (integrating again)

$\mathbf{r}(t) = 100t\mathbf{i} + \left(100\sqrt{3}\,t - 4.9t^2\right)\mathbf{j} + \mathbf{D}$ where $\mathbf{0} = \mathbf{r}(0) = \mathbf{D}$. Thus the position function of the projectile is

$\mathbf{r}(t) = 100t\mathbf{i} + \left(100\sqrt{3}\,t - 4.9t^2\right)\mathbf{j}$.

(a) Parametric equations for the projectile are $x(t) = 100t$, $y(t) = 100\sqrt{3}\,t - 4.9t^2$. The projectile reaches the ground when

$y(t) = 0$ (and $t > 0$) $\Rightarrow$ $100\sqrt{3}\,t - 4.9t^2 = t(100\sqrt{3} - 4.9t) = 0$ $\Rightarrow$ $t = \frac{100\sqrt{3}}{4.9} \approx 35.3$ s. So the range is

$x\left(\frac{100\sqrt{3}}{4.9}\right) = 100\left(\frac{100\sqrt{3}}{4.9}\right) \approx 3535$ m.

(b) The maximum height is reached when $y(t)$ has a critical number (or equivalently, when the vertical component

of velocity is 0): $y'(t) = 0$ $\Rightarrow$ $100\sqrt{3} - 9.8t = 0$ $\Rightarrow$ $t = \frac{100\sqrt{3}}{9.8} \approx 17.7$ s. Thus the maximum height is

$y\left(\frac{100\sqrt{3}}{9.8}\right) = 100\sqrt{3}\left(\frac{100\sqrt{3}}{9.8}\right) - 4.9\left(\frac{100\sqrt{3}}{9.8}\right)^2 \approx 1531$ m.

(c) From part (a), impact occurs at $t = \frac{100\sqrt{3}}{4.9}$ s. Thus, the velocity at impact is

$\mathbf{v}\left(\frac{100\sqrt{3}}{4.9}\right) = 100\mathbf{i} + \left[100\sqrt{3} - 9.8\left(\frac{100\sqrt{3}}{4.9}\right)\right]\mathbf{j} = 100\mathbf{i} - 100\sqrt{3}\mathbf{j}$ and the speed is

$\left|\mathbf{v}\left(\frac{100\sqrt{3}}{4.9}\right)\right| = \sqrt{10{,}000 + 30{,}000} = 200$ m/s.

25. As in Example 5, $\mathbf{r}(t) = (v_0\cos 45°)t\,\mathbf{i} + \left[(v_0\sin 45°)t - \frac{1}{2}gt^2\right]\mathbf{j} = \frac{1}{2}\left[v_0\sqrt{2}\,t\,\mathbf{i} + \left(v_0\sqrt{2}\,t - gt^2\right)\mathbf{j}\right]$. The ball lands when

$y = 0$ (and $t > 0$) $\Rightarrow$ $t = \frac{v_0\sqrt{2}}{g}$ s. Now since it lands 90 m away, $90 = x = \frac{1}{2}v_0\sqrt{2}\,\frac{v_0\sqrt{2}}{g}$ or $v_0^2 = 90g$ and the initial

velocity is $v_0 = \sqrt{90g} \approx 30$ m/s.

27. Let α be the angle of elevation. Then $v_0 = 150$ m/s and from Example 5, the horizontal distance traveled by the projectile is

$d = \dfrac{v_0^2\sin 2\alpha}{g}$. Thus $\dfrac{150^2\sin 2\alpha}{g} = 800$ $\Rightarrow$ $\sin 2\alpha = \dfrac{800g}{150^2} \approx 0.3484$ $\Rightarrow$ $2\alpha \approx 20.4°$ or $180 - 20.4 = 159.6°$.

Two angles of elevation then are $\alpha \approx 10.2°$ and $\alpha \approx 79.8°$.

29. Place the catapult at the origin and assume the catapult is 100 meters from the city, so the city lies between $(100, 0)$

and $(600, 0)$. The initial speed is $v_0 = 80$ m/s and let θ be the angle the catapult is set at. As in Example 5, the trajectory of

the catapulted rock is given by $\mathbf{r}(t) = (80\cos\theta)t\,\mathbf{i} + \left[(80\sin\theta)t - 4.9t^2\right]\mathbf{j}$. The top of the near city wall is at $(100, 15)$,

which the rock will hit when $(80\cos\theta)t = 100$ $\Rightarrow$ $t = \dfrac{5}{4\cos\theta}$ and $(80\sin\theta)t - 4.9t^2 = 15$ $\Rightarrow$

© 2012 Cengage Learning. All Rights Reserved. May not be scanned, copied, or duplicated, or posted to a publicly accessible website, in whole or in part.

$80 \sin\theta \cdot \dfrac{5}{4\cos\theta} - 4.9\left(\dfrac{5}{4\cos\theta}\right)^2 = 15 \quad\Rightarrow\quad 100\tan\theta - 7.65625\sec^2\theta = 15$. Replacing $\sec^2\theta$ with $\tan^2\theta + 1$ gives

$7.65625\tan^2\theta - 100\tan\theta + 22.65625 = 0$. Using the quadratic formula, we have $\tan\theta \approx 0.230635,\ 12.8306 \quad\Rightarrow$

$\theta \approx 13.0°,\ 85.5°$. So for $13.0° < \theta < 85.5°$, the rock will land beyond the near city wall. The base of the far wall is

located at $(600, 0)$ which the rock hits if $(80\cos\theta)t = 600 \quad\Rightarrow\quad t = \dfrac{15}{2\cos\theta}$ and $(80\sin\theta)t - 4.9t^2 = 0 \quad\Rightarrow$

$80\sin\theta \cdot \dfrac{15}{2\cos\theta} - 4.9\left(\dfrac{15}{2\cos\theta}\right)^2 = 0 \quad\Rightarrow\quad 600\tan\theta - 275.625\sec^2\theta = 0 \quad\Rightarrow$

$275.625\tan^2\theta - 600\tan\theta + 275.625 = 0$. Solutions are $\tan\theta \approx 0.658678,\ 1.51819 \quad\Rightarrow\quad \theta \approx 33.4°,\ 56.6°$. Thus the

rock lands beyond the enclosed city ground for $33.4° < \theta < 56.6°$, and the angles that allow the rock to land on city ground

are $13.0° < \theta < 33.4°,\ 56.6° < \theta < 85.5°$. If you consider that the rock can hit the far wall and bounce back into the city, we

calculate the angles that cause the rock to hit the top of the wall at $(600, 15)$: $(80\cos\theta)t = 600 \quad\Rightarrow\quad t = \dfrac{15}{2\cos\theta}$ and

$(80\sin\theta)t - 4.9t^2 = 15 \quad\Rightarrow\quad 600\tan\theta - 275.625\sec^2\theta = 15 \quad\Rightarrow\quad 275.625\tan^2\theta - 600\tan\theta + 290.625 = 0$.

Solutions are $\tan\theta \approx 0.727506,\ 1.44936 \quad\Rightarrow\quad \theta \approx 36.0°,\ 55.4°$, so the catapult should be set with angle θ where

$13.0° < \theta < 36.0°,\ 55.4° < \theta < 85.5°$.

31. Here $\mathbf{a}(t) = -4\,\mathbf{j} - 32\,\mathbf{k}$ so $\mathbf{v}(t) = -4t\,\mathbf{j} - 32t\,\mathbf{k} + \mathbf{v}_0 = -4t\,\mathbf{j} - 32t\,\mathbf{k} + 50\,\mathbf{i} + 80\,\mathbf{k} = 50\,\mathbf{i} - 4t\,\mathbf{j} + (80 - 32t)\,\mathbf{k}$ and

$\mathbf{r}(t) = 50t\,\mathbf{i} - 2t^2\,\mathbf{j} + (80t - 16t^2)\,\mathbf{k}$ (note that $\mathbf{r}_0 = \mathbf{0}$). The ball lands when the z-component of $\mathbf{r}(t)$ is zero

and $t > 0$: $80t - 16t^2 = 16t(5 - t) = 0 \quad\Rightarrow\quad t = 5$. The position of the ball then is

$\mathbf{r}(5) = 50(5)\,\mathbf{i} - 2(5)^2\,\mathbf{j} + [80(5) - 16(5)^2]\,\mathbf{k} = 250\,\mathbf{i} - 50\,\mathbf{j}$ or equivalently the point $(250, -50, 0)$. This is a distance of

$\sqrt{250^2 + (-50)^2 + 0^2} = \sqrt{65{,}000} \approx 255$ ft from the origin at an angle of $\tan^{-1}\left(\frac{50}{250}\right) \approx 11.3°$ from the eastern direction

toward the south. The speed of the ball is $|\mathbf{v}(5)| = |50\,\mathbf{i} - 20\,\mathbf{j} - 80\,\mathbf{k}| = \sqrt{50^2 + (-20)^2 + (-80)^2} = \sqrt{9300} \approx 96.4$ ft/s.

33. (a) After t seconds, the boat will be $5t$ meters west of point A. The velocity

of the water at that location is $\frac{3}{400}(5t)(40 - 5t)\,\mathbf{j}$. The velocity of the

boat in still water is $5\,\mathbf{i}$, so the resultant velocity of the boat is

$\mathbf{v}(t) = 5\,\mathbf{i} + \frac{3}{400}(5t)(40 - 5t)\,\mathbf{j} = 5\,\mathbf{i} + \left(\frac{3}{2}t - \frac{3}{16}t^2\right)\mathbf{j}$. Integrating, we obtain

$\mathbf{r}(t) = 5t\,\mathbf{i} + \left(\frac{3}{4}t^2 - \frac{1}{16}t^3\right)\mathbf{j} + \mathbf{C}$. If we place the origin at A (and consider $\mathbf{j}$

to coincide with the northern direction) then $\mathbf{r}(0) = \mathbf{0} \quad\Rightarrow\quad \mathbf{C} = \mathbf{0}$ and we have $\mathbf{r}(t) = 5t\,\mathbf{i} + \left(\frac{3}{4}t^2 - \frac{1}{16}t^3\right)\mathbf{j}$. The boat

reaches the east bank after 8 s, and it is located at $\mathbf{r}(8) = 5(8)\,\mathbf{i} + \left(\frac{3}{4}(8)^2 - \frac{1}{16}(8)^3\right)\mathbf{j} = 40\,\mathbf{i} + 16\,\mathbf{j}$. Thus the boat is 16 m

downstream.

(b) Let α be the angle north of east that the boat heads. Then the velocity of the boat in still water is given by

$5(\cos\alpha)\,\mathbf{i} + 5(\sin\alpha)\,\mathbf{j}$. At t seconds, the boat is $5(\cos\alpha)t$ meters from the west bank, at which point the velocity

of the water is $\frac{3}{400}[5(\cos\alpha)t][40 - 5(\cos\alpha)t]\,\mathbf{j}$. The resultant velocity of the boat is given by

© 2012 Cengage Learning. All Rights Reserved. May not be scanned, copied, or duplicated, or posted to a publicly accessible website, in whole or in part.

$$\mathbf{v}(t) = 5(\cos\alpha)\,\mathbf{i} + \left[5\sin\alpha + \tfrac{3}{400}(5t\cos\alpha)(40 - 5t\cos\alpha)\right]\mathbf{j} = (5\cos\alpha)\,\mathbf{i} + \left(5\sin\alpha + \tfrac{3}{2}t\cos\alpha - \tfrac{3}{16}t^2\cos^2\alpha\right)\mathbf{j}.$$

Integrating, $\mathbf{r}(t) = (5t\cos\alpha)\,\mathbf{i} + \left(5t\sin\alpha + \tfrac{3}{4}t^2\cos\alpha - \tfrac{1}{16}t^3\cos^2\alpha\right)\mathbf{j}$ (where we have again placed

the origin at A). The boat will reach the east bank when $5t\cos\alpha = 40 \Rightarrow t = \dfrac{40}{5\cos\alpha} = \dfrac{8}{\cos\alpha}$.

In order to land at point $B(40, 0)$ we need $5t\sin\alpha + \tfrac{3}{4}t^2\cos\alpha - \tfrac{1}{16}t^3\cos^2\alpha = 0 \Rightarrow$

$$5\left(\frac{8}{\cos\alpha}\right)\sin\alpha + \tfrac{3}{4}\left(\frac{8}{\cos\alpha}\right)^2\cos\alpha - \tfrac{1}{16}\left(\frac{8}{\cos\alpha}\right)^3\cos^2\alpha = 0 \Rightarrow \frac{1}{\cos\alpha}(40\sin\alpha + 48 - 32) = 0 \Rightarrow$$

$40\sin\alpha + 16 = 0 \Rightarrow \sin\alpha = -\tfrac{2}{5}$. Thus $\alpha = \sin^{-1}\left(-\tfrac{2}{5}\right) \approx -23.6°$, so the boat should head $23.6°$ south of

east (upstream). The path does seem realistic. The boat initially heads

upstream to counteract the effect of the current. Near the center of the river,

the current is stronger and the boat is pushed downstream. When the boat

nears the eastern bank, the current is slower and the boat is able to progress

upstream to arrive at point B.

35. If $\mathbf{r}'(t) = \mathbf{c} \times \mathbf{r}(t)$ then $\mathbf{r}'(t)$ is perpendicular to both $\mathbf{c}$ and $\mathbf{r}(t)$. Remember that $\mathbf{r}'(t)$ points in the direction of motion, so if

$\mathbf{r}'(t)$ is always perpendicular to $\mathbf{c}$, the path of the particle must lie in a plane perpendicular to $\mathbf{c}$. But $\mathbf{r}'(t)$ is also perpendicular

to the position vector $\mathbf{r}(t)$ which confines the path to a sphere centered at the origin. Considering both restrictions, the path

must be contained in a circle that lies in a plane perpendicular to $\mathbf{c}$, and the circle is centered on a line through the origin in the

direction of $\mathbf{c}$.

37. $\mathbf{r}(t) = (3t - t^3)\,\mathbf{i} + 3t^2\,\mathbf{j} \Rightarrow \mathbf{r}'(t) = (3 - 3t^2)\,\mathbf{i} + 6t\,\mathbf{j}$,

$|\mathbf{r}'(t)| = \sqrt{(3 - 3t^2)^2 + (6t)^2} = \sqrt{9 + 18t^2 + 9t^4} = \sqrt{(3 - 3t^2)^2} = 3 + 3t^2$,

$\mathbf{r}''(t) = -6t\,\mathbf{i} + 6\,\mathbf{j}$, $\mathbf{r}'(t) \times \mathbf{r}''(t) = (18 + 18t^2)\,\mathbf{k}$. Then Equation 9 gives

$$a_T = \frac{\mathbf{r}'(t) \cdot \mathbf{r}''(t)}{|\mathbf{r}'(t)|} = \frac{(3 - 3t^2)(-6t) + (6t)(6)}{3 + 3t^2} = \frac{18t + 18t^3}{3 + 3t^2} = \frac{18t(1 + t^2)}{3(1 + t^2)} = 6t \quad \Big[\text{or by Equation 8,}$$

$$a_T = v' = \frac{d}{dt}\left[3 + 3t^2\right] = 6t\Big] \quad \text{and Equation 10 gives } a_N = \frac{|\mathbf{r}'(t) \times \mathbf{r}''(t)|}{|\mathbf{r}'(t)|} = \frac{18 + 18t^2}{3 + 3t^2} = \frac{18(1 + t^2)}{3(1 + t^2)} = 6.$$

39. $\mathbf{r}(t) = \cos t\,\mathbf{i} + \sin t\,\mathbf{j} + t\,\mathbf{k} \Rightarrow \mathbf{r}'(t) = -\sin t\,\mathbf{i} + \cos t\,\mathbf{j} + \mathbf{k}$, $|\mathbf{r}'(t)| = \sqrt{\sin^2 t + \cos^2 t + 1} = \sqrt{2}$,

$\mathbf{r}''(t) = -\cos t\,\mathbf{i} - \sin t\,\mathbf{j}$, $\mathbf{r}'(t) \times \mathbf{r}''(t) = \sin t\,\mathbf{i} - \cos t\,\mathbf{j} + \mathbf{k}$.

Then $a_T = \dfrac{\mathbf{r}'(t) \cdot \mathbf{r}''(t)}{|\mathbf{r}'(t)|} = \dfrac{\sin t\cos t - \sin t\cos t}{\sqrt{2}} = 0$ and $a_N = \dfrac{|\mathbf{r}'(t) \times \mathbf{r}''(t)|}{|\mathbf{r}'(t)|} = \dfrac{\sqrt{\sin^2 t + \cos^2 t + 1}}{\sqrt{2}} = \dfrac{\sqrt{2}}{\sqrt{2}} = 1$.

41. $\mathbf{r}(t) = e^t\,\mathbf{i} + \sqrt{2}\,t\,\mathbf{j} + e^{-t}\,\mathbf{k} \Rightarrow \mathbf{r}'(t) = e^t\,\mathbf{i} + \sqrt{2}\,\mathbf{j} - e^{-t}\,\mathbf{k}$, $|\mathbf{r}(t)| = \sqrt{e^{2t} + 2 + e^{-2t}} = \sqrt{(e^t + e^{-t})^2} = e^t + e^{-t}$,

$\mathbf{r}''(t) = e^t\,\mathbf{i} + e^{-t}\,\mathbf{k}$. Then $a_T = \dfrac{e^{2t} - e^{-2t}}{e^t + e^{-t}} = \dfrac{(e^t + e^{-t})(e^t - e^{-t})}{e^t + e^{-t}} = e^t - e^{-t} = 2\sinh t$

and $a_N = \dfrac{\left|\sqrt{2}\,e^{-t}\,\mathbf{i} - 2\,\mathbf{j} - \sqrt{2}\,e^t\,\mathbf{k}\right|}{e^t + e^{-t}} = \dfrac{\sqrt{2(e^{-2t} + 2 + e^{2t})}}{e^t + e^{-t}} = \sqrt{2}\,\dfrac{e^t + e^{-t}}{e^t + e^{-t}} = \sqrt{2}.$

© 2012 Cengage Learning. All Rights Reserved. May not be scanned, copied, or duplicated, or posted to a publicly accessible website, in whole or in part.

43. The tangential component of **a** is the length of the projection of **a** onto **T**, so we sketch

the scalar projection of **a** in the tangential direction to the curve and estimate its length to

be 4.5 (using the fact that **a** has length 10 as a guide). Similarly, the normal component of

a is the length of the projection of **a** onto **N**, so we sketch the scalar projection of **a** in the

normal direction to the curve and estimate its length to be 9.0. Thus $a_T \approx 4.5 \text{ cm/s}^2$ and

$a_N \approx 9.0 \text{ cm/s}^2$.

45. If the engines are turned off at time t, then the spacecraft will continue to travel in the direction of $\mathbf{v}(t)$, so we need a t such

that for some scalar $s > 0$, $\mathbf{r}(t) + s\,\mathbf{v}(t) = \langle 6, 4, 9 \rangle$. $\mathbf{v}(t) = \mathbf{r}'(t) = \mathbf{i} + \dfrac{1}{t}\mathbf{j} + \dfrac{8t}{(t^2+1)^2}\mathbf{k} \;\Rightarrow$

$\mathbf{r}(t) + s\,\mathbf{v}(t) = \left\langle 3 + t + s,\, 2 + \ln t + \dfrac{s}{t},\, 7 - \dfrac{4}{t^2+1} + \dfrac{8st}{(t^2+1)^2} \right\rangle \;\Rightarrow\; 3 + t + s = 6 \;\Rightarrow\; s = 3 - t,$

so $7 - \dfrac{4}{t^2+1} + \dfrac{8(3-t)t}{(t^2+1)^2} = 9 \;\Leftrightarrow\; \dfrac{24t - 12t^2 - 4}{(t^2+1)^2} = 2 \;\Leftrightarrow\; t^4 + 8t^2 - 12t + 3 = 0.$

It is easily seen that $t = 1$ is a root of this polynomial. Also $2 + \ln 1 + \dfrac{3-1}{1} = 4$, so $t = 1$ is the desired solution.

13   Review

CONCEPT CHECK

1. A vector function is a function whose domain is a set of real numbers and whose range is a set of vectors. To find the derivative
or integral, we can differentiate or integrate each component of the vector function.

2. The tip of the moving vector $\mathbf{r}(t)$ of a continuous vector function traces out a space curve.

3. The tangent vector to a smooth curve at a point P with position vector $\mathbf{r}(t)$ is the vector $\mathbf{r}'(t)$. The tangent line at P is the line

through P parallel to the tangent vector $\mathbf{r}'(t)$. The unit tangent vector is $\mathbf{T}(t) = \dfrac{\mathbf{r}'(t)}{|\mathbf{r}'(t)|}$.

4. (a) (a)–(f) See Theorem 13.2.3.

5. Use Formula 13.3.2, or equivalently, 13.3.3.

6. (a) The curvature of a curve is $\kappa = \left|\dfrac{d\mathbf{T}}{ds}\right|$ where $\mathbf{T}$ is the unit tangent vector.

(b) $\kappa(t) = \left|\dfrac{\mathbf{T}'(t)}{\mathbf{r}'(t)}\right|$ 　　　　(c) $\kappa(t) = \dfrac{|\mathbf{r}'(t) \times \mathbf{r}''(t)|}{|\mathbf{r}'(t)|^3}$ 　　　　(d) $\kappa(x) = \dfrac{|f''(x)|}{[1 + (f'(x))^2]^{3/2}}$

7. (a) The unit normal vector: $\mathbf{N}(t) = \dfrac{\mathbf{T}'(t)}{|\mathbf{T}'(t)|}$.   The binormal vector: $\mathbf{B}(t) = \mathbf{T}(t) \times \mathbf{N}(t)$.

(b) See the discussion preceding Example 7 in Section 13.3.

8. (a) If $\mathbf{r}(t)$ is the position vector of the particle on the space curve, the velocity $\mathbf{v}(t) = \mathbf{r}'(t)$, the speed is given by $|\mathbf{v}(t)|$,
and the acceleration $\mathbf{a}(t) = \mathbf{v}'(t) = \mathbf{r}''(t)$.

© 2012 Cengage Learning. All Rights Reserved. May not be scanned, copied, or duplicated, or posted to a publicly accessible website, in whole or in part.

(b) $\mathbf{a} = a_T\mathbf{T} + a_N\mathbf{N}$ where $a_T = v'$ and $a_N = \kappa v^2$.

9. See the statement of Kepler's Laws on page 892 [ET 868].

TRUE-FALSE QUIZ

1. True. If we reparametrize the curve by replacing $u = t^3$, we have $\mathbf{r}(u) = u\mathbf{i} + 2u\mathbf{j} + 3u\mathbf{k}$, which is a line through the origin with direction vector $\mathbf{i} + 2\mathbf{j} + 3\mathbf{k}$.

3. False. The vector function represents a line, but the line does not pass through the origin; the x-component is 0 only for $t = 0$ which corresponds to the point $(0, 3, 0)$ not $(0, 0, 0)$.

5. False. By Formula 5 of Theorem 13.2.3, $\dfrac{d}{dt}[\mathbf{u}(t) \times \mathbf{v}(t)] = \mathbf{u}'(t) \times \mathbf{v}(t) + \mathbf{u}(t) \times \mathbf{v}'(t)$.

7. False. κ is the magnitude of the rate of change of the unit tangent vector $\mathbf{T}$ with respect to arc length s, not with respect to t.

9. True. At an inflection point where f is twice continuously differentiable we must have $f''(x) = 0$, and by Equation 13.3.11, the curvature is 0 there.

11. False. If $\mathbf{r}(t)$ is the position of a moving particle at time t and $|\mathbf{r}(t)| = 1$ then the particle lies on the unit circle or the unit sphere, but this does not mean that the speed $|\mathbf{r}'(t)|$ must be constant. As a counterexample, let $\mathbf{r}(t) = \langle t, \sqrt{1 - t^2}\rangle$, then $\mathbf{r}'(t) = \langle 1, -t/\sqrt{1 - t^2}\rangle$ and $|\mathbf{r}(t)| = \sqrt{t^2 + 1 - t^2} = 1$ but $|\mathbf{r}'(t)| = \sqrt{1 + t^2/(1 - t^2)} = 1/\sqrt{1 - t^2}$ which is not constant.

13. True. See the discussion preceding Example 7 in Section 13.3.

EXERCISES

1. (a) The corresponding parametric equations for the curve are $x = t$, $y = \cos \pi t$, $z = \sin \pi t$. Since $y^2 + z^2 = 1$, the curve is contained in a circular cylinder with axis the x-axis. Since $x = t$, the curve is a helix.

(b) $\mathbf{r}(t) = t\mathbf{i} + \cos \pi t\,\mathbf{j} + \sin \pi t\,\mathbf{k} \Rightarrow$
$\mathbf{r}'(t) = \mathbf{i} - \pi\sin \pi t\,\mathbf{j} + \pi\cos \pi t\,\mathbf{k} \Rightarrow$
$\mathbf{r}''(t) = -\pi^2\cos \pi t\,\mathbf{j} - \pi^2\sin \pi t\,\mathbf{k}$

3. The projection of the curve C of intersection onto the xy-plane is the circle $x^2 + y^2 = 16$, $z = 0$. So we can write $x = 4\cos t$, $y = 4\sin t$, $0 \le t \le 2\pi$. From the equation of the plane, we have $z = 5 - x = 5 - 4\cos t$, so parametric equations for C are $x = 4\cos t$, $y = 4\sin t$, $z = 5 - 4\cos t$, $0 \le t \le 2\pi$, and the corresponding vector function is $\mathbf{r}(t) = 4\cos t\,\mathbf{i} + 4\sin t\,\mathbf{j} + (5 - 4\cos t)\,\mathbf{k}$, $0 \le t \le 2\pi$.

© 2012 Cengage Learning. All Rights Reserved. May not be scanned, copied, or duplicated, or posted to a publicly accessible website, in whole or in part.

5. $\int_0^1 (t^2\,\mathbf{i} + t\cos\pi t\,\mathbf{j} + \sin\pi t\,\mathbf{k})\,dt = \left(\int_0^1 t^2\,dt\right)\mathbf{i} + \left(\int_0^1 t\cos\pi t\,dt\right)\mathbf{j} + \left(\int_0^1 \sin\pi t\,dt\right)\mathbf{k}$

$$= \left[\tfrac{1}{3}t^3\right]_0^1\mathbf{i} + \left(\tfrac{t}{\pi}\sin\pi t\Big]_0^1 - \int_0^1 \tfrac{1}{\pi}\sin\pi t\,dt\right)\mathbf{j} + \left[-\tfrac{1}{\pi}\cos\pi t\right]_0^1\mathbf{k}$$

$$= \tfrac{1}{3}\mathbf{i} + \left[\tfrac{1}{\pi^2}\cos\pi t\right]_0^1\mathbf{j} + \tfrac{2}{\pi}\mathbf{k} = \tfrac{1}{3}\mathbf{i} - \tfrac{2}{\pi^2}\mathbf{j} + \tfrac{2}{\pi}\mathbf{k}$$

where we integrated by parts in the y-component.

7. $\mathbf{r}(t) = \langle t^2, t^3, t^4\rangle \Rightarrow \mathbf{r}'(t) = \langle 2t, 3t^2, 4t^3\rangle \Rightarrow |\mathbf{r}'(t)| = \sqrt{4t^2 + 9t^4 + 16t^6}$ and

$L = \int_0^3 |\mathbf{r}'(t)|\,dt = \int_0^3 \sqrt{4t^2 + 9t^4 + 16t^6}\,dt$. Using Simpson's Rule with $f(t) = \sqrt{4t^2 + 9t^4 + 16t^6}$ and $n = 6$ we

have $\Delta t = \frac{3-0}{6} = \frac{1}{2}$ and

$$L \approx \tfrac{\Delta t}{3}\left[f(0) + 4f\left(\tfrac{1}{2}\right) + 2f(1) + 4f\left(\tfrac{3}{2}\right) + 2f(2) + 4f\left(\tfrac{5}{2}\right) + f(3)\right]$$

$$= \tfrac{1}{6}\left[\sqrt{0 + 0 + 0} + 4\cdot\sqrt{4\left(\tfrac{1}{2}\right)^2 + 9\left(\tfrac{1}{2}\right)^4 + 16\left(\tfrac{1}{2}\right)^6} + 2\cdot\sqrt{4(1)^2 + 9(1)^4 + 16(1)^6}\right.$$

$$+ 4\cdot\sqrt{4\left(\tfrac{3}{2}\right)^2 + 9\left(\tfrac{3}{2}\right)^4 + 16\left(\tfrac{3}{2}\right)^6} + 2\cdot\sqrt{4(2)^2 + 9(2)^4 + 16(2)^6}$$

$$\left. + 4\cdot\sqrt{4\left(\tfrac{5}{2}\right)^2 + 9\left(\tfrac{5}{2}\right)^4 + 16\left(\tfrac{5}{2}\right)^6} + \sqrt{4(3)^2 + 9(3)^4 + 16(3)^6}\right]$$

$$\approx 86.631$$

9. The angle of intersection of the two curves, θ, is the angle between their respective tangents at the point of intersection.

For both curves the point $(1, 0, 0)$ occurs when $t = 0$.

$\mathbf{r}_1'(t) = -\sin t\,\mathbf{i} + \cos t\,\mathbf{j} + \mathbf{k} \Rightarrow \mathbf{r}_1'(0) = \mathbf{j} + \mathbf{k}$ and $\mathbf{r}_2'(t) = \mathbf{i} + 2t\,\mathbf{j} + 3t^2\,\mathbf{k} \Rightarrow \mathbf{r}_2'(0) = \mathbf{i}$.

$\mathbf{r}_1'(0)\cdot\mathbf{r}_2'(0) = (\mathbf{j} + \mathbf{k})\cdot\mathbf{i} = 0$. Therefore, the curves intersect in a right angle, that is, $\theta = \frac{\pi}{2}$.

11. (a) $\mathbf{T}(t) = \dfrac{\mathbf{r}'(t)}{|\mathbf{r}'(t)|} = \dfrac{\langle t^2, t, 1\rangle}{|\langle t^2, t, 1\rangle|} = \dfrac{\langle t^2, t, 1\rangle}{\sqrt{t^4 + t^2 + 1}}$

(b) $\mathbf{T}'(t) = -\tfrac{1}{2}(t^4 + t^2 + 1)^{-3/2}(4t^3 + 2t)\langle t^2, t, 1\rangle + (t^4 + t^2 + 1)^{-1/2}\langle 2t, 1, 0\rangle$

$$= \frac{-2t^3 - t}{(t^4 + t^2 + 1)^{3/2}}\langle t^2, t, 1\rangle + \frac{1}{(t^4 + t^2 + 1)^{1/2}}\langle 2t, 1, 0\rangle$$

$$= \frac{\langle -2t^5 - t^3, -2t^4 - t^2, -2t^3 - t\rangle + \langle 2t^5 + 2t^3 + 2t, t^4 + t^2 + 1, 0\rangle}{(t^4 + t^2 + 1)^{3/2}} = \frac{\langle t^3 + 2t, -t^4 + 1, -2t^3 - t\rangle}{(t^4 + t^2 + 1)^{3/2}}$$

$$|\mathbf{T}'(t)| = \frac{\sqrt{t^6 + 4t^4 + 4t^2 + t^8 - 2t^4 + 1 + 4t^6 + 4t^4 + t^2}}{(t^4 + t^2 + 1)^{3/2}} = \frac{\sqrt{t^8 + 5t^6 + 6t^4 + 5t^2 + 1}}{(t^4 + t^2 + 1)^{3/2}} \quad \text{and}$$

$$\mathbf{N}(t) = \frac{\langle t^3 + 2t, 1 - t^4, -2t^3 - t\rangle}{\sqrt{t^8 + 5t^6 + 6t^4 + 5t^2 + 1}}.$$

(c) $\kappa(t) = \dfrac{|\mathbf{T}'(t)|}{|\mathbf{r}'(t)|} = \dfrac{\sqrt{t^8 + 5t^6 + 6t^4 + 5t^2 + 1}}{(t^4 + t^2 + 1)^2}$ or $\dfrac{\sqrt{t^4 + 4t^2 + 1}}{(t^4 + t^2 + 1)^{3/2}}$

13. $y' = 4x^3$, $y'' = 12x^2$ and $\kappa(x) = \dfrac{|y''|}{[1 + (y')^2]^{3/2}} = \dfrac{|12x^2|}{(1 + 16x^6)^{3/2}}$, so $\kappa(1) = \dfrac{12}{17^{3/2}}$.

© 2012 Cengage Learning. All Rights Reserved. May not be scanned, copied, or duplicated, or posted to a publicly accessible website, in whole or in part.

15. $\mathbf{r}(t) = \langle \sin 2t, t, \cos 2t \rangle \quad \Rightarrow \quad \mathbf{r}'(t) = \langle 2\cos 2t, 1, -2\sin 2t \rangle \quad \Rightarrow \quad \mathbf{T}(t) = \frac{1}{\sqrt{5}} \langle 2\cos 2t, 1, -2\sin 2t \rangle \quad \Rightarrow$

$\mathbf{T}'(t) = \frac{1}{\sqrt{5}} \langle -4\sin 2t, 0, -4\cos 2t \rangle \quad \Rightarrow \quad \mathbf{N}(t) = \langle -\sin 2t, 0, -\cos 2t \rangle$. So $\mathbf{N} = \mathbf{N}(\pi) = \langle 0, 0, -1 \rangle$ and

$\mathbf{B} = \mathbf{T} \times \mathbf{N} = \frac{1}{\sqrt{5}} \langle -1, 2, 0 \rangle$. So a normal to the osculating plane is $\langle -1, 2, 0 \rangle$ and an equation is

$-1(x - 0) + 2(y - \pi) + 0(z - 1) = 0$ or $x - 2y + 2\pi = 0$.

17. $\mathbf{r}(t) = t\ln t \, \mathbf{i} + t\mathbf{j} + e^{-t}\mathbf{k}, \quad \mathbf{v}(t) = \mathbf{r}'(t) = (1 + \ln t)\mathbf{i} + \mathbf{j} - e^{-t}\mathbf{k},$

$|\mathbf{v}(t)| = \sqrt{(1 + \ln t)^2 + 1^2 + (-e^{-t})^2} = \sqrt{2 + 2\ln t + (\ln t)^2 + e^{-2t}}, \quad \mathbf{a}(t) = \mathbf{v}'(t) = \frac{1}{t}\mathbf{i} + e^{-t}\mathbf{k}$

19. We set up the axes so that the shot leaves the athlete's hand 7 ft above the origin. Then we are given $\mathbf{r}(0) = 7\mathbf{j}$,

$|\mathbf{v}(0)| = 43$ ft/s, and $\mathbf{v}(0)$ has direction given by a $45°$ angle of elevation. Then a unit vector in the direction of $\mathbf{v}(0)$ is

$\frac{1}{\sqrt{2}}(\mathbf{i} + \mathbf{j}) \quad \Rightarrow \quad \mathbf{v}(0) = \frac{43}{\sqrt{2}}(\mathbf{i} + \mathbf{j})$. Assuming air resistance is negligible, the only external force is due to gravity, so as in

Example 13.4.5 we have $\mathbf{a} = -g\mathbf{j}$ where here $g \approx 32$ ft/s^2. Since $\mathbf{v}'(t) = \mathbf{a}(t)$, we integrate, giving $\mathbf{v}(t) = -gt\mathbf{j} + \mathbf{C}$

where $\mathbf{C} = \mathbf{v}(0) = \frac{43}{\sqrt{2}}(\mathbf{i} + \mathbf{j}) \quad \Rightarrow \quad \mathbf{v}(t) = \frac{43}{\sqrt{2}}\mathbf{i} + \left(\frac{43}{\sqrt{2}} - gt\right)\mathbf{j}$. Since $\mathbf{r}'(t) = \mathbf{v}(t)$ we integrate again, so

$\mathbf{r}(t) = \frac{43}{\sqrt{2}}t\mathbf{i} + \left(\frac{43}{\sqrt{2}}t - \frac{1}{2}gt^2\right)\mathbf{j} + \mathbf{D}$. But $\mathbf{D} = \mathbf{r}(0) = 7\mathbf{j} \quad \Rightarrow \quad \mathbf{r}(t) = \frac{43}{\sqrt{2}}t\mathbf{i} + \left(\frac{43}{\sqrt{2}}t - \frac{1}{2}gt^2 + 7\right)\mathbf{j}$.

(a) At 2 seconds, the shot is at $\mathbf{r}(2) = \frac{43}{\sqrt{2}}(2)\mathbf{i} + \left(\frac{43}{\sqrt{2}}(2) - \frac{1}{2}g(2)^2 + 7\right)\mathbf{j} \approx 60.8\mathbf{i} + 3.8\mathbf{j}$, so the shot is about 3.8 ft above

the ground, at a horizontal distance of 60.8 ft from the athlete.

(b) The shot reaches its maximum height when the vertical component of velocity is 0: $\frac{43}{\sqrt{2}} - gt = 0 \quad \Rightarrow$

$t = \frac{43}{\sqrt{2}\,g} \approx 0.95$ s. Then $\mathbf{r}(0.95) \approx 28.9\mathbf{i} + 21.4\mathbf{j}$, so the maximum height is approximately 21.4 ft.

(c) The shot hits the ground when the vertical component of $\mathbf{r}(t)$ is 0, so $\frac{43}{\sqrt{2}}t - \frac{1}{2}gt^2 + 7 = 0 \quad \Rightarrow$

$-16t^2 + \frac{43}{\sqrt{2}}t + 7 = 0 \quad \Rightarrow \quad t \approx 2.11$ s. $\quad \mathbf{r}(2.11) \approx 64.2\mathbf{i} - 0.08\mathbf{j}$, thus the shot lands approximately 64.2 ft from the

athlete.

21. (a) Instead of proceeding directly, we use Formula 3 of Theorem 13.2.3: $\mathbf{r}(t) = t\mathbf{R}(t) \quad \Rightarrow$

$\mathbf{v} = \mathbf{r}'(t) = \mathbf{R}(t) + t\mathbf{R}'(t) = \cos\omega t\,\mathbf{i} + \sin\omega t\,\mathbf{j} + t\mathbf{v}_d.$

(b) Using the same method as in part (a) and starting with $\mathbf{v} = \mathbf{R}(t) + t\mathbf{R}'(t)$, we have

$\mathbf{a} = \mathbf{v}' = \mathbf{R}'(t) + \mathbf{R}'(t) + t\mathbf{R}''(t) = 2\mathbf{R}'(t) + t\mathbf{R}''(t) = 2\mathbf{v}_d + t\mathbf{a}_d.$

(c) Here we have $\mathbf{r}(t) = e^{-t}\cos\omega t\,\mathbf{i} + e^{-t}\sin\omega t\,\mathbf{j} = e^{-t}\mathbf{R}(t)$. So, as in parts (a) and (b),

$\mathbf{v} = \mathbf{r}'(t) = e^{-t}\mathbf{R}'(t) - e^{-t}\mathbf{R}(t) = e^{-t}[\mathbf{R}'(t) - \mathbf{R}(t)] \quad \Rightarrow$

$\mathbf{a} = \mathbf{v}' = e^{-t}[\mathbf{R}''(t) - \mathbf{R}'(t)] - e^{-t}[\mathbf{R}'(t) - \mathbf{R}(t)] = e^{-t}[\mathbf{R}''(t) - 2\mathbf{R}'(t) + \mathbf{R}(t)]$

$= e^{-t}\mathbf{a}_d - 2e^{-t}\mathbf{v}_d + e^{-t}\mathbf{R}$

Thus, the Coriolis acceleration (the sum of the "extra" terms not involving $\mathbf{a}_d$) is $-2e^{-t}\mathbf{v}_d + e^{-t}\mathbf{R}$.

© 2012 Cengage Learning. All Rights Reserved. May not be scanned, copied, or duplicated, or posted to a publicly accessible website, in whole or in part.

23. (a) $\mathbf{r}(t) = R\cos\omega t\,\mathbf{i} + R\sin\omega t\,\mathbf{j} \;\Rightarrow\; \mathbf{v} = \mathbf{r}'(t) = -\omega R\sin\omega t\,\mathbf{i} + \omega R\cos\omega t\,\mathbf{j}$, so $\mathbf{r} = R(\cos\omega t\,\mathbf{i} + \sin\omega t\,\mathbf{j})$ and

$\mathbf{v} = \omega R(-\sin\omega t\,\mathbf{i} + \cos\omega t\,\mathbf{j})$. $\mathbf{v}\cdot\mathbf{r} = \omega R^2(-\cos\omega t\sin\omega t + \sin\omega t\cos\omega t) = 0$, so $\mathbf{v}\perp\mathbf{r}$. Since $\mathbf{r}$ points along a

radius of the circle, and $\mathbf{v}\perp\mathbf{r}$, $\mathbf{v}$ is tangent to the circle. Because it is a velocity vector, $\mathbf{v}$ points in the direction of motion.

(b) In (a), we wrote $\mathbf{v}$ in the form $\omega R\,\mathbf{u}$, where $\mathbf{u}$ is the unit vector $-\sin\omega t\,\mathbf{i} + \cos\omega t\,\mathbf{j}$. Clearly $|\mathbf{v}| = \omega R\,|\mathbf{u}| = \omega R$. At

speed ωR, the particle completes one revolution, a distance $2\pi R$, in time $T = \dfrac{2\pi R}{\omega R} = \dfrac{2\pi}{\omega}$.

(c) $\mathbf{a} = \dfrac{d\mathbf{v}}{dt} = -\omega^2 R\cos\omega t\,\mathbf{i} - \omega^2 R\sin\omega t\,\mathbf{j} = -\omega^2 R(\cos\omega t\,\mathbf{i} + \sin\omega t\,\mathbf{j})$, so $\mathbf{a} = -\omega^2\mathbf{r}$. This shows that $\mathbf{a}$ is proportional

to $\mathbf{r}$ and points in the opposite direction (toward the origin). Also, $|\mathbf{a}| = \omega^2\,|\mathbf{r}| = \omega^2 R$.

(d) By Newton's Second Law (see Section 13.4), $\mathbf{F} = m\mathbf{a}$, so $|\mathbf{F}| = m\,|\mathbf{a}| = mR\omega^2 = \dfrac{m\,(\omega R)^2}{R} = \dfrac{m\,|\mathbf{v}|^2}{R}$.

© 2012 Cengage Learning. All Rights Reserved. May not be scanned, copied, or duplicated, or posted to a publicly accessible website, in whole or in part.

□ PROBLEMS PLUS

1. (a) The projectile reaches maximum height when $0 = \dfrac{dy}{dt} = \dfrac{d}{dt}\left[(v_0\sin\alpha)t - \tfrac{1}{2}gt^2\right] = v_0\sin\alpha - gt$; that is, when

$t = \dfrac{v_0\sin\alpha}{g}$ and $y = (v_0\sin\alpha)\left(\dfrac{v_0\sin\alpha}{g}\right) - \dfrac{1}{2}g\left(\dfrac{v_0\sin\alpha}{g}\right)^2 = \dfrac{v_0^2\sin^2\alpha}{2g}$. This is the maximum height attained when

the projectile is fired with an angle of elevation α. This maximum height is largest when $\alpha = \frac{\pi}{2}$. In that case, $\sin\alpha = 1$

and the maximum height is $\dfrac{v_0^2}{2g}$.

(b) Let $R = v_0^2/g$. We are asked to consider the parabola $x^2 + 2Ry - R^2 = 0$ which can be rewritten as $y = -\dfrac{1}{2R}x^2 + \dfrac{R}{2}$.

The points on or inside this parabola are those for which $-R \le x \le R$ and $0 \le y \le \dfrac{-1}{2R}x^2 + \dfrac{R}{2}$. When the projectile is

fired at angle of elevation α, the points (x, y) along its path satisfy the relations $x = (v_0\cos\alpha)\,t$ and

$y = (v_0\sin\alpha)t - \tfrac{1}{2}gt^2$, where $0 \le t \le (2v_0\sin\alpha)/g$ (as in Example 13.4.5). Thus

$|x| \le \left| v_0\cos\alpha\left(\dfrac{2v_0\sin\alpha}{g}\right)\right| = \left|\dfrac{v_0^2}{g}\sin 2\alpha\right| \le \left|\dfrac{v_0^2}{g}\right| = |R|$. This shows that $-R \le x \le R$.

For t in the specified range, we also have $y = t\left(v_0\sin\alpha - \tfrac{1}{2}gt\right) = \tfrac{1}{2}gt\left(\dfrac{2v_0\sin\alpha}{g} - t\right) \ge 0$ and

$y = (v_0\sin\alpha)\dfrac{x}{v_0\cos\alpha} - \dfrac{g}{2}\left(\dfrac{x}{v_0\cos\alpha}\right)^2 = (\tan\alpha)\,x - \dfrac{g}{2v_0^2\cos^2\alpha}x^2 = -\dfrac{1}{2R\cos^2\alpha}x^2 + (\tan\alpha)\,x$. Thus

$$y - \left(\dfrac{-1}{2R}x^2 + \dfrac{R}{2}\right) = \dfrac{-1}{2R\cos^2\alpha}x^2 + \dfrac{1}{2R}x^2 + (\tan\alpha)\,x - \dfrac{R}{2}$$

$$= \dfrac{x^2}{2R}\left(1 - \dfrac{1}{\cos^2\alpha}\right) + (\tan\alpha)\,x - \dfrac{R}{2} = \dfrac{x^2(1 - \sec^2\alpha) + 2R(\tan\alpha)\,x - R^2}{2R}$$

$$= \dfrac{-(\tan^2\alpha)\,x^2 + 2R(\tan\alpha)\,x - R^2}{2R} = \dfrac{-[(\tan\alpha)\,x - R]^2}{2R} \le 0$$

We have shown that every target that can be hit by the projectile lies on or inside the parabola $y = -\dfrac{1}{2R}x^2 + \dfrac{R}{2}$.

Now let (a, b) be any point on or inside the parabola $y = -\dfrac{1}{2R}x^2 + \dfrac{R}{2}$. Then $-R \le a \le R$ and $0 \le b \le -\dfrac{1}{2R}a^2 + \dfrac{R}{2}$.

We seek an angle α such that (a, b) lies in the path of the projectile; that is, we wish to find an angle α such that

$b = -\dfrac{1}{2R\cos^2\alpha}a^2 + (\tan\alpha)\,a$ or equivalently $b = \dfrac{-1}{2R}(\tan^2\alpha + 1)a^2 + (\tan\alpha)\,a$. Rearranging this equation we get

$\dfrac{a^2}{2R}\tan^2\alpha - a\tan\alpha + \left(\dfrac{a^2}{2R} + b\right) = 0$ or $a^2(\tan\alpha)^2 - 2aR(\tan\alpha) + (a^2 + 2bR) = 0$ $(\star)$. This quadratic equation

for $\tan\alpha$ has real solutions exactly when the discriminant is nonnegative. Now $B^2 - 4AC \ge 0$ $\Leftrightarrow$

© 2012 Cengage Learning. All Rights Reserved. May not be scanned, copied, or duplicated, or posted to a publicly accessible website, in whole or in part.

$(-2aR)^2 - 4a^2(a^2 + 2bR) \geq 0 \iff 4a^2(R^2 - a^2 - 2bR) \geq 0 \iff -a^2 - 2bR + R^2 \geq 0 \iff$

$b \leq \dfrac{1}{2R}(R^2 - a^2) \iff b \leq \dfrac{-1}{2R}a^2 + \dfrac{R}{2}$. This condition is satisfied since (a, b) is on or inside the parabola

$y = -\dfrac{1}{2R}x^2 + \dfrac{R}{2}$. It follows that (a, b) lies in the path of the projectile when $\tan \alpha$ satisfies $(\star)$, that is, when

$\tan \alpha = \dfrac{2aR \pm \sqrt{4a^2(R^2 - a^2 - 2bR)}}{2a^2} = \dfrac{R \pm \sqrt{R^2 - 2bR - a^2}}{a}.$

(c)

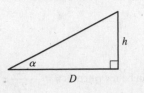

If the gun is pointed at a target with height h at a distance D downrange, then $\tan \alpha = h/D$. When the projectile reaches a distance D downrange (remember we are assuming that it doesn't hit the ground first), we have $D = x = (v_0 \cos \alpha)t$,

so $t = \dfrac{D}{v_0 \cos \alpha}$ and $y = (v_0 \sin \alpha)t - \frac{1}{2}gt^2 = D \tan \alpha - \dfrac{gD^2}{2v_0^2 \cos^2 \alpha}.$

Meanwhile, the target, whose x-coordinate is also D, has fallen from height h to height

$h - \frac{1}{2}gt^2 = D \tan \alpha - \dfrac{gD^2}{2v_0^2 \cos^2 \alpha}$. Thus the projectile hits the target.

3. (a) $\mathbf{a} = -g\mathbf{j} \implies \mathbf{v} = \mathbf{v}_0 - gt\mathbf{j} = 2\mathbf{i} - gt\mathbf{j} \implies \mathbf{s} = \mathbf{s}_0 + 2t\mathbf{i} - \frac{1}{2}gt^2\mathbf{j} = 3.5\mathbf{j} + 2t\mathbf{i} - \frac{1}{2}gt^2\mathbf{j} \implies$

$\mathbf{s} = 2t\mathbf{i} + \left(3.5 - \frac{1}{2}gt^2\right)\mathbf{j}$. Therefore $y = 0$ when $t = \sqrt{7/g}$ seconds. At that instant, the ball is $2\sqrt{7/g} \approx 0.94$ ft to the

right of the table top. Its coordinates (relative to an origin on the floor directly under the table's edge) are $(0.94, 0)$. At

impact, the velocity is $\mathbf{v} = 2\mathbf{i} - \sqrt{7g}\,\mathbf{j}$, so the speed is $|\mathbf{v}| = \sqrt{4 + 7g} \approx 15$ ft/s.

(b) The slope of the curve when $t = \sqrt{\dfrac{7}{g}}$ is $\dfrac{dy}{dx} = \dfrac{dy/dt}{dx/dt} = \dfrac{-gt}{2} = \dfrac{-g\sqrt{7/g}}{2} = \dfrac{-\sqrt{7g}}{2}$. Thus $\cot \theta = \dfrac{\sqrt{7g}}{2}$

and $\theta \approx 7.6°$.

(c) From (a), $|\mathbf{v}| = \sqrt{4 + 7g}$. So the ball rebounds with speed $0.8\sqrt{4 + 7g} \approx 12.08$ ft/s at angle of inclination

$90° - \theta \approx 82.3886°$. By Example 13.4.5, the horizontal distance traveled between bounces is $d = \dfrac{v_0^2 \sin 2\alpha}{g}$, where

$v_0 \approx 12.08$ ft/s and $\alpha \approx 82.3886°$. Therefore, $d \approx 1.197$ ft. So the ball strikes the floor at about

$2\sqrt{7/g} + 1.197 \approx 2.13$ ft to the right of the table's edge.

5. The trajectory of the projectile is given by $\mathbf{r}(t) = (v \cos \alpha)t\,\mathbf{i} + \left[(v \sin \alpha)t - \frac{1}{2}gt^2\right]\mathbf{j}$, so

$\mathbf{v}(t) = \mathbf{r}'(t) = v \cos \alpha\,\mathbf{i} + (v \sin \alpha - gt)\,\mathbf{j}$ and

$$|\mathbf{v}(t)| = \sqrt{(v \cos \alpha)^2 + (v \sin \alpha - gt)^2} = \sqrt{v^2 - (2vg \sin \alpha)\,t + g^2t^2} = \sqrt{g^2\left(t^2 - \frac{2v}{g}(\sin \alpha)\,t + \frac{v^2}{g^2}\right)}$$

$$= g\sqrt{\left(t - \frac{v}{g}\sin \alpha\right)^2 + \frac{v^2}{g^2} - \frac{v^2}{g^2}\sin^2 \alpha} = g\sqrt{\left(t - \frac{v}{g}\sin \alpha\right)^2 + \frac{v^2}{g^2}\cos^2 \alpha}$$

© 2012 Cengage Learning. All Rights Reserved. May not be scanned, copied, or duplicated, or posted to a publicly accessible website, in whole or in part.

The projectile hits the ground when $(v\sin\alpha)t - \frac{1}{2}gt^2 = 0 \Rightarrow t = \frac{2v}{g}\sin\alpha$, so the distance traveled by the projectile is

$$L(\alpha) = \int_0^{(2v/g)\sin\alpha} |\mathbf{v}(t)|\, dt = \int_0^{(2v/g)\sin\alpha} g\sqrt{\left(t - \frac{v}{g}\sin\alpha\right)^2 + \frac{v^2}{g^2}\cos^2\alpha}\, dt$$

$$= g\left[\frac{t - (v/g)\sin\alpha}{2}\sqrt{\left(t - \frac{v}{g}\sin\alpha\right)^2 + \left(\frac{v}{g}\cos\alpha\right)^2}\right.$$

$$\left. + \frac{[(v/g)\cos\alpha]^2}{2}\ln\left(t - \frac{v}{g}\sin\alpha + \sqrt{\left(t - \frac{v}{g}\sin\alpha\right)^2 + \left(\frac{v}{g}\cos\alpha\right)^2}\right)\right]_0^{(2v/g)\sin\alpha}$$

[using Formula 21 in the Table of Integrals]

$$= \frac{g}{2}\left[\frac{v}{g}\sin\alpha\sqrt{\left(\frac{v}{g}\sin\alpha\right)^2 + \left(\frac{v}{g}\cos\alpha\right)^2} + \left(\frac{v}{g}\cos\alpha\right)^2\ln\left(\frac{v}{g}\sin\alpha + \sqrt{\left(\frac{v}{g}\sin\alpha\right)^2 + \left(\frac{v}{g}\cos\alpha\right)^2}\right)\right.$$

$$\left. + \frac{v}{g}\sin\alpha\sqrt{\left(\frac{v}{g}\sin\alpha\right)^2 + \left(\frac{v}{g}\cos\alpha\right)^2} - \left(\frac{v}{g}\cos\alpha\right)^2\ln\left(-\frac{v}{g}\sin\alpha + \sqrt{\left(\frac{v}{g}\sin\alpha\right)^2 + \left(\frac{v}{g}\cos\alpha\right)^2}\right)\right]$$

$$= \frac{g}{2}\left[\frac{v}{g}\sin\alpha\cdot\frac{v}{g} + \frac{v^2}{g^2}\cos^2\alpha\ln\left(\frac{v}{g}\sin\alpha + \frac{v}{g}\right) + \frac{v}{g}\sin\alpha\cdot\frac{v}{g} - \frac{v^2}{g^2}\cos^2\alpha\ln\left(-\frac{v}{g}\sin\alpha + \frac{v}{g}\right)\right]$$

$$= \frac{v^2}{g}\sin\alpha + \frac{v^2}{2g}\cos^2\alpha\ln\left(\frac{(v/g)\sin\alpha + v/g}{-(v/g)\sin\alpha + v/g}\right) = \frac{v^2}{g}\sin\alpha + \frac{v^2}{2g}\cos^2\alpha\ln\left(\frac{1 + \sin\alpha}{1 - \sin\alpha}\right)$$

We want to maximize $L(\alpha)$ for $0 \le \alpha \le \pi/2$.

$$L'(\alpha) = \frac{v^2}{g}\cos\alpha + \frac{v^2}{2g}\left[\cos^2\alpha\cdot\frac{1 - \sin\alpha}{1 + \sin\alpha}\cdot\frac{2\cos\alpha}{(1 - \sin\alpha)^2} - 2\cos\alpha\sin\alpha\ln\left(\frac{1 + \sin\alpha}{1 - \sin\alpha}\right)\right]$$

$$= \frac{v^2}{g}\cos\alpha + \frac{v^2}{2g}\left[\cos^2\alpha\cdot\frac{2}{\cos\alpha} - 2\cos\alpha\sin\alpha\ln\left(\frac{1 + \sin\alpha}{1 - \sin\alpha}\right)\right]$$

$$= \frac{v^2}{g}\cos\alpha + \frac{v^2}{g}\cos\alpha\left[1 - \sin\alpha\ln\left(\frac{1 + \sin\alpha}{1 - \sin\alpha}\right)\right] = \frac{v^2}{g}\cos\alpha\left[2 - \sin\alpha\ln\left(\frac{1 + \sin\alpha}{1 - \sin\alpha}\right)\right]$$

$L(\alpha)$ has critical points for $0 < \alpha < \pi/2$ when $L'(\alpha) = 0 \Rightarrow 2 - \sin\alpha\ln\left(\dfrac{1 + \sin\alpha}{1 - \sin\alpha}\right) = 0$ [since $\cos\alpha \ne 0$].

Solving by graphing (or using a CAS) gives $\alpha \approx 0.9855$. Compare values at the critical point and the endpoints:

$L(0) = 0$, $L(\pi/2) = v^2/g$, and $L(0.9855) \approx 1.20v^2/g$. Thus the distance traveled by the projectile is maximized for $\alpha \approx 0.9855$ or $\approx 56°$.

7. We can write the vector equation as $\mathbf{r}(t) = \mathbf{a}t^2 + \mathbf{b}t + \mathbf{c}$ where $\mathbf{a} = \langle a_1, a_2, a_3\rangle$, $\mathbf{b} = \langle b_1, b_2, b_3\rangle$, and $\mathbf{c} = \langle c_1, c_2, c_3\rangle$.

Then $\mathbf{r}'(t) = 2t\,\mathbf{a} + \mathbf{b}$ which says that each tangent vector is the sum of a scalar multiple of $\mathbf{a}$ and the vector $\mathbf{b}$. Thus the tangent vectors are all parallel to the plane determined by $\mathbf{a}$ and $\mathbf{b}$ so the curve must be parallel to this plane. [Here we assume that $\mathbf{a}$ and $\mathbf{b}$ are nonparallel. Otherwise the tangent vectors are all parallel and the curve lies along a single line.] A normal

© 2012 Cengage Learning. All Rights Reserved. May not be scanned, copied, or duplicated, or posted to a publicly accessible website, in whole or in part.

vector for the plane is $\mathbf{a} \times \mathbf{b} = \langle a_2b_3 - a_3b_2, a_3b_1 - a_1b_3, a_1b_2 - a_2b_1 \rangle$. The point (c_1, c_2, c_3) lies on the plane (when $t = 0$), so an equation of the plane is

$$(a_2b_3 - a_3b_2)(x - c_1) + (a_3b_1 - a_1b_3)(y - c_2) + (a_1b_2 - a_2b_1)(z - c_3) = 0$$

or

$$(a_2b_3 - a_3b_2)x + (a_3b_1 - a_1b_3)y + (a_1b_2 - a_2b_1)z = a_2b_3c_1 - a_3b_2c_1 + a_3b_1c_2 - a_1b_3c_2 + a_1b_2c_3 - a_2b_1c_3$$

© 2012 Cengage Learning. All Rights Reserved. May not be scanned, copied, or duplicated, or posted to a publicly accessible website, in whole or in part.

14 ☐ PARTIAL DERIVATIVES

14.1 Functions of Several Variables

1. (a) From Table 1, $f(-15, 40) = -27$, which means that if the temperature is $-15°C$ and the wind speed is 40 km/h, then the air would feel equivalent to approximately $-27°C$ without wind.

   (b) The question is asking: when the temperature is $-20°C$, what wind speed gives a wind-chill index of $-30°C$? From Table 1, the speed is 20 km/h.

   (c) The question is asking: when the wind speed is 20 km/h, what temperature gives a wind-chill index of $-49°C$? From Table 1, the temperature is $-35°C$.

   (d) The function $W = f(-5, v)$ means that we fix T at -5 and allow v to vary, resulting in a function of one variable. In other words, the function gives wind-chill index values for different wind speeds when the temperature is $-5°C$. From Table 1 (look at the row corresponding to $T = -5$), the function decreases and appears to approach a constant value as v increases.

   (e) The function $W = f(T, 50)$ means that we fix v at 50 and allow T to vary, again giving a function of one variable. In other words, the function gives wind-chill index values for different temperatures when the wind speed is 50 km/h. From Table 1 (look at the column corresponding to $v = 50$), the function increases almost linearly as T increases.

3. $P(120, 20) = 1.47(120)^{0.65}(20)^{0.35} \approx 94.2$, so when the manufacturer invests $20 million in capital and 120,000 hours of labor are completed yearly, the monetary value of the production is about $94.2 million.

5. (a) $f(160, 70) = 0.1091(160)^{0.425}(70)^{0.725} \approx 20.5$, which means that the surface area of a person 70 inches (5 feet 10 inches) tall who weighs 160 pounds is approximately 20.5 square feet.

   (b) Answers will vary depending on the height and weight of the reader.

7. (a) According to Table 4, $f(40, 15) = 25$, which means that if a 40-knot wind has been blowing in the open sea for 15 hours, it will create waves with estimated heights of 25 feet.

   (b) $h = f(30, t)$ means we fix v at 30 and allow t to vary, resulting in a function of one variable. Thus here, $h = f(30, t)$ gives the wave heights produced by 30-knot winds blowing for t hours. From the table (look at the row corresponding to $v = 30$), the function increases but at a declining rate as t increases. In fact, the function values appear to be approaching a limiting value of approximately 19, which suggests that 30-knot winds cannot produce waves higher than about 19 feet.

   (c) $h = f(v, 30)$ means we fix t at 30, again giving a function of one variable. So, $h = f(v, 30)$ gives the wave heights produced by winds of speed v blowing for 30 hours. From the table (look at the column corresponding to $t = 30$), the function appears to increase at an increasing rate, with no apparent limiting value. This suggests that faster winds (lasting 30 hours) always create higher waves.

© 2012 Cengage Learning. All Rights Reserved. May not be scanned, copied, or duplicated, or posted to a publicly accessible website, in whole or in part.

9. (a) $g(2, -1) = \cos(2 + 2(-1)) = \cos(0) = 1$

(b) $x + 2y$ is defined for all choices of values for x and y and the cosine function is defined for all input values, so the domain of g is $\mathbb{R}^2$.

(c) The range of the cosine function is $[-1, 1]$ and $x + 2y$ generates all possible input values for the cosine function, so the range of $\cos(x + 2y)$ is $[-1, 1]$.

11. (a) $f(1, 1, 1) = \sqrt{1} + \sqrt{1} + \sqrt{1} + \ln(4 - 1^2 - 1^2 - 1^2) = 3 + \ln 1 = 3$

(b) $\sqrt{x}$, $\sqrt{y}$, $\sqrt{z}$ are defined only when $x \geq 0$, $y \geq 0$, $z \geq 0$, and $\ln(4 - x^2 - y^2 - z^2)$ is defined when

$4 - x^2 - y^2 - z^2 > 0 \quad \Leftrightarrow \quad x^2 + y^2 + z^2 < 4$, thus the domain is

$\{(x, y, z) \mid x^2 + y^2 + z^2 < 4,\ x \geq 0,\ y \geq 0,\ z \geq 0\}$, the portion of the interior of a sphere of radius 2, centered at the origin, that is in the first octant.

13. $\sqrt{2x - y}$ is defined only when $2x - y \geq 0$, or $y \leq 2x$.
So the domain of f is $\{(x, y) \mid y \leq 2x\}$.

15. $\ln(9 - x^2 - 9y^2)$ is defined only when
$9 - x^2 - 9y^2 > 0$, or $\frac{1}{9}x^2 + y^2 < 1$. So the domain of f is $\{(x, y) \mid \frac{1}{9}x^2 + y^2 < 1\}$, the interior of an ellipse.

17. $\sqrt{1 - x^2}$ is defined only when $1 - x^2 \geq 0$, or
$x^2 \leq 1 \quad \Leftrightarrow \quad -1 \leq x \leq 1$, and $\sqrt{1 - y^2}$ is defined only when $1 - y^2 \geq 0$, or $y^2 \leq 1 \quad \Leftrightarrow \quad -1 \leq y \leq 1$.
Thus the domain of f is
$\{(x, y) \mid -1 \leq x \leq 1,\ -1 \leq y \leq 1\}$.

19. $\sqrt{y - x^2}$ is defined only when $y - x^2 \geq 0$, or $y \geq x^2$.
In addition, f is not defined if $1 - x^2 = 0 \quad \Leftrightarrow \quad x = \pm 1$. Thus the domain of f is
$\{(x, y) \mid y \geq x^2,\ x \neq \pm 1\}$.

© 2012 Cengage Learning. All Rights Reserved. May not be scanned, copied, or duplicated, or posted to a publicly accessible website, in whole or in part.

21. We need $1 - x^2 - y^2 - z^2 \geq 0$ or $x^2 + y^2 + z^2 \leq 1$, so $D = \{(x, y, z) \mid x^2 + y^2 + z^2 \leq 1\}$ (the points inside or on the sphere of radius 1, center the origin).

23. $z = 1 + y$, a plane which intersects the yz-plane in the line $z = 1 + y$, $x = 0$. The portion of this plane for $x \geq 0$, $z \geq 0$ is shown.

25. $z = 10 - 4x - 5y$ or $4x + 5y + z = 10$, a plane with intercepts 2.5, 2, and 10.

27. $z = y^2 + 1$, a parabolic cylinder

29. $z = 9 - x^2 - 9y^2$, an elliptic paraboloid opening downward with vertex at $(0, 0, 9)$.

31. $z = \sqrt{4 - 4x^2 - y^2}$ so $4x^2 + y^2 + z^2 = 4$ or $x^2 + \dfrac{y^2}{4} + \dfrac{z^2}{4} = 1$ and $z \geq 0$, the top half of an ellipsoid.

33. The point $(-3, 3)$ lies between the level curves with z-values 50 and 60. Since the point is a little closer to the level curve with $z = 60$, we estimate that $f(-3, 3) \approx 56$. The point $(3, -2)$ appears to be just about halfway between the level curves with z-values 30 and 40, so we estimate $f(3, -2) \approx 35$. The graph rises as we approach the origin, gradually from above, steeply from below.

© 2012 Cengage Learning. All Rights Reserved. May not be scanned, copied, or duplicated, or posted to a publicly accessible website, in whole or in part.

35. The point $(160, 10)$, corresponding to day 160 and a depth of 10 m, lies between the isothermals with temperature values of 8 and $12°C$. Since the point appears to be located about three-fourths the distance from the $8°C$ isothermal to the $12°C$ isothermal, we estimate the temperature at that point to be approximately $11°C$. The point $(180, 5)$ lies between the 16 and $20°C$ isothermals, very close to the $20°C$ level curve, so we estimate the temperature there to be about $19.5°C$.

37. Near A, the level curves are very close together, indicating that the terrain is quite steep. At B, the level curves are much farther apart, so we would expect the terrain to be much less steep than near A, perhaps almost flat.

39.

41.

43. The level curves are $(y - 2x)^2 = k$ or $y = 2x \pm \sqrt{k}$, $k \geq 0$, a family of pairs of parallel lines.

45. The level curves are $\sqrt{x} + y = k$ or $y = -\sqrt{x} + k$, a family of vertical translations of the graph of the root function $y = -\sqrt{x}$.

47. The level curves are $ye^x = k$ or $y = ke^{-x}$, a family of exponential curves.

49. The level curves are $\sqrt{y^2 - x^2} = k$ or $y^2 - x^2 = k^2$, $k \geq 0$. When $k = 0$ the level curve is the pair of lines $y = \pm x$. For $k > 0$, the level curves are hyperbolas with axis the y-axis.

© 2012 Cengage Learning. All Rights Reserved. May not be scanned, copied, or duplicated, or posted to a publicly accessible website, in whole or in part.

51. The contour map consists of the level curves $k = x^2 + 9y^2$, a family of ellipses with major axis the x-axis. (Or, if $k = 0$, the origin.)

The graph of $f(x, y)$ is the surface $z = x^2 + 9y^2$, an elliptic paraboloid.

If we visualize lifting each ellipse $k = x^2 + 9y^2$ of the contour map to the plane $z = k$, we have horizontal traces that indicate the shape of the graph of f.

53. The isothermals are given by $k = 100/(1 + x^2 + 2y^2)$ or $x^2 + 2y^2 = (100 - k)/k$ $[0 < k \leq 100]$, a family of ellipses.

55. $f(x, y) = xy^2 - x^3$

The traces parallel to the yz-plane (such as the left-front trace in the graph above) are parabolas; those parallel to the xz-plane (such as the right-front trace) are cubic curves. The surface is called a monkey saddle because a monkey sitting on the surface near the origin has places for both legs and tail to rest.

57. $f(x, y) = e^{-(x^2 + y^2)/3} \left(\sin(x^2) + \cos(y^2) \right)$

© 2012 Cengage Learning. All Rights Reserved. May not be scanned, copied, or duplicated, or posted to a publicly accessible website, in whole or in part.

59. $z = \sin(xy)$ (a) C   (b) II

Reasons: This function is periodic in both x and y, and the function is the same when x is interchanged with y, so its graph is

symmetric about the plane $y = x$. In addition, the function is 0 along the x- and y-axes. These conditions are satisfied only by

C and II.

61. $z = \sin(x - y)$ (a) F   (b) I

Reasons: This function is periodic in both x and y but is constant along the lines $y = x + k$, a condition satisfied only

by F and I.

63. $z = (1 - x^2)(1 - y^2)$ (a) B   (b) VI

Reasons: This function is 0 along the lines $x = \pm 1$ and $y = \pm 1$. The only contour map in which this could occur is VI. Also

note that the trace in the xz-plane is the parabola $z = 1 - x^2$ and the trace in the yz-plane is the parabola $z = 1 - y^2$, so the

graph is B.

65. $k = x + 3y + 5z$ is a family of parallel planes with normal vector $\langle 1, 3, 5 \rangle$.

67. Equations for the level surfaces are $k = y^2 + z^2$. For $k > 0$, we have a family of circular cylinders with axis the x-axis and

radius $\sqrt{k}$. When $k = 0$ the level surface is the x-axis. (There are no level surfaces for $k < 0$.)

69. (a) The graph of g is the graph of f shifted upward 2 units.

(b) The graph of g is the graph of f stretched vertically by a factor of 2.

(c) The graph of g is the graph of f reflected about the xy-plane.

(d) The graph of $g(x, y) = -f(x, y) + 2$ is the graph of f reflected about the xy-plane and then shifted upward 2 units.

71. $f(x, y) = 3x - x^4 - 4y^2 - 10xy$

Three-dimensional view

Front view

It does appear that the function has a maximum value, at the higher of the two "hilltops." From the front view graph, the

maximum value appears to be approximately 15. Both hilltops could be considered local maximum points, as the values of f

there are larger than at the neighboring points. There does not appear to be any local minimum point; although the valley shape

between the two peaks looks like a minimum of some kind, some neighboring points have lower function values.

© 2012 Cengage Learning. All Rights Reserved. May not be scanned, copied, or duplicated, or posted to a publicly accessible website, in whole or in part.

73. $f(x, y) = \dfrac{x + y}{x^2 + y^2}$. As both x and y become large, the function values appear to approach 0, regardless of which direction is considered. As (x, y) approaches the origin, the graph exhibits asymptotic behavior. From some directions, $f(x, y) \to \infty$, while in others $f(x, y) \to -\infty$. (These are the vertical spikes visible in the graph.) If the graph is examined carefully, however, one can see that $f(x, y)$ approaches 0 along the line $y = -x$.

75. $f(x, y) = e^{cx^2 + y^2}$. First, if $c = 0$, the graph is the cylindrical surface

$z = e^{y^2}$ (whose level curves are parallel lines). When $c > 0$, the vertical trace above the y-axis remains fixed while the sides of the surface in the x-direction "curl" upward, giving the graph a shape resembling an elliptic paraboloid. The level curves of the surface are ellipses centered at the origin.

$c = 0$

For $0 < c < 1$, the ellipses have major axis the x-axis and the eccentricity increases as $c \to 0$.

$c = 0.5$ (level curves in increments of 1)

For $c = 1$ the level curves are circles centered at the origin.

$c = 1$ (level curves in increments of 1)

[continued]

© 2012 Cengage Learning. All Rights Reserved. May not be scanned, copied, or duplicated, or posted to a publicly accessible website, in whole or in part.

When $c > 1$, the level curves are ellipses with major axis the y-axis, and the eccentricity increases as c increases.

$c = 2$ (level curves in increments of 4)

For values of $c < 0$, the sides of the surface in the x-direction curl downward and approach the xy-plane (while the vertical trace $x = 0$ remains fixed), giving a saddle-shaped appearance to the graph near the point $(0, 0, 1)$. The level curves consist of a family of hyperbolas. As c decreases, the surface becomes flatter in the x-direction and the surface's approach to the curve in the trace $x = 0$ becomes steeper, as the graphs demonstrate.

$c = -0.5$ (level curves in increments of 0.25)

$c = -2$ (level curves in increments of 0.25)

77. $z = x^2 + y^2 + cxy$. When $c < -2$, the surface intersects the plane $z = k \neq 0$ in a hyperbola. (See the following graph.) It intersects the plane $x = y$ in the parabola $z = (2 + c)x^2$, and the plane $x = -y$ in the parabola $z = (2 - c)x^2$. These parabolas open in opposite directions, so the surface is a hyperbolic paraboloid.

When $c = -2$ the surface is $z = x^2 + y^2 - 2xy = (x - y)^2$. So the surface is constant along each line $x - y = k$. That is, the surface is a cylinder with axis $x - y = 0$, $z = 0$. The shape of the cylinder is determined by its intersection with the plane $x + y = 0$, where $z = 4x^2$, and hence the cylinder is parabolic with minima of 0 on the line $y = x$.

© 2012 Cengage Learning. All Rights Reserved. May not be scanned, copied, or duplicated, or posted to a publicly accessible website, in whole or in part.

$c = -5, z = 2$ $c = -10$ $c = -2$

When $-2 < c \leq 0$, $z \geq 0$ for all x and y. If x and y have the same sign, then

$x^2 + y^2 + cxy \geq x^2 + y^2 - 2xy = (x - y)^2 \geq 0$. If they have opposite signs, then $cxy \geq 0$. The intersection with the

surface and the plane $z = k > 0$ is an ellipse (see graph below). The intersection with the surface and the planes $x = 0$ and

$y = 0$ are parabolas $z = y^2$ and $z = x^2$ respectively, so the surface is an elliptic paraboloid.

When $c > 0$ the graphs have the same shape, but are reflected in the plane $x = 0$, because

$x^2 + y^2 + cxy = (-x)^2 + y^2 + (-c)(-x)y$. That is, the value of z is the same for c at (x, y) as it is for $-c$ at $(-x, y)$.

$c = -1, z = 2$ $c = 0$ $c = 10$

So the surface is an elliptic paraboloid for $0 < c < 2$, a parabolic cylinder for $c = 2$, and a hyperbolic paraboloid for $c > 2$.

79. (a) $P = bL^\alpha K^{1-\alpha}$ $\Rightarrow$ $\dfrac{P}{K} = bL^\alpha K^{-\alpha}$ $\Rightarrow$ $\dfrac{P}{K} = b\left(\dfrac{L}{K}\right)^\alpha$ $\Rightarrow$ $\ln \dfrac{P}{K} = \ln\left(b\left(\dfrac{L}{K}\right)^\alpha\right)$ $\Rightarrow$

$\ln \dfrac{P}{K} = \ln b + \alpha \ln\left(\dfrac{L}{K}\right)$

(b) We list the values for $\ln(L/K)$ and $\ln(P/K)$ for the years 1899–1922. (Historically, these values were rounded to

2 decimal places.)

Year	$x = \ln(L/K)$	$y = \ln(P/K)$		Year	$x = \ln(L/K)$	$y = \ln(P/K)$
1899	0	0		1911	−0.38	−0.34
1900	−0.02	−0.06		1912	−0.38	−0.24
1901	−0.04	−0.02		1913	−0.41	−0.25
1902	−0.04	0		1914	−0.47	−0.37
1903	−0.07	−0.05		1915	−0.53	−0.34
1904	−0.13	−0.12		1916	−0.49	−0.28
1905	−0.18	−0.04		1917	−0.53	−0.39
1906	−0.20	−0.07		1918	−0.60	−0.50
1907	−0.23	−0.15		1919	−0.68	−0.57
1908	−0.41	−0.38		1920	−0.74	−0.57
1909	−0.33	−0.24		1921	−1.05	−0.85
1910	−0.35	−0.27		1922	−0.98	−0.59

© 2012 Cengage Learning. All Rights Reserved. May not be scanned, copied, or duplicated, or posted to a publicly accessible website, in whole or in part.

After entering the (x, y) pairs into a calculator or CAS, the resulting least squares regression line through the points is approximately $y = 0.75136x + 0.01053$, which we round to $y = 0.75x + 0.01$.

(c) Comparing the regression line from part (b) to the equation $y = \ln b + \alpha x$ with $x = \ln(L/K)$ and $y = \ln(P/K)$, we have

$\alpha = 0.75$ and $\ln b = 0.01 \Rightarrow b = e^{0.01} \approx 1.01$. Thus, the Cobb-Douglas production function is

$P = bL^{\alpha}K^{1-\alpha} = 1.01L^{0.75}K^{0.25}$.

14.2   Limits and Continuity

1. In general, we can't say anything about $f(3, 1)$! $\lim\limits_{(x,y)\to(3,1)} f(x, y) = 6$ means that the values of $f(x, y)$ approach 6 as

(x, y) approaches, but is not equal to, $(3, 1)$. If f is continuous, we know that $\lim\limits_{(x,y)\to(a,b)} f(x, y) = f(a, b)$, so

$\lim\limits_{(x,y)\to(3,1)} f(x, y) = f(3, 1) = 6$.

3. We make a table of values of

$f(x, y) = \dfrac{x^2 y^3 + x^3 y^2 - 5}{2 - xy}$ for a set

of (x, y) points near the origin.

x \ y	−0.2	−0.1	−0.05	0	0.05	0.1	0.2
−0.2	−2.551	−2.525	−2.513	−2.500	−2.488	−2.475	−2.451
−0.1	−2.525	−2.513	−2.506	−2.500	−2.494	−2.488	−2.475
−0.05	−2.513	−2.506	−2.503	−2.500	−2.497	−2.494	−2.488
0	−2.500	−2.500	−2.500		−2.500	−2.500	−2.500
0.05	−2.488	−2.494	−2.497	−2.500	−2.503	−2.506	−2.513
0.1	−2.475	−2.488	−2.494	−2.500	−2.506	−2.513	−2.525
0.2	−2.451	−2.475	−2.488	−2.500	−2.513	−2.525	−2.551

As the table shows, the values of $f(x, y)$ seem to approach -2.5 as (x, y) approaches the origin from a variety of different

directions. This suggests that $\lim\limits_{(x,y)\to(0,0)} f(x, y) = -2.5$. Since f is a rational function, it is continuous on its domain. f is

defined at $(0, 0)$, so we can use direct substitution to establish that $\lim\limits_{(x,y)\to(0,0)} f(x, y) = \dfrac{0^2 0^3 + 0^3 0^2 - 5}{2 - 0 \cdot 0} = -\dfrac{5}{2}$, verifying

our guess.

5. $f(x, y) = 5x^3 - x^2 y^2$ is a polynomial, and hence continuous, so $\lim\limits_{(x,y)\to(1,2)} f(x, y) = f(1, 2) = 5(1)^3 - (1)^2(2)^2 = 1$.

7. $f(x, y) = \dfrac{4 - xy}{x^2 + 3y^2}$ is a rational function and hence continuous on its domain.

$(2, 1)$ is in the domain of f, so f is continuous there and $\lim\limits_{(x,y)\to(2,1)} f(x, y) = f(2, 1) = \dfrac{4 - (2)(1)}{(2)^2 + 3(1)^2} = \dfrac{2}{7}$.

9. $f(x, y) = (x^4 - 4y^2)/(x^2 + 2y^2)$.   First approach $(0, 0)$ along the x-axis. Then $f(x, 0) = x^4/x^2 = x^2$ for $x \neq 0$, so

$f(x, y) \to 0$. Now approach $(0, 0)$ along the y-axis. For $y \neq 0$, $f(0, y) = -4y^2/2y^2 = -2$, so $f(x, y) \to -2$. Since f has

two different limits along two different lines, the limit does not exist.

© 2012 Cengage Learning. All Rights Reserved. May not be scanned, copied, or duplicated, or posted to a publicly accessible website, in whole or in part.

11. $f(x, y) = (y^2 \sin^2 x)/(x^4 + y^4)$.   On the x-axis, $f(x, 0) = 0$ for $x \neq 0$, so $f(x, y) \to 0$ as $(x, y) \to (0, 0)$ along the

x-axis. Approaching $(0, 0)$ along the line $y = x$, $f(x, x) = \dfrac{x^2 \sin^2 x}{x^4 + x^4} = \dfrac{\sin^2 x}{2x^2} = \dfrac{1}{2}\left(\dfrac{\sin x}{x}\right)^2$ for $x \neq 0$ and

$\displaystyle\lim_{x \to 0} \dfrac{\sin x}{x} = 1$, so $f(x, y) \to \frac{1}{2}$. Since f has two different limits along two different lines, the limit does not exist.

13. $f(x, y) = \dfrac{xy}{\sqrt{x^2 + y^2}}$.   We can see that the limit along any line through $(0, 0)$ is 0, as well as along other paths through

$(0, 0)$ such as $x = y^2$ and $y = x^2$. So we suspect that the limit exists and equals 0; we use the Squeeze Theorem to prove our

assertion. $0 \leq \left| \dfrac{xy}{\sqrt{x^2 + y^2}} \right| \leq |x|$ since $|y| \leq \sqrt{x^2 + y^2}$, and $|x| \to 0$ as $(x, y) \to (0, 0)$. So $\displaystyle\lim_{(x,y)\to(0,0)} f(x, y) = 0$.

15. Let $f(x, y) = \dfrac{x^2 y e^y}{x^4 + 4y^2}$. Then $f(x, 0) = 0$ for $x \neq 0$, so $f(x, y) \to 0$ as $(x, y) \to (0, 0)$ along the x-axis. Approaching

$(0, 0)$ along the y-axis or the line $y = x$ also gives a limit of 0. But $f(x, x^2) = \dfrac{x^2 x^2 e^{x^2}}{x^4 + 4(x^2)^2} = \dfrac{x^4 e^{x^2}}{5x^4} = \dfrac{e^{x^2}}{5}$ for $x \neq 0$, so

$f(x, y) \to e^0/5 = \frac{1}{5}$ as $(x, y) \to (0, 0)$ along the parabola $y = x^2$. Thus the limit doesn't exist.

17. $\displaystyle\lim_{(x,y)\to(0,0)} \dfrac{x^2 + y^2}{\sqrt{x^2 + y^2 + 1} - 1} = \lim_{(x,y)\to(0,0)} \dfrac{x^2 + y^2}{\sqrt{x^2 + y^2 + 1} - 1} \cdot \dfrac{\sqrt{x^2 + y^2 + 1} + 1}{\sqrt{x^2 + y^2 + 1} + 1}$

$$= \lim_{(x,y)\to(0,0)} \dfrac{(x^2 + y^2)\left(\sqrt{x^2 + y^2 + 1} + 1\right)}{x^2 + y^2} = \lim_{(x,y)\to(0,0)} \left(\sqrt{x^2 + y^2 + 1} + 1\right) = 2$$

19. e^{y^2} is a composition of continuous functions and hence continuous. xz is a continuous function and $\tan t$ is continuous for

$t \neq \frac{\pi}{2} + n\pi$ (n an integer), so the composition $\tan(xz)$ is continuous for $xz \neq \frac{\pi}{2} + n\pi$. Thus the product

$f(x, y, z) = e^{y^2} \tan(xz)$ is a continuous function for $xz \neq \frac{\pi}{2} + n\pi$. If $x = \pi$ and $z = \frac{1}{3}$ then $xz \neq \frac{\pi}{2} + n\pi$, so

$$\lim_{(x,y,z)\to(\pi,0,1/3)} f(x, y, z) = f(\pi, 0, 1/3) = e^{0^2} \tan(\pi \cdot 1/3) = 1 \cdot \tan(\pi/3) = \sqrt{3}.$$

21. $f(x, y, z) = \dfrac{xy + yz^2 + xz^2}{x^2 + y^2 + z^4}$.   Then $f(x, 0, 0) = 0/x^2 = 0$ for $x \neq 0$, so as $(x, y, z) \to (0, 0, 0)$ along the x-axis,

$f(x, y, z) \to 0$. But $f(x, x, 0) = x^2/(2x^2) = \frac{1}{2}$ for $x \neq 0$, so as $(x, y, z) \to (0, 0, 0)$ along the line $y = x$, $z = 0$,

$f(x, y, z) \to \frac{1}{2}$. Thus the limit doesn't exist.

23.

From the ridges on the graph, we see that as $(x, y) \to (0, 0)$ along the lines under the two ridges, $f(x, y)$ approaches different values. So the limit does not exist.

© 2012 Cengage Learning. All Rights Reserved. May not be scanned, copied, or duplicated, or posted to a publicly accessible website, in whole or in part.

25. $h(x,y) = g(f(x,y)) = (2x+3y-6)^2 + \sqrt{2x+3y-6}$. Since f is a polynomial, it is continuous on $\mathbb{R}^2$ and g is

continuous on its domain $\{t \mid t \geq 0\}$. Thus h is continuous on its domain.

$D = \{(x,y) \mid 2x+3y-6 \geq 0\} = \{(x,y) \mid y \geq -\frac{2}{3}x+2\}$, which consists of all points on or above the line $y = -\frac{2}{3}x+2$.

27.

From the graph, it appears that f is discontinuous along the line $y = x$.

If we consider $f(x,y) = e^{1/(x-y)}$ as a composition of functions,

$g(x,y) = 1/(x-y)$ is a rational function and therefore continuous except

where $x - y = 0 \;\Rightarrow\; y = x$. Since the function $h(t) = e^t$ is continuous

everywhere, the composition $h(g(x,y)) = e^{1/(x-y)} = f(x,y)$ is

continuous except along the line $y = x$, as we suspected.

29. The functions xy and $1 + e^{x-y}$ are continuous everywhere, and $1 + e^{x-y}$ is never zero, so $F(x,y) = \dfrac{xy}{1+e^{x-y}}$ is continuous

on its domain $\mathbb{R}^2$.

31. $F(x,y) = \dfrac{1+x^2+y^2}{1-x^2-y^2}$ is a rational function and thus is continuous on its domain

$\{(x,y) \mid 1 - x^2 - y^2 \neq 0\} = \{(x,y) \mid x^2 + y^2 \neq 1\}$.

33. $G(x,y) = \ln(x^2+y^2-4) = g(f(x,y))$ where $f(x,y) = x^2+y^2-4$, continuous on $\mathbb{R}^2$, and $g(t) = \ln t$, continuous on its

domain $\{t \mid t > 0\}$. Thus G is continuous on its domain $\{(x,y) \mid x^2+y^2-4 > 0\} = \{(x,y) \mid x^2+y^2 > 4\}$, the exterior

of the circle $x^2 + y^2 = 4$.

35. $f(x,y,z) = h(g(x,y,z))$ where $g(x,y,z) = x^2+y^2+z^2$, a polynomial that is continuous

everywhere, and $h(t) = \arcsin t$, continuous on $[-1,1]$. Thus f is continuous on its domain

$\{(x,y,z) \mid -1 \leq x^2+y^2+z^2 \leq 1\} = \{(x,y,z) \mid x^2+y^2+z^2 \leq 1\}$, so f is continuous on the unit ball.

37. $f(x,y) = \begin{cases} \dfrac{x^2 y^3}{2x^2+y^2} & \text{if } (x,y) \neq (0,0) \\ 1 & \text{if } (x,y) = (0,0) \end{cases}$ The first piece of f is a rational function defined everywhere except at the

origin, so f is continuous on $\mathbb{R}^2$ except possibly at the origin. Since $x^2 \leq 2x^2 + y^2$, we have $\left| x^2 y^3 /(2x^2+y^2) \right| \leq |y^3|$. We

know that $|y^3| \to 0$ as $(x,y) \to (0,0)$. So, by the Squeeze Theorem, $\displaystyle\lim_{(x,y)\to(0,0)} f(x,y) = \lim_{(x,y)\to(0,0)} \dfrac{x^2 y^3}{2x^2+y^2} = 0$.

But $f(0,0) = 1$, so f is discontinuous at $(0,0)$. Therefore, f is continuous on the set $\{(x,y) \mid (x,y) \neq (0,0)\}$.

39. $\displaystyle\lim_{(x,y)\to(0,0)} \dfrac{x^3+y^3}{x^2+y^2} = \lim_{r\to 0^+} \dfrac{(r\cos\theta)^3 + (r\sin\theta)^3}{r^2} = \lim_{r\to 0^+} (r\cos^3\theta + r\sin^3\theta) = 0$

41. $\displaystyle\lim_{(x,y)\to(0,0)} \dfrac{e^{-x^2-y^2}-1}{x^2+y^2} = \lim_{r\to 0^+} \dfrac{e^{-r^2}-1}{r^2} = \lim_{r\to 0^+} \dfrac{e^{-r^2}(-2r)}{2r}$ [using l'Hospital's Rule]

$\qquad\qquad = \displaystyle\lim_{r\to 0^+} -e^{-r^2} = -e^0 = -1$

© 2012 Cengage Learning. All Rights Reserved. May not be scanned, copied, or duplicated, or posted to a publicly accessible website, in whole or in part.

43. $f(x, y) = \begin{cases} \dfrac{\sin(xy)}{xy} & \text{if } (x, y) \neq (0, 0) \\ 1 & \text{if } (x, y) = (0, 0) \end{cases}$

From the graph, it appears that f is continuous everywhere. We know

xy is continuous on $\mathbb{R}^2$ and $\sin t$ is continuous everywhere, so

$\sin(xy)$ is continuous on $\mathbb{R}^2$ and $\dfrac{\sin(xy)}{xy}$ is continuous on $\mathbb{R}^2$

except possibly where $xy = 0$. To show that f is continuous at those points, consider any point (a, b) in $\mathbb{R}^2$ where $ab = 0$.

Because xy is continuous, $xy \to ab = 0$ as $(x, y) \to (a, b)$. If we let $t = xy$, then $t \to 0$ as $(x, y) \to (a, b)$ and

$$\lim_{(x,y)\to(a,b)} \frac{\sin(xy)}{xy} = \lim_{t\to 0} \frac{\sin(t)}{t} = 1 \text{ by Equation 2.4.2 [ET 3.3.2]. Thus } \lim_{(x,y)\to(a,b)} f(x, y) = f(a, b) \text{ and } f \text{ is continuous}$$

on $\mathbb{R}^2$.

45. Since $|\mathbf{x} - \mathbf{a}|^2 = |\mathbf{x}|^2 + |\mathbf{a}|^2 - 2|\mathbf{x}||\mathbf{a}|\cos\theta \geq |\mathbf{x}|^2 + |\mathbf{a}|^2 - 2|\mathbf{x}||\mathbf{a}| = (|\mathbf{x}| - |\mathbf{a}|)^2$, we have $\big||\mathbf{x}| - |\mathbf{a}|\big| \leq |\mathbf{x} - \mathbf{a}|$. Let

$\epsilon > 0$ be given and set $\delta = \epsilon$. Then if $0 < |\mathbf{x} - \mathbf{a}| < \delta$, $\big||\mathbf{x}| - |\mathbf{a}|\big| \leq |\mathbf{x} - \mathbf{a}| < \delta = \epsilon$. Hence $\lim_{\mathbf{x}\to\mathbf{a}} |\mathbf{x}| = |\mathbf{a}|$ and

$f(\mathbf{x}) = |\mathbf{x}|$ is continuous on $\mathbb{R}^n$.

14.3   Partial Derivatives

1. (a) $\partial T/\partial x$ represents the rate of change of T when we fix y and t and consider T as a function of the single variable x, which

describes how quickly the temperature changes when longitude changes but latitude and time are constant. $\partial T/\partial y$

represents the rate of change of T when we fix x and t and consider T as a function of y, which describes how quickly the

temperature changes when latitude changes but longitude and time are constant. $\partial T/\partial t$ represents the rate of change of T

when we fix x and y and consider T as a function of t, which describes how quickly the temperature changes over time for

a constant longitude and latitude.

(b) $f_x(158, 21, 9)$ represents the rate of change of temperature at longitude $158°$W, latitude $21°$N at 9:00 AM when only

longitude varies. Since the air is warmer to the west than to the east, increasing longitude results in an increased air

temperature, so we would expect $f_x(158, 21, 9)$ to be positive. $f_y(158, 21, 9)$ represents the rate of change of temperature

at the same time and location when only latitude varies. Since the air is warmer to the south and cooler to the north,

increasing latitude results in a decreased air temperature, so we would expect $f_y(158, 21, 9)$ to be negative. $f_t(158, 21, 9)$

represents the rate of change of temperature at the same time and location when only time varies. Since typically air

temperature increases from the morning to the afternoon as the sun warms it, we would expect $f_t(158, 21, 9)$ to be

positive.

3. (a) By Definition 4, $f_T(-15, 30) = \lim_{h\to 0} \dfrac{f(-15 + h, 30) - f(-15, 30)}{h}$, which we can approximate by considering $h = 5$

and $h = -5$ and using the values given in the table:

© 2012 Cengage Learning. All Rights Reserved. May not be scanned, copied, or duplicated, or posted to a publicly accessible website, in whole or in part.

$$f_T(-15, 30) \approx \frac{f(-10, 30) - f(-15, 30)}{5} = \frac{-20 - (-26)}{5} = \frac{6}{5} = 1.2,$$

$$f_T(-15, 30) \approx \frac{f(-20, 30) - f(-15, 30)}{-5} = \frac{-33 - (-26)}{-5} = \frac{-7}{-5} = 1.4. \text{ Averaging these values, we estimate}$$

$f_T(-15, 30)$ to be approximately 1.3. Thus, when the actual temperature is $-15°$C and the wind speed is 30 km/h, the

apparent temperature rises by about $1.3°$C for every degree that the actual temperature rises.

Similarly, $f_v(-15, 30) = \lim\limits_{h \to 0} \dfrac{f(-15, 30 + h) - f(-15, 30)}{h}$ which we can approximate by considering $h = 10$

and $h = -10$: $f_v(-15, 30) \approx \dfrac{f(-15, 40) - f(-15, 30)}{10} = \dfrac{-27 - (-26)}{10} = \dfrac{-1}{10} = -0.1,$

$$f_v(-15, 30) \approx \frac{f(-15, 20) - f(-15, 30)}{-10} = \frac{-24 - (-26)}{-10} = \frac{2}{-10} = -0.2. \text{ Averaging these values, we estimate}$$

$f_v(-15, 30)$ to be approximately -0.15. Thus, when the actual temperature is $-15°$C and the wind speed is 30 km/h, the

apparent temperature decreases by about $0.15°$C for every km/h that the wind speed increases.

(b) For a fixed wind speed v, the values of the wind-chill index W increase as temperature T increases (look at a column of

the table), so $\dfrac{\partial W}{\partial T}$ is positive. For a fixed temperature T, the values of W decrease (or remain constant) as v increases

(look at a row of the table), so $\dfrac{\partial W}{\partial v}$ is negative (or perhaps 0).

(c) For fixed values of T, the function values $f(T, v)$ appear to become constant (or nearly constant) as v increases, so the

corresponding rate of change is 0 or near 0 as v increases. This suggests that $\lim\limits_{v \to \infty} (\partial W / \partial v) = 0$.

5. (a) If we start at $(1, 2)$ and move in the positive x-direction, the graph of f increases. Thus $f_x(1, 2)$ is positive.

(b) If we start at $(1, 2)$ and move in the positive y-direction, the graph of f decreases. Thus $f_y(1, 2)$ is negative.

7. (a) $f_{xx} = \frac{\partial}{\partial x}(f_x)$, so f_{xx} is the rate of change of f_x in the x-direction. f_x is negative at $(-1, 2)$ and if we move in the

positive x-direction, the surface becomes less steep. Thus the values of f_x are increasing and $f_{xx}(-1, 2)$ is positive.

(b) f_{yy} is the rate of change of f_y in the y-direction. f_y is negative at $(-1, 2)$ and if we move in the positive y-direction, the

surface becomes steeper. Thus the values of f_y are decreasing, and $f_{yy}(-1, 2)$ is negative.

9. First of all, if we start at the point $(3, -3)$ and move in the positive y-direction, we see that both b and c decrease, while a

increases. Both b and c have a low point at about $(3, -1.5)$, while a is 0 at this point. So a is definitely the graph of f_y, and

one of b and c is the graph of f. To see which is which, we start at the point $(-3, -1.5)$ and move in the positive x-direction.

b traces out a line with negative slope, while c traces out a parabola opening downward. This tells us that b is the x-derivative

of c. So c is the graph of f, b is the graph of f_x, and a is the graph of f_y.

© 2012 Cengage Learning. All Rights Reserved. May not be scanned, copied, or duplicated, or posted to a publicly accessible website, in whole or in part.

11. $f(x, y) = 16 - 4x^2 - y^2 \Rightarrow f_x(x, y) = -8x$ and $f_y(x, y) = -2y \Rightarrow f_x(1, 2) = -8$ and $f_y(1, 2) = -4$. The graph

of f is the paraboloid $z = 16 - 4x^2 - y^2$ and the vertical plane $y = 2$ intersects it in the parabola $z = 12 - 4x^2$, $y = 2$

(the curve C_1 in the first figure). The slope of the tangent line

to this parabola at $(1, 2, 8)$ is $f_x(1, 2) = -8$. Similarly the

plane $x = 1$ intersects the paraboloid in the parabola

$z = 12 - y^2$, $x = 1$ (the curve C_2 in the second figure) and

the slope of the tangent line at $(1, 2, 8)$ is $f_y(1, 2) = -4$.

13. $f(x, y) = x^2 y^3 \Rightarrow f_x = 2xy^3, \quad f_y = 3x^2 y^2$

Note that traces of f in planes parallel to the xz-plane are parabolas which open downward for $y < 0$ and upward for $y > 0$,

and the traces of f_x in these planes are straight lines, which have negative slopes for $y < 0$ and positive slopes for $y > 0$. The

traces of f in planes parallel to the yz-plane are cubic curves, and the traces of f_y in these planes are parabolas.

15. $f(x, y) = y^5 - 3xy \Rightarrow f_x(x, y) = 0 - 3y = -3y, f_y(x, y) = 5y^4 - 3x$

17. $f(x, t) = e^{-t} \cos \pi x \Rightarrow f_x(x, t) = e^{-t}(-\sin \pi x)(\pi) = -\pi e^{-t} \sin \pi x, f_t(x, t) = e^{-t}(-1) \cos \pi x = -e^{-t} \cos \pi x$

19. $z = (2x + 3y)^{10} \Rightarrow \dfrac{\partial z}{\partial x} = 10(2x + 3y)^9 \cdot 2 = 20(2x + 3y)^9, \dfrac{\partial z}{\partial y} = 10(2x + 3y)^9 \cdot 3 = 30(2x + 3y)^9$

21. $f(x, y) = x/y = xy^{-1} \Rightarrow f_x(x, y) = y^{-1} = 1/y, f_y(x, y) = -xy^{-2} = -x/y^2$

23. $f(x, y) = \dfrac{ax + by}{cx + dy} \Rightarrow f_x(x, y) = \dfrac{(cx + dy)(a) - (ax + by)(c)}{(cx + dy)^2} = \dfrac{(ad - bc)y}{(cx + dy)^2},$

$f_y(x, y) = \dfrac{(cx + dy)(b) - (ax + by)(d)}{(cx + dy)^2} = \dfrac{(bc - ad)x}{(cx + dy)^2}$

© 2012 Cengage Learning. All Rights Reserved. May not be scanned, copied, or duplicated, or posted to a publicly accessible website, in whole or in part.

25. $g(u,v) = (u^2v - v^3)^5 \Rightarrow g_u(u,v) = 5(u^2v - v^3)^4 \cdot 2uv = 10uv(u^2v - v^3)^4,$

$g_v(u,v) = 5(u^2v - v^3)^4(u^2 - 3v^2) = 5(u^2 - 3v^2)(u^2v - v^3)^4$

27. $R(p,q) = \tan^{-1}(pq^2) \Rightarrow R_p(p,q) = \dfrac{1}{1 + (pq^2)^2} \cdot q^2 = \dfrac{q^2}{1 + p^2q^4}, R_q(p,q) = \dfrac{1}{1 + (pq^2)^2} \cdot 2pq = \dfrac{2pq}{1 + p^2q^4}$

29. $F(x,y) = \displaystyle\int_y^x \cos(e^t)\,dt \Rightarrow F_x(x,y) = \dfrac{\partial}{\partial x}\displaystyle\int_y^x \cos(e^t)\,dt = \cos(e^x)$ by the Fundamental Theorem of Calculus, Part 1;

$F_y(x,y) = \dfrac{\partial}{\partial y}\displaystyle\int_y^x \cos(e^t)\,dt = \dfrac{\partial}{\partial y}\left[-\displaystyle\int_x^y \cos(e^t)\,dt\right] = -\dfrac{\partial}{\partial y}\displaystyle\int_x^y \cos(e^t)\,dt = -\cos(e^y).$

31. $f(x,y,z) = xz - 5x^2y^3z^4 \Rightarrow f_x(x,y,z) = z - 10xy^3z^4, f_y(x,y,z) = -15x^2y^2z^4, f_z(x,y,z) = x - 20x^2y^3z^3$

33. $w = \ln(x + 2y + 3z) \Rightarrow \dfrac{\partial w}{\partial x} = \dfrac{1}{x + 2y + 3z}, \dfrac{\partial w}{\partial y} = \dfrac{2}{x + 2y + 3z}, \dfrac{\partial w}{\partial z} = \dfrac{3}{x + 2y + 3z}$

35. $u = xy\sin^{-1}(yz) \Rightarrow \dfrac{\partial u}{\partial x} = y\sin^{-1}(yz), \dfrac{\partial u}{\partial y} = xy \cdot \dfrac{1}{\sqrt{1 - (yz)^2}}(z) + \sin^{-1}(yz) \cdot x = \dfrac{xyz}{\sqrt{1 - y^2z^2}} + x\sin^{-1}(yz),$

$\dfrac{\partial u}{\partial z} = xy \cdot \dfrac{1}{\sqrt{1 - (yz)^2}}(y) = \dfrac{xy^2}{\sqrt{1 - y^2z^2}}$

37. $h(x,y,z,t) = x^2y\cos(z/t) \Rightarrow h_x(x,y,z,t) = 2xy\cos(z/t), h_y(x,y,z,t) = x^2\cos(z/t),$

$h_z(x,y,z,t) = -x^2y\sin(z/t)(1/t) = (-x^2y/t)\sin(z/t), h_t(x,y,z,t) = -x^2y\sin(z/t)(-zt^{-2}) = (x^2yz/t^2)\sin(z/t)$

39. $u = \sqrt{x_1^2 + x_2^2 + \cdots + x_n^2}.$ For each $i = 1, \ldots, n, u_{x_i} = \frac{1}{2}(x_1^2 + x_2^2 + \cdots + x_n^2)^{-1/2}(2x_i) = \dfrac{x_i}{\sqrt{x_1^2 + x_2^2 + \cdots + x_n^2}}.$

41. $f(x,y) = \ln\left(x + \sqrt{x^2 + y^2}\right) \Rightarrow$

$f_x(x,y) = \dfrac{1}{x + \sqrt{x^2 + y^2}}\left[1 + \frac{1}{2}(x^2 + y^2)^{-1/2}(2x)\right] = \dfrac{1}{x + \sqrt{x^2 + y^2}}\left(1 + \dfrac{x}{\sqrt{x^2 + y^2}}\right),$

so $f_x(3,4) = \dfrac{1}{3 + \sqrt{3^2 + 4^2}}\left(1 + \dfrac{3}{\sqrt{3^2 + 4^2}}\right) = \frac{1}{8}\left(1 + \frac{3}{5}\right) = \frac{1}{5}.$

43. $f(x,y,z) = \dfrac{y}{x + y + z} \Rightarrow f_y(x,y,z) = \dfrac{1(x + y + z) - y(1)}{(x + y + z)^2} = \dfrac{x + z}{(x + y + z)^2},$

so $f_y(2,1,-1) = \dfrac{2 + (-1)}{(2 + 1 + (-1))^2} = \dfrac{1}{4}.$

45. $f(x,y) = xy^2 - x^3y \Rightarrow$

$f_x(x,y) = \displaystyle\lim_{h\to0}\dfrac{f(x+h,y) - f(x,y)}{h} = \lim_{h\to0}\dfrac{(x+h)y^2 - (x+h)^3y - (xy^2 - x^3y)}{h}$

$= \displaystyle\lim_{h\to0}\dfrac{h(y^2 - 3x^2y - 3xyh - yh^2)}{h} = \lim_{h\to0}(y^2 - 3x^2y - 3xyh - yh^2) = y^2 - 3x^2y$

$f_y(x,y) = \displaystyle\lim_{h\to0}\dfrac{f(x,y+h) - f(x,y)}{h} = \lim_{h\to0}\dfrac{x(y+h)^2 - x^3(y+h) - (xy^2 - x^3y)}{h} = \lim_{h\to0}\dfrac{h(2xy + xh - x^3)}{h}$

$= \displaystyle\lim_{h\to0}(2xy + xh - x^3) = 2xy - x^3$

© 2012 Cengage Learning. All Rights Reserved. May not be scanned, copied, or duplicated, or posted to a publicly accessible website, in whole or in part.

47. $x^2 + 2y^2 + 3z^2 = 1$ $\Rightarrow$ $\dfrac{\partial}{\partial x}(x^2 + 2y^2 + 3z^2) = \dfrac{\partial}{\partial x}(1)$ $\Rightarrow$ $2x + 0 + 6z\dfrac{\partial z}{\partial x} = 0$ $\Rightarrow$ $6z\dfrac{\partial z}{\partial x} = -2x$ $\Rightarrow$

$\dfrac{\partial z}{\partial x} = \dfrac{-2x}{6z} = -\dfrac{x}{3z}$, and $\dfrac{\partial}{\partial y}(x^2 + 2y^2 + 3z^2) = \dfrac{\partial}{\partial y}(1)$ $\Rightarrow$ $0 + 4y + 6z\dfrac{\partial z}{\partial y} = 0$ $\Rightarrow$ $6z\dfrac{\partial z}{\partial y} = -4y$ $\Rightarrow$

$\dfrac{\partial z}{\partial y} = \dfrac{-4y}{6z} = -\dfrac{2y}{3z}$.

49. $e^z = xyz$ $\Rightarrow$ $\dfrac{\partial}{\partial x}(e^z) = \dfrac{\partial}{\partial x}(xyz)$ $\Rightarrow$ $e^z\dfrac{\partial z}{\partial x} = y\left(x\dfrac{\partial z}{\partial x} + z \cdot 1\right)$ $\Rightarrow$ $e^z\dfrac{\partial z}{\partial x} - xy\dfrac{\partial z}{\partial x} = yz$ $\Rightarrow$

$(e^z - xy)\dfrac{\partial z}{\partial x} = yz$, so $\dfrac{\partial z}{\partial x} = \dfrac{yz}{e^z - xy}$.

$\dfrac{\partial}{\partial y}(e^z) = \dfrac{\partial}{\partial y}(xyz)$ $\Rightarrow$ $e^z\dfrac{\partial z}{\partial y} = x\left(y\dfrac{\partial z}{\partial x} + z \cdot 1\right)$ $\Rightarrow$ $e^z\dfrac{\partial z}{\partial y} - xy\dfrac{\partial z}{\partial y} = xz$ $\Rightarrow$ $(e^z - xy)\dfrac{\partial z}{\partial y} = xz$, so

$\dfrac{\partial z}{\partial y} = \dfrac{xz}{e^z - xy}$.

51. (a) $z = f(x) + g(y)$ $\Rightarrow$ $\dfrac{\partial z}{\partial x} = f'(x)$, $\dfrac{\partial z}{\partial y} = g'(y)$

(b) $z = f(x+y)$. Let $u = x + y$. Then $\dfrac{\partial z}{\partial x} = \dfrac{df}{du}\dfrac{\partial u}{\partial x} = \dfrac{df}{du}(1) = f'(u) = f'(x+y)$,

$\dfrac{\partial z}{\partial y} = \dfrac{df}{du}\dfrac{\partial u}{\partial y} = \dfrac{df}{du}(1) = f'(u) = f'(x+y)$.

53. $f(x,y) = x^3y^5 + 2x^4y$ $\Rightarrow$ $f_x(x,y) = 3x^2y^5 + 8x^3y$, $f_y(x,y) = 5x^3y^4 + 2x^4$. Then $f_{xx}(x,y) = 6xy^5 + 24x^2y$,

$f_{xy}(x,y) = 15x^2y^4 + 8x^3$, $f_{yx}(x,y) = 15x^2y^4 + 8x^3$, and $f_{yy}(x,y) = 20x^3y^3$.

55. $w = \sqrt{u^2 + v^2}$ $\Rightarrow$ $w_u = \frac{1}{2}(u^2 + v^2)^{-1/2} \cdot 2u = \dfrac{u}{\sqrt{u^2 + v^2}}$, $w_v = \frac{1}{2}(u^2 + v^2)^{-1/2} \cdot 2v = \dfrac{v}{\sqrt{u^2 + v^2}}$. Then

$w_{uu} = \dfrac{1 \cdot \sqrt{u^2 + v^2} - u \cdot \frac{1}{2}(u^2 + v^2)^{-1/2}(2u)}{\left(\sqrt{u^2 + v^2}\right)^2} = \dfrac{\sqrt{u^2 + v^2} - u^2/\sqrt{u^2 + v^2}}{u^2 + v^2} = \dfrac{u^2 + v^2 - u^2}{(u^2 + v^2)^{3/2}} = \dfrac{v^2}{(u^2 + v^2)^{3/2}}$,

$w_{uv} = u\left(-\frac{1}{2}\right)(u^2 + v^2)^{-3/2}(2v) = -\dfrac{uv}{(u^2 + v^2)^{3/2}}$, $w_{vu} = v\left(-\frac{1}{2}\right)(u^2 + v^2)^{-3/2}(2u) = -\dfrac{uv}{(u^2 + v^2)^{3/2}}$,

$w_{vv} = \dfrac{1 \cdot \sqrt{u^2 + v^2} - v \cdot \frac{1}{2}(u^2 + v^2)^{-1/2}(2v)}{\left(\sqrt{u^2 + v^2}\right)^2} = \dfrac{\sqrt{u^2 + v^2} - v^2/\sqrt{u^2 + v^2}}{u^2 + v^2} = \dfrac{u^2 + v^2 - v^2}{(u^2 + v^2)^{3/2}} = \dfrac{u^2}{(u^2 + v^2)^{3/2}}$.

57. $z = \arctan\dfrac{x+y}{1-xy}$ $\Rightarrow$

$z_x = \dfrac{1}{1 + \left(\frac{x+y}{1-xy}\right)^2} \cdot \dfrac{(1)(1-xy) - (x+y)(-y)}{(1-xy)^2} = \dfrac{1+y^2}{(1-xy)^2 + (x+y)^2} = \dfrac{1+y^2}{1 + x^2 + y^2 + x^2y^2}$

$= \dfrac{1+y^2}{(1+x^2)(1+y^2)} = \dfrac{1}{1+x^2}$,

© 2012 Cengage Learning. All Rights Reserved. May not be scanned, copied, or duplicated, or posted to a publicly accessible website, in whole or in part.

$$z_y = \frac{1}{1 + \left(\frac{x+y}{1-xy}\right)^2} \cdot \frac{(1)(1-xy) - (x+y)(-x)}{(1-xy)^2} = \frac{1+x^2}{(1-xy)^2 + (x+y)^2} = \frac{1+x^2}{(1+x^2)(1+y^2)} = \frac{1}{1+y^2}.$$

Then $z_{xx} = -(1+x^2)^{-2} \cdot 2x = -\dfrac{2x}{(1+x^2)^2}$, $z_{xy} = 0$, $z_{yx} = 0$, $z_{yy} = -(1+y^2)^{-2} \cdot 2y = -\dfrac{2y}{(1+y^2)^2}$.

59. $u = x^4 y^3 - y^4 \;\Rightarrow\; u_x = 4x^3 y^3$, $u_{xy} = 12x^3 y^2$ and $u_y = 3x^4 y^2 - 4y^3$, $u_{yx} = 12x^3 y^2$.

Thus $u_{xy} = u_{yx}$.

61. $u = \cos(x^2 y) \;\Rightarrow\; u_x = -\sin(x^2 y) \cdot 2xy = -2xy \sin(x^2 y)$,

$u_{xy} = -2xy \cdot \cos(x^2 y) \cdot x^2 + \sin(x^2 y) \cdot (-2x) = -2x^3 y \cos(x^2 y) - 2x \sin(x^2 y)$ and

$u_y = -\sin(x^2 y) \cdot x^2 = -x^2 \sin(x^2 y)$, $u_{yx} = -x^2 \cdot \cos(x^2 y) \cdot 2xy + \sin(x^2 y) \cdot (-2x) = -2x^3 y \cos(x^2 y) - 2x \sin(x^2 y)$.

Thus $u_{xy} = u_{yx}$.

63. $f(x,y) = x^4 y^2 - x^3 y \;\Rightarrow\; f_x = 4x^3 y^2 - 3x^2 y$, $f_{xx} = 12x^2 y^2 - 6xy$, $f_{xxx} = 24xy^2 - 6y$ and

$f_{xy} = 8x^3 y - 3x^2$, $f_{xyx} = 24x^2 y - 6x$.

65. $f(x,y,z) = e^{xyz^2} \;\Rightarrow\; f_x = e^{xyz^2} \cdot yz^2 = yz^2 e^{xyz^2}$, $f_{xy} = yz^2 \cdot e^{xyz^2}(xz^2) + e^{xyz^2} \cdot z^2 = (xyz^4 + z^2)e^{xyz^2}$,

$f_{xyz} = (xyz^4 + z^2) \cdot e^{xyz^2}(2xyz) + e^{xyz^2} \cdot (4xyz^3 + 2z) = (2x^2 y^2 z^5 + 6xyz^3 + 2z)e^{xyz^2}$.

67. $u = e^{r\theta} \sin\theta \;\Rightarrow\; \dfrac{\partial u}{\partial \theta} = e^{r\theta} \cos\theta + \sin\theta \cdot e^{r\theta}(r) = e^{r\theta}(\cos\theta + r\sin\theta)$,

$\dfrac{\partial^2 u}{\partial r\, \partial \theta} = e^{r\theta}(\sin\theta) + (\cos\theta + r\sin\theta)e^{r\theta}(\theta) = e^{r\theta}(\sin\theta + \theta\cos\theta + r\theta\sin\theta)$,

$\dfrac{\partial^3 u}{\partial r^2\, \partial \theta} = e^{r\theta}(\theta\sin\theta) + (\sin\theta + \theta\cos\theta + r\theta\sin\theta) \cdot e^{r\theta}(\theta) = \theta e^{r\theta}(2\sin\theta + \theta\cos\theta + r\theta\sin\theta)$.

69. $w = \dfrac{x}{y+2z} = x(y+2z)^{-1} \;\Rightarrow\; \dfrac{\partial w}{\partial x} = (y+2z)^{-1}$, $\dfrac{\partial^2 w}{\partial y\, \partial x} = -(y+2z)^{-2}(1) = -(y+2z)^{-2}$,

$\dfrac{\partial^3 w}{\partial z\, \partial y\, \partial x} = -(-2)(y+2z)^{-3}(2) = 4(y+2z)^{-3} = \dfrac{4}{(y+2z)^3}$ and $\dfrac{\partial w}{\partial y} = x(-1)(y+2z)^{-2}(1) = -x(y+2z)^{-2}$,

$\dfrac{\partial^2 w}{\partial x\, \partial y} = -(y+2z)^{-2}$, $\dfrac{\partial^3 w}{\partial x^2\, \partial y} = 0$.

71. Assuming that the third partial derivatives of f are continuous (easily verified), we can write $f_{xzy} = f_{yxz}$. Then

$f(x,y,z) = xy^2 z^3 + \arcsin\left(x\sqrt{z}\right) \;\Rightarrow\; f_y = 2xyz^3 + 0$, $f_{yx} = 2yz^3$, and $f_{yxz} = 6yz^2 = f_{xzy}$.

73. By Definition 4, $f_x(3,2) = \lim\limits_{h \to 0} \dfrac{f(3+h, 2) - f(3,2)}{h}$ which we can approximate by considering $h = 0.5$ and $h = -0.5$:

$f_x(3,2) \approx \dfrac{f(3.5, 2) - f(3,2)}{0.5} = \dfrac{22.4 - 17.5}{0.5} = 9.8$, $f_x(3,2) \approx \dfrac{f(2.5, 2) - f(3,2)}{-0.5} = \dfrac{10.2 - 17.5}{-0.5} = 14.6$. Averaging

these values, we estimate $f_x(3,2)$ to be approximately 12.2. Similarly, $f_x(3, 2.2) = \lim\limits_{h \to 0} \dfrac{f(3+h, 2.2) - f(3, 2.2)}{h}$ which

© 2012 Cengage Learning. All Rights Reserved. May not be scanned, copied, or duplicated, or posted to a publicly accessible website, in whole or in part.

we can approximate by considering $h = 0.5$ and $h = -0.5$: $f_x(3, 2.2) \approx \dfrac{f(3.5, 2.2) - f(3, 2.2)}{0.5} = \dfrac{26.1 - 15.9}{0.5} = 20.4$,

$f_x(3, 2.2) \approx \dfrac{f(2.5, 2.2) - f(3, 2.2)}{-0.5} = \dfrac{9.3 - 15.9}{-0.5} = 13.2$. Averaging these values, we have $f_x(3, 2.2) \approx 16.8$.

To estimate $f_{xy}(3, 2)$, we first need an estimate for $f_x(3, 1.8)$:

$f_x(3, 1.8) \approx \dfrac{f(3.5, 1.8) - f(3, 1.8)}{0.5} = \dfrac{20.0 - 18.1}{0.5} = 3.8$, $f_x(3, 1.8) \approx \dfrac{f(2.5, 1.8) - f(3, 1.8)}{-0.5} = \dfrac{12.5 - 18.1}{-0.5} = 11.2$.

Averaging these values, we get $f_x(3, 1.8) \approx 7.5$. Now $f_{xy}(x, y) = \dfrac{\partial}{\partial y}[f_x(x, y)]$ and $f_x(x, y)$ is itself a function of two

variables, so Definition 4 says that $f_{xy}(x, y) = \dfrac{\partial}{\partial y}[f_x(x, y)] = \lim\limits_{h \to 0} \dfrac{f_x(x, y + h) - f_x(x, y)}{h}$ $\Rightarrow$

$f_{xy}(3, 2) = \lim\limits_{h \to 0} \dfrac{f_x(3, 2 + h) - f_x(3, 2)}{h}$. We can estimate this value using our previous work with $h = 0.2$ and $h = -0.2$:

$f_{xy}(3, 2) \approx \dfrac{f_x(3, 2.2) - f_x(3, 2)}{0.2} = \dfrac{16.8 - 12.2}{0.2} = 23$, $f_{xy}(3, 2) \approx \dfrac{f_x(3, 1.8) - f_x(3, 2)}{-0.2} = \dfrac{7.5 - 12.2}{-0.2} = 23.5$.

Averaging these values, we estimate $f_{xy}(3, 2)$ to be approximately 23.25.

75. $u = e^{-\alpha^2 k^2 t} \sin kx \Rightarrow u_x = k e^{-\alpha^2 k^2 t} \cos kx$, $u_{xx} = -k^2 e^{-\alpha^2 k^2 t} \sin kx$, and $u_t = -\alpha^2 k^2 e^{-\alpha^2 k^2 t} \sin kx$.
Thus $\alpha^2 u_{xx} = u_t$.

77. $u = \dfrac{1}{\sqrt{x^2 + y^2 + z^2}} \Rightarrow u_x = \left(-\tfrac{1}{2}\right)(x^2 + y^2 + z^2)^{-3/2}(2x) = -x(x^2 + y^2 + z^2)^{-3/2}$ and

$u_{xx} = -(x^2 + y^2 + z^2)^{-3/2} - x\left(-\tfrac{3}{2}\right)(x^2 + y^2 + z^2)^{-5/2}(2x) = \dfrac{2x^2 - y^2 - z^2}{(x^2 + y^2 + z^2)^{5/2}}$.

By symmetry, $u_{yy} = \dfrac{2y^2 - x^2 - z^2}{(x^2 + y^2 + z^2)^{5/2}}$ and $u_{zz} = \dfrac{2z^2 - x^2 - y^2}{(x^2 + y^2 + z^2)^{5/2}}$.

Thus $u_{xx} + u_{yy} + u_{zz} = \dfrac{2x^2 - y^2 - z^2 + 2y^2 - x^2 - z^2 + 2z^2 - x^2 - y^2}{(x^2 + y^2 + z^2)^{5/2}} = 0$.

79. Let $v = x + at$, $w = x - at$. Then $u_t = \dfrac{\partial[f(v) + g(w)]}{\partial t} = \dfrac{df(v)}{dv}\dfrac{\partial v}{\partial t} + \dfrac{dg(w)}{dw}\dfrac{\partial w}{\partial t} = af'(v) - ag'(w)$ and

$u_{tt} = \dfrac{\partial[af'(v) - ag'(w)]}{\partial t} = a[af''(v) + ag''(w)] = a^2[f''(v) + g''(w)]$. Similarly, by using the Chain Rule we have

$u_x = f'(v) + g'(w)$ and $u_{xx} = f''(v) + g''(w)$. Thus $u_{tt} = a^2 u_{xx}$.

81. $z = \ln(e^x + e^y) \Rightarrow \dfrac{\partial z}{\partial x} = \dfrac{e^x}{e^x + e^y}$ and $\dfrac{\partial z}{\partial y} = \dfrac{e^y}{e^x + e^y}$, so $\dfrac{\partial z}{\partial x} + \dfrac{\partial z}{\partial y} = \dfrac{e^x}{e^x + e^y} + \dfrac{e^y}{e^x + e^y} = \dfrac{e^x + e^y}{e^x + e^y} = 1$.

$\dfrac{\partial^2 z}{\partial x^2} = \dfrac{e^x(e^x + e^y) - e^x(e^x)}{(e^x + e^y)^2} = \dfrac{e^{x+y}}{(e^x + e^y)^2}$, $\dfrac{\partial^2 z}{\partial x \, \partial y} = \dfrac{0 - e^y(e^x)}{(e^x + e^y)^2} = -\dfrac{e^{x+y}}{(e^x + e^y)^2}$, and

$\dfrac{\partial^2 z}{\partial y^2} = \dfrac{e^y(e^x + e^y) - e^y(e^y)}{(e^x + e^y)^2} = \dfrac{e^{x+y}}{(e^x + e^y)^2}$. Thus

$$\dfrac{\partial^2 z}{\partial x^2} \dfrac{\partial^2 z}{\partial y^2} - \left(\dfrac{\partial^2 z}{\partial x \, \partial y}\right)^2 = \dfrac{e^{x+y}}{(e^x + e^y)^2} \cdot \dfrac{e^{x+y}}{(e^x + e^y)^2} - \left(-\dfrac{e^{x+y}}{(e^x + e^y)^2}\right)^2 = \dfrac{(e^{x+y})^2}{(e^x + e^y)^4} - \dfrac{(e^{x+y})^2}{(e^x + e^y)^4} = 0$$

© 2012 Cengage Learning. All Rights Reserved. May not be scanned, copied, or duplicated, or posted to a publicly accessible website, in whole or in part.

83. By the Chain Rule, taking the partial derivative of both sides with respect to R_1 gives

$$\frac{\partial R^{-1}}{\partial R}\frac{\partial R}{\partial R_1} = \frac{\partial\left[(1/R_1) + (1/R_2) + (1/R_3)\right]}{\partial R_1} \quad \text{or} \quad -R^{-2}\frac{\partial R}{\partial R_1} = -R_1^{-2}. \quad \text{Thus } \frac{\partial R}{\partial R_1} = \frac{R^2}{R_1^2}.$$

85. If we fix $K = K_0$, $P(L, K_0)$ is a function of a single variable L, and $\dfrac{dP}{dL} = \alpha\,\dfrac{P}{L}$ is a separable differential equation. Then

$$\frac{dP}{P} = \alpha\,\frac{dL}{L} \;\Rightarrow\; \int\frac{dP}{P} = \int\alpha\,\frac{dL}{L} \;\Rightarrow\; \ln|P| = \alpha\ln|L| + C\,(K_0), \text{ where } C(K_0) \text{ can depend on } K_0. \text{ Then}$$

$$|P| = e^{\alpha\ln|L| + C(K_0)}, \text{ and since } P > 0 \text{ and } L > 0, \text{ we have } P = e^{\alpha\ln L}e^{C(K_0)} = e^{C(K_0)}e^{\ln L^\alpha} = C_1(K_0)L^\alpha \text{ where}$$

$$C_1(K_0) = e^{C(K_0)}.$$

87. $\left(P + \dfrac{n^2a}{V^2}\right)(V - nb) = nRT \;\Rightarrow\; T = \dfrac{1}{nR}\left(P + \dfrac{n^2a}{V^2}\right)(V - nb), \text{ so } \dfrac{\partial T}{\partial P} = \dfrac{1}{nR}\,(1)(V - nb) = \dfrac{V - nb}{nR}.$

We can also write $P + \dfrac{n^2a}{V^2} = \dfrac{nRT}{V - nb} \;\Rightarrow\; P = \dfrac{nRT}{V - nb} - \dfrac{n^2a}{V^2} = nRT(V - nb)^{-1} - n^2aV^{-2},$ so

$$\frac{\partial P}{\partial V} = -nRT(V - nb)^{-2}(1) + 2n^2aV^{-3} = \frac{2n^2a}{V^3} - \frac{nRT}{(V - nb)^2}.$$

89. By Exercise 88, $PV = mRT \;\Rightarrow\; P = \dfrac{mRT}{V}, \text{ so } \dfrac{\partial P}{\partial T} = \dfrac{mR}{V}.$ Also, $PV = mRT \;\Rightarrow\; V = \dfrac{mRT}{P} \text{ and } \dfrac{\partial V}{\partial T} = \dfrac{mR}{P}.$

Since $T = \dfrac{PV}{mR},$ we have $T\dfrac{\partial P}{\partial T}\dfrac{\partial V}{\partial T} = \dfrac{PV}{mR}\cdot\dfrac{mR}{V}\cdot\dfrac{mR}{P} = mR.$

91. $\dfrac{\partial K}{\partial m} = \frac{1}{2}v^2,$ $\dfrac{\partial K}{\partial v} = mv,$ $\dfrac{\partial^2 K}{\partial v^2} = m.$ Thus $\dfrac{\partial K}{\partial m}\cdot\dfrac{\partial^2 K}{\partial v^2} = \frac{1}{2}v^2m = K.$

93. $f_x(x, y) = x + 4y \;\Rightarrow\; f_{xy}(x, y) = 4$ and $f_y(x, y) = 3x - y \;\Rightarrow\; f_{yx}(x, y) = 3.$ Since f_{xy} and f_{yx} are continuous everywhere but $f_{xy}(x, y) \neq f_{yx}(x, y),$ Clairaut's Theorem implies that such a function $f(x, y)$ does not exist.

95. By the geometry of partial derivatives, the slope of the tangent line is $f_x(1, 2).$ By implicit differentiation of

$4x^2 + 2y^2 + z^2 = 16,$ we get $8x + 2z\,(\partial z/\partial x) = 0 \;\Rightarrow\; \partial z/\partial x = -4x/z,$ so when $x = 1$ and $z = 2$ we have

$\partial z/\partial x = -2.$ So the slope is $f_x(1, 2) = -2.$ Thus the tangent line is given by $z - 2 = -2(x - 1),$ $y = 2.$ Taking the

parameter to be $t = x - 1,$ we can write parametric equations for this line: $x = 1 + t,$ $y = 2,$ $z = 2 - 2t.$

97. By Clairaut's Theorem, $f_{xyy} = (f_{xy})_y = (f_{yx})_y = f_{yxy} = (f_y)_{xy} = (f_y)_{yx} = f_{yyx}.$

99. Let $g(x) = f(x, 0) = x(x^2)^{-3/2}e^0 = x\,|x|^{-3}.$ But we are using the point $(1, 0),$ so near $(1, 0),$ $g(x) = x^{-2}.$ Then

$g'(x) = -2x^{-3}$ and $g'(1) = -2,$ so using (1) we have $f_x(1, 0) = g'(1) = -2.$

101. **(a)**

(b) For $(x, y) \neq (0, 0),$

$$f_x(x, y) = \frac{(3x^2y - y^3)(x^2 + y^2) - (x^3y - xy^3)(2x)}{(x^2 + y^2)^2}$$

$$= \frac{x^4y + 4x^2y^3 - y^5}{(x^2 + y^2)^2}$$

and by symmetry $f_y(x, y) = \dfrac{x^5 - 4x^3y^2 - xy^4}{(x^2 + y^2)^2}.$

© 2012 Cengage Learning. All Rights Reserved. May not be scanned, copied, or duplicated, or posted to a publicly accessible website, in whole or in part.

(c) $f_x(0,0) = \lim\limits_{h \to 0} \dfrac{f(h,0) - f(0,0)}{h} = \lim\limits_{h \to 0} \dfrac{(0/h^2) - 0}{h} = 0$ and $f_y(0,0) = \lim\limits_{h \to 0} \dfrac{f(0,h) - f(0,0)}{h} = 0$.

(d) By (3), $f_{xy}(0,0) = \dfrac{\partial f_x}{\partial y} = \lim\limits_{h \to 0} \dfrac{f_x(0,h) - f_x(0,0)}{h} = \lim\limits_{h \to 0} \dfrac{(-h^5 - 0)/h^4}{h} = -1$ while by (2),

$f_{yx}(0,0) = \dfrac{\partial f_y}{\partial x} = \lim\limits_{h \to 0} \dfrac{f_y(h,0) - f_y(0,0)}{h} = \lim\limits_{h \to 0} \dfrac{h^5/h^4}{h} = 1$.

(e) For $(x,y) \neq (0,0)$, we use a CAS to compute

$$f_{xy}(x,y) = \frac{x^6 + 9x^4 y^2 - 9x^2 y^4 - y^6}{(x^2 + y^2)^3}$$

Now as $(x,y) \to (0,0)$ along the x-axis, $f_{xy}(x,y) \to 1$ while as

$(x,y) \to (0,0)$ along the y-axis, $f_{xy}(x,y) \to -1$. Thus f_{xy} isn't

continuous at $(0,0)$ and Clairaut's Theorem doesn't apply, so there is

no contradiction. The graphs of f_{xy} and f_{yx} are identical except at the

origin, where we observe the discontinuity.

14.4   Tangent Planes and Linear Approximations

1. $z = f(x,y) = 3y^2 - 2x^2 + x \Rightarrow f_x(x,y) = -4x + 1,\ f_y(x,y) = 6y$, so $f_x(2,-1) = -7,\ f_y(2,-1) = -6$.

By Equation 2, an equation of the tangent plane is $z - (-3) = f_x(2,-1)(x - 2) + f_y(2,-1)[y - (-1)] \Rightarrow$

$z + 3 = -7(x - 2) - 6(y + 1)$ or $z = -7x - 6y + 5$.

3. $z = f(x,y) = \sqrt{xy} \Rightarrow f_x(x,y) = \frac{1}{2}(xy)^{-1/2} \cdot y = \frac{1}{2}\sqrt{y/x},\ f_y(x,y) = \frac{1}{2}(xy)^{-1/2} \cdot x = \frac{1}{2}\sqrt{x/y}$, so $f_x(1,1) = \frac{1}{2}$

and $f_y(1,1) = \frac{1}{2}$. Thus an equation of the tangent plane is $z - 1 = f_x(1,1)(x - 1) + f_y(1,1)(y - 1) \Rightarrow$

$z - 1 = \frac{1}{2}(x - 1) + \frac{1}{2}(y - 1)$ or $x + y - 2z = 0$.

5. $z = f(x,y) = x\sin(x + y) \Rightarrow f_x(x,y) = x \cdot \cos(x + y) + \sin(x + y) \cdot 1 = x\cos(x + y) + \sin(x + y)$,

$f_y(x,y) = x\cos(x + y)$, so $f_x(-1,1) = (-1)\cos 0 + \sin 0 = -1,\ f_y(-1,1) = (-1)\cos 0 = -1$ and an equation of the

tangent plane is $z - 0 = (-1)(x + 1) + (-1)(y - 1)$ or $x + y + z = 0$.

7. $z = f(x,y) = x^2 + xy + 3y^2$, so $f_x(x,y) = 2x + y \Rightarrow f_x(1,1) = 3,\ f_y(x,y) = x + 6y \Rightarrow f_y(1,1) = 7$ and an

equation of the tangent plane is $z - 5 = 3(x - 1) + 7(y - 1)$ or $z = 3x + 7y - 5$. After zooming in, the surface and the

© 2012 Cengage Learning. All Rights Reserved. May not be scanned, copied, or duplicated, or posted to a publicly accessible website, in whole or in part.

tangent plane become almost indistinguishable. (Here, the tangent plane is below the surface.) If we zoom in farther, the surface and the tangent plane will appear to coincide.

9. $f(x, y) = \dfrac{xy \sin (x - y)}{1 + x^2 + y^2}$.   A CAS gives $f_x(x, y) = \dfrac{y \sin (x - y) + xy \cos (x - y)}{1 + x^2 + y^2} - \dfrac{2x^2 y \sin (x - y)}{(1 + x^2 + y^2)^2}$ and

$f_y(x, y) = \dfrac{x \sin (x - y) - xy \cos (x - y)}{1 + x^2 + y^2} - \dfrac{2xy^2 \sin (x - y)}{(1 + x^2 + y^2)^2}$. We use the CAS to evaluate these at $(1, 1)$, and then

substitute the results into Equation 2 to compute an equation of the tangent plane: $z = \frac{1}{3}x - \frac{1}{3}y$. The surface and tangent

plane are shown in the first graph below. After zooming in, the surface and the tangent plane become almost indistinguishable,

as shown in the second graph. (Here, the tangent plane is shown with fewer traces than the surface.) If we zoom in farther, the

surface and the tangent plane will appear to coincide.

11. $f(x, y) = 1 + x \ln(xy - 5)$.   The partial derivatives are $f_x(x, y) = x \cdot \dfrac{1}{xy - 5} (y) + \ln(xy - 5) \cdot 1 = \dfrac{xy}{xy - 5} + \ln(xy - 5)$

and $f_y(x, y) = x \cdot \dfrac{1}{xy - 5} (x) = \dfrac{x^2}{xy - 5}$, so $f_x(2, 3) = 6$ and $f_y(2, 3) = 4$. Both f_x and f_y are continuous functions for

$xy > 5$, so by Theorem 8, f is differentiable at $(2, 3)$. By Equation 3, the linearization of f at $(2, 3)$ is given by

$L(x, y) = f(2, 3) + f_x(2, 3)(x - 2) + f_y(2, 3)(y - 3) = 1 + 6(x - 2) + 4(y - 3) = 6x + 4y - 23$.

13. $f(x, y) = \dfrac{x}{x + y}$.   The partial derivatives are $f_x(x, y) = \dfrac{1(x + y) - x(1)}{(x + y)^2} = y/(x + y)^2$ and

$f_y(x, y) = x(-1)(x + y)^{-2} \cdot 1 = -x/(x + y)^2$, so $f_x(2, 1) = \frac{1}{9}$ and $f_y(2, 1) = -\frac{2}{9}$. Both f_x and f_y are continuous

functions for $y \neq -x$, so f is differentiable at $(2, 1)$ by Theorem 8. The linearization of f at $(2, 1)$ is given by

$L(x, y) = f(2, 1) + f_x(2, 1)(x - 2) + f_y(2, 1)(y - 1) = \frac{2}{3} + \frac{1}{9}(x - 2) - \frac{2}{9}(y - 1) = \frac{1}{9}x - \frac{2}{9}y + \frac{2}{3}$.

© 2012 Cengage Learning. All Rights Reserved. May not be scanned, copied, or duplicated, or posted to a publicly accessible website, in whole or in part.

15. $f(x, y) = e^{-xy} \cos y$.   The partial derivatives are $f_x(x, y) = e^{-xy}(-y) \cos y = -ye^{-xy} \cos y$ and

$f_y(x, y) = e^{-xy}(-\sin y) + (\cos y)e^{-xy}(-x) = -e^{-xy}(\sin y + x \cos y)$, so $f_x(\pi, 0) = 0$ and $f_y(\pi, 0) = -\pi$.

Both f_x and f_y are continuous functions, so f is differentiable at $(\pi, 0)$, and the linearization of f at $(\pi, 0)$ is

$L(x, y) = f(\pi, 0) + f_x(\pi, 0)(x - \pi) + f_y(\pi, 0)(y - 0) = 1 + 0(x - \pi) - \pi(y - 0) = 1 - \pi y.$

17. Let $f(x, y) = \dfrac{2x + 3}{4y + 1}$. Then $f_x(x, y) = \dfrac{2}{4y + 1}$ and $f_y(x, y) = (2x + 3)(-1)(4y + 1)^{-2}(4) = \dfrac{-8x - 12}{(4y + 1)^2}$. Both f_x and f_y

are continuous functions for $y \neq -\frac{1}{4}$, so by Theorem 8, f is differentiable at $(0, 0)$. We have $f_x(0, 0) = 2$, $f_y(0, 0) = -12$

and the linear approximation of f at $(0, 0)$ is $f(x, y) \approx f(0, 0) + f_x(0, 0)(x - 0) + f_y(0, 0)(y - 0) = 3 + 2x - 12y.$

19. We can estimate $f(2.2, 4.9)$ using a linear approximation of f at $(2, 5)$, given by

$f(x, y) \approx f(2, 5) + f_x(2, 5)(x - 2) + f_y(2, 5)(y - 5) = 6 + 1(x - 2) + (-1)(y - 5) = x - y + 9$. Thus

$f(2.2, 4.9) \approx 2.2 - 4.9 + 9 = 6.3.$

21. $f(x, y, z) = \sqrt{x^2 + y^2 + z^2}$ $\Rightarrow$ $f_x(x, y, z) = \dfrac{x}{\sqrt{x^2 + y^2 + z^2}}$, $f_y(x, y, z) = \dfrac{y}{\sqrt{x^2 + y^2 + z^2}}$, and

$f_z(x, y, z) = \dfrac{z}{\sqrt{x^2 + y^2 + z^2}}$, so $f_x(3, 2, 6) = \frac{3}{7}$, $f_y(3, 2, 6) = \frac{2}{7}$, $f_z(3, 2, 6) = \frac{6}{7}$. Then the linear approximation of f

at $(3, 2, 6)$ is given by

$$f(x, y, z) \approx f(3, 2, 6) + f_x(3, 2, 6)(x - 3) + f_y(3, 2, 6)(y - 2) + f_z(3, 2, 6)(z - 6)$$
$$= 7 + \tfrac{3}{7}(x - 3) + \tfrac{2}{7}(y - 2) + \tfrac{6}{7}(z - 6) = \tfrac{3}{7}x + \tfrac{2}{7}y + \tfrac{6}{7}z.$$

Thus $\sqrt{(3.02)^2 + (1.97)^2 + (5.99)^2} = f(3.02, 1.97, 5.99) \approx \frac{3}{7}(3.02) + \frac{2}{7}(1.97) + \frac{6}{7}(5.99) \approx 6.9914.$

23. From the table, $f(94, 80) = 127$. To estimate $f_T(94, 80)$ and $f_H(94, 80)$ we follow the procedure used in Section 14.3. Since

$f_T(94, 80) = \lim\limits_{h \to 0} \dfrac{f(94 + h, 80) - f(94, 80)}{h}$, we approximate this quantity with $h = \pm 2$ and use the values given in the

table:

$f_T(94, 80) \approx \dfrac{f(96, 80) - f(94, 80)}{2} = \dfrac{135 - 127}{2} = 4,$ $f_T(94, 80) \approx \dfrac{f(92, 80) - f(94, 80)}{-2} = \dfrac{119 - 127}{-2} = 4$

Averaging these values gives $f_T(94, 80) \approx 4$. Similarly, $f_H(94, 80) = \lim\limits_{h \to 0} \dfrac{f(94, 80 + h) - f(94, 80)}{h}$, so we use $h = \pm 5$:

$f_H(94, 80) \approx \dfrac{f(94, 85) - f(94, 80)}{5} = \dfrac{132 - 127}{5} = 1,$ $f_H(94, 80) \approx \dfrac{f(94, 75) - f(94, 80)}{-5} = \dfrac{122 - 127}{-5} = 1$

Averaging these values gives $f_H(94, 80) \approx 1$. The linear approximation, then, is

$$f(T, H) \approx f(94, 80) + f_T(94, 80)(T - 94) + f_H(94, 80)(H - 80)$$

$$\approx 127 + 4(T - 94) + 1(H - 80) \quad \text{[or } 4T + H - 329]$$

© 2012 Cengage Learning. All Rights Reserved. May not be scanned, copied, or duplicated, or posted to a publicly accessible website, in whole or in part.

Thus when $T = 95$ and $H = 78$, $f(95, 78) \approx 127 + 4(95 - 94) + 1(78 - 80) = 129$, so we estimate the heat index to be approximately $129°F$.

25. $z = e^{-2x} \cos 2\pi t \implies$

$$dz = \frac{\partial z}{\partial x} dx + \frac{\partial z}{\partial t} dt = e^{-2x}(-2) \cos 2\pi t \, dx + e^{-2x}(-\sin 2\pi t)(2\pi) \, dt = -2e^{-2x} \cos 2\pi t \, dx - 2\pi e^{-2x} \sin 2\pi t \, dt$$

27. $m = p^5 q^3 \implies dm = \frac{\partial m}{\partial p} dp + \frac{\partial m}{\partial q} dq = 5p^4 q^3 \, dp + 3p^5 q^2 \, dq$

29. $R = \alpha \beta^2 \cos \gamma \implies dR = \frac{\partial R}{\partial \alpha} d\alpha + \frac{\partial R}{\partial \beta} d\beta + \frac{\partial R}{\partial \gamma} d\gamma = \beta^2 \cos \gamma \, d\alpha + 2\alpha\beta \cos \gamma \, d\beta - \alpha\beta^2 \sin \gamma \, d\gamma$

31. $dx = \Delta x = 0.05$, $dy = \Delta y = 0.1$, $z = 5x^2 + y^2$, $z_x = 10x$, $z_y = 2y$. Thus when $x = 1$ and $y = 2$,

$dz = z_x(1, 2) \, dx + z_y(1, 2) \, dy = (10)(0.05) + (4)(0.1) = 0.9$ while

$\Delta z = f(1.05, 2.1) - f(1, 2) = 5(1.05)^2 + (2.1)^2 - 5 - 4 = 0.9225.$

33. $dA = \frac{\partial A}{\partial x} dx + \frac{\partial A}{\partial y} dy = y \, dx + x \, dy$ and $|\Delta x| \le 0.1$, $|\Delta y| \le 0.1$. We use $dx = 0.1$, $dy = 0.1$ with $x = 30$, $y = 24$; then

the maximum error in the area is about $dA = 24(0.1) + 30(0.1) = 5.4 \text{ cm}^2$.

35. The volume of a can is $V = \pi r^2 h$ and $\Delta V \approx dV$ is an estimate of the amount of tin. Here $dV = 2\pi r h \, dr + \pi r^2 \, dh$, so put

$dr = 0.04$, $dh = 0.08$ (0.04 on top, 0.04 on bottom) and then $\Delta V \approx dV = 2\pi(48)(0.04) + \pi(16)(0.08) \approx 16.08 \text{ cm}^3$.

Thus the amount of tin is about 16 cm^3.

37. $T = \dfrac{mgR}{2r^2 + R^2}$, so the differential of T is

$$dT = \frac{\partial T}{\partial R} dR + \frac{\partial T}{\partial r} dr = \frac{(2r^2 + R^2)(mg) - mgR(2R)}{(2r^2 + R^2)^2} dR + \frac{(2r^2 + R^2)(0) - mgR(4r)}{(2r^2 + R^2)^2} dr$$

$$= \frac{mg(2r^2 - R^2)}{(2r^2 + R^2)^2} dR - \frac{4mgRr}{(2r^2 + R^2)^2} dr$$

Here we have $\Delta R = 0.1$ and $\Delta r = 0.1$, so we take $dR = 0.1$, $dr = 0.1$ with $R = 3$, $r = 0.7$. Then the change in the

tension T is approximately

$$dT = \frac{mg[2(0.7)^2 - (3)^2]}{[2(0.7)^2 + (3)^2]^2} (0.1) - \frac{4mg(3)(0.7)}{[2(0.7)^2 + (3)^2]^2} (0.1)$$

$$= -\frac{0.802mg}{(9.98)^2} - \frac{0.84mg}{(9.98)^2} = -\frac{1.642}{99.6004} mg \approx -0.0165mg$$

Because the change is negative, tension decreases.

© 2012 Cengage Learning. All Rights Reserved. May not be scanned, copied, or duplicated, or posted to a publicly accessible website, in whole or in part.

39. First we find $\dfrac{\partial R}{\partial R_1}$ implicitly by taking partial derivatives of both sides with respect to R_1:

$$\frac{\partial}{\partial R_1}\left(\frac{1}{R}\right) = \frac{\partial\left[(1/R_1)+(1/R_2)+(1/R_3)\right]}{\partial R_1} \;\Rightarrow\; -R^{-2}\frac{\partial R}{\partial R_1} = -R_1^{-2} \;\Rightarrow\; \frac{\partial R}{\partial R_1} = \frac{R^2}{R_1^2}. \text{ Then by symmetry,}$$

$\dfrac{\partial R}{\partial R_2} = \dfrac{R^2}{R_2^2}$, $\dfrac{\partial R}{\partial R_3} = \dfrac{R^2}{R_3^2}$. When $R_1 = 25$, $R_2 = 40$ and $R_3 = 50$, $\dfrac{1}{R} = \dfrac{17}{200}$ $\Leftrightarrow$ $R = \dfrac{200}{17}\ \Omega$. Since the possible error

for each R_i is 0.5%, the maximum error of R is attained by setting $\Delta R_i = 0.005 R_i$. So

$$\Delta R \approx dR = \frac{\partial R}{\partial R_1}\Delta R_1 + \frac{\partial R}{\partial R_2}\Delta R_2 + \frac{\partial R}{\partial R_3}\Delta R_3 = (0.005)R^2\left(\frac{1}{R_1}+\frac{1}{R_2}+\frac{1}{R_3}\right) = (0.005)R = \tfrac{1}{17} \approx 0.059\ \Omega.$$

41. The errors in measurement are at most 2%, so $\left|\dfrac{\Delta w}{w}\right| \leq 0.02$ and $\left|\dfrac{\Delta h}{h}\right| \leq 0.02$. The relative error in the calculated surface

area is

$$\frac{\Delta S}{S} \approx \frac{dS}{S} = \frac{0.1091(0.425w^{0.425-1})h^{0.725}\,dw + 0.1091w^{0.425}(0.725h^{0.725-1})\,dh}{0.1091w^{0.425}h^{0.725}} = 0.425\frac{dw}{w} + 0.725\frac{dh}{h}$$

To estimate the maximum relative error, we use $\dfrac{dw}{w} = \left|\dfrac{\Delta w}{w}\right| = 0.02$ and $\dfrac{dh}{h} = \left|\dfrac{\Delta h}{h}\right| = 0.02$ $\Rightarrow$

$\dfrac{dS}{S} = 0.425\,(0.02) + 0.725\,(0.02) = 0.023$. Thus the maximum percentage error is approximately 2.3%.

43. $\Delta z = f(a + \Delta x, b + \Delta y) - f(a,b) = (a+\Delta x)^2 + (b+\Delta y)^2 - (a^2+b^2)$

$ = a^2 + 2a\,\Delta x + (\Delta x)^2 + b^2 + 2b\,\Delta y + (\Delta y)^2 - a^2 - b^2 = 2a\,\Delta x + (\Delta x)^2 + 2b\,\Delta y + (\Delta y)^2$

But $f_x(a,b) = 2a$ and $f_y(a,b) = 2b$ and so $\Delta z = f_x(a,b)\,\Delta x + f_y(a,b)\,\Delta y + \Delta x\,\Delta x + \Delta y\,\Delta y$, which is Definition 7

with $\varepsilon_1 = \Delta x$ and $\varepsilon_2 = \Delta y$. Hence f is differentiable.

45. To show that f is continuous at (a,b) we need to show that $\displaystyle\lim_{(x,y)\to(a,b)} f(x,y) = f(a,b)$ or

equivalently $\displaystyle\lim_{(\Delta x,\Delta y)\to(0,0)} f(a + \Delta x, b + \Delta y) = f(a,b)$. Since f is differentiable at (a,b),

$f(a+\Delta x, b+\Delta y) - f(a,b) = \Delta z = f_x(a,b)\,\Delta x + f_y(a,b)\,\Delta y + \varepsilon_1\,\Delta x + \varepsilon_2\,\Delta y$, where ε_1 and $\varepsilon_2 \to 0$ as

$(\Delta x, \Delta y) \to (0,0)$. Thus $f(a+\Delta x, b+\Delta y) = f(a,b) + f_x(a,b)\,\Delta x + f_y(a,b)\,\Delta y + \varepsilon_1\,\Delta x + \varepsilon_2\,\Delta y$. Taking the limit of

both sides as $(\Delta x, \Delta y) \to (0,0)$ gives $\displaystyle\lim_{(\Delta x,\Delta y)\to(0,0)} f(a+\Delta x, b+\Delta y) = f(a,b)$. Thus f is continuous at (a,b).

14.5   The Chain Rule

1. $z = x^2 + y^2 + xy$, $x = \sin t$, $y = e^t$ $\Rightarrow$ $\dfrac{dz}{dt} = \dfrac{\partial z}{\partial x}\dfrac{dx}{dt} + \dfrac{\partial z}{\partial y}\dfrac{dy}{dt} = (2x+y)\cos t + (2y+x)e^t$

3. $z = \sqrt{1+x^2+y^2}$, $x = \ln t$, $y = \cos t$ $\Rightarrow$

$\dfrac{dz}{dt} = \dfrac{\partial z}{\partial x}\dfrac{dx}{dt} + \dfrac{\partial z}{\partial y}\dfrac{dy}{dt} = \tfrac{1}{2}(1+x^2+y^2)^{-1/2}(2x)\cdot\dfrac{1}{t} + \tfrac{1}{2}(1+x^2+y^2)^{-1/2}(2y)(-\sin t) = \dfrac{1}{\sqrt{1+x^2+y^2}}\left(\dfrac{x}{t} - y\sin t\right)$

© 2012 Cengage Learning. All Rights Reserved. May not be scanned, copied, or duplicated, or posted to a publicly accessible website, in whole or in part.

5. $w = xe^{y/z}$, $x = t^2$, $y = 1 - t$, $z = 1 + 2t$ $\Rightarrow$

$$\frac{dw}{dt} = \frac{\partial w}{\partial x}\frac{dx}{dt} + \frac{\partial w}{\partial y}\frac{dy}{dt} + \frac{\partial w}{\partial z}\frac{dz}{dt} = e^{y/z} \cdot 2t + xe^{y/z}\left(\frac{1}{z}\right) \cdot (-1) + xe^{y/z}\left(-\frac{y}{z^2}\right) \cdot 2 = e^{y/z}\left(2t - \frac{x}{z} - \frac{2xy}{z^2}\right)$$

7. $z = x^2 y^3$, $x = s\cos t$, $y = s\sin t$ $\Rightarrow$

$$\frac{\partial z}{\partial s} = \frac{\partial z}{\partial x}\frac{\partial x}{\partial s} + \frac{\partial z}{\partial y}\frac{\partial y}{\partial s} = 2xy^3\cos t + 3x^2 y^2\sin t$$

$$\frac{\partial z}{\partial t} = \frac{\partial z}{\partial x}\frac{\partial x}{\partial t} + \frac{\partial z}{\partial y}\frac{\partial y}{\partial t} = (2xy^3)(-s\sin t) + (3x^2 y^2)(s\cos t) = -2sxy^3\sin t + 3sx^2 y^2\cos t$$

9. $z = \sin\theta\cos\phi$, $\theta = st^2$, $\phi = s^2 t$ $\Rightarrow$

$$\frac{\partial z}{\partial s} = \frac{\partial z}{\partial \theta}\frac{\partial \theta}{\partial s} + \frac{\partial z}{\partial \phi}\frac{\partial \phi}{\partial s} = (\cos\theta\cos\phi)(t^2) + (-\sin\theta\sin\phi)(2st) = t^2\cos\theta\cos\phi - 2st\sin\theta\sin\phi$$

$$\frac{\partial z}{\partial t} = \frac{\partial z}{\partial \theta}\frac{\partial \theta}{\partial t} + \frac{\partial z}{\partial \phi}\frac{\partial \phi}{\partial t} = (\cos\theta\cos\phi)(2st) + (-\sin\theta\sin\phi)(s^2) = 2st\cos\theta\cos\phi - s^2\sin\theta\sin\phi$$

11. $z = e^r\cos\theta$, $r = st$, $\theta = \sqrt{s^2 + t^2}$ $\Rightarrow$

$$\frac{\partial z}{\partial s} = \frac{\partial z}{\partial r}\frac{\partial r}{\partial s} + \frac{\partial z}{\partial \theta}\frac{\partial \theta}{\partial s} = e^r\cos\theta \cdot t + e^r(-\sin\theta) \cdot \tfrac{1}{2}(s^2 + t^2)^{-1/2}(2s) = te^r\cos\theta - e^r\sin\theta \cdot \frac{s}{\sqrt{s^2 + t^2}}$$

$$= e^r\left(t\cos\theta - \frac{s}{\sqrt{s^2 + t^2}}\sin\theta\right)$$

$$\frac{\partial z}{\partial t} = \frac{\partial z}{\partial r}\frac{\partial r}{\partial t} + \frac{\partial z}{\partial \theta}\frac{\partial \theta}{\partial t} = e^r\cos\theta \cdot s + e^r(-\sin\theta) \cdot \tfrac{1}{2}(s^2 + t^2)^{-1/2}(2t) = se^r\cos\theta - e^r\sin\theta \cdot \frac{t}{\sqrt{s^2 + t^2}}$$

$$= e^r\left(s\cos\theta - \frac{t}{\sqrt{s^2 + t^2}}\sin\theta\right)$$

13. When $t = 3$, $x = g(3) = 2$ and $y = h(3) = 7$. By the Chain Rule (2),

$$\frac{dz}{dt} = \frac{\partial f}{\partial x}\frac{dx}{dt} + \frac{\partial f}{\partial y}\frac{dy}{dt} = f_x(2, 7)g'(3) + f_y(2, 7)\,h'(3) = (6)(5) + (-8)(-4) = 62.$$

15. $g(u, v) = f(x(u, v), y(u, v))$ where $x = e^u + \sin v$, $y = e^u + \cos v$ $\Rightarrow$

$\dfrac{\partial x}{\partial u} = e^u$, $\dfrac{\partial x}{\partial v} = \cos v$, $\dfrac{\partial y}{\partial u} = e^u$, $\dfrac{\partial y}{\partial v} = -\sin v$. By the Chain Rule (3), $\dfrac{\partial g}{\partial u} = \dfrac{\partial f}{\partial x}\dfrac{\partial x}{\partial u} + \dfrac{\partial f}{\partial y}\dfrac{\partial y}{\partial u}$. Then

$$g_u(0, 0) = f_x(x(0, 0), y(0, 0))\,x_u(0, 0) + f_y(x(0, 0), y(0, 0))\,y_u(0, 0) = f_x(1, 2)(e^0) + f_y(1, 2)(e^0) = 2(1) + 5(1) = 7.$$

Similarly, $\dfrac{\partial g}{\partial v} = \dfrac{\partial f}{\partial x}\dfrac{\partial x}{\partial v} + \dfrac{\partial f}{\partial y}\dfrac{\partial y}{\partial v}$. Then

$$g_v(0, 0) = f_x(x(0, 0), y(0, 0))\,x_v(0, 0) + f_y(x(0, 0), y(0, 0))\,y_v(0, 0) = f_x(1, 2)(\cos 0) + f_y(1, 2)(-\sin 0)$$

$$= 2(1) + 5(0) = 2$$

17.

$u = f(x, y)$, $x = x(r, s, t)$, $y = y(r, s, t)$ $\Rightarrow$

$$\frac{\partial u}{\partial r} = \frac{\partial u}{\partial x}\frac{\partial x}{\partial r} + \frac{\partial u}{\partial y}\frac{\partial y}{\partial r}, \quad \frac{\partial u}{\partial s} = \frac{\partial u}{\partial x}\frac{\partial x}{\partial s} + \frac{\partial u}{\partial y}\frac{\partial y}{\partial s},$$

$$\frac{\partial u}{\partial t} = \frac{\partial u}{\partial x}\frac{\partial x}{\partial t} + \frac{\partial u}{\partial y}\frac{\partial y}{\partial t}$$

© 2012 Cengage Learning. All Rights Reserved. May not be scanned, copied, or duplicated, or posted to a publicly accessible website, in whole or in part.

19.

$w = f(r, s, t), \quad r = r(x, y), \quad s = s(x, y), \quad t = t(x, y) \quad \Rightarrow$

$$\frac{\partial w}{\partial x} = \frac{\partial w}{\partial r}\frac{\partial r}{\partial x} + \frac{\partial w}{\partial s}\frac{\partial s}{\partial x} + \frac{\partial w}{\partial t}\frac{\partial t}{\partial x}, \quad \frac{\partial w}{\partial y} = \frac{\partial w}{\partial r}\frac{\partial r}{\partial y} + \frac{\partial w}{\partial s}\frac{\partial s}{\partial y} + \frac{\partial w}{\partial t}\frac{\partial t}{\partial y}$$

21. $z = x^4 + x^2 y, \ x = s + 2t - u, \ y = stu^2 \quad \Rightarrow$

$$\frac{\partial z}{\partial s} = \frac{\partial z}{\partial x}\frac{\partial x}{\partial s} + \frac{\partial z}{\partial y}\frac{\partial y}{\partial s} = (4x^3 + 2xy)(1) + (x^2)(tu^2),$$

$$\frac{\partial z}{\partial t} = \frac{\partial z}{\partial x}\frac{\partial x}{\partial t} + \frac{\partial z}{\partial y}\frac{\partial y}{\partial t} = (4x^3 + 2xy)(2) + (x^2)(su^2),$$

$$\frac{\partial z}{\partial u} = \frac{\partial z}{\partial x}\frac{\partial x}{\partial u} + \frac{\partial z}{\partial y}\frac{\partial y}{\partial u} = (4x^3 + 2xy)(-1) + (x^2)(2stu).$$

When $s = 4$, $t = 2$, and $u = 1$ we have $x = 7$ and $y = 8$,

so $\dfrac{\partial z}{\partial s} = (1484)(1) + (49)(2) = 1582, \quad \dfrac{\partial z}{\partial t} = (1484)(2) + (49)(4) = 3164, \quad \dfrac{\partial z}{\partial u} = (1484)(-1) + (49)(16) = -700.$

23. $w = xy + yz + zx, \ x = r\cos\theta, \ y = r\sin\theta, \ z = r\theta \quad \Rightarrow$

$$\frac{\partial w}{\partial r} = \frac{\partial w}{\partial x}\frac{\partial x}{\partial r} + \frac{\partial w}{\partial y}\frac{\partial y}{\partial r} + \frac{\partial w}{\partial z}\frac{\partial z}{\partial r} = (y + z)(\cos\theta) + (x + z)(\sin\theta) + (y + x)(\theta),$$

$$\frac{\partial w}{\partial\theta} = \frac{\partial w}{\partial x}\frac{\partial x}{\partial\theta} + \frac{\partial w}{\partial y}\frac{\partial y}{\partial\theta} + \frac{\partial w}{\partial z}\frac{\partial z}{\partial\theta} = (y + z)(-r\sin\theta) + (x + z)(r\cos\theta) + (y + x)(r).$$

When $r = 2$ and $\theta = \pi/2$ we have $x = 0$, $y = 2$, and $z = \pi$, so $\dfrac{\partial w}{\partial r} = (2 + \pi)(0) + (0 + \pi)(1) + (2 + 0)(\pi/2) = 2\pi$ and

$$\frac{\partial w}{\partial\theta} = (2 + \pi)(-2) + (0 + \pi)(0) + (2 + 0)(2) = -2\pi.$$

25. $N = \dfrac{p + q}{p + r}, \ p = u + vw, \ q = v + uw, \ r = w + uv \quad \Rightarrow$

$$\frac{\partial N}{\partial u} = \frac{\partial N}{\partial p}\frac{\partial p}{\partial u} + \frac{\partial N}{\partial q}\frac{\partial q}{\partial u} + \frac{\partial N}{\partial r}\frac{\partial r}{\partial u}$$

$$= \frac{(p + r)(1) - (p + q)(1)}{(p + r)^2}(1) + \frac{(p + r)(1) - (p + q)(0)}{(p + r)^2}(w) + \frac{(p + r)(0) - (p + q)(1)}{(p + r)^2}(v)$$

$$= \frac{(r - q) + (p + r)w - (p + q)v}{(p + r)^2},$$

$$\frac{\partial N}{\partial v} = \frac{\partial N}{\partial p}\frac{\partial p}{\partial v} + \frac{\partial N}{\partial q}\frac{\partial q}{\partial v} + \frac{\partial N}{\partial r}\frac{\partial r}{\partial v} = \frac{r - q}{(p + r)^2}(w) + \frac{p + r}{(p + r)^2}(1) + \frac{-(p + q)}{(p + r)^2}(u) = \frac{(r - q)w + (p + r) - (p + q)u}{(p + r)^2},$$

$$\frac{\partial N}{\partial w} = \frac{\partial N}{\partial p}\frac{\partial p}{\partial w} + \frac{\partial N}{\partial q}\frac{\partial q}{\partial w} + \frac{\partial N}{\partial r}\frac{\partial r}{\partial w} = \frac{r - q}{(p + r)^2}(v) + \frac{p + r}{(p + r)^2}(u) + \frac{-(p + q)}{(p + r)^2}(1) = \frac{(r - q)v + (p + r)u - (p + q)}{(p + r)^2}.$$

When $u = 2$, $v = 3$, and $w = 4$ we have $p = 14$, $q = 11$, and $r = 10$, so $\dfrac{\partial N}{\partial u} = \dfrac{-1 + (24)(4) - (25)(3)}{(24)^2} = \dfrac{20}{576} = \dfrac{5}{144},$

$$\frac{\partial N}{\partial v} = \frac{(-1)(4) + 24 - (25)(2)}{(24)^2} = \frac{-30}{576} = -\frac{5}{96}, \text{ and } \frac{\partial N}{\partial w} = \frac{(-1)(3) + (24)(2) - 25}{(24)^2} = \frac{20}{576} = \frac{5}{144}.$$

© 2012 Cengage Learning. All Rights Reserved. May not be scanned, copied, or duplicated, or posted to a publicly accessible website, in whole or in part.

27. $y \cos x = x^2 + y^2$, so let $F(x, y) = y \cos x - x^2 - y^2 = 0$. Then by Equation 6

$$\frac{dy}{dx} = -\frac{F_x}{F_y} = -\frac{-y \sin x - 2x}{\cos x - 2y} = \frac{2x + y \sin x}{\cos x - 2y}.$$

29. $\tan^{-1}(x^2 y) = x + xy^2$, so let $F(x, y) = \tan^{-1}(x^2 y) - x - xy^2 = 0$. Then

$$F_x(x, y) = \frac{1}{1 + (x^2 y)^2}(2xy) - 1 - y^2 = \frac{2xy}{1 + x^4 y^2} - 1 - y^2 = \frac{2xy - (1 + y^2)(1 + x^4 y^2)}{1 + x^4 y^2},$$

$$F_y(x, y) = \frac{1}{1 + (x^2 y)^2}(x^2) - 2xy = \frac{x^2}{1 + x^4 y^2} - 2xy = \frac{x^2 - 2xy(1 + x^4 y^2)}{1 + x^4 y^2}$$

and $\qquad \dfrac{dy}{dx} = -\dfrac{F_x}{F_y} = -\dfrac{[2xy - (1 + y^2)(1 + x^4 y^2)]/(1 + x^4 y^2)}{[x^2 - 2xy(1 + x^4 y^2)]/(1 + x^4 y^2)} = \dfrac{(1 + y^2)(1 + x^4 y^2) - 2xy}{x^2 - 2xy(1 + x^4 y^2)}$

$$= \frac{1 + x^4 y^2 + y^2 + x^4 y^4 - 2xy}{x^2 - 2xy - 2x^5 y^3}$$

31. $x^2 + 2y^2 + 3z^2 = 1$, so let $F(x, y, z) = x^2 + 2y^2 + 3z^2 - 1 = 0$. Then by Equations 7

$$\frac{\partial z}{\partial x} = -\frac{F_x}{F_z} = -\frac{2x}{6z} = -\frac{x}{3z} \quad \text{and} \quad \frac{\partial z}{\partial y} = -\frac{F_y}{F_z} = -\frac{4y}{6z} = -\frac{2y}{3z}.$$

33. $e^z = xyz$, so let $F(x, y, z) = e^z - xyz = 0$. Then $\dfrac{\partial z}{\partial x} = -\dfrac{F_x}{F_z} = -\dfrac{-yz}{e^z - xy} = \dfrac{yz}{e^z - xy}$ and

$$\frac{\partial z}{\partial y} = -\frac{F_y}{F_z} = -\frac{-xz}{e^z - xy} = \frac{xz}{e^z - xy}.$$

35. Since x and y are each functions of t, $T(x, y)$ is a function of t, so by the Chain Rule, $\dfrac{dT}{dt} = \dfrac{\partial T}{\partial x}\dfrac{dx}{dt} + \dfrac{\partial T}{\partial y}\dfrac{dy}{dt}$. After

3 seconds, $x = \sqrt{1 + t} = \sqrt{1 + 3} = 2$, $y = 2 + \frac{1}{3}t = 2 + \frac{1}{3}(3) = 3$, $\dfrac{dx}{dt} = \dfrac{1}{2\sqrt{1 + t}} = \dfrac{1}{2\sqrt{1 + 3}} = \dfrac{1}{4}$, and $\dfrac{dy}{dt} = \dfrac{1}{3}$.

Then $\dfrac{dT}{dt} = T_x(2, 3)\dfrac{dx}{dt} + T_y(2, 3)\dfrac{dy}{dt} = 4\left(\frac{1}{4}\right) + 3\left(\frac{1}{3}\right) = 2$. Thus the temperature is rising at a rate of $2°\text{C/s}$.

37. $C = 1449.2 + 4.6T - 0.055T^2 + 0.00029T^3 + 0.016D$, so $\dfrac{\partial C}{\partial T} = 4.6 - 0.11T + 0.00087T^2$ and $\dfrac{\partial C}{\partial D} = 0.016$.

According to the graph, the diver is experiencing a temperature of approximately $12.5°\text{C}$ at $t = 20$ minutes, so

$\dfrac{\partial C}{\partial T} = 4.6 - 0.11(12.5) + 0.00087(12.5)^2 \approx 3.36$. By sketching tangent lines at $t = 20$ to the graphs given, we estimate

$\dfrac{dD}{dt} \approx \dfrac{1}{2}$ and $\dfrac{dT}{dt} \approx -\dfrac{1}{10}$. Then, by the Chain Rule, $\dfrac{dC}{dt} = \dfrac{\partial C}{\partial T}\dfrac{dT}{dt} + \dfrac{\partial C}{\partial D}\dfrac{dD}{dt} \approx (3.36)\left(-\frac{1}{10}\right) + (0.016)\left(\frac{1}{2}\right) \approx -0.33$.

Thus the speed of sound experienced by the diver is decreasing at a rate of approximately 0.33 m/s per minute.

39. (a) $V = \ell w h$, so by the Chain Rule,

$$\frac{dV}{dt} = \frac{\partial V}{\partial \ell}\frac{d\ell}{dt} + \frac{\partial V}{\partial w}\frac{dw}{dt} + \frac{\partial V}{\partial h}\frac{dh}{dt} = wh\frac{d\ell}{dt} + \ell h\frac{dw}{dt} + \ell w\frac{dh}{dt} = 2 \cdot 2 \cdot 2 + 1 \cdot 2 \cdot 2 + 1 \cdot 2 \cdot (-3) = 6 \text{ m}^3/\text{s}.$$

© 2012 Cengage Learning. All Rights Reserved. May not be scanned, copied, or duplicated, or posted to a publicly accessible website, in whole or in part.

(b) $S = 2(\ell w + \ell h + wh)$, so by the Chain Rule,

$$\frac{dS}{dt} = \frac{\partial S}{\partial \ell}\frac{d\ell}{dt} + \frac{\partial S}{\partial w}\frac{dw}{dt} + \frac{\partial S}{\partial h}\frac{dh}{dt} = 2(w+h)\frac{d\ell}{dt} + 2(\ell + h)\frac{dw}{dt} + 2(\ell + w)\frac{dh}{dt}$$
$$= 2(2+2)2 + 2(1+2)2 + 2(1+2)(-3) = 10 \text{ m}^2/\text{s}$$

(c) $L^2 = \ell^2 + w^2 + h^2 \;\Rightarrow\; 2L\frac{dL}{dt} = 2\ell\frac{d\ell}{dt} + 2w\frac{dw}{dt} + 2h\frac{dh}{dt} = 2(1)(2) + 2(2)(2) + 2(2)(-3) = 0 \;\Rightarrow$

$dL/dt = 0$ m/s.

41. $\dfrac{dP}{dt} = 0.05$, $\dfrac{dT}{dt} = 0.15$, $V = 8.31\dfrac{T}{P}$ and $\dfrac{dV}{dt} = \dfrac{8.31}{P}\dfrac{dT}{dt} - 8.31\dfrac{T}{P^2}\dfrac{dP}{dt}$. Thus when $P = 20$ and $T = 320$,

$$\frac{dV}{dt} = 8.31\left[\frac{0.15}{20} - \frac{(0.05)(320)}{400}\right] \approx -0.27 \text{ L/s}.$$

43. Let x be the length of the first side of the triangle and y the length of the second side. The area A of the triangle is given by

$A = \frac{1}{2}xy\sin\theta$ where θ is the angle between the two sides. Thus A is a function of x, y, and θ, and x, y, and θ are each in turn

functions of time t. We are given that $\dfrac{dx}{dt} = 3$, $\dfrac{dy}{dt} = -2$, and because A is constant, $\dfrac{dA}{dt} = 0$. By the Chain Rule,

$$\frac{dA}{dt} = \frac{\partial A}{\partial x}\frac{dx}{dt} + \frac{\partial A}{\partial y}\frac{dy}{dt} + \frac{\partial A}{\partial \theta}\frac{d\theta}{dt} \;\Rightarrow\; \frac{dA}{dt} = \frac{1}{2}y\sin\theta\cdot\frac{dx}{dt} + \frac{1}{2}x\sin\theta\cdot\frac{dy}{dt} + \frac{1}{2}xy\cos\theta\cdot\frac{d\theta}{dt}.$$ When $x = 20$, $y = 30$,

and $\theta = \pi/6$ we have

$$0 = \frac{1}{2}(30)\left(\sin\frac{\pi}{6}\right)(3) + \frac{1}{2}(20)\left(\sin\frac{\pi}{6}\right)(-2) + \frac{1}{2}(20)(30)\left(\cos\frac{\pi}{6}\right)\frac{d\theta}{dt}$$

$$= 45\cdot\frac{1}{2} - 20\cdot\frac{1}{2} + 300\cdot\frac{\sqrt{3}}{2}\cdot\frac{d\theta}{dt} = \frac{25}{2} + 150\sqrt{3}\frac{d\theta}{dt}$$

Solving for $\dfrac{d\theta}{dt}$ gives $\dfrac{d\theta}{dt} = \dfrac{-25/2}{150\sqrt{3}} = -\dfrac{1}{12\sqrt{3}}$, so the angle between the sides is decreasing at a rate of

$1/(12\sqrt{3}) \approx 0.048$ rad/s.

45. (a) By the Chain Rule, $\dfrac{\partial z}{\partial r} = \dfrac{\partial z}{\partial x}\cos\theta + \dfrac{\partial z}{\partial y}\sin\theta$, $\dfrac{\partial z}{\partial \theta} = \dfrac{\partial z}{\partial x}(-r\sin\theta) + \dfrac{\partial z}{\partial y}r\cos\theta$.

(b) $\left(\dfrac{\partial z}{\partial r}\right)^2 = \left(\dfrac{\partial z}{\partial x}\right)^2\cos^2\theta + 2\dfrac{\partial z}{\partial x}\dfrac{\partial z}{\partial y}\cos\theta\sin\theta + \left(\dfrac{\partial z}{\partial y}\right)^2\sin^2\theta,$

$\left(\dfrac{\partial z}{\partial \theta}\right)^2 = \left(\dfrac{\partial z}{\partial x}\right)^2 r^2\sin^2\theta - 2\dfrac{\partial z}{\partial x}\dfrac{\partial z}{\partial y}r^2\cos\theta\sin\theta + \left(\dfrac{\partial z}{\partial y}\right)^2 r^2\cos^2\theta.$ Thus

$$\left(\frac{\partial z}{\partial r}\right)^2 + \frac{1}{r^2}\left(\frac{\partial z}{\partial \theta}\right)^2 = \left[\left(\frac{\partial z}{\partial x}\right)^2 + \left(\frac{\partial z}{\partial y}\right)^2\right](\cos^2\theta + \sin^2\theta) = \left(\frac{\partial z}{\partial x}\right)^2 + \left(\frac{\partial z}{\partial y}\right)^2.$$

47. Let $u = x - y$. Then $\dfrac{\partial z}{\partial x} = \dfrac{dz}{du}\dfrac{\partial u}{\partial x} = \dfrac{dz}{du}$ and $\dfrac{\partial z}{\partial y} = \dfrac{dz}{du}(-1)$. Thus $\dfrac{\partial z}{\partial x} + \dfrac{\partial z}{\partial y} = 0.$

© 2012 Cengage Learning. All Rights Reserved. May not be scanned, copied, or duplicated, or posted to a publicly accessible website, in whole or in part.

49. Let $u = x + at$, $v = x - at$. Then $z = f(u) + g(v)$, so $\partial z/\partial u = f'(u)$ and $\partial z/\partial v = g'(v)$.

Thus $\dfrac{\partial z}{\partial t} = \dfrac{\partial z}{\partial u}\dfrac{\partial u}{\partial t} + \dfrac{\partial z}{\partial v}\dfrac{\partial v}{\partial t} = af'(u) - ag'(v)$ and

$$\frac{\partial^2 z}{\partial t^2} = a\frac{\partial}{\partial t}\left[f'(u) - g'(v)\right] = a\left(\frac{df'(u)}{du}\frac{\partial u}{\partial t} - \frac{dg'(v)}{dv}\frac{\partial v}{\partial t}\right) = a^2 f''(u) + a^2 g''(v).$$

Similarly $\dfrac{\partial z}{\partial x} = f'(u) + g'(v)$ and $\dfrac{\partial^2 z}{\partial x^2} = f''(u) + g''(v)$. Thus $\dfrac{\partial^2 z}{\partial t^2} = a^2\dfrac{\partial^2 z}{\partial x^2}$.

51. $\dfrac{\partial z}{\partial s} = \dfrac{\partial z}{\partial x}\,2s + \dfrac{\partial z}{\partial y}\,2r$. Then

$$\frac{\partial^2 z}{\partial r\,\partial s} = \frac{\partial}{\partial r}\left(\frac{\partial z}{\partial x}\,2s\right) + \frac{\partial}{\partial r}\left(\frac{\partial z}{\partial y}\,2r\right)$$

$$= \frac{\partial^2 z}{\partial x^2}\frac{\partial x}{\partial r}\,2s + \frac{\partial}{\partial y}\left(\frac{\partial z}{\partial x}\right)\frac{\partial y}{\partial r}\,2s + \frac{\partial z}{\partial x}\frac{\partial}{\partial r}\,2s + \frac{\partial^2 z}{\partial y^2}\frac{\partial y}{\partial r}\,2r + \frac{\partial}{\partial x}\left(\frac{\partial z}{\partial y}\right)\frac{\partial x}{\partial r}\,2r + \frac{\partial z}{\partial y}\,2$$

$$= 4rs\frac{\partial^2 z}{\partial x^2} + \frac{\partial^2 z}{\partial y\,\partial x}\,4s^2 + 0 + 4rs\frac{\partial^2 z}{\partial y^2} + \frac{\partial^2 z}{\partial x\,\partial y}\,4r^2 + 2\frac{\partial z}{\partial y}$$

By the continuity of the partials, $\dfrac{\partial^2 z}{\partial r\,\partial s} = 4rs\dfrac{\partial^2 z}{\partial x^2} + 4rs\dfrac{\partial^2 z}{\partial y^2} + (4r^2 + 4s^2)\dfrac{\partial^2 z}{\partial x\,\partial y} + 2\dfrac{\partial z}{\partial y}$.

53. $\dfrac{\partial z}{\partial r} = \dfrac{\partial z}{\partial x}\cos\theta + \dfrac{\partial z}{\partial y}\sin\theta$ and $\dfrac{\partial z}{\partial\theta} = -\dfrac{\partial z}{\partial x}\,r\sin\theta + \dfrac{\partial z}{\partial y}\,r\cos\theta$. Then

$$\frac{\partial^2 z}{\partial r^2} = \cos\theta\left(\frac{\partial^2 z}{\partial x^2}\cos\theta + \frac{\partial^2 z}{\partial y\,\partial x}\sin\theta\right) + \sin\theta\left(\frac{\partial^2 z}{\partial y^2}\sin\theta + \frac{\partial^2 z}{\partial x\,\partial y}\cos\theta\right)$$

$$= \cos^2\theta\frac{\partial^2 z}{\partial x^2} + 2\cos\theta\sin\theta\frac{\partial^2 z}{\partial x\,\partial y} + \sin^2\theta\frac{\partial^2 z}{\partial y^2}$$

and

$$\frac{\partial^2 z}{\partial\theta^2} = -r\cos\theta\frac{\partial z}{\partial x} + (-r\sin\theta)\left(\frac{\partial^2 z}{\partial x^2}(-r\sin\theta) + \frac{\partial^2 z}{\partial y\,\partial x}\,r\cos\theta\right)$$

$$-r\sin\theta\frac{\partial z}{\partial y} + r\cos\theta\left(\frac{\partial^2 z}{\partial y^2}\,r\cos\theta + \frac{\partial^2 z}{\partial x\,\partial y}(-r\sin\theta)\right)$$

$$= -r\cos\theta\frac{\partial z}{\partial x} - r\sin\theta\frac{\partial z}{\partial y} + r^2\sin^2\theta\frac{\partial^2 z}{\partial x^2} - 2r^2\cos\theta\sin\theta\frac{\partial^2 z}{\partial x\,\partial y} + r^2\cos^2\theta\frac{\partial^2 z}{\partial y^2}$$

Thus

$$\frac{\partial^2 z}{\partial r^2} + \frac{1}{r^2}\frac{\partial^2 z}{\partial\theta^2} + \frac{1}{r}\frac{\partial z}{\partial r} = (\cos^2\theta + \sin^2\theta)\frac{\partial^2 z}{\partial x^2} + (\sin^2\theta + \cos^2\theta)\frac{\partial^2 z}{\partial y^2}$$

$$-\frac{1}{r}\cos\theta\frac{\partial z}{\partial x} - \frac{1}{r}\sin\theta\frac{\partial z}{\partial y} + \frac{1}{r}\left(\cos\theta\frac{\partial z}{\partial x} + \sin\theta\frac{\partial z}{\partial y}\right)$$

$$= \frac{\partial^2 z}{\partial x^2} + \frac{\partial^2 z}{\partial y^2} \text{ as desired.}$$

55. (a) Since f is a polynomial, it has continuous second-order partial derivatives, and

$$f(tx, ty) = (tx)^2(ty) + 2(tx)(ty)^2 + 5(ty)^3 = t^3 x^2 y + 2t^3 xy^2 + 5t^3 y^3 = t^3(x^2 y + 2xy^2 + 5y^3) = t^3 f(x, y).$$

Thus, f is homogeneous of degree 3.

© 2012 Cengage Learning. All Rights Reserved. May not be scanned, copied, or duplicated, or posted to a publicly accessible website, in whole or in part.

(b) Differentiating both sides of $f(tx, ty) = t^n f(x, y)$ with respect to t using the Chain Rule, we get

$$\frac{\partial}{\partial t} f(tx, ty) = \frac{\partial}{\partial t} [t^n f(x, y)] \quad \Leftrightarrow$$

$$\frac{\partial}{\partial (tx)} f(tx, ty) \cdot \frac{\partial (tx)}{\partial t} + \frac{\partial}{\partial (ty)} f(tx, ty) \cdot \frac{\partial (ty)}{\partial t} = x \frac{\partial}{\partial (tx)} f(tx, ty) + y \frac{\partial}{\partial (ty)} f(tx, ty) = nt^{n-1} f(x, y).$$

Setting $t = 1$: $x \frac{\partial}{\partial x} f(x, y) + y \frac{\partial}{\partial y} f(x, y) = nf(x, y).$

57. Differentiating both sides of $f(tx, ty) = t^n f(x, y)$ with respect to x using the Chain Rule, we get

$$\frac{\partial}{\partial x} f(tx, ty) = \frac{\partial}{\partial x} [t^n f(x, y)] \quad \Leftrightarrow$$

$$\frac{\partial}{\partial (tx)} f(tx, ty) \cdot \frac{\partial (tx)}{\partial x} + \frac{\partial}{\partial (ty)} f(tx, ty) \cdot \frac{\partial (ty)}{\partial x} = t^n \frac{\partial}{\partial x} f(x, y) \quad \Leftrightarrow \quad tf_x(tx, ty) = t^n f_x(x, y).$$

Thus $f_x(tx, ty) = t^{n-1} f_x(x, y).$

59. Given a function defined implicitly by $F(x, y) = 0$, where F is differentiable and $F_y \neq 0$, we know that $\frac{dy}{dx} = -\frac{F_x}{F_y}$. Let

$G(x, y) = -\frac{F_x}{F_y}$ so $\frac{dy}{dx} = G(x, y)$. Differentiating both sides with respect to x and using the Chain Rule gives

$$\frac{d^2 y}{dx^2} = \frac{\partial G}{\partial x} \frac{dx}{dx} + \frac{\partial G}{\partial y} \frac{dy}{dx} \text{ where } \frac{\partial G}{\partial x} = \frac{\partial}{\partial x} \left(-\frac{F_x}{F_y} \right) = -\frac{F_y F_{xx} - F_x F_{yx}}{F_y^2}, \frac{\partial G}{\partial y} = \frac{\partial}{\partial y} \left(-\frac{F_x}{F_y} \right) = -\frac{F_y F_{xy} - F_x F_{yy}}{F_y^2}.$$

Thus

$$\frac{d^2 y}{dx^2} = \left(-\frac{F_y F_{xx} - F_x F_{yx}}{F_y^2} \right) (1) + \left(-\frac{F_y F_{xy} - F_x F_{yy}}{F_y^2} \right) \left(-\frac{F_x}{F_y} \right)$$

$$= -\frac{F_{xx} F_y^2 - F_{yx} F_x F_y - F_{xy} F_y F_x + F_{yy} F_x^2}{F_y^3}$$

But F has continuous second derivatives, so by Clauraut's Theorem, $F_{yx} = F_{xy}$ and we have

$$\frac{d^2 y}{dx^2} = -\frac{F_{xx} F_y^2 - 2F_{xy} F_x F_y + F_{yy} F_x^2}{F_y^3} \text{ as desired.}$$

14.6   Directional Derivatives and the Gradient Vector

1. We can approximate the directional derivative of the pressure function at K in the direction of S by the average rate of change of pressure between the points where the red line intersects the contour lines closest to K (extend the red line slightly at the left). In the direction of S, the pressure changes from 1000 millibars to 996 millibars and we estimate the distance between these two points to be approximately 50 km (using the fact that the distance from K to S is 300 km). Then the rate of change of pressure in the direction given is approximately $\frac{996 - 1000}{50} = -0.08$ millibar/km.

3. $D_{\mathbf{u}} f(-20, 30) = \nabla f(-20, 30) \cdot \mathbf{u} = f_T(-20, 30) \left(\frac{1}{\sqrt{2}} \right) + f_v(-20, 30) \left(\frac{1}{\sqrt{2}} \right).$

$f_T(-20, 30) = \lim\limits_{h \to 0} \frac{f(-20 + h, 30) - f(-20, 30)}{h}$, so we can approximate $f_T(-20, 30)$ by considering $h = \pm 5$ and

© 2012 Cengage Learning. All Rights Reserved. May not be scanned, copied, or duplicated, or posted to a publicly accessible website, in whole or in part.

using the values given in the table: $f_T(-20, 30) \approx \dfrac{f(-15, 30) - f(-20, 30)}{5} = \dfrac{-26 - (-33)}{5} = 1.4$,

$f_T(-20, 30) \approx \dfrac{f(-25, 30) - f(-20, 30)}{-5} = \dfrac{-39 - (-33)}{-5} = 1.2$. Averaging these values gives $f_T(-20, 30) \approx 1.3$.

Similarly, $f_v(-20, 30) = \lim\limits_{h \to 0} \dfrac{f(-20, 30 + h) - f(-20, 30)}{h}$, so we can approximate $f_v(-20, 30)$ with $h = \pm 10$:

$f_v(-20, 30) \approx \dfrac{f(-20, 40) - f(-20, 30)}{10} = \dfrac{-34 - (-33)}{10} = -0.1$,

$f_v(-20, 30) \approx \dfrac{f(-20, 20) - f(-20, 30)}{-10} = \dfrac{-30 - (-33)}{-10} = -0.3$. Averaging these values gives $f_v(-20, 30) \approx -0.2$.

Then $D_{\mathbf{u}} f(-20, 30) \approx 1.3\left(\frac{1}{\sqrt{2}}\right) + (-0.2)\left(\frac{1}{\sqrt{2}}\right) \approx 0.778$.

5. $f(x, y) = ye^{-x} \Rightarrow f_x(x, y) = -ye^{-x}$ and $f_y(x, y) = e^{-x}$. If $\mathbf{u}$ is a unit vector in the direction of $\theta = 2\pi/3$, then

from Equation 6, $D_{\mathbf{u}} f(0, 4) = f_x(0, 4) \cos\left(\frac{2\pi}{3}\right) + f_y(0, 4) \sin\left(\frac{2\pi}{3}\right) = -4 \cdot \left(-\frac{1}{2}\right) + 1 \cdot \frac{\sqrt{3}}{2} = 2 + \frac{\sqrt{3}}{2}$.

7. $f(x, y) = \sin(2x + 3y)$

(a) $\nabla f(x, y) = \dfrac{\partial f}{\partial x}\mathbf{i} + \dfrac{\partial f}{\partial y}\mathbf{j} = [\cos(2x + 3y) \cdot 2]\mathbf{i} + [\cos(2x + 3y) \cdot 3]\mathbf{j} = 2\cos(2x + 3y)\mathbf{i} + 3\cos(2x + 3y)\mathbf{j}$

(b) $\nabla f(-6, 4) = (2\cos 0)\mathbf{i} + (3\cos 0)\mathbf{j} = 2\mathbf{i} + 3\mathbf{j}$

(c) By Equation 9, $D_{\mathbf{u}} f(-6, 4) = \nabla f(-6, 4) \cdot \mathbf{u} = (2\mathbf{i} + 3\mathbf{j}) \cdot \frac{1}{2}(\sqrt{3}\mathbf{i} - \mathbf{j}) = \frac{1}{2}(2\sqrt{3} - 3) = \sqrt{3} - \frac{3}{2}$.

9. $f(x, y, z) = x^2yz - xyz^3$

(a) $\nabla f(x, y, z) = \langle f_x(x, y, z), f_y(x, y, z), f_z(x, y, z)\rangle = \langle 2xyz - yz^3, x^2z - xz^3, x^2y - 3xyz^2\rangle$

(b) $\nabla f(2, -1, 1) = \langle -4 + 1, 4 - 2, -4 + 6\rangle = \langle -3, 2, 2\rangle$

(c) By Equation 14, $D_{\mathbf{u}} f(2, -1, 1) = \nabla f(2, -1, 1) \cdot \mathbf{u} = \langle -3, 2, 2\rangle \cdot \langle 0, \frac{4}{5}, -\frac{3}{5}\rangle = 0 + \frac{8}{5} - \frac{6}{5} = \frac{2}{5}$.

11. $f(x, y) = e^x \sin y \Rightarrow \nabla f(x, y) = \langle e^x \sin y, e^x \cos y\rangle$, $\nabla f(0, \pi/3) = \left\langle \frac{\sqrt{3}}{2}, \frac{1}{2}\right\rangle$, and a

unit vector in the direction of $\mathbf{v}$ is $\mathbf{u} = \dfrac{1}{\sqrt{(-6)^2 + 8^2}}\langle -6, 8\rangle = \frac{1}{10}\langle -6, 8\rangle = \langle -\frac{3}{5}, \frac{4}{5}\rangle$, so

$D_{\mathbf{u}} f(0, \pi/3) = \nabla f(0, \pi/3) \cdot \mathbf{u} = \left\langle \frac{\sqrt{3}}{2}, \frac{1}{2}\right\rangle \cdot \langle -\frac{3}{5}, \frac{4}{5}\rangle = -\frac{3\sqrt{3}}{10} + \frac{4}{10} = \frac{4 - 3\sqrt{3}}{10}$.

13. $g(p, q) = p^4 - p^2q^3 \Rightarrow \nabla g(p, q) = (4p^3 - 2pq^3)\mathbf{i} + (-3p^2q^2)\mathbf{j}$, $\nabla g(2, 1) = 28\mathbf{i} - 12\mathbf{j}$, and a unit

vector in the direction of $\mathbf{v}$ is $\mathbf{u} = \dfrac{1}{\sqrt{1^2 + 3^2}}(\mathbf{i} + 3\mathbf{j}) = \frac{1}{\sqrt{10}}(\mathbf{i} + 3\mathbf{j})$, so

$D_{\mathbf{u}} g(2, 1) = \nabla g(2, 1) \cdot \mathbf{u} = (28\mathbf{i} - 12\mathbf{j}) \cdot \frac{1}{\sqrt{10}}(\mathbf{i} + 3\mathbf{j}) = \frac{1}{\sqrt{10}}(28 - 36) = -\frac{8}{\sqrt{10}}$ or $-\frac{4\sqrt{10}}{5}$.

15. $f(x, y, z) = xe^y + ye^z + ze^x \Rightarrow \nabla f(x, y, z) = \langle e^y + ze^x, xe^y + e^z, ye^z + e^x\rangle$, $\nabla f(0, 0, 0) = \langle 1, 1, 1\rangle$, and a unit

vector in the direction of $\mathbf{v}$ is $\mathbf{u} = \dfrac{1}{\sqrt{25 + 1 + 4}}\langle 5, 1, -2\rangle = \frac{1}{\sqrt{30}}\langle 5, 1, -2\rangle$, so

$D_{\mathbf{u}} f(0, 0, 0) = \nabla f(0, 0, 0) \cdot \mathbf{u} = \langle 1, 1, 1\rangle \cdot \frac{1}{\sqrt{30}}\langle 5, 1, -2\rangle = \frac{4}{\sqrt{30}}$.

© 2012 Cengage Learning. All Rights Reserved. May not be scanned, copied, or duplicated, or posted to a publicly accessible website, in whole or in part.

17. $h(r, s, t) = \ln(3r + 6s + 9t) \Rightarrow \nabla h(r, s, t) = \langle 3/(3r + 6s + 9t), 6/(3r + 6s + 9t), 9/(3r + 6s + 9t) \rangle$,

$\nabla h(1, 1, 1) = \langle \frac{1}{6}, \frac{1}{3}, \frac{1}{2} \rangle$, and a unit vector in the direction of $\mathbf{v} = 4\mathbf{i} + 12\mathbf{j} + 6\mathbf{k}$

is $\mathbf{u} = \frac{1}{\sqrt{16+144+36}} (4\mathbf{i} + 12\mathbf{j} + 6\mathbf{k}) = \frac{2}{7}\mathbf{i} + \frac{6}{7}\mathbf{j} + \frac{3}{7}\mathbf{k}$, so

$D_\mathbf{u} h(1, 1, 1) = \nabla h(1, 1, 1) \cdot \mathbf{u} = \langle \frac{1}{6}, \frac{1}{3}, \frac{1}{2} \rangle \cdot \langle \frac{2}{7}, \frac{6}{7}, \frac{3}{7} \rangle = \frac{1}{21} + \frac{2}{7} + \frac{3}{14} = \frac{23}{42}$.

19. $f(x, y) = \sqrt{xy} \Rightarrow \nabla f(x, y) = \left\langle \frac{1}{2}(xy)^{-1/2}(y), \frac{1}{2}(xy)^{-1/2}(x) \right\rangle = \left\langle \dfrac{y}{2\sqrt{xy}}, \dfrac{x}{2\sqrt{xy}} \right\rangle$, so $\nabla f(2, 8) = \langle 1, \frac{1}{4} \rangle$.

The unit vector in the direction of $\overrightarrow{PQ} = \langle 5 - 2, 4 - 8 \rangle = \langle 3, -4 \rangle$ is $\mathbf{u} = \langle \frac{3}{5}, -\frac{4}{5} \rangle$, so

$D_\mathbf{u} f(2, 8) = \nabla f(2, 8) \cdot \mathbf{u} = \langle 1, \frac{1}{4} \rangle \cdot \langle \frac{3}{5}, -\frac{4}{5} \rangle = \frac{2}{5}$.

21. $f(x, y) = 4y\sqrt{x} \Rightarrow \nabla f(x, y) = \left\langle 4y \cdot \frac{1}{2}x^{-1/2}, 4\sqrt{x} \right\rangle = \langle 2y/\sqrt{x}, 4\sqrt{x} \rangle$.

$\nabla f(4, 1) = \langle 1, 8 \rangle$ is the direction of maximum rate of change, and the maximum rate is $|\nabla f(4, 1)| = \sqrt{1 + 64} = \sqrt{65}$.

23. $f(x, y) = \sin(xy) \Rightarrow \nabla f(x, y) = \langle y \cos(xy), x \cos(xy) \rangle$, $\nabla f(1, 0) = \langle 0, 1 \rangle$. Thus the maximum rate of change is

$|\nabla f(1, 0)| = 1$ in the direction $\langle 0, 1 \rangle$.

25. $f(x, y, z) = \sqrt{x^2 + y^2 + z^2} \Rightarrow$

$\nabla f(x, y, z) = \left\langle \frac{1}{2}(x^2 + y^2 + z^2)^{-1/2} \cdot 2x, \frac{1}{2}(x^2 + y^2 + z^2)^{-1/2} \cdot 2y, \frac{1}{2}(x^2 + y^2 + z^2)^{-1/2} \cdot 2z \right\rangle$

$= \left\langle \dfrac{x}{\sqrt{x^2 + y^2 + z^2}}, \dfrac{y}{\sqrt{x^2 + y^2 + z^2}}, \dfrac{z}{\sqrt{x^2 + y^2 + z^2}} \right\rangle$,

$\nabla f(3, 6, -2) = \left\langle \frac{3}{\sqrt{49}}, \frac{6}{\sqrt{49}}, \frac{-2}{\sqrt{49}} \right\rangle = \langle \frac{3}{7}, \frac{6}{7}, -\frac{2}{7} \rangle$. Thus the maximum rate of change is

$|\nabla f(3, 6, -2)| = \sqrt{\left(\frac{3}{7}\right)^2 + \left(\frac{6}{7}\right)^2 + \left(-\frac{2}{7}\right)^2} = \sqrt{\frac{9 + 36 + 4}{49}} = 1$ in the direction $\langle \frac{3}{7}, \frac{6}{7}, -\frac{2}{7} \rangle$ or equivalently $\langle 3, 6, -2 \rangle$.

27. (a) As in the proof of Theorem 15, $D_\mathbf{u} f = |\nabla f| \cos \theta$. Since the minimum value of $\cos \theta$ is -1 occurring when $\theta = \pi$, the

minimum value of $D_\mathbf{u} f$ is $-|\nabla f|$ occurring when $\theta = \pi$, that is when $\mathbf{u}$ is in the opposite direction of ∇f

(assuming $\nabla f \neq \mathbf{0}$).

(b) $f(x, y) = x^4 y - x^2 y^3 \Rightarrow \nabla f(x, y) = \langle 4x^3 y - 2xy^3, x^4 - 3x^2 y^2 \rangle$, so f decreases fastest at the point $(2, -3)$ in the

direction $-\nabla f(2, -3) = -\langle 12, -92 \rangle = \langle -12, 92 \rangle$.

29. The direction of fastest change is $\nabla f(x, y) = (2x - 2)\mathbf{i} + (2y - 4)\mathbf{j}$, so we need to find all points (x, y) where $\nabla f(x, y)$ is

parallel to $\mathbf{i} + \mathbf{j} \Leftrightarrow (2x - 2)\mathbf{i} + (2y - 4)\mathbf{j} = k(\mathbf{i} + \mathbf{j}) \Leftrightarrow k = 2x - 2$ and $k = 2y - 4$. Then $2x - 2 = 2y - 4 \Rightarrow$

$y = x + 1$, so the direction of fastest change is $\mathbf{i} + \mathbf{j}$ at all points on the line $y = x + 1$.

© 2012 Cengage Learning. All Rights Reserved. May not be scanned, copied, or duplicated, or posted to a publicly accessible website, in whole or in part.

31. $T = \dfrac{k}{\sqrt{x^2 + y^2 + z^2}}$ and $120 = T(1, 2, 2) = \dfrac{k}{3}$ so $k = 360$.

(a) $\mathbf{u} = \dfrac{\langle 1, -1, 1 \rangle}{\sqrt{3}}$,

$$D_{\mathbf{u}}T(1, 2, 2) = \nabla T(1, 2, 2) \cdot \mathbf{u} = \left[-360(x^2 + y^2 + z^2)^{-3/2} \langle x, y, z \rangle \right]_{(1,2,2)} \cdot \mathbf{u} = -\tfrac{40}{3} \langle 1, 2, 2 \rangle \cdot \tfrac{1}{\sqrt{3}} \langle 1, -1, 1 \rangle = -\tfrac{40}{3\sqrt{3}}$$

(b) From (a), $\nabla T = -360(x^2 + y^2 + z^2)^{-3/2} \langle x, y, z \rangle$, and since $\langle x, y, z \rangle$ is the position vector of the point (x, y, z), the

vector $- \langle x, y, z \rangle$, and thus ∇T, always points toward the origin.

33. $\nabla V(x, y, z) = \langle 10x - 3y + yz, xz - 3x, xy \rangle$, $\nabla V(3, 4, 5) = \langle 38, 6, 12 \rangle$

(a) $D_{\mathbf{u}} V(3, 4, 5) = \langle 38, 6, 12 \rangle \cdot \tfrac{1}{\sqrt{3}} \langle 1, 1, -1 \rangle = \dfrac{32}{\sqrt{3}}$

(b) $\nabla V(3, 4, 5) = \langle 38, 6, 12 \rangle$, or equivalently, $\langle 19, 3, 6 \rangle$.

(c) $|\nabla V(3, 4, 5)| = \sqrt{38^2 + 6^2 + 12^2} = \sqrt{1624} = 2\sqrt{406}$

35. A unit vector in the direction of $\overrightarrow{AB}$ is $\mathbf{i}$ and a unit vector in the direction of $\overrightarrow{AC}$ is $\mathbf{j}$. Thus $D_{\overrightarrow{AB}} f(1, 3) = f_x(1, 3) = 3$ and

$D_{\overrightarrow{AC}} f(1, 3) = f_y(1, 3) = 26$. Therefore $\nabla f(1, 3) = \langle f_x(1, 3), f_y(1, 3) \rangle = \langle 3, 26 \rangle$, and by definition,

$D_{\overrightarrow{AD}} f(1, 3) = \nabla f \cdot \mathbf{u}$ where $\mathbf{u}$ is a unit vector in the direction of $\overrightarrow{AD}$, which is $\left\langle \tfrac{5}{13}, \tfrac{12}{13} \right\rangle$. Therefore,

$D_{\overrightarrow{AD}} f(1, 3) = \langle 3, 26 \rangle \cdot \left\langle \tfrac{5}{13}, \tfrac{12}{13} \right\rangle = 3 \cdot \tfrac{5}{13} + 26 \cdot \tfrac{12}{13} = \dfrac{327}{13}$.

37. (a) $\nabla(au + bv) = \left\langle \dfrac{\partial(au + bv)}{\partial x}, \dfrac{\partial(au + bv)}{\partial y} \right\rangle = \left\langle a\dfrac{\partial u}{\partial x} + b\dfrac{\partial v}{\partial x}, a\dfrac{\partial u}{\partial y} + b\dfrac{\partial v}{\partial y} \right\rangle = a\left\langle \dfrac{\partial u}{\partial x}, \dfrac{\partial u}{\partial y} \right\rangle + b\left\langle \dfrac{\partial v}{\partial x}, \dfrac{\partial v}{\partial y} \right\rangle$

$\qquad = a\,\nabla u + b\,\nabla v$

(b) $\nabla(uv) = \left\langle v\dfrac{\partial u}{\partial x} + u\dfrac{\partial v}{\partial x}, v\dfrac{\partial u}{\partial y} + u\dfrac{\partial v}{\partial y} \right\rangle = v\left\langle \dfrac{\partial u}{\partial x}, \dfrac{\partial u}{\partial y} \right\rangle + u\left\langle \dfrac{\partial v}{\partial x}, \dfrac{\partial v}{\partial y} \right\rangle = v\,\nabla u + u\,\nabla v$

(c) $\nabla\left(\dfrac{u}{v}\right) = \left\langle \dfrac{v\dfrac{\partial u}{\partial x} - u\dfrac{\partial v}{\partial x}}{v^2}, \dfrac{v\dfrac{\partial u}{\partial y} - u\dfrac{\partial v}{\partial y}}{v^2} \right\rangle = \dfrac{v\left\langle \dfrac{\partial u}{\partial x}, \dfrac{\partial u}{\partial y} \right\rangle - u\left\langle \dfrac{\partial v}{\partial x}, \dfrac{\partial v}{\partial y} \right\rangle}{v^2} = \dfrac{v\,\nabla u - u\,\nabla v}{v^2}$

(d) $\nabla u^n = \left\langle \dfrac{\partial(u^n)}{\partial x}, \dfrac{\partial(u^n)}{\partial y} \right\rangle = \left\langle nu^{n-1}\dfrac{\partial u}{\partial x}, nu^{n-1}\dfrac{\partial u}{\partial y} \right\rangle = nu^{n-1}\,\nabla u$

39. $f(x, y) = x^3 + 5x^2 y + y^3 \quad \Rightarrow$

$D_{\mathbf{u}} f(x, y) = \nabla f(x, y) \cdot \mathbf{u} = \langle 3x^2 + 10xy, 5x^2 + 3y^2 \rangle \cdot \langle \tfrac{3}{5}, \tfrac{4}{5} \rangle = \tfrac{9}{5}x^2 + 6xy + 4x^2 + \tfrac{12}{5}y^2 = \tfrac{29}{5}x^2 + 6xy + \tfrac{12}{5}y^2$. Then

$D_{\mathbf{u}}^2 f(x, y) = D_{\mathbf{u}}[D_{\mathbf{u}} f(x, y)] = \nabla[D_{\mathbf{u}} f(x, y)] \cdot \mathbf{u} = \langle \tfrac{58}{5}x + 6y, 6x + \tfrac{24}{5}y \rangle \cdot \langle \tfrac{3}{5}, \tfrac{4}{5} \rangle$

$\qquad = \tfrac{174}{25}x + \tfrac{18}{5}y + \tfrac{24}{5}x + \tfrac{96}{25}y = \tfrac{294}{25}x + \tfrac{186}{25}y$

and $D_{\mathbf{u}}^2 f(2, 1) = \tfrac{294}{25}(2) + \tfrac{186}{25}(1) = \dfrac{774}{25}$.

© 2012 Cengage Learning. All Rights Reserved. May not be scanned, copied, or duplicated, or posted to a publicly accessible website, in whole or in part.

41. Let $F(x, y, z) = 2(x - 2)^2 + (y - 1)^2 + (z - 3)^2$. Then $2(x - 2)^2 + (y - 1)^2 + (z - 3)^2 = 10$ is a level surface of F.

$F_x(x, y, z) = 4(x - 2)$ $\Rightarrow$ $F_x(3, 3, 5) = 4$, $F_y(x, y, z) = 2(y - 1)$ $\Rightarrow$ $F_y(3, 3, 5) = 4$, and

$F_z(x, y, z) = 2(z - 3)$ $\Rightarrow$ $F_z(3, 3, 5) = 4$.

(a) Equation 19 gives an equation of the tangent plane at $(3, 3, 5)$ as $4(x - 3) + 4(y - 3) + 4(z - 5) = 0$ $\Leftrightarrow$

$4x + 4y + 4z = 44$ or equivalently $x + y + z = 11$.

(b) By Equation 20, the normal line has symmetric equations $\dfrac{x - 3}{4} = \dfrac{y - 3}{4} = \dfrac{z - 5}{4}$ or equivalently

$x - 3 = y - 3 = z - 5$. Corresponding parametric equations are $x = 3 + t$, $y = 3 + t$, $z = 5 + t$.

43. Let $F(x, y, z) = xyz^2$. Then $xyz^2 = 6$ is a level surface of F and $\nabla F(x, y, z) = \langle yz^2, xz^2, 2xyz \rangle$.

(a) $\nabla F(3, 2, 1) = \langle 2, 3, 12 \rangle$ is a normal vector for the tangent plane at $(3, 2, 1)$, so an equation of the tangent plane

is $2(x - 3) + 3(y - 2) + 12(z - 1) = 0$ or $2x + 3y + 12z = 24$.

(b) The normal line has direction $\langle 2, 3, 12 \rangle$, so parametric equations are $x = 3 + 2t$, $y = 2 + 3t$, $z = 1 + 12t$, and

symmetric equations are $\dfrac{x - 3}{2} = \dfrac{y - 2}{3} = \dfrac{z - 1}{12}$.

45. Let $F(x, y, z) = x + y + z - e^{xyz}$. Then $x + y + z = e^{xyz}$ is the level surface $F(x, y, z) = 0$,

and $\nabla F(x, y, z) = \langle 1 - yze^{xyz}, 1 - xze^{xyz}, 1 - xye^{xyz} \rangle$.

(a) $\nabla F(0, 0, 1) = \langle 1, 1, 1 \rangle$ is a normal vector for the tangent plane at $(0, 0, 1)$, so an equation of the tangent plane

is $1(x - 0) + 1(y - 0) + 1(z - 1) = 0$ or $x + y + z = 1$.

(b) The normal line has direction $\langle 1, 1, 1 \rangle$, so parametric equations are $x = t$, $y = t$, $z = 1 + t$, and symmetric equations are

$x = y = z - 1$.

47. $F(x, y, z) = xy + yz + zx$, $\nabla F(x, y, z) = \langle y + z, x + z, y + x \rangle$, $\nabla F(1, 1, 1) = \langle 2, 2, 2 \rangle$, so an equation of the tangent

plane is $2x + 2y + 2z = 6$ or $x + y + z = 3$, and the normal line is given by $x - 1 = y - 1 - z$, 1 or $x - y = z$. To graph

the surface we solve for z: $z = \dfrac{3 - xy}{x + y}$.

© 2012 Cengage Learning. All Rights Reserved. May not be scanned, copied, or duplicated, or posted to a publicly accessible website, in whole or in part.

49. $f(x, y) = xy \Rightarrow \nabla f(x, y) = \langle y, x \rangle$, $\nabla f(3, 2) = \langle 2, 3 \rangle$. $\nabla f(3, 2)$

is perpendicular to the tangent line, so the tangent line has equation

$\nabla f(3, 2) \cdot \langle x - 3, y - 2 \rangle = 0 \Rightarrow \langle 2, 3 \rangle \cdot \langle x - 3, x - 2 \rangle = 0 \Rightarrow$

$2(x - 3) + 3(y - 2) = 0$ or $2x + 3y = 12$.

51. $\nabla F(x_0, y_0, z_0) = \left\langle \dfrac{2x_0}{a^2}, \dfrac{2y_0}{b^2}, \dfrac{2z_0}{c^2} \right\rangle$. Thus an equation of the tangent plane at (x_0, y_0, z_0) is

$\dfrac{2x_0}{a^2} x + \dfrac{2y_0}{b^2} y + \dfrac{2z_0}{c^2} z = 2 \left(\dfrac{x_0^2}{a^2} + \dfrac{y_0^2}{b^2} + \dfrac{z_0^2}{c^2} \right) = 2(1) = 2$ since (x_0, y_0, z_0) is a point on the ellipsoid. Hence

$\dfrac{x_0}{a^2} x + \dfrac{y_0}{b^2} y + \dfrac{z_0}{c^2} z = 1$ is an equation of the tangent plane.

53. $\nabla F(x_0, y_0, z_0) = \left\langle \dfrac{2x_0}{a^2}, \dfrac{2y_0}{b^2}, \dfrac{-1}{c} \right\rangle$, so an equation of the tangent plane is $\dfrac{2x_0}{a^2} x + \dfrac{2y_0}{b^2} y - \dfrac{1}{c} z = \dfrac{2x_0^2}{a^2} + \dfrac{2y_0^2}{b^2} - \dfrac{z_0}{c}$

or $\dfrac{2x_0}{a^2} x + \dfrac{2y_0}{b^2} y = \dfrac{z}{c} + 2 \left(\dfrac{x_0^2}{a^2} + \dfrac{y_0^2}{b^2} \right) - \dfrac{z_0}{c}$. But $\dfrac{z_0}{c} = \dfrac{x_0^2}{a^2} + \dfrac{y_0^2}{b^2}$, so the equation can be written as

$\dfrac{2x_0}{a^2} x + \dfrac{2y_0}{b^2} y = \dfrac{z + z_0}{c}$.

55. The hyperboloid $x^2 - y^2 - z^2 = 1$ is a level surface of $F(x, y, z) = x^2 - y^2 - z^2$ and $\nabla F(x, y, z) = \langle 2x, -2y, -2z \rangle$ is a

normal vector to the surface and hence a normal vector for the tangent plane at (x, y, z). The tangent plane is parallel to the

plane $z = x + y$ or $x + y - z = 0$ if and only if the corresponding normal vectors are parallel, so we need a point (x_0, y_0, z_0)

on the hyperboloid where $\langle 2x_0, -2y_0, -2z_0 \rangle = c \langle 1, 1, -1 \rangle$ or equivalently $\langle x_0, -y_0, -z_0 \rangle = k \langle 1, 1, -1 \rangle$ for some $k \neq 0$.

Then we must have $x_0 = k$, $y_0 = -k$, $z_0 = k$ and substituting into the equation of the hyperboloid gives

$k^2 - (-k)^2 - k^2 = 1 \Leftrightarrow -k^2 = 1$, an impossibility. Thus there is no such point on the hyperboloid.

57. Let (x_0, y_0, z_0) be a point on the cone [other than $(0, 0, 0)$]. The cone is a level surface of $F(x, y, z) = x^2 + y^2 - z^2$ and

$\nabla F(x, y, z) = \langle 2x, 2y, -2z \rangle$, so $\nabla F(x_0, y_0, z_0) = \langle 2x_0, 2y_0, -2z_0 \rangle$ is a normal vector to the cone at this point and an

equation of the tangent plane there is $2x_0 (x - x_0) + 2y_0 (y - y_0) - 2z_0 (z - z_0) = 0$ or $x_0 x + y_0 y - z_0 z = x_0^2 + y_0^2 - z_0^2$.

But $x_0^2 + y_0^2 = z_0^2$ so the tangent plane is given by $x_0 x + y_0 y - z_0 z = 0$, a plane which always contains the origin.

59. Let $F(x, y, z) = x^2 + y^2 - z$. Then the paraboloid is the level surface $F(x, y, z) = 0$ and $\nabla F(x, y, z) = \langle 2x, 2y, -1 \rangle$, so

$\nabla F(1, 1, 2) = \langle 2, 2, -1 \rangle$ is a normal vector to the surface. Thus the normal line at $(1, 1, 2)$ is given by $x = 1 + 2t$,

$y = 1 + 2t$, $z = 2 - t$. Substitution into the equation of the paraboloid $z = x^2 + y^2$ gives $2 - t = (1 + 2t)^2 + (1 + 2t)^2 \Leftrightarrow$

$2 - t = 2 + 8t + 8t^2 \Leftrightarrow 8t^2 + 9t = 0 \Leftrightarrow t(8t + 9) = 0$. Thus the line intersects the paraboloid when $t = 0$,

corresponding to the given point $(1, 1, 2)$, or when $t = -\dfrac{9}{8}$, corresponding to the point $\left(-\dfrac{5}{4}, -\dfrac{5}{4}, \dfrac{25}{8} \right)$.

© 2012 Cengage Learning. All Rights Reserved. May not be scanned, copied, or duplicated, or posted to a publicly accessible website, in whole or in part.

61. Let (x_0, y_0, z_0) be a point on the surface. Then an equation of the tangent plane at the point is

$$\frac{x}{2\sqrt{x_0}} + \frac{y}{2\sqrt{y_0}} + \frac{z}{2\sqrt{z_0}} = \frac{\sqrt{x_0} + \sqrt{y_0} + \sqrt{z_0}}{2}. \text{ But } \sqrt{x_0} + \sqrt{y_0} + \sqrt{z_0} = \sqrt{c}, \text{ so the equation is}$$

$$\frac{x}{\sqrt{x_0}} + \frac{y}{\sqrt{y_0}} + \frac{z}{\sqrt{z_0}} = \sqrt{c}.$$ The x-, y-, and z-intercepts are $\sqrt{cx_0}$, $\sqrt{cy_0}$ and $\sqrt{cz_0}$ respectively. (The x-intercept is found by

setting $y = z = 0$ and solving the resulting equation for x, and the y- and z-intercepts are found similarly.) So the sum of the

intercepts is $\sqrt{c}\left(\sqrt{x_0} + \sqrt{y_0} + \sqrt{z_0}\right) = c$, a constant.

63. If $f(x, y, z) = z - x^2 - y^2$ and $g(x, y, z) = 4x^2 + y^2 + z^2$, then the tangent line is perpendicular to both ∇f and ∇g

at $(-1, 1, 2)$. The vector $\mathbf{v} = \nabla f \times \nabla g$ will therefore be parallel to the tangent line.

We have $\nabla f(x, y, z) = \langle -2x, -2y, 1 \rangle \;\Rightarrow\; \nabla f(-1, 1, 2) = \langle 2, -2, 1 \rangle$, and $\nabla g(x, y, z) = \langle 8x, 2y, 2z \rangle \;\Rightarrow\;$

$$\nabla g(-1, 1, 2) = \langle -8, 2, 4 \rangle. \text{ Hence } \mathbf{v} = \nabla f \times \nabla g = \begin{vmatrix} \mathbf{i} & \mathbf{j} & \mathbf{k} \\ 2 & -2 & 1 \\ -8 & 2 & 4 \end{vmatrix} = -10\,\mathbf{i} - 16\,\mathbf{j} - 12\,\mathbf{k}.$$

Parametric equations are: $x = -1 - 10t$, $y = 1 - 16t$, $z = 2 - 12t$.

65. (a) The direction of the normal line of F is given by ∇F, and that of G by ∇G. Assuming that

$\nabla F \neq 0 \neq \nabla G$, the two normal lines are perpendicular at P if $\nabla F \cdot \nabla G = 0$ at P $\Leftrightarrow$

$\langle \partial F/\partial x, \partial F/\partial y, \partial F/\partial z \rangle \cdot \langle \partial G/\partial x, \partial G/\partial y, \partial G/\partial z \rangle = 0$ at P $\Leftrightarrow$ $F_x G_x + F_y G_y + F_z G_z = 0$ at P.

(b) Here $F = x^2 + y^2 - z^2$ and $G = x^2 + y^2 + z^2 - r^2$, so

$\nabla F \cdot \nabla G = \langle 2x, 2y, -2z \rangle \cdot \langle 2x, 2y, 2z \rangle = 4x^2 + 4y^2 - 4z^2 = 4F = 0$, since the point (x, y, z) lies on the graph of

$F = 0$. To see that this is true without using calculus, note that $G = 0$ is the equation of a sphere centered at the origin and

$F = 0$ is the equation of a right circular cone with vertex at the origin (which is generated by lines through the origin). At

any point of intersection, the sphere's normal line (which passes through the origin) lies on the cone, and thus is

perpendicular to the cone's normal line. So the surfaces with equations $F = 0$ and $G = 0$ are everywhere orthogonal.

67. Let $\mathbf{u} = \langle a, b \rangle$ and $\mathbf{v} = \langle c, d \rangle$. Then we know that at the given point, $D_{\mathbf{u}} f = \nabla f \cdot \mathbf{u} = a f_x + b f_y$ and

$D_{\mathbf{v}} f = \nabla f \cdot \mathbf{v} = c f_x + d f_y$. But these are just two linear equations in the two unknowns f_x and f_y, and since $\mathbf{u}$ and $\mathbf{v}$ are

not parallel, we can solve the equations to find $\nabla f = \langle f_x, f_y \rangle$ at the given point. In fact,

$$\nabla f = \left\langle \frac{d\,D_{\mathbf{u}} f - b\,D_{\mathbf{v}} f}{ad - bc}, \frac{a\,D_{\mathbf{v}} f - c\,D_{\mathbf{u}} f}{ad - bc} \right\rangle.$$

© 2012 Cengage Learning. All Rights Reserved. May not be scanned, copied, or duplicated, or posted to a publicly accessible website, in whole or in part.

14.7 Maximum and Minimum Values

1. (a) First we compute $D(1, 1) = f_{xx}(1, 1) f_{yy}(1, 1) - [f_{xy}(1, 1)]^2 = (4)(2) - (1)^2 = 7$. Since $D(1, 1) > 0$ and

$f_{xx}(1, 1) > 0$, f has a local minimum at $(1, 1)$ by the Second Derivatives Test.

(b) $D(1, 1) = f_{xx}(1, 1) f_{yy}(1, 1) - [f_{xy}(1, 1)]^2 = (4)(2) - (3)^2 = -1$. Since $D(1, 1) < 0$, f has a saddle point at $(1, 1)$ by

the Second Derivatives Test.

3. In the figure, a point at approximately $(1, 1)$ is enclosed by level curves which are oval in shape and indicate that as we move

away from the point in any direction the values of f are increasing. Hence we would expect a local minimum at or near $(1, 1)$.

The level curves near $(0, 0)$ resemble hyperbolas, and as we move away from the origin, the values of f increase in some

directions and decrease in others, so we would expect to find a saddle point there.

To verify our predictions, we have $f(x, y) = 4 + x^3 + y^3 - 3xy \Rightarrow f_x(x, y) = 3x^2 - 3y$, $f_y(x, y) = 3y^2 - 3x$. We

have critical points where these partial derivatives are equal to 0: $3x^2 - 3y = 0$, $3y^2 - 3x = 0$. Substituting $y = x^2$ from the

first equation into the second equation gives $3(x^2)^2 - 3x = 0 \Rightarrow 3x(x^3 - 1) = 0 \Rightarrow x = 0$ or $x = 1$. Then we have

two critical points, $(0, 0)$ and $(1, 1)$. The second partial derivatives are $f_{xx}(x, y) = 6x$, $f_{xy}(x, y) = -3$, and $f_{yy}(x, y) = 6y$,

so $D(x, y) = f_{xx}(x, y) f_{yy}(x, y) - [f_{xy}(x, y)]^2 = (6x)(6y) - (-3)^2 = 36xy - 9$. Then $D(0, 0) = 36(0)(0) - 9 = -9$,

and $D(1, 1) = 36(1)(1) - 9 = 27$. Since $D(0, 0) < 0$, f has a saddle point at $(0, 0)$ by the Second Derivatives Test. Since

$D(1, 1) > 0$ and $f_{xx}(1, 1) > 0$, f has a local minimum at $(1, 1)$.

5. $f(x, y) = x^2 + xy + y^2 + y \Rightarrow f_x = 2x + y$, $f_y = x + 2y + 1$, $f_{xx} = 2$, $f_{xy} = 1$, $f_{yy} = 2$. Then $f_x = 0$ implies

$y = -2x$, and substitution into $f_y = x + 2y + 1 = 0$ gives $x + 2(-2x) + 1 = 0 \Rightarrow -3x = -1 \Rightarrow x = \frac{1}{3}$.

Then $y = -\frac{2}{3}$ and the only critical point is $\left(\frac{1}{3}, -\frac{2}{3}\right)$.

$D(x, y) = f_{xx} f_{yy} - (f_{xy})^2 = (2)(2) - (1)^2 = 3$, and since

$D\left(\frac{1}{3}, -\frac{2}{3}\right) = 3 > 0$ and $f_{xx}\left(\frac{1}{3}, -\frac{2}{3}\right) = 2 > 0$, $f\left(\frac{1}{3}, -\frac{2}{3}\right) = -\frac{1}{3}$ is a local

minimum by the Second Derivatives Test.

7. $f(x, y) = (x - y)(1 - xy) = x - y - x^2y + xy^2 \Rightarrow f_x = 1 - 2xy + y^2$, $f_y = -1 - x^2 + 2xy$, $f_{xx} = -2y$,

$f_{xy} = -2x + 2y$, $f_{yy} = 2x$. Then $f_x = 0$ implies $1 - 2xy + y^2 = 0$ and $f_y = 0$ implies $-1 - x^2 + 2xy = 0$. Adding the

two equations gives $1 + y^2 - 1 - x^2 = 0 \Rightarrow y^2 = x^2 \Rightarrow y = \pm x$, but if $y = -x$ then $f_x = 0$ implies

© 2012 Cengage Learning. All Rights Reserved. May not be scanned, copied, or duplicated, or posted to a publicly accessible website, in whole or in part.

$1 + 2x^2 + x^2 = 0 \Rightarrow 3x^2 = -1$ which has no real solution. If $y = x$

then substitution into $f_x = 0$ gives $1 - 2x^2 + x^2 = 0 \Rightarrow x^2 = 1 \Rightarrow$

$x = \pm 1$, so the critical points are $(1, 1)$ and $(-1, -1)$. Now

$D(1, 1) = (-2)(2) - 0^2 = -4 < 0$ and

$D(-1, -1) = (2)(-2) - 0^2 = -4 < 0$, so $(1, 1)$ and $(-1, -1)$ are

saddle points.

9. $f(x, y) = y^3 + 3x^2 y - 6x^2 - 6y^2 + 2 \Rightarrow f_x = 6xy - 12x$, $f_y = 3y^2 + 3x^2 - 12y$, $f_{xx} = 6y - 12$, $f_{xy} = 6x$,

$f_{yy} = 6y - 12$. Then $f_x = 0$ implies $6x(y - 2) = 0$, so $x = 0$ or $y = 2$. If $x = 0$ then substitution into $f_y = 0$ gives

$3y^2 - 12y = 0 \Rightarrow 3y(y - 4) = 0 \Rightarrow y = 0$ or $y = 4$, so we have critical points $(0, 0)$ and $(0, 4)$. If $y = 2$,

substitution into $f_y = 0$ gives $12 + 3x^2 - 24 = 0 \Rightarrow x^2 = 4 \Rightarrow$

$x = \pm 2$, so we have critical points $(\pm 2, 2)$.

$D(0, 0) = (-12)(-12) - 0^2 = 144 > 0$ and $f_{xx}(0, 0) = -12 < 0$, so

$f(0, 0) = 2$ is a local maximum. $D(0, 4) = (12)(12) - 0^2 = 144 > 0$

and $f_{xx}(0, 4) = 12 > 0$, so $f(0, 4) = -30$ is a local minimum.

$D(\pm 2, 2) = (0)(0) - (\pm 12)^2 = -144 < 0$, so $(\pm 2, 2)$ are saddle points.

11. $f(x, y) = x^3 - 12xy + 8y^3 \Rightarrow f_x = 3x^2 - 12y$, $f_y = -12x + 24y^2$, $f_{xx} = 6x$, $f_{xy} = -12$, $f_{yy} = 48y$. Then $f_x = 0$

implies $x^2 = 4y$ and $f_y = 0$ implies $x = 2y^2$. Substituting the second equation into the first gives $(2y^2)^2 = 4y \Rightarrow$

$4y^4 = 4y \Rightarrow 4y(y^3 - 1) = 0 \Rightarrow y = 0$ or $y = 1$. If $y = 0$ then

$x = 0$ and if $y = 1$ then $x = 2$, so the critical points are $(0, 0)$ and $(2, 1)$.

$D(0, 0) = (0)(0) - (-12)^2 = -144 < 0$, so $(0, 0)$ is a saddle point.

$D(2, 1) = (12)(48) - (-12)^2 = 432 > 0$ and $f_{xx}(2, 1) = 12 > 0$ so

$f(2, 1) = -8$ is a local minimum.

13. $f(x, y) = e^x \cos y \Rightarrow f_x = e^x \cos y$, $f_y = -e^x \sin y$.

Now $f_x = 0$ implies $\cos y = 0$ or $y = \frac{\pi}{2} + n\pi$ for n an integer.

But $\sin\left(\frac{\pi}{2} + n\pi\right) \neq 0$, so there are no critical points.

© 2012 Cengage Learning. All Rights Reserved. May not be scanned, copied, or duplicated, or posted to a publicly accessible website, in whole or in part.

15. $f(x, y) = (x^2 + y^2)e^{y^2 - x^2}$ $\Rightarrow$

$f_x = (x^2 + y^2)e^{y^2 - x^2}(-2x) + 2xe^{y^2 - x^2} = 2xe^{y^2 - x^2}(1 - x^2 - y^2)$,

$f_y = (x^2 + y^2)e^{y^2 - x^2}(2y) + 2ye^{y^2 - x^2} = 2ye^{y^2 - x^2}(1 + x^2 + y^2)$,

$f_{xx} = 2xe^{y^2 - x^2}(-2x) + (1 - x^2 - y^2)\left(2x\left(-2xe^{y^2 - x^2}\right) + 2e^{y^2 - x^2}\right) = 2e^{y^2 - x^2}\left((1 - x^2 - y^2)(1 - 2x^2) - 2x^2\right)$,

$f_{xy} = 2xe^{y^2 - x^2}(-2y) + 2x(2y)e^{y^2 - x^2}(1 - x^2 - y^2) = -4xye^{y^2 - x^2}(x^2 + y^2)$,

$f_{yy} = 2ye^{y^2 - x^2}(2y) + (1 + x^2 + y^2)\left(2y\left(2ye^{y^2 - x^2}\right) + 2e^{y^2 - x^2}\right) = 2e^{y^2 - x^2}\left((1 + x^2 + y^2)(1 + 2y^2) + 2y^2\right)$.

$f_y = 0$ implies $y = 0$, and substituting into $f_x = 0$ gives

$2xe^{-x^2}(1 - x^2) = 0$ $\Rightarrow$ $x = 0$ or $x = \pm 1$. Thus the critical points are

$(0, 0)$ and $(\pm 1, 0)$. Now $D(0, 0) = (2)(2) - 0 > 0$ and $f_{xx}(0, 0) = 2 > 0$,

so $f(0, 0) = 0$ is a local minimum. $D(\pm 1, 0) = (-4e^{-1})(4e^{-1}) - 0 < 0$

so $(\pm 1, 0)$ are saddle points.

17. $f(x, y) = y^2 - 2y \cos x$ $\Rightarrow$ $f_x = 2y \sin x$, $f_y = 2y - 2\cos x$,

$f_{xx} = 2y \cos x$, $f_{xy} = 2\sin x$, $f_{yy} = 2$. Then $f_x = 0$ implies $y = 0$ or

$\sin x = 0$ $\Rightarrow$ $x = 0, \pi$, or 2π for $-1 \leq x \leq 7$. Substituting $y = 0$ into

$f_y = 0$ gives $\cos x = 0$ $\Rightarrow$ $x = \frac{\pi}{2}$ or $\frac{3\pi}{2}$, substituting $x = 0$ or $x = 2\pi$

into $f_y = 0$ gives $y = 1$, and substituting $x = \pi$ into $f_y = 0$ gives $y = -1$.

Thus the critical points are $(0, 1)$, $\left(\frac{\pi}{2}, 0\right)$, $(\pi, -1)$, $\left(\frac{3\pi}{2}, 0\right)$, and $(2\pi, 1)$.

$D\left(\frac{\pi}{2}, 0\right) = D\left(\frac{3\pi}{2}, 0\right) = -4 < 0$ so $\left(\frac{\pi}{2}, 0\right)$ and $\left(\frac{3\pi}{2}, 0\right)$ are saddle points. $D(0, 1) = D(\pi, -1) = D(2\pi, 1) = 4 > 0$ and

$f_{xx}(0, 1) = f_{xx}(\pi, -1) = f_{xx}(2\pi, 1) = 2 > 0$, so $f(0, 1) = f(\pi, -1) = f(2\pi, 1) = -1$ are local minima.

19. $f(x, y) = x^2 + 4y^2 - 4xy + 2$ $\Rightarrow$ $f_x = 2x - 4y$, $f_y = 8y - 4x$, $f_{xx} = 2$, $f_{xy} = -4$, $f_{yy} = 8$. Then $f_x = 0$

and $f_y = 0$ each implies $y = \frac{1}{2}x$, so all points of the form $\left(x_0, \frac{1}{2}x_0\right)$ are critical points and for each of these we have

$D\left(x_0, \frac{1}{2}x_0\right) = (2)(8) - (-4)^2 = 0$. The Second Derivatives Test gives no information, but

$f(x, y) = x^2 + 4y^2 - 4xy + 2 = (x - 2y)^2 + 2 \geq 2$ with equality if and only if $y = \frac{1}{2}x$. Thus $f\left(x_0, \frac{1}{2}x_0\right) = 2$ are all local

(and absolute) minima.

© 2012 Cengage Learning. All Rights Reserved. May not be scanned, copied, or duplicated, or posted to a publicly accessible website, in whole or in part.

21. $f(x, y) = x^2 + y^2 + x^{-2}y^{-2}$

From the graphs, there appear to be local minima of about $f(1, \pm1) = f(-1, \pm1) \approx 3$ (and no local maxima or saddle

points). $f_x = 2x - 2x^{-3}y^{-2}$, $f_y = 2y - 2x^{-2}y^{-3}$, $f_{xx} = 2 + 6x^{-4}y^{-2}$, $f_{xy} = 4x^{-3}y^{-3}$, $f_{yy} = 2 + 6x^{-2}y^{-4}$. Then

$f_x = 0$ implies $2x^4y^2 - 2 = 0$ or $x^4y^2 = 1$ or $y^2 = x^{-4}$. Note that neither x nor y can be zero. Now $f_y = 0$ implies

$2x^2y^4 - 2 = 0$, and with $y^2 = x^{-4}$ this implies $2x^{-6} - 2 = 0$ or $x^6 = 1$. Thus $x = \pm1$ and if $x = 1$, $y = \pm1$; if $x = -1$,

$y = \pm1$. So the critical points are $(1, 1)$, $(1, -1)$, $(-1, 1)$ and $(-1, -1)$. Now $D(1, \pm1) = D(-1, \pm1) = 64 - 16 > 0$ and

$f_{xx} > 0$ always, so $f(1, \pm1) = f(-1, \pm1) = 3$ are local minima.

23. $f(x, y) = \sin x + \sin y + \sin(x + y)$, $0 \le x \le 2\pi$, $0 \le y \le 2\pi$

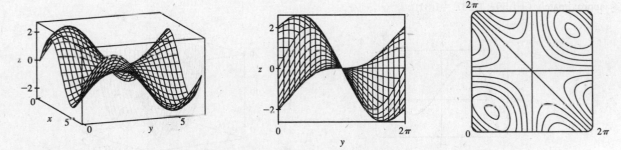

From the graphs it appears that f has a local maximum at about $(1, 1)$ with value approximately 2.6, a local minimum

at about $(5, 5)$ with value approximately -2.6, and a saddle point at about $(3, 3)$.

$f_x = \cos x + \cos(x + y)$, $f_y = \cos y + \cos(x + y)$, $f_{xx} = -\sin x - \sin(x + y)$, $f_{yy} = -\sin y - \sin(x + y)$,

$f_{xy} = -\sin(x + y)$. Setting $f_x = 0$ and $f_y = 0$ and subtracting gives $\cos x - \cos y = 0$ or $\cos x = \cos y$. Thus $x = y$

or $x = 2\pi - y$. If $x = y$, $f_x = 0$ becomes $\cos x + \cos 2x = 0$ or $2\cos^2 x + \cos x - 1 = 0$, a quadratic in $\cos x$. Thus

$\cos x = -1$ or $\frac{1}{2}$ and $x = \pi$, $\frac{\pi}{3}$, or $\frac{5\pi}{3}$, giving the critical points (π, π), $\left(\frac{\pi}{3}, \frac{\pi}{3}\right)$ and $\left(\frac{5\pi}{3}, \frac{5\pi}{3}\right)$. Similarly if

$x = 2\pi - y$, $f_x = 0$ becomes $(\cos x) + 1 = 0$ and the resulting critical point is (π, π). Now

$D(x, y) = \sin x \sin y + \sin x \sin(x + y) + \sin y \sin(x + y)$. So $D(\pi, \pi) = 0$ and the Second Derivatives Test doesn't apply.

However, along the line $y = x$ we have $f(x, x) = 2\sin x + \sin 2x = 2\sin x + 2\sin x \cos x = 2\sin x(1 + \cos x)$, and

© 2012 Cengage Learning. All Rights Reserved. May not be scanned, copied, or duplicated, or posted to a publicly accessible website, in whole or in part.

$f(x, x) > 0$ for $0 < x < \pi$ while $f(x, x) < 0$ for $\pi < x < 2\pi$. Thus every disk with center (π, π) contains points where f is positive as well as points where f is negative, so the graph crosses its tangent plane $(z = 0)$ there and (π, π) is a saddle point.

$D\left(\frac{\pi}{3}, \frac{\pi}{3}\right) = \frac{9}{4} > 0$ and $f_{xx}\left(\frac{\pi}{3}, \frac{\pi}{3}\right) < 0$ so $f\left(\frac{\pi}{3}, \frac{\pi}{3}\right) = \frac{3\sqrt{3}}{2}$ is a local maximum while $D\left(\frac{5\pi}{3}, \frac{5\pi}{3}\right) = \frac{9}{4} > 0$ and

$f_{xx}\left(\frac{5\pi}{3}, \frac{5\pi}{3}\right) > 0$, so $f\left(\frac{5\pi}{3}, \frac{5\pi}{3}\right) = -\frac{3\sqrt{3}}{2}$ is a local minimum.

25. $f(x, y) = x^4 + y^4 - 4x^2y + 2y \Rightarrow f_x(x, y) = 4x^3 - 8xy$ and $f_y(x, y) = 4y^3 - 4x^2 + 2$. $f_x = 0 \Rightarrow$

$4x(x^2 - 2y) = 0$, so $x = 0$ or $x^2 = 2y$. If $x = 0$ then substitution into $f_y = 0$ gives $4y^3 = -2 \Rightarrow y = -\frac{1}{\sqrt[3]{2}}$, so

$\left(0, -\frac{1}{\sqrt[3]{2}}\right)$ is a critical point. Substituting $x^2 = 2y$ into $f_y = 0$ gives $4y^3 - 8y + 2 = 0$. Using a graph, solutions are

approximately $y = -1.526$, 0.259, and 1.267. (Alternatively, we could have used a calculator or a CAS to find these roots.)

We have $x^2 = 2y \Rightarrow x = \pm\sqrt{2y}$, so $y = -1.526$ gives no real-valued solution for x, but

$y = 0.259 \Rightarrow x \approx \pm 0.720$ and $y = 1.267 \Rightarrow x \approx \pm 1.592$. Thus to three decimal places, the critical points are

$\left(0, -\frac{1}{\sqrt[3]{2}}\right) \approx (0, -0.794)$, $(\pm 0.720, 0.259)$, and $(\pm 1.592, 1.267)$. Now since $f_{xx} = 12x^2 - 8y$, $f_{xy} = -8x$, $f_{yy} = 12y^2$,

and $D = (12x^2 - 8y)(12y^2) - 64x^2$, we have $D(0, -0.794) > 0$, $f_{xx}(0, -0.794) > 0$, $D(\pm 0.720, 0.259) < 0$,

$D(\pm 1.592, 1.267) > 0$, and $f_{xx}(\pm 1.592, 1.267) > 0$. Therefore $f(0, -0.794) \approx -1.191$ and $f(\pm 1.592, 1.267) \approx -1.310$

are local minima, and $(\pm 0.720, 0.259)$ are saddle points. There is no highest point on the graph, but the lowest points are

approximately $(\pm 1.592, 1.267, -1.310)$.

27. $f(x, y) = x^4 + y^3 - 3x^2 + y^2 + x - 2y + 1 \Rightarrow f_x(x, y) = 4x^3 - 6x + 1$ and $f_y(x, y) = 3y^2 + 2y - 2$. From the

graphs, we see that to three decimal places, $f_x = 0$ when $x \approx -1.301$, 0.170, or 1.131, and $f_y = 0$ when $y \approx -1.215$ or

0.549. (Alternatively, we could have used a calculator or a CAS to find these roots. We could also use the quadratic formula to

find the solutions of $f_y = 0$.) So, to three decimal places, f has critical points at $(-1.301, -1.215)$, $(-1.301, 0.549)$,

$(0.170, -1.215)$, $(0.170, 0.549)$, $(1.131, -1.215)$, and $(1.131, 0.549)$. Now since $f_{xx} = 12x^2 - 6$, $f_{xy} = 0$, $f_{yy} = 6y + 2$,

and $D = (12x^2 - 6)(6y + 2)$, we have $D(-1.301, -1.215) < 0$, $D(-1.301, 0.549) > 0$, $f_{xx}(-1.301, 0.549) > 0$,

$D(0.170, -1.215) > 0$, $f_{xx}(0.170, -1.215) < 0$, $D(0.170, 0.549) < 0$, $D(1.131, -1.215) < 0$, $D(1.131, 0.549) > 0$, and

$f_{xx}(1.131, 0.549) > 0$. Therefore, to three decimal places, $f(-1.301, 0.549) \approx -3.145$ and $f(1.131, 0.549) \approx -0.701$ are

© 2012 Cengage Learning. All Rights Reserved. May not be scanned, copied, or duplicated, or posted to a publicly accessible website, in whole or in part.

local minima, $f(0.170, -1.215) \approx 3.197$ is a local maximum, and $(-1.301, -1.215)$, $(0.170, 0.549)$, and $(1.131, -1.215)$ are saddle points. There is no highest or lowest point on the graph.

29. Since f is a polynomial it is continuous on D, so an absolute maximum and minimum exist. Here $f_x = 2x - 2$, $f_y = 2y$, and

setting $f_x = f_y = 0$ gives $(1, 0)$ as the only critical point (which is inside D), where $f(1, 0) = -1$. Along L_1: $x = 0$ and

$f(0, y) = y^2$ for $-2 \leq y \leq 2$, a quadratic function which attains its minimum at $y = 0$, where $f(0, 0) = 0$, and its maximum

at $y = \pm 2$, where $f(0, \pm 2) = 4$. Along L_2: $y = x - 2$ for $0 \leq x \leq 2$, and $f(x, x - 2) = 2x^2 - 6x + 4 = 2\left(x - \frac{3}{2}\right)^2 - \frac{1}{2}$,

a quadratic which attains its minimum at $x = \frac{3}{2}$, where $f\left(\frac{3}{2}, -\frac{1}{2}\right) = -\frac{1}{2}$, and its maximum at $x = 0$, where $f(0, -2) = 4$.

Along L_3: $y = 2 - x$ for $0 \leq x \leq 2$, and

$f(x, 2 - x) = 2x^2 - 6x + 4 = 2\left(x - \frac{3}{2}\right)^2 - \frac{1}{2}$, a quadratic which attains

its minimum at $x - \frac{3}{2}$, where $f\left(\frac{3}{2}, \frac{1}{2}\right) = -\frac{1}{2}$, and its maximum at $x = 0$,

where $f(0, 2) = 4$. Thus the absolute maximum of f on D is $f(0, \pm 2) = 4$

and the absolute minimum is $f(1, 0) = -1$.

31. $f_x(x, y) = 2x + 2xy$, $f_y(x, y) = 2y + x^2$, and setting $f_x = f_y = 0$

gives $(0, 0)$ as the only critical point in D, with $f(0, 0) = 4$.

On L_1: $y = -1$, $f(x, -1) = 5$, a constant.

On L_2: $x = 1$, $f(1, y) = y^2 + y + 5$, a quadratic in y which attains its

maximum at $(1, 1)$, $f(1, 1) = 7$ and its minimum at $\left(1, -\frac{1}{2}\right)$, $f\left(1, -\frac{1}{2}\right) = \frac{19}{4}$.

On L_3: $f(x, 1) = 2x^2 + 5$ which attains its maximum at $(-1, 1)$ and $(1, 1)$

with $f(\pm 1, 1) = 7$ and its minimum at $(0, 1)$, $f(0, 1) = 5$.

© 2012 Cengage Learning. All Rights Reserved. May not be scanned, copied, or duplicated, or posted to a publicly accessible website, in whole or in part.

On L_4: $f(-1, y) = y^2 + y + 5$ with maximum at $(-1, 1)$, $f(-1, 1) = 7$ and minimum at $\left(-1, -\frac{1}{2}\right)$, $f\left(-1, -\frac{1}{2}\right) = \frac{19}{4}$.

Thus the absolute maximum is attained at both $(\pm 1, 1)$ with $f(\pm 1, 1) = 7$ and the absolute minimum on D is attained at

$(0, 0)$ with $f(0, 0) = 4$.

33. $f(x, y) = x^4 + y^4 - 4xy + 2$ is a polynomial and hence continuous on D, so

it has an absolute maximum and minimum on D. $f_x(x, y) = 4x^3 - 4y$ and

$f_y(x, y) = 4y^3 - 4x$; then $f_x = 0$ implies $y = x^3$, and substitution into

$f_y = 0 \Rightarrow x = y^3$ gives $x^9 - x = 0 \Rightarrow x(x^8 - 1) = 0 \Rightarrow x = 0$

or $x = \pm 1$. Thus the critical points are $(0, 0)$, $(1, 1)$, and $(-1, -1)$, but only

$(1, 1)$ with $f(1, 1) = 0$ is inside D. On L_1: $y = 0$, $f(x, 0) = x^4 + 2$,

$0 \le x \le 3$, a polynomial in x which attains its maximum at $x = 3$, $f(3, 0) = 83$, and its minimum at $x = 0$, $f(0, 0) = 2$.

On L_2: $x = 3$, $f(3, y) = y^4 - 12y + 83$, $0 \le y \le 2$, a polynomial in y which attains its minimum at $y = \sqrt[3]{3}$,

$f(3, \sqrt[3]{3}) = 83 - 9\sqrt[3]{3} \approx 70.0$, and its maximum at $y = 0$, $f(3, 0) = 83$.

On L_3: $y = 2$, $f(x, 2) = x^4 - 8x + 18$, $0 \le x \le 3$, a polynomial in x which attains its minimum at $x = \sqrt[3]{2}$,

$f(\sqrt[3]{2}, 2) = 18 - 6\sqrt[3]{2} \approx 10.4$, and its maximum at $x = 3$, $f(3, 2) = 75$. On L_4: $x = 0$, $f(0, y) = y^4 + 2$, $0 \le y \le 2$, a

polynomial in y which attains its maximum at $y = 2$, $f(0, 2) = 18$, and its minimum at $y = 0$, $f(0, 0) = 2$. Thus the absolute

maximum of f on D is $f(3, 0) = 83$ and the absolute minimum is $f(1, 1) = 0$.

35. $f_x(x, y) = 6x^2$ and $f_y(x, y) = 4y^3$. And so $f_x = 0$ and $f_y = 0$ only occur when $x = y = 0$. Hence, the only critical point

inside the disk is at $x = y = 0$ where $f(0, 0) = 0$. Now on the circle $x^2 + y^2 = 1$, $y^2 = 1 - x^2$ so let

$g(x) = f(x, y) = 2x^3 + (1 - x^2)^2 = x^4 + 2x^3 - 2x^2 + 1, -1 \le x \le 1$. Then $g'(x) = 4x^3 + 6x^2 - 4x = 0 \Rightarrow x = 0$,

-2, or $\frac{1}{2}$. $f(0, \pm 1) = g(0) = 1$, $f\left(\frac{1}{2}, \pm \frac{\sqrt{3}}{2}\right) = g\left(\frac{1}{2}\right) = \frac{13}{16}$, and $(-2, -3)$ is not in D. Checking the endpoints, we get

$f(-1, 0) = g(-1) = -2$ and $f(1, 0) = g(1) = 2$. Thus the absolute maximum and minimum of f on D are $f(1, 0) = 2$ and

$f(-1, 0) = -2$.

Another method: On the boundary $x^2 + y^2 = 1$ we can write $x = \cos \theta$, $y = \sin \theta$, so $f(\cos \theta, \sin \theta) = 2 \cos^3 \theta + \sin^4 \theta$,

$0 \le \theta \le 2\pi$.

37. $f(x, y) = -(x^2 - 1)^2 - (x^2 y - x - 1)^2 \Rightarrow f_x(x, y) = -2(x^2 - 1)(2x) - 2(x^2 y - x - 1)(2xy - 1)$ and

$f_y(x, y) = -2(x^2 y - x - 1)x^2$. Setting $f_y(x, y) = 0$ gives either $x = 0$ or $x^2 y - x - 1 = 0$.

There are no critical points for $x = 0$, since $f_x(0, y) = -2$, so we set $x^2 y - x - 1 = 0 \Leftrightarrow y = \dfrac{x + 1}{x^2}$ $[x \ne 0]$,

so $f_x\left(x, \dfrac{x + 1}{x^2}\right) = -2(x^2 - 1)(2x) - 2\left(x^2 \dfrac{x + 1}{x^2} - x - 1\right)\left(2x \dfrac{x + 1}{x^2} - 1\right) = -4x(x^2 - 1)$. Therefore

$f_x(x, y) = f_y(x, y) = 0$ at the points $(1, 2)$ and $(-1, 0)$. To classify these critical points, we calculate

© 2012 Cengage Learning. All Rights Reserved. May not be scanned, copied, or duplicated, or posted to a publicly accessible website, in whole or in part.

$f_{xx}(x, y) = -12x^2 - 12x^2 y^2 + 12xy + 4y + 2$, $f_{yy}(x, y) = -2x^4$,

and $f_{xy}(x, y) = -8x^3 y + 6x^2 + 4x$. In order to use the Second Derivatives

Test we calculate

$D(-1, 0) = f_{xx}(-1, 0) f_{yy}(-1, 0) - [f_{xy}(-1, 0)]^2 = 16 > 0$,

$f_{xx}(-1, 0) = -10 < 0$, $D(1, 2) = 16 > 0$, and $f_{xx}(1, 2) = -26 < 0$, so

both $(-1, 0)$ and $(1, 2)$ give local maxima.

39. Let d be the distance from $(2, 0, -3)$ to any point (x, y, z) on the plane $x + y + z = 1$, so $d = \sqrt{(x - 2)^2 + y^2 + (z + 3)^2}$

where $z = 1 - x - y$, and we minimize $d^2 = f(x, y) = (x - 2)^2 + y^2 + (4 - x - y)^2$. Then

$f_x(x, y) = 2(x - 2) + 2(4 - x - y)(-1) = 4x + 2y - 12$, $f_y(x, y) = 2y + 2(4 - x - y)(-1) = 2x + 4y - 8$. Solving

$4x + 2y - 12 = 0$ and $2x + 4y - 8 = 0$ simultaneously gives $x = \frac{8}{3}$, $y = \frac{2}{3}$, so the only critical point is $\left(\frac{8}{3}, \frac{2}{3}\right)$. An absolute

minimum exists (since there is a minimum distance from the point to the plane) and it must occur at a critical point, so the

shortest distance occurs for $x = \frac{8}{3}$, $y = \frac{2}{3}$ for which $d = \sqrt{\left(\frac{8}{3} - 2\right)^2 + \left(\frac{2}{3}\right)^2 + \left(4 - \frac{8}{3} - \frac{2}{3}\right)^2} = \sqrt{\frac{4}{3}} = \frac{2}{\sqrt{3}}$.

41. Let d be the distance from the point $(4, 2, 0)$ to any point (x, y, z) on the cone, so $d = \sqrt{(x - 4)^2 + (y - 2)^2 + z^2}$ where

$z^2 = x^2 + y^2$, and we minimize $d^2 = (x - 4)^2 + (y - 2)^2 + x^2 + y^2 = f(x, y)$. Then

$f_x(x, y) = 2(x - 4) + 2x = 4x - 8$, $f_y(x, y) = 2(y - 2) + 2y = 4y - 4$, and the critical points occur when

$f_x = 0 \Rightarrow x = 2$, $f_y = 0 \Rightarrow y = 1$. Thus the only critical point is $(2, 1)$. An absolute minimum exists (since there is a

minimum distance from the cone to the point) which must occur at a critical point, so the points on the cone closest

to $(4, 2, 0)$ are $\left(2, 1, \pm\sqrt{5}\right)$.

43. $x + y + z = 100$, so maximize $f(x, y) = xy(100 - x - y)$. $f_x = 100y - 2xy - y^2$, $f_y = 100x - x^2 - 2xy$,

$f_{xx} = -2y$, $f_{yy} = -2x$, $f_{xy} = 100 - 2x - 2y$. Then $f_x = 0$ implies $y = 0$ or $y = 100 - 2x$. Substituting $y = 0$ into

$f_y = 0$ gives $x = 0$ or $x = 100$ and substituting $y = 100 - 2x$ into $f_y = 0$ gives $3x^2 - 100x = 0$ so $x = 0$ or $\frac{100}{3}$.

Thus the critical points are $(0, 0)$, $(100, 0)$, $(0, 100)$ and $\left(\frac{100}{3}, \frac{100}{3}\right)$.

$D(0, 0) = D(100, 0) = D(0, 100) = -10{,}000$ while $D\left(\frac{100}{3}, \frac{100}{3}\right) = \frac{10{,}000}{3}$ and $f_{xx}\left(\frac{100}{3}, \frac{100}{3}\right) = -\frac{200}{3} < 0$. Thus $(0, 0)$,

$(100, 0)$ and $(0, 100)$ are saddle points whereas $f\left(\frac{100}{3}, \frac{100}{3}\right)$ is a local maximum. Thus the numbers are $x = y = z = \frac{100}{3}$.

45. Center the sphere at the origin so that its equation is $x^2 + y^2 + z^2 = r^2$, and orient the inscribed rectangular box so that its

edges are parallel to the coordinate axes. Any vertex of the box satisfies $x^2 + y^2 + z^2 = r^2$, so take (x, y, z) to be the vertex

in the first octant. Then the box has length $2x$, width $2y$, and height $2z = 2\sqrt{r^2 - x^2 - y^2}$ with volume given by

$V(x, y) = (2x)(2y)\left(2\sqrt{r^2 - x^2 - y^2}\right) = 8xy\sqrt{r^2 - x^2 - y^2}$ for $0 < x < r$, $0 < y < r$. Then

$V_x = (8xy) \cdot \frac{1}{2}(r^2 - x^2 - y^2)^{-1/2}(-2x) + \sqrt{r^2 - x^2 - y^2} \cdot 8y = \frac{8y(r^2 - 2x^2 - y^2)}{\sqrt{r^2 - x^2 - y^2}}$ and $V_y = \frac{8x(r^2 - x^2 - 2y^2)}{\sqrt{r^2 - x^2 - y^2}}$.

Setting $V_x = 0$ gives $y = 0$ or $2x^2 + y^2 = r^2$, but $y > 0$ so only the latter solution applies. Similarly, $V_y = 0$ with $x > 0$

© 2012 Cengage Learning. All Rights Reserved. May not be scanned, copied, or duplicated, or posted to a publicly accessible website, in whole or in part.

implies $x^2 + 2y^2 = r^2$. Substituting, we have $2x^2 + y^2 = x^2 + 2y^2 \;\Rightarrow\; x^2 = y^2 \;\Rightarrow\; y = x$. Then $x^2 + 2y^2 = r^2 \;\Rightarrow\;$

$3x^2 = r^2 \;\Rightarrow\; x = \sqrt{r^2/3} = r/\sqrt{3} = y$. Thus the only critical point is $\left(r/\sqrt{3},\, r/\sqrt{3}\right)$. There must be a maximum

volume and here it must occur at a critical point, so the maximum volume occurs when $x = y = r/\sqrt{3}$ and the maximum

volume is $V\left(\frac{r}{\sqrt{3}},\, \frac{r}{\sqrt{3}}\right) = 8\left(\frac{r}{\sqrt{3}}\right)\left(\frac{r}{\sqrt{3}}\right)\sqrt{r^2 - \left(\frac{r}{\sqrt{3}}\right)^2 - \left(\frac{r}{\sqrt{3}}\right)^2} = \dfrac{8}{3\sqrt{3}}\, r^3$.

47. Maximize $f(x, y) = \dfrac{xy}{3}(6 - x - 2y)$, then the maximum volume is $V = xyz$.

$f_x = \frac{1}{3}(6y - 2xy - y^2) = \frac{1}{3}y(6 - 2x - 2y)$ and $f_y = \frac{1}{3}x(6 - x - 4y)$. Setting $f_x = 0$ and $f_y = 0$ gives the critical point

$(2, 1)$ which geometrically must give a maximum. Thus the volume of the largest such box is $V = (2)(1)\left(\frac{2}{3}\right) = \frac{4}{3}$.

49. Let the dimensions be x, y, and z; then $4x + 4y + 4z = c$ and the volume is

$V = xyz = xy\left(\frac{1}{4}c - x - y\right) = \frac{1}{4}cxy - x^2y - xy^2$, $x > 0$, $y > 0$. Then $V_x = \frac{1}{4}cy - 2xy - y^2$ and $V_y = \frac{1}{4}cx - x^2 - 2xy$,

so $V_x = 0 = V_y$ when $2x + y = \frac{1}{4}c$ and $x + 2y = \frac{1}{4}c$. Solving, we get $x = \frac{1}{12}c$, $y = \frac{1}{12}c$ and $z = \frac{1}{4}c - x - y = \frac{1}{12}c$. From

the geometrical nature of the problem, this critical point must give an absolute maximum. Thus the box is a cube with edge

length $\frac{1}{12}c$.

51. Let the dimensions be x, y and z, then minimize $xy + 2(xz + yz)$ if $xyz = 32{,}000 \text{ cm}^3$. Then

$f(x, y) = xy + [64{,}000(x + y)/xy] = xy + 64{,}000(x^{-1} + y^{-1})$, $f_x = y - 64{,}000x^{-2}$, $f_y = x - 64{,}000y^{-2}$.

And $f_x = 0$ implies $y = 64{,}000/x^2$; substituting into $f_y = 0$ implies $x^3 = 64{,}000$ or $x = 40$ and then $y = 40$. Now

$D(x, y) = [(2)(64{,}000)]^2 x^{-3} y^{-3} - 1 > 0$ for $(40, 40)$ and $f_{xx}(40, 40) > 0$ so this is indeed a minimum. Thus the

dimensions of the box are $x = y = 40$ cm, $z = 20$ cm.

53. Let x, y, z be the dimensions of the rectangular box. Then the volume of the box is xyz and

$L = \sqrt{x^2 + y^2 + z^2} \;\Rightarrow\; L^2 = x^2 + y^2 + z^2 \;\Rightarrow\; z = \sqrt{L^2 - x^2 - y^2}$.

Substituting, we have volume $V(x, y) = xy\sqrt{L^2 - x^2 - y^2}$ $(x, y > 0)$.

$V_x = xy \cdot \frac{1}{2}(L^2 - x^2 - y^2)^{-1/2}(-2x) + y\sqrt{L^2 - x^2 - y^2} = y\sqrt{L^2 - x^2 - y^2} - \dfrac{x^2 y}{\sqrt{L^2 - x^2 - y^2}}$,

$V_y = x\sqrt{L^2 - x^2 - y^2} - \dfrac{xy^2}{\sqrt{L^2 - x^2 - y^2}}$. $V_x = 0$ implies $y(L^2 - x^2 - y^2) = x^2 y \;\Rightarrow\; y(L^2 - 2x^2 - y^2) = 0 \;\Rightarrow\;$

$2x^2 + y^2 = L^2$ (since $y > 0$), and $V_y = 0$ implies $x(L^2 - x^2 - y^2) = xy^2 \;\Rightarrow\; x(L^2 - x^2 - 2y^2) = 0 \;\Rightarrow\;$

$x^2 + 2y^2 = L^2$ (since $x > 0$). Substituting $y^2 = L^2 - 2x^2$ into $x^2 + 2y^2 = L^2$ gives $x^2 + 2L^2 - 4x^2 = L^2 \;\Rightarrow\;$

$3x^2 = L^2 \;\Rightarrow\; x = L/\sqrt{3}$ (since $x > 0$) and then $y = \sqrt{L^2 - 2\left(L/\sqrt{3}\right)^2} = L/\sqrt{3}$.

So the only critical point is $\left(L/\sqrt{3},\, L/\sqrt{3}\right)$ which, from the geometrical nature of the problem, must give an absolute

maximum. Thus the maximum volume is $V\left(L/\sqrt{3},\, L/\sqrt{3}\right) = \left(L/\sqrt{3}\right)^2 \sqrt{L^2 - \left(L/\sqrt{3}\right)^2 - \left(L/\sqrt{3}\right)^2} = L^3/(3\sqrt{3})$

cubic units.

© 2012 Cengage Learning. All Rights Reserved. May not be scanned, copied, or duplicated, or posted to a publicly accessible website, in whole or in part.

55. Note that here the variables are m and b, and $f(m, b) = \sum\limits_{i=1}^{n} [y_i - (mx_i + b)]^2$. Then $f_m = \sum\limits_{i=1}^{n} -2x_i[y_i - (mx_i + b)] = 0$

implies $\sum\limits_{i=1}^{n} (x_iy_i - mx_i^2 - bx_i) = 0$ or $\sum\limits_{i=1}^{n} x_iy_i = m\sum\limits_{i=1}^{n} x_i^2 + b\sum\limits_{i=1}^{n} x_i$ and $f_b = \sum\limits_{i=1}^{n} -2[y_i - (mx_i + b)] = 0$ implies

$\sum\limits_{i=1}^{n} y_i = m\sum\limits_{i=1}^{n} x_i + \sum\limits_{i=1}^{n} b = m\left(\sum\limits_{i=1}^{n} x_i\right) + nb$. Thus we have the two desired equations.

Now $f_{mm} = \sum\limits_{i=1}^{n} 2x_i^2$, $f_{bb} = \sum\limits_{i=1}^{n} 2 = 2n$ and $f_{mb} = \sum\limits_{i=1}^{n} 2x_i$. And $f_{mm}(m, b) > 0$ always and

$D(m, b) = 4n\left(\sum\limits_{i=1}^{n} x_i^2\right) - 4\left(\sum\limits_{i=1}^{n} x_i\right)^2 = 4\left[n\left(\sum\limits_{i=1}^{n} x_i^2\right) - \left(\sum\limits_{i=1}^{n} x_i\right)^2\right] > 0$ always so the solutions of these two

equations do indeed minimize $\sum\limits_{i=1}^{n} d_i^2$.

14.8 Lagrange Multipliers

1. At the extreme values of f, the level curves of f just touch the curve $g(x, y) = 8$ with a common tangent line. (See Figure 1

and the accompanying discussion.) We can observe several such occurrences on the contour map, but the level curve

$f(x, y) = c$ with the largest value of c which still intersects the curve $g(x, y) = 8$ is approximately $c = 59$, and the smallest

value of c corresponding to a level curve which intersects $g(x, y) = 8$ appears to be $c = 30$. Thus we estimate the maximum

value of f subject to the constraint $g(x, y) = 8$ to be about 59 and the minimum to be 30.

3. $f(x, y) = x^2 + y^2$, $g(x, y) = xy = 1$, and $\nabla f = \lambda \nabla g \Rightarrow \langle 2x, 2y \rangle = \langle \lambda y, \lambda x \rangle$, so $2x = \lambda y$, $2y = \lambda x$, and $xy = 1$.

From the last equation, $x \neq 0$ and $y \neq 0$, so $2x = \lambda y \Rightarrow \lambda = 2x/y$. Substituting, we have $2y = (2x/y)x \Rightarrow$

$y^2 = x^2 \Rightarrow y = \pm x$. But $xy = 1$, so $x = y = \pm 1$ and the possible points for the extreme values of f are $(1, 1)$ and

$(-1, -1)$. Here there is no maximum value, since the constraint $xy = 1$ allows x or y to become arbitrarily large, and hence

$f(x, y) = x^2 + y^2$ can be made arbitrarily large. The minimum value is $f(1, 1) = f(-1, -1) = 2$.

5. $f(x, y) = y^2 - x^2$, $g(x, y) = \frac{1}{4}x^2 + y^2 = 1$, and $\nabla f = \lambda \nabla g \Rightarrow \langle -2x, 2y \rangle = \langle \frac{1}{2}\lambda x, 2\lambda y \rangle$, so $-2x = \frac{1}{2}\lambda x$, $2y = 2\lambda y$,

and $\frac{1}{4}x^2 + y^2 = 1$. From the first equation we have $x(4 + \lambda) = 0 \Rightarrow x = 0$ or $\lambda = -4$. If $x = 0$ then the third equation

gives $y = \pm 1$. If $\lambda = -4$ then the second equation gives $2y = -8y \Rightarrow y = 0$, and substituting into the third equation,

we have $x = \pm 2$. Thus the possible extreme values of f occur at the points $(0, \pm 1)$ and $(\pm 2, 0)$. Evaluating f at these points,

we see that the maximum value is $f(0, \pm 1) = 1$ and the minimum is $f(\pm 2, 0) = -4$.

7. $f(x, y, z) = 2x + 2y + z$, $g(x, y, z) = x^2 + y^2 + z^2 = 9$, and $\nabla f = \lambda \nabla g \Rightarrow \langle 2, 2, 1 \rangle = \langle 2\lambda x, 2\lambda y, 2\lambda z \rangle$, so $2\lambda x = 2$,

$2\lambda y = 2$, $2\lambda z = 1$, and $x^2 + y^2 + z^2 = 9$. The first three equations imply $x = \dfrac{1}{\lambda}$, $y = \dfrac{1}{\lambda}$, and $z = \dfrac{1}{2\lambda}$. But substitution into

the fourth equation gives $\left(\dfrac{1}{\lambda}\right)^2 + \left(\dfrac{1}{\lambda}\right)^2 + \left(\dfrac{1}{2\lambda}\right)^2 = 9 \Rightarrow \dfrac{9}{4\lambda^2} = 9 \Rightarrow \lambda = \pm\frac{1}{2}$, so f has possible extreme values at

© 2012 Cengage Learning. All Rights Reserved. May not be scanned, copied, or duplicated, or posted to a publicly accessible website, in whole or in part.

the points $(2, 2, 1)$ and $(-2, -2, -1)$. The maximum value of f on $x^2 + y^2 + z^2 = 9$ is $f(2, 2, 1) = 9$, and the minimum is $f(-2, -2, -1) = -9$.

9. $f(x, y, z) = xyz$, $g(x, y, z) = x^2 + 2y^2 + 3z^2 = 6$. $\nabla f = \lambda \nabla g \Rightarrow \langle yz, xz, xy \rangle = \lambda \langle 2x, 4y, 6z \rangle$. If any of x, y, or z is

zero then $x = y = z = 0$ which contradicts $x^2 + 2y^2 + 3z^2 = 6$. Then $\lambda = (yz)/(2x) = (xz)/(4y) = (xy)/(6z)$ or

$x^2 = 2y^2$ and $z^2 = \frac{2}{3}y^2$. Thus $x^2 + 2y^2 + 3z^2 = 6$ implies $6y^2 = 6$ or $y = \pm 1$. Then the possible points are

$\left(\sqrt{2}, \pm 1, \sqrt{\frac{2}{3}} \right)$, $\left(\sqrt{2}, \pm 1, -\sqrt{\frac{2}{3}} \right)$, $\left(-\sqrt{2}, \pm 1, \sqrt{\frac{2}{3}} \right)$, $\left(-\sqrt{2}, \pm 1, -\sqrt{\frac{2}{3}} \right)$. The maximum value of f on the ellipsoid is

$\frac{2}{\sqrt{3}}$, occurring when all coordinates are positive or exactly two are negative and the minimum is $-\frac{2}{\sqrt{3}}$ occurring when 1 or 3 of

the coordinates are negative.

11. $f(x, y, z) = x^2 + y^2 + z^2$, $g(x, y, z) = x^4 + y^4 + z^4 = 1 \Rightarrow \nabla f = \langle 2x, 2y, 2z \rangle$, $\lambda \nabla g = \langle 4\lambda x^3, 4\lambda y^3, 4\lambda z^3 \rangle$.

Case 1: If $x \neq 0$, $y \neq 0$ and $z \neq 0$, then $\nabla f = \lambda \nabla g$ implies $\lambda = 1/(2x^2) = 1/(2y^2) = 1/(2z^2)$ or $x^2 = y^2 = z^2$ and

$3x^4 = 1$ or $x = \pm \frac{1}{\sqrt[4]{3}}$ giving the points $\left(\pm \frac{1}{\sqrt[4]{3}}, \frac{1}{\sqrt[4]{3}}, \frac{1}{\sqrt[4]{3}} \right)$, $\left(\pm \frac{1}{\sqrt[4]{3}}, -\frac{1}{\sqrt[4]{3}}, \frac{1}{\sqrt[4]{3}} \right)$, $\left(\pm \frac{1}{\sqrt[4]{3}}, \frac{1}{\sqrt[4]{3}}, -\frac{1}{\sqrt[4]{3}} \right)$, $\left(\pm \frac{1}{\sqrt[4]{3}}, -\frac{1}{\sqrt[4]{3}}, -\frac{1}{\sqrt[4]{3}} \right)$

all with an f-value of $\sqrt{3}$.

Case 2: If one of the variables equals zero and the other two are not zero, then the squares of the two nonzero coordinates are

equal with common value $\frac{1}{\sqrt{2}}$ and corresponding f value of $\sqrt{2}$.

Case 3: If exactly two of the variables are zero, then the third variable has value ± 1 with the corresponding f value of 1. Thus

on $x^4 + y^4 + z^4 = 1$, the maximum value of f is $\sqrt{3}$ and the minimum value is 1.

13. $f(x, y, z, t) = x + y + z + t$, $g(x, y, z, t) = x^2 + y^2 + z^2 + t^2 = 1 \Rightarrow \langle 1, 1, 1, 1 \rangle = \langle 2\lambda x, 2\lambda y, 2\lambda z, 2\lambda t \rangle$, so

$\lambda = 1/(2x) = 1/(2y) = 1/(2z) = 1/(2t)$ and $x = y = z = t$. But $x^2 + y^2 + z^2 + t^2 = 1$, so the possible points are

$\left(\pm \frac{1}{2}, \pm \frac{1}{2}, \pm \frac{1}{2}, \pm \frac{1}{2} \right)$. Thus the maximum value of f is $f\left(\frac{1}{2}, \frac{1}{2}, \frac{1}{2}, \frac{1}{2} \right) = 2$ and the minimum value is

$f\left(-\frac{1}{2}, -\frac{1}{2}, -\frac{1}{2}, -\frac{1}{2} \right) = -2$.

15. $f(x, y, z) = x + 2y$, $g(x, y, z) = x + y + z = 1$, $h(x, y, z) = y^2 + z^2 = 4 \Rightarrow \nabla f = \langle 1, 2, 0 \rangle$, $\lambda \nabla g = \langle \lambda, \lambda, \lambda \rangle$

and $\mu \nabla h = \langle 0, 2\mu y, 2\mu z \rangle$. Then $1 = \lambda$, $2 = \lambda + 2\mu y$ and $0 = \lambda + 2\mu z$ so $\mu y = \frac{1}{2} = -\mu z$ or $y = 1/(2\mu)$, $z = -1/(2\mu)$.

Thus $x + y + z = 1$ implies $x = 1$ and $y^2 + z^2 = 4$ implies $\mu = \pm \frac{1}{2\sqrt{2}}$. Then the possible points are $\left(1, \pm \sqrt{2}, \mp \sqrt{2} \right)$

and the maximum value is $f\left(1, \sqrt{2}, -\sqrt{2} \right) = 1 + 2\sqrt{2}$ and the minimum value is $f\left(1, -\sqrt{2}, \sqrt{2} \right) = 1 - 2\sqrt{2}$.

17. $f(x, y, z) = yz + xy$, $g(x, y, z) = xy = 1$, $h(x, y, z) = y^2 + z^2 = 1 \Rightarrow \nabla f = \langle y, x + z, y \rangle$, $\lambda \nabla g = \langle \lambda y, \lambda x, 0 \rangle$,

$\mu \nabla h = \langle 0, 2\mu y, 2\mu z \rangle$. Then $y = \lambda y$ implies $\lambda = 1$ [$y \neq 0$ since $g(x, y, z) = 1$], $x + z = \lambda x + 2\mu y$ and $y = 2\mu z$. Thus

$\mu = z/(2y) = y/(2y)$ or $y^2 = z^2$, and so $y^2 + z^2 = 1$ implies $y = \pm \frac{1}{\sqrt{2}}$, $z = \pm \frac{1}{\sqrt{2}}$. Then $xy = 1$ implies $x = \pm \sqrt{2}$ and

© 2012 Cengage Learning. All Rights Reserved. May not be scanned, copied, or duplicated, or posted to a publicly accessible website, in whole or in part.

the possible points are $\left(\pm\sqrt{2}, \pm\frac{1}{\sqrt{2}}, \frac{1}{\sqrt{2}}\right)$, $\left(\pm\sqrt{2}, \pm\frac{1}{\sqrt{2}}, -\frac{1}{\sqrt{2}}\right)$. Hence the maximum of f subject to the constraints is

$f\left(\pm\sqrt{2}, \pm\frac{1}{\sqrt{2}}, \pm\frac{1}{\sqrt{2}}\right) = \frac{3}{2}$ and the minimum is $f\left(\pm\sqrt{2}, \pm\frac{1}{\sqrt{2}}, \mp\frac{1}{\sqrt{2}}\right) = \frac{1}{2}$.

Note: Since $xy = 1$ is one of the constraints we could have solved the problem by solving $f(y, z) = yz + 1$ subject to

$y^2 + z^2 = 1$.

19. $f(x, y) = x^2 + y^2 + 4x - 4y$. For the interior of the region, we find the critical points: $f_x = 2x + 4$, $f_y = 2y - 4$, so the

only critical point is $(-2, 2)$ (which is inside the region) and $f(-2, 2) = -8$. For the boundary, we use Lagrange multipliers.

$g(x, y) = x^2 + y^2 = 9$, so $\nabla f = \lambda \nabla g$ $\Rightarrow$ $\langle 2x + 4, 2y - 4 \rangle = \langle 2\lambda x, 2\lambda y \rangle$. Thus $2x + 4 = 2\lambda x$ and $2y - 4 = 2\lambda y$.

Adding the two equations gives $2x + 2y = 2\lambda x + 2\lambda y$ $\Rightarrow$ $x + y = \lambda(x + y)$ $\Rightarrow$ $(x + y)(\lambda - 1) = 0$, so

$x + y = 0$ $\Rightarrow$ $y = -x$ or $\lambda - 1 = 0$ $\Rightarrow$ $\lambda = 1$. But $\lambda = 1$ leads to a contradiction in $2x + 4 = 2\lambda x$, so $y = -x$ and

$x^2 + y^2 = 9$ implies $2y^2 = 9$ $\Rightarrow$ $y = \pm\frac{3}{\sqrt{2}}$. We have $f\left(\frac{3}{\sqrt{2}}, -\frac{3}{\sqrt{2}}\right) = 9 + 12\sqrt{2} \approx 25.97$ and

$f\left(-\frac{3}{\sqrt{2}}, \frac{3}{\sqrt{2}}\right) = 9 - 12\sqrt{2} \approx -7.97$, so the maximum value of f on the disk $x^2 + y^2 \leq 9$ is $f\left(\frac{3}{\sqrt{2}}, -\frac{3}{\sqrt{2}}\right) = 9 + 12\sqrt{2}$

and the minimum is $f(-2, 2) = -8$.

21. $f(x, y) = e^{-xy}$.   For the interior of the region, we find the critical points: $f_x = -ye^{-xy}$, $f_y = -xe^{-xy}$, so the only

critical point is $(0, 0)$, and $f(0, 0) = 1$. For the boundary, we use Lagrange multipliers. $g(x, y) = x^2 + 4y^2 = 1$ $\Rightarrow$

$\lambda \nabla g = \langle 2\lambda x, 8\lambda y \rangle$, so setting $\nabla f = \lambda \nabla g$ we get $-ye^{-xy} = 2\lambda x$ and $-xe^{-xy} = 8\lambda y$. The first of these gives

$e^{-xy} = -2\lambda x / y$, and then the second gives $-x(-2\lambda x / y) = 8\lambda y$ $\Rightarrow$ $x^2 = 4y^2$. Solving this last equation with the

constraint $x^2 + 4y^2 = 1$ gives $x = \pm\frac{1}{\sqrt{2}}$ and $y = \pm\frac{1}{2\sqrt{2}}$. Now $f\left(\pm\frac{1}{\sqrt{2}}, \mp\frac{1}{2\sqrt{2}}\right) = e^{1/4} \approx 1.284$ and

$f\left(\pm\frac{1}{\sqrt{2}}, \pm\frac{1}{2\sqrt{2}}\right) = e^{-1/4} \approx 0.779$. The former are the maxima on the region and the latter are the minima.

23. (a) $f(x, y) = x$, $g(x, y) = y^2 + x^4 - x^3 = 0$ $\Rightarrow$ $\nabla f = \langle 1, 0 \rangle = \lambda \nabla g = \lambda \langle 4x^3 - 3x^2, 2y \rangle$. Then

$1 = \lambda(4x^3 - 3x^2)$ **(1)** and $0 = 2\lambda y$ **(2)**. We have $\lambda \neq 0$ from **(1)**, so **(2)** gives $y = 0$. Then, from the constraint equation,

$x^4 - x^3 = 0$ $\Rightarrow$ $x^3(x - 1) = 0$ $\Rightarrow$ $x = 0$ or $x = 1$. But $x = 0$ contradicts **(1)**, so the only possible extreme value

subject to the constraint is $f(1, 0) = 1$. (The question remains whether this is indeed the minimum of f.)

(b) The constraint is $y^2 + x^4 - x^3 = 0$ $\Leftrightarrow$ $y^2 = x^3 - x^4$. The left side is non-negative, so we must have $x^3 - x^4 \geq 0$

which is true only for $0 \leq x \leq 1$. Therefore the minimum possible value for $f(x, y) = x$ is 0 which occurs for $x = y = 0$.

However, $\lambda \nabla g(0, 0) = \lambda \langle 0 - 0, 0 \rangle = \langle 0, 0 \rangle$ and $\nabla f(0, 0) = \langle 1, 0 \rangle$, so $\nabla f(0, 0) \neq \lambda \nabla g(0, 0)$ for all values of λ.

(c) Here $\nabla g(0, 0) = \mathbf{0}$ but the method of Lagrange multipliers requires that $\nabla g \neq \mathbf{0}$ everywhere on the constraint curve.

© 2012 Cengage Learning. All Rights Reserved. May not be scanned, copied, or duplicated, or posted to a publicly accessible website, in whole or in part.

25. $P(L, K) = bL^\alpha K^{1-\alpha}$, $g(L, K) = mL + nK = p \Rightarrow \nabla P = \langle \alpha b L^{\alpha-1} K^{1-\alpha}, (1-\alpha)bL^\alpha K^{-\alpha} \rangle$, $\lambda \nabla g = \langle \lambda m, \lambda n \rangle$.

Then $\alpha b(K/L)^{1-\alpha} = \lambda m$ and $(1-\alpha)b(L/K)^\alpha = \lambda n$ and $mL + nK = p$, so $\alpha b(K/L)^{1-\alpha}/m = (1-\alpha)b(L/K)^\alpha/n$ or

$n\alpha/[m(1-\alpha)] = (L/K)^\alpha (L/K)^{1-\alpha}$ or $L = Kn\alpha/[m(1-\alpha)]$. Substituting into $mL + nK = p$ gives $K = (1-\alpha)p/n$

and $L = \alpha p/m$ for the maximum production.

27. Let the sides of the rectangle be x and y. Then $f(x, y) = xy$, $g(x, y) = 2x + 2y = p \Rightarrow \nabla f(x, y) = \langle y, x \rangle$,

$\lambda \nabla g = \langle 2\lambda, 2\lambda \rangle$. Then $\lambda = \frac{1}{2}y = \frac{1}{2}x$ implies $x = y$ and the rectangle with maximum area is a square with side length $\frac{1}{4}p$.

29. The distance from $(2, 0, -3)$ to a point (x, y, z) on the plane is $d = \sqrt{(x-2)^2 + y^2 + (z+3)^2}$, so we seek to minimize

$d^2 = f(x, y, z) = (x-2)^2 + y^2 + (z+3)^2$ subject to the constraint that (x, y, z) lies on the plane $x + y + z = 1$, that is,

that $g(x, y, z) = x + y + z = 1$. Then $\nabla f = \lambda \nabla g \Rightarrow \langle 2(x-2), 2y, 2(z+3) \rangle = \langle \lambda, \lambda, \lambda \rangle$, so $x = (\lambda + 4)/2$,

$y = \lambda/2$, $z = (\lambda - 6)/2$. Substituting into the constraint equation gives $\dfrac{\lambda+4}{2} + \dfrac{\lambda}{2} + \dfrac{\lambda-6}{2} = 1 \Rightarrow 3\lambda - 2 = 2 \Rightarrow$

$\lambda = \frac{4}{3}$, so $x = \frac{8}{3}$, $y = \frac{2}{3}$, and $z = -\frac{7}{3}$. This must correspond to a minimum, so the shortest distance is

$d = \sqrt{\left(\frac{8}{3} - 2\right)^2 + \left(\frac{2}{3}\right)^2 + \left(-\frac{7}{3} + 3\right)^2} = \sqrt{\frac{4}{3}} = \frac{2}{\sqrt{3}}$.

31. Let $f(x, y, z) = d^2 = (x-4)^2 + (y-2)^2 + z^2$. Then we want to minimize f subject to the constraint

$g(x, y, z) = x^2 + y^2 - z^2 = 0$. $\nabla f = \lambda \nabla g \Rightarrow \langle 2(x-4), 2(y-2), 2z \rangle = \langle 2\lambda x, 2\lambda y, -2\lambda z \rangle$, so $x - 4 = \lambda x$,

$y - 2 = \lambda y$, and $z = -\lambda z$. From the last equation we have $z + \lambda z = 0 \Rightarrow z(1 + \lambda) = 0$, so either $z = 0$ or $\lambda = -1$.

But from the constraint equation we have $z = 0 \Rightarrow x^2 + y^2 = 0 \Rightarrow x = y = 0$ which is not possible from the first

two equations. So $\lambda = -1$ and $x - 4 = \lambda x \Rightarrow x = 2$, $y - 2 = \lambda y \Rightarrow y = 1$, and $x^2 + y^2 - z^2 = 0 \Rightarrow$

$4 + 1 - z^2 = 0 \Rightarrow z = \pm\sqrt{5}$. This must correspond to a minimum, so the points on the cone closest to $(4, 2, 0)$

are $\left(2, 1, \pm\sqrt{5}\right)$.

33. $f(x, y, z) = xyz$, $g(x, y, z) = x + y + z = 100 \Rightarrow \nabla f = \langle yz, xz, xy \rangle = \lambda \nabla g = \langle \lambda, \lambda, \lambda \rangle$. Then $\lambda = yz = xz = xy$

implies $x = y = z = \frac{100}{3}$.

35. If the dimensions are $2x$, $2y$, and $2z$, then maximize $f(x, y, z) = (2x)(2y)(2z) = 8xyz$ subject to

$g(x, y, z) = x^2 + y^2 + z^2 = r^2$ $(x > 0, y > 0, z > 0)$. Then $\nabla f = \lambda \nabla g \Rightarrow \langle 8yz, 8xz, 8xy \rangle = \lambda \langle 2x, 2y, 2z \rangle \Rightarrow$

$8yz = 2\lambda x$, $8xz = 2\lambda y$, and $8xy = 2\lambda z$, so $\lambda = \dfrac{4yz}{x} = \dfrac{4xz}{y} = \dfrac{4xy}{z}$. This gives $x^2 z = y^2 z \Rightarrow x^2 = y^2$ (since $z \neq 0$)

and $xy^2 = xz^2 \Rightarrow z^2 = y^2$, so $x^2 = y^2 = z^2 \Rightarrow x = y = z$, and substituting into the constraint

equation gives $3x^2 = r^2 \Rightarrow x = r/\sqrt{3} = y = z$. Thus the largest volume of such a box is

$f\left(\dfrac{r}{\sqrt{3}}, \dfrac{r}{\sqrt{3}}, \dfrac{r}{\sqrt{3}}\right) = 8\left(\dfrac{r}{\sqrt{3}}\right)\left(\dfrac{r}{\sqrt{3}}\right)\left(\dfrac{r}{\sqrt{3}}\right) = \dfrac{8}{3\sqrt{3}}r^3$.

© 2012 Cengage Learning. All Rights Reserved. May not be scanned, copied, or duplicated, or posted to a publicly accessible website, in whole or in part.

37. $f(x, y, z) = xyz$, $g(x, y, z) = x + 2y + 3z = 6$ $\Rightarrow$ $\nabla f = \langle yz, xz, xy \rangle = \lambda \nabla g = \langle \lambda, 2\lambda, 3\lambda \rangle$.

Then $\lambda = yz = \frac{1}{2}xz = \frac{1}{3}xy$ implies $x = 2y$, $z = \frac{2}{3}y$. But $2y + 2y + 2y = 6$ so $y = 1$, $x = 2$, $z = \frac{2}{3}$ and the volume

is $V = \frac{4}{3}$.

39. $f(x, y, z) = xyz$, $g(x, y, z) = 4(x + y + z) = c$ $\Rightarrow$ $\nabla f = \langle yz, xz, xy \rangle$, $\lambda \nabla g = \langle 4\lambda, 4\lambda, 4\lambda \rangle$. Thus

$4\lambda = yz = xz = xy$ or $x = y = z = \frac{1}{12}c$ are the dimensions giving the maximum volume.

41. If the dimensions of the box are given by x, y, and z, then we need to find the maximum value of $f(x, y, z) = xyz$

$[x, y, z > 0]$ subject to the constraint $L = \sqrt{x^2 + y^2 + z^2}$ or $g(x, y, z) = x^2 + y^2 + z^2 = L^2$. $\nabla f = \lambda \nabla g$ $\Rightarrow$

$\langle yz, xz, xy \rangle = \lambda \langle 2x, 2y, 2z \rangle$, so $yz = 2\lambda x$ $\Rightarrow$ $\lambda = \dfrac{yz}{2x}$, $xz = 2\lambda y$ $\Rightarrow$ $\lambda = \dfrac{xz}{2y}$, and $xy = 2\lambda z$ $\Rightarrow$ $\lambda = \dfrac{xy}{2z}$.

Thus $\lambda = \dfrac{yz}{2x} = \dfrac{xz}{2y}$ $\Rightarrow$ $x^2 = y^2$ [since $z \neq 0$] $\Rightarrow$ $x = y$ and $\lambda = \dfrac{yz}{2x} = \dfrac{xy}{2z}$ $\Rightarrow$ $x = z$ [since $y \neq 0$].

Substituting into the constraint equation gives $x^2 + x^2 + x^2 = L^2$ $\Rightarrow$ $x^2 = L^2/3$ $\Rightarrow$ $x = L/\sqrt{3} = y = z$ and the

maximum volume is $\left(L/\sqrt{3}\right)^3 = L^3/(3\sqrt{3})$.

43. We need to find the extreme values of $f(x, y, z) = x^2 + y^2 + z^2$ subject to the two constraints $g(x, y, z) = x + y + 2z = 2$

and $h(x, y, z) = x^2 + y^2 - z = 0$. $\nabla f = \langle 2x, 2y, 2z \rangle$, $\lambda \nabla g = \langle \lambda, \lambda, 2\lambda \rangle$ and $\mu \nabla h = \langle 2\mu x, 2\mu y, -\mu \rangle$. Thus we need

$2x = \lambda + 2\mu x$ **(1)**, $2y = \lambda + 2\mu y$ **(2)**, $2z = 2\lambda - \mu$ **(3)**, $x + y + 2z = 2$ **(4)**, and $x^2 + y^2 - z = 0$ **(5)**.

From **(1)** and **(2)**, $2(x - y) = 2\mu(x - y)$, so if $x \neq y$, $\mu = 1$. Putting this in **(3)** gives $2z = 2\lambda - 1$ or $\lambda = z + \frac{1}{2}$, but putting

$\mu = 1$ into **(1)** says $\lambda = 0$. Hence $z + \frac{1}{2} = 0$ or $z = -\frac{1}{2}$. Then **(4)** and **(5)** become $x + y - 3 = 0$ and $x^2 + y^2 + \frac{1}{2} = 0$. The

last equation cannot be true, so this case gives no solution. So we must have $x = y$. Then **(4)** and **(5)** become $2x + 2z = 2$ and

$2x^2 - z = 0$ which imply $z = 1 - x$ and $z = 2x^2$. Thus $2x^2 = 1 - x$ or $2x^2 + x - 1 = (2x - 1)(x + 1) = 0$ so $x = \frac{1}{2}$ or

$x = -1$. The two points to check are $\left(\frac{1}{2}, \frac{1}{2}, \frac{1}{2}\right)$ and $(-1, -1, 2)$: $f\left(\frac{1}{2}, \frac{1}{2}, \frac{1}{2}\right) = \frac{3}{4}$ and $f(-1, -1, 2) = 6$. Thus $\left(\frac{1}{2}, \frac{1}{2}, \frac{1}{2}\right)$ is

the point on the ellipse nearest the origin and $(-1, -1, 2)$ is the one farthest from the origin.

45. $f(x, y, z) = ye^{x-z}$, $g(x, y, z) = 9x^2 + 4y^2 + 36z^2 = 36$, $h(x, y, z) = xy + yz = 1$. $\nabla f = \lambda \nabla g + \mu \nabla h$ $\Rightarrow$

$\langle ye^{x-z}, e^{x-z}, -ye^{x-z} \rangle = \lambda \langle 18x, 8y, 72z \rangle + \mu \langle y, x + z, y \rangle$, so $ye^{x-z} = 18\lambda x + \mu y$, $e^{x-z} = 8\lambda y + \mu(x + z)$,

$-ye^{x-z} = 72\lambda z + \mu y$, $9x^2 + 4y^2 + 36z^2 = 36$, $xy + yz = 1$. Using a CAS to solve these 5 equations simultaneously for x,

y, z, λ, and μ (in Maple, use the `allvalues` command), we get 4 real-valued solutions:

$$x \approx 0.222444, \qquad y \approx -2.157012, \qquad z \approx -0.686049, \qquad \lambda \approx -0.200401, \qquad \mu \approx 2.108584$$

$$x \approx -1.951921, \qquad y \approx -0.545867, \qquad z \approx 0.119973, \qquad \lambda \approx 0.003141, \qquad \mu \approx -0.076238$$

$$x \approx 0.155142, \qquad y \approx 0.904622, \qquad z \approx 0.950293, \qquad \lambda \approx -0.012447, \qquad \mu \approx 0.489938$$

$$x \approx 1.138731, \qquad y \approx 1.768057, \qquad z \approx -0.573138, \qquad \lambda \approx 0.317141, \qquad \mu \approx 1.862675$$

Substituting these values into f gives $f(0.222444, -2.157012, -0.686049) \approx -5.3506$,

© 2012 Cengage Learning. All Rights Reserved. May not be scanned, copied, or duplicated, or posted to a publicly accessible website, in whole or in part.

$f(-1.951921, -0.545867, 0.119973) \approx -0.0688$, $f(0.155142, 0.904622, 0.950293) \approx 0.4084$,

$f(1.138731, 1.768057, -0.573138) \approx 9.7938$. Thus the maximum is approximately 9.7938, and the minimum is approximately -5.3506.

47. (a) We wish to maximize $f(x_1, x_2, \ldots, x_n) = \sqrt[n]{x_1 x_2 \cdots x_n}$ subject to

$g(x_1, x_2, \ldots, x_n) = x_1 + x_2 + \cdots + x_n = c$ and $x_i > 0$.

$\nabla f = \left\langle \frac{1}{n}(x_1 x_2 \cdots x_n)^{\frac{1}{n}-1}(x_2 \cdots x_n), \frac{1}{n}(x_1 x_2 \cdots x_n)^{\frac{1}{n}-1}(x_1 x_3 \cdots x_n), \ldots, \frac{1}{n}(x_1 x_2 \cdots x_n)^{\frac{1}{n}-1}(x_1 \cdots x_{n-1}) \right\rangle$

and $\lambda \nabla g = \langle \lambda, \lambda, \ldots, \lambda \rangle$, so we need to solve the system of equations

$$\frac{1}{n}(x_1 x_2 \cdots x_n)^{\frac{1}{n}-1}(x_2 \cdots x_n) = \lambda \quad \Rightarrow \quad x_1^{1/n} x_2^{1/n} \cdots x_n^{1/n} = n\lambda x_1$$

$$\frac{1}{n}(x_1 x_2 \cdots x_n)^{\frac{1}{n}-1}(x_1 x_3 \cdots x_n) = \lambda \quad \Rightarrow \quad x_1^{1/n} x_2^{1/n} \cdots x_n^{1/n} = n\lambda x_2$$

$$\vdots$$

$$\frac{1}{n}(x_1 x_2 \cdots x_n)^{\frac{1}{n}-1}(x_1 \cdots x_{n-1}) = \lambda \quad \Rightarrow \quad x_1^{1/n} x_2^{1/n} \cdots x_n^{1/n} = n\lambda x_n$$

This implies $n\lambda x_1 = n\lambda x_2 = \cdots = n\lambda x_n$. Note $\lambda \neq 0$, otherwise we can't have all $x_i > 0$. Thus $x_1 = x_2 = \cdots = x_n$. But $x_1 + x_2 + \cdots + x_n = c \Rightarrow nx_1 = c \Rightarrow x_1 = \frac{c}{n} = x_2 = x_3 = \cdots = x_n$. Then the only point where f can have an extreme value is $\left(\frac{c}{n}, \frac{c}{n}, \ldots, \frac{c}{n}\right)$. Since we can choose values for $(x_1, x_2, \ldots, x_n)$ that make f as close to zero (but not equal) as we like, f has no minimum value. Thus the maximum value is

$f\left(\frac{c}{n}, \frac{c}{n}, \ldots, \frac{c}{n}\right) = \sqrt[n]{\frac{c}{n} \cdot \frac{c}{n} \cdot \cdots \cdot \frac{c}{n}} = \frac{c}{n}$.

(b) From part (a), $\frac{c}{n}$ is the maximum value of f. Thus $f(x_1, x_2, \ldots, x_n) = \sqrt[n]{x_1 x_2 \cdots x_n} \leq \frac{c}{n}$. But

$x_1 + x_2 + \cdots + x_n = c$, so $\sqrt[n]{x_1 x_2 \cdots x_n} \leq \dfrac{x_1 + x_2 + \cdots + x_n}{n}$. These two means are equal when f attains its

maximum value $\frac{c}{n}$, but this can occur only at the point $\left(\frac{c}{n}, \frac{c}{n}, \ldots, \frac{c}{n}\right)$ we found in part (a). So the means are equal only

when $x_1 = x_2 = x_3 = \cdots = x_n = \dfrac{c}{n}$.

14 Review

CONCEPT CHECK

1. (a) A function f of two variables is a rule that assigns to each ordered pair (x, y) of real numbers in its domain a unique real number denoted by $f(x, y)$.

(b) One way to visualize a function of two variables is by graphing it, resulting in the surface $z = f(x, y)$. Another method for visualizing a function of two variables is a contour map. The contour map consists of level curves of the function which are horizontal traces of the graph of the function projected onto the xy-plane. Also, we can use an arrow diagram such as Figure 1 in Section 14.1.

© 2012 Cengage Learning. All Rights Reserved. May not be scanned, copied, or duplicated, or posted to a publicly accessible website, in whole or in part.

2. A function f of three variables is a rule that assigns to each ordered triple (x, y, z) in its domain a unique real number $f(x, y, z)$. We can visualize a function of three variables by examining its level surfaces $f(x, y, z) = k$, where k is a constant.

3. $\lim\limits_{(x,y)\to(a,b)} f(x, y) = L$ means the values of $f(x, y)$ approach the number L as the point (x, y) approaches the point (a, b) along any path that is within the domain of f. We can show that a limit at a point does not exist by finding two different paths approaching the point along which $f(x, y)$ has different limits.

4. (a) See Definition 14.2.4.

   (b) If f is continuous on $\mathbb{R}^2$, its graph will appear as a surface without holes or breaks.

5. (a) See (2) and (3) in Section 14.3.

   (b) See "Interpretations of Partial Derivatives" on page 927 [ET 903].

   (c) To find f_x, regard y as a constant and differentiate $f(x, y)$ with respect to x. To find f_y, regard x as a constant and differentiate $f(x, y)$ with respect to y.

6. See the statement of Clairaut's Theorem on page 931 [ET 907].

7. (a) See (2) in Section 14.4.

   (b) See (19) and the preceding discussion in Section 14.6.

8. See (3) and (4) and the accompanying discussion in Section 14.4. We can interpret the linearization of f at (a, b) geometrically as the linear function whose graph is the tangent plane to the graph of f at (a, b). Thus it is the linear function which best approximates f near (a, b).

9. (a) See Definition 14.4.7.

   (b) Use Theorem 14.4.8.

10. See (10) and the associated discussion in Section 14.4.

11. See (2) and (3) in Section 14.5.

12. See (7) and the preceding discussion in Section 14.5.

13. (a) See Definition 14.6.2. We can interpret it as the rate of change of f at (x_0, y_0) in the direction of $\mathbf{u}$. Geometrically, if P is the point $(x_0, y_0, f(x_0, y_0))$ on the graph of f and C is the curve of intersection of the graph of f with the vertical plane that passes through P in the direction $\mathbf{u}$, the directional derivative of f at (x_0, y_0) in the direction of $\mathbf{u}$ is the slope of the tangent line to C at P. (See Figure 5 in Section 14.6.)

   (b) See Theorem 14.6.3.

14. (a) See (8) and (13) in Section 14.6.

   (b) $D_{\mathbf{u}} f(x, y) = \nabla f(x, y) \cdot \mathbf{u}$ or $D_{\mathbf{u}} f(x, y, z) = \nabla f(x, y, z) \cdot \mathbf{u}$

© 2012 Cengage Learning. All Rights Reserved. May not be scanned, copied, or duplicated, or posted to a publicly accessible website, in whole or in part.

(c) The gradient vector of a function points in the direction of maximum rate of increase of the function. On a graph of the function, the gradient points in the direction of steepest ascent.

15. (a) f has a local maximum at (a, b) if $f(x, y) \leq f(a, b)$ when (x, y) is near (a, b).

   (b) f has an absolute maximum at (a, b) if $f(x, y) \leq f(a, b)$ for all points (x, y) in the domain of f.

   (c) f has a local minimum at (a, b) if $f(x, y) \geq f(a, b)$ when (x, y) is near (a, b).

   (d) f has an absolute minimum at (a, b) if $f(x, y) \geq f(a, b)$ for all points (x, y) in the domain of f.

   (e) f has a saddle point at (a, b) if $f(a, b)$ is a local maximum in one direction but a local minimum in another.

16. (a) By Theorem 14.7.2, if f has a local maximum at (a, b) and the first-order partial derivatives of f exist there, then

$$f_x(a, b) = 0 \text{ and } f_y(a, b) = 0.$$

   (b) A critical point of f is a point (a, b) such that $f_x(a, b) = 0$ and $f_y(a, b) = 0$ or one of these partial derivatives does not exist.

17. See (3) in Section 14.7.

18. (a) See Figure 11 and the accompanying discussion in Section 14.7.

   (b) See Theorem 14.7.8.

   (c) See the procedure outlined in (9) in Section 14.7.

19. See the discussion beginning on page 981 [ET 957]; see "Two Constraints" on page 985 [ET 961].

TRUE-FALSE QUIZ

1. True. $f_y(a, b) = \lim\limits_{h \to 0} \dfrac{f(a, b + h) - f(a, b)}{h}$ from Equation 14.3.3. Let $h = y - b$. As $h \to 0$, $y \to b$. Then by substituting,

we get $f_y(a, b) = \lim\limits_{y \to b} \dfrac{f(a, y) - f(a, b)}{y - b}$.

3. False. $f_{xy} = \dfrac{\partial^2 f}{\partial y \, \partial x}$.

5. False. See Example 14.2.3.

7. True. If f has a local minimum and f is differentiable at (a, b) then by Theorem 14.7.2, $f_x(a, b) = 0$ and $f_y(a, b) = 0$, so

$$\nabla f(a, b) = \langle f_x(a, b), f_y(a, b) \rangle = \langle 0, 0 \rangle = \mathbf{0}.$$

9. False. $\nabla f(x, y) = \langle 0, 1/y \rangle$.

11. True. $\nabla f = \langle \cos x, \cos y \rangle$, so $|\nabla f| = \sqrt{\cos^2 x + \cos^2 y}$. But $|\cos \theta| \leq 1$, so $|\nabla f| \leq \sqrt{2}$. Now

$$D_{\mathbf{u}} f(x, y) = \nabla f \cdot \mathbf{u} = |\nabla f| \, |\mathbf{u}| \cos \theta, \text{ but } \mathbf{u} \text{ is a unit vector, so } |D_{\mathbf{u}} f(x, y)| \leq \sqrt{2} \cdot 1 \cdot 1 = \sqrt{2}.$$

© 2012 Cengage Learning. All Rights Reserved. May not be scanned, copied, or duplicated, or posted to a publicly accessible website, in whole or in part.

EXERCISES

1. $\ln(x + y + 1)$ is defined only when $x + y + 1 > 0$ $\Leftrightarrow$ $y > -x - 1$, so the domain of f is $\{(x, y) \mid y > -x - 1\}$, all those points above the line $y = -x - 1$.

3. $z = f(x, y) = 1 - y^2$, a parabolic cylinder

5. The level curves are $\sqrt{4x^2 + y^2} = k$ or $4x^2 + y^2 = k^2$, $k \geq 0$, a family of ellipses.

7.

9. f is a rational function, so it is continuous on its domain. Since f is defined at $(1, 1)$, we use direct substitution to evaluate the limit: $\displaystyle\lim_{(x,y) \to (1,1)} \frac{2xy}{x^2 + 2y^2} = \frac{2(1)(1)}{1^2 + 2(1)^2} = \frac{2}{3}$.

11. (a) $T_x(6, 4) = \displaystyle\lim_{h \to 0} \frac{T(6 + h, 4) - T(6, 4)}{h}$, so we can approximate $T_x(6, 4)$ by considering $h = \pm 2$ and

using the values given in the table: $T_x(6, 4) \approx \dfrac{T(8, 4) - T(6, 4)}{2} = \dfrac{86 - 80}{2} = 3$,

$T_x(6, 4) \approx \dfrac{T(4, 4) - T(6, 4)}{-2} = \dfrac{72 - 80}{-2} = 4$. Averaging these values, we estimate $T_x(6, 4)$ to be approximately

$3.5°\text{C/m}$. Similarly, $T_y(6, 4) = \displaystyle\lim_{h \to 0} \frac{T(6, 4 + h) - T(6, 4)}{h}$, which we can approximate with $h = \pm 2$:

$T_y(6, 4) \approx \dfrac{T(6, 6) - T(6, 4)}{2} = \dfrac{75 - 80}{2} = -2.5$, $T_y(6, 4) \approx \dfrac{T(6, 2) - T(6, 4)}{-2} = \dfrac{87 - 80}{-2} = -3.5$. Averaging these

values, we estimate $T_y(6, 4)$ to be approximately $-3.0°\text{C/m}$.

© 2012 Cengage Learning. All Rights Reserved. May not be scanned, copied, or duplicated, or posted to a publicly accessible website, in whole or in part.

(b) Here $\mathbf{u} = \left\langle \frac{1}{\sqrt{2}}, \frac{1}{\sqrt{2}} \right\rangle$, so by Equation 14.6.9, $D_{\mathbf{u}} T(6,4) = \nabla T(6,4) \cdot \mathbf{u} = T_x(6,4) \frac{1}{\sqrt{2}} + T_y(6,4) \frac{1}{\sqrt{2}}$. Using our

estimates from part (a), we have $D_{\mathbf{u}} T(6,4) \approx (3.5) \frac{1}{\sqrt{2}} + (-3.0) \frac{1}{\sqrt{2}} = \frac{1}{2\sqrt{2}} \approx 0.35$. This means that as we move

through the point $(6,4)$ in the direction of $\mathbf{u}$, the temperature increases at a rate of approximately $0.35°\text{C/m}$.

Alternatively, we can use Definition 14.6.2: $D_{\mathbf{u}} T(6,4) = \lim\limits_{h \to 0} \dfrac{T\left(6 + h \frac{1}{\sqrt{2}}, 4 + h \frac{1}{\sqrt{2}}\right) - T(6,4)}{h}$,

which we can estimate with $h = \pm 2\sqrt{2}$. Then $D_{\mathbf{u}} T(6,4) \approx \dfrac{T(8,6) - T(6,4)}{2\sqrt{2}} = \dfrac{80 - 80}{2\sqrt{2}} = 0$,

$D_{\mathbf{u}} T(6,4) \approx \dfrac{T(4,2) - T(6,4)}{-2\sqrt{2}} = \dfrac{74 - 80}{-2\sqrt{2}} = \dfrac{3}{\sqrt{2}}$. Averaging these values, we have $D_{\mathbf{u}} T(6,4) \approx \frac{3}{2\sqrt{2}} \approx 1.1°\text{C/m}$.

(c) $T_{xy}(x,y) = \dfrac{\partial}{\partial y}\left[T_x(x,y)\right] = \lim\limits_{h \to 0} \dfrac{T_x(x, y+h) - T_x(x,y)}{h}$, so $T_{xy}(6,4) = \lim\limits_{h \to 0} \dfrac{T_x(6, 4+h) - T_x(6,4)}{h}$ which we can

estimate with $h = \pm 2$. We have $T_x(6,4) \approx 3.5$ from part (a), but we will also need values for $T_x(6,6)$ and $T_x(6,2)$. If we

use $h = \pm 2$ and the values given in the table, we have

$T_x(6,6) \approx \dfrac{T(8,6) - T(6,6)}{2} = \dfrac{80 - 75}{2} = 2.5, \; T_x(6,6) \approx \dfrac{T(4,6) - T(6,6)}{-2} = \dfrac{68 - 75}{-2} = 3.5$.

Averaging these values, we estimate $T_x(6,6) \approx 3.0$. Similarly,

$T_x(6,2) \approx \dfrac{T(8,2) - T_x(6,2)}{2} = \dfrac{90 - 87}{2} = 1.5, \; T_x(6,2) \approx \dfrac{T(4,2) - T(6,2)}{-2} = \dfrac{74 - 87}{-2} = 6.5$.

Averaging these values, we estimate $T_x(6,2) \approx 4.0$. Finally, we estimate $T_{xy}(6,4)$:

$T_{xy}(6,4) \approx \dfrac{T_x(6,6) - T_x(6,4)}{2} = \dfrac{3.0 - 3.5}{2} = -0.25, \; T_{xy}(6,4) \approx \dfrac{T_x(6,2) - T_x(6,4)}{-2} = \dfrac{4.0 - 3.5}{-2} = -0.25$.

Averaging these values, we have $T_{xy}(6,4) \approx -0.25$.

13. $f(x,y) = (5y^3 + 2x^2 y)^8 \quad \Rightarrow \quad f_x = 8(5y^3 + 2x^2 y)^7 (4xy) = 32xy(5y^3 + 2x^2 y)^7$,

$f_y = 8(5y^3 + 2x^2 y)^7 (15y^2 + 2x^2) = (16x^2 + 120y^2)(5y^3 + 2x^2 y)^7$

15. $F(\alpha, \beta) = \alpha^2 \ln(\alpha^2 + \beta^2) \quad \Rightarrow \quad F_\alpha = \alpha^2 \cdot \dfrac{1}{\alpha^2 + \beta^2}\,(2\alpha) + \ln(\alpha^2 + \beta^2) \cdot 2\alpha = \dfrac{2\alpha^3}{\alpha^2 + \beta^2} + 2\alpha \ln(\alpha^2 + \beta^2)$,

$F_\beta = \alpha^2 \cdot \dfrac{1}{\alpha^2 + \beta^2}\,(2\beta) = \dfrac{2\alpha^2 \beta}{\alpha^2 + \beta^2}$

17. $S(u,v,w) = u \arctan(v\sqrt{w}) \quad \Rightarrow \quad S_u = \arctan(v\sqrt{w}), \; S_v = u \cdot \dfrac{1}{1 + (v\sqrt{w})^2}\,(\sqrt{w}) = \dfrac{u\sqrt{w}}{1 + v^2 w}$,

$S_w = u \cdot \dfrac{1}{1 + (v\sqrt{w})^2}\left(v \cdot \tfrac{1}{2} w^{-1/2}\right) = \dfrac{uv}{2\sqrt{w}\,(1 + v^2 w)}$

19. $f(x,y) = 4x^3 - xy^2 \quad \Rightarrow \quad f_x = 12x^2 - y^2, \; f_y = -2xy, \; f_{xx} = 24x, \; f_{yy} = -2x, \; f_{xy} = f_{yx} = -2y$

© 2012 Cengage Learning. All Rights Reserved. May not be scanned, copied, or duplicated, or posted to a publicly accessible website, in whole or in part.

21. $f(x, y, z) = x^k y^l z^m \quad \Rightarrow \quad f_x = kx^{k-1} y^l z^m, \; f_y = lx^k y^{l-1} z^m, \; f_z = mx^k y^l z^{m-1}, \; f_{xx} = k(k-1)x^{k-2} y^l z^m,$

$f_{yy} = l(l-1)x^k y^{l-2} z^m, \; f_{zz} = m(m-1)x^k y^l z^{m-2}, \; f_{xy} = f_{yx} = klx^{k-1} y^{l-1} z^m, \; f_{xz} = f_{zx} = kmx^{k-1} y^l z^{m-1},$

$f_{yz} = f_{zy} = lmx^k y^{l-1} z^{m-1}$

23. $z = xy + xe^{y/x} \quad \Rightarrow \quad \dfrac{\partial z}{\partial x} = y - \dfrac{y}{x}e^{y/x} + e^{y/x}, \; \dfrac{\partial z}{\partial y} = x + e^{y/x}$ and

$$x\dfrac{\partial z}{\partial x} + y\dfrac{\partial z}{\partial y} = x\left(y - \dfrac{y}{x}e^{y/x} + e^{y/x}\right) + y\left(x + e^{y/x}\right) = xy - ye^{y/x} + xe^{y/x} + xy + ye^{y/x} = xy + xy + xe^{y/x} = xy + z.$$

25. (a) $z_x = 6x + 2 \quad \Rightarrow \quad z_x(1, -2) = 8$ and $z_y = -2y \quad \Rightarrow \quad z_y(1, -2) = 4$, so an equation of the tangent plane is

$z - 1 = 8(x - 1) + 4(y + 2)$ or $z = 8x + 4y + 1$.

(b) A normal vector to the tangent plane (and the surface) at $(1, -2, 1)$ is $\langle 8, 4, -1 \rangle$. Then parametric equations for the normal

line there are $x = 1 + 8t, \; y = -2 + 4t, \; z = 1 - t$, and symmetric equations are $\dfrac{x - 1}{8} = \dfrac{y + 2}{4} = \dfrac{z - 1}{-1}$.

27. (a) Let $F(x, y, z) = x^2 + 2y^2 - 3z^2$. Then $F_x = 2x, \; F_y = 4y, \; F_z = -6z$, so $F_x(2, -1, 1) = 4, \; F_y(2, -1, 1) = -4,$

$F_z(2, -1, 1) = -6$. From Equation 14.6.19, an equation of the tangent plane is $4(x - 2) - 4(y + 1) - 6(z - 1) = 0$

or, equivalently, $2x - 2y - 3z = 3$.

(b) From Equations 14.6.20, symmetric equations for the normal line are $\dfrac{x - 2}{4} = \dfrac{y + 1}{-4} = \dfrac{z - 1}{-6}$.

29. (a) Let $F(x, y, z) = x + 2y + 3z - \sin(xyz)$. Then $F_x = 1 - yz\cos(xyz), \; F_y = 2 - xz\cos(xyz), \; F_z = 3 - xy\cos(xyz),$

so $F_x(2, -1, 0) = 1, \; F_y(2, -1, 0) = 2, \; F_z(2, -1, 0) = 5$. From Equation 14.6.19, an equation of the tangent plane is

$1(x - 2) + 2(y + 1) + 5(z - 0) = 0$ or $x + 2y + 5z = 0$.

(b) From Equations 14.6.20, symmetric equations for the normal line are $\dfrac{x - 2}{1} = \dfrac{y + 1}{2} = \dfrac{z}{5}$ or $x - 2 = \dfrac{y + 1}{2} = \dfrac{z}{5}$.

Parametric equations are $x = 2 + t, \; y = -1 + 2t, \; z = 5t$.

31. The hyperboloid is a level surface of the function $F(x, y, z) = x^2 + 4y^2 - z^2$, so a normal vector to the surface at (x_0, y_0, z_0)

is $\nabla F(x_0, y_0, z_0) = \langle 2x_0, 8y_0, -2z_0 \rangle$. A normal vector for the plane $2x + 2y + z = 5$ is $\langle 2, 2, 1 \rangle$. For the planes to be

parallel, we need the normal vectors to be parallel, so $\langle 2x_0, 8y_0, -2z_0 \rangle = k\langle 2, 2, 1 \rangle$, or $x_0 = k, \; y_0 = \frac{1}{4}k$, and $z_0 = -\frac{1}{2}k$.

But $x_0^2 + 4y_0^2 - z_0^2 = 4 \quad \Rightarrow \quad k^2 + \frac{1}{4}k^2 - \frac{1}{4}k^2 = 4 \quad \Rightarrow \quad k^2 = 4 \quad \Rightarrow \quad k = \pm 2$. So there are two such points:

$\left(2, \frac{1}{2}, -1\right)$ and $\left(-2, -\frac{1}{2}, 1\right)$.

33. $f(x, y, z) = x^3 \sqrt{y^2 + z^2} \quad \Rightarrow \quad f_x(x, y, z) = 3x^2 \sqrt{y^2 + z^2}, \; f_y(x, y, z) = \dfrac{yx^3}{\sqrt{y^2 + z^2}}, \; f_z(x, y, z) = \dfrac{zx^3}{\sqrt{y^2 + z^2}},$

so $f(2, 3, 4) = 8(5) = 40, \; f_x(2, 3, 4) = 3(4)\sqrt{25} = 60, \; f_y(2, 3, 4) = \dfrac{3(8)}{\sqrt{25}} = \dfrac{24}{5}$, and $f_z(2, 3, 4) = \dfrac{4(8)}{\sqrt{25}} = \dfrac{32}{5}$. Then the

© 2012 Cengage Learning. All Rights Reserved. May not be scanned, copied, or duplicated, or posted to a publicly accessible website, in whole or in part.

linear approximation of f at $(2, 3, 4)$ is

$$f(x, y, z) \approx f(2, 3, 4) + f_x(2, 3, 4)(x - 2) + f_y(2, 3, 4)(y - 3) + f_z(2, 3, 4)(z - 4)$$

$$= 40 + 60(x - 2) + \tfrac{24}{5}(y - 3) + \tfrac{32}{5}(z - 4) = 60x + \tfrac{24}{5}y + \tfrac{32}{5}z - 120$$

Then $(1.98)^3 \sqrt{(3.01)^2 + (3.97)^2} = f(1.98, 3.01, 3.97) \approx 60(1.98) + \tfrac{24}{5}(3.01) + \tfrac{32}{5}(3.97) - 120 = 38.656$.

35. $\dfrac{du}{dp} = \dfrac{\partial u}{\partial x}\dfrac{dx}{dp} + \dfrac{\partial u}{\partial y}\dfrac{dy}{dp} + \dfrac{\partial u}{\partial z}\dfrac{dz}{dp} = 2xy^3(1 + 6p) + 3x^2y^2(pe^p + e^p) + 4z^3(p\cos p + \sin p)$

37. By the Chain Rule, $\dfrac{\partial z}{\partial s} = \dfrac{\partial z}{\partial x}\dfrac{\partial x}{\partial s} + \dfrac{\partial z}{\partial y}\dfrac{\partial y}{\partial s}$. When $s = 1$ and $t = 2$, $x = g(1, 2) = 3$ and $y = h(1, 2) = 6$, so

$\dfrac{\partial z}{\partial s} = f_x(3, 6)g_s(1, 2) + f_y(3, 6)h_s(1, 2) = (7)(-1) + (8)(-5) = -47$. Similarly, $\dfrac{\partial z}{\partial t} = \dfrac{\partial z}{\partial x}\dfrac{\partial x}{\partial t} + \dfrac{\partial z}{\partial y}\dfrac{\partial y}{\partial t}$, so

$\dfrac{\partial z}{\partial t} = f_x(3, 6)g_t(1, 2) + f_y(3, 6)h_t(1, 2) = (7)(4) + (8)(10) = 108$.

39. $\dfrac{\partial z}{\partial x} = 2xf'(x^2 - y^2)$, $\quad \dfrac{\partial z}{\partial y} = 1 - 2yf'(x^2 - y^2)$ $\quad \left[\text{where } f' = \dfrac{df}{d(x^2 - y^2)} \right]$. Then

$y\,\dfrac{\partial z}{\partial x} + x\,\dfrac{\partial z}{\partial y} = 2xyf'(x^2 - y^2) + x - 2xyf'(x^2 - y^2) = x$.

41. $\dfrac{\partial z}{\partial x} = \dfrac{\partial z}{\partial u}y + \dfrac{\partial z}{\partial v}\dfrac{-y}{x^2}$ and

$$\dfrac{\partial^2 z}{\partial x^2} = y\,\dfrac{\partial}{\partial x}\left(\dfrac{\partial z}{\partial u}\right) + \dfrac{2y}{x^3}\dfrac{\partial z}{\partial v} + \dfrac{-y}{x^2}\dfrac{\partial}{\partial x}\left(\dfrac{\partial z}{\partial v}\right) = \dfrac{2y}{x^3}\dfrac{\partial z}{\partial v} + y\left(\dfrac{\partial^2 z}{\partial u^2}y + \dfrac{\partial^2 z}{\partial v\,\partial u}\dfrac{-y}{x^2}\right) + \dfrac{-y}{x^2}\left(\dfrac{\partial^2 z}{\partial v^2}\dfrac{-y}{x^2} + \dfrac{\partial^2 z}{\partial u\,\partial v}y\right)$$

$$= \dfrac{2y}{x^3}\dfrac{\partial z}{\partial v} + y^2\,\dfrac{\partial^2 z}{\partial u^2} - \dfrac{2y^2}{x^2}\dfrac{\partial^2 z}{\partial u\,\partial v} + \dfrac{y^2}{x^4}\dfrac{\partial^2 z}{\partial v^2}$$

Also $\dfrac{\partial z}{\partial y} = x\,\dfrac{\partial z}{\partial u} + \dfrac{1}{x}\dfrac{\partial z}{\partial v}$ and

$$\dfrac{\partial^2 z}{\partial y^2} = x\,\dfrac{\partial}{\partial y}\left(\dfrac{\partial z}{\partial u}\right) + \dfrac{1}{x}\dfrac{\partial}{\partial y}\left(\dfrac{\partial z}{\partial v}\right) = x\left(\dfrac{\partial^2 z}{\partial u^2}x + \dfrac{\partial^2 z}{\partial v\,\partial u}\dfrac{1}{x}\right) + \dfrac{1}{x}\left(\dfrac{\partial^2 z}{\partial v^2}\dfrac{1}{x} + \dfrac{\partial^2 z}{\partial u\,\partial v}x\right) = x^2\,\dfrac{\partial^2 z}{\partial u^2} + 2\dfrac{\partial^2 z}{\partial u\,\partial v} + \dfrac{1}{x^2}\dfrac{\partial^2 z}{\partial v^2}$$

Thus

$$x^2\,\dfrac{\partial^2 z}{\partial x^2} - y^2\,\dfrac{\partial^2 z}{\partial y^2} = \dfrac{2y}{x}\dfrac{\partial z}{\partial v} + x^2y^2\,\dfrac{\partial^2 z}{\partial u^2} - 2y^2\,\dfrac{\partial^2 z}{\partial u\,\partial v} + \dfrac{y^2}{x^2}\dfrac{\partial^2 z}{\partial v^2} - x^2y^2\,\dfrac{\partial^2 z}{\partial u^2} - 2y^2\,\dfrac{\partial^2 z}{\partial u\,\partial v} - \dfrac{y^2}{x^2}\dfrac{\partial^2 z}{\partial v^2}$$

$$= \dfrac{2y}{x}\dfrac{\partial z}{\partial v} - 4y^2\,\dfrac{\partial^2 z}{\partial u\,\partial v} = 2v\,\dfrac{\partial z}{\partial v} - 4uv\,\dfrac{\partial^2 z}{\partial u\,\partial v}$$

since $y = xv = \dfrac{uv}{y}$ or $y^2 = uv$.

43. $f(x, y, z) = x^2e^{yz^2} \quad \Rightarrow \quad \nabla f = \langle f_x, f_y, f_z \rangle = \left\langle 2xe^{yz^2},\ x^2e^{yz^2} \cdot z^2,\ x^2e^{yz^2} \cdot 2yz \right\rangle = \left\langle 2xe^{yz^2},\ x^2z^2e^{yz^2},\ 2x^2yze^{yz^2} \right\rangle$

© 2012 Cengage Learning. All Rights Reserved. May not be scanned, copied, or duplicated, or posted to a publicly accessible website, in whole or in part.

45. $f(x,y) = x^2 e^{-y} \Rightarrow \nabla f = \langle 2xe^{-y}, -x^2 e^{-y} \rangle$, $\nabla f(-2,0) = \langle -4, -4 \rangle$. The direction is given by $\langle 4, -3 \rangle$, so

$\mathbf{u} = \frac{1}{\sqrt{4^2 + (-3)^2}} \langle 4, -3 \rangle = \frac{1}{5} \langle 4, -3 \rangle$ and $D_{\mathbf{u}} f(-2,0) = \nabla f(-2,0) \cdot \mathbf{u} = \langle -4, -4 \rangle \cdot \frac{1}{5} \langle 4, -3 \rangle = \frac{1}{5}(-16 + 12) = -\frac{4}{5}$.

47. $\nabla f = \langle 2xy, x^2 + 1/(2\sqrt{y}) \rangle$, $|\nabla f(2,1)| = |\langle 4, \frac{9}{2} \rangle|$. Thus the maximum rate of change of f at $(2,1)$ is $\frac{\sqrt{145}}{2}$ in the

direction $\langle 4, \frac{9}{2} \rangle$.

49. First we draw a line passing through Homestead and the eye of the hurricane. We can approximate the directional derivative at

Homestead in the direction of the eye of the hurricane by the average rate of change of wind speed between the points where

this line intersects the contour lines closest to Homestead. In the direction of the eye of the hurricane, the wind speed changes

from 45 to 50 knots. We estimate the distance between these two points to be approximately 8 miles, so the rate of change of

wind speed in the direction given is approximately $\frac{50 - 45}{8} = \frac{5}{8} = 0.625$ knot/mi.

51. $f(x,y) = x^2 - xy + y^2 + 9x - 6y + 10 \Rightarrow f_x = 2x - y + 9$,

$f_y = -x + 2y - 6$, $f_{xx} = 2 = f_{yy}$, $f_{xy} = -1$. Then $f_x = 0$ and $f_y = 0$ imply

$y = 1$, $x = -4$. Thus the only critical point is $(-4, 1)$ and $f_{xx}(-4, 1) > 0$,

$D(-4, 1) = 3 > 0$, so $f(-4, 1) = -11$ is a local minimum.

53. $f(x,y) = 3xy - x^2 y - xy^2 \Rightarrow f_x = 3y - 2xy - y^2$, $f_y = 3x - x^2 - 2xy$,

$f_{xx} = -2y$, $f_{yy} = -2x$, $f_{xy} = 3 - 2x - 2y$. Then $f_x = 0$ implies

$y(3 - 2x - y) = 0$ so $y = 0$ or $y = 3 - 2x$. Substituting into $f_y = 0$ implies

$x(3 - x) = 0$ or $3x(-1 + x) = 0$. Hence the critical points are $(0,0)$, $(3,0)$,

$(0,3)$ and $(1,1)$. $D(0,0) = D(3,0) = D(0,3) = -9 < 0$ so $(0,0)$, $(3,0)$, and

$(0,3)$ are saddle points. $D(1,1) = 3 > 0$ and $f_{xx}(1,1) = -2 < 0$, so

$f(1,1) - 1$ is a local maximum.

55. First solve inside D. Here $f_x = 4y^2 - 2xy^2 - y^3$, $f_y = 8xy - 2x^2 y - 3xy^2$.

Then $f_x = 0$ implies $y = 0$ or $y = 4 - 2x$, but $y = 0$ isn't inside D. Substituting

$y = 4 - 2x$ into $f_y = 0$ implies $x = 0$, $x = 2$ or $x = 1$, but $x = 0$ isn't inside D,

and when $x = 2$, $y = 0$ but $(2,0)$ isn't inside D. Thus the only critical point inside

D is $(1,2)$ and $f(1,2) = 4$. Secondly we consider the boundary of D.

On L_1: $f(x,0) = 0$ and so $f = 0$ on L_1. On L_2: $x = -y + 6$ and

$f(-y + 6, y) = y^2(6 - y)(-2) = -2(6y^2 - y^3)$ which has critical points

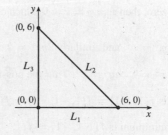

at $y = 0$ and $y = 4$. Then $f(6,0) = 0$ while $f(2,4) = -64$. On L_3: $f(0,y) = 0$, so $f = 0$ on L_3. Thus on D the absolute

maximum of f is $f(1,2) = 4$ while the absolute minimum is $f(2,4) = -64$.

© 2012 Cengage Learning. All Rights Reserved. May not be scanned, copied, or duplicated, or posted to a publicly accessible website, in whole or in part.

57. $f(x,y) = x^3 - 3x + y^4 - 2y^2$

From the graphs, it appears that f has a local maximum $f(-1, 0) \approx 2$, local minima $f(1, \pm1) \approx -3$, and saddle points at

$(-1, \pm1)$ and $(1, 0)$.

To find the exact quantities, we calculate $f_x = 3x^2 - 3 = 0 \iff x = \pm1$ and $f_y = 4y^3 - 4y = 0 \iff$

$y = 0, \pm1$, giving the critical points estimated above. Also $f_{xx} = 6x$, $f_{xy} = 0$, $f_{yy} = 12y^2 - 4$, so using the Second

Derivatives Test, $D(-1, 0) = 24 > 0$ and $f_{xx}(-1, 0) = -6 < 0$ indicating a local maximum $f(-1, 0) = 2$;

$D(1, \pm1) = 48 > 0$ and $f_{xx}(1, \pm1) = 6 > 0$ indicating local minima $f(1, \pm1) = -3$; and $D(-1, \pm1) = -48$ and

$D(1, 0) = -24$, indicating saddle points.

59. $f(x,y) = x^2 y$, $g(x,y) = x^2 + y^2 = 1 \implies \nabla f = \langle 2xy, x^2 \rangle = \lambda \nabla g = \langle 2\lambda x, 2\lambda y \rangle$. Then $2xy = 2\lambda x$ implies $x = 0$ or

$y = \lambda$. If $x = 0$ then $x^2 + y^2 = 1$ gives $y = \pm1$ and we have possible points $(0, \pm1)$ where $f(0, \pm1) = 0$. If $y = \lambda$ then

$x^2 = 2\lambda y$ implies $x^2 = 2y^2$ and substitution into $x^2 + y^2 = 1$ gives $3y^2 = 1 \implies y = \pm\frac{1}{\sqrt{3}}$ and $x = \pm\sqrt{\frac{2}{3}}$. The

corresponding possible points are $\left(\pm\sqrt{\frac{2}{3}}, \pm\frac{1}{\sqrt{3}}\right)$. The absolute maximum is $f\left(\pm\sqrt{\frac{2}{3}}, \frac{1}{\sqrt{3}}\right) = \frac{2}{3\sqrt{3}}$ while the absolute

minimum is $f\left(\pm\sqrt{\frac{2}{3}}, -\frac{1}{\sqrt{3}}\right) = -\frac{2}{3\sqrt{3}}$.

61. $f(x,y,z) = xyz$, $g(x,y,z) = x^2 + y^2 + z^2 = 3$. $\nabla f = \lambda \nabla g \implies \langle yz, xz, xy \rangle = \lambda\langle 2x, 2y, 2z \rangle$. If any of x, y, or z is

zero, then $x = y = z = 0$ which contradicts $x^2 + y^2 + z^2 = 3$. Then $\lambda = \frac{yz}{2x} = \frac{xz}{2y} = \frac{xy}{2z} \implies 2y^2z = 2x^2z \implies$

$y^2 = x^2$, and similarly $2yz^2 = 2x^2y \implies z^2 = x^2$. Substituting into the constraint equation gives $x^2 + x^2 + x^2 = 3 \implies$

$x^2 = 1 = y^2 = z^2$. Thus the possible points are $(1, 1, \pm1)$, $(1, -1, \pm1)$, $(-1, 1, \pm1)$, $(-1, -1, \pm1)$. The absolute maximum

is $f(1, 1, 1) = f(1, -1, -1) = f(-1, 1, -1) = f(-1, -1, 1) = 1$ and the absolute

minimum is $f(1, 1, -1) = f(1, -1, 1) = f(-1, 1, 1) = f(-1, -1, -1) = -1$.

63. $f(x,y,z) = x^2 + y^2 + z^2$, $g(x,y,z) = xy^2z^3 = 2 \implies \nabla f = \langle 2x, 2y, 2z \rangle = \lambda \nabla g = \langle \lambda y^2z^3, 2\lambda xyz^3, 3\lambda xy^2z^2 \rangle$.

Since $xy^2z^3 = 2$, $x \neq 0$, $y \neq 0$ and $z \neq 0$, so $2x = \lambda y^2z^3$ **(1)**, $1 = \lambda xz^3$ **(2)**, $2 = 3\lambda xy^2z$ **(3)**. Then **(2)** and **(3)** imply

$\frac{1}{xz^3} = \frac{2}{3xy^2z}$ or $y^2 = \frac{2}{3}z^2$ so $y = \pm z\sqrt{\frac{2}{3}}$. Similarly **(1)** and **(3)** imply $\frac{2x}{y^2z^3} = \frac{2}{3xy^2z}$ or $3x^2 = z^2$ so $x = \pm\frac{1}{\sqrt{3}}z$. But

© 2012 Cengage Learning. All Rights Reserved. May not be scanned, copied, or duplicated, or posted to a publicly accessible website, in whole or in part.

CHAPTER 14 REVIEW □ **243**

$xy^2z^3 = 2$ so x and z must have the same sign, that is, $x = \frac{1}{\sqrt{3}}z$. Thus $g(x, y, z) = 2$ implies $\frac{1}{\sqrt{3}}z\left(\frac{2}{3}z^2\right)z^3 = 2$ or

$z = \pm 3^{1/4}$ and the possible points are $(\pm 3^{-1/4}, 3^{-1/4}\sqrt{2}, \pm 3^{1/4})$, $(\pm 3^{-1/4}, -3^{-1/4}\sqrt{2}, \pm 3^{1/4})$. However at each of these

points f takes on the same value, $2\sqrt{3}$. But $(2, 1, 1)$ also satisfies $g(x, y, z) = 2$ and $f(2, 1, 1) = 6 > 2\sqrt{3}$. Thus f has an

absolute minimum value of $2\sqrt{3}$ and no absolute maximum subject to the constraint $xy^2z^3 = 2$.

Alternate solution: $g(x, y, z) = xy^2z^3 = 2$ implies $y^2 = \dfrac{2}{xz^3}$, so minimize $f(x, z) = x^2 + \dfrac{2}{xz^3} + z^2$. Then

$f_x = 2x - \dfrac{2}{x^2z^3}$, $f_z = -\dfrac{6}{xz^4} + 2z$, $f_{xx} = 2 + \dfrac{4}{x^3z^3}$, $f_{zz} = \dfrac{24}{xz^5} + 2$ and $f_{xz} = \dfrac{6}{x^2z^4}$. Now $f_x = 0$ implies

$2x^3z^3 - 2 = 0$ or $z = 1/x$. Substituting into $f_y = 0$ implies $-6x^3 + 2x^{-1} = 0$ or $x = \dfrac{1}{\sqrt[4]{3}}$, so the two critical points are

$\left(\pm\dfrac{1}{\sqrt[4]{3}}, \pm\sqrt[4]{3}\right)$. Then $D\left(\pm\dfrac{1}{\sqrt[4]{3}}, \pm\sqrt[4]{3}\right) = (2 + 4)\left(2 + \frac{24}{3}\right) - \left(\frac{6}{\sqrt{3}}\right)^2 > 0$ and $f_{xx}\left(\pm\dfrac{1}{\sqrt[4]{3}}, \pm\sqrt[4]{3}\right) = 6 > 0$, so each point

is a minimum. Finally, $y^2 = \dfrac{2}{xz^3}$, so the four points closest to the origin are $\left(\pm\dfrac{1}{\sqrt[4]{3}}, \dfrac{\sqrt{2}}{\sqrt[4]{3}}, \pm\sqrt[4]{3}\right)$, $\left(\pm\dfrac{1}{\sqrt[4]{3}}, -\dfrac{\sqrt{2}}{\sqrt[4]{3}}, \pm\sqrt[4]{3}\right)$.

65.

The area of the triangle is $\frac{1}{2}ca\sin\theta$ and the area of the rectangle is bc. Thus, the area of the whole object is $f(a, b, c) = \frac{1}{2}ca\sin\theta + bc$. The perimeter of the object is $g(a, b, c) = 2a + 2b + c = P$. To simplify $\sin\theta$ in terms of a, b, and c notice that $a^2\sin^2\theta + \left(\frac{1}{2}c\right)^2 = a^2 \Rightarrow \sin\theta = \dfrac{1}{2a}\sqrt{4a^2 - c^2}$.

Thus $f(a, b, c) = \dfrac{c}{4}\sqrt{4a^2 - c^2} + bc$. (Instead of using θ, we could just have

used the Pythagorean Theorem.) As a result, by Lagrange's method, we must find a, b, c, and λ by solving $\nabla f = \lambda\nabla g$ which

gives the following equations: $ca(4a^2 - c^2)^{-1/2} = 2\lambda$ **(1)**, $c = 2\lambda$ **(2)**, $\frac{1}{4}(4a^2 - c^2)^{1/2} - \frac{1}{4}c^2(4a^2 - c^2)^{-1/2} + b = \lambda$

(3), and $2a + 2b + c = P$ **(4)**. From **(2)**, $\lambda = \frac{1}{2}c$ and so **(1)** produces $ca(4a^2 - c^2)^{-1/2} = c \Rightarrow (4a^2 - c^2)^{1/2} = a \Rightarrow$

$4a^2 - c^2 = a^2 \Rightarrow c = \sqrt{3}\,a$ **(5)**. Similarly, since $(4a^2 - c^2)^{1/2} = a$ and $\lambda = \frac{1}{2}c$, **(3)** gives $\dfrac{a}{4} - \dfrac{c^2}{4a} + b = \dfrac{c}{2}$, so from

(5), $\dfrac{a}{4} - \dfrac{3a}{4} + b = \dfrac{\sqrt{3}\,a}{2} \Rightarrow -\dfrac{a}{2} - \dfrac{\sqrt{3}\,a}{2} = -b \Rightarrow b = \dfrac{a}{2}\left(1 + \sqrt{3}\right)$ **(6)**. Substituting **(5)** and **(6)** into **(4)** we get:

$2a + a\left(1 + \sqrt{3}\right) + \sqrt{3}\,a = P \Rightarrow 3a + 2\sqrt{3}\,a = P \Rightarrow a = \dfrac{P}{3 + 2\sqrt{3}} = \dfrac{2\sqrt{3} - 3}{3}P$ and thus

$b = \dfrac{\left(2\sqrt{3} - 3\right)\left(1 + \sqrt{3}\right)}{6}P = \dfrac{3 - \sqrt{3}}{6}P$ and $c = \left(2 - \sqrt{3}\right)P$.

© 2012 Cengage Learning. All Rights Reserved. May not be scanned, copied, or duplicated, or posted to a publicly accessible website, in whole or in part.

☐ PROBLEMS PLUS

1. The areas of the smaller rectangles are $A_1 = xy$, $A_2 = (L - x)y$,

$A_3 = (L - x)(W - y)$, $A_4 = x(W - y)$. For $0 \le x \le L$, $0 \le y \le W$, let

$$f(x, y) = A_1^2 + A_2^2 + A_3^2 + A_4^2$$
$$= x^2 y^2 + (L - x)^2 y^2 + (L - x)^2 (W - y)^2 + x^2 (W - y)^2$$
$$= [x^2 + (L - x)^2][y^2 + (W - y)^2]$$

Then we need to find the maximum and minimum values of $f(x, y)$. Here

$f_x(x, y) = [2x - 2(L - x)][y^2 + (W - y)^2] = 0 \Rightarrow 4x - 2L = 0$ or $x = \frac{1}{2}L$, and

$f_y(x, y) = [x^2 + (L - x)^2][2y - 2(W - y)] = 0 \Rightarrow 4y - 2W = 0$ or $y = W/2$. Also

$f_{xx} = 4[y^2 + (W - y)^2]$, $f_{yy} = 4[x^2 + (L - x)^2]$, and $f_{xy} = (4x - 2L)(4y - 2W)$. Then

$D = 16[y^2 + (W - y)^2][x^2 + (L - x)^2] - (4x - 2L)^2(4y - 2W)^2$. Thus when $x = \frac{1}{2}L$ and $y = \frac{1}{2}W$, $D > 0$ and

$f_{xx} = 2W^2 > 0$. Thus a minimum of f occurs at $\left(\frac{1}{2}L, \frac{1}{2}W\right)$ and this minimum value is $f\left(\frac{1}{2}L, \frac{1}{2}W\right) = \frac{1}{4}L^2W^2$.

There are no other critical points, so the maximum must occur on the boundary. Now along the width of the rectangle let

$g(y) = f(0, y) = f(L, y) = L^2[y^2 + (W - y)^2]$, $0 \le y \le W$. Then $g'(y) = L^2[2y - 2(W - y)] = 0 \Leftrightarrow y = \frac{1}{2}W$.

And $g\left(\frac{1}{2}\right) = \frac{1}{2}L^2W^2$. Checking the endpoints, we get $g(0) = g(W) = L^2W^2$. Along the length of the rectangle let

$h(x) = f(x, 0) = f(x, W) = W^2[x^2 + (L - x)^2]$, $0 \le x \le L$. By symmetry $h'(x) = 0 \Leftrightarrow x = \frac{1}{2}L$ and

$h\left(\frac{1}{2}L\right) = \frac{1}{2}L^2W^2$. At the endpoints we have $h(0) = h(L) = L^2W^2$. Therefore L^2W^2 is the maximum value of f.

This maximum value of f occurs when the "cutting" lines correspond to sides of the rectangle.

3. (a) The area of a trapezoid is $\frac{1}{2}h(b_1 + b_2)$, where h is the height (the distance between the two parallel sides) and b_1, b_2 are

the lengths of the bases (the parallel sides). From the figure in the text, we see that $h = x \sin\theta$, $b_1 = w - 2x$, and

$b_2 = w - 2x + 2x \cos\theta$. Therefore the cross-sectional area of the rain gutter is

$$A(x, \theta) = \frac{1}{2}x \sin\theta \left[(w - 2x) + (w - 2x + 2x \cos\theta)\right] = (x \sin\theta)(w - 2x + x \cos\theta)$$
$$= wx \sin\theta - 2x^2 \sin\theta + x^2 \sin\theta \cos\theta, \quad 0 < x \le \frac{1}{2}w, 0 < \theta \le \frac{\pi}{2}$$

We look for the critical points of A: $\partial A / \partial x = w \sin\theta - 4x \sin\theta + 2x \sin\theta \cos\theta$ and

$\partial A / \partial \theta = wx \cos\theta - 2x^2 \cos\theta + x^2(\cos^2\theta - \sin^2\theta)$, so $\partial A / \partial x = 0 \Leftrightarrow \sin\theta (w - 4x + 2x \cos\theta) = 0 \Leftrightarrow$

$\cos\theta = \dfrac{4x - w}{2x} = 2 - \dfrac{w}{2x}$ $\left(0 < \theta \le \frac{\pi}{2} \Rightarrow \sin\theta > 0\right)$. If, in addition, $\partial A / \partial \theta = 0$, then

$$0 = wx \cos\theta - 2x^2 \cos\theta + x^2(2\cos^2\theta - 1)$$
$$= wx\left(2 - \frac{w}{2x}\right) - 2x^2\left(2 - \frac{w}{2x}\right) + x^2\left[2\left(2 - \frac{w}{2x}\right)^2 - 1\right]$$
$$= 2wx - \frac{1}{2}w^2 - 4x^2 + wx + x^2\left[8 - \frac{4w}{x} + \frac{w^2}{2x^2} - 1\right] = -wx + 3x^2 = x(3x - w)$$

© 2012 Cengage Learning. All Rights Reserved. May not be scanned, copied, or duplicated, or posted to a publicly accessible website, in whole or in part.

Since $x > 0$, we must have $x = \frac{1}{3}w$, in which case $\cos\theta = \frac{1}{2}$, so $\theta = \frac{\pi}{3}$, $\sin\theta = \frac{\sqrt{3}}{2}$, $k = \frac{\sqrt{3}}{6}w$, $b_1 = \frac{1}{3}w$, $b_2 = \frac{2}{3}w$,

and $A = \frac{\sqrt{3}}{12}w^2$. As in Example 14.7.6, we can argue from the physical nature of this problem that we have found a local

maximum of A. Now checking the boundary of A, let

$$g(\theta) = A(w/2, \theta) = \frac{1}{2}w^2\sin\theta - \frac{1}{2}w^2\sin\theta + \frac{1}{4}w^2\sin\theta\cos\theta = \frac{1}{8}w^2\sin 2\theta, \; 0 < \theta \le \frac{\pi}{2}.$$ Clearly g is maximized when

$\sin 2\theta = 1$ in which case $A = \frac{1}{8}w^2$. Also along the line $\theta = \frac{\pi}{2}$, let $h(x) = A\left(x, \frac{\pi}{2}\right) = wx - 2x^2, \; 0 < x < \frac{1}{2}w \;\; \Rightarrow$

$h'(x) = w - 4x = 0 \;\; \Leftrightarrow \;\; x = \frac{1}{4}w$, and $h\left(\frac{1}{4}w\right) = w\left(\frac{1}{4}w\right) - 2\left(\frac{1}{4}w\right)^2 = \frac{1}{8}w^2$. Since $\frac{1}{8}w^2 < \frac{\sqrt{3}}{12}w^2$, we conclude that

the local maximum found earlier was an absolute maximum.

(b) If the metal were bent into a semi-circular gutter of radius r, we would have $w = \pi r$ and $A = \frac{1}{2}\pi r^2 = \frac{1}{2}\pi\left(\frac{w}{\pi}\right)^2 = \frac{w^2}{2\pi}$.

Since $\dfrac{w^2}{2\pi} > \dfrac{\sqrt{3}\,w^2}{12}$, it *would* be better to bend the metal into a gutter with a semicircular cross-section.

5. Let $g(x, y) = xf\left(\dfrac{y}{x}\right)$. Then $g_x(x, y) = f\left(\dfrac{y}{x}\right) + xf'\left(\dfrac{y}{x}\right)\left(-\dfrac{y}{x^2}\right) = f\left(\dfrac{y}{x}\right) - \dfrac{y}{x}\,f'\left(\dfrac{y}{x}\right)$ and

$g_y(x, y) = xf'\left(\dfrac{y}{x}\right)\left(\dfrac{1}{x}\right) = f'\left(\dfrac{y}{x}\right)$. Thus the tangent plane at (x_0, y_0, z_0) on the surface has equation

$$z - x_0 f\left(\frac{y_0}{x_0}\right) = \left[f\left(\frac{y_0}{x_0}\right) - y_0 x_0^{-1} f'\left(\frac{y_0}{x_0}\right)\right](x - x_0) + f'\left(\frac{y_0}{x_0}\right)(y - y_0) \;\; \Rightarrow$$

$\left[f\left(\dfrac{y_0}{x_0}\right) - y_0 x_0^{-1} f'\left(\dfrac{y_0}{x_0}\right)\right]x + \left[f'\left(\dfrac{y_0}{x_0}\right)\right]y - z = 0$. But any plane whose equation is of the form $ax + by + cz = 0$

passes through the origin. Thus the origin is the common point of intersection.

7. Since we are minimizing the area of the ellipse, and the circle lies above the x-axis,

the ellipse will intersect the circle for only one value of y. This y-value must

satisfy both the equation of the circle and the equation of the ellipse. Now

$\dfrac{x^2}{a^2} + \dfrac{y^2}{b^2} = 1 \;\; \Rightarrow \;\; x^2 = \dfrac{a^2}{b^2}\left(b^2 - y^2\right)$. Substituting into the equation of the

circle gives $\dfrac{a^2}{b^2}\left(b^2 - y^2\right) + y^2 - 2y = 0 \;\; \Rightarrow \;\; \left(\dfrac{b^2 - a^2}{b^2}\right)y^2 - 2y + a^2 = 0$.

In order for there to be only one solution to this quadratic equation, the discriminant must be 0, so $4 - 4a^2\,\dfrac{b^2 - a^2}{b^2} = 0 \;\; \Rightarrow$

$b^2 - a^2 b^2 + a^4 = 0$. The area of the ellipse is $A(a, b) = \pi ab$, and we minimize this function subject to the constraint

$g(a, b) = b^2 - a^2 b^2 + a^4 = 0$.

Now $\nabla A = \lambda\nabla g \;\; \Leftrightarrow \;\; \pi b = \lambda(4a^3 - 2ab^2), \; \pi a = \lambda(2b - 2ba^2) \;\; \Rightarrow \;\; \lambda = \dfrac{\pi b}{2a(2a^2 - b^2)}$ **(1)**,

$\lambda = \dfrac{\pi a}{2b(1 - a^2)}$ **(2)**, $b^2 - a^2 b^2 + a^4 = 0$ **(3)**. Comparing **(1)** and **(2)** gives $\dfrac{\pi b}{2a(2a^2 - b^2)} = \dfrac{\pi a}{2b(1 - a^2)} \;\; \Rightarrow$

$2\pi b^2 = 4\pi a^4 \;\; \Leftrightarrow \;\; a^2 = \frac{1}{\sqrt{2}}\,b$. Substitute this into **(3)** to get $b = \frac{3}{\sqrt{2}} \;\; \Rightarrow \;\; a = \sqrt{\frac{3}{2}}$.

© 2012 Cengage Learning. All Rights Reserved. May not be scanned, copied, or duplicated, or posted to a publicly accessible website, in whole or in part.

15 ☐ MULTIPLE INTEGRALS

15.1 Double Integrals over Rectangles

1. (a) The subrectangles are shown in the figure.

The surface is the graph of $f(x, y) = xy$ and $\Delta A = 4$, so we estimate

$$V \approx \sum_{i=1}^{3} \sum_{j=1}^{2} f(x_i, y_j) \, \Delta A$$

$$= f(2,2) \, \Delta A + f(2,4) \, \Delta A + f(4,2) \, \Delta A + f(4,4) \, \Delta A + f(6,2) \, \Delta A + f(6,4) \, \Delta A$$

$$= 4(4) + 8(4) + 8(4) + 16(4) + 12(4) + 24(4) = 288$$

(b) $V \approx \sum_{i=1}^{3} \sum_{j=1}^{2} f(\overline{x}_i, \overline{y}_j) \, \Delta A = f(1,1) \, \Delta A + f(1,3) \, \Delta A + f(3,1) \, \Delta A + f(3,3) \, \Delta A + f(5,1) \, \Delta A + f(5,3) \, \Delta A$

$$= 1(4) + 3(4) + 3(4) + 9(4) + 5(4) + 15(4) = 144$$

3. (a) The subrectangles are shown in the figure. Since $\Delta A = 1 \cdot \frac{1}{2} = \frac{1}{2}$, we estimate

$$\iint_R xe^{-xy} \, dA \approx \sum_{i=1}^{2} \sum_{j=1}^{2} f(x_{ij}^*, y_{ij}^*) \, \Delta A$$

$$= f\left(1, \tfrac{1}{2}\right) \Delta A + f(1,1) \, \Delta A + f\left(2, \tfrac{1}{2}\right) \Delta A + f(2,1) \, \Delta A$$

$$= e^{-1/2}\left(\tfrac{1}{2}\right) + e^{-1}\left(\tfrac{1}{2}\right) + 2e^{-1}\left(\tfrac{1}{2}\right) + 2e^{-2}\left(\tfrac{1}{2}\right) \approx 0.990$$

(b) $\iint_R xe^{-xy} \, dA \approx \sum_{i=1}^{2} \sum_{j=1}^{2} f(\overline{x}_i, \overline{y}_j) \, \Delta A$

$$= f\left(\tfrac{1}{2}, \tfrac{1}{4}\right) \Delta A + f\left(\tfrac{1}{2}, \tfrac{3}{4}\right) \Delta A + f\left(\tfrac{3}{2}, \tfrac{1}{4}\right) \Delta A + f\left(\tfrac{3}{2}, \tfrac{3}{4}\right) \Delta A$$

$$= \tfrac{1}{2}e^{-1/8}\left(\tfrac{1}{2}\right) + \tfrac{1}{2}e^{-3/8}\left(\tfrac{1}{2}\right) + \tfrac{3}{2}e^{-3/8}\left(\tfrac{1}{2}\right) + \tfrac{3}{2}e^{-9/8}\left(\tfrac{1}{2}\right) \approx 1.151$$

5. (a) Each subrectangle and its midpoint are shown in the figure.

The area of each subrectangle is $\Delta A = 2$, so we evaluate f

at each midpoint and estimate

$$\iint_R f(x, y) \, dA \approx \sum_{i=1}^{2} \sum_{j=1}^{2} f(\overline{x}_i, \overline{y}_j) \, \Delta A$$

$$= f(1, 2.5) \, \Delta A + f(1, 3.5) \, \Delta A$$

$$+ f(3, 2.5) \, \Delta A + f(3, 3.5) \, \Delta A$$

$$= -2(2) + (-1)(2) + 2(2) + 3(2) = 4$$

© 2012 Cengage Learning. All Rights Reserved. May not be scanned, copied, or duplicated, or posted to a publicly accessible website, in whole or in part.

(b) The subrectangles are shown in the figure.

In each subrectangle, the sample point closest to the origin

is the lower left corner, and the area of each subrectangle is $\Delta A = \frac{1}{2}$.

Thus we estimate

$$\iint_R f(x,y)\, dA \approx \sum_{i=1}^{4} \sum_{j=1}^{4} f\left(x_{ij}^*, y_{ij}^*\right) \Delta A$$

$$= f(0,2)\,\Delta A + f(0,2.5)\,\Delta A + f(0,3)\,\Delta A + f(0,3.5)\,\Delta A$$

$$+ f(1,2)\,\Delta A + f(1,2.5)\,\Delta A + f(1,3)\,\Delta A + f(1,3.5)\,\Delta A$$

$$+ f(2,2)\,\Delta A + f(2,2.5)\,\Delta A + f(2,3)\,\Delta A + f(2,3.5)\,\Delta A$$

$$+ f(3,2)\,\Delta A + f(3,2.5)\,\Delta A + f(3,3)\,\Delta A + f(3,3.5)\,\Delta A$$

$$= -3\left(\tfrac{1}{2}\right) + (-5)\left(\tfrac{1}{2}\right) + (-6)\left(\tfrac{1}{2}\right) + (-4)\left(\tfrac{1}{2}\right) + (-1)\left(\tfrac{1}{2}\right) + (-2)\left(\tfrac{1}{2}\right) + (-3)\left(\tfrac{1}{2}\right) + (-1)\left(\tfrac{1}{2}\right)$$

$$+ 1\left(\tfrac{1}{2}\right) + 0\left(\tfrac{1}{2}\right) + (-1)\left(\tfrac{1}{2}\right) + 1\left(\tfrac{1}{2}\right) + 2\left(\tfrac{1}{2}\right) + 2\left(\tfrac{1}{2}\right) + 1\left(\tfrac{1}{2}\right) + 3\left(\tfrac{1}{2}\right)$$

$$= -8$$

7. The values of $f(x,y) = \sqrt{52 - x^2 - y^2}$ get smaller as we move farther from the origin, so on any of the subrectangles in the

problem, the function will have its largest value at the lower left corner of the subrectangle and its smallest value at the upper

right corner, and any other value will lie between these two. So using these subrectangles we have $U < V < L$. (Note that this

is true no matter how R is divided into subrectangles.)

9. (a) With $m = n = 2$, we have $\Delta A = 4$. Using the contour map to estimate the value of f at the center of each subrectangle,

we have

$$\iint_R f(x,y)\, dA \approx \sum_{i=1}^{2} \sum_{j=1}^{2} f\left(\overline{x}_i, \overline{y}_j\right) \Delta A = \Delta A[f(1,1) + f(1,3) + f(3,1) + f(3,3)] \approx 4(27 + 4 + 14 + 17) = 248$$

(b) $f_{\text{ave}} = \frac{1}{A(R)} \iint_R f(x,y)\, dA \approx \frac{1}{16}(248) = 15.5$

11. $z = 3 > 0$, so we can interpret the integral as the volume of the solid S that lies below the plane $z = 3$ and above the

rectangle $[-2, 2] \times [1, 6]$. S is a rectangular solid, thus $\iint_R 3\, dA = 4 \cdot 5 \cdot 3 = 60$.

13. $z = f(x,y) = 4 - 2y \geq 0$ for $0 \leq y \leq 1$. Thus the integral represents the volume of that

part of the rectangular solid $[0, 1] \times [0, 1] \times [0, 4]$ which lies below the plane $z = 4 - 2y$.

So

$$\iint_R (4 - 2y)\, dA = (1)(1)(2) + \tfrac{1}{2}(1)(1)(2) = 3$$

© 2012 Cengage Learning. All Rights Reserved. May not be scanned, copied, or duplicated, or posted to a publicly accessible website, in whole or in part.

15. To calculate the estimates using a programmable calculator, we can use an algorithm
similar to that of Exercise 4.1.9 [ET 5.1.9]. In Maple, we can define the function

$f(x, y) = \sqrt{1 + xe^{-y}}$ (calling it f), load the `student` package, and then use the
command

$$\text{middlesum}(\text{middlesum}(f, x=0..1, m),$$
$$y=0..1, m);$$

to get the estimate with $n = m^2$ squares of equal size. Mathematica has no special
Riemann sum command, but we can define f and then use nested `Sum` commands to
calculate the estimates.

n	estimate
1	1.141606
4	1.143191
16	1.143535
64	1.143617
256	1.143637
1024	1.143642

17. If we divide R into mn subrectangles, $\iint_R k \, dA \approx \sum_{i=1}^{m} \sum_{j=1}^{n} f\left(x_{ij}^*, y_{ij}^*\right) \Delta A$ for any choice of sample points $\left(x_{ij}^*, y_{ij}^*\right)$.

But $f\left(x_{ij}^*, y_{ij}^*\right) = k$ always and $\sum_{i=1}^{m} \sum_{j=1}^{n} \Delta A = $ area of $R = (b-a)(d-c)$. Thus, no matter how we choose the sample

points, $\sum_{i=1}^{m} \sum_{j=1}^{n} f\left(x_{ij}^*, y_{ij}^*\right) \Delta A = k \sum_{i=1}^{m} \sum_{j=1}^{n} \Delta A = k(b-a)(d-c)$ and so

$$\iint_R k \, dA = \lim_{m,n\to\infty} \sum_{i=1}^{m} \sum_{j=1}^{n} f\left(x_{ij}^*, y_{ij}^*\right) \Delta A = \lim_{m,n\to\infty} k \sum_{i=1}^{m} \sum_{j=1}^{n} \Delta A = \lim_{m,n\to\infty} k(b-a)(d-c) = k(b-a)(d-c).$$

15.2   Iterated Integrals

1. $\int_0^5 12x^2y^3 \, dx = \left[12\frac{x^3}{3}y^3\right]_{x=0}^{x=5} = 4x^3y^3\Big]_{x=0}^{x=5} = 4(5)^3 y^3 - 4(0)^3 y^3 = 500y^3,$

$\int_0^1 12x^2y^3 \, dy = \left[12x^2\frac{y^4}{4}\right]_{y=0}^{y=1} = 3x^2y^4\Big]_{y=0}^{y=1} = 3x^2(1)^4 - 3x^2(0)^4 = 3x^2$

3. $\int_1^4 \int_0^2 (6x^2y - 2x) \, dy \, dx = \int_1^4 \left[3x^2y^2 - 2xy\right]_{y=0}^{y=2} dx = \int_1^4 (12x^2 - 4x) \, dx = \left[4x^3 - 2x^2\right]_1^4 = (256 - 32) - (4 - 2) = 222$

5. $\int_0^2 \int_0^4 y^3 e^{2x} \, dy \, dx = \int_0^2 e^{2x} \, dx \int_0^4 y^3 \, dy$ [as in Example 5] $= \left[\frac{1}{2}e^{2x}\right]_0^2 \left[\frac{1}{4}y^4\right]_0^4 = \frac{1}{2}(e^4 - 1)(64 - 0) = 32(e^4 - 1)$

7. $\int_{-3}^3 \int_0^{\pi/2} (y + y^2 \cos x) \, dx \, dy = \int_{-3}^3 \left[xy + y^2 \sin x\right]_{x=0}^{x=\pi/2} dy$

$= \int_{-3}^3 \left(\frac{\pi}{2}y + y^2\right) dy = \left[\frac{\pi}{4}y^2 + \frac{1}{3}y^3\right]_{-3}^3$

$= \left[\frac{9\pi}{4} + 9 - \left(\frac{9\pi}{4} - 9\right)\right] = 18$

9. $\int_1^4 \int_1^2 \left(\frac{x}{y} + \frac{y}{x}\right) dy \, dx = \int_1^4 \left[x\ln|y| + \frac{1}{x} \cdot \frac{1}{2}y^2\right]_{y=1}^{y=2} dx = \int_1^4 \left(x\ln 2 + \frac{3}{2x}\right) dx = \left[\frac{1}{2}x^2\ln 2 + \frac{3}{2}\ln|x|\right]_1^4$

$= 8\ln 2 + \frac{3}{2}\ln 4 - \frac{1}{2}\ln 2 = \frac{15}{2}\ln 2 + 3\ln 4^{1/2} = \frac{21}{2}\ln 2$

11. $\int_0^1 \int_0^1 v(u + v^2)^4 \, du \, dv = \int_0^1 \left[\frac{1}{5}v(u + v^2)^5\right]_{u=0}^{u=1} dv = \frac{1}{5}\int_0^1 v\left[(1 + v^2)^5 - (0 + v^2)^5\right] dv$

$= \frac{1}{5}\int_0^1 \left[v(1 + v^2)^5 - v^{11}\right] dv = \frac{1}{5}\left[\frac{1}{2} \cdot \frac{1}{6}(1 + v^2)^6 - \frac{1}{12}v^{12}\right]_0^1$

[substitute $t = 1 + v^2 \Rightarrow dt = 2v \, dv$ in the first term]

$= \frac{1}{60}\left[(2^6 - 1) - (1 - 0)\right] = \frac{1}{60}(63 - 1) = \frac{31}{30}$

© 2012 Cengage Learning. All Rights Reserved. May not be scanned, copied, or duplicated, or posted to a publicly accessible website, in whole or in part.

13. $\int_0^2 \int_0^\pi r\sin^2\theta \, d\theta \, dr = \int_0^2 r \, dr \int_0^\pi \sin^2\theta \, d\theta$ [as in Example 5] $= \int_0^2 r \, dr \int_0^\pi \frac{1}{2}(1-\cos 2\theta) \, d\theta$

$$= \left[\tfrac{1}{2}r^2\right]_0^2 \cdot \tfrac{1}{2}\left[\theta - \tfrac{1}{2}\sin 2\theta\right]_0^\pi = (2-0)\cdot\tfrac{1}{2}\left[\left(\pi - \tfrac{1}{2}\sin 2\pi\right) - \left(0 - \tfrac{1}{2}\sin 0\right)\right]$$

$$= 2\cdot\tfrac{1}{2}[(\pi - 0) - (0-0)] = \pi$$

15. $\iint_R \sin(x-y)\,dA = \int_0^{\pi/2}\int_0^{\pi/2}\sin(x-y)\,dy\,dx = \int_0^{\pi/2}\left[\cos(x-y)\right]_{y=0}^{y=\pi/2}\,dx = \int_0^{\pi/2}\left[\cos(x-\tfrac{\pi}{2}) - \cos x\right]\,dx$

$$= \left[\sin(x-\tfrac{\pi}{2}) - \sin x\right]_0^{\pi/2} = \sin 0 - \sin\tfrac{\pi}{2} - \left[\sin(-\tfrac{\pi}{2}) - \sin 0\right]$$

$$= 0 - 1 - (-1-0) = 0$$

17. $\iint_R \dfrac{xy^2}{x^2+1}\,dA = \int_0^1\int_{-3}^3 \dfrac{xy^2}{x^2+1}\,dy\,dx = \int_0^1 \dfrac{x}{x^2+1}\,dx \int_{-3}^3 y^2\,dy = \left[\tfrac{1}{2}\ln(x^2+1)\right]_0^1 \left[\tfrac{1}{3}y^3\right]_{-3}^3$

$$= \tfrac{1}{2}(\ln 2 - \ln 1)\cdot \tfrac{1}{3}(27+27) = 9\ln 2$$

19. $\int_0^{\pi/6}\int_0^{\pi/3} x\sin(x+y)\,dy\,dx$

$$= \int_0^{\pi/6}\left[-x\cos(x+y)\right]_{y=0}^{y=\pi/3}\,dx = \int_0^{\pi/6}\left[x\cos x - x\cos(x+\tfrac{\pi}{3})\right]\,dx$$

$$= x\left[\sin x - \sin(x+\tfrac{\pi}{3})\right]_0^{\pi/6} - \int_0^{\pi/6}\left[\sin x - \sin(x+\tfrac{\pi}{3})\right]\,dx \qquad \text{[by integrating by parts separately for each term]}$$

$$= \tfrac{\pi}{6}\left[\tfrac{1}{2} - 1\right] - \left[-\cos x + \cos(x+\tfrac{\pi}{3})\right]_0^{\pi/6} = -\tfrac{\pi}{12} - \left[-\tfrac{\sqrt{3}}{2} + 0 - (-1+\tfrac{1}{2})\right] = \tfrac{\sqrt{3}-1}{2} - \tfrac{\pi}{12}$$

21. $\iint_R ye^{-xy}\,dA = \int_0^3\int_0^2 ye^{-xy}\,dx\,dy = \int_0^3 \left[-e^{-xy}\right]_{x=0}^{x=2}\,dy = \int_0^3(-e^{-2y}+1)\,dy = \left[\tfrac{1}{2}e^{-2y}+y\right]_0^3$

$$= \tfrac{1}{2}e^{-6} + 3 - (\tfrac{1}{2}+0) = \tfrac{1}{2}e^{-6} + \tfrac{5}{2}$$

23. $z = f(x,y) = 4 - x - 2y \ge 0$ for $0 \le x \le 1$ and $0 \le y \le 1$. So the solid

is the region in the first octant which lies below the plane $z = 4 - x - 2y$

and above $[0,1]\times[0,1]$.

25. The solid lies under the plane $4x + 6y - 2z + 15 = 0$ or $z = 2x + 3y + \tfrac{15}{2}$ so

$$V = \iint_R (2x + 3y + \tfrac{15}{2})\,dA = \int_{-1}^1\int_{-1}^2 (2x+3y+\tfrac{15}{2})\,dx\,dy = \int_{-1}^1 \left[x^2 + 3xy + \tfrac{15}{2}x\right]_{x=-1}^{x=2}\,dy$$

$$= \int_{-1}^1 \left[(19+6y) - (-\tfrac{13}{2} - 3y)\right]\,dy = \int_{-1}^1 (\tfrac{51}{2} + 9y)\,dy = \left[\tfrac{51}{2}y + \tfrac{9}{2}y^2\right]_{-1}^1 = 30 - (-21) = 51$$

27. $V = \int_{-2}^2\int_{-1}^1 \left(1 - \tfrac{1}{4}x^2 - \tfrac{1}{9}y^2\right)\,dx\,dy = 4\int_0^2\int_0^1\left(1 - \tfrac{1}{4}x^2 - \tfrac{1}{9}y^2\right)\,dx\,dy$

$$= 4\int_0^2 \left[x - \tfrac{1}{12}x^3 - \tfrac{1}{9}y^2 x\right]_{x=0}^{x=1}\,dy = 4\int_0^2\left(\tfrac{11}{12} - \tfrac{1}{9}y^2\right)\,dy = 4\left[\tfrac{11}{12}y - \tfrac{1}{27}y^3\right]_0^2 = 4\cdot\tfrac{83}{54} = \tfrac{166}{27}$$

29. Here we need the volume of the solid lying under the surface $z = x\sec^2 y$ and above the rectangle $R = [0,2]\times[0,\pi/4]$ in the xy-plane.

$$V = \int_0^2\int_0^{\pi/4} x\sec^2 y \, dy\,dx = \int_0^2 x\,dx \int_0^{\pi/4}\sec^2 y \, dy = \left[\tfrac{1}{2}x^2\right]_0^2 \left[\tan y\right]_0^{\pi/4}$$

$$= (2-0)(\tan\tfrac{\pi}{4} - \tan 0) = 2(1-0) = 2$$

© 2012 Cengage Learning. All Rights Reserved. May not be scanned, copied, or duplicated, or posted to a publicly accessible website, in whole or in part.

31. The solid lies below the surface $z = 2 + x^2 + (y-2)^2$ and above the plane $z = 1$ for $-1 \le x \le 1$, $0 \le y \le 4$. The volume

of the solid is the difference in volumes between the solid that lies under $z = 2 + x^2 + (y-2)^2$ over the rectangle

$R = [-1, 1] \times [0, 4]$ and the solid that lies under $z = 1$ over R.

$$V = \int_0^4 \int_{-1}^1 [2 + x^2 + (y-2)^2] \, dx \, dy - \int_0^4 \int_{-1}^1 (1) \, dx \, dy = \int_0^4 \left[2x + \tfrac{1}{3}x^3 + x(y-2)^2 \right]_{x=-1}^{x=1} dy - \int_{-1}^1 dx \int_0^4 dy$$

$$= \int_0^4 \left[(2 + \tfrac{1}{3} + (y-2)^2) - (-2 - \tfrac{1}{3} - (y-2)^2) \right] dy - [x]_{-1}^1 \, [y]_0^4$$

$$= \int_0^4 \left[\tfrac{14}{3} + 2(y-2)^2 \right] dy - [1 - (-1)][4 - 0] = \left[\tfrac{14}{3}y + \tfrac{2}{3}(y-2)^3 \right]_0^4 - (2)(4)$$

$$= \left[\left(\tfrac{56}{3} + \tfrac{16}{3} \right) - \left(0 - \tfrac{16}{3} \right) \right] - 8 = \tfrac{88}{3} - 8 = \tfrac{64}{3}$$

33. In Maple, we can calculate the integral by defining the integrand as `f`

and then using the command `int(int(f,x=0..1),y=0..1);`.

In Mathematica, we can use the command

$$\texttt{Integrate[f,\{x,0,1\},\{y,0,1\}]}$$

We find that $\iint_R x^5 y^3 e^{xy} \, dA = 21e - 57 \approx 0.0839$. We can use `plot3d`

(in Maple) or `Plot3D` (in Mathematica) to graph the function.

35. R is the rectangle $[-1, 1] \times [0, 5]$. Thus, $A(R) = 2 \cdot 5 = 10$ and

$$f_{\text{ave}} = \frac{1}{A(R)} \iint_R f(x, y) \, dA = \tfrac{1}{10} \int_0^5 \int_{-1}^1 x^2 y \, dx \, dy = \tfrac{1}{10} \int_0^5 \left[\tfrac{1}{3}x^3 y \right]_{x=-1}^{x=1} dy = \tfrac{1}{10} \int_0^5 \tfrac{2}{3}y \, dy = \tfrac{1}{10} \left[\tfrac{1}{3}y^2 \right]_0^5 = \tfrac{5}{6}.$$

37. $\displaystyle\iint_R \frac{xy}{1 + x^4} \, dA = \int_{-1}^1 \int_0^1 \frac{xy}{1 + x^4} \, dy \, dx = \int_{-1}^1 \frac{x}{1 + x^4} \, dx \int_0^1 y \, dy$ [by Equation 5] but $f(x) = \dfrac{x}{1 + x^4}$ is an odd

function so $\displaystyle\int_{-1}^1 f(x) \, dx = 0$ by (6) in Section 4.5 [ET (7) in Section 5.5]. Thus $\displaystyle\iint_R \frac{xy}{1 + x^4} \, dA = 0 \cdot \int_0^1 y \, dy = 0.$

39. Let $f(x, y) = \dfrac{x - y}{(x + y)^3}$. Then a CAS gives $\int_0^1 \int_0^1 f(x, y) \, dy \, dx = \tfrac{1}{2}$ and $\int_0^1 \int_0^1 f(x, y) \, dx \, dy = -\tfrac{1}{2}$.

To explain the seeming violation of Fubini's Theorem, note that f has an infinite discontinuity at $(0, 0)$ and thus does not

satisfy the conditions of Fubini's Theorem. In fact, both iterated integrals involve improper integrals which diverge at their

lower limits of integration.

15.3 Double Integrals over General Regions

1. $\int_0^4 \int_0^{\sqrt{y}} xy^2 \, dx \, dy = \int_0^4 \left[\tfrac{1}{2}x^2 y^2 \right]_{x=0}^{x=\sqrt{y}} dy = \int_0^4 \tfrac{1}{2}y^2 [(\sqrt{y})^2 - 0^2] dy = \tfrac{1}{2} \int_0^4 y^3 \, dy = \tfrac{1}{2} \left[\tfrac{1}{4}y^4 \right]_0^4 = \tfrac{1}{2}(64 - 0) = 32$

3. $\int_0^1 \int_{x^2}^x (1 + 2y) \, dy \, dx = \int_0^1 \left[y + y^2 \right]_{y=x^2}^{y=x} dx = \int_0^1 \left[x + x^2 - x^2 - (x^2)^2 \right] dx$

$$= \int_0^1 (x - x^4) \, dx = \left[\tfrac{1}{2}x^2 - \tfrac{1}{5}x^5 \right]_0^1 = \tfrac{1}{2} - \tfrac{1}{5} - 0 + 0 = \tfrac{3}{10}$$

5. $\int_0^1 \int_0^{s^2} \cos(s^3) \, dt \, ds = \int_0^1 \left[t \cos(s^3) \right]_{t=0}^{t=s^2} ds = \int_0^1 s^2 \cos(s^3) \, ds = \tfrac{1}{3} \sin(s^3) \Big|_0^1 = \tfrac{1}{3}(\sin 1 - \sin 0) = \tfrac{1}{3} \sin 1$

© 2012 Cengage Learning. All Rights Reserved. May not be scanned, copied, or duplicated, or posted to a publicly accessible website, in whole or in part.

7. $\iint_D y^2\,dA = \int_{-1}^{1}\int_{-y-2}^{y} y^2\,dx\,dy = \int_{-1}^{1}\left[xy^2\right]_{x=-y-2}^{x=y}\,dy = \int_{-1}^{1} y^2\left[y - (-y-2)\right]\,dy$

$\qquad = \int_{-1}^{1}(2y^3 + 2y^2)\,dy = \left[\tfrac{1}{2}y^4 + \tfrac{2}{3}y^3\right]_{-1}^{1} = \tfrac{1}{2} + \tfrac{2}{3} - \tfrac{1}{2} + \tfrac{2}{3} = \tfrac{4}{3}$

9. $\iint_D x\,dA = \int_0^{\pi}\int_0^{\sin x} x\,dy\,dx = \int_0^{\pi}\left[xy\right]_{y=0}^{y=\sin x}\,dx = \int_0^{\pi} x\sin x\,dx \quad \begin{bmatrix} \text{integrate by parts} \\ \text{with } u = x,\, dv = \sin x\,dx \end{bmatrix}$

$\qquad = \left[-x\cos x + \sin x\right]_0^{\pi} = -\pi\cos\pi + \sin\pi + 0 - \sin 0 = \pi$

11. (a) At the right we sketch an example of a region D that can be described as lying

between the graphs of two continuous functions of x (a type I region) but not as

lying between graphs of two continuous functions of y (a type II region). The

regions shown in Figures 6 and 8 in the text are additional examples.

(b) Now we sketch an example of a region D that can be described as lying between

the graphs of two continuous functions of y but not as lying between graphs of two

continuous functions of x. The first region shown in Figure 7 is another example.

13.

As a type I region, D lies between the lower boundary $y = 0$ and the upper

boundary $y = x$ for $0 \le x \le 1$, so $D = \{(x,y) \mid 0 \le x \le 1, 0 \le y \le x\}$. If we

describe D as a type II region, D lies between the left boundary $x = y$ and the

right boundary $x = 1$ for $0 \le y \le 1$, so $D = \{(x,y) \mid 0 \le y \le 1, y \le x \le 1\}$.

Thus $\iint_D x\,dA = \int_0^1\int_0^x x\,dy\,dx = \int_0^1\left[xy\right]_{y=0}^{y=x}\,dx = \int_0^1 x^2\,dx = \tfrac{1}{3}x^3\Big]_0^1 = \tfrac{1}{3}(1-0) = \tfrac{1}{3}$ or

$\iint_D x\,dA = \int_0^1\int_y^1 x\,dx\,dy = \int_0^1\left[\tfrac{1}{2}x^2\right]_{x=y}^{x=1}\,dy = \tfrac{1}{2}\int_0^1(1 - y^2)\,dy = \tfrac{1}{2}\left[y - \tfrac{1}{3}y^3\right]_0^1 = \tfrac{1}{2}\left[(1 - \tfrac{1}{3}) - 0\right] = \tfrac{1}{3}$.

15.

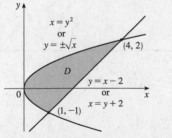

The curves $y = x - 2$ or $x = y + 2$ and $x = y^2$ intersect when $y + 2 = y^2 \;\Leftrightarrow\;$

$y^2 - y - 2 = 0 \;\Leftrightarrow\; (y-2)(y+1) = 0 \;\Leftrightarrow\; y = -1, y = 2$, so the points of

intersection are $(1, -1)$ and $(4, 2)$. If we describe D as a type I region, the upper

boundary curve is $y = \sqrt{x}$ but the lower boundary curve consists of two parts,

$y = -\sqrt{x}$ for $0 \le x \le 1$ and $y = x - 2$ for $1 \le x \le 4$.

Thus $D = \{(x,y) \mid 0 \le x \le 1,\ -\sqrt{x} \le y \le \sqrt{x}\} \cup \{(x,y) \mid 1 \le x \le 4,\ x - 2 \le y \le \sqrt{x}\}$ and

$\iint_D y\,dA = \int_0^1\int_{-\sqrt{x}}^{\sqrt{x}} y\,dy\,dx + \int_1^4\int_{x-2}^{\sqrt{x}} y\,dy\,dx$. If we describe D as a type II region, D is enclosed by the left boundary

$x = y^2$ and the right boundary $x = y + 2$ for $-1 \le y \le 2$, so $D = \{(x,y) \mid -1 \le y \le 2,\ y^2 \le x \le y + 2\}$ and

© 2012 Cengage Learning. All Rights Reserved. May not be scanned, copied, or duplicated, or posted to a publicly accessible website, in whole or in part.

$\iint_D y \, dA = \int_{-1}^2 \int_{y^2}^{y+2} y \, dx \, dy$. In either case, the resulting iterated integrals are not difficult to evaluate but the region D is more simply described as a type II region, giving one iterated integral rather than a sum of two, so we evaluate the latter integral:

$$\iint_D y \, dA = \int_{-1}^2 \int_{y^2}^{y+2} y \, dx \, dy = \int_{-1}^2 [xy]_{x=y^2}^{x=y+2} \, dy = \int_{-1}^2 (y+2-y^2)y \, dy = \int_{-1}^2 (y^2 + 2y - y^3) \, dy$$

$$= \left[\tfrac{1}{3}y^3 + y^2 - \tfrac{1}{4}y^4\right]_{-1}^2 = \left(\tfrac{8}{3} + 4 - 4\right) - \left(-\tfrac{1}{3} + 1 - \tfrac{1}{4}\right) = \tfrac{9}{4}$$

17. $\int_0^1 \int_0^{x^2} x \cos y \, dy \, dx = \int_0^1 [x \sin y]_{y=0}^{y=x^2} \, dx = \int_0^1 x \sin x^2 \, dx = -\tfrac{1}{2}\cos x^2\big]_0^1 = \tfrac{1}{2}(1 - \cos 1)$

19.

$$\iint_D y^2 \, dA = \int_1^2 \int_{y-1}^{7-3y} y^2 \, dx \, dy = \int_1^2 [xy^2]_{x=y-1}^{x=7-3y} \, dy$$

$$= \int_1^2 [(7-3y) - (y-1)] y^2 \, dy = \int_1^2 (8y^2 - 4y^3) \, dy$$

$$= \left[\tfrac{8}{3}y^3 - y^4\right]_1^2 = \tfrac{64}{3} - 16 - \tfrac{8}{3} + 1 = \tfrac{11}{3}$$

21.

$$\int_{-2}^2 \int_{-\sqrt{4-x^2}}^{\sqrt{4-x^2}} (2x - y) \, dy \, dx$$

$$= \int_{-2}^2 \left[2xy - \tfrac{1}{2}y^2\right]_{y=-\sqrt{4-x^2}}^{y=\sqrt{4-x^2}} \, dx$$

$$= \int_{-2}^2 \left[2x\sqrt{4-x^2} - \tfrac{1}{2}(4-x^2) + 2x\sqrt{4-x^2} + \tfrac{1}{2}(4-x^2)\right] dx$$

$$= \int_{-2}^2 4x\sqrt{4-x^2} \, dx = -\tfrac{4}{3}(4-x^2)^{3/2}\Big]_{-2}^2 = 0$$

[Or, note that $4x\sqrt{4-x^2}$ is an odd function, so $\int_{-2}^2 4x\sqrt{4-x^2} \, dx = 0$.]

23.

$$V = \int_0^1 \int_{1-x}^{1-x^2} (1 - x + 2y) \, dy \, dx = \int_0^1 \left[y - xy + y^2\right]_{y=1-x}^{y=1-x^2} dx$$

$$= \int_0^1 \left[\left((1-x^2) - x(1-x^2) + (1-x^2)^2\right)\right.$$

$$\left. - \left((1-x) - x(1-x) + (1-x)^2\right)\right] dx$$

$$= \int_0^1 \left[(x^4 + x^3 - 3x^2 - x + 2) - (2x^2 - 4x + 2)\right] dx$$

$$= \int_0^1 (x^4 + x^3 - 5x^2 + 3x) \, dx = \left[\tfrac{1}{5}x^5 + \tfrac{1}{4}x^4 - \tfrac{5}{3}x^3 + \tfrac{3}{2}x^2\right]_0^1$$

$$= \tfrac{1}{5} + \tfrac{1}{4} - \tfrac{5}{3} + \tfrac{3}{2} = \tfrac{17}{60}$$

25.

$$V = \int_1^2 \int_1^{7-3y} xy \, dx \, dy = \int_1^2 \left[\tfrac{1}{2}x^2 y\right]_{x=1}^{x=7-3y} dy$$

$$= \tfrac{1}{2}\int_1^2 (48y - 42y^2 + 9y^3) \, dy$$

$$= \tfrac{1}{2}\left[24y^2 - 14y^3 + \tfrac{9}{4}y^4\right]_1^2 = \tfrac{31}{8}$$

© 2012 Cengage Learning. All Rights Reserved. May not be scanned, copied, or duplicated, or posted to a publicly accessible website, in whole or in part.

27.

$$V = \int_0^2 \int_0^{3-\frac{3}{2}x} (6 - 3x - 2y)\, dy\, dx$$

$$= \int_0^2 \left[6y - 3xy - y^2 \right]_{y=0}^{y=3-\frac{3}{2}x} dx$$

$$= \int_0^2 \left[6(3 - \tfrac{3}{2}x) - 3x(3 - \tfrac{3}{2}x) - (3 - \tfrac{3}{2}x)^2 \right] dx$$

$$= \int_0^2 \left(\tfrac{9}{4}x^2 - 9x + 9 \right) dx = \left[\tfrac{3}{4}x^3 - \tfrac{9}{2}x^2 + 9x \right]_0^2 = 6 - 0 = 6$$

29.

$$V = \int_{-2}^2 \int_{x^2}^4 x^2\, dy\, dx$$

$$= \int_{-2}^2 x^2 \left[y \right]_{y=x^2}^{y=4} dx = \int_{-2}^2 (4x^2 - x^4)\, dx$$

$$= \left[\tfrac{4}{3}x^3 - \tfrac{1}{5}x^5 \right]_{-2}^2 = \tfrac{32}{3} - \tfrac{32}{5} + \tfrac{32}{3} - \tfrac{32}{5} = \tfrac{128}{15}$$

31.

$$V = \int_0^1 \int_0^{\sqrt{1-x^2}} y\, dy\, dx = \int_0^1 \left[\frac{y^2}{2} \right]_{y=0}^{y=\sqrt{1-x^2}} dx$$

$$= \int_0^1 \frac{1-x^2}{2}\, dx = \tfrac{1}{2}\left[x - \tfrac{1}{3}x^3 \right]_0^1 = \tfrac{1}{3}$$

33.

From the graph, it appears that the two curves intersect at $x = 0$ and at $x \approx 1.213$. Thus the desired integral is

$$\iint_D x\, dA \approx \int_0^{1.213} \int_{x^4}^{3x-x^2} x\, dy\, dx = \int_0^{1.213} \left[xy \right]_{y=x^4}^{y=3x-x^2} dx$$

$$= \int_0^{1.213} (3x^2 - x^3 - x^5)\, dx = \left[x^3 - \tfrac{1}{4}x^4 - \tfrac{1}{6}x^6 \right]_0^{1.213}$$

$$\approx 0.713$$

35. The two bounding curves $y = 1 - x^2$ and $y = x^2 - 1$ intersect at $(\pm 1, 0)$ with $1 - x^2 \geq x^2 - 1$ on $[-1, 1]$. Within this region, the plane $z = 2x + 2y + 10$ is above the plane $z = 2 - x - y$, so

$$V = \int_{-1}^1 \int_{x^2-1}^{1-x^2} (2x + 2y + 10)\, dy\, dx - \int_{-1}^1 \int_{x^2-1}^{1-x^2} (2 - x - y)\, dy\, dx$$

$$= \int_{-1}^1 \int_{x^2-1}^{1-x^2} (2x + 2y + 10 - (2 - x - y))\, dy\, dx$$

$$= \int_{-1}^1 \int_{x^2-1}^{1-x^2} (3x + 3y + 8)\, dy\, dx = \int_{-1}^1 \left[3xy + \tfrac{3}{2}y^2 + 8y \right]_{y=x^2-1}^{y=1-x^2} dx$$

$$= \int_{-1}^1 \left[3x(1 - x^2) + \tfrac{3}{2}(1 - x^2)^2 + 8(1 - x^2) - 3x(x^2 - 1) - \tfrac{3}{2}(x^2 - 1)^2 - 8(x^2 - 1) \right] dx$$

$$= \int_{-1}^1 (-6x^3 - 16x^2 + 6x + 16)\, dx = \left[-\tfrac{3}{2}x^4 - \tfrac{16}{3}x^3 + 3x^2 + 16x \right]_{-1}^1$$

$$= -\tfrac{3}{2} - \tfrac{16}{3} + 3 + 16 + \tfrac{3}{2} - \tfrac{16}{3} - 3 + 16 = \tfrac{64}{3}$$

© 2012 Cengage Learning. All Rights Reserved. May not be scanned, copied, or duplicated, or posted to a publicly accessible website, in whole or in part.

37. The solid lies below the plane $z = 1 - x - y$
or $x + y + z = 1$ and above the region
$D = \{(x, y) \mid 0 \le x \le 1, 0 \le y \le 1 - x\}$
in the xy-plane. The solid is a tetrahedron.

39. The two bounding curves $y = x^3 - x$ and $y = x^2 + x$ intersect at the origin and at $x = 2$, with $x^2 + x > x^3 - x$ on $(0, 2)$.
Using a CAS, we find that the volume is

$$V = \int_0^2 \int_{x^3 - x}^{x^2 + x} z \, dy \, dx = \int_0^2 \int_{x^3 - x}^{x^2 + x} (x^3 y^4 + xy^2) \, dy \, dx = \frac{13{,}984{,}735{,}616}{14{,}549{,}535}$$

41. The two surfaces intersect in the circle $x^2 + y^2 = 1$, $z = 0$ and the region of integration is the disk D: $x^2 + y^2 \le 1$.

Using a CAS, the volume is $\iint_D (1 - x^2 - y^2) \, dA = \int_{-1}^1 \int_{-\sqrt{1-x^2}}^{\sqrt{1-x^2}} (1 - x^2 - y^2) \, dy \, dx = \frac{\pi}{2}$.

43.

Because the region of integration is

$D = \{(x, y) \mid 0 \le x \le y, 0 \le y \le 1\} = \{(x, y) \mid x \le y \le 1, 0 \le x \le 1\}$

we have $\int_0^1 \int_0^y f(x, y) \, dx \, dy = \iint_D f(x, y) \, dA = \int_0^1 \int_x^1 f(x, y) \, dy \, dx$.

45.

Because the region of integration is

$$D = \{(x, y) \mid 0 \le y \le \cos x, 0 \le x \le \pi/2\}$$
$$= \{(x, y) \mid 0 \le x \le \cos^{-1} y, 0 \le y \le 1\}$$

we have

$\int_0^{\pi/2} \int_0^{\cos x} f(x, y) \, dy \, dx = \iint_D f(x, y) \, dA = \int_0^1 \int_0^{\cos^{-1} y} f(x, y) \, dx \, dy$.

47.

Because the region of integration is

$D = \{(x, y) \mid 0 \le y \le \ln x, 1 \le x \le 2\} = \{(x, y) \mid e^y \le x \le 2, 0 \le y \le \ln 2\}$

we have

$$\int_1^2 \int_0^{\ln x} f(x, y) \, dy \, dx = \iint_D f(x, y) \, dA = \int_0^{\ln 2} \int_{e^y}^2 f(x, y) \, dx \, dy$$

49.

$$\int_0^1 \int_{3y}^3 e^{x^2} \, dx \, dy = \int_0^3 \int_0^{x/3} e^{x^2} \, dy \, dx = \int_0^3 \left[e^{x^2} y \right]_{y=0}^{y=x/3} dx$$

$$= \int_0^3 \left(\frac{x}{3} \right) e^{x^2} \, dx = \tfrac{1}{6} e^{x^2} \Big]_0^3 = \frac{e^9 - 1}{6}$$

© 2012 Cengage Learning. All Rights Reserved. May not be scanned, copied, or duplicated, or posted to a publicly accessible website, in whole or in part.

51.

$$\int_0^4 \int_{\sqrt{x}}^2 \frac{1}{y^3+1} \, dy \, dx = \int_0^2 \int_0^{y^2} \frac{1}{y^3+1} \, dx \, dy$$

$$= \int_0^2 \frac{1}{y^3+1} \left[x \right]_{x=0}^{x=y^2} dy = \int_0^2 \frac{y^2}{y^3+1} \, dy$$

$$= \tfrac{1}{3} \ln|y^3+1| \Big]_0^2 = \tfrac{1}{3}(\ln 9 - \ln 1) = \tfrac{1}{3} \ln 9$$

53.

$$\int_0^1 \int_{\arcsin y}^{\pi/2} \cos x \sqrt{1+\cos^2 x} \, dx \, dy$$

$$= \int_0^{\pi/2} \int_0^{\sin x} \cos x \sqrt{1+\cos^2 x} \, dy \, dx$$

$$= \int_0^{\pi/2} \cos x \sqrt{1+\cos^2 x} \left[y \right]_{y=0}^{y=\sin x} dx$$

$$= \int_0^{\pi/2} \cos x \sqrt{1+\cos^2 x} \sin x \, dx \qquad \begin{bmatrix} \text{Let } u = \cos x, \, du = -\sin x \, dx, \\ dx = du/(-\sin x) \end{bmatrix}$$

$$= \int_1^0 -u \sqrt{1+u^2} \, du = -\tfrac{1}{3}\left(1+u^2\right)^{3/2} \Big]_1^0$$

$$= \tfrac{1}{3}\left(\sqrt{8}-1\right) = \tfrac{1}{3}\left(2\sqrt{2}-1\right)$$

55. $D = \{(x,y) \mid 0 \le x \le 1, \, -x+1 \le y \le 1\} \cup \{(x,y) \mid -1 \le x \le 0, \, x+1 \le y \le 1\}$

$\cup \{(x,y) \mid 0 \le x \le 1, \, -1 \le y \le x-1\} \cup \{(x,y) \mid -1 \le x \le 0, \, -1 \le y \le -x-1\}$, all type I.

$$\iint_D x^2 \, dA = \int_0^1 \int_{1-x}^1 x^2 \, dy \, dx + \int_{-1}^0 \int_{x+1}^1 x^2 \, dy \, dx + \int_0^1 \int_{-1}^{x-1} x^2 \, dy \, dx + \int_{-1}^0 \int_{-1}^{-x-1} x^2 \, dy \, dx$$

$$= 4 \int_0^1 \int_{1-x}^1 x^2 \, dy \, dx \qquad \text{[by symmetry of the regions and because } f(x,y) = x^2 \ge 0]$$

$$= 4 \int_0^1 x^3 \, dx = 4\left[\tfrac{1}{4}x^4\right]_0^1 = 1$$

57. Here $Q = \{(x,y) \mid x^2+y^2 \le \tfrac{1}{4}, \, x \ge 0, \, y \ge 0\}$, and $0 \le (x^2+y^2)^2 \le \left(\tfrac{1}{4}\right)^2 \quad \Rightarrow \quad -\tfrac{1}{16} \le -(x^2+y^2)^2 \le 0$ so

$e^{-1/16} \le e^{-(x^2+y^2)^2} \le e^0 = 1$ since e^t is an increasing function. We have $A(Q) = \tfrac{1}{4}\pi\left(\tfrac{1}{2}\right)^2 = \tfrac{\pi}{16}$, so by Property 11,

$e^{-1/16} A(Q) \le \iint_Q e^{-(x^2+y^2)^2} dA \le 1 \cdot A(Q) \quad \Rightarrow \quad \tfrac{\pi}{16}e^{-1/16} \le \iint_Q e^{-(x^2+y^2)^2} dA \le \tfrac{\pi}{16}$ or we can say

$0.1844 < \iint_Q e^{-(x^2+y^2)^2} dA < 0.1964$. (We have rounded the lower bound down and the upper bound up to preserve the

inequalities.)

59. The average value of a function f of two variables defined on a rectangle R was

defined in Section 15.1 as $f_{\text{ave}} = \frac{1}{A(R)} \iint_R f(x,y) dA$. Extending this definition

to general regions D, we have $f_{\text{ave}} = \frac{1}{A(D)} \iint_D f(x,y) dA$.

Here $D = \{(x,y) \mid 0 \le x \le 1, \, 0 \le y \le 3x\}$, so $A(D) = \tfrac{1}{2}(1)(3) = \tfrac{3}{2}$ and

$$f_{\text{ave}} = \frac{1}{A(D)} \iint_D f(x,y) dA = \frac{1}{3/2} \int_0^1 \int_0^{3x} xy \, dy \, dx$$

$$= \tfrac{2}{3} \int_0^1 \left[\tfrac{1}{2}xy^2\right]_{y=0}^{y=3x} dx = \tfrac{1}{3} \int_0^1 9x^3 \, dx = \tfrac{3}{4}x^4 \Big]_0^1 = \tfrac{3}{4}$$

© 2012 Cengage Learning. All Rights Reserved. May not be scanned, copied, or duplicated, or posted to a publicly accessible website, in whole or in part.

61. Since $m \leq f(x, y) \leq M$, $\iint_D m \, dA \leq \iint_D f(x, y) \, dA \leq \iint_D M \, dA$ by (8) $\Rightarrow$

$m \iint_D 1 \, dA \leq \iint_D f(x, y) \, dA \leq M \iint_D 1 \, dA$ by (7) $\Rightarrow$ $mA(D) \leq \iint_D f(x, y) \, dA \leq MA(D)$ by (10).

63.

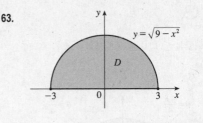

First we can write $\iint_D (x + 2) \, dA = \iint_D x \, dA + \iint_D 2 \, dA$. But $f(x, y) = x$ is

an odd function with respect to x [that is, $f(-x, y) = -f(x, y)$] and D is

symmetric with respect to x. Consequently, the volume above D and below the

graph of f is the same as the volume below D and above the graph of f, so

$\iint_D x \, dA = 0$. Also, $\iint_D 2 \, dA = 2 \cdot A(D) = 2 \cdot \frac{1}{2}\pi(3)^2 = 9\pi$ since D is a half

disk of radius 3. Thus $\iint_D (x + 2) \, dA = 0 + 9\pi = 9\pi$.

65. We can write $\iint_D (2x + 3y) \, dA = \iint_D 2x \, dA + \iint_D 3y \, dA$. $\iint_D 2x \, dA$ represents the volume of the solid lying under the

plane $z = 2x$ and above the rectangle D. This solid region is a triangular cylinder with length b and whose cross-section is a

triangle with width a and height $2a$. (See the first figure.)

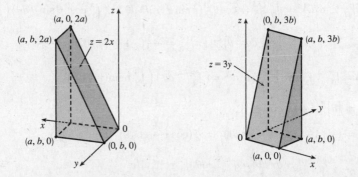

Thus its volume is $\frac{1}{2} \cdot a \cdot 2a \cdot b = a^2 b$. Similarly, $\iint_D 3y \, dA$ represents the volume of a triangular cylinder with length a,

triangular cross-section with width b and height $3b$, and volume $\frac{1}{2} \cdot b \cdot 3b \cdot a = \frac{3}{2}ab^2$. (See the second figure.) Thus

$$\iint_D (2x + 3y) \, dA = a^2 b + \frac{3}{2}ab^2$$

67. $\iint_D \left(ax^3 + by^3 + \sqrt{a^2 - x^2}\right) dA = \iint_D ax^3 \, dA + \iint_D by^3 \, dA + \iint_D \sqrt{a^2 - x^2} \, dA$. Now ax^3 is odd with respect

to x and by^3 is odd with respect to y, and the region of integration is symmetric with respect to both x and y,

so $\iint_D ax^3 \, dA = \iint_D by^3 \, dA = 0$.

$\iint_D \sqrt{a^2 - x^2} \, dA$ represents the volume of the solid region under the

graph of $z = \sqrt{a^2 - x^2}$ and above the rectangle D, namely a half circular

cylinder with radius a and length $2b$ (see the figure) whose volume is

$\frac{1}{2} \cdot \pi r^2 h = \frac{1}{2}\pi a^2 (2b) = \pi a^2 b$. Thus

$$\iint_D \left(ax^3 + by^3 + \sqrt{a^2 - x^2}\right) dA = 0 + 0 + \pi a^2 b = \pi a^2 b.$$

© 2012 Cengage Learning. All Rights Reserved. May not be scanned, copied, or duplicated, or posted to a publicly accessible website, in whole or in part.

15.4   Double Integrals in Polar Coordinates

1. The region R is more easily described by polar coordinates: $R = \{(r, \theta) \mid 0 \leq r \leq 4, 0 \leq \theta \leq \frac{3\pi}{2}\}$.

Thus $\iint_R f(x, y)\, dA = \int_0^{3\pi/2} \int_0^4 f(r\cos\theta, r\sin\theta)\, r\, dr\, d\theta$.

3. The region R is more easily described by rectangular coordinates: $R = \{(x, y) \mid -1 \leq x \leq 1, 0 \leq y \leq \frac{1}{2}x + \frac{1}{2}\}$.

Thus $\iint_R f(x, y)\, dA = \int_{-1}^1 \int_0^{(x+1)/2} f(x, y)\, dy\, dx$.

5. The integral $\int_{\pi/4}^{3\pi/4} \int_1^2 r\, dr\, d\theta$ represents the area of the region

$R = \{(r, \theta) \mid 1 \leq r \leq 2, \pi/4 \leq \theta \leq 3\pi/4\}$, the top quarter portion of a

ring (annulus).

$\int_{\pi/4}^{3\pi/4} \int_1^2 r\, dr\, d\theta = \left(\int_{\pi/4}^{3\pi/4} d\theta \right) \left(\int_1^2 r\, dr \right)$

$= \left[\theta \right]_{\pi/4}^{3\pi/4} \left[\frac{1}{2}r^2 \right]_1^2 = \left(\frac{3\pi}{4} - \frac{\pi}{4} \right) \cdot \frac{1}{2}(4 - 1) = \frac{\pi}{2} \cdot \frac{3}{2} = \frac{3\pi}{4}$

7. The half disk D can be described in polar coordinates as $D = \{(r, \theta) \mid 0 \leq r \leq 5, 0 \leq \theta \leq \pi\}$. Then

$$\iint_D x^2 y\, dA = \int_0^\pi \int_0^5 (r\cos\theta)^2 (r\sin\theta)\, r\, dr\, d\theta = \left(\int_0^\pi \cos^2\theta \sin\theta\, d\theta \right) \left(\int_0^5 r^4\, dr \right)$$

$$= \left[-\frac{1}{3}\cos^3\theta \right]_0^\pi \left[\frac{1}{5}r^5 \right]_0^5 = -\frac{1}{3}(-1 - 1) \cdot 625 = \frac{1250}{3}$$

9. $\iint_R \sin(x^2 + y^2)\, dA = \int_0^{\pi/2} \int_1^3 \sin(r^2)\, r\, dr\, d\theta = \left(\int_0^{\pi/2} d\theta \right) \left(\int_1^3 r\sin(r^2)\, dr \right)$

$= \left[\theta \right]_0^{\pi/2} \left[-\frac{1}{2}\cos(r^2) \right]_1^3$

$= \left(\frac{\pi}{2} \right) \left[-\frac{1}{2}(\cos 9 - \cos 1) \right] = \frac{\pi}{4}(\cos 1 - \cos 9)$

11. $\iint_D e^{-x^2 - y^2}\, dA = \int_{-\pi/2}^{\pi/2} \int_0^2 e^{-r^2} r\, dr\, d\theta = \int_{-\pi/2}^{\pi/2} d\theta \int_0^2 re^{-r^2}\, dr$

$= \left[\theta \right]_{-\pi/2}^{\pi/2} \left[-\frac{1}{2}e^{-r^2} \right]_0^2 = \pi \left(-\frac{1}{2} \right)(e^{-4} - e^0) = \frac{\pi}{2}(1 - e^{-4})$

13. R is the region shown in the figure, and can be described

by $R = \{(r, \theta) \mid 0 \leq \theta \leq \pi/4, 1 \leq r \leq 2\}$. Thus

$\iint_R \arctan(y/x)\, dA = \int_0^{\pi/4} \int_1^2 \arctan(\tan\theta)\, r\, dr\, d\theta$ since $y/x = \tan\theta$.

Also, $\arctan(\tan\theta) = \theta$ for $0 \leq \theta \leq \pi/4$, so the integral becomes

$\int_0^{\pi/4} \int_1^2 \theta\, r\, dr\, d\theta = \int_0^{\pi/4} \theta\, d\theta \int_1^2 r\, dr = \left[\frac{1}{2}\theta^2 \right]_0^{\pi/4} \left[\frac{1}{2}r^2 \right]_1^2 = \frac{\pi^2}{32} \cdot \frac{3}{2} = \frac{3}{64}\pi^2$.

15. One loop is given by the region

$D = \{(r, \theta) \mid -\pi/6 \leq \theta \leq \pi/6, 0 \leq r \leq \cos 3\theta\}$, so the area is

$\iint_D dA = \int_{-\pi/6}^{\pi/6} \int_0^{\cos 3\theta} r\, dr\, d\theta = \int_{-\pi/6}^{\pi/6} \left[\frac{1}{2}r^2 \right]_{r=0}^{r=\cos 3\theta} d\theta$

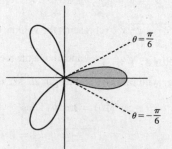

$= \int_{-\pi/6}^{\pi/6} \frac{1}{2}\cos^2 3\theta\, d\theta = 2\int_0^{\pi/6} \frac{1}{2}\left(\frac{1 + \cos 6\theta}{2} \right) d\theta$

$= \frac{1}{2}\left[\theta + \frac{1}{6}\sin 6\theta \right]_0^{\pi/6} = \frac{\pi}{12}$

© 2012 Cengage Learning. All Rights Reserved. May not be scanned, copied, or duplicated, or posted to a publicly accessible website, in whole or in part.

17. In polar coordinates the circle $(x-1)^2 + y^2 = 1 \iff x^2 + y^2 = 2x$ is $r^2 = 2r\cos\theta \Rightarrow r = 2\cos\theta$,

and the circle $x^2 + y^2 = 1$ is $r = 1$. The curves intersect in the first quadrant when

$2\cos\theta = 1 \Rightarrow \cos\theta = \frac{1}{2} \Rightarrow \theta = \pi/3$, so the portion of the region in the first quadrant is given by

$D = \{(r, \theta) \mid 1 \le r \le 2\cos\theta, 0 \le \theta \le \pi/2\}$. By symmetry, the total area

is twice the area of D:

$2A(D) = 2\iint_D dA = 2\int_0^{\pi/3} \int_1^{2\cos\theta} r\,dr\,d\theta = 2\int_0^{\pi/3} \left[\frac{1}{2}r^2\right]_{r=1}^{r=2\cos\theta} d\theta$

$\qquad = \int_0^{\pi/3} (4\cos^2\theta - 1)\,d\theta = \int_0^{\pi/3} \left[4 \cdot \frac{1}{2}(1 + \cos 2\theta) - 1\right] d\theta$

$\qquad = \int_0^{\pi/3}(1 + 2\cos 2\theta)\,d\theta = [\theta + \sin 2\theta]_0^{\pi/3} = \frac{\pi}{3} + \frac{\sqrt{3}}{2}$

19. $V = \iint_{x^2+y^2\le 4} \sqrt{x^2+y^2}\,dA = \int_0^{2\pi}\int_0^2 \sqrt{r^2}\, r\,dr\,d\theta = \int_0^{2\pi} d\theta \int_0^2 r^2\,dr = [\theta]_0^{2\pi}\left[\frac{1}{3}r^3\right]_0^2 = 2\pi\left(\frac{8}{3}\right) = \frac{16}{3}\pi$

21. The hyperboloid of two sheets $-x^2 - y^2 + z^2 = 1$ intersects the plane $z = 2$ when $-x^2 - y^2 + 4 = 1$ or $x^2 + y^2 = 3$. So the

solid region lies above the surface $z = \sqrt{1 + x^2 + y^2}$ and below the plane $z = 2$ for $x^2 + y^2 \le 3$, and its volume is

$$V = \iint\limits_{x^2+y^2\le 3} \left(2 - \sqrt{1 + x^2 + y^2}\right) dA = \int_0^{2\pi}\int_0^{\sqrt{3}} \left(2 - \sqrt{1 + r^2}\right) r\,dr\,d\theta$$

$$= \int_0^{2\pi} d\theta \int_0^{\sqrt{3}} \left(2r - r\sqrt{1 + r^2}\right) dr = [\theta]_0^{2\pi}\left[r^2 - \frac{1}{3}(1 + r^2)^{3/2}\right]_0^{\sqrt{3}}$$

$$= 2\pi\left(3 - \frac{8}{3} - 0 + \frac{1}{3}\right) = \frac{4}{3}\pi$$

23. By symmetry,

$$V = 2\iint\limits_{x^2+y^2\le a^2} \sqrt{a^2 - x^2 - y^2}\,dA = 2\int_0^{2\pi}\int_0^a \sqrt{a^2 - r^2}\, r\,dr\,d\theta = 2\int_0^{2\pi} d\theta \int_0^a r\sqrt{a^2 - r^2}\,dr$$

$$= 2[\theta]_0^{2\pi}\left[-\frac{1}{3}(a^2 - r^2)^{3/2}\right]_0^a = 2(2\pi)\left(0 + \frac{1}{3}a^3\right) = \frac{4\pi}{3}a^3$$

25. The cone $z = \sqrt{x^2 + y^2}$ intersects the sphere $x^2 + y^2 + z^2 = 1$ when $x^2 + y^2 + \left(\sqrt{x^2+y^2}\right)^2 = 1$ or $x^2 + y^2 = \frac{1}{2}$. So

$$V = \iint\limits_{x^2+y^2\le 1/2} \left(\sqrt{1 - x^2 - y^2} - \sqrt{x^2 + y^2}\right) dA = \int_0^{2\pi}\int_0^{1/\sqrt{2}} \left(\sqrt{1 - r^2} - r\right) r\,dr\,d\theta$$

$$= \int_0^{2\pi} d\theta \int_0^{1/\sqrt{2}} \left(r\sqrt{1 - r^2} - r^2\right) dr = [\theta]_0^{2\pi}\left[-\frac{1}{3}(1 - r^2)^{3/2} - \frac{1}{3}r^3\right]_0^{1/\sqrt{2}} = 2\pi\left(-\frac{1}{3}\right)\left(\frac{1}{\sqrt{2}} - 1\right) = \frac{\pi}{3}\left(2 - \sqrt{2}\right)$$

27. The given solid is the region inside the cylinder $x^2 + y^2 = 4$ between the surfaces $z = \sqrt{64 - 4x^2 - 4y^2}$

and $z = -\sqrt{64 - 4x^2 - 4y^2}$. So

$$V = \iint\limits_{x^2+y^2\le 4} \left[\sqrt{64 - 4x^2 - 4y^2} - \left(-\sqrt{64 - 4x^2 - 4y^2}\right)\right] dA = \iint\limits_{x^2+y^2\le 4} 2\sqrt{64 - 4x^2 - 4y^2}\,dA$$

$$= 4\int_0^{2\pi}\int_0^2 \sqrt{16 - r^2}\, r\,dr\,d\theta = 4\int_0^{2\pi} d\theta \int_0^2 r\sqrt{16 - r^2}\,dr = 4[\theta]_0^{2\pi}\left[-\frac{1}{3}(16 - r^2)^{3/2}\right]_0^2$$

$$= 8\pi\left(-\frac{1}{3}\right)(12^{3/2} - 16^{2/3}) = \frac{8\pi}{3}\left(64 - 24\sqrt{3}\right)$$

© 2012 Cengage Learning. All Rights Reserved. May not be scanned, copied, or duplicated, or posted to a publicly accessible website, in whole or in part.

29.

$$\int_{-3}^{3}\int_{0}^{\sqrt{9-x^2}} \sin(x^2+y^2)\,dy\,dx = \int_{0}^{\pi}\int_{0}^{3} \sin\left(r^2\right) r\,dr\,d\theta$$

$$= \int_{0}^{\pi} d\theta \int_{0}^{3} r\sin\left(r^2\right) dr = [\theta]_{0}^{\pi} \left[-\tfrac{1}{2}\cos\left(r^2\right)\right]_{0}^{3}$$

$$= \pi\left(-\tfrac{1}{2}\right)(\cos 9 - 1) = \tfrac{\pi}{2}(1 - \cos 9)$$

31.

$$\int_{0}^{\pi/4}\int_{0}^{\sqrt{2}} (r\cos\theta + r\sin\theta) r\,dr\,d\theta = \int_{0}^{\pi/4} (\cos\theta+\sin\theta)\,d\theta \int_{0}^{\sqrt{2}} r^2\,dr$$

$$= [\sin\theta - \cos\theta]_{0}^{\pi/4} \left[\tfrac{1}{3}r^3\right]_{0}^{\sqrt{2}}$$

$$= \left[\tfrac{\sqrt{2}}{2} - \tfrac{\sqrt{2}}{2} - 0 + 1\right] \cdot \tfrac{1}{3}\left(2\sqrt{2}-0\right) = \tfrac{2\sqrt{2}}{3}$$

33. $D = \{(r,\theta) \mid 0 \le r \le 1, 0 \le \theta \le 2\pi\}$, so

$$\iint_{D} e^{(x^2+y^2)^2}\,dA = \int_{0}^{2\pi}\int_{0}^{1} e^{(r^2)^2} r\,dr\,d\theta = \int_{0}^{2\pi} d\theta \int_{0}^{1} re^{r^4}\,dr = 2\pi\int_{0}^{1} re^{r^4}\,dr.$$ Using a calculator, we estimate

$2\pi\int_{0}^{1} re^{r^4}\,dr \approx 4.5951$.

35. The surface of the water in the pool is a circular disk D with radius 20 ft. If we place D on coordinate axes with the origin at the center of D and define $f(x,y)$ to be the depth of the water at (x,y), then the volume of water in the pool is the volume of the solid that lies above $D = \{(x,y) \mid x^2 + y^2 \le 400\}$ and below the graph of $f(x,y)$. We can associate north with the positive y-direction, so we are given that the depth is constant in the x-direction and the depth increases linearly in the y-direction from $f(0,-20) = 2$ to $f(0,20) = 7$. The trace in the yz-plane is a line segment from $(0,-20,2)$ to $(0,20,7)$. The slope of this line is $\frac{7-2}{20-(-20)} = \frac{1}{8}$, so an equation of the line is $z - 7 = \frac{1}{8}(y-20) \;\Rightarrow\; z = \frac{1}{8}y + \frac{9}{2}$. Since $f(x,y)$ is independent of x, $f(x,y) = \frac{1}{8}y + \frac{9}{2}$. Thus the volume is given by $\iint_{D} f(x,y)\,dA$, which is most conveniently evaluated using polar coordinates. Then $D = \{(r,\theta) \mid 0 \le r \le 20, 0 \le \theta \le 2\pi\}$ and substituting $x = r\cos\theta$, $y = r\sin\theta$ the integral becomes

$$\int_{0}^{2\pi}\int_{0}^{20} \left(\tfrac{1}{8}r\sin\theta + \tfrac{9}{2}\right) r\,dr\,d\theta = \int_{0}^{2\pi} \left[\tfrac{1}{24}r^3\sin\theta + \tfrac{9}{4}r^2\right]_{r=0}^{r=20} d\theta = \int_{0}^{2\pi} \left(\tfrac{1000}{3}\sin\theta + 900\right) d\theta$$

$$= \left[-\tfrac{1000}{3}\cos\theta + 900\theta\right]_{0}^{2\pi} = 1800\pi$$

Thus the pool contains $1800\pi \approx 5655 \text{ ft}^3$ of water.

37. As in Exercise 15.3.59, $f_{\text{ave}} = \frac{1}{A(D)} \iint_{D} f(x,y)\,dA$. Here $D = \{(r,\theta) \mid a \le r \le b, 0 \le \theta \le 2\pi\}$,

so $A(D) = \pi b^2 - \pi a^2 = \pi(b^2 - a^2)$ and

$$f_{\text{ave}} = \frac{1}{A(D)} \iint_{D} \frac{1}{\sqrt{x^2+y^2}}\,dA = \frac{1}{\pi(b^2-a^2)} \int_{0}^{2\pi}\int_{a}^{b} \frac{1}{\sqrt{r^2}} r\,dr\,d\theta = \frac{1}{\pi(b^2-a^2)} \int_{0}^{2\pi} d\theta \int_{a}^{b} dr$$

$$= \frac{1}{\pi(b^2-a^2)} [\theta]_{0}^{2\pi} [r]_{a}^{b} = \frac{1}{\pi(b^2-a^2)} (2\pi)(b-a) = \frac{2(b-a)}{(b+a)(b-a)} = \frac{2}{a+b}$$

© 2012 Cengage Learning. All Rights Reserved. May not be scanned, copied, or duplicated, or posted to a publicly accessible website, in whole or in part.

39. $\int_{1/\sqrt{2}}^{1} \int_{\sqrt{1-x^2}}^{x} xy \, dy \, dx + \int_{1}^{\sqrt{2}} \int_{0}^{x} xy \, dy \, dx + \int_{\sqrt{2}}^{2} \int_{0}^{\sqrt{4-x^2}} xy \, dy \, dx$

$\qquad = \int_{0}^{\pi/4} \int_{1}^{2} r^3 \cos\theta \sin\theta \, dr \, d\theta = \int_{0}^{\pi/4} \left[\frac{r^4}{4} \cos\theta \sin\theta\right]_{r=1}^{r=2} d\theta$

$\qquad = \frac{15}{4} \int_{0}^{\pi/4} \sin\theta \cos\theta \, d\theta = \frac{15}{4} \left[\frac{\sin^2\theta}{2}\right]_{0}^{\pi/4} = \frac{15}{16}$

41. (a) We integrate by parts with $u = x$ and $dv = xe^{-x^2} dx$. Then $du = dx$ and $v = -\frac{1}{2}e^{-x^2}$, so

$$\int_{0}^{\infty} x^2 e^{-x^2} dx = \lim_{t \to \infty} \int_{0}^{t} x^2 e^{-x^2} dx = \lim_{t \to \infty} \left(-\frac{1}{2}xe^{-x^2}\Big]_{0}^{t} + \int_{0}^{t} \frac{1}{2}e^{-x^2} dx\right)$$

$$= \lim_{t \to \infty} \left(-\frac{1}{2}te^{-t^2}\right) + \frac{1}{2}\int_{0}^{\infty} e^{-x^2} dx = 0 + \frac{1}{2}\int_{0}^{\infty} e^{-x^2} dx \qquad \text{[by l'Hospital's Rule]}$$

$$= \frac{1}{4}\int_{-\infty}^{\infty} e^{-x^2} dx \qquad \text{[since } e^{-x^2} \text{ is an even function]}$$

$$= \frac{1}{4}\sqrt{\pi} \qquad \text{[by Exercise 40(c)]}$$

(b) Let $u = \sqrt{x}$. Then $u^2 = x \Rightarrow dx = 2u \, du \Rightarrow$

$$\int_{0}^{\infty} \sqrt{x}e^{-x} dx = \lim_{t \to \infty} \int_{0}^{t} \sqrt{x}e^{-x} dx = \lim_{t \to \infty} \int_{0}^{\sqrt{t}} ue^{-u^2} 2u \, du = 2\int_{0}^{\infty} u^2 e^{-u^2} du = 2\left(\frac{1}{4}\sqrt{\pi}\right) \quad \text{[by part(a)]} = \frac{1}{2}\sqrt{\pi}.$$

15.5 Applications of Double Integrals

1. $Q = \iint_D \sigma(x, y) \, dA = \int_{0}^{5} \int_{2}^{5} (2x + 4y) \, dy \, dx = \int_{0}^{5} \left[2xy + 2y^2\right]_{y=2}^{y=5} dx$

$\qquad = \int_{0}^{5} (10x + 50 - 4x - 8) \, dx = \int_{0}^{5} (6x + 42) \, dx = \left[3x^2 + 42x\right]_{0}^{5} = 75 + 210 = 285$ C

3. $m = \iint_D \rho(x, y) \, dA = \int_{1}^{3} \int_{1}^{4} ky^2 \, dy \, dx = k\int_{1}^{3} dx \int_{1}^{4} y^2 \, dy = k[x]_{1}^{3} \left[\frac{1}{3}y^3\right]_{1}^{4} = k(2)(21) = 42k,$

$\bar{x} = \frac{1}{m} \iint_D x\rho(x, y) \, dA = \frac{1}{42k} \int_{1}^{3} \int_{1}^{4} kxy^2 \, dy \, dx = \frac{1}{42} \int_{1}^{3} x \, dx \int_{1}^{4} y^2 \, dy = \frac{1}{42} \left[\frac{1}{2}x^2\right]_{1}^{3} \left[\frac{1}{3}y^3\right]_{1}^{4} = \frac{1}{42}(4)(21) = 2,$

$\bar{y} = \frac{1}{m} \iint_D y\rho(x, y) \, dA = \frac{1}{42k} \int_{1}^{3} \int_{1}^{4} ky^3 \, dy \, dx = \frac{1}{42} \int_{1}^{3} dx \int_{1}^{4} y^3 \, dy = \frac{1}{42} [x]_{1}^{3} \left[\frac{1}{4}y^4\right]_{1}^{4} = \frac{1}{42}(2)\left(\frac{255}{4}\right) = \frac{85}{28}$

Hence $m = 42k$, $(\bar{x}, \bar{y}) = \left(2, \frac{85}{28}\right)$.

5. $m = \int_{0}^{2} \int_{x/2}^{3-x} (x + y) \, dy \, dx = \int_{0}^{2} \left[xy + \frac{1}{2}y^2\right]_{y=x/2}^{y=3-x} dx = \int_{0}^{2} \left[x\left(3 - \frac{3}{2}x\right) + \frac{1}{2}(3 - x)^2 - \frac{1}{8}x^2\right] dx$

$\qquad = \int_{0}^{2} \left(-\frac{9}{8}x^2 + \frac{9}{2}\right) dx = \left[-\frac{9}{8}\left(\frac{1}{3}x^3\right) + \frac{9}{2}x\right]_{0}^{2} = 6,$

$M_y = \int_{0}^{2} \int_{x/2}^{3-x} (x^2 + xy) \, dy \, dx = \int_{0}^{2} \left[x^2y + \frac{1}{2}xy^2\right]_{y=x/2}^{y=3-x} dx = \int_{0}^{2} \left(\frac{9}{2}x - \frac{9}{8}x^3\right) dx = \frac{9}{2},$

$M_x = \int_{0}^{2} \int_{x/2}^{3-y} (xy + y^2) \, dy \, dx = \int_{0}^{2} \left[\frac{1}{2}xy^2 + \frac{1}{3}y^3\right]_{y=x/2}^{y=3-x} dx = \int_{0}^{2} \left(9 - \frac{9}{2}x\right) dx = 9.$

Hence $m = 6$, $(\bar{x}, \bar{y}) = \left(\frac{M_y}{m}, \frac{M_x}{m}\right) = \left(\frac{3}{4}, \frac{3}{2}\right)$.

7. $m = \int_{-1}^{1} \int_{0}^{1-x^2} ky \, dy \, dx = k\int_{-1}^{1} \left[\frac{1}{2}y^2\right]_{y=0}^{y=1-x^2} dx = \frac{1}{2}k\int_{-1}^{1} (1 - x^2)^2 \, dx = \frac{1}{2}k\int_{-1}^{1} (1 - 2x^2 + x^4) \, dx$

$\qquad = \frac{1}{2}k\left[x - \frac{2}{3}x^3 + \frac{1}{5}x^5\right]_{-1}^{1} = \frac{1}{2}k\left(1 - \frac{2}{3} + \frac{1}{5} + 1 - \frac{2}{3} + \frac{1}{5}\right) = \frac{8}{15}k,$

© 2012 Cengage Learning. All Rights Reserved. May not be scanned, copied, or duplicated, or posted to a publicly accessible website, in whole or in part.

$M_y = \int_{-1}^{1} \int_0^{1-x^2} kxy \, dy \, dx = k \int_{-1}^{1} \left[\frac{1}{2} xy^2 \right]_{y=0}^{y=1-x^2} dx = \frac{1}{2} k \int_{-1}^{1} x (1-x^2)^2 \, dx = \frac{1}{2} k \int_{-1}^{1} (x - 2x^3 + x^5) \, dx$

$\quad = \frac{1}{2} k \left[\frac{1}{2} x^2 - \frac{1}{2} x^4 + \frac{1}{6} x^6 \right]_{-1}^{1} = \frac{1}{2} k \left(\frac{1}{2} - \frac{1}{2} + \frac{1}{6} - \frac{1}{2} + \frac{1}{2} - \frac{1}{6} \right) = 0,$

$M_x = \int_{-1}^{1} \int_0^{1-x^2} ky^2 \, dy \, dx = k \int_{-1}^{1} \left[\frac{1}{3} y^3 \right]_{y=0}^{y=1-x^2} dx = \frac{1}{3} k \int_{-1}^{1} (1-x^2)^3 \, dx = \frac{1}{3} k \int_{-1}^{1} (1 - 3x^2 + 3x^4 - x^6) \, dx$

$\quad = \frac{1}{3} k \left[x - x^3 + \frac{3}{5} x^5 - \frac{1}{7} x^7 \right]_{-1}^{1} = \frac{1}{3} k \left(1 - 1 + \frac{3}{5} - \frac{1}{7} + 1 - 1 + \frac{3}{5} - \frac{1}{7} \right) = \frac{32}{105} k.$

Hence $m = \frac{8}{15} k$, $(\overline{x}, \overline{y}) = \left(0, \frac{32k/105}{8k/15} \right) = \left(0, \frac{4}{7} \right).$

9. Note that $\sin(\pi x / L) \geq 0$ for $0 \leq x \leq L$.

$m = \int_0^L \int_0^{\sin(\pi x / L)} y \, dy \, dx = \int_0^L \frac{1}{2} \sin^2(\pi x / L) \, dx = \frac{1}{2} \left[\frac{1}{2} x - \frac{L}{4\pi} \sin(2\pi x / L) \right]_0^L = \frac{1}{4} L,$

$\quad M_y = \int_0^L \int_0^{\sin(\pi x / L)} x \cdot y \, dy \, dx = \frac{1}{2} \int_0^L x \sin^2(\pi x / L) \, dx \quad \begin{bmatrix} \text{integrate by parts with} \\ u = x, \, dv = \sin^2(\pi x / L) \, dx \end{bmatrix}$

$\quad\quad = \frac{1}{2} \cdot x \left(\frac{1}{2} x - \frac{L}{4\pi} \sin(2\pi x / L) \right) \Big]_0^L - \frac{1}{2} \int_0^L \left[\frac{1}{2} x - \frac{L}{4\pi} \sin(2\pi x / L) \right] dx$

$\quad\quad = \frac{1}{4} L^2 - \frac{1}{2} \left[\frac{1}{4} x^2 + \frac{L^2}{4\pi^2} \cos(2\pi x / L) \right]_0^L = \frac{1}{4} L^2 - \frac{1}{2} \left(\frac{1}{4} L^2 + \frac{L^2}{4\pi^2} - \frac{L^2}{4\pi^2} \right) = \frac{1}{8} L^2,$

$\quad M_x = \int_0^L \int_0^{\sin(\pi x / L)} y \cdot y \, dy \, dx = \int_0^L \frac{1}{3} \sin^3(\pi x / L) \, dx = \frac{1}{3} \int_0^L \left[1 - \cos^2(\pi x / L) \right] \sin(\pi x / L) \, dx$

$\quad\quad\quad\quad [\text{substitute } u = \cos(\pi x / L)] \quad \Rightarrow \quad du = -\frac{\pi}{L} \sin(\pi x / L)]$

$\quad\quad = \frac{1}{3} \left(-\frac{L}{\pi} \right) \left[\cos(\pi x / L) - \frac{1}{3} \cos^3(\pi x / L) \right]_0^L = -\frac{L}{3\pi} \left(-1 + \frac{1}{3} - 1 + \frac{1}{3} \right) = \frac{4}{9\pi} L.$

Hence $m = \frac{L}{4}$, $(\overline{x}, \overline{y}) = \left(\frac{L^2/8}{L/4}, \frac{4L/(9\pi)}{L/4} \right) = \left(\frac{L}{2}, \frac{16}{9\pi} \right).$

11. $\rho(x, y) = ky = kr \sin \theta$, $m = \int_0^{\pi/2} \int_0^1 kr^2 \sin \theta \, dr \, d\theta = \frac{1}{3} k \int_0^{\pi/2} \sin \theta \, d\theta = \frac{1}{3} k \left[-\cos \theta \right]_0^{\pi/2} = \frac{1}{3} k,$

$M_y = \int_0^{\pi/2} \int_0^1 kr^3 \sin \theta \cos \theta \, dr \, d\theta = \frac{1}{4} k \int_0^{\pi/2} \sin \theta \cos \theta \, d\theta = \frac{1}{8} k \left[-\cos 2\theta \right]_0^{\pi/2} = \frac{1}{8} k,$

$M_x = \int_0^{\pi/2} \int_0^1 kr^3 \sin^2 \theta \, dr \, d\theta = \frac{1}{4} k \int_0^{\pi/2} \sin^2 \theta \, d\theta = \frac{1}{8} k \left[\theta + \sin 2\theta \right]_0^{\pi/2} = \frac{\pi}{16} k.$

Hence $(\overline{x}, \overline{y}) = \left(\frac{3}{8}, \frac{3\pi}{16} \right).$

13.

$\rho(x, y) = k \sqrt{x^2 + y^2} = kr,$

$m = \iint_D \rho(x, y) \, dA = \int_0^\pi \int_1^2 kr \cdot r \, dr \, d\theta$

$\quad = k \int_0^\pi d\theta \int_1^2 r^2 \, dr = k(\pi) \left[\frac{1}{3} r^3 \right]_1^2 = \frac{7}{3} \pi k,$

$M_y = \iint_D x \rho(x, y) \, dA = \int_0^\pi \int_1^2 (r \cos \theta)(kr) \, r \, dr \, d\theta = k \int_0^\pi \cos \theta \, d\theta \int_1^2 r^3 \, dr$

$\quad = k \left[\sin \theta \right]_0^\pi \left[\frac{1}{4} r^4 \right]_1^2 = k(0) \left(\frac{15}{4} \right) = 0 \quad$ [this is to be expected as the region and density function are symmetric about the y-axis]

$M_x = \iint_D y \rho(x, y) \, dA = \int_0^\pi \int_1^2 (r \sin \theta)(kr) \, r \, dr \, d\theta = k \int_0^\pi \sin \theta \, d\theta \int_1^2 r^3 \, dr$

$\quad = k \left[-\cos \theta \right]_0^\pi \left[\frac{1}{4} r^4 \right]_1^2 = k(1+1) \left(\frac{15}{4} \right) = \frac{15}{2} k.$

Hence $(\overline{x}, \overline{y}) = \left(0, \frac{15k/2}{7\pi k/3} \right) = \left(0, \frac{45}{14\pi} \right).$

© 2012 Cengage Learning. All Rights Reserved. May not be scanned, copied, or duplicated, or posted to a publicly accessible website, in whole or in part.

15. Placing the vertex opposite the hypotenuse at $(0,0)$, $\rho(x,y) = k(x^2 + y^2)$. Then

$$m = \int_0^a \int_0^{a-x} k(x^2 + y^2)\, dy\, dx = k\int_0^a \left[ax^2 - x^3 + \tfrac{1}{3}(a-x)^3 \right] dx = k\left[\tfrac{1}{3}ax^3 - \tfrac{1}{4}x^4 - \tfrac{1}{12}(a-x)^4 \right]_0^a = \tfrac{1}{6}ka^4.$$

By symmetry,

$$M_y = M_x = \int_0^a \int_0^{a-x} ky(x^2 + y^2)\, dy\, dx = k\int_0^a \left[\tfrac{1}{2}(a-x)^2 x^2 + \tfrac{1}{4}(a-x)^4 \right] dx$$

$$= k\left[\tfrac{1}{6}a^2 x^3 - \tfrac{1}{4}ax^4 + \tfrac{1}{10}x^5 - \tfrac{1}{20}(a-x)^5 \right]_0^a = \tfrac{1}{15}ka^5$$

Hence $(\overline{x}, \overline{y}) = \left(\tfrac{2}{5}a, \tfrac{2}{5}a \right)$.

17. $I_x = \iint_D y^2 \rho(x,y)\, dA = \int_{-1}^1 \int_0^{1-x^2} y^2 \cdot ky\, dy\, dx = k\int_{-1}^1 \left[\tfrac{1}{4}y^4 \right]_{y=0}^{y=1-x^2} dx = \tfrac{1}{4}k\int_{-1}^1 (1-x^2)^4\, dx$

$\quad = \tfrac{1}{4}k\int_{-1}^1 (x^8 - 4x^6 + 6x^4 - 4x^2 + 1)\, dx = \tfrac{1}{4}k\left[\tfrac{1}{9}x^9 - \tfrac{4}{7}x^7 + \tfrac{6}{5}x^5 - \tfrac{4}{3}x^3 + x \right]_{-1}^1 = \tfrac{64}{315}k,$

$I_y = \iint_D x^2 \rho(x,y)\, dA = \int_{-1}^1 \int_0^{1-x^2} kx^2 y\, dy\, dx = k\int_{-1}^1 \left[\tfrac{1}{2}x^2 y^2 \right]_{y=0}^{y=1-x^2} dx = \tfrac{1}{2}k\int_{-1}^1 x^2(1-x^2)^2\, dx$

$\quad = \tfrac{1}{2}k\int_{-1}^1 (x^2 - 2x^4 + x^6)\, dx = \tfrac{1}{2}k\left[\tfrac{1}{3}x^3 - \tfrac{2}{5}x^5 + \tfrac{1}{7}x^7 \right]_{-1}^1 = \tfrac{8}{105}k,$

and $I_0 = I_x + I_y = \tfrac{64}{315}k + \tfrac{8}{105}k = \tfrac{88}{315}k.$

19. As in Exercise 15, we place the vertex opposite the hypotenuse at $(0,0)$ and the equal sides along the positive axes.

$I_x = \int_0^a \int_0^{a-x} y^2 k(x^2 + y^2)\, dy\, dx = k\int_0^a \int_0^{a-x} (x^2 y^2 + y^4)\, dy\, dx = k\int_0^a \left[\tfrac{1}{3}x^2 y^3 + \tfrac{1}{5}y^5 \right]_{y=0}^{y=a-x} dx$

$\quad = k\int_0^a \left[\tfrac{1}{3}x^2(a-x)^3 + \tfrac{1}{5}(a-x)^5 \right] dx = k\left[\tfrac{1}{3}\left(\tfrac{1}{3}a^3 x^3 - \tfrac{3}{4}a^2 x^4 + \tfrac{3}{5}ax^5 - \tfrac{1}{6}x^6 \right) - \tfrac{1}{30}(a-x)^6 \right]_0^a = \tfrac{7}{180}ka^6,$

$I_y = \int_0^a \int_0^{a-x} x^2 k(x^2 + y^2)\, dy\, dx = k\int_0^a \int_0^{a-x} (x^4 + x^2 y^2)\, dy\, dx = k\int_0^a \left[x^4 y + \tfrac{1}{3}x^2 y^3 \right]_{y=0}^{y=a-x} dx$

$\quad = k\int_0^a \left[x^4(a-x) + \tfrac{1}{3}x^2(a-x)^3 \right] dx = k\left[\tfrac{1}{5}ax^5 - \tfrac{1}{6}x^6 + \tfrac{1}{3}\left(\tfrac{1}{3}a^3 x^3 - \tfrac{3}{4}a^2 x^4 + \tfrac{3}{5}ax^5 - \tfrac{1}{6}x^6 \right) \right]_0^a = \tfrac{7}{180}ka^6,$

and $I_0 = I_x + I_y = \tfrac{7}{90}ka^6.$

21. $I_x = \iint_D y^2 \rho(x,y)\, dA = \int_0^h \int_0^b \rho y^2\, dx\, dy = \rho \int_0^b dx \int_0^h y^2\, dy = \rho\left[x \right]_0^b \left[\tfrac{1}{3}y^3 \right]_0^h = \rho b\left(\tfrac{1}{3}h^3 \right) = \tfrac{1}{3}\rho bh^3,$

$I_y = \iint_D x^2 \rho(x,y)\, dA = \int_0^h \int_0^b \rho x^2\, dx\, dy = \rho \int_0^b x^2\, dx \int_0^h dy = \rho\left[\tfrac{1}{3}x^3 \right]_0^b \left[y \right]_0^h = \tfrac{1}{3}\rho b^3 h,$

and $m = \rho\,(\text{area of rectangle}) = \rho bh$ since the lamina is homogeneous. Hence $\overline{\overline{x}}^2 = \dfrac{I_y}{m} = \dfrac{\tfrac{1}{3}\rho b^3 h}{\rho bh} = \dfrac{b^2}{3} \quad \Rightarrow \quad \overline{\overline{x}} = \dfrac{b}{\sqrt{3}}$

and $\overline{\overline{y}}^2 = \dfrac{I_x}{m} = \dfrac{\tfrac{1}{3}\rho bh^3}{\rho bh} = \dfrac{h^2}{3} \quad \Rightarrow \quad \overline{\overline{y}} = \dfrac{h}{\sqrt{3}}.$

23. In polar coordinates, the region is $D = \left\{ (r,\theta) \mid 0 \le r \le a,\, 0 \le \theta \le \tfrac{\pi}{2} \right\}$, so

$I_x = \iint_D y^2 \rho\, dA = \int_0^{\pi/2} \int_0^a \rho(r\sin\theta)^2\, r\, dr\, d\theta = \rho \int_0^{\pi/2} \sin^2\theta\, d\theta \int_0^a r^3\, dr$

$\quad = \rho\left[\tfrac{1}{2}\theta - \tfrac{1}{4}\sin 2\theta \right]_0^{\pi/2} \left[\tfrac{1}{4}r^4 \right]_0^a = \rho\left(\tfrac{\pi}{4} \right)\left(\tfrac{1}{4}a^4 \right) = \tfrac{1}{16}\rho a^4 \pi,$

$I_y = \iint_D x^2 \rho\, dA = \int_0^{\pi/2} \int_0^a \rho(r\cos\theta)^2\, r\, dr\, d\theta = \rho \int_0^{\pi/2} \cos^2\theta\, d\theta \int_0^a r^3\, dr$

$\quad = \rho\left[\tfrac{1}{2}\theta + \tfrac{1}{4}\sin 2\theta \right]_0^{\pi/2} \left[\tfrac{1}{4}r^4 \right]_0^a = \rho\left(\tfrac{\pi}{4} \right)\left(\tfrac{1}{4}a^4 \right) = \tfrac{1}{16}\rho a^4 \pi,$

and $m = \rho \cdot A(D) = \rho \cdot \tfrac{1}{4}\pi a^2$ since the lamina is homogeneous. Hence $\overline{\overline{x}}^2 = \overline{\overline{y}}^2 = \dfrac{\tfrac{1}{16}\rho a^4 \pi}{\tfrac{1}{4}\rho a^2 \pi} = \dfrac{a^2}{4} \quad \Rightarrow \quad \overline{\overline{x}} = \overline{\overline{y}} = \dfrac{a}{2}.$

© 2012 Cengage Learning. All Rights Reserved. May not be scanned, copied, or duplicated, or posted to a publicly accessible website, in whole or in part.

25. The right loop of the curve is given by $D = \{(r, \theta) \mid 0 \le r \le \cos 2\theta,\ -\pi/4 \le \theta \le \pi/4\}$. Using a CAS, we

find $m = \iint_D \rho(x, y)\, dA = \iint_D (x^2 + y^2)\, dA = \int_{-\pi/4}^{\pi/4} \int_0^{\cos 2\theta} r^2\, r\, dr\, d\theta = \dfrac{3\pi}{64}$. Then

$\bar{x} = \dfrac{1}{m} \iint_D x\rho(x, y)\, dA = \dfrac{64}{3\pi} \int_{-\pi/4}^{\pi/4} \int_0^{\cos 2\theta} (r \cos \theta)\, r^2\, r\, dr\, d\theta = \dfrac{64}{3\pi} \int_{-\pi/4}^{\pi/4} \int_0^{\cos 2\theta} r^4 \cos \theta\, dr\, d\theta = \dfrac{16384\sqrt{2}}{10395\pi}$ and

$\bar{y} = \dfrac{1}{m} \iint_D y\rho(x, y)\, dA = \dfrac{64}{3\pi} \int_{-\pi/4}^{\pi/4} \int_0^{\cos 2\theta} (r \sin \theta)\, r^2\, r\, dr\, d\theta = \dfrac{64}{3\pi} \int_{-\pi/4}^{\pi/4} \int_0^{\cos 2\theta} r^4 \sin \theta\, dr\, d\theta = 0$, so

$(\bar{x}, \bar{y}) = \left(\dfrac{16384\sqrt{2}}{10395\pi}, 0 \right)$.

The moments of inertia are

$I_x = \iint_D y^2 \rho(x, y)\, dA = \int_{-\pi/4}^{\pi/4} \int_0^{\cos 2\theta} (r \sin \theta)^2\, r^2\, r\, dr\, d\theta = \int_{-\pi/4}^{\pi/4} \int_0^{\cos 2\theta} r^5 \sin^2 \theta\, dr\, d\theta = \dfrac{5\pi}{384} - \dfrac{4}{105}$,

$I_y = \iint_D x^2 \rho(x, y)\, dA = \int_{-\pi/4}^{\pi/4} \int_0^{\cos 2\theta} (r \cos \theta)^2\, r^2\, r\, dr\, d\theta = \int_{-\pi/4}^{\pi/4} \int_0^{\cos 2\theta} r^5 \cos^2 \theta\, dr\, d\theta = \dfrac{5\pi}{384} + \dfrac{4}{105}$, and

$I_0 = I_x + I_y = \dfrac{5\pi}{192}$.

27. (a) $f(x, y)$ is a joint density function, so we know $\iint_{\mathbb{R}^2} f(x, y)\, dA = 1$. Since $f(x, y) = 0$ outside the

rectangle $[0, 1] \times [0, 2]$, we can say

$$\iint_{\mathbb{R}^2} f(x, y)\, dA = \int_{-\infty}^{\infty} \int_{-\infty}^{\infty} f(x, y)\, dy\, dx = \int_0^1 \int_0^2 Cx(1 + y)\, dy\, dx$$

$$= C \int_0^1 x\left[y + \tfrac{1}{2}y^2 \right]_{y=0}^{y=2} dx = C \int_0^1 4x\, dx = C\left[2x^2 \right]_0^1 = 2C$$

Then $2C = 1 \Rightarrow C = \tfrac{1}{2}$.

(b) $P(X \le 1, Y \le 1) = \int_{-\infty}^1 \int_{-\infty}^1 f(x, y)\, dy\, dx = \int_0^1 \int_0^1 \tfrac{1}{2} x(1 + y)\, dy\, dx$

$$= \int_0^1 \tfrac{1}{2} x\left[y + \tfrac{1}{2}y^2 \right]_{y=0}^{y=1} dx = \int_0^1 \tfrac{1}{2} x\left(\tfrac{3}{2} \right) dx = \tfrac{3}{4}\left[\tfrac{1}{2} x^2 \right]_0^1 = \tfrac{3}{8} \text{ or } 0.375$$

(c) $P(X + Y \le 1) = P((X, Y) \in D)$ where D is the triangular region shown in

the figure. Thus

$P(X + Y \le 1) = \iint_D f(x, y)\, dA = \int_0^1 \int_0^{1-x} \tfrac{1}{2} x(1 + y)\, dy\, dx$

$$= \int_0^1 \tfrac{1}{2} x\left[y + \tfrac{1}{2}y^2 \right]_{y=0}^{y=1-x} dx = \int_0^1 \tfrac{1}{2} x\left(\tfrac{1}{2}x^2 - 2x + \tfrac{3}{2} \right) dx$$

$$= \tfrac{1}{4} \int_0^1 (x^3 - 4x^2 + 3x)\, dx = \tfrac{1}{4}\left[\tfrac{x^4}{4} - 4\tfrac{x^3}{3} + 3\tfrac{x^2}{2} \right]_0^1$$

$$= \tfrac{5}{48} \approx 0.1042$$

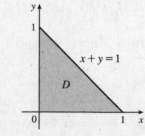

29. (a) $f(x, y) \ge 0$, so f is a joint density function if $\iint_{\mathbb{R}^2} f(x, y)\, dA = 1$. Here, $f(x, y) = 0$ outside the first quadrant, so

$$\iint_{\mathbb{R}^2} f(x, y)\, dA = \int_0^\infty \int_0^\infty 0.1e^{-(0.5x + 0.2y)}\, dy\, dx = 0.1 \int_0^\infty \int_0^\infty e^{-0.5x} e^{-0.2y}\, dy\, dx = 0.1 \int_0^\infty e^{-0.5x}\, dx \int_0^\infty e^{-0.2y}\, dy$$

$$= 0.1 \lim_{t \to \infty} \int_0^t e^{-0.5x}\, dx\ \lim_{t \to \infty} \int_0^t e^{-0.2y}\, dy = 0.1 \lim_{t \to \infty} \left[-2e^{-0.5x} \right]_0^t\ \lim_{t \to \infty} \left[-5e^{-0.2y} \right]_0^t$$

$$= 0.1 \lim_{t \to \infty} \left[-2(e^{-0.5t} - 1) \right] \lim_{t \to \infty} \left[-5(e^{-0.2t} - 1) \right] = (0.1) \cdot (-2)(0 - 1) \cdot (-5)(0 - 1) = 1$$

Thus $f(x, y)$ is a joint density function.

© 2012 Cengage Learning. All Rights Reserved. May not be scanned, copied, or duplicated, or posted to a publicly accessible website, in whole or in part.

(b) (i) No restriction is placed on X, so

$$P(Y \geq 1) = \int_{-\infty}^{\infty} \int_{1}^{\infty} f(x,y) \, dy \, dx = \int_{0}^{\infty} \int_{1}^{\infty} 0.1 e^{-(0.5x+0.2y)} \, dy \, dx$$

$$= 0.1 \int_{0}^{\infty} e^{-0.5x} \, dx \int_{1}^{\infty} e^{-0.2y} \, dy = 0.1 \lim_{t \to \infty} \int_{0}^{t} e^{-0.5x} \, dx \lim_{t \to \infty} \int_{1}^{t} e^{-0.2y} \, dy$$

$$= 0.1 \lim_{t \to \infty} \left[-2e^{-0.5x}\right]_{0}^{t} \lim_{t \to \infty} \left[-5e^{-0.2y}\right]_{1}^{t} = 0.1 \lim_{t \to \infty} \left[-2(e^{-0.5t} - 1)\right] \lim_{t \to \infty} \left[-5(e^{-0.2t} - e^{-0.2})\right]$$

$$(0.1) \cdot (-2)(0-1) \cdot (-5)(0 - e^{-0.2}) = e^{-0.2} \approx 0.8187$$

(ii) $P(X \leq 2, Y \leq 4) = \int_{-\infty}^{2} \int_{-\infty}^{4} f(x,y) \, dy \, dx = \int_{0}^{2} \int_{0}^{4} 0.1 e^{-(0.5x+0.2y)} \, dy \, dx$

$$= 0.1 \int_{0}^{2} e^{-0.5x} \, dx \int_{0}^{4} e^{-0.2y} \, dy = 0.1 \left[-2e^{-0.5x}\right]_{0}^{2} \left[-5e^{-0.2y}\right]_{0}^{4}$$

$$= (0.1) \cdot (-2)(e^{-1} - 1) \cdot (-5)(e^{-0.8} - 1)$$

$$= (e^{-1} - 1)(e^{-0.8} - 1) = 1 + e^{-1.8} - e^{-0.8} - e^{-1} \approx 0.3481$$

(c) The expected value of X is given by

$$\mu_1 = \iint_{\mathbb{R}^2} x \, f(x,y) \, dA = \int_{0}^{\infty} \int_{0}^{\infty} x \left[0.1 e^{-(0.5x+0.2y)}\right] dy \, dx$$

$$= 0.1 \int_{0}^{\infty} x e^{-0.5x} \, dx \int_{0}^{\infty} e^{-0.2y} \, dy = 0.1 \lim_{t \to \infty} \int_{0}^{t} x e^{-0.5x} \, dx \lim_{t \to \infty} \int_{0}^{t} e^{-0.2y} \, dy$$

To evaluate the first integral, we integrate by parts with $u = x$ and $dv = e^{-0.5x} \, dx$ (or we can use Formula 96 in the Table of Integrals): $\int x e^{-0.5x} \, dx = -2x e^{-0.5x} - \int -2 e^{-0.5x} \, dx = -2x e^{-0.5x} - 4 e^{-0.5x} = -2(x+2) e^{-0.5x}$. Thus

$$\mu_1 = 0.1 \lim_{t \to \infty} \left[-2(x+2)e^{-0.5x}\right]_{0}^{t} \lim_{t \to \infty} \left[-5e^{-0.2y}\right]_{0}^{t}$$

$$= 0.1 \lim_{t \to \infty} (-2)\left[(t+2)e^{-0.5t} - 2\right] \lim_{t \to \infty} (-5)\left[e^{-0.2t} - 1\right]$$

$$= 0.1(-2)\left(\lim_{t \to \infty} \frac{t+2}{e^{0.5t}} - 2\right)(-5)(-1) = 2 \qquad \text{[by l'Hospital's Rule]}$$

The expected value of Y is given by

$$\mu_2 = \iint_{\mathbb{R}^2} y \, f(x,y) \, dA = \int_{0}^{\infty} \int_{0}^{\infty} y \left[0.1 e^{-(0.5+0.2y)}\right] dy \, dx$$

$$= 0.1 \int_{0}^{\infty} e^{-0.5x} \, dx \int_{0}^{\infty} y e^{-0.2y} \, dy = 0.1 \lim_{t \to \infty} \int_{0}^{t} e^{-0.5x} \, dx \lim_{t \to \infty} \int_{0}^{t} y e^{-0.2y} \, dy$$

To evaluate the second integral, we integrate by parts with $u = y$ and $dv = e^{-0.2y} \, dy$ (or again we can use Formula 96 in the Table of Integrals) which gives $\int y e^{-0.2y} \, dy = -5y e^{-0.2y} + \int 5 e^{-0.2y} \, dy = -5(y+5)e^{-0.2y}$. Then

$$\mu_2 = 0.1 \lim_{t \to \infty} \left[-2e^{-0.5x}\right]_{0}^{t} \lim_{t \to \infty} \left[-5(y+5)e^{-0.2y}\right]_{0}^{t}$$

$$= 0.1 \lim_{t \to \infty} \left[-2(e^{-0.5t} - 1)\right] \lim_{t \to \infty} \left(-5\left[(t+5)e^{-0.2t} - 5\right]\right)$$

$$= 0.1(-2)(-1) \cdot (-5)\left(\lim_{t \to \infty} \frac{t+5}{e^{0.2t}} - 5\right) = 5 \qquad \text{[by l'Hospital's Rule]}$$

31. (a) The random variables X and Y are normally distributed with $\mu_1 = 45$, $\mu_2 = 20$, $\sigma_1 = 0.5$, and $\sigma_2 = 0.1$.

The individual density functions for X and Y, then, are $f_1(x) = \dfrac{1}{0.5\sqrt{2\pi}} e^{-(x-45)^2/0.5}$ and

$f_2(y) = \dfrac{1}{0.1\sqrt{2\pi}} e^{-(y-20)^2/0.02}$. Since X and Y are independent, the joint density function is the product

© 2012 Cengage Learning. All Rights Reserved. May not be scanned, copied, or duplicated, or posted to a publicly accessible website, in whole or in part.

$$f(x,y) = f_1(x)f_2(y) = \frac{1}{0.5\sqrt{2\pi}}e^{-(x-45)^2/0.5}\frac{1}{0.1\sqrt{2\pi}}e^{-(y-20)^2/0.02} = \frac{10}{\pi}e^{-2(x-45)^2-50(y-20)^2}.$$

Then $P(40 \le X \le 50, 20 \le Y \le 25) = \int_{40}^{50}\int_{20}^{25} f(x,y)\,dy\,dx = \frac{10}{\pi}\int_{40}^{50}\int_{20}^{25} e^{-2(x-45)^2-50(y-20)^2}\,dy\,dx.$

Using a CAS or calculator to evaluate the integral, we get $P(40 \le X \le 50, 20 \le Y \le 25) \approx 0.500.$

(b) $P(4(X-45)^2 + 100(Y-20)^2 \le 2) = \iint_D \frac{10}{\pi}e^{-2(x-45)^2-50(y-20)^2}\,dA$, where D is the region enclosed by the ellipse

$4(x-45)^2 + 100(y-20)^2 = 2$. Solving for y gives $y = 20 \pm \frac{1}{10}\sqrt{2-4(x-45)^2}$, the upper and lower halves of the

ellipse, and these two halves meet where $y = 20$ [since the ellipse is centered at $(45, 20)$] $\Rightarrow$ $4(x-45)^2 = 2$ $\Rightarrow$

$x = 45 \pm \frac{1}{\sqrt{2}}$. Thus

$$\iint_D \frac{10}{\pi}e^{-2(x-45)^2-50(y-20)^2}\,dA = \frac{10}{\pi}\int_{45-1/\sqrt{2}}^{45+1/\sqrt{2}}\int_{20-\frac{1}{10}\sqrt{2-4(x-45)^2}}^{20+\frac{1}{10}\sqrt{2-4(x-45)^2}} e^{-2(x-45)^2-50(y-20)^2}\,dy\,dx.$$

Using a CAS or calculator to evaluate the integral, we get $P(4(X-45)^2 + 100(Y-20)^2 \le 2) \approx 0.632.$

33. (a) If $f(P, A)$ is the probability that an individual at A will be infected by an individual at P, and $k\,dA$ is the number of

infected individuals in an element of area dA, then $f(P, A)k\,dA$ is the number of infections that should result from

exposure of the individual at A to infected people in the element of area dA. Integration over D gives the number of

infections of the person at A due to all the infected people in D. In rectangular coordinates (with the origin at the city's

center), the exposure of a person at A is

$$E = \iint_D kf(P, A)\,dA = k\iint_D \frac{1}{20}[20 - d(P, A)]\,dA = k\iint_D \left[1 - \frac{1}{20}\sqrt{(x-x_0)^2 + (y-y_0)^2}\right]dA$$

(b) If $A = (0, 0)$, then

$$E = k\iint_D \left[1 - \frac{1}{20}\sqrt{x^2 + y^2}\right]dA$$

$$= k\int_0^{2\pi}\int_0^{10}\left(1 - \frac{1}{20}r\right)r\,dr\,d\theta = 2\pi k\left[\frac{1}{2}r^2 - \frac{1}{60}r^3\right]_0^{10}$$

$$= 2\pi k\left(50 - \frac{50}{3}\right) = \frac{200}{3}\pi k \approx 209k$$

For A at the edge of the city, it is convenient to use a polar coordinate system centered at A. Then the polar equation for

the circular boundary of the city becomes $r = 20\cos\theta$ instead of $r = 10$, and the distance from A to a point P in the city

is again r (see the figure). So

$$E = k\int_{-\pi/2}^{\pi/2}\int_0^{20\cos\theta}\left(1 - \frac{1}{20}r\right)r\,dr\,d\theta = k\int_{-\pi/2}^{\pi/2}\left[\frac{1}{2}r^2 - \frac{1}{60}r^3\right]_{r=0}^{r=20\cos\theta}d\theta$$

$$= k\int_{-\pi/2}^{\pi/2}\left(200\cos^2\theta - \frac{400}{3}\cos^3\theta\right)d\theta = 200k\int_{-\pi/2}^{\pi/2}\left[\frac{1}{2} + \frac{1}{2}\cos 2\theta - \frac{2}{3}(1 - \sin^2\theta)\cos\theta\right]d\theta$$

$$= 200k\left[\frac{1}{2}\theta + \frac{1}{4}\sin 2\theta - \frac{2}{3}\sin\theta + \frac{2}{3}\cdot\frac{1}{3}\sin^3\theta\right]_{-\pi/2}^{\pi/2} = 200k\left[\frac{\pi}{4} + 0 - \frac{2}{3} + \frac{2}{9} + \frac{\pi}{4} + 0 - \frac{2}{3} + \frac{2}{9}\right]$$

$$= 200k\left(\frac{\pi}{2} - \frac{8}{9}\right) \approx 136k$$

Therefore the risk of infection is much lower at the edge of the city than in the middle, so it is better to live at the edge.

© 2012 Cengage Learning. All Rights Reserved. May not be scanned, copied, or duplicated, or posted to a publicly accessible website, in whole or in part.

15.6 Surface Area

1. Here $z = f(x, y) = 2 + 3x + 4y$ and D is the rectangle $[0, 5] \times [1, 4]$, so by Formula 2 the area of the surface is

$$A(S) = \iint_D \sqrt{[f_x(x, y)]^2 + [f_y(x, y)]^2 + 1} \, dA = \iint_D \sqrt{3^2 + 4^2 + 1} \, dA = \sqrt{26} \iint_D dA$$
$$= \sqrt{26} \, A(D) = \sqrt{26} \, (5)(3) = 15\sqrt{26}$$

3. $z = f(x, y) = 6 - 3x - 2y$ which intersects the xy-plane in the line $3x + 2y = 6$, so D is the triangular region given by $\{(x, y) \mid 0 \le x \le 2, 0 \le y \le 3 - \frac{3}{2}x\}$. Thus

$$A(S) = \iint_D \sqrt{(-3)^2 + (-2)^2 + 1} \, dA = \sqrt{14} \iint_D dA = \sqrt{14} \, A(D) = \sqrt{14} \left(\frac{1}{2} \cdot 2 \cdot 3\right) = 3\sqrt{14}$$

5. $y^2 + z^2 = 9 \implies z = \sqrt{9 - y^2}$. $f_x = 0$, $f_y = -y(9 - y^2)^{-1/2} \implies$

$$A(S) = \int_0^4 \int_0^2 \sqrt{0^2 + [-y(9 - y^2)^{-1/2}]^2 + 1} \, dy \, dx = \int_0^4 \int_0^2 \sqrt{\frac{y^2}{9 - y^2} + 1} \, dy \, dx$$
$$= \int_0^4 \int_0^2 \frac{3}{\sqrt{9 - y^2}} \, dy \, dx = 3 \int_0^4 \left[\sin^{-1} \frac{y}{3}\right]_{y=0}^{y=2} dx = 3 \left[(\sin^{-1}(\tfrac{2}{3}))x\right]_0^4 = 12 \sin^{-1}(\tfrac{2}{3})$$

7. $z = f(x, y) = y^2 - x^2$ with $1 \le x^2 + y^2 \le 4$. Then

$$A(S) = \iint_D \sqrt{1 + 4x^2 + 4y^2} \, dA = \int_0^{2\pi} \int_1^2 \sqrt{1 + 4r^2} \, r \, dr \, d\theta = \int_0^{2\pi} d\theta \int_1^2 r\sqrt{1 + 4r^2} \, dr$$
$$= \left[\theta\right]_0^{2\pi} \left[\tfrac{1}{12}(1 + 4r^2)^{3/2}\right]_1^2 = \frac{\pi}{6}\left(17\sqrt{17} - 5\sqrt{5}\right)$$

9. $z = f(x, y) = xy$ with $x^2 + y^2 \le 1$, so $f_x = y$, $f_y = x \implies$

$$A(S) = \iint_D \sqrt{y^2 + x^2 + 1} \, dA = \int_0^{2\pi} \int_0^1 \sqrt{r^2 + 1} \, r \, dr \, d\theta = \int_0^{2\pi} \left[\tfrac{1}{3}(r^2 + 1)^{3/2}\right]_{r=0}^{r=1} d\theta$$
$$= \int_0^{2\pi} \tfrac{1}{3}\left(2\sqrt{2} - 1\right) d\theta = \frac{2\pi}{3}\left(2\sqrt{2} - 1\right)$$

11. $z = \sqrt{a^2 - x^2 - y^2}$, $z_x = -x(a^2 - x^2 - y^2)^{-1/2}$, $z_y = -y(a^2 - x^2 - y^2)^{-1/2}$,

$$A(S) = \iint_D \sqrt{\frac{x^2 + y^2}{a^2 - x^2 - y^2} + 1} \, dA$$
$$= \int_{-\pi/2}^{\pi/2} \int_0^{a\cos\theta} \sqrt{\frac{r^2}{a^2 - r^2} + 1} \, r \, dr \, d\theta$$
$$= \int_{-\pi/2}^{\pi/2} \int_0^{a\cos\theta} \frac{ar}{\sqrt{a^2 - r^2}} \, dr \, d\theta$$
$$= \int_{-\pi/2}^{\pi/2} \left[-a\sqrt{a^2 - r^2}\right]_{r=0}^{r=a\cos\theta} d\theta$$
$$= \int_{-\pi/2}^{\pi/2} -a\left(\sqrt{a^2 - a^2\cos^2\theta} - a\right) d\theta = 2a^2 \int_0^{\pi/2} \left(1 - \sqrt{1 - \cos^2\theta}\right) d\theta$$
$$= 2a^2 \int_0^{\pi/2} d\theta - 2a^2 \int_0^{\pi/2} \sqrt{\sin^2\theta} \, d\theta = a^2\pi - 2a^2 \int_0^{\pi/2} \sin\theta \, d\theta = a^2(\pi - 2)$$

$r = a\cos\theta$

© 2012 Cengage Learning. All Rights Reserved. May not be scanned, copied, or duplicated, or posted to a publicly accessible website, in whole or in part.

13. $z = f(x, y) = e^{-x^2 - y^2}$, $f_x = -2xe^{-x^2 - y^2}$, $f_y = -2ye^{-x^2 - y^2}$. Then

$$A(S) = \iint\limits_{x^2 + y^2 \leq 4} \sqrt{(-2xe^{-x^2 - y^2})^2 + (-2ye^{-x^2 - y^2})^2 + 1}\, dA = \iint\limits_{x^2 + y^2 \leq 4} \sqrt{4(x^2 + y^2)e^{-2(x^2 + y^2)} + 1}\, dA.$$

Converting to polar coordinates we have

$$A(S) = \int_0^{2\pi} \int_0^2 \sqrt{4r^2 e^{-2r^2} + 1}\, r\, dr\, d\theta = \int_0^{2\pi} d\theta \int_0^2 r\sqrt{4r^2 e^{-2r^2} + 1}\, dr$$

$$= 2\pi \int_0^2 r\sqrt{4r^2 e^{-2r^2} + 1}\, dr \approx 13.9783 \text{ using a calculator.}$$

15. (a) The midpoints of the four squares are $\left(\frac{1}{4}, \frac{1}{4}\right)$, $\left(\frac{1}{4}, \frac{3}{4}\right)$, $\left(\frac{3}{4}, \frac{1}{4}\right)$, and $\left(\frac{3}{4}, \frac{3}{4}\right)$. Here $f(x, y) = x^2 + y^2$, so the Midpoint Rule

gives

$$A(S) = \iint_D \sqrt{[f_x(x, y)]^2 + [f_y(x, y)]^2 + 1}\, dA = \iint_D \sqrt{(2x)^2 + (2y)^2 + 1}\, dA$$

$$\approx \frac{1}{4}\left(\sqrt{\left[2\left(\frac{1}{4}\right)\right]^2 + \left[2\left(\frac{1}{4}\right)\right]^2 + 1} + \sqrt{\left[2\left(\frac{1}{4}\right)\right]^2 + \left[2\left(\frac{3}{4}\right)\right]^2 + 1}\right.$$

$$\left. + \sqrt{\left[2\left(\frac{3}{4}\right)\right]^2 + \left[2\left(\frac{1}{4}\right)\right]^2 + 1} + \sqrt{\left[2\left(\frac{3}{4}\right)\right]^2 + \left[2\left(\frac{3}{4}\right)\right]^2 + 1}\right)$$

$$= \frac{1}{4}\left(\sqrt{\frac{3}{2}} + 2\sqrt{\frac{7}{2}} + \sqrt{\frac{11}{2}}\right) \approx 1.8279$$

(b) A CAS estimates the integral to be $A(S) = \iint_D \sqrt{1 + (2x)^2 + (2y)^2}\, dA = \int_0^1 \int_0^1 \sqrt{1 + 4x^2 + 4y^2}\, dy\, dx \approx 1.8616$.

This agrees with the Midpoint estimate only in the first decimal place.

17. $z = 1 + 2x + 3y + 4y^2$, so

$$A(S) = \iint_D \sqrt{1 + \left(\frac{\partial z}{\partial x}\right)^2 + \left(\frac{\partial z}{\partial y}\right)^2}\, dA = \int_1^4 \int_0^1 \sqrt{1 + 4 + (3 + 8y)^2}\, dy\, dx = \int_1^4 \int_0^1 \sqrt{14 + 48y + 64y^2}\, dy\, dx.$$

Using a CAS, we have $\int_1^4 \int_0^1 \sqrt{14 + 48y + 64y^2}\, dy\, dx = \frac{45}{8}\sqrt{14} + \frac{15}{16}\ln\left(11\sqrt{5} + 3\sqrt{14}\sqrt{5}\right) - \frac{15}{16}\ln\left(3\sqrt{5} + \sqrt{14}\sqrt{5}\right)$

or $\frac{45}{8}\sqrt{14} + \frac{15}{16}\ln\dfrac{11\sqrt{5} + 3\sqrt{70}}{3\sqrt{5} + \sqrt{70}}$.

19. $f(x, y) = 1 + x^2 y^2 \implies f_x = 2xy^2$, $f_y = 2x^2 y$. We use a CAS (with precision reduced to five significant digits, to speed

up the calculation) to estimate the integral

$$A(S) = \int_{-1}^1 \int_{-\sqrt{1-x^2}}^{\sqrt{1-x^2}} \sqrt{f_x^2 + f_y^2 + 1}\, dy\, dx = \int_{-1}^1 \int_{-\sqrt{1-x^2}}^{\sqrt{1-x^2}} \sqrt{4x^2 y^4 + 4x^4 y^2 + 1}\, dy\, dx, \text{ and find that } A(S) \approx 3.3213.$$

21. Here $z = f(x, y) = ax + by + c$, $f_x(x, y) = a$, $f_y(x, y) = b$, so

$$A(S) = \iint_D \sqrt{a^2 + b^2 + 1}\, dA = \sqrt{a^2 + b^2 + 1} \iint_D dA = \sqrt{a^2 + b^2 + 1}\, A(D).$$

23. If we project the surface onto the xz-plane, then the surface lies "above" the disk $x^2 + z^2 \leq 25$ in the xz-plane.

We have $y = f(x, z) = x^2 + z^2$ and, adapting Formula 2, the area of the surface is

$$A(S) = \iint\limits_{x^2 + z^2 \leq 25} \sqrt{[f_x(x, z)]^2 + [f_z(x, z)]^2 + 1}\, dA = \iint\limits_{x^2 + z^2 \leq 25} \sqrt{4x^2 + 4z^2 + 1}\, dA$$

© 2012 Cengage Learning. All Rights Reserved. May not be scanned, copied, or duplicated, or posted to a publicly accessible website, in whole or in part.

Converting to polar coordinates $x = r\cos\theta$, $z = r\sin\theta$ we have

$$A(S) = \int_0^{2\pi} \int_0^5 \sqrt{4r^2+1}\, r\, dr\, d\theta = \int_0^{2\pi} d\theta \int_0^5 r(4r^2+1)^{1/2}\, dr = \left[\theta\right]_0^{2\pi} \left[\frac{1}{12}(4r^2+1)^{3/2}\right]_0^5 = \frac{\pi}{6}\left(101\sqrt{101}-1\right)$$

15.7 Triple Integrals

1. $\iiint_B xyz^2\, dV = \int_0^1 \int_0^3 \int_{-1}^2 xyz^2\, dy\, dz\, dx = \int_0^1 \int_0^3 \left[\frac{1}{2}xy^2z^2\right]_{y=-1}^{y=2} dz\, dx = \int_0^1 \int_0^3 \frac{3}{2}xz^2\, dz\, dx$

$$= \int_0^1 \left[\frac{1}{2}xz^3\right]_{z=0}^{z=3} dx = \int_0^1 \frac{27}{2}x\, dx = \frac{27}{4}x^2\Big]_0^1 = \frac{27}{4}$$

3. $\int_0^2 \int_0^{z^2} \int_0^{y-z}(2x-y)\, dx\, dy\, dz = \int_0^2 \int_0^{z^2}\left[x^2-xy\right]_{x=0}^{x=y-z} dy\, dz = \int_0^2 \int_0^{z^2}\left[(y-z)^2-(y-z)y\right] dy\, dz$

$$= \int_0^2 \int_0^{z^2}\left(z^2-yz\right) dy\, dz = \int_0^2 \left[yz^2-\frac{1}{2}y^2z\right]_{y=0}^{y=z^2} dz = \int_0^2 \left(z^4-\frac{1}{2}z^5\right) dz$$

$$= \left[\frac{1}{5}z^5-\frac{1}{12}z^6\right]_0^2 = \frac{32}{5}-\frac{64}{12} = \frac{16}{15}$$

5. $\int_1^2 \int_0^{2z} \int_0^{\ln x} xe^{-y}\, dy\, dx\, dz = \int_1^2 \int_0^{2z}\left[-xe^{-y}\right]_{y=0}^{y=\ln x} dx\, dz = \int_1^2 \int_0^{2z}\left(-xe^{-\ln x}+xe^0\right) dx\, dz$

$$= \int_1^2 \int_0^{2z}(-1+x)\, dx\, dz = \int_1^2 \left[-x+\frac{1}{2}x^2\right]_{x=0}^{x=2z} dz$$

$$= \int_1^2 \left(-2z+2z^2\right) dz = \left[-z^2+\frac{2}{3}z^3\right]_1^2 = -4+\frac{16}{3}+1-\frac{2}{3} = \frac{5}{3}$$

7. $\int_0^{\pi/2} \int_0^y \int_0^x \cos(x+y+z)\, dz\, dx\, dy = \int_0^{\pi/2} \int_0^y \left[\sin(x+y+z)\right]_{z=0}^{z=x} dx\, dy$

$$= \int_0^{\pi/2} \int_0^y \left[\sin(2x+y)-\sin(x+y)\right] dx\, dy$$

$$= \int_0^{\pi/2} \left[-\frac{1}{2}\cos(2x+y)+\cos(x+y)\right]_{x=0}^{x=y} dy$$

$$= \int_0^{\pi/2} \left[-\frac{1}{2}\cos 3y+\cos 2y+\frac{1}{2}\cos y-\cos y\right] dy$$

$$= \left[-\frac{1}{6}\sin 3y+\frac{1}{2}\sin 2y-\frac{1}{2}\sin y\right]_0^{\pi/2} = \frac{1}{6}-\frac{1}{2} = -\frac{1}{3}$$

9. $\iiint_E y\, dV - \int_0^3 \int_0^x \int_{x-y}^{x+y} y\, dz\, dy\, dx = \int_0^3 \int_0^x \left[yz\right]_{z=x-y}^{z=x+y} dy\, dx = \int_0^3 \int_0^x 2y^2\, dy\, dx$

$$= \int_0^3 \left[\frac{2}{3}y^3\right]_{y=0}^{y=x} dx = \int_0^3 \frac{2}{3}x^3\, dx = \frac{1}{6}x^4\Big]_0^3 = \frac{81}{6} = \frac{27}{2}$$

11. $\iiint_E \frac{z}{x^2+z^2}\, dV = \int_1^4 \int_y^4 \int_0^z \frac{z}{x^2+z^2}\, dx\, dz\, dy = \int_1^4 \int_y^4 \left[z\cdot\frac{1}{z}\tan^{-1}\frac{x}{z}\right]_{x=0}^{x=z} dz\, dy$

$$= \int_1^4 \int_y^4 \left[\tan^{-1}(1)-\tan^{-1}(0)\right] dz\, dy = \int_1^4 \int_y^4 \left(\frac{\pi}{4}-0\right) dz\, dy = \frac{\pi}{4} \int_1^4 \left[z\right]_{z=y}^{z=4} dy$$

$$= \frac{\pi}{4} \int_1^4 (4-y)\, dy = \frac{\pi}{4}\left[4y-\frac{1}{2}y^2\right]_1^4 = \frac{\pi}{4}\left(16-8-4+\frac{1}{2}\right) = \frac{9\pi}{8}$$

13. Here $E = \{(x,y,z) \mid 0 \le x \le 1, 0 \le y \le \sqrt{x}, 0 \le z \le 1+x+y\}$, so

$$\iiint_E 6xy\, dV = \int_0^1 \int_0^{\sqrt{x}} \int_0^{1+x+y} 6xy\, dz\, dy\, dx = \int_0^1 \int_0^{\sqrt{x}} \left[6xyz\right]_{z=0}^{z=1+x+y} dy\, dx$$

$$= \int_0^1 \int_0^{\sqrt{x}} 6xy(1+x+y)\, dy\, dx = \int_0^1 \left[3xy^2+3x^2y^2+2xy^3\right]_{y=0}^{y=\sqrt{x}} dx$$

$$= \int_0^1 \left(3x^2+3x^3+2x^{5/2}\right) dx = \left[x^3+\frac{3}{4}x^4+\frac{4}{7}x^{7/2}\right]_0^1 = \frac{65}{28}$$

© 2012 Cengage Learning. All Rights Reserved. May not be scanned, copied, or duplicated, or posted to a publicly accessible website, in whole or in part.

15.

Here $T = \{(x, y, z) \mid 0 \le x \le 1, 0 \le y \le 1 - x, 0 \le z \le 1 - x - y\}$, so

$$\iiint_T x^2\, dV = \int_0^1 \int_0^{1-x} \int_0^{1-x-y} x^2\, dz\, dy\, dx = \int_0^1 \int_0^{1-x} x^2(1 - x - y)\, dy\, dx$$

$$= \int_0^1 \int_0^{1-x} (x^2 - x^3 - x^2 y)\, dy\, dx = \int_0^1 \left[x^2 y - x^3 y - \tfrac{1}{2} x^2 y^2 \right]_{y=0}^{y=1-x} dx$$

$$= \int_0^1 \left[x^2(1 - x) - x^3(1 - x) - \tfrac{1}{2} x^2 (1 - x)^2 \right] dx$$

$$= \int_0^1 \left(\tfrac{1}{2} x^4 - x^3 + \tfrac{1}{2} x^2 \right) dx = \left[\tfrac{1}{10} x^5 - \tfrac{1}{4} x^4 + \tfrac{1}{6} x^3 \right]_0^1$$

$$= \tfrac{1}{10} - \tfrac{1}{4} + \tfrac{1}{6} = \tfrac{1}{60}$$

17.

The projection of E on the yz-plane is the disk $y^2 + z^2 \le 1$. Using polar coordinates $y = r \cos\theta$ and $z = r \sin\theta$, we get

$$\iiint_E x\, dV = \iint_D \left[\int_{4y^2 + 4z^2}^4 x\, dx \right] dA = \tfrac{1}{2} \iint_D \left[4^2 - (4y^2 + 4z^2)^2 \right] dA$$

$$= 8 \int_0^{2\pi} \int_0^1 (1 - r^4)\, r\, dr\, d\theta = 8 \int_0^{2\pi} d\theta \int_0^1 (r - r^5)\, dr$$

$$= 8(2\pi) \left[\tfrac{1}{2} r^2 - \tfrac{1}{6} r^6 \right]_0^1 = \tfrac{16\pi}{3}$$

19. The plane $2x + y + z = 4$ intersects the xy-plane when

$2x + y + 0 = 4 \;\Rightarrow\; y = 4 - 2x$, so

$E = \{(x, y, z) \mid 0 \le x \le 2, 0 \le y \le 4 - 2x, 0 \le z \le 4 - 2x - y\}$ and

$$V = \int_0^2 \int_0^{4-2x} \int_0^{4-2x-y} dz\, dy\, dx = \int_0^2 \int_0^{4-2x} (4 - 2x - y)\, dy\, dx$$

$$= \int_0^2 \left[4y - 2xy - \tfrac{1}{2} y^2 \right]_{y=0}^{y=4-2x} dx$$

$$= \int_0^2 \left[4(4 - 2x) - 2x(4 - 2x) - \tfrac{1}{2}(4 - 2x)^2 \right] dx$$

$$= \int_0^2 (2x^2 - 8x + 8)\, dx = \left[\tfrac{2}{3} x^3 - 4x^2 + 8x \right]_0^2 = \tfrac{16}{3}$$

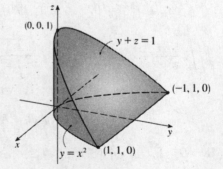

21. The plane $y + z = 1$ intersects the xy-plane in the line $y = 1$, so

$E = \{(x, y, z) \mid -1 \le x \le 1, x^2 \le y \le 1, 0 \le z \le 1 - y\}$ and

$$V = \iiint_E dV = \int_{-1}^1 \int_{x^2}^1 \int_0^{1-y} dz\, dy\, dx = \int_{-1}^1 \int_{x^2}^1 (1 - y)\, dy\, dx$$

$$= \int_{-1}^1 \left[y - \tfrac{1}{2} y^2 \right]_{y=x^2}^{y=1} dx = \int_{-1}^1 \left(\tfrac{1}{2} - x^2 + \tfrac{1}{2} x^4 \right) dx$$

$$= \left[\tfrac{1}{2} x - \tfrac{1}{3} x^3 + \tfrac{1}{10} x^5 \right]_{-1}^1 = \tfrac{1}{2} - \tfrac{1}{3} + \tfrac{1}{10} + \tfrac{1}{2} - \tfrac{1}{3} + \tfrac{1}{10} = \tfrac{8}{15}$$

© 2012 Cengage Learning. All Rights Reserved. May not be scanned, copied, or duplicated, or posted to a publicly accessible website, in whole or in part.

23. (a) The wedge can be described as the region

$$D = \{(x,y,z) \mid y^2 + z^2 \le 1, 0 \le x \le 1, 0 \le y \le x\}$$
$$= \{(x,y,z) \mid 0 \le x \le 1, 0 \le y \le x, 0 \le z \le \sqrt{1-y^2}\}$$

So the integral expressing the volume of the wedge is

$$\iiint_D dV = \int_0^1 \int_0^x \int_0^{\sqrt{1-y^2}} dz\, dy\, dx.$$

(b) A CAS gives $\int_0^1 \int_0^x \int_0^{\sqrt{1-y^2}} dz\, dy\, dx = \frac{\pi}{4} - \frac{1}{3}$.

(Or use Formulas 30 and 87 from the Table of Integrals.)

25. Here $f(x,y,z) = \cos(xyz)$ and $\Delta V = \frac{1}{2} \cdot \frac{1}{2} \cdot \frac{1}{2} = \frac{1}{8}$, so the Midpoint Rule gives

$$\iiint_B f(x,y,z)\, dV \approx \sum_{i=1}^{l} \sum_{j=1}^{m} \sum_{k=1}^{n} f(\overline{x}_i, \overline{y}_j, \overline{z}_k)\, \Delta V$$

$$= \frac{1}{8}\left[f\left(\tfrac{1}{4},\tfrac{1}{4},\tfrac{1}{4}\right) + f\left(\tfrac{1}{4},\tfrac{1}{4},\tfrac{3}{4}\right) + f\left(\tfrac{1}{4},\tfrac{3}{4},\tfrac{1}{4}\right) + f\left(\tfrac{1}{4},\tfrac{3}{4},\tfrac{3}{4}\right)\right.$$
$$\left. + f\left(\tfrac{3}{4},\tfrac{1}{4},\tfrac{1}{4}\right) + f\left(\tfrac{3}{4},\tfrac{1}{4},\tfrac{3}{4}\right) + f\left(\tfrac{3}{4},\tfrac{3}{4},\tfrac{1}{4}\right) + f\left(\tfrac{3}{4},\tfrac{3}{4},\tfrac{3}{4}\right)\right]$$

$$= \frac{1}{8}\left[\cos\tfrac{1}{64} + \cos\tfrac{3}{64} + \cos\tfrac{3}{64} + \cos\tfrac{9}{64} + \cos\tfrac{3}{64} + \cos\tfrac{9}{64} + \cos\tfrac{9}{64} + \cos\tfrac{27}{64}\right] \approx 0.985$$

27. $E = \{(x,y,z) \mid 0 \le x \le 1, 0 \le z \le 1-x, 0 \le y \le 2-2z\}$,

the solid bounded by the three coordinate planes and the planes

$z = 1-x$, $y = 2-2z$.

29.

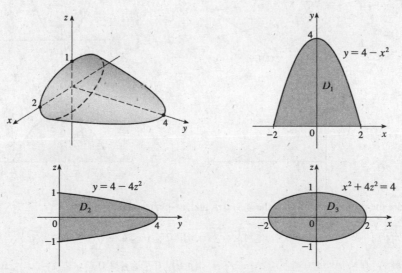

If D_1, D_2, D_3 are the projections of E on the xy-, yz-, and xz-planes, then

$$D_1 = \{(x,y) \mid -2 \le x \le 2, 0 \le y \le 4-x^2\} = \{(x,y) \mid 0 \le y \le 4,\ -\sqrt{4-y} \le x \le \sqrt{4-y}\}$$

$$D_2 = \{(y,z) \mid 0 \le y \le 4,\ -\tfrac{1}{2}\sqrt{4-y} \le z \le \tfrac{1}{2}\sqrt{4-y}\} = \{(y,z) \mid -1 \le z \le 1, 0 \le y \le 4-4z^2\}$$

$$D_3 = \{(x,z) \mid x^2 + 4z^2 \le 4\}$$

[continued]

© 2012 Cengage Learning. All Rights Reserved. May not be scanned, copied, or duplicated, or posted to a publicly accessible website, in whole or in part.

Therefore

$$E = \left\{ (x,y,z) \mid -2 \le x \le 2, 0 \le y \le 4 - x^2, -\tfrac{1}{2}\sqrt{4 - x^2 - y} \le z \le \tfrac{1}{2}\sqrt{4 - x^2 - y} \right\}$$

$$= \left\{ (x,y,z) \mid 0 \le y \le 4, -\sqrt{4 - y} \le x \le \sqrt{4 - y}, -\tfrac{1}{2}\sqrt{4 - x^2 - y} \le z \le \tfrac{1}{2}\sqrt{4 - x^2 - y} \right\}$$

$$= \left\{ (x,y,z) \mid -1 \le z \le 1, 0 \le y \le 4 - 4z^2, -\sqrt{4 - y - 4z^2} \le x \le \sqrt{4 - y - 4z^2} \right\}$$

$$= \left\{ (x,y,z) \mid 0 \le y \le 4, -\tfrac{1}{2}\sqrt{4 - y} \le z \le \tfrac{1}{2}\sqrt{4 - y}, -\sqrt{4 - y - 4z^2} \le x \le \sqrt{4 - y - 4z^2} \right\}$$

$$= \left\{ (x,y,z) \mid -2 \le x \le 2, -\tfrac{1}{2}\sqrt{4 - x^2} \le z \le \tfrac{1}{2}\sqrt{4 - x^2}, 0 \le y \le 4 - x^2 - 4z^2 \right\}$$

$$= \left\{ (x,y,z) \mid -1 \le z \le 1, -\sqrt{4 - 4z^2} \le x \le \sqrt{4 - 4z^2}, 0 \le y \le 4 - x^2 - 4z^2 \right\}$$

Then

$$\iiint_E f(x,y,z)\,dV = \int_{-2}^{2} \int_0^{4-x^2} \int_{-\sqrt{4-x^2-y}/2}^{\sqrt{4-x^2-y}/2} f(x,y,z)\,dz\,dy\,dx = \int_0^4 \int_{-\sqrt{4-y}}^{\sqrt{4-y}} \int_{-\sqrt{4-x^2-y}/2}^{\sqrt{4-x^2-y}/2} f(x,y,z)\,dz\,dx\,dy$$

$$= \int_{-1}^{1} \int_0^{4-4z^2} \int_{-\sqrt{4-y-4z^2}}^{\sqrt{4-y-4z^2}} f(x,y,z)\,dx\,dy\,dz = \int_0^4 \int_{-\sqrt{4-y}/2}^{\sqrt{4-y}/2} \int_{-\sqrt{4-y-4z^2}}^{\sqrt{4-y-4z^2}} f(x,y,z)\,dx\,dz\,dy$$

$$= \int_{-2}^{2} \int_{-\sqrt{4-x^2}/2}^{\sqrt{4-x^2}/2} \int_0^{4-x^2-4z^2} f(x,y,z)\,dy\,dz\,dx = \int_{-1}^{1} \int_{-\sqrt{4-4z^2}}^{\sqrt{4-4z^2}} \int_0^{4-x^2-4z^2} f(x,y,z)\,dy\,dx\,dz$$

31.

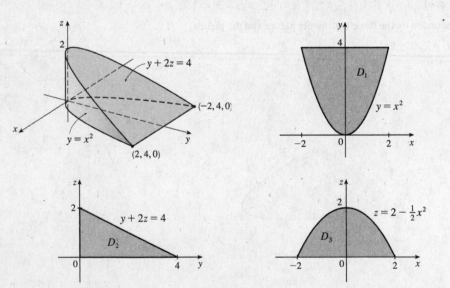

If D_1, D_2, and D_3 are the projections of E on the xy-, yz-, and xz-planes, then

$$D_1 = \left\{ (x,y) \mid -2 \le x \le 2, x^2 \le y \le 4 \right\} = \left\{ (x,y) \mid 0 \le y \le 4, -\sqrt{y} \le x \le \sqrt{y} \right\},$$

$$D_2 = \left\{ (y,z) \mid 0 \le y \le 4, 0 \le z \le 2 - \tfrac{1}{2}y \right\} = \left\{ (y,z) \mid 0 \le z \le 2, 0 \le y \le 4 - 2z \right\}, \text{ and}$$

$$D_3 = \left\{ (x,z) \mid -2 \le x \le 2, 0 \le z \le 2 - \tfrac{1}{2}x^2 \right\} = \left\{ (x,z) \mid 0 \le z \le 2, -\sqrt{4 - 2z} \le x \le \sqrt{4 - 2z} \right\}$$

© 2012 Cengage Learning. All Rights Reserved. May not be scanned, copied, or duplicated, or posted to a publicly accessible website, in whole or in part.

Therefore
$$E = \left\{ (x,y,z) \mid -2 \le x \le 2,\, x^2 \le y \le 4,\, 0 \le z \le 2 - \tfrac{1}{2}y \right\}$$
$$= \left\{ (x,y,z) \mid 0 \le y \le 4,\, -\sqrt{y} \le x \le \sqrt{y},\, 0 \le z \le 2 - \tfrac{1}{2}y \right\}$$
$$= \left\{ (x,y,z) \mid 0 \le y \le 4,\, 0 \le z \le 2 - \tfrac{1}{2}y,\, -\sqrt{y} \le x \le \sqrt{y} \right\}$$
$$= \left\{ (x,y,z) \mid 0 \le z \le 2,\, 0 \le y \le 4 - 2z,\, -\sqrt{y} \le x \le \sqrt{y} \right\}$$
$$= \left\{ (x,y,z) \mid -2 \le x \le 2,\, 0 \le z \le 2 - \tfrac{1}{2}x^2,\, x^2 \le y \le 4 - 2z \right\}$$
$$= \left\{ (x,y,z) \mid 0 \le z \le 2,\, -\sqrt{4-2z} \le x \le \sqrt{4-2z},\, x^2 \le y \le 4 - 2z \right\}$$

Then
$$\iiint_E f(x,y,z)\,dV = \int_{-2}^{2} \int_{x^2}^{4} \int_0^{2-y/2} f(x,y,z)\,dz\,dy\,dx = \int_0^4 \int_{-\sqrt{y}}^{\sqrt{y}} \int_0^{2-y/2} f(x,y,z)\,dz\,dx\,dy$$

$$= \int_0^4 \int_0^{2-y/2} \int_{-\sqrt{y}}^{\sqrt{y}} f(x,y,z)\,dx\,dz\,dy = \int_0^2 \int_0^{4-2z} \int_{-\sqrt{y}}^{\sqrt{y}} f(x,y,z)\,dx\,dy\,dz$$

$$= \int_{-2}^2 \int_0^{2-x^2/2} \int_{x^2}^{4-2z} f(x,y,z)\,dy\,dz\,dx = \int_0^2 \int_{-\sqrt{4-2z}}^{\sqrt{4-2z}} \int_{x^2}^{4-2z} f(x,y,z)\,dy\,dx\,dz$$

33.

The diagrams show the projections of E on the xy-, yz-, and xz-planes.

Therefore

$$\int_0^1 \int_{\sqrt{x}}^1 \int_0^{1-y} f(x,y,z)\,dz\,dy\,dx = \int_0^1 \int_0^{y^2} \int_0^{1-y} f(x,y,z)\,dz\,dx\,dy = \int_0^1 \int_0^{1-z} \int_0^{y^2} f(x,y,z)\,dx\,dy\,dz$$

$$= \int_0^1 \int_0^{1-y} \int_0^{y^2} f(x,y,z)\,dx\,dz\,dy = \int_0^1 \int_0^{1-\sqrt{x}} \int_{\sqrt{x}}^{1-z} f(x,y,z)\,dy\,dz\,dx$$

$$= \int_0^1 \int_0^{(1-z)^2} \int_{\sqrt{x}}^{1-z} f(x,y,z)\,dy\,dx\,dz$$

35.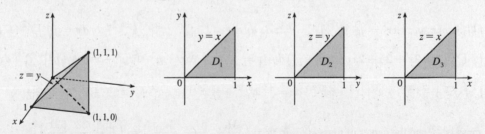

$$\int_0^1 \int_y^1 \int_0^y f(x,y,z)\,dz\,dx\,dy = \iiint_E f(x,y,z)\,dV \text{ where } E = \{(x,y,z) \mid 0 \le z \le y,\, y \le x \le 1,\, 0 \le y \le 1\}.$$

If D_1, D_2, and D_3 are the projections of E on the xy-, yz- and xz-planes then

$$D_1 = \{(x,y) \mid 0 \le y \le 1,\, y \le x \le 1\} = \{(x,y) \mid 0 \le x \le 1,\, 0 \le y \le x\},$$

$$D_2 = \{(y,z) \mid 0 \le y \le 1,\, 0 \le z \le y\} = \{(y,z) \mid 0 \le z \le 1,\, z \le y \le 1\}, \text{ and}$$

$$D_3 = \{(x,z) \mid 0 \le x \le 1,\, 0 \le z \le x\} = \{(x,z) \mid 0 \le z \le 1,\, z \le x \le 1\}.$$

[continued]

© 2012 Cengage Learning. All Rights Reserved. May not be scanned, copied, or duplicated, or posted to a publicly accessible website, in whole or in part.

Thus we also have

$$E = \{(x,y,z) \mid 0 \le x \le 1, 0 \le y \le x, 0 \le z \le y\} = \{(x,y,z) \mid 0 \le y \le 1, 0 \le z \le y, y \le x \le 1\}$$

$$= \{(x,y,z) \mid 0 \le z \le 1, z \le y \le 1, y \le x \le 1\} = \{(x,y,z) \mid 0 \le x \le 1, 0 \le z \le x, z \le y \le x\}$$

$$= \{(x,y,z) \mid 0 \le z \le 1, z \le x \le 1, z \le y \le x\}.$$

Then

$$\int_0^1 \int_y^1 \int_0^y f(x,y,z)\,dz\,dx\,dy = \int_0^1 \int_0^x \int_0^y f(x,y,z)\,dz\,dy\,dx = \int_0^1 \int_0^y \int_y^1 f(x,y,z)\,dx\,dz\,dy$$

$$= \int_0^1 \int_z^1 \int_y^1 f(x,y,z)\,dx\,dy\,dz = \int_0^1 \int_0^x \int_z^x f(x,y,z)\,dy\,dz\,dx$$

$$= \int_0^1 \int_z^1 \int_z^x f(x,y,z)\,dy\,dx\,dz$$

37. The region C is the solid bounded by a circular cylinder of radius 2 with axis the z-axis for $-2 \le z \le 2$. We can write $\iiint_C (4 + 5x^2yz^2)\,dV = \iiint_C 4\,dV + \iiint_C 5x^2yz^2\,dV$, but $f(x,y,z) = 5x^2yz^2$ is an odd function with respect to y. Since C is symmetrical about the xz-plane, we have $\iiint_C 5x^2yz^2\,dV = 0$. Thus $\iiint_C (4 + 5x^2yz^2)\,dV = \iiint_C 4\,dV = 4 \cdot V(E) = 4 \cdot \pi(2)^2(4) = 64\pi$.

39. $m = \iiint_E \rho(x,y,z)\,dV = \int_0^1 \int_0^{\sqrt{x}} \int_0^{1+x+y} 2\,dz\,dy\,dx = \int_0^1 \int_0^{\sqrt{x}} 2(1 + x + y)\,dy\,dx$

$= \int_0^1 \left[2y + 2xy + y^2\right]_{y=0}^{y=\sqrt{x}}\,dx = \int_0^1 \left(2\sqrt{x} + 2x^{3/2} + x\right)\,dx = \left[\frac{4}{3}x^{3/2} + \frac{4}{5}x^{5/2} + \frac{1}{2}x^2\right]_0^1 = \frac{79}{30}$

$M_{yz} = \iiint_E x\rho(x,y,z)\,dV = \int_0^1 \int_0^{\sqrt{x}} \int_0^{1+x+y} 2x\,dz\,dy\,dx = \int_0^1 \int_0^{\sqrt{x}} 2x(1 + x + y)\,dy\,dx$

$= \int_0^1 \left[2xy + 2x^2y + xy^2\right]_{y=0}^{y=\sqrt{x}}\,dx = \int_0^1 (2x^{3/2} + 2x^{5/2} + x^2)\,dx = \left[\frac{4}{5}x^{5/2} + \frac{4}{7}x^{7/2} + \frac{1}{3}x^3\right]_0^1 = \frac{179}{105}$

$M_{xz} = \iiint_E y\rho(x,y,z)\,dV = \int_0^1 \int_0^{\sqrt{x}} \int_0^{1+x+y} 2y\,dz\,dy\,dx = \int_0^1 \int_0^{\sqrt{x}} 2y(1 + x + y)\,dy\,dx$

$= \int_0^1 \left[y^2 + xy^2 + \frac{2}{3}y^3\right]_{y=0}^{y=\sqrt{x}}\,dx = \int_0^1 \left(x + x^2 + \frac{2}{3}x^{3/2}\right)\,dx = \left[\frac{1}{2}x^2 + \frac{1}{3}x^3 + \frac{4}{15}x^{5/2}\right]_0^1 = \frac{11}{10}$

$M_{xy} = \iiint_E z\rho(x,y,z)\,dV = \int_0^1 \int_0^{\sqrt{x}} \int_0^{1+x+y} 2z\,dz\,dy\,dx = \int_0^1 \int_0^{\sqrt{x}} \left[z^2\right]_{z=0}^{z=1+x+y}\,dy\,dx = \int_0^1 \int_0^{\sqrt{x}} (1 + x + y)^2\,dy\,dx$

$= \int_0^1 \int_0^{\sqrt{x}} (1 + 2x + 2y + 2xy + x^2 + y^2)\,dy\,dx = \int_0^1 \left[y + 2xy + y^2 + xy^2 + x^2y + \frac{1}{3}y^3\right]_{y=0}^{y=\sqrt{x}}\,dx$

$= \int_0^1 \left(\sqrt{x} + \frac{7}{3}x^{3/2} + x + x^2 + x^{5/2}\right)\,dx = \left[\frac{2}{3}x^{3/2} + \frac{14}{15}x^{5/2} + \frac{1}{2}x^2 + \frac{1}{3}x^3 + \frac{2}{7}x^{7/2}\right]_0^1 = \frac{571}{210}$

Thus the mass is $\frac{79}{30}$ and the center of mass is $(\overline{x}, \overline{y}, \overline{z}) = \left(\dfrac{M_{yz}}{m}, \dfrac{M_{xz}}{m}, \dfrac{M_{xy}}{m}\right) = \left(\dfrac{358}{553}, \dfrac{33}{79}, \dfrac{571}{553}\right)$.

41. $m = \int_0^a \int_0^a \int_0^a (x^2 + y^2 + z^2)\,dx\,dy\,dz = \int_0^a \int_0^a \left[\frac{1}{3}x^3 + xy^2 + xz^2\right]_{x=0}^{x=a}\,dy\,dz = \int_0^a \int_0^a \left(\frac{1}{3}a^3 + ay^2 + az^2\right)\,dy\,dz$

$= \int_0^a \left[\frac{1}{3}a^3y + \frac{1}{3}ay^3 + ayz^2\right]_{y=0}^{y=a}\,dz = \int_0^a \left(\frac{2}{3}a^4 + a^2z^2\right)\,dz = \left[\frac{2}{3}a^4z + \frac{1}{3}a^2z^3\right]_0^a = \frac{2}{3}a^5 + \frac{1}{3}a^5 = a^5$

$M_{yz} = \int_0^a \int_0^a \int_0^a \left[x^3 + x(y^2 + z^2)\right]\,dx\,dy\,dz = \int_0^a \int_0^a \left[\frac{1}{4}a^4 + \frac{1}{2}a^2(y^2 + z^2)\right]\,dy\,dz$

$= \int_0^a \left(\frac{1}{4}a^5 + \frac{1}{6}a^5 + \frac{1}{2}a^3z^2\right)\,dz = \frac{1}{4}a^6 + \frac{1}{3}a^6 = \frac{7}{12}a^6 = M_{xz} = M_{xy}$ by symmetry of E and $\rho(x,y,z)$

Hence $(\overline{x}, \overline{y}, \overline{z}) = \left(\frac{7}{12}a, \frac{7}{12}a, \frac{7}{12}a\right)$.

© 2012 Cengage Learning. All Rights Reserved. May not be scanned, copied, or duplicated, or posted to a publicly accessible website, in whole or in part.

43. $I_x = \int_0^L \int_0^L \int_0^L k(y^2 + z^2)\, dz\, dy\, dx = k \int_0^L \int_0^L \left(Ly^2 + \frac{1}{3}L^3\right) dy\, dx = k \int_0^L \frac{2}{3}L^4\, dx = \frac{2}{3}kL^5.$

By symmetry, $I_x = I_y = I_z = \frac{2}{3}kL^5.$

45. $I_z = \iiint_E (x^2 + y^2)\, \rho(x, y, z)\, dV = \iint_{x^2+y^2 \leq a^2} \left[\int_0^h k(x^2 + y^2)\, dz\right] dA = \iint_{x^2+y^2 \leq a^2} k(x^2 + y^2)h\, dA$

$= kh \int_0^{2\pi} \int_0^a (r^2)\, r\, dr\, d\theta = kh \int_0^{2\pi} d\theta \int_0^a r^3\, dr = kh(2\pi)\left[\frac{1}{4}r^4\right]_0^a = 2\pi kh \cdot \frac{1}{4}a^4 = \frac{1}{2}\pi kha^4$

47. (a) $m = \int_{-1}^1 \int_{x^2}^1 \int_0^{1-y} \sqrt{x^2 + y^2}\, dz\, dy\, dx$

(b) $(\overline{x}, \overline{y}, \overline{z})$ where $\overline{x} = \frac{1}{m} \int_{-1}^1 \int_{x^2}^1 \int_0^{1-y} x\sqrt{x^2 + y^2}\, dz\, dy\, dx,\ \overline{y} = \frac{1}{m} \int_{-1}^1 \int_{x^2}^1 \int_0^{1-y} y\sqrt{x^2 + y^2}\, dz\, dy\, dx$, and

$\overline{z} = \frac{1}{m} \int_{-1}^1 \int_{x^2}^1 \int_0^{1-y} z\sqrt{x^2 + y^2}\, dz\, dy\, dx.$

(c) $I_z = \int_{-1}^1 \int_{x^2}^1 \int_0^{1-y} (x^2 + y^2)\sqrt{x^2 + y^2}\, dz\, dy\, dx = \int_{-1}^1 \int_{x^2}^1 \int_0^{1-y} (x^2 + y^2)^{3/2}\, dz\, dy\, dx$

49. (a) $m = \int_0^1 \int_0^{\sqrt{1-x^2}} \int_0^y (1 + x + y + z)\, dz\, dy\, dx = \frac{3\pi}{32} + \frac{11}{24}$

(b) $(\overline{x}, \overline{y}, \overline{z}) = \left(m^{-1} \int_0^1 \int_0^{\sqrt{1-x^2}} \int_0^y x(1 + x + y + z)\, dz\, dy\, dx,\right.$

$m^{-1} \int_0^1 \int_0^{\sqrt{1-x^2}} \int_0^y y(1 + x + y + z)\, dz\, dy\, dx,$

$\left. m^{-1} \int_0^1 \int_0^{\sqrt{1-x^2}} \int_0^y z(1 + x + y + z)\, dz\, dy\, dx\right)$

$= \left(\frac{28}{9\pi + 44},\ \frac{30\pi + 128}{45\pi + 220},\ \frac{45\pi + 208}{135\pi + 660}\right)$

(c) $I_z = \int_0^1 \int_0^{\sqrt{1-x^2}} \int_0^y (x^2 + y^2)(1 + x + y + z)\, dz\, dy\, dx = \frac{68 + 15\pi}{240}$

51. (a) $f(x, y, z)$ is a joint density function, so we know $\iiint_{\mathbb{R}^3} f(x, y, z)\, dV = 1$. Here we have

$\iiint_{\mathbb{R}^3} f(x, y, z)\, dV = \int_{-\infty}^{\infty} \int_{-\infty}^{\infty} \int_{-\infty}^{\infty} f(x, y, z)\, dz\, dy\, dx = \int_0^2 \int_0^2 \int_0^2 Cxyz\, dz\, dy\, dx$

$= C \int_0^2 x\, dx \int_0^2 y\, dy \int_0^2 z\, dz = C\left[\frac{1}{2}x^2\right]_0^2 \left[\frac{1}{2}y^2\right]_0^2 \left[\frac{1}{2}z^2\right]_0^2 = 8C$

Then we must have $8C = 1 \Rightarrow C = \frac{1}{8}.$

(b) $P(X \leq 1, Y \leq 1, Z \leq 1) = \int_{-\infty}^1 \int_{-\infty}^1 \int_{-\infty}^1 f(x, y, z)\, dz\, dy\, dx = \int_0^1 \int_0^1 \int_0^1 \frac{1}{8}xyz\, dz\, dy\, dx$

$= \frac{1}{8} \int_0^1 x\, dx \int_0^1 y\, dy \int_0^1 z\, dz = \frac{1}{8}\left[\frac{1}{2}x^2\right]_0^1 \left[\frac{1}{2}y^2\right]_0^1 \left[\frac{1}{2}z^2\right]_0^1 = \frac{1}{8}\left(\frac{1}{2}\right)^3 = \frac{1}{64}$

(c) $P(X + Y + Z \leq 1) = P((X, Y, Z) \in E)$ where E is the solid region in the first octant bounded by the coordinate planes and the plane $x + y + z = 1$. The plane $x + y + z = 1$ meets the xy-plane in the line $x + y = 1$, so we have

$P(X + Y + Z \leq 1) = \iiint_E f(x, y, z)\, dV = \int_0^1 \int_0^{1-x} \int_0^{1-x-y} \frac{1}{8}xyz\, dz\, dy\, dx$

$= \frac{1}{8} \int_0^1 \int_0^{1-x} xy\left[\frac{1}{2}z^2\right]_{z=0}^{z=1-x-y} dy\, dx = \frac{1}{16} \int_0^1 \int_0^{1-x} xy(1 - x - y)^2\, dy\, dx$

$= \frac{1}{16} \int_0^1 \int_0^{1-x} [(x^3 - 2x^2 + x)y + (2x^2 - 2x)y^2 + xy^3]\, dy\, dx$

$= \frac{1}{16} \int_0^1 \left[(x^3 - 2x^2 + x)\frac{1}{2}y^2 + (2x^2 - 2x)\frac{1}{3}y^3 + x\left(\frac{1}{4}y^4\right)\right]_{y=0}^{y=1-x} dx$

$= \frac{1}{192} \int_0^1 (x - 4x^2 + 6x^3 - 4x^4 + x^5)\, dx = \frac{1}{192}\left(\frac{1}{30}\right) = \frac{1}{5760}$

© 2012 Cengage Learning. All Rights Reserved. May not be scanned, copied, or duplicated, or posted to a publicly accessible website, in whole or in part.

53. $V(E) = L^3 \quad \Rightarrow \quad f_{\text{ave}} = \dfrac{1}{L^3} \displaystyle\int_0^L \int_0^L \int_0^L xyz \, dx \, dy \, dz = \dfrac{1}{L^3} \int_0^L x \, dx \int_0^L y \, dy \int_0^L z \, dz$

$$= \dfrac{1}{L^3} \left[\dfrac{x^2}{2}\right]_0^L \left[\dfrac{y^2}{2}\right]_0^L \left[\dfrac{z^2}{2}\right]_0^L = \dfrac{1}{L^3} \dfrac{L^2}{2} \dfrac{L^2}{2} \dfrac{L^2}{2} = \dfrac{L^3}{8}$$

55. (a) The triple integral will attain its maximum when the integrand $1 - x^2 - 2y^2 - 3z^2$ is positive in the region E and negative everywhere else. For if E contains some region F where the integrand is negative, the integral could be increased by excluding F from E, and if E fails to contain some part G of the region where the integrand is positive, the integral could be increased by including G in E. So we require that $x^2 + 2y^2 + 3z^2 \leq 1$. This describes the region bounded by the ellipsoid $x^2 + 2y^2 + 3z^2 = 1$.

(b) The maximum value of $\iiint_E (1 - x^2 - 2y^2 - 3z^2) \, dV$ occurs when E is the solid region bounded by the ellipsoid $x^2 + 2y^2 + 3z^2 = 1$. The projection of E on the xy-plane is the planar region bounded by the ellipse $x^2 + 2y^2 = 1$, so

$$E = \left\{ (x, y, z) \mid -1 \leq x \leq 1, -\sqrt{\tfrac{1}{2}(1 - x^2)} \leq y \leq \sqrt{\tfrac{1}{2}(1 - x^2)}, -\sqrt{\tfrac{1}{3}(1 - x^2 - 2y^2)} \leq z \leq \sqrt{\tfrac{1}{3}(1 - x^2 - 2y^2)} \right\}$$

and

$$\iiint_E (1 - x^2 - 2y^2 - 3z^2) \, dV = \int_{-1}^{1} \int_{-\sqrt{\frac{1}{2}(1-x^2)}}^{\sqrt{\frac{1}{2}(1-x^2)}} \int_{-\sqrt{\frac{1}{3}(1-x^2-2y^2)}}^{\sqrt{\frac{1}{3}(1-x^2-2y^2)}} (1 - x^2 - 2y^2 - 3z^2) \, dz \, dy \, dx = \dfrac{4\sqrt{6}}{45}\pi$$

using a CAS.

15.8 Triple Integrals in Cylindrical Coordinates

1. (a)

From Equations 1, $x = r\cos\theta = 4\cos\dfrac{\pi}{3} = 4 \cdot \dfrac{1}{2} = 2$,

$y = r\sin\theta = 4\sin\dfrac{\pi}{3} = 4 \cdot \dfrac{\sqrt{3}}{2} = 2\sqrt{3}$, $z = -2$, so the point is

$\left(2, 2\sqrt{3}, -2\right)$ in rectangular coordinates.

(b)

$x = 2\cos\left(-\tfrac{\pi}{2}\right) = 0$, $y = 2\sin\left(-\tfrac{\pi}{2}\right) = -2$,

and $z = 1$, so the point is $(0, -2, 1)$ in rectangular coordinates.

© 2012 Cengage Learning. All Rights Reserved. May not be scanned, copied, or duplicated, or posted to a publicly accessible website, in whole or in part.

3. (a) From Equations 2 we have $r^2 = (-1)^2 + 1^2 = 2$ so $r = \sqrt{2}$; $\tan\theta = \frac{1}{-1} = -1$ and the point $(-1, 1)$ is in the second

quadrant of the xy-plane, so $\theta = \frac{3\pi}{4} + 2n\pi$; $z = 1$. Thus, one set of cylindrical coordinates is $\left(\sqrt{2}, \frac{3\pi}{4}, 1\right)$.

(b) $r^2 = (-2)^2 + (2\sqrt{3})^2 = 16$ so $r = 4$; $\tan\theta = \frac{2\sqrt{3}}{-2} = -\sqrt{3}$ and the point $(-2, 2\sqrt{3})$ is in the second quadrant of the

xy-plane, so $\theta = \frac{2\pi}{3} + 2n\pi$; $z = 3$. Thus, one set of cylindrical coordinates is $\left(4, \frac{2\pi}{3}, 3\right)$.

5. Since $\theta = \frac{\pi}{4}$ but r and z may vary, the surface is a vertical half-plane including the z-axis and intersecting the xy-plane in the

half-line $y = x$, $x \geq 0$.

7. $z = 4 - r^2 = 4 - (x^2 + y^2)$ or $4 - x^2 - y^2$, so the surface is a circular paraboloid with vertex $(0, 0, 4)$, axis the z-axis, and

opening downward.

9. (a) Substituting $x^2 + y^2 = r^2$ and $x = r\cos\theta$, the equation $x^2 - x + y^2 + z^2 = 1$ becomes $r^2 - r\cos\theta + z^2 = 1$ or

$z^2 = 1 + r\cos\theta - r^2$.

(b) Substituting $x = r\cos\theta$ and $y = r\sin\theta$, the equation $z = x^2 - y^2$ becomes

$z = (r\cos\theta)^2 - (r\sin\theta)^2 = r^2(\cos^2\theta - \sin^2\theta)$ or $z = r^2\cos 2\theta$.

11.

$0 \leq r \leq 2$ and $0 \leq z \leq 1$ describe a solid circular cylinder with

radius 2, axis the z-axis, and height 1, but $-\pi/2 \leq \theta \leq \pi/2$ restricts

the solid to the first and fourth quadrants of the xy-plane, so we have

a half-cylinder.

13. We can position the cylindrical shell vertically so that its axis coincides with the z-axis and its base lies in the xy-plane. If we

use centimeters as the unit of measurement, then cylindrical coordinates conveniently describe the shell as $6 \leq r \leq 7$,

$0 \leq \theta \leq 2\pi$, $0 \leq z \leq 20$.

15.

The region of integration is given in cylindrical coordinates by

$E = \left\{(r, \theta, z) \mid -\pi/2 \leq \theta \leq \pi/2, 0 \leq r \leq 2, 0 \leq z \leq r^2\right\}$. This

represents the solid region above quadrants I and IV of the xy-plane enclosed

by the circular cylinder $r = 2$, bounded above by the circular paraboloid

$z = r^2$ ($z = x^2 + y^2$), and bounded below by the xy-plane ($z = 0$).

$\int_{-\pi/2}^{\pi/2} \int_0^2 \int_0^{r^2} r\, dz\, dr\, d\theta = \int_{-\pi/2}^{\pi/2} \int_0^2 \left[rz\right]_{z=0}^{z=r^2} dr\, d\theta = \int_{-\pi/2}^{\pi/2} \int_0^2 r^3\, dr\, d\theta$

$= \int_{-\pi/2}^{\pi/2} d\theta \int_0^2 r^3\, dr = \left[\theta\right]_{-\pi/2}^{\pi/2} \left[\frac{1}{4}r^4\right]_0^2$

$= \pi(4 - 0) = 4\pi$

© 2012 Cengage Learning. All Rights Reserved. May not be scanned, copied, or duplicated, or posted to a publicly accessible website, in whole or in part.

17. In cylindrical coordinates, E is given by $\{(r, \theta, z) \mid 0 \le \theta \le 2\pi, 0 \le r \le 4, -5 \le z \le 4\}$. So

$$\iiint_E \sqrt{x^2 + y^2}\, dV = \int_0^{2\pi} \int_0^4 \int_{-5}^4 \sqrt{r^2}\, r\, dz\, dr\, d\theta = \int_0^{2\pi} d\theta \int_0^4 r^2\, dr \int_{-5}^4 dz$$

$$= \left[\theta\right]_0^{2\pi} \left[\tfrac{1}{3}r^3\right]_0^4 \left[z\right]_{-5}^4 = (2\pi)\left(\tfrac{64}{3}\right)(9) = 384\pi$$

19. The paraboloid $z = 4 - x^2 - y^2 = 4 - r^2$ intersects the xy-plane in the circle $x^2 + y^2 = 4$ or $r^2 = 4 \Rightarrow r = 2$, so in

cylindrical coordinates, E is given by $\{(r, \theta, z) \mid 0 \le \theta \le \pi/2, 0 \le r \le 2, 0 \le z \le 4 - r^2\}$. Thus

$$\iiint_E (x + y + z)\, dV = \int_0^{\pi/2} \int_0^2 \int_0^{4-r^2} (r\cos\theta + r\sin\theta + z)\, r\, dz\, dr\, d\theta = \int_0^{\pi/2} \int_0^2 \left[r^2(\cos\theta + \sin\theta)z + \tfrac{1}{2}rz^2\right]_{z=0}^{z=4-r^2} dr\, d\theta$$

$$= \int_0^{\pi/2} \int_0^2 \left[(4r^2 - r^4)(\cos\theta + \sin\theta) + \tfrac{1}{2}r(4 - r^2)^2\right] dr\, d\theta$$

$$= \int_0^{\pi/2} \left[\left(\tfrac{4}{3}r^3 - \tfrac{1}{5}r^5\right)(\cos\theta + \sin\theta) - \tfrac{1}{12}(4 - r^2)^3\right]_{r=0}^{r=2} d\theta$$

$$= \int_0^{\pi/2} \left[\tfrac{64}{15}(\cos\theta + \sin\theta) + \tfrac{16}{3}\right] d\theta = \left[\tfrac{64}{15}(\sin\theta - \cos\theta) + \tfrac{16}{3}\theta\right]_0^{\pi/2}$$

$$= \tfrac{64}{15}(1 - 0) + \tfrac{16}{3} \cdot \tfrac{\pi}{2} - \tfrac{64}{15}(0 - 1) - 0 = \tfrac{8}{3}\pi + \tfrac{128}{15}$$

21. In cylindrical coordinates, E is bounded by the cylinder $r = 1$, the plane $z = 0$, and the cone $z = 2r$. So

$E = \{(r, \theta, z) \mid 0 \le \theta \le 2\pi, 0 \le r \le 1, 0 \le z \le 2r\}$ and

$$\iiint_E x^2\, dV = \int_0^{2\pi} \int_0^1 \int_0^{2r} r^2\cos^2\theta\, r\, dz\, dr\, d\theta = \int_0^{2\pi} \int_0^1 \left[r^3\cos^2\theta\, z\right]_{z=0}^{z=2r} dr\, d\theta = \int_0^{2\pi} \int_0^1 2r^4\cos^2\theta\, dr\, d\theta$$

$$= \int_0^{2\pi} \left[\tfrac{2}{5}r^5\cos^2\theta\right]_{r=0}^{r=1} d\theta = \tfrac{2}{5}\int_0^{2\pi} \cos^2\theta\, d\theta = \tfrac{2}{5}\int_0^{2\pi} \tfrac{1}{2}(1 + \cos 2\theta)\, d\theta = \tfrac{1}{5}\left[\theta + \tfrac{1}{2}\sin 2\theta\right]_0^{2\pi} = \tfrac{2\pi}{5}$$

23. In cylindrical coordinates, E is bounded below by the cone $z = r$ and above by the sphere $r^2 + z^2 = 2$ or $z = \sqrt{2 - r^2}$. The

cone and the sphere intersect when $2r^2 = 2 \Rightarrow r = 1$, so $E = \{(r, \theta, z) \mid 0 \le \theta \le 2\pi, 0 \le r \le 1, r \le z \le \sqrt{2 - r^2}\}$

and the volume is

$$\iiint_E dV = \int_0^{2\pi} \int_0^1 \int_r^{\sqrt{2-r^2}} r\, dz\, dr\, d\theta = \int_0^{2\pi} \int_0^1 \left[rz\right]_{z=r}^{z=\sqrt{2-r^2}} dr\, d\theta = \int_0^{2\pi} \int_0^1 \left(r\sqrt{2 - r^2} - r^2\right) dr\, d\theta$$

$$= \int_0^{2\pi} d\theta \int_0^1 \left(r\sqrt{2 - r^2} - r^2\right) dr = 2\pi \left[-\tfrac{1}{3}(2 - r^2)^{3/2} - \tfrac{1}{3}r^3\right]_0^1$$

$$= 2\pi\left(-\tfrac{1}{3}\right)\left(1 + 1 - 2^{3/2}\right) = -\tfrac{2}{3}\pi\left(2 - 2\sqrt{2}\right) = \tfrac{4}{3}\pi\left(\sqrt{2} - 1\right)$$

25. (a) The paraboloids intersect when $x^2 + y^2 = 36 - 3x^2 - 3y^2 \Rightarrow x^2 + y^2 = 9$, so the region of integration

is $D = \{(x, y) \mid x^2 + y^2 \le 9\}$. Then, in cylindrical coordinates,

$E = \{(r, \theta, z) \mid r^2 \le z \le 36 - 3r^2, 0 \le r \le 3, 0 \le \theta \le 2\pi\}$ and

$$V = \int_0^{2\pi} \int_0^3 \int_{r^2}^{36-3r^2} r\, dz\, dr\, d\theta = \int_0^{2\pi} \int_0^3 (36r - 4r^3)\, dr\, d\theta = \int_0^{2\pi} \left[18r^2 - r^4\right]_{r=0}^{r=3} d\theta = \int_0^{2\pi} 81\, d\theta = 162\pi.$$

(b) For constant density K, $m = KV = 162\pi K$ from part (a). Since the region is homogeneous and symmetric,

$M_{yz} = M_{xz} = 0$ and

© 2012 Cengage Learning. All Rights Reserved. May not be scanned, copied, or duplicated, or posted to a publicly accessible website, in whole or in part.

$$M_{xy} = \int_0^{2\pi} \int_0^3 \int_{r^2}^{36-3r^2} (zK)\, r\, dz\, dr\, d\theta = K \int_0^{2\pi} \int_0^3 r \left[\tfrac{1}{2}z^2\right]_{z=r^2}^{z=36-3r^2} dr\, d\theta$$

$$= \tfrac{K}{2} \int_0^{2\pi} \int_0^3 r\left((36-3r^2)^2 - r^4\right) dr\, d\theta = \tfrac{K}{2} \int_0^{2\pi} d\theta \int_0^3 (8r^5 - 216r^3 + 1296r)\, dr$$

$$= \tfrac{K}{2}(2\pi)\left[\tfrac{8}{6}r^6 - \tfrac{216}{4}r^4 + \tfrac{1296}{2}r^2\right]_0^3 = \pi K(2430) = 2430\pi K$$

Thus $(\overline{x}, \overline{y}, \overline{z}) = \left(\dfrac{M_{yz}}{m}, \dfrac{M_{xz}}{m}, \dfrac{M_{xy}}{m}\right) = \left(0, 0, \dfrac{2430\pi K}{162\pi K}\right) = (0, 0, 15)$.

27. The paraboloid $z = 4x^2 + 4y^2$ intersects the plane $z = a$ when $a = 4x^2 + 4y^2$ or $x^2 + y^2 = \tfrac{1}{4}a$. So, in cylindrical

coordinates, $E = \left\{(r, \theta, z) \mid 0 \le r \le \tfrac{1}{2}\sqrt{a}, 0 \le \theta \le 2\pi, 4r^2 \le z \le a\right\}$. Thus

$$m = \int_0^{2\pi} \int_0^{\sqrt{a}/2} \int_{4r^2}^a Kr\, dz\, dr\, d\theta = K \int_0^{2\pi} \int_0^{\sqrt{a}/2} (ar - 4r^3)\, dr\, d\theta$$

$$= K \int_0^{2\pi} \left[\tfrac{1}{2}ar^2 - r^4\right]_{r=0}^{r=\sqrt{a}/2} d\theta = K \int_0^{2\pi} \tfrac{1}{16}a^2\, d\theta = \tfrac{1}{8}a^2\pi K$$

Since the region is homogeneous and symmetric, $M_{yz} = M_{xz} = 0$ and

$$M_{xy} = \int_0^{2\pi} \int_0^{\sqrt{a}/2} \int_{4r^2}^a Krz\, dz\, dr\, d\theta = K \int_0^{2\pi} \int_0^{\sqrt{a}/2} \left(\tfrac{1}{2}a^2 r - 8r^5\right) dr\, d\theta$$

$$= K \int_0^{2\pi} \left[\tfrac{1}{4}a^2 r^2 - \tfrac{4}{3}r^6\right]_{r=0}^{r=\sqrt{a}/2} d\theta = K \int_0^{2\pi} \tfrac{1}{24}a^3\, d\theta = \tfrac{1}{12}a^3\pi K$$

Hence $(\overline{x}, \overline{y}, \overline{z}) = \left(0, 0, \tfrac{2}{3}a\right)$.

29. The region of integration is the region above the cone $z = \sqrt{x^2 + y^2}$, or $z = r$, and below the plane $z = 2$. Also, we have

$-2 \le y \le 2$ with $-\sqrt{4-y^2} \le x \le \sqrt{4-y^2}$ which describes a circle of radius 2 in the xy-plane centered at $(0,0)$. Thus,

$$\int_{-2}^2 \int_{-\sqrt{4-y^2}}^{\sqrt{4-y^2}} \int_{\sqrt{x^2+y^2}}^2 xz\, dz\, dx\, dy = \int_0^{2\pi} \int_0^2 \int_r^2 (r\cos\theta)\, z\, r\, dz\, dr\, d\theta = \int_0^{2\pi} \int_0^2 \int_r^2 r^2(\cos\theta)\, z\, dz\, dr\, d\theta$$

$$= \int_0^{2\pi} \int_0^2 r^2(\cos\theta) \left[\tfrac{1}{2}z^2\right]_{z=r}^{z=2} dr\, d\theta = \tfrac{1}{2} \int_0^{2\pi} \int_0^2 r^2(\cos\theta)\left(4 - r^2\right) dr\, d\theta$$

$$= \tfrac{1}{2} \int_0^{2\pi} \cos\theta\, d\theta \int_0^2 \left(4r^2 - r^4\right) dr = \tfrac{1}{2} \left[\sin\theta\right]_0^{2\pi} \left[\tfrac{4}{3}r^3 - \tfrac{1}{5}r^5\right]_0^2 = 0$$

31. (a) The mountain comprises a solid conical region C. The work done in lifting a small volume of material ΔV with density

$g(P)$ to a height $h(P)$ above sea level is $h(P)g(P)\,\Delta V$. Summing over the whole mountain we get

$W = \iiint_C h(P)g(P)\, dV$.

(b) Here C is a solid right circular cone with radius $R = 62{,}000$ ft, height $H = 12{,}400$ ft,

and density $g(P) = 200$ lb/ft^3 at all points P in C. We use cylindrical coordinates:

$$W = \int_0^{2\pi} \int_0^H \int_0^{R(1-z/H)} z \cdot 200r\, dr\, dz\, d\theta = 2\pi \int_0^H 200z \left[\tfrac{1}{2}r^2\right]_{r=0}^{r=R(1-z/H)} dz$$

$$= 400\pi \int_0^H z\,\frac{R^2}{2}\left(1 - \frac{z}{H}\right)^2 dz = 200\pi R^2 \int_0^H \left(z - \frac{2z^2}{H} + \frac{z^3}{H^2}\right) dz$$

$$= 200\pi R^2 \left[\frac{z^2}{2} - \frac{2z^3}{3H} + \frac{z^4}{4H^2}\right]_0^H = 200\pi R^2 \left(\frac{H^2}{2} - \frac{2H^2}{3} + \frac{H^2}{4}\right)$$

$$= \tfrac{50}{3}\pi R^2 H^2 = \tfrac{50}{3}\pi(62{,}000)^2(12{,}400)^2 \approx 3.1 \times 10^{19}\text{ ft-lb}$$

$$\frac{r}{R} = \frac{H-z}{H} = 1 - \frac{z}{H}$$

© 2012 Cengage Learning. All Rights Reserved. May not be scanned, copied, or duplicated, or posted to a publicly accessible website, in whole or in part.

15.9 Triple Integrals in Spherical Coordinates

1. (a)

From Equations 1, $x = \rho \sin\phi\cos\theta = 6\sin\frac{\pi}{6}\cos\frac{\pi}{3} = 6 \cdot \frac{1}{2} \cdot \frac{1}{2} = \frac{3}{2}$,

$y = \rho\sin\phi\sin\theta = 6\sin\frac{\pi}{6}\sin\frac{\pi}{3} = 6 \cdot \frac{1}{2} \cdot \frac{\sqrt{3}}{2} = \frac{3\sqrt{3}}{2}$, and

$z = \rho\cos\phi = 6\cos\frac{\pi}{6} = 6 \cdot \frac{\sqrt{3}}{2} = 3\sqrt{3}$, so the point is $\left(\frac{3}{2}, \frac{3\sqrt{3}}{2}, 3\sqrt{3}\right)$ in

rectangular coordinates.

(b)

$x = 3\sin\frac{3\pi}{4}\cos\frac{\pi}{2} = 3 \cdot \frac{\sqrt{2}}{2} \cdot 0 = 0$,

$y = 3\sin\frac{3\pi}{4}\sin\frac{\pi}{2} = 3 \cdot \frac{\sqrt{2}}{2} \cdot 1 = \frac{3\sqrt{2}}{2}$, and

$z = 3\cos\frac{3\pi}{4} = 3\left(-\frac{\sqrt{2}}{2}\right) = -\frac{3\sqrt{2}}{2}$, so the point is $\left(0, \frac{3\sqrt{2}}{2}, -\frac{3\sqrt{2}}{2}\right)$ in

rectangular coordinates.

3. (a) From Equations 1 and 2, $\rho = \sqrt{x^2 + y^2 + z^2} = \sqrt{0^2 + (-2)^2 + 0^2} = 2$, $\cos\phi = \frac{z}{\rho} = \frac{0}{2} = 0 \implies \phi = \frac{\pi}{2}$, and

$\cos\theta = \frac{x}{\rho\sin\phi} = \frac{0}{2\sin(\pi/2)} = 0 \implies \theta = \frac{3\pi}{2}$ [since $y < 0$]. Thus spherical coordinates are $\left(2, \frac{3\pi}{2}, \frac{\pi}{2}\right)$.

(b) $\rho = \sqrt{1 + 1 + 2} = 2$, $\cos\phi = \frac{z}{\rho} = \frac{-\sqrt{2}}{2} \implies \phi = \frac{3\pi}{4}$, and

$\cos\theta = \frac{x}{\rho\sin\phi} = \frac{-1}{2\sin(3\pi/4)} = \frac{-1}{2\left(\sqrt{2}/2\right)} = -\frac{1}{\sqrt{2}} \implies \theta = \frac{3\pi}{4}$ [since $y > 0$]. Thus spherical coordinates

are $\left(2, \frac{3\pi}{4}, \frac{3\pi}{4}\right)$.

5. Since $\phi = \frac{\pi}{3}$, the surface is the top half of the right circular cone with vertex at the origin and axis the positive z-axis.

7. $\rho = \sin\theta\sin\phi \implies \rho^2 = \rho\sin\theta\sin\phi \iff x^2 + y^2 + z^2 = y \iff x^2 + y^2 - y + \frac{1}{4} + z^2 = \frac{1}{4} \iff$

$x^2 + \left(y - \frac{1}{2}\right)^2 + z^2 = \frac{1}{4}$. Therefore, the surface is a sphere of radius $\frac{1}{2}$ centered at $\left(0, \frac{1}{2}, 0\right)$.

9. (a) $x = \rho\sin\phi\cos\theta$, $y = \rho\sin\phi\sin\theta$, and $z = \rho\cos\phi$, so the equation $z^2 = x^2 + y^2$ becomes

$(\rho\cos\phi)^2 = (\rho\sin\phi\cos\theta)^2 + (\rho\sin\phi\sin\theta)^2$ or $\rho^2\cos^2\phi = \rho^2\sin^2\phi$. If $\rho \neq 0$, this becomes $\cos^2\phi = \sin^2\phi$. ($\rho = 0$

corresponds to the origin which is included in the surface.) There are many equivalent equations in spherical coordinates,

such as $\tan^2\phi = 1$, $2\cos^2\phi = 1$, $\cos 2\phi = 0$, or even $\phi = \frac{\pi}{4}$, $\phi = \frac{3\pi}{4}$.

(b) $x^2 + z^2 = 9 \iff (\rho\sin\phi\cos\theta)^2 + (\rho\cos\phi)^2 = 9 \iff \rho^2\sin^2\phi\cos^2\theta + \rho^2\cos^2\phi = 9$ or

$\rho^2\left(\sin^2\phi\cos^2\theta + \cos^2\phi\right) = 9$.

© 2012 Cengage Learning. All Rights Reserved. May not be scanned, copied, or duplicated, or posted to a publicly accessible website, in whole or in part.

11. $2 \le \rho \le 4$ represents the solid region between and including the spheres of radii 2 and 4, centered at the origin. $0 \le \phi \le \frac{\pi}{3}$ restricts the solid to that portion on or above the cone $\phi = \frac{\pi}{3}$, and $0 \le \theta \le \pi$ further restricts the solid to that portion on or to the right of the xz-plane.

13. $\rho \le 1$ represents the solid sphere of radius 1 centered at the origin.

$\frac{3\pi}{4} \le \phi \le \pi$ restricts the solid to that portion on or below the cone $\phi = \frac{3\pi}{4}$.

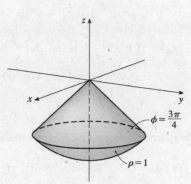

15. $z \ge \sqrt{x^2 + y^2}$ because the solid lies above the cone. Squaring both sides of this inequality gives $z^2 \ge x^2 + y^2 \Rightarrow$

$2z^2 \ge x^2 + y^2 + z^2 = \rho^2 \Rightarrow z^2 = \rho^2 \cos^2 \phi \ge \frac{1}{2}\rho^2 \Rightarrow \cos^2 \phi \ge \frac{1}{2}$. The cone opens upward so that the inequality is

$\cos \phi \ge \frac{1}{\sqrt{2}}$, or equivalently $0 \le \phi \le \frac{\pi}{4}$. In spherical coordinates the sphere $z = x^2 + y^2 + z^2$ is $\rho \cos \phi = \rho^2 \Rightarrow$

$\rho = \cos \phi$. $0 \le \rho \le \cos \phi$ because the solid lies below the sphere. The solid can therefore be described as the region in

spherical coordinates satisfying $0 \le \rho \le \cos \phi$, $0 \le \phi \le \frac{\pi}{4}$.

17.

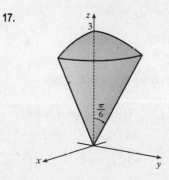

The region of integration is given in spherical coordinates by

$E = \{(\rho, \theta, \phi) \mid 0 \le \rho \le 3, 0 \le \theta \le \pi/2, 0 \le \phi \le \pi/6\}$. This represents the solid

region in the first octant bounded above by the sphere $\rho = 3$ and below by the cone

$\phi = \pi/6$.

$$\int_0^{\pi/6} \int_0^{\pi/2} \int_0^3 \rho^2 \sin \phi \, d\rho \, d\theta \, d\phi = \int_0^{\pi/6} \sin \phi \, d\phi \int_0^{\pi/2} d\theta \int_0^3 \rho^2 \, d\rho$$

$$= \left[-\cos \phi \right]_0^{\pi/6} \left[\theta \right]_0^{\pi/2} \left[\tfrac{1}{3}\rho^3 \right]_0^3$$

$$= \left(1 - \frac{\sqrt{3}}{2} \right) \left(\frac{\pi}{2} \right) (9) = \frac{9\pi}{4} \left(2 - \sqrt{3} \right)$$

19. The solid E is most conveniently described if we use cylindrical coordinates:

$E = \left\{ (r, \theta, z) \mid 0 \le \theta \le \frac{\pi}{2}, 0 \le r \le 3, 0 \le z \le 2 \right\}$. Then

$\iiint_E f(x, y, z) \, dV = \int_0^{\pi/2} \int_0^3 \int_0^2 f(r \cos \theta, r \sin \theta, z) \, r \, dz \, dr \, d\theta$.

© 2012 Cengage Learning. All Rights Reserved. May not be scanned, copied, or duplicated, or posted to a publicly accessible website, in whole or in part.

21. In spherical coordinates, B is represented by $\{(\rho, \theta, \phi) \mid 0 \le \rho \le 5, 0 \le \theta \le 2\pi, 0 \le \phi \le \pi\}$. Thus

$$\iiint_B (x^2 + y^2 + z^2)^2 \, dV = \int_0^\pi \int_0^{2\pi} \int_0^5 (\rho^2)^2 \rho^2 \sin\phi \, d\rho \, d\theta \, d\phi = \int_0^\pi \sin\phi \, d\phi \int_0^{2\pi} d\theta \int_0^5 \rho^6 \, d\rho$$

$$= \left[-\cos\phi \right]_0^\pi \left[\theta \right]_0^{2\pi} \left[\tfrac{1}{7}\rho^7 \right]_0^5 = (2)(2\pi)\left(\tfrac{78,125}{7} \right)$$

$$= \tfrac{312,500}{7}\pi \approx 140{,}249.7$$

23. In spherical coordinates, E is represented by $\{(\rho, \theta, \phi) \mid 2 \le \rho \le 3, 0 \le \theta \le 2\pi, 0 \le \phi \le \pi\}$ and

$$x^2 + y^2 = \rho^2 \sin^2\phi \cos^2\theta + \rho^2 \sin^2\phi \sin^2\theta = \rho^2 \sin^2\phi \left(\cos^2\theta + \sin^2\theta\right) = \rho^2 \sin^2\phi. \text{ Thus}$$

$$\iiint_E (x^2 + y^2) \, dV = \int_0^\pi \int_0^{2\pi} \int_2^3 (\rho^2 \sin^2\phi) \rho^2 \sin\phi \, d\rho \, d\theta \, d\phi = \int_0^\pi \sin^3\phi \, d\phi \int_0^{2\pi} d\theta \int_2^3 \rho^4 \, d\rho$$

$$= \int_0^\pi (1 - \cos^2\phi) \sin\phi \, d\phi \left[\theta \right]_0^{2\pi} \left[\tfrac{1}{5}\rho^5 \right]_2^3 = \left[-\cos\phi + \tfrac{1}{3}\cos^3\phi \right]_0^\pi (2\pi) \cdot \tfrac{1}{5}(243 - 32)$$

$$= \left(1 - \tfrac{1}{3} + 1 - \tfrac{1}{3}\right)(2\pi)\left(\tfrac{211}{5}\right) = \tfrac{1688\pi}{15}$$

25. In spherical coordinates, E is represented by $\{(\rho, \theta, \phi) \mid 0 \le \rho \le 1, 0 \le \theta \le \tfrac{\pi}{2}, 0 \le \phi \le \tfrac{\pi}{2}\}$. Thus

$$\iiint_E x e^{x^2 + y^2 + z^2} \, dV = \int_0^{\pi/2} \int_0^{\pi/2} \int_0^1 (\rho \sin\phi \cos\theta) e^{\rho^2} \rho^2 \sin\phi \, d\rho \, d\theta \, d\phi = \int_0^{\pi/2} \sin^2\phi \, d\phi \int_0^{\pi/2} \cos\theta \, d\theta \int_0^1 \rho^3 e^{\rho^2} \, d\rho$$

$$= \int_0^{\pi/2} \tfrac{1}{2}(1 - \cos 2\phi) \, d\phi \int_0^{\pi/2} \cos\theta \, d\theta \left(\left. \tfrac{1}{2}\rho^2 e^{\rho^2} \right|_0^1 - \int_0^1 \rho e^{\rho^2} \, d\rho \right)$$

$$\left[\text{integrate by parts with } u = \rho^2, \, dv = \rho e^{\rho^2} \, d\rho \right]$$

$$= \left[\tfrac{1}{2}\phi - \tfrac{1}{4}\sin 2\phi \right]_0^{\pi/2} \left[\sin\theta \right]_0^{\pi/2} \left[\tfrac{1}{2}\rho^2 e^{\rho^2} - \tfrac{1}{2}e^{\rho^2} \right]_0^1 = \left(\tfrac{\pi}{4} - 0 \right)(1 - 0)\left(0 + \tfrac{1}{2} \right) = \tfrac{\pi}{8}$$

27. The solid region is given by $E = \{(\rho, \theta, \phi) \mid 0 \le \rho \le a, 0 \le \theta \le 2\pi, \tfrac{\pi}{6} \le \phi \le \tfrac{\pi}{3}\}$ and its volume is

$$V = \iiint_E dV = \int_{\pi/6}^{\pi/3} \int_0^{2\pi} \int_0^a \rho^2 \sin\phi \, d\rho \, d\theta \, d\phi = \int_{\pi/6}^{\pi/3} \sin\phi \, d\phi \int_0^{2\pi} d\theta \int_0^a \rho^2 \, d\rho$$

$$= \left[-\cos\phi \right]_{\pi/6}^{\pi/3} \left[\theta \right]_0^{2\pi} \left[\tfrac{1}{3}\rho^3 \right]_0^a = \left(-\tfrac{1}{2} + \tfrac{\sqrt{3}}{2} \right)(2\pi)\left(\tfrac{1}{3}a^3 \right) = \tfrac{\sqrt{3}-1}{3}\pi a^3$$

29. (a) Since $\rho = 4\cos\phi$ implies $\rho^2 = 4\rho\cos\phi$, the equation is that of a sphere of radius 2 with center at $(0, 0, 2)$. Thus

$$V = \int_0^{2\pi} \int_0^{\pi/3} \int_0^{4\cos\phi} \rho^2 \sin\phi \, d\rho \, d\phi \, d\theta = \int_0^{2\pi} \int_0^{\pi/3} \left[\tfrac{1}{3}\rho^3 \right]_{\rho=0}^{\rho=4\cos\phi} \sin\phi \, d\phi \, d\theta = \int_0^{2\pi} \int_0^{\pi/3} \left(\tfrac{64}{3}\cos^3\phi \right) \sin\phi \, d\phi \, d\theta$$

$$= \int_0^{2\pi} \left[-\tfrac{16}{3}\cos^4\phi \right]_{\phi=0}^{\phi=\pi/3} d\theta = \int_0^{2\pi} -\tfrac{16}{3}\left(\tfrac{1}{16} - 1 \right) d\theta = 5\theta \Big]_0^{2\pi} = 10\pi$$

(b) By the symmetry of the problem $M_{yz} = M_{xz} = 0$. Then

$$M_{xy} = \int_0^{2\pi} \int_0^{\pi/3} \int_0^{4\cos\phi} \rho^3 \cos\phi \sin\phi \, d\rho \, d\phi \, d\theta = \int_0^{2\pi} \int_0^{\pi/3} \cos\phi \sin\phi \left(64\cos^4\phi \right) d\phi \, d\theta$$

$$= \int_0^{2\pi} 64 \left[-\tfrac{1}{6}\cos^6\phi \right]_{\phi=0}^{\phi=\pi/3} d\theta = \int_0^{2\pi} \tfrac{21}{2} \, d\theta = 21\pi$$

Hence $(\overline{x}, \overline{y}, \overline{z}) = (0, 0, 2.1)$.

© 2012 Cengage Learning. All Rights Reserved. May not be scanned, copied, or duplicated, or posted to a publicly accessible website, in whole or in part.

31. (a) By the symmetry of the region, $M_{yz} = 0$ and $M_{xz} = 0$. Assuming constant density K,

$$m = \iiint_E K\, dV = K \iiint_E dV = \tfrac{\pi}{8} K \text{ (from Example 4). Then}$$

$$M_{xy} = \iiint_E z K\, dV = K \int_0^{2\pi} \int_0^{\pi/4} \int_0^{\cos\phi} (\rho\cos\phi)\, \rho^2 \sin\phi\, d\rho\, d\phi\, d\theta = K \int_0^{2\pi} \int_0^{\pi/4} \sin\phi\cos\phi \left[\tfrac14 \rho^4\right]_{\rho=0}^{\rho=\cos\phi} d\phi\, d\theta$$

$$= \tfrac14 K \int_0^{2\pi} \int_0^{\pi/4} \sin\phi\cos\phi \left(\cos^4\phi\right) d\phi\, d\theta = \tfrac14 K \int_0^{2\pi} d\theta \int_0^{\pi/4} \cos^5\phi \sin\phi\, d\phi$$

$$= \tfrac14 K \left[\theta\right]_0^{2\pi} \left[-\tfrac16 \cos^6\phi\right]_0^{\pi/4} = \tfrac14 K(2\pi)\left(-\tfrac16\right)\left[\left(\tfrac{\sqrt2}{2}\right)^6 - 1\right] = -\tfrac{\pi}{12}K\left(-\tfrac78\right) = \tfrac{7\pi}{96}K$$

Thus the centroid is $(\overline{x}, \overline{y}, \overline{z}) = \left(\dfrac{M_{yz}}{m}, \dfrac{M_{xz}}{m}, \dfrac{M_{xy}}{m}\right) = \left(0, 0, \dfrac{7\pi K/96}{\pi K/8}\right) = \left(0, 0, \tfrac{7}{12}\right)$.

(b) As in Exercise 23, $x^2 + y^2 = \rho^2 \sin^2\phi$ and

$$I_z = \iiint_E (x^2+y^2)\, K\, dV = K \int_0^{2\pi} \int_0^{\pi/4} \int_0^{\cos\phi} (\rho^2 \sin^2\phi)\, \rho^2 \sin\phi\, d\rho\, d\phi\, d\theta = K \int_0^{2\pi} \int_0^{\pi/4} \sin^3\phi \left[\tfrac15 \rho^5\right]_{\rho=0}^{\rho=\cos\phi} d\phi\, d\theta$$

$$= \tfrac15 K \int_0^{2\pi} \int_0^{\pi/4} \sin^3\phi \cos^5\phi\, d\phi\, d\theta = \tfrac15 K \int_0^{2\pi} d\theta \int_0^{\pi/4} \cos^5\phi \left(1 - \cos^2\phi\right) \sin\phi\, d\phi$$

$$= \tfrac15 K \left[\theta\right]_0^{2\pi} \left[-\tfrac16 \cos^6\phi + \tfrac18 \cos^8\phi\right]_0^{\pi/4}$$

$$= \tfrac15 K(2\pi)\left[-\tfrac16 \left(\tfrac{\sqrt2}{2}\right)^6 + \tfrac18 \left(\tfrac{\sqrt2}{2}\right)^8 + \tfrac16 - \tfrac18\right] = \tfrac{2\pi}{5}K\left(\tfrac{11}{384}\right) = \tfrac{11\pi}{960}K$$

33. (a) The density function is $\rho(x, y, z) = K$, a constant, and by the symmetry of the problem $M_{xz} = M_{yz} = 0$. Then

$$M_{xy} = \int_0^{2\pi} \int_0^{\pi/2} \int_0^a K\rho^3 \sin\phi \cos\phi\, d\rho\, d\phi\, d\theta = \tfrac12 \pi K a^4 \int_0^{\pi/2} \sin\phi \cos\phi\, d\phi = \tfrac18 \pi K a^4. \text{ But the mass is } K(\text{volume of}$$

the hemisphere) $= \tfrac23 \pi K a^3$, so the centroid is $\left(0, 0, \tfrac38 a\right)$.

(b) Place the center of the base at $(0, 0, 0)$; the density function is $\rho(x, y, z) = K$. By symmetry, the moments of inertia about any two such diameters will be equal, so we just need to find I_x:

$$I_x = \int_0^{2\pi} \int_0^{\pi/2} \int_0^a (K\rho^2 \sin\phi)\, \rho^2 \left(\sin^2\phi \sin^2\theta + \cos^2\phi\right) d\rho\, d\phi\, d\theta$$

$$= K \int_0^{2\pi} \int_0^{\pi/2} (\sin^3\phi \sin^2\theta + \sin\phi \cos^2\phi)\left(\tfrac15 a^5\right) d\phi\, d\theta$$

$$= \tfrac15 K a^5 \int_0^{2\pi} \left[\sin^2\theta \left(-\cos\phi + \tfrac13 \cos^3\phi\right) + \left(-\tfrac13 \cos^3\phi\right)\right]_{\phi=0}^{\phi=\pi/2} d\theta = \tfrac15 K a^5 \int_0^{2\pi} \left[\tfrac23 \sin^2\theta + \tfrac13\right] d\theta$$

$$= \tfrac15 K a^5 \left[\tfrac23 \left(\tfrac12 \theta - \tfrac14 \sin 2\theta\right) + \tfrac13 \theta\right]_0^{2\pi} = \tfrac15 K a^5 \left[\tfrac23 (\pi - 0) + \tfrac13 (2\pi - 0)\right] = \tfrac{4}{15} K a^5 \pi$$

35. In spherical coordinates $z = \sqrt{x^2 + y^2}$ becomes $\cos\phi = \sin\phi$ or $\phi = \tfrac{\pi}{4}$. Then

$$V = \int_0^{2\pi} \int_0^{\pi/4} \int_0^1 \rho^2 \sin\phi\, d\rho\, d\phi\, d\theta = \int_0^{2\pi} d\theta \int_0^{\pi/4} \sin\phi\, d\phi \int_0^1 \rho^2\, d\rho = 2\pi\left(-\tfrac{\sqrt2}{2} + 1\right)\left(\tfrac13\right) = \tfrac13 \pi (2 - \sqrt2),$$

$$M_{xy} = \int_0^{2\pi} \int_0^{\pi/4} \int_0^1 \rho^3 \sin\phi \cos\phi\, d\rho\, d\phi\, d\theta = 2\pi \left[-\tfrac14 \cos 2\phi\right]_0^{\pi/4} \left(\tfrac14\right) = \tfrac{\pi}{8} \text{ and by symmetry } M_{yz} = M_{xz} = 0.$$

Hence $(\overline{x}, \overline{y}, \overline{z}) = \left(0, 0, \dfrac{3}{8(2 - \sqrt2)}\right)$.

37. In cylindrical coordinates the paraboloid is given by $z = r^2$ and the plane by $z = 2r\sin\theta$ and they intersect in the circle

$r = 2\sin\theta$. Then $\iiint_E z\, dV = \int_0^\pi \int_0^{2\sin\theta} \int_{r^2}^{2r\sin\theta} rz\, dz\, dr\, d\theta = \tfrac{5\pi}{6}$ [using a CAS].

© 2012 Cengage Learning. All Rights Reserved. May not be scanned, copied, or duplicated, or posted to a publicly accessible website, in whole or in part.

39. The region E of integration is the region above the cone $z = \sqrt{x^2 + y^2}$ and below the sphere $x^2 + y^2 + z^2 = 2$ in the first

octant. Because E is in the first octant we have $0 \le \theta \le \frac{\pi}{2}$. The cone has equation $\phi = \frac{\pi}{4}$ (as in Example 4), so $0 \le \phi \le \frac{\pi}{4}$,

and $0 \le \rho \le \sqrt{2}$. So the integral becomes

$$\int_0^{\pi/4} \int_0^{\pi/2} \int_0^{\sqrt{2}} (\rho \sin \phi \cos \theta)(\rho \sin \phi \sin \theta) \rho^2 \sin \phi \, d\rho \, d\theta \, d\phi$$

$$= \int_0^{\pi/4} \sin^3 \phi \, d\phi \int_0^{\pi/2} \sin \theta \cos \theta \, d\theta \int_0^{\sqrt{2}} \rho^4 \, d\rho = \left(\int_0^{\pi/4} (1 - \cos^2 \phi) \sin \phi \, d\phi \right) \left[\tfrac{1}{2} \sin^2 \theta \right]_0^{\pi/2} \left[\tfrac{1}{5} \rho^5 \right]_0^{\sqrt{2}}$$

$$= \left[\tfrac{1}{3} \cos^3 \phi - \cos \phi \right]_0^{\pi/4} \cdot \tfrac{1}{2} \cdot \tfrac{1}{5} (\sqrt{2})^5 = \left[\tfrac{\sqrt{2}}{12} - \tfrac{\sqrt{2}}{2} - (\tfrac{1}{3} - 1) \right] \cdot \tfrac{2\sqrt{2}}{5} = \tfrac{4\sqrt{2}-5}{15}$$

41. The region of integration is the solid sphere $x^2 + y^2 + (z-2)^2 \le 4$ or equivalently

$$\rho^2 \sin^2 \phi + (\rho \cos \phi - 2)^2 = \rho^2 - 4\rho \cos \phi + 4 \le 4 \quad \Rightarrow \quad \rho \le 4 \cos \phi, \text{ so } 0 \le \theta \le 2\pi, 0 \le \phi \le \frac{\pi}{2}, \text{ and}$$

$0 \le \rho \le 4 \cos \phi$. Also $(x^2 + y^2 + z^2)^{3/2} = (\rho^2)^{3/2} = \rho^3$, so the integral becomes

$$\int_0^{\pi/2} \int_0^{2\pi} \int_0^{4 \cos \phi} (\rho^3) \rho^2 \sin \phi \, d\rho \, d\theta \, d\phi = \int_0^{\pi/2} \int_0^{2\pi} \sin \phi \left[\tfrac{1}{6} \rho^6 \right]_{\rho=0}^{\rho=4 \cos \phi} d\theta \, d\phi = \tfrac{1}{6} \int_0^{\pi/2} \int_0^{2\pi} \sin \phi \left(4096 \cos^6 \phi \right) d\theta \, d\phi$$

$$= \tfrac{1}{6} (4096) \int_0^{\pi/2} \cos^6 \phi \sin \phi \, d\phi \int_0^{2\pi} d\theta = \tfrac{2048}{3} \left[-\tfrac{1}{7} \cos^7 \phi \right]_0^{\pi/2} \left[\theta \right]_0^{2\pi}$$

$$= \tfrac{2048}{3} \left(\tfrac{1}{7} \right) (2\pi) = \tfrac{4096\pi}{21}$$

43. In cylindrical coordinates, the equation of the cylinder is $r = 3, 0 \le z \le 10$.

The hemisphere is the upper part of the sphere radius 3, center $(0, 0, 10)$, equation

$r^2 + (z - 10)^2 = 3^2, z \ge 10$. In Maple, we can use the `coords=cylindrical` option

in a regular `plot3d` command. In Mathematica, we can use `ParametricPlot3D`.

45. If E is the solid enclosed by the surface $\rho = 1 + \frac{1}{5} \sin 6\theta \sin 5\phi$, it can be described in spherical coordinates as

$E = \{ (\rho, \theta, \phi) \mid 0 \le \rho \le 1 + \frac{1}{5} \sin 6\theta \sin 5\phi, 0 \le \theta \le 2\pi, 0 \le \phi \le \pi \}$. Its volume is given by

$$V(E) = \iiint_E dV = \int_0^\pi \int_0^{2\pi} \int_0^{1 + (\sin 6\theta \sin 5\phi)/5} \rho^2 \sin \phi \, d\rho \, d\theta \, d\phi = \tfrac{136\pi}{99} \quad \text{[using a CAS]}.$$

47. (a) From the diagram, $z = r \cot \phi_0$ to $z = \sqrt{a^2 - r^2}, r = 0$

to $r = a \sin \phi_0$ (or use $a^2 - r^2 = r^2 \cot^2 \phi_0$). Thus

$$V = \int_0^{2\pi} \int_0^{a \sin \phi_0} \int_{r \cot \phi_0}^{\sqrt{a^2 - r^2}} r \, dz \, dr \, d\theta$$

$$= 2\pi \int_0^{a \sin \phi_0} \left(r \sqrt{a^2 - r^2} - r^2 \cot \phi_0 \right) dr$$

$$= \tfrac{2\pi}{3} \left[-(a^2 - r^2)^{3/2} - r^3 \cot \phi_0 \right]_0^{a \sin \phi_0}$$

$$= \tfrac{2\pi}{3} \left[-\left(a^2 - a^2 \sin^2 \phi_0 \right)^{3/2} - a^3 \sin^3 \phi_0 \cot \phi_0 + a^3 \right]$$

$$= \tfrac{2}{3} \pi a^3 \left[1 - (\cos^3 \phi_0 + \sin^2 \phi_0 \cos \phi_0) \right] = \tfrac{2}{3} \pi a^3 (1 - \cos \phi_0)$$

On the right: a quarter-circle diagram with z-axis vertical, labelled $r^2 + z^2 = a^2$, with r, a, angle ϕ_0, and $z = r \cot \phi_0$.

© 2012 Cengage Learning. All Rights Reserved. May not be scanned, copied, or duplicated, or posted to a publicly accessible website, in whole or in part.

(b) The wedge in question is the shaded area rotated from $\theta = \theta_1$ to $\theta = \theta_2$.
Letting

$V_{ij} = $ volume of the region bounded by the sphere of radius ρ_i

and the cone with angle ϕ_j ($\theta = \theta_1$ to θ_2)

and letting V be the volume of the wedge, we have

$V = (V_{22} - V_{21}) - (V_{12} - V_{11})$

$= \frac{1}{3}(\theta_2 - \theta_1)\left[\rho_2^3(1 - \cos\phi_2) - \rho_2^3(1 - \cos\phi_1) - \rho_1^3(1 - \cos\phi_2) + \rho_1^3(1 - \cos\phi_1)\right]$

$= \frac{1}{3}(\theta_2 - \theta_1)\left[(\rho_2^3 - \rho_1^3)(1 - \cos\phi_2) - (\rho_2^3 - \rho_1^3)(1 - \cos\phi_1)\right] = \frac{1}{3}(\theta_2 - \theta_1)\left[(\rho_2^3 - \rho_1^3)(\cos\phi_1 - \cos\phi_2)\right]$

Or: Show that $V = \displaystyle\int_{\theta_1}^{\theta_2}\int_{\rho_1\sin\phi_1}^{\rho_2\sin\phi_2}\int_{r\cot\phi_2}^{r\cot\phi_1} r\,dz\,dr\,d\theta$.

(c) By the Mean Value Theorem with $f(\rho) = \rho^3$ there exists some $\tilde{\rho}$ with $\rho_1 \leq \tilde{\rho} \leq \rho_2$ such that

$f(\rho_2) - f(\rho_1) = f'(\tilde{\rho})(\rho_2 - \rho_1)$ or $\rho_1^3 - \rho_2^3 = 3\tilde{\rho}^2\Delta\rho$. Similarly there exists ϕ with $\phi_1 \leq \tilde{\phi} \leq \phi_2$

such that $\cos\phi_2 - \cos\phi_1 = \left(-\sin\tilde{\phi}\right)\Delta\phi$. Substituting into the result from (b) gives

$\Delta V = (\tilde{\rho}^2\,\Delta\rho)(\theta_2 - \theta_1)(\sin\tilde{\phi})\,\Delta\phi = \tilde{\rho}^2\sin\tilde{\phi}\,\Delta\rho\,\Delta\phi\,\Delta\theta$.

15.10   Change of Variables in Multiple Integrals

1. $x = 5u - v,\ y = u + 3v$.

The Jacobian is $\dfrac{\partial(x, y)}{\partial(u, v)} = \begin{vmatrix} \partial x/\partial u & \partial x/\partial v \\ \partial y/\partial u & \partial y/\partial v \end{vmatrix} = \begin{vmatrix} 5 & -1 \\ 1 & 3 \end{vmatrix} = 5(3) - (-1)(1) = 16$.

3. $x = e^{-r}\sin\theta,\ y = e^r\cos\theta$.

$\dfrac{\partial(x, y)}{\partial(r, \theta)} = \begin{vmatrix} \partial x/\partial r & \partial x/\partial\theta \\ \partial y/\partial r & \partial y/\partial\theta \end{vmatrix} = \begin{vmatrix} -e^{-r}\sin\theta & e^{-r}\cos\theta \\ e^r\cos\theta & -e^r\sin\theta \end{vmatrix} = e^{-r}e^r\sin^2\theta - e^{-r}e^r\cos^2\theta = \sin^2\theta - \cos^2\theta$ or $-\cos 2\theta$

5. $x = u/v,\ y = v/w,\ z = w/u$.

$\dfrac{\partial(x, y, z)}{\partial(u, v, w)} = \begin{vmatrix} \partial x/\partial u & \partial x/\partial v & \partial x/\partial w \\ \partial y/\partial u & \partial y/\partial v & \partial y/\partial w \\ \partial z/\partial u & \partial z/\partial v & \partial z/\partial w \end{vmatrix} = \begin{vmatrix} 1/v & -u/v^2 & 0 \\ 0 & 1/w & -v/w^2 \\ -w/u^2 & 0 & 1/u \end{vmatrix}$

$= \dfrac{1}{v}\begin{vmatrix} 1/w & -v/w^2 \\ 0 & 1/u \end{vmatrix} - \left(-\dfrac{u}{v^2}\right)\begin{vmatrix} 0 & -v/w^2 \\ -w/u^2 & 1/u \end{vmatrix} + 0\begin{vmatrix} 0 & 1/w \\ -w/u^2 & 0 \end{vmatrix}$

$= \dfrac{1}{v}\left(\dfrac{1}{uw} - 0\right) + \dfrac{u}{v^2}\left(0 - \dfrac{v}{u^2w}\right) + 0 = \dfrac{1}{uvw} - \dfrac{1}{uvw} = 0$

7. The transformation maps the boundary of S to the boundary of the image R, so we first look at side S_1 in the uv-plane. S_1 is
described by $v = 0, 0 \leq u \leq 3$, so $x = 2u + 3v = 2u$ and $y = u - v = u$. Eliminating u, we have $x = 2y, 0 \leq x \leq 6$. S_2 is

© 2012 Cengage Learning. All Rights Reserved. May not be scanned, copied, or duplicated, or posted to a publicly accessible website, in whole or in part.

the line segment $u = 3$, $0 \leq v \leq 2$, so $x = 6 + 3v$ and $y = 3 - v$. Then $v = 3 - y$ $\Rightarrow$ $x = 6 + 3(3 - y) = 15 - 3y$, $6 \leq x \leq 12$. S_3 is the line segment $v = 2$, $0 \leq u \leq 3$, so $x = 2u + 6$ and $y = u - 2$, giving $u = y + 2$ $\Rightarrow$ $x = 2y + 10$, $6 \leq x \leq 12$. Finally, S_4 is the segment $u = 0$, $0 \leq v \leq 2$, so $x = 3v$ and $y = -v$ $\Rightarrow$ $x = -3y$, $0 \leq x \leq 6$. The image of set S is the region R shown in the xy-plane, a parallelogram bounded by these four segments.

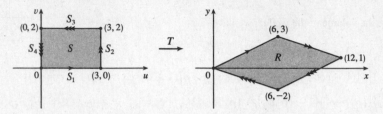

9. S_1 is the line segment $u = v$, $0 \leq u \leq 1$, so $y = v = u$ and $x = u^2 = y^2$. Since $0 \leq u \leq 1$, the image is the portion of the parabola $x = y^2$, $0 \leq y \leq 1$. S_2 is the segment $v = 1$, $0 \leq u \leq 1$, thus $y = v = 1$ and $x = u^2$, so $0 \leq x \leq 1$. The image is the line segment $y = 1$, $0 \leq x \leq 1$. S_3 is the segment $u = 0$, $0 \leq v \leq 1$, so $x = u^2 = 0$ and $y = v$ $\Rightarrow$ $0 \leq y \leq 1$. The image is the segment $x = 0$, $0 \leq y \leq 1$. Thus, the image of S is the region R in the first quadrant bounded by the parabola $x = y^2$, the y-axis, and the line $y = 1$.

11. R is a parallelogram enclosed by the parallel lines $y = 2x - 1$, $y = 2x + 1$ and the parallel lines $y = 1 - x$, $y = 3 - x$. The first pair of equations can be written as $y - 2x = -1$, $y - 2x = 1$. If we let $u = y - 2x$ then these lines are mapped to the vertical lines $u = -1$, $u = 1$ in the uv-plane. Similarly, the second pair of equations can be written as $x + y = 1$, $x + y = 3$, and setting $v = x + y$ maps these lines to the horizontal lines $v = 1$, $v = 3$ in the uv-plane. Boundary curves are mapped to boundary curves under a transformation, so here the equations $u = y - 2x$, $v = x + y$ define a transformation T^{-1} that maps R in the xy-plane to the square S enclosed by the lines $u = -1$, $u = 1$, $v = 1$, $v = 3$ in the uv-plane. To find the transformation T that maps S to R we solve $u = y - 2x$, $v = x + y$ for x, y: Subtracting the first equation from the second gives $v - u = 3x$ $\Rightarrow$ $x = \frac{1}{3}(v - u)$ and adding twice the second equation to the first gives $u + 2v = 3y$ $\Rightarrow$ $y = \frac{1}{3}(u + 2v)$. Thus one possible transformation T (there are many) is given by $x = \frac{1}{3}(v - u)$, $y = \frac{1}{3}(u + 2v)$.

© 2012 Cengage Learning. All Rights Reserved. May not be scanned, copied, or duplicated, or posted to a publicly accessible website, in whole or in part.

13. R is a portion of an annular region (see the figure) that is easily described in polar coordinates as

$R = \{(r,\theta) \mid 1 \le r \le \sqrt{2}, 0 \le \theta \le \pi/2\}$. If we converted a double integral over R to polar coordinates the resulting region of integration is a rectangle (in the $r\theta$-plane), so we can create a transformation T here by letting u play the role of r and v the role of θ. Thus T is defined by $x = u\cos v$, $y = u\sin v$ and T maps the rectangle $S = \{(u,v) \mid 1 \le u \le \sqrt{2}, 0 \le v \le \pi/2\}$ in the uv-plane to R in the xy-plane.

15. $\dfrac{\partial(x,y)}{\partial(u,v)} = \begin{vmatrix} 2 & 1 \\ 1 & 2 \end{vmatrix} = 3$ and $x - 3y = (2u+v) - 3(u+2v) = -u - 5v$. To find the region S in the uv-plane that

corresponds to R we first find the corresponding boundary under the given transformation. The line through $(0,0)$ and $(2,1)$ is $y = \frac{1}{2}x$ which is the image of $u + 2v = \frac{1}{2}(2u+v) \Rightarrow v = 0$; the line through $(2,1)$ and $(1,2)$ is $x + y = 3$ which is the image of $(2u+v) + (u+2v) = 3 \Rightarrow u + v = 1$; the line through $(0,0)$ and $(1,2)$ is $y = 2x$ which is the image of $u + 2v = 2(2u+v) \Rightarrow u = 0$. Thus S is the triangle $0 \le v \le 1 - u$, $0 \le u \le 1$ in the uv-plane and

$$\iint_R (x - 3y)\, dA = \int_0^1 \int_0^{1-u} (-u - 5v) \, |3| \, dv\, du = -3 \int_0^1 \left[uv + \tfrac{5}{2}v^2 \right]_{v=0}^{v=1-u} du$$

$$= -3 \int_0^1 \left(u - u^2 + \tfrac{5}{2}(1-u)^2 \right) du = -3\left[\tfrac{1}{2}u^2 - \tfrac{1}{3}u^3 - \tfrac{5}{6}(1-u)^3 \right]_0^1 = -3\left(\tfrac{1}{2} - \tfrac{1}{3} + \tfrac{5}{6} \right) = -3$$

17. $\dfrac{\partial(x,y)}{\partial(u,v)} = \begin{vmatrix} 2 & 0 \\ 0 & 3 \end{vmatrix} = 6$, $x^2 = 4u^2$ and the planar ellipse $9x^2 + 4y^2 \le 36$ is the image of the disk $u^2 + v^2 \le 1$. Thus

$$\iint_R x^2\, dA = \iint_{u^2+v^2 \le 1} (4u^2)(6)\, du\, dv = \int_0^{2\pi} \int_0^1 (24 r^2 \cos^2\theta)\, r\, dr\, d\theta = 24 \int_0^{2\pi} \cos^2\theta \, d\theta \int_0^1 r^3\, dr$$

$$= 24 \left[\tfrac{1}{2}x + \tfrac{1}{4}\sin 2x \right]_0^{2\pi} \left[\tfrac{1}{4}r^4 \right]_0^1 = 24(\pi)\left(\tfrac{1}{4} \right) = 6\pi$$

19. $\dfrac{\partial(x,y)}{\partial(u,v)} = \begin{vmatrix} 1/v & -u/v^2 \\ 0 & 1 \end{vmatrix} = \dfrac{1}{v}$, $xy = u$, $y = x$ is the image of the parabola $v^2 = u$, $y = 3x$ is the image of the parabola

$v^2 = 3u$, and the hyperbolas $xy = 1$, $xy = 3$ are the images of the lines $u = 1$ and $u = 3$ respectively. Thus

$$\iint_R xy\, dA = \int_1^3 \int_{\sqrt{u}}^{\sqrt{3u}} u\left(\tfrac{1}{v} \right) dv\, du = \int_1^3 u\left(\ln\sqrt{3u} - \ln\sqrt{u} \right) du = \int_1^3 u \ln\sqrt{3}\, du = 4\ln\sqrt{3} = 2\ln 3.$$

21. (a) $\dfrac{\partial(x,y,z)}{\partial(u,v,w)} = \begin{vmatrix} a & 0 & 0 \\ 0 & b & 0 \\ 0 & 0 & c \end{vmatrix} = abc$ and since $u = \dfrac{x}{a}$, $v = \dfrac{y}{b}$, $w = \dfrac{z}{c}$ the solid enclosed by the ellipsoid is the image of the

ball $u^2 + v^2 + w^2 \le 1$. So

$$\iiint_E dV = \iiint_{u^2+v^2+w^2 \le 1} abc\, du\, dv\, dw = (abc)(\text{volume of the ball}) = \tfrac{4}{3}\pi abc$$

© 2012 Cengage Learning. All Rights Reserved. May not be scanned, copied, or duplicated, or posted to a publicly accessible website, in whole or in part.

(b) If we approximate the surface of the earth by the ellipsoid $\dfrac{x^2}{6378^2} + \dfrac{y^2}{6378^2} + \dfrac{z^2}{6356^2} = 1$, then we can estimate

the volume of the earth by finding the volume of the solid E enclosed by the ellipsoid. From part (a), this is

$$\iiint_E dV = \tfrac{4}{3}\pi(6378)(6378)(6356) \approx 1.083 \times 10^{12} \text{ km}^3.$$

(c) The moment of intertia about the z-axis is $I_z = \iiint_E (x^2 + y^2)\,\rho(x, y, z)\,dV$, where E is the solid enclosed by

$\dfrac{x^2}{a^2} + \dfrac{y^2}{b^2} + \dfrac{z^2}{c^2} = 1$. As in part (a), we use the transformation $x = au$, $y = bv$, $z = cw$, so $\left|\dfrac{\partial(x, y, z)}{\partial(u, v, w)}\right| = abc$ and

$$I_z = \iiint_E (x^2 + y^2)\,k\,dV = \iiint_{u^2+v^2+w^2 \le 1} k(a^2u^2 + b^2v^2)(abc)\,du\,dv\,dw$$

$$= abck \int_0^\pi \int_0^{2\pi} \int_0^1 (a^2\rho^2 \sin^2\phi\cos^2\theta + b^2\rho^2\sin^2\phi\sin^2\theta)\,\rho^2\sin\phi\,d\rho\,d\theta\,d\phi$$

$$= abck \left[a^2 \int_0^\pi \int_0^{2\pi} \int_0^1 (\rho^2\sin^2\phi\cos^2\theta)\,\rho^2\sin\phi\,d\rho\,d\theta\,d\phi + b^2 \int_0^\pi \int_0^{2\pi} \int_0^1 (\rho^2\sin^2\phi\sin^2\theta)\,\rho^2\sin\phi\,d\rho\,d\theta\,d\phi \right]$$

$$= a^3bck \int_0^\pi \sin^3\phi\,d\phi \int_0^{2\pi} \cos^2\theta\,d\theta \int_0^1 \rho^4\,d\rho + ab^3ck \int_0^\pi \sin^3\phi\,d\phi \int_0^{2\pi} \sin^2\theta\,d\theta \int_0^1 \rho^4\,d\rho$$

$$= a^3bck \left[\tfrac{1}{3}\cos^3\phi - \cos\phi\right]_0^\pi \left[\tfrac{1}{2}\theta + \tfrac{1}{4}\sin 2\theta\right]_0^{2\pi} \left[\tfrac{1}{5}\rho^5\right]_0^1 + ab^3ck \left[\tfrac{1}{3}\cos^3\phi - \cos\phi\right]_0^\pi \left[\tfrac{1}{2}\theta - \tfrac{1}{4}\sin 2\theta\right]_0^{2\pi} \left[\tfrac{1}{5}\rho^5\right]_0^1$$

$$= a^3bck \left(\tfrac{4}{3}\right)(\pi)\left(\tfrac{1}{5}\right) + ab^3ck \left(\tfrac{4}{3}\right)(\pi)\left(\tfrac{1}{5}\right) = \tfrac{4}{15}\pi\,(a^2 + b^2)\,abck$$

23. Letting $u = x - 2y$ and $v = 3x - y$, we have $x = \tfrac{1}{5}(2v - u)$ and $y = \tfrac{1}{5}(v - 3u)$. Then $\dfrac{\partial(x, y)}{\partial(u, v)} = \begin{vmatrix} -1/5 & 2/5 \\ -3/5 & 1/5 \end{vmatrix} = \dfrac{1}{5}$

and R is the image of the rectangle enclosed by the lines $u = 0$, $u = 4$, $v = 1$, and $v = 8$. Thus

$$\iint_R \frac{x - 2y}{3x - y}\,dA = \int_0^4 \int_1^8 \frac{u}{v}\left|\frac{1}{5}\right|\,dv\,du = \frac{1}{5}\int_0^4 u\,du \int_1^8 \frac{1}{v}\,dv = \frac{1}{5}\left[\tfrac{1}{2}u^2\right]_0^4 \left[\ln|v|\right]_1^8 = \tfrac{8}{5}\ln 8.$$

25. Letting $u = y - x$, $v = y + x$, we have $y = \tfrac{1}{2}(u + v)$, $x = \tfrac{1}{2}(v - u)$. Then $\dfrac{\partial(x, y)}{\partial(u, v)} = \begin{vmatrix} -1/2 & 1/2 \\ 1/2 & 1/2 \end{vmatrix} = -\dfrac{1}{2}$ and R is the

image of the trapezoidal region with vertices $(-1, 1)$, $(-2, 2)$, $(2, 2)$, and $(1, 1)$. Thus

$$\iint_R \cos\frac{y - x}{y + x}\,dA = \int_1^2 \int_{-v}^v \cos\frac{u}{v}\left|-\frac{1}{2}\right|\,du\,dv = \frac{1}{2}\int_1^2 \left[v\sin\frac{u}{v}\right]_{u=-v}^{u=v}\,dv = \frac{1}{2}\int_1^2 2v\sin(1)\,dv = \tfrac{3}{2}\sin 1$$

27. Let $u = x + y$ and $v = -x + y$. Then $u + v = 2y \ \Rightarrow\ y = \tfrac{1}{2}(u + v)$ and $u - v = 2x \ \Rightarrow\ x = \tfrac{1}{2}(u - v)$.

$\dfrac{\partial(x, y)}{\partial(u, v)} = \begin{vmatrix} 1/2 & -1/2 \\ 1/2 & 1/2 \end{vmatrix} = \dfrac{1}{2}$. Now $|u| = |x + y| \le |x| + |y| \le 1 \ \Rightarrow\ -1 \le u \le 1$, and

$|v| = |-x + y| \le |x| + |y| \le 1 \ \Rightarrow\ -1 \le v \le 1$. R is the image of the square

region with vertices $(1, 1)$, $(1, -1)$, $(-1, -1)$, and $(-1, 1)$.

So $\iint_R e^{x+y}\,dA = \tfrac{1}{2}\int_{-1}^1 \int_{-1}^1 e^u\,du\,dv = \tfrac{1}{2}\left[e^u\right]_{-1}^1 \left[v\right]_{-1}^1 = e - e^{-1}$.

© 2012 Cengage Learning. All Rights Reserved. May not be scanned, copied, or duplicated, or posted to a publicly accessible website, in whole or in part.

15 Review

CONCEPT CHECK

1. (a) A double Riemann sum of f is $\sum_{i=1}^{m} \sum_{j=1}^{n} f\left(x_{ij}^*, y_{ij}^*\right) \Delta A$, where ΔA is the area of each subrectangle and $\left(x_{ij}^*, y_{ij}^*\right)$ is a sample point in each subrectangle. If $f(x, y) \geq 0$, this sum represents an approximation to the volume of the solid that lies above the rectangle R and below the graph of f.

   (b) $\iint_R f(x, y)\, dA = \lim_{m,n \to \infty} \sum_{i=1}^{m} \sum_{j=1}^{n} f\left(x_{ij}^*, y_{ij}^*\right) \Delta A$

   (c) If $f(x, y) \geq 0$, $\iint_R f(x, y)\, dA$ represents the volume of the solid that lies above the rectangle R and below the surface $z = f(x, y)$. If f takes on both positive and negative values, $\iint_R f(x, y)\, dA$ is the difference of the volume above R but below the surface $z = f(x, y)$ and the volume below R but above the surface $z = f(x, y)$.

   (d) We usually evaluate $\iint_R f(x, y)\, dA$ as an iterated integral according to Fubini's Theorem (see Theorem 15.2.4).

   (e) The Midpoint Rule for Double Integrals says that we approximate the double integral $\iint_R f(x, y)\, dA$ by the double Riemann sum $\sum_{i=1}^{m} \sum_{j=1}^{n} f\left(\overline{x}_i, \overline{y}_j\right) \Delta A$ where the sample points $\left(\overline{x}_i, \overline{y}_j\right)$ are the centers of the subrectangles.

   (f) $f_{\text{ave}} = \dfrac{1}{A(R)} \iint_R f(x, y)\, dA$ where $A(R)$ is the area of R.

2. (a) See (1) and (2) and the accompanying discussion in Section 15.3.

   (b) See (3) and the accompanying discussion in Section 15.3.

   (c) See (5) and the preceding discussion in Section 15.3.

   (d) See (6)–(11) in Section 15.3.

3. We may want to change from rectangular to polar coordinates in a double integral if the region R of integration is more easily described in polar coordinates. To accomplish this, we use $\iint_R f(x, y)\, dA = \int_\alpha^\beta \int_a^b f(r \cos \theta, r \sin \theta)\, r\, dr\, d\theta$ where R is given by $0 \leq a \leq r \leq b$, $\alpha \leq \theta \leq \beta$.

4. (a) $m = \iint_D \rho(x, y)\, dA$

   (b) $M_x = \iint_D y\rho(x, y)\, dA$, $M_y = \iint_D x\rho(x, y)\, dA$

   (c) The center of mass is $(\overline{x}, \overline{y})$ where $\overline{x} = \dfrac{M_y}{m}$ and $\overline{y} = \dfrac{M_x}{m}$.

   (d) $I_x = \iint_D y^2 \rho(x, y)\, dA$, $I_y = \iint_D x^2 \rho(x, y)\, dA$, $I_0 = \iint_D (x^2 + y^2)\rho(x, y)\, dA$

5. (a) $P(a \leq X \leq b, c \leq Y \leq d) = \int_a^b \int_c^d f(x, y)\, dy\, dx$

   (b) $f(x, y) \geq 0$ and $\iint_{\mathbb{R}^2} f(x, y)\, dA = 1$.

   (c) The expected value of X is $\mu_1 = \iint_{\mathbb{R}^2} x f(x, y)\, dA$; the expected value of Y is $\mu_2 = \iint_{\mathbb{R}^2} y f(x, y)\, dA$.

© 2012 Cengage Learning. All Rights Reserved. May not be scanned, copied, or duplicated, or posted to a publicly accessible website, in whole or in part.

6. $A(S) = \iint_D \sqrt{[f_x(x,y)]^2 + [f_y(x,y)]^2 + 1}\, dA$

7. (a) $\iiint_B f(x,y,z)\, dV = \lim\limits_{l,m,n\to\infty} \sum\limits_{i=1}^{l} \sum\limits_{j=1}^{m} \sum\limits_{k=1}^{n} f\left(x_{ijk}^*, y_{ijk}^*, z_{ijk}^*\right) \Delta V$

   (b) We usually evaluate $\iiint_B f(x,y,z)\, dV$ as an iterated integral according to Fubini's Theorem for Triple Integrals (see Theorem 15.7.4).

   (c) See the paragraph following Example 15.7.1.

   (d) See (5) and (6) and the accompanying discussion in Section 15.7.

   (e) See (10) and the accompanying discussion in Section 15.7.

   (f) See (11) and the preceding discussion in Section 15.7.

8. (a) $m = \iiint_E \rho(x,y,z)\, dV$

   (b) $M_{yz} = \iiint_E x\rho(x,y,z)\, dV,\ M_{xz} = \iiint_E y\rho(x,y,z)\, dV,\ M_{xy} = \iiint_E z\rho(x,y,z)\, dV.$

   (c) The center of mass is $(\overline{x}, \overline{y}, \overline{z})$ where $\overline{x} = \dfrac{M_{yz}}{m}, \overline{y} = \dfrac{M_{xz}}{m}$, and $\overline{z} = \dfrac{M_{xy}}{m}$.

   (d) $I_x = \iiint_E (y^2 + z^2)\rho(x,y,z)\, dV,\ I_y = \iiint_E (x^2 + z^2)\rho(x,y,z)\, dV,\ I_z = \iiint_E (x^2 + y^2)\rho(x,y,z)\, dV.$

9. (a) See Formula 15.8.4 and the accompanying discussion.

   (b) See Formula 15.9.3 and the accompanying discussion.

   (c) We may want to change from rectangular to cylindrical or spherical coordinates in a triple integral if the region E of integration is more easily described in cylindrical or spherical coordinates or if the triple integral is easier to evaluate using cylindrical or spherical coordinates.

10. (a) $\dfrac{\partial(x,y)}{\partial(u,v)} = \begin{vmatrix} \partial x/\partial u & \partial x/\partial v \\ \partial y/\partial u & \partial y/\partial v \end{vmatrix} = \dfrac{\partial x}{\partial u}\dfrac{\partial y}{\partial v} - \dfrac{\partial x}{\partial v}\dfrac{\partial y}{\partial u}$

    (b) See (9) and the accompanying discussion in Section 15.10.

    (c) See (13) and the accompanying discussion in Section 15.10.

TRUE-FALSE QUIZ

1. This is true by Fubini's Theorem.

3. True by Equation 15.2.5.

5. True. By Equation 15.2.5 we can write $\int_0^1 \int_0^1 f(x)\, f(y)\, dy\, dx = \int_0^1 f(x)\, dx \int_0^1 f(y)\, dy$. But $\int_0^1 f(y)\, dy = \int_0^1 f(x)\, dx$ so this becomes $\int_0^1 f(x)\, dx \int_0^1 f(x)\, dx = \left[\int_0^1 f(x)\, dx\right]^2$.

7. True: $\iint_D \sqrt{4 - x^2 - y^2}\, dA = $ the volume under the surface $x^2 + y^2 + z^2 = 4$ and above the xy-plane
   $= \frac{1}{2}$ (the volume of the sphere $x^2 + y^2 + z^2 = 4$) $= \frac{1}{2} \cdot \frac{4}{3}\pi(2)^3 = \frac{16}{3}\pi$

© 2012 Cengage Learning. All Rights Reserved. May not be scanned, copied, or duplicated, or posted to a publicly accessible website, in whole or in part.

9. The volume enclosed by the cone $z = \sqrt{x^2 + y^2}$ and the plane $z = 2$ is, in cylindrical coordinates,

$V = \int_0^{2\pi} \int_0^2 \int_r^2 r \, dz \, dr \, d\theta \neq \int_0^{2\pi} \int_0^2 \int_r^2 dz \, dr \, d\theta$, so the assertion is false.

EXERCISES

1. As shown in the contour map, we divide R into 9 equally sized subsquares, each with area $\Delta A = 1$. Then we approximate

$\iint_R f(x, y) \, dA$ by a Riemann sum with $m = n = 3$ and the sample points the upper right corners of each square, so

$$\iint_R f(x, y) \, dA \approx \sum_{i=1}^{3} \sum_{j=1}^{3} f(x_i, y_j) \, \Delta A$$

$$= \Delta A \left[f(1,1) + f(1,2) + f(1,3) + f(2,1) + f(2,2) + f(2,3) + f(3,1) + f(3,2) + f(3,3) \right]$$

Using the contour lines to estimate the function values, we have

$$\iint_R f(x, y) \, dA \approx 1[2.7 + 4.7 + 8.0 + 4.7 + 6.7 + 10.0 + 6.7 + 8.6 + 11.9] \approx 64.0$$

3. $\int_1^2 \int_0^2 (y + 2xe^y) \, dx \, dy = \int_1^2 \left[xy + x^2 e^y \right]_{x=0}^{x=2} dy = \int_1^2 (2y + 4e^y) \, dy = \left[y^2 + 4e^y \right]_1^2$

$$= 4 + 4e^2 - 1 - 4e = 4e^2 - 4e + 3$$

5. $\int_0^1 \int_0^x \cos(x^2) \, dy \, dx = \int_0^1 \left[\cos(x^2) y \right]_{y=0}^{y=x} dx = \int_0^1 x \cos(x^2) \, dx = \frac{1}{2} \sin(x^2) \Big]_0^1 = \frac{1}{2} \sin 1$

7. $\int_0^{\pi} \int_0^1 \int_0^{\sqrt{1-y^2}} y \sin x \, dz \, dy \, dx = \int_0^{\pi} \int_0^1 \left[(y \sin x) z \right]_{z=0}^{z=\sqrt{1-y^2}} dy \, dx = \int_0^{\pi} \int_0^1 y \sqrt{1-y^2} \sin x \, dy \, dx$

$$= \int_0^{\pi} \left[-\frac{1}{3}(1 - y^2)^{3/2} \sin x \right]_{y=0}^{y=1} dx = \int_0^{\pi} \frac{1}{3} \sin x \, dx = -\frac{1}{3} \cos x \Big]_0^{\pi} = \frac{2}{3}$

9. The region R is more easily described by polar coordinates: $R = \{(r, \theta) \mid 2 \leq r \leq 4, 0 \leq \theta \leq \pi\}$. Thus

$\iint_R f(x, y) \, dA = \int_0^{\pi} \int_2^4 f(r \cos \theta, r \sin \theta) \, r \, dr \, d\theta$.

11.

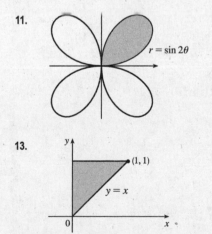

$r = \sin 2\theta$

The region whose area is given by $\int_0^{\pi/2} \int_0^{\sin 2\theta} r \, dr \, d\theta$ is

$\{(r, \theta) \mid 0 \leq \theta \leq \frac{\pi}{2}, 0 \leq r \leq \sin 2\theta\}$, which is the region contained in the

loop in the first quadrant of the four-leaved rose $r = \sin 2\theta$.

13.

$\int_0^1 \int_x^1 \cos(y^2) \, dy \, dx = \int_0^1 \int_0^y \cos(y^2) \, dx \, dy$

$$= \int_0^1 \cos(y^2) \left[x \right]_{x=0}^{x=y} dy = \int_0^1 y \cos(y^2) \, dy$$

$$= \left[\frac{1}{2} \sin(y^2) \right]_0^1 = \frac{1}{2} \sin 1$$

$y = x$, point $(1,1)$

15. $\iint_R ye^{xy} \, dA = \int_0^3 \int_0^2 ye^{xy} \, dx \, dy = \int_0^3 \left[e^{xy} \right]_{x=0}^{x=2} dy = \int_0^3 (e^{2y} - 1) \, dy = \left[\frac{1}{2}e^{2y} - y \right]_0^3 = \frac{1}{2}e^6 - 3 - \frac{1}{2} = \frac{1}{2}e^6 - \frac{7}{2}$

© 2012 Cengage Learning. All Rights Reserved. May not be scanned, copied, or duplicated, or posted to a publicly accessible website, in whole or in part.

17.

$$\iint_D \frac{y}{1+x^2}\, dA = \int_0^1 \int_0^{\sqrt{x}} \frac{y}{1+x^2}\, dy\, dx = \int_0^1 \frac{1}{1+x^2} \left[\tfrac{1}{2}y^2\right]_{y=0}^{y=\sqrt{x}} dx$$

$$= \tfrac{1}{2}\int_0^1 \frac{x}{1+x^2}\, dx = \left[\tfrac{1}{4}\ln(1+x^2)\right]_0^1 = \tfrac{1}{4}\ln 2$$

19.

$$\iint_D y\, dA = \int_0^2 \int_{y^2}^{8-y^2} y\, dx\, dy$$

$$= \int_0^2 y\left[x\right]_{x=y^2}^{x=8-y^2} dy = \int_0^2 y(8-y^2-y^2)\, dy$$

$$= \int_0^2 (8y - 2y^3)\, dy = \left[4y^2 - \tfrac{1}{2}y^4\right]_0^2 = 8$$

21.

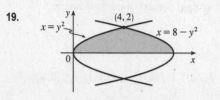

$$\iint_D \left(x^2+y^2\right)^{3/2} dA = \int_0^{\pi/3} \int_0^3 (r^2)^{3/2}\, r\, dr\, d\theta$$

$$= \int_0^{\pi/3} d\theta \int_0^3 r^4\, dr = \left[\theta\right]_0^{\pi/3} \left[\tfrac{1}{5}r^5\right]_0^3$$

$$= \frac{\pi}{3}\frac{3^5}{5} = \frac{81\pi}{5}$$

23. $\iiint_E xy\, dV = \int_0^3 \int_0^x \int_0^{x+y} xy\, dz\, dy\, dx = \int_0^3 \int_0^x xy\left[z\right]_{z=0}^{z=x+y} dy\, dx = \int_0^3 \int_0^x xy(x+y)\, dy\, dx$

$= \int_0^3 \int_0^x (x^2 y + xy^2)\, dy\, dx = \int_0^3 \left[\tfrac{1}{2}x^2 y^2 + \tfrac{1}{3}xy^3\right]_{y=0}^{y=x} dx = \int_0^3 \left(\tfrac{1}{2}x^4 + \tfrac{1}{3}x^4\right) dx$

$= \tfrac{5}{6}\int_0^3 x^4\, dx = \left[\tfrac{1}{6}x^5\right]_0^3 = \tfrac{81}{2} = 40.5$

25. $\iiint_E y^2 z^2\, dV = \int_{-1}^1 \int_{-\sqrt{1-y^2}}^{\sqrt{1-y^2}} \int_0^{1-y^2-z^2} y^2 z^2\, dx\, dz\, dy = \int_{-1}^1 \int_{-\sqrt{1-y^2}}^{\sqrt{1-y^2}} y^2 z^2 (1-y^2-z^2)\, dz\, dy$

$= \int_0^{2\pi} \int_0^1 (r^2 \cos^2\theta)(r^2 \sin^2\theta)(1-r^2)\, r\, dr\, d\theta = \int_0^{2\pi} \int_0^1 \tfrac{1}{4}\sin^2 2\theta\, (r^5 - r^7)\, dr\, d\theta$

$= \int_0^{2\pi} \tfrac{1}{8}(1-\cos 4\theta)\left[\tfrac{1}{6}r^6 - \tfrac{1}{8}r^8\right]_{r=0}^{r=1} d\theta = \tfrac{1}{192}\left[\theta - \tfrac{1}{4}\sin 4\theta\right]_0^{2\pi} = \tfrac{2\pi}{192} = \tfrac{\pi}{96}$

27. $\iiint_E yz\, dV = \int_{-2}^2 \int_0^{\sqrt{4-x^2}} \int_0^y yz\, dz\, dy\, dx = \int_{-2}^2 \int_0^{\sqrt{4-x^2}} \tfrac{1}{2}y^3\, dy\, dx = \int_0^\pi \int_0^2 \tfrac{1}{2}r^3(\sin^3\theta)\, r\, dr\, d\theta$

$= \tfrac{16}{5}\int_0^\pi \sin^3\theta\, d\theta = \tfrac{16}{5}\left[-\cos\theta + \tfrac{1}{3}\cos^3\theta\right]_0^\pi = \tfrac{64}{15}$

29. $V = \int_0^2 \int_1^4 (x^2 + 4y^2)\, dy\, dx = \int_0^2 \left[x^2 y + \tfrac{4}{3}y^3\right]_{y=1}^{y=4} dx = \int_0^2 (3x^2 + 84)\, dx = 176$

31.

$$V = \int_0^2 \int_0^y \int_0^{(2-y)/2} dz\, dx\, dy = \int_0^2 \int_0^y \left(1 - \tfrac{1}{2}y\right) dx\, dy$$

$$= \int_0^2 \left(y - \tfrac{1}{2}y^2\right) dy = \tfrac{2}{3}$$

© 2012 Cengage Learning. All Rights Reserved. May not be scanned, copied, or duplicated, or posted to a publicly accessible website, in whole or in part.

33. Using the wedge above the plane $z = 0$ and below the plane $z = mx$ and noting that we have the same volume for $m < 0$ as for $m > 0$ (so use $m > 0$), we have

$$V = 2 \int_0^{a/3} \int_0^{\sqrt{a^2 - 9y^2}} mx \, dx \, dy = 2 \int_0^{a/3} \tfrac{1}{2} m(a^2 - 9y^2) \, dy = m\big[a^2 y - 3y^3\big]_0^{a/3} = m\big(\tfrac{1}{3}a^3 - \tfrac{1}{9}a^3\big) = \tfrac{2}{9}ma^3.$$

35. (a) $m = \int_0^1 \int_0^{1-y^2} y \, dx \, dy = \int_0^1 (y - y^3) \, dy = \tfrac{1}{2} - \tfrac{1}{4} = \tfrac{1}{4}$

(b) $M_y = \int_0^1 \int_0^{1-y^2} xy \, dx \, dy = \int_0^1 \tfrac{1}{2} y(1 - y^2)^2 \, dy = -\tfrac{1}{12}(1 - y^2)^3\big]_0^1 = \tfrac{1}{12}$,

$M_x = \int_0^1 \int_0^{1-y^2} y^2 \, dx \, dy = \int_0^1 (y^2 - y^4) \, dy = \tfrac{2}{15}$. Hence $(\bar{x}, \bar{y}) = \big(\tfrac{1}{3}, \tfrac{8}{15}\big)$.

(c) $I_x = \int_0^1 \int_0^{1-y^2} y^3 \, dx \, dy = \int_0^1 (y^3 - y^5) \, dy = \tfrac{1}{12}$,

$I_y = \int_0^1 \int_0^{1-y^2} yx^2 \, dx \, dy = \int_0^1 \tfrac{1}{3} y(1 - y^2)^3 \, dy = -\tfrac{1}{24}(1 - y^2)^4\big]_0^1 = \tfrac{1}{24}$,

$I_0 = I_x + I_y = \tfrac{1}{8}, \overline{\overline{y}}^2 = \tfrac{1/12}{1/4} = \tfrac{1}{3} \Rightarrow \overline{\overline{y}} = \tfrac{1}{\sqrt{3}}$, and $\overline{\overline{x}}^2 = \tfrac{1/24}{1/4} = \tfrac{1}{6} \Rightarrow \overline{\overline{x}} = \tfrac{1}{\sqrt{6}}$.

37. (a) The equation of the cone with the suggested orientation is $(h - z) = \tfrac{h}{a}\sqrt{x^2 + y^2}, 0 \le z \le h$. Then $V = \tfrac{1}{3}\pi a^2 h$ is the volume of one frustum of a cone; by symmetry $M_{yz} = M_{xz} = 0$; and

$$M_{xy} = \iint\limits_{x^2 + y^2 \le a^2} \int_0^{h - (h/a)\sqrt{x^2 + y^2}} z \, dz \, dA = \int_0^{2\pi} \int_0^a \int_0^{(h/a)(a-r)} rz \, dz \, dr \, d\theta = \pi \int_0^a r \frac{h^2}{a^2}(a - r)^2 \, dr$$

$$= \frac{\pi h^2}{a^2} \int_0^a (a^2 r - 2ar^2 + r^3) \, dr = \frac{\pi h^2}{a^2}\left(\frac{a^4}{2} - \frac{2a^4}{3} + \frac{a^4}{4}\right) = \frac{\pi h^2 a^2}{12}$$

Hence the centroid is $(\bar{x}, \bar{y}, \bar{z}) = \big(0, 0, \tfrac{1}{4}h\big)$.

(b) $I_z = \int_0^{2\pi} \int_0^a \int_0^{(h/a)(a-r)} r^3 \, dz \, dr \, d\theta = 2\pi \int_0^a \frac{h}{a}(ar^3 - r^4) \, dr = \frac{2\pi h}{a}\left(\frac{a^5}{4} - \frac{a^5}{5}\right) = \frac{\pi a^4 h}{10}$

39. Let D represent the given triangle; then D can be described as the area enclosed by the x- and y-axes and the line $y = 2 - 2x$, or equivalently $D = \{(x, y) \mid 0 \le x \le 1, 0 \le y \le 2 - 2x\}$. We want to find the surface area of the part of the graph of $z = x^2 + y$ that lies over D, so using Equation 15.6.3 we have

$$A(S) = \iint_D \sqrt{1 + \left(\frac{\partial z}{\partial x}\right)^2 + \left(\frac{\partial z}{\partial y}\right)^2} \, dA = \iint_D \sqrt{1 + (2x)^2 + (1)^2} \, dA = \int_0^1 \int_0^{2-2x} \sqrt{2 + 4x^2} \, dy \, dx$$

$$= \int_0^1 \sqrt{2 + 4x^2} \, [y]_{y=0}^{y=2-2x} \, dx = \int_0^1 (2 - 2x)\sqrt{2 + 4x^2} \, dx = \int_0^1 2\sqrt{2 + 4x^2} \, dx - \int_0^1 2x\sqrt{2 + 4x^2} \, dx$$

Using Formula 21 in the Table of Integrals with $a = \sqrt{2}$, $u = 2x$, and $du = 2 \, dx$, we have

$\int 2\sqrt{2 + 4x^2} \, dx = x\sqrt{2 + 4x^2} + \ln\big(2x + \sqrt{2 + 4x^2}\big)$. If we substitute $u = 2 + 4x^2$ in the second integral, then

$du = 8x \, dx$ and $\int 2x\sqrt{2 + 4x^2} \, dx = \tfrac{1}{4} \int \sqrt{u} \, du = \tfrac{1}{4} \cdot \tfrac{2}{3} u^{3/2} = \tfrac{1}{6}(2 + 4x^2)^{3/2}$. Thus

$$A(S) = \left[x\sqrt{2 + 4x^2} + \ln\big(2x + \sqrt{2 + 4x^2}\big) - \tfrac{1}{6}(2 + 4x^2)^{3/2}\right]_0^1$$

$$= \sqrt{6} + \ln\big(2 + \sqrt{6}\big) - \tfrac{1}{6}(6)^{3/2} - \ln\sqrt{2} + \frac{\sqrt{2}}{3} = \ln\frac{2 + \sqrt{6}}{\sqrt{2}} + \frac{\sqrt{2}}{3}$$

$$= \ln\big(\sqrt{2} + \sqrt{3}\big) + \frac{\sqrt{2}}{3} \approx 1.6176$$

© 2012 Cengage Learning. All Rights Reserved. May not be scanned, copied, or duplicated, or posted to a publicly accessible website, in whole or in part.

41.

$$\int_0^3 \int_{-\sqrt{9-x^2}}^{\sqrt{9-x^2}} (x^3 + xy^2)\, dy\, dx = \int_0^3 \int_{-\sqrt{9-x^2}}^{\sqrt{9-x^2}} x(x^2 + y^2)\, dy\, dx$$

$$= \int_{-\pi/2}^{\pi/2} \int_0^3 (r\cos\theta)(r^2)\, r\, dr\, d\theta$$

$$= \int_{-\pi/2}^{\pi/2} \cos\theta\, d\theta \int_0^3 r^4\, dr$$

$$= \left[\sin\theta\right]_{-\pi/2}^{\pi/2} \left[\tfrac{1}{5}r^5\right]_0^3 = 2\cdot\tfrac{1}{5}(243) = \tfrac{486}{5} = 97.2$$

43. From the graph, it appears that $1 - x^2 = e^x$ at $x \approx -0.71$ and at

$x = 0$, with $1 - x^2 > e^x$ on $(-0.71, 0)$. So the desired integral is

$$\iint_D y^2\, dA \approx \int_{-0.71}^0 \int_{e^x}^{1-x^2} y^2\, dy\, dx$$

$$= \tfrac{1}{3} \int_{-0.71}^0 [(1 - x^2)^3 - e^{3x}]\, dx$$

$$= \tfrac{1}{3} \left[x - x^3 + \tfrac{3}{5}x^5 - \tfrac{1}{7}x^7 - \tfrac{1}{3}e^{3x}\right]_{-0.71}^0 \approx 0.0512$$

45. (a) $f(x, y)$ is a joint density function, so we know that $\iint_{\mathbb{R}^2} f(x, y)\, dA = 1$. Since $f(x, y) = 0$ outside the rectangle

$[0, 3] \times [0, 2]$, we can say

$$\iint_{\mathbb{R}^2} f(x, y)\, dA = \int_{-\infty}^{\infty} \int_{-\infty}^{\infty} f(x, y)\, dy\, dx = \int_0^3 \int_0^2 C(x + y)\, dy\, dx$$

$$= C \int_0^3 \left[xy + \tfrac{1}{2}y^2\right]_{y=0}^{y=2} dx = C \int_0^3 (2x + 2)\, dx = C\left[x^2 + 2x\right]_0^3 = 15C$$

Then $15C = 1 \;\Rightarrow\; C = \tfrac{1}{15}$.

(b) $P(X \le 2, Y \ge 1) = \int_{-\infty}^2 \int_1^{\infty} f(x, y)\, dy\, dx = \int_0^2 \int_1^2 \tfrac{1}{15}(x, y)\, dy\, dx = \tfrac{1}{15} \int_0^2 \left[xy + \tfrac{1}{2}y^2\right]_{y=1}^{y=2} dx$

$$= \tfrac{1}{15} \int_0^2 \left(x + \tfrac{3}{2}\right) dx = \tfrac{1}{15} \left[\tfrac{1}{2}x^2 + \tfrac{3}{2}x\right]_0^2 = \tfrac{1}{3}$$

(c) $P(X + Y \le 1) = P((X, Y) \in D)$ where D is the triangular region shown in

the figure. Thus

$$P(X + Y \le 1) = \iint_D f(x, y)\, dA = \int_0^1 \int_0^{1-x} \tfrac{1}{15}(x + y)\, dy\, dx$$

$$= \tfrac{1}{15} \int_0^1 \left[xy + \tfrac{1}{2}y^2\right]_{y=0}^{y=1-x} dx$$

$$= \tfrac{1}{15} \int_0^1 \left[x(1 - x) + \tfrac{1}{2}(1 - x)^2\right] dx$$

$$= \tfrac{1}{30} \int_0^1 (1 - x^2)\, dx = \tfrac{1}{30} \left[x - \tfrac{1}{3}x^3\right]_0^1 = \tfrac{1}{45}$$

47.

$$\int_{-1}^1 \int_{x^2}^1 \int_0^{1-y} f(x, y, z)\, dz\, dy\, dx = \int_0^1 \int_0^{1-z} \int_{-\sqrt{y}}^{\sqrt{y}} f(x, y, z)\, dx\, dy\, dz$$

© 2012 Cengage Learning. All Rights Reserved. May not be scanned, copied, or duplicated, or posted to a publicly accessible website, in whole or in part.

49. Since $u = x - y$ and $v = x + y$, $x = \frac{1}{2}(u + v)$ and $y = \frac{1}{2}(v - u)$.

Thus $\dfrac{\partial(x, y)}{\partial(u, v)} = \begin{vmatrix} 1/2 & 1/2 \\ -1/2 & 1/2 \end{vmatrix} = \dfrac{1}{2}$ and $\displaystyle\iint_R \dfrac{x - y}{x + y}\, dA = \int_2^4 \int_{-2}^0 \dfrac{u}{v}\left(\dfrac{1}{2}\right) du\, dv = -\int_2^4 \dfrac{dv}{v} = -\ln 2$.

51. Let $u = y - x$ and $v = y + x$ so $x = y - u = (v - x) - u \Rightarrow x = \frac{1}{2}(v - u)$ and $y = v - \frac{1}{2}(v - u) = \frac{1}{2}(v + u)$.

$\left|\dfrac{\partial(x, y)}{\partial(u, v)}\right| = \left|\dfrac{\partial x}{\partial u}\dfrac{\partial y}{\partial v} - \dfrac{\partial x}{\partial v}\dfrac{\partial y}{\partial u}\right| = \left|-\frac{1}{2}\left(\frac{1}{2}\right) - \frac{1}{2}\left(\frac{1}{2}\right)\right| = \left|-\frac{1}{2}\right| = \frac{1}{2}$. R is the image under this transformation of the square

with vertices $(u, v) = (0, 0)$, $(-2, 0)$, $(0, 2)$, and $(-2, 2)$. So

$$\iint_R xy\, dA = \int_0^2 \int_{-2}^0 \dfrac{v^2 - u^2}{4}\left(\dfrac{1}{2}\right) du\, dv = \frac{1}{8}\int_0^2 \left[v^2 u - \frac{1}{3}u^3\right]_{u=-2}^{u=0} dv = \frac{1}{8}\int_0^2 \left(2v^2 - \frac{8}{3}\right) dv = \frac{1}{8}\left[\frac{2}{3}v^3 - \frac{8}{3}v\right]_0^2 = 0$$

This result could have been anticipated by symmetry, since the integrand is an odd function of y and R is symmetric about the x-axis.

53. For each r such that D_r lies within the domain, $A(D_r) = \pi r^2$, and by the Mean Value Theorem for Double Integrals there

exists (x_r, y_r) in D_r such that $f(x_r, y_r) = \dfrac{1}{\pi r^2}\displaystyle\iint_{D_r} f(x, y)\, dA$. But $\displaystyle\lim_{r \to 0^+}(x_r, y_r) = (a, b)$,

so $\displaystyle\lim_{r \to 0^+}\dfrac{1}{\pi r^2}\iint_{D_r} f(x, y)\, dA = \lim_{r \to 0^+} f(x_r, y_r) = f(a, b)$ by the continuity of f.

© 2012 Cengage Learning. All Rights Reserved. May not be scanned, copied, or duplicated, or posted to a publicly accessible website, in whole or in part.

□ PROBLEMS PLUS

1.

Let $R = \bigcup_{i=1}^{5} R_i$, where

$$R_i = \{(x,y) \mid x + y \geq i + 2, \, x + y < i + 3, \, 1 \leq x \leq 3, \, 2 \leq y \leq 5\}.$$

$\iint_R [\![x+y]\!] \, dA = \sum_{i=1}^{5} \iint_{R_i} [\![x+y]\!] \, dA = \sum_{i=1}^{5} [\![x+y]\!] \iint_{R_i} dA$, since

$[\![x+y]\!] = \text{constant} = i + 2$ for $(x,y) \in R_i$. Therefore

$$\iint_R [\![x+y]\!] \, dA = \sum_{i=1}^{5} (i+2) \, [A(R_i)]$$
$$= 3A(R_1) + 4A(R_2) + 5A(R_3) + 6A(R_4) + 7A(R_5)$$
$$= 3\left(\tfrac{1}{2}\right) + 4\left(\tfrac{3}{2}\right) + 5(2) + 6\left(\tfrac{3}{2}\right) + 7\left(\tfrac{1}{2}\right) = 30$$

3. $f_{\text{ave}} = \dfrac{1}{b-a} \displaystyle\int_a^b f(x)\,dx = \dfrac{1}{1-0}\int_0^1 \left[\int_x^1 \cos(t^2)\,dt\right] dx$

$= \int_0^1 \int_x^1 \cos(t^2)\,dt\,dx = \int_0^1 \int_0^t \cos(t^2)\,dx\,dt$ [changing the order of integration]

$= \int_0^1 t\cos(t^2)\,dt = \tfrac{1}{2}\sin(t^2)\big]_0^1 = \tfrac{1}{2}\sin 1$

5. Since $|xy| < 1$, except at $(1,1)$, the formula for the sum of a geometric series gives $\dfrac{1}{1-xy} = \displaystyle\sum_{n=0}^{\infty} (xy)^n$, so

$$\int_0^1\int_0^1 \frac{1}{1-xy}\,dx\,dy = \int_0^1\int_0^1 \sum_{n=0}^{\infty}(xy)^n\,dx\,dy = \sum_{n=0}^{\infty}\int_0^1\int_0^1 (xy)^n\,dx\,dy = \sum_{n=0}^{\infty}\left[\int_0^1 x^n\,dx\right]\left[\int_0^1 y^n\,dy\right]$$

$$= \sum_{n=0}^{\infty} \frac{1}{n+1}\cdot\frac{1}{n+1} = \sum_{n=0}^{\infty}\frac{1}{(n+1)^2} = \frac{1}{1^2}+\frac{1}{2^2}+\frac{1}{3^2}+\cdots = \sum_{n=1}^{\infty}\frac{1}{n^2}$$

7. (a) Since $|xyz| < 1$ except at $(1,1,1)$, the formula for the sum of a geometric series gives $\dfrac{1}{1-xyz} = \displaystyle\sum_{n=0}^{\infty}(xyz)^n$, so

$$\int_0^1\int_0^1\int_0^1 \frac{1}{1-xyz}\,dx\,dy\,dz = \int_0^1\int_0^1\int_0^1 \sum_{n=0}^{\infty}(xyz)^n\,dx\,dy\,dz = \sum_{n=0}^{\infty}\int_0^1\int_0^1\int_0^1 (xyz)^n\,dx\,dy\,dz$$

$$= \sum_{n=0}^{\infty}\left[\int_0^1 x^n\,dx\right]\left[\int_0^1 y^n\,dy\right]\left[\int_0^1 z^n\,dz\right] = \sum_{n=0}^{\infty}\frac{1}{n+1}\cdot\frac{1}{n+1}\cdot\frac{1}{n+1}$$

$$= \sum_{n=0}^{\infty}\frac{1}{(n+1)^3} = \frac{1}{1^3}+\frac{1}{2^3}+\frac{1}{3^3}+\cdots = \sum_{n=1}^{\infty}\frac{1}{n^3}$$

(b) Since $|-xyz| < 1$, except at $(1,1,1)$, the formula for the sum of a geometric series gives $\dfrac{1}{1+xyz} = \displaystyle\sum_{n=0}^{\infty}(-xyz)^n$, so

$$\int_0^1\int_0^1\int_0^1 \frac{1}{1+xyz}\,dx\,dy\,dz = \int_0^1\int_0^1\int_0^1 \sum_{n=0}^{\infty}(-xyz)^n\,dx\,dy\,dz = \sum_{n=0}^{\infty}\int_0^1\int_0^1\int_0^1 (-xyz)^n\,dx\,dy\,dz$$

$$= \sum_{n=0}^{\infty}(-1)^n\left[\int_0^1 x^n\,dx\right]\left[\int_0^1 y^n\,dy\right]\left[\int_0^1 z^n\,dz\right] = \sum_{n=0}^{\infty}(-1)^n\frac{1}{n+1}\cdot\frac{1}{n+1}\cdot\frac{1}{n+1}$$

$$= \sum_{n=0}^{\infty}\frac{(-1)^n}{(n+1)^3} = \frac{1}{1^3}-\frac{1}{2^3}+\frac{1}{3^3}-\cdots = \sum_{n=0}^{\infty}\frac{(-1)^{n-1}}{n^3}$$

[continued]

© 2012 Cengage Learning. All Rights Reserved. May not be scanned, copied, or duplicated, or posted to a publicly accessible website, in whole or in part.

To evaluate this sum, we first write out a few terms: $s = 1 - \dfrac{1}{2^3} + \dfrac{1}{3^3} - \dfrac{1}{4^3} + \dfrac{1}{5^3} - \dfrac{1}{6^3} \approx 0.8998$. Notice that

$a_7 = \dfrac{1}{7^3} < 0.003$. By the Alternating Series Estimation Theorem from Section 11.5, we have $|s - s_6| \le a_7 < 0.003$.

This error of 0.003 will not affect the second decimal place, so we have $s \approx 0.90$.

9. (a) $x = r\cos\theta$, $y = r\sin\theta$, $z = z$. Then $\dfrac{\partial u}{\partial r} = \dfrac{\partial u}{\partial x}\dfrac{\partial x}{\partial r} + \dfrac{\partial u}{\partial y}\dfrac{\partial y}{\partial r} + \dfrac{\partial u}{\partial z}\dfrac{\partial z}{\partial r} = \dfrac{\partial u}{\partial x}\cos\theta + \dfrac{\partial u}{\partial y}\sin\theta$ and

$$\frac{\partial^2 u}{\partial r^2} = \cos\theta\left[\frac{\partial^2 u}{\partial x^2}\frac{\partial x}{\partial r} + \frac{\partial^2 u}{\partial y\,\partial x}\frac{\partial y}{\partial r} + \frac{\partial^2 u}{\partial z\,\partial x}\frac{\partial z}{\partial r}\right] + \sin\theta\left[\frac{\partial^2 u}{\partial y^2}\frac{\partial y}{\partial r} + \frac{\partial^2 u}{\partial x\,\partial y}\frac{\partial x}{\partial r} + \frac{\partial^2 u}{\partial z\,\partial y}\frac{\partial z}{\partial r}\right]$$

$$= \frac{\partial^2 u}{\partial x^2}\cos^2\theta + \frac{\partial^2 u}{\partial y^2}\sin^2\theta + 2\frac{\partial^2 u}{\partial y\,\partial x}\cos\theta\sin\theta$$

Similarly $\dfrac{\partial u}{\partial \theta} = -\dfrac{\partial u}{\partial x}r\sin\theta + \dfrac{\partial u}{\partial y}r\cos\theta$ and

$$\frac{\partial^2 u}{\partial \theta^2} = \frac{\partial^2 u}{\partial x^2}r^2\sin^2\theta + \frac{\partial^2 u}{\partial y^2}r^2\cos^2\theta - 2\frac{\partial^2 u}{\partial y\,\partial x}r^2\sin\theta\cos\theta - \frac{\partial u}{\partial x}r\cos\theta - \frac{\partial u}{\partial y}r\sin\theta. \text{ So}$$

$$\frac{\partial^2 u}{\partial r^2} + \frac{1}{r}\frac{\partial u}{\partial r} + \frac{1}{r^2}\frac{\partial^2 u}{\partial \theta^2} + \frac{\partial^2 u}{\partial z^2} = \frac{\partial^2 u}{\partial x^2}\cos^2\theta + \frac{\partial^2 u}{\partial y^2}\sin^2\theta + 2\frac{\partial^2 u}{\partial y\,\partial x}\cos\theta\sin\theta + \frac{\partial u}{\partial x}\frac{\cos\theta}{r} + \frac{\partial u}{\partial y}\frac{\sin\theta}{r}$$

$$+ \frac{\partial^2 u}{\partial x^2}\sin^2\theta + \frac{\partial^2 u}{\partial y^2}\cos^2\theta - 2\frac{\partial^2 u}{\partial y\,\partial x}\sin\theta\cos\theta$$

$$- \frac{\partial u}{\partial x}\frac{\cos\theta}{r} - \frac{\partial u}{\partial y}\frac{\sin\theta}{r} + \frac{\partial^2 u}{\partial z^2}$$

$$= \frac{\partial^2 u}{\partial x^2} + \frac{\partial^2 u}{\partial y^2} + \frac{\partial^2 u}{\partial z^2}$$

(b) $x = \rho\sin\phi\cos\theta$, $y = \rho\sin\phi\sin\theta$, $z = \rho\cos\phi$. Then

$$\frac{\partial u}{\partial \rho} = \frac{\partial u}{\partial x}\frac{\partial x}{\partial \rho} + \frac{\partial u}{\partial y}\frac{\partial y}{\partial \rho} + \frac{\partial u}{\partial z}\frac{\partial z}{\partial \rho} = \frac{\partial u}{\partial x}\sin\phi\cos\theta + \frac{\partial u}{\partial y}\sin\phi\sin\theta + \frac{\partial u}{\partial z}\cos\phi, \text{ and}$$

$$\frac{\partial^2 u}{\partial \rho^2} = \sin\phi\cos\theta\left[\frac{\partial^2 u}{\partial x^2}\frac{\partial x}{\partial \rho} + \frac{\partial^2 u}{\partial y\,\partial x}\frac{\partial y}{\partial \rho} + \frac{\partial^2 u}{\partial z\,\partial x}\frac{\partial z}{\partial \rho}\right]$$

$$+ \sin\phi\sin\theta\left[\frac{\partial^2 u}{\partial y^2}\frac{\partial y}{\partial \rho} + \frac{\partial^2 u}{\partial x\,\partial y}\frac{\partial x}{\partial \rho} + \frac{\partial^2 u}{\partial z\,\partial y}\frac{\partial z}{\partial \rho}\right]$$

$$+ \cos\phi\left[\frac{\partial^2 u}{\partial z^2}\frac{\partial z}{\partial \rho} + \frac{\partial^2 u}{\partial x\,\partial z}\frac{\partial x}{\partial \rho} + \frac{\partial^2 u}{\partial y\,\partial z}\frac{\partial y}{\partial \rho}\right]$$

$$= 2\frac{\partial^2 u}{\partial y\,\partial x}\sin^2\phi\sin\theta\cos\theta + 2\frac{\partial^2 u}{\partial z\,\partial x}\sin\phi\cos\phi\cos\theta + 2\frac{\partial^2 u}{\partial y\,\partial z}\sin\phi\cos\phi\sin\theta$$

$$+ \frac{\partial^2 u}{\partial x^2}\sin^2\phi\cos^2\theta + \frac{\partial^2 u}{\partial y^2}\sin^2\phi\sin^2\theta + \frac{\partial^2 u}{\partial z^2}\cos^2\phi$$

Similarly $\dfrac{\partial u}{\partial \phi} = \dfrac{\partial u}{\partial x}\rho\cos\phi\cos\theta + \dfrac{\partial u}{\partial y}\rho\cos\phi\sin\theta - \dfrac{\partial u}{\partial z}\rho\sin\phi$, and

© 2012 Cengage Learning. All Rights Reserved. May not be scanned, copied, or duplicated, or posted to a publicly accessible website, in whole or in part.

$$\frac{\partial^2 u}{\partial \phi^2} = 2 \frac{\partial^2 u}{\partial y\, \partial x} \rho^2 \cos^2 \phi \sin \theta \cos \theta - 2 \frac{\partial^2 u}{\partial x\, \partial z} \rho^2 \sin \phi \cos \phi \cos \theta$$

$$- 2 \frac{\partial^2 u}{\partial y\, \partial z} \rho^2 \sin \phi \cos \phi \sin \theta + \frac{\partial^2 u}{\partial x^2} \rho^2 \cos^2 \phi \cos^2 \theta + \frac{\partial^2 u}{\partial y^2} \rho^2 \cos^2 \phi \sin^2 \theta$$

$$+ \frac{\partial^2 u}{\partial z^2} \rho^2 \sin^2 \phi - \frac{\partial u}{\partial x} \rho \sin \phi \cos \theta - \frac{\partial u}{\partial y} \rho \sin \phi \sin \theta - \frac{\partial u}{\partial z} \rho \cos \phi$$

And $\dfrac{\partial u}{\partial \theta} = -\dfrac{\partial u}{\partial x} \rho \sin \phi \sin \theta + \dfrac{\partial u}{\partial y} \rho \sin \phi \cos \theta$, while

$$\frac{\partial^2 u}{\partial \theta^2} = -2 \frac{\partial^2 u}{\partial y\, \partial x} \rho^2 \sin^2 \phi \cos \theta \sin \theta + \frac{\partial^2 u}{\partial x^2} \rho^2 \sin^2 \phi \sin^2 \theta$$

$$+ \frac{\partial^2 u}{\partial y^2} \rho^2 \sin^2 \phi \cos^2 \theta - \frac{\partial u}{\partial x} \rho \sin \phi \cos \theta - \frac{\partial u}{\partial y} \rho \sin \phi \sin \theta$$

Therefore

$$\frac{\partial^2 u}{\partial \rho^2} + \frac{2}{\rho} \frac{\partial u}{\partial \rho} + \frac{\cot \phi}{\rho^2} \frac{\partial u}{\partial \phi} + \frac{1}{\rho^2} \frac{\partial^2 u}{\partial \phi^2} + \frac{1}{\rho^2 \sin^2 \phi} \frac{\partial^2 u}{\partial \theta^2}$$

$$= \frac{\partial^2 u}{\partial x^2} \left[(\sin^2 \phi \cos^2 \theta) + (\cos^2 \phi \cos^2 \theta) + \sin^2 \theta \right]$$

$$+ \frac{\partial^2 u}{\partial y^2} \left[(\sin^2 \phi \sin^2 \theta) + (\cos^2 \phi \sin^2 \theta) + \cos^2 \theta \right] + \frac{\partial^2 u}{\partial z^2} \left[\cos^2 \phi + \sin^2 \phi \right]$$

$$+ \frac{\partial u}{\partial x} \left[\frac{2 \sin^2 \phi \cos \theta + \cos^2 \phi \cos \theta - \sin^2 \phi \cos \theta - \cos \theta}{\rho \sin \phi} \right]$$

$$+ \frac{\partial u}{\partial y} \left[\frac{2 \sin^2 \phi \sin \theta + \cos^2 \phi \sin \theta - \sin^2 \phi \sin \theta - \sin \theta}{\rho \sin \phi} \right]$$

But $2 \sin^2 \phi \cos \theta + \cos^2 \phi \cos \theta - \sin^2 \phi \cos \theta - \cos \theta = (\sin^2 \phi + \cos^2 \phi - 1) \cos \theta = 0$ and similarly the coefficient of $\partial u/\partial y$ is 0. Also $\sin^2 \phi \cos^2 \theta + \cos^2 \phi \cos^2 \theta + \sin^2 \theta = \cos^2 \theta (\sin^2 \phi + \cos^2 \phi) + \sin^2 \theta = 1$, and similarly the coefficient of $\partial^2 u/\partial y^2$ is 1. So Laplace's Equation in spherical coordinates is as stated.

11. $\int_0^x \int_0^y \int_0^z f(t)\, dt\, dz\, dy = \iiint_E f(t)\, dV$, where

$E = \{(t, z, y) \mid 0 \le t \le z,\, 0 \le z \le y,\, 0 \le y \le x\}$.

If we let D be the projection of E on the yt-plane then

$D = \{(y, t) \mid 0 \le t \le x,\, t \le y \le x\}$. And we see from the diagram

that $E = \{(t, z, y) \mid t \le z \le y,\, t \le y \le x,\, 0 \le t \le x\}$. So

$$\int_0^x \int_0^y \int_0^z f(t)\, dt\, dz\, dy = \int_0^x \int_t^x \int_t^y f(t)\, dz\, dy\, dt = \int_0^x \left[\int_t^x (y - t) f(t)\, dy \right] dt$$

$$= \int_0^x \left[\left(\tfrac{1}{2} y^2 - ty \right) f(t) \right]_{y=t}^{y=x} dt = \int_0^x \left[\tfrac{1}{2} x^2 - tx - \tfrac{1}{2} t^2 + t^2 \right] f(t)\, dt$$

$$= \int_0^x \left[\tfrac{1}{2} x^2 - tx + \tfrac{1}{2} t^2 \right] f(t)\, dt = \int_0^x \left(\tfrac{1}{2} x^2 - 2tx + t^2 \right) f(t)\, dt$$

$$= \tfrac{1}{2} \int_0^x (x - t)^2 f(t)\, dt$$

© 2012 Cengage Learning. All Rights Reserved. May not be scanned, copied, or duplicated, or posted to a publicly accessible website, in whole or in part.

13. The volume is $V = \iiint_R dV$ where R is the solid region given. From Exercise 15.10.21(a), the transformation $x = au$,

$y = bv$, $z = cw$ maps the unit ball $u^2 + v^2 + w^2 \leq 1$ to the solid ellipsoid

$\dfrac{x^2}{a^2} + \dfrac{y^2}{b^2} + \dfrac{z^2}{c^2} \leq 1$ with $\dfrac{\partial(x, y, z)}{\partial(u, v, w)} = abc$. The same transformation maps the

plane $u + v + w = 1$ to $\dfrac{x}{a} + \dfrac{y}{b} + \dfrac{z}{c} = 1$. Thus the region R in xyz-space

corresponds to the region S in uvw-space consisting of the smaller piece of the

unit ball cut off by the plane $u + v + w = 1$, a "cap of a sphere" (see the figure).

We will need to compute the volume of S, but first consider the general case

where a horizontal plane slices the upper portion of a sphere of radius r to produce

a cap of height h. We use spherical coordinates. From the figure, a line through the

origin at angle ϕ from the z-axis intersects the plane when $\cos \phi = (r - h)/a \;\Rightarrow$

$a = (r - h)/\cos \phi$, and the line passes through the outer rim of the cap when

$a = r \;\Rightarrow\; \cos \phi = (r - h)/r \;\Rightarrow\; \phi = \cos^{-1}((r - h)/r)$. Thus the cap

is described by $\left\{ (\rho, \theta, \phi) \mid (r - h)/\cos \phi \leq \rho \leq r, 0 \leq \theta \leq 2\pi, 0 \leq \phi \leq \cos^{-1}((r - h)/r) \right\}$ and its volume is

$$V = \int_0^{2\pi} \int_0^{\cos^{-1}((r-h)/r)} \int_{(r-h)/\cos \phi}^{r} \rho^2 \sin \phi \, d\rho \, d\phi \, d\theta$$

$$= \int_0^{2\pi} \int_0^{\cos^{-1}((r-h)/r)} \left[\tfrac{1}{3}\rho^3 \sin \phi \right]_{\rho=(r-h)/\cos \phi}^{\rho=r} d\phi \, d\theta$$

$$= \frac{1}{3} \int_0^{2\pi} \int_0^{\cos^{-1}((r-h)/r)} \left[r^3 \sin \phi - \frac{(r-h)^3}{\cos^3 \phi} \sin \phi \right] d\phi \, d\theta$$

$$= \frac{1}{3} \int_0^{2\pi} \left[-r^3 \cos \phi - \tfrac{1}{2}(r - h)^3 \cos^{-2} \phi \right]_{\phi=0}^{\phi=\cos^{-1}((r-h)/r)} d\theta$$

$$= \frac{1}{3} \int_0^{2\pi} \left[-r^3 \left(\frac{r-h}{r} \right) - \frac{1}{2}(r - h)^3 \left(\frac{r-h}{r} \right)^{-2} + r^3 + \frac{1}{2}(r - h)^3 \right] d\theta$$

$$= \tfrac{1}{3} \int_0^{2\pi} (\tfrac{3}{2}rh^2 - \tfrac{1}{2}h^3) \, d\theta = \tfrac{1}{3}(\tfrac{3}{2}rh^2 - \tfrac{1}{2}h^3)(2\pi) = \pi h^2(r - \tfrac{1}{3}h)$$

(This volume can also be computed by treating the cap as a solid of revolution and using the single variable disk method;
see Exercise 5.2.49 [ET 6.2.49].)

To determine the height h of the cap cut from the unit ball by the plane

$u + v + w = 1$, note that the line $u = v = w$ passes through the origin with

direction vector $\langle 1, 1, 1 \rangle$ which is perpendicular to the plane. Therefore this line

coincides with a radius of the sphere that passes through the center of the cap and

h is measured along this line. The line intersects the plane at $\left(\tfrac{1}{3}, \tfrac{1}{3}, \tfrac{1}{3} \right)$ and the

sphere at $\left(\tfrac{1}{\sqrt{3}}, \tfrac{1}{\sqrt{3}}, \tfrac{1}{\sqrt{3}} \right)$. (See the figure.)

© 2012 Cengage Learning. All Rights Reserved. May not be scanned, copied, or duplicated, or posted to a publicly accessible website, in whole or in part.

The distance between these points is $h = \sqrt{3\left(\frac{1}{\sqrt{3}} - \frac{1}{3}\right)^2} = \sqrt{3}\left(\frac{1}{\sqrt{3}} - \frac{1}{3}\right) = 1 - \frac{1}{\sqrt{3}}$. Thus the volume of R is

$$V = \iiint_R dV = \iiint_S \left|\frac{\partial(x, y, z)}{\partial(u, v, w)}\right| dV = abc \iiint_S dV = abc\, V(S)$$

$$= abc \cdot \pi h^2\left(r - \tfrac{1}{3}h\right) = abc \cdot \pi\left(1 - \tfrac{1}{\sqrt{3}}\right)^2\left[1 - \tfrac{1}{3}\left(1 - \tfrac{1}{\sqrt{3}}\right)\right]$$

$$= abc\pi\left(\tfrac{4}{3} - \tfrac{2}{\sqrt{3}}\right)\left(\tfrac{2}{3} + \tfrac{1}{3\sqrt{3}}\right) = abc\pi\left(\tfrac{2}{3} - \tfrac{8}{9\sqrt{3}}\right) \approx 0.482abc$$

© 2012 Cengage Learning. All Rights Reserved. May not be scanned, copied, or duplicated, or posted to a publicly accessible website, in whole or in part.

16 □ VECTOR CALCULUS

16.1 Vector Fields

1. $\mathbf{F}(x, y) = 0.3\,\mathbf{i} - 0.4\,\mathbf{j}$

All vectors in this field are identical, with length 0.5 and

parallel to $\langle 3, -4 \rangle$.

3. $\mathbf{F}(x, y) = -\frac{1}{2}\,\mathbf{i} + (y - x)\,\mathbf{j}$

The length of the vector $-\frac{1}{2}\,\mathbf{i} + (y - x)\,\mathbf{j}$ is

$\sqrt{\frac{1}{4} + (y - x)^2}$. Vectors along the line $y = x$ are

horizontal with length $\frac{1}{2}$.

5. $\mathbf{F}(x, y) = \dfrac{y\,\mathbf{i} + x\,\mathbf{j}}{\sqrt{x^2 + y^2}}$

The length of the vector $\dfrac{y\,\mathbf{i} + x\,\mathbf{j}}{\sqrt{x^2 + y^2}}$ is 1.

7. $\mathbf{F}(x, y, z) = \mathbf{k}$

All vectors in this field are parallel to the z-axis and have

length 1.

9. $\mathbf{F}(x, y, z) = x\,\mathbf{k}$

At each point (x, y, z), $\mathbf{F}(x, y, z)$ is a vector of length $|x|$.

For $x > 0$, all point in the direction of the positive z-axis,

while for $x < 0$, all are in the direction of the negative

z-axis. In each plane $x = k$, all the vectors are identical.

© 2012 Cengage Learning. All Rights Reserved. May not be scanned, copied, or duplicated, or posted to a publicly accessible website, in whole or in part.

11. $\mathbf{F}(x, y) = \langle x, -y \rangle$ corresponds to graph IV. In the first quadrant all the vectors have positive x-components and negative y-components, in the second quadrant all vectors have negative x- and y-components, in the third quadrant all vectors have negative x-components and positive y-components, and in the fourth quadrant all vectors have positive x- and y-components. In addition, the vectors get shorter as we approach the origin.

13. $\mathbf{F}(x, y) = \langle y, y + 2 \rangle$ corresponds to graph I. As in Exercise 12, all vectors in quadrants I and II have positive x-components while all vectors in quadrants III and IV have negative x-components. Vectors along the line $y = -2$ are horizontal, and the vectors are independent of x (vectors along horizontal lines are identical).

15. $\mathbf{F}(x, y, z) = \mathbf{i} + 2\mathbf{j} + 3\mathbf{k}$ corresponds to graph IV, since all vectors have identical length and direction.

17. $\mathbf{F}(x, y, z) = x\mathbf{i} + y\mathbf{j} + 3\mathbf{k}$ corresponds to graph III; the projection of each vector onto the xy-plane is $x\mathbf{i} + y\mathbf{j}$, which points away from the origin, and the vectors point generally upward because their z-components are all 3.

19.

The vector field seems to have very short vectors near the line $y = 2x$.

For $\mathbf{F}(x, y) = \langle 0, 0 \rangle$ we must have $y^2 - 2xy = 0$ and $3xy - 6x^2 = 0$.

The first equation holds if $y = 0$ or $y = 2x$, and the second holds if $x = 0$ or $y = 2x$. So both equations hold [and thus $\mathbf{F}(x, y) = \mathbf{0}$] along the line $y = 2x$.

21. $f(x, y) = xe^{xy} \quad \Rightarrow$

$\nabla f(x, y) = f_x(x, y)\,\mathbf{i} + f_y(x, y)\,\mathbf{j} = (xe^{xy} \cdot y + e^{xy})\,\mathbf{i} + (xe^{xy} \cdot x)\,\mathbf{j} = (xy + 1)e^{xy}\,\mathbf{i} + x^2 e^{xy}\,\mathbf{j}$

23. $\nabla f(x, y, z) = f_x(x, y, z)\,\mathbf{i} + f_y(x, y, z)\,\mathbf{j} + f_z(x, y, z)\,\mathbf{k} = \dfrac{x}{\sqrt{x^2 + y^2 + z^2}}\,\mathbf{i} + \dfrac{y}{\sqrt{x^2 + y^2 + z^2}}\,\mathbf{j} + \dfrac{z}{\sqrt{x^2 + y^2 + z^2}}\,\mathbf{k}$

25. $f(x, y) = x^2 - y \quad \Rightarrow \quad \nabla f(x, y) = 2x\,\mathbf{i} - \mathbf{j}$.

The length of $\nabla f(x, y)$ is $\sqrt{4x^2 + 1}$. When $x \neq 0$, the vectors point away from the y-axis in a slightly downward direction with length that increases as the distance from the y-axis increases.

27. We graph $\nabla f(x, y) = \dfrac{2x}{1 + x^2 + 2y^2}\,\mathbf{i} + \dfrac{4y}{1 + x^2 + 2y^2}\,\mathbf{j}$ along with a contour map of f.

The graph shows that the gradient vectors are perpendicular to the level curves. Also, the gradient vectors point in the direction in which f is increasing and are longer where the level curves are closer together.

© 2012 Cengage Learning. All Rights Reserved. May not be scanned, copied, or duplicated, or posted to a publicly accessible website, in whole or in part.

29. $f(x,y) = x^2 + y^2 \Rightarrow \nabla f(x,y) = 2x\,\mathbf{i} + 2y\,\mathbf{j}$. Thus, each vector $\nabla f(x,y)$ has the same direction and twice the length of the position vector of the point (x,y), so the vectors all point directly away from the origin and their lengths increase as we move away from the origin. Hence, ∇f is graph III.

31. $f(x,y) = (x+y)^2 \Rightarrow \nabla f(x,y) = 2(x+y)\,\mathbf{i} + 2(x+y)\,\mathbf{j}$. The x- and y-components of each vector are equal, so all vectors are parallel to the line $y = x$. The vectors are $\mathbf{0}$ along the line $y = -x$ and their length increases as the distance from this line increases. Thus, ∇f is graph II.

33. At $t = 3$ the particle is at $(2,1)$ so its velocity is $\mathbf{V}(2,1) = \langle 4,3\rangle$. After 0.01 units of time, the particle's change in location should be approximately $0.01\,\mathbf{V}(2,1) = 0.01\,\langle 4,3\rangle = \langle 0.04, 0.03\rangle$, so the particle should be approximately at the point $(2.04, 1.03)$.

35. (a) We sketch the vector field $\mathbf{F}(x,y) = x\,\mathbf{i} - y\,\mathbf{j}$ along with several approximate flow lines. The flow lines appear to be hyperbolas with shape similar to the graph of $y = \pm 1/x$, so we might guess that the flow lines have equations $y = C/x$.

(b) If $x = x(t)$ and $y = y(t)$ are parametric equations of a flow line, then the velocity vector of the flow line at the point (x,y) is $x'(t)\,\mathbf{i} + y'(t)\,\mathbf{j}$. Since the velocity vectors coincide with the vectors in the vector field, we have

$x'(t)\,\mathbf{i} + y'(t)\,\mathbf{j} = x\,\mathbf{i} - y\,\mathbf{j} \Rightarrow dx/dt = x, dy/dt = -y$. To solve these differential equations, we know

$dx/dt = x \Rightarrow dx/x = dt \Rightarrow \ln|x| = t + C \Rightarrow x = \pm e^{t+C} = Ae^t$ for some constant A, and

$dy/dt = -y \Rightarrow dy/y = -dt \Rightarrow \ln|y| = -t + K \Rightarrow y = \pm e^{-t+K} = Be^{-t}$ for some constant B. Therefore

$xy = Ae^t Be^{-t} = AB =$ constant. If the flow line passes through $(1,1)$ then $(1)(1) -$ constant $= 1 \Rightarrow xy = 1 \Rightarrow y = 1/x, x > 0$.

16.2 Line Integrals

1. $x = t^3$ and $y = t$, $0 \le t \le 2$, so by Formula 3

$$\int_C y^3\,ds = \int_0^2 t^3 \sqrt{\left(\frac{dx}{dt}\right)^2 + \left(\frac{dy}{dt}\right)^2}\,dt = \int_0^2 t^3 \sqrt{(3t^2)^2 + (1)^2}\,dt = \int_0^2 t^3 \sqrt{9t^4 + 1}\,dt$$

$$= \frac{1}{36} \cdot \frac{2}{3}\left(9t^4 + 1\right)^{3/2}\Big]_0^2 = \frac{1}{54}(145^{3/2} - 1) \text{ or } \frac{1}{54}\left(145\sqrt{145} - 1\right)$$

3. Parametric equations for C are $x = 4\cos t$, $y = 4\sin t$, $-\frac{\pi}{2} \le t \le \frac{\pi}{2}$. Then

$$\int_C xy^4\,ds = \int_{-\pi/2}^{\pi/2}(4\cos t)(4\sin t)^4 \sqrt{(-4\sin t)^2 + (4\cos t)^2}\,dt = \int_{-\pi/2}^{\pi/2} 4^5 \cos t \sin^4 t \sqrt{16(\sin^2 t + \cos^2 t)}\,dt$$

$$= 4^5 \int_{-\pi/2}^{\pi/2}(\sin^4 t \cos t)(4)\,dt = (4)^6\left[\tfrac{1}{5}\sin^5 t\right]_{-\pi/2}^{\pi/2} = \frac{2\cdot 4^6}{5} = 1638.4$$

© 2012 Cengage Learning. All Rights Reserved. May not be scanned, copied, or duplicated, or posted to a publicly accessible website, in whole or in part.

5. If we choose x as the parameter, parametric equations for C are $x = x$, $y = \sqrt{x}$ for $1 \le x \le 4$ and

$$\int_C \left(x^2 y^3 - \sqrt{x} \right) dy = \int_1^4 \left[x^2 \cdot (\sqrt{x})^3 - \sqrt{x} \right] \frac{1}{2\sqrt{x}} \, dx = \tfrac{1}{2} \int_1^4 \left(x^3 - 1 \right) dx$$

$$= \tfrac{1}{2} \left[\tfrac{1}{4} x^4 - x \right]_1^4 = \tfrac{1}{2} \left(64 - 4 - \tfrac{1}{4} + 1 \right) = \tfrac{243}{8}$$

7.

$C = C_1 + C_2$

On C_1: $x = x, y = \tfrac{1}{2}x$ $\Rightarrow$ $dy = \tfrac{1}{2} dx$, $0 \le x \le 2$.

On C_2: $x = x, y = 3 - x$ $\Rightarrow$ $dy = -dx$, $2 \le x \le 3$.

Then

$$\int_C (x + 2y) \, dx + x^2 \, dy = \int_{C_1} (x + 2y) \, dx + x^2 \, dy + \int_{C_2} (x + 2y) \, dx + x^2 \, dy$$

$$= \int_0^2 \left[x + 2 \left(\tfrac{1}{2} x \right) + x^2 \left(\tfrac{1}{2} \right) \right] dx + \int_2^3 \left[x + 2(3 - x) + x^2(-1) \right] dx$$

$$= \int_0^2 \left(2x + \tfrac{1}{2} x^2 \right) dx + \int_2^3 \left(6 - x - x^2 \right) dx$$

$$= \left[x^2 + \tfrac{1}{6} x^3 \right]_0^2 + \left[6x - \tfrac{1}{2} x^2 - \tfrac{1}{3} x^3 \right]_2^3 = \tfrac{16}{3} - 0 + \tfrac{9}{2} - \tfrac{22}{3} = \tfrac{5}{2}$$

9. $x = 2 \sin t$, $y = t$, $z = -2 \cos t$, $0 \le t \le \pi$. Then by Formula 9,

$$\int_C xyz \, ds = \int_0^\pi (2 \sin t)(t)(-2 \cos t) \sqrt{\left(\tfrac{dx}{dt} \right)^2 + \left(\tfrac{dy}{dt} \right)^2 + \left(\tfrac{dz}{dt} \right)^2} \, dt$$

$$= \int_0^\pi -4t \sin t \, \cos t \, \sqrt{(2 \cos t)^2 + (1)^2 + (2 \sin t)^2} \, dt = \int_0^\pi -2t \sin 2t \, \sqrt{4(\cos^2 t + \sin^2 t) + 1} \, dt$$

$$= -2\sqrt{5} \int_0^\pi t \sin 2t \, dt = -2\sqrt{5} \left[-\tfrac{1}{2} t \cos 2t + \tfrac{1}{4} \sin 2t \right]_0^\pi \qquad \begin{bmatrix} \text{integrate by parts with} \\ u = t, \, dv = \sin 2t \, dt \end{bmatrix}$$

$$= -2\sqrt{5} \left(-\tfrac{\pi}{2} - 0 \right) = \sqrt{5} \, \pi$$

11. Parametric equations for C are $x = t$, $y = 2t$, $z = 3t$, $0 \le t \le 1$. Then

$$\int_C x e^{yz} \, ds = \int_0^1 t e^{(2t)(3t)} \sqrt{1^2 + 2^2 + 3^2} \, dt = \sqrt{14} \int_0^1 t e^{6t^2} \, dt = \sqrt{14} \left[\tfrac{1}{12} e^{6t^2} \right]_0^1 = \tfrac{\sqrt{14}}{12} (e^6 - 1).$$

13. $\int_C xy e^{yz} \, dy = \int_0^1 (t)(t^2) e^{(t^2)(t^3)} \cdot 2t \, dt = \int_0^1 2t^4 e^{t^5} \, dt = \tfrac{2}{5} e^{t^5} \Big]_0^1 = \tfrac{2}{5} (e^1 - e^0) = \tfrac{2}{5}(e - 1)$

15. Parametric equations for C are $x = 1 + 3t$, $y = t$, $z = 2t$, $0 \le t \le 1$. Then

$$\int_C z^2 \, dx + x^2 \, dy + y^2 \, dz = \int_0^1 (2t)^2 \cdot 3 \, dt + (1 + 3t)^2 \, dt + t^2 \cdot 2 \, dt = \int_0^1 \left(23t^2 + 6t + 1 \right) dt$$

$$= \left[\tfrac{23}{3} t^3 + 3t^2 + t \right]_0^1 = \tfrac{23}{3} + 3 + 1 = \tfrac{35}{3}$$

17. (a) Along the line $x = -3$, the vectors of $\mathbf{F}$ have positive y-components, so since the path goes upward, the integrand $\mathbf{F} \cdot \mathbf{T}$ is

always positive. Therefore $\int_{C_1} \mathbf{F} \cdot d\mathbf{r} = \int_{C_1} \mathbf{F} \cdot \mathbf{T} \, ds$ is positive.

© 2012 Cengage Learning. All Rights Reserved. May not be scanned, copied, or duplicated, or posted to a publicly accessible website, in whole or in part.

(b) All of the (nonzero) field vectors along the circle with radius 3 are pointed in the clockwise direction, that is, opposite the direction to the path. So $\mathbf{F} \cdot \mathbf{T}$ is negative, and therefore $\int_{C_2} \mathbf{F} \cdot d\mathbf{r} = \int_{C_2} \mathbf{F} \cdot \mathbf{T} \, ds$ is negative.

19. $\mathbf{r}(t) = 11t^4 \, \mathbf{i} + t^3 \, \mathbf{j}$, so $\mathbf{F}(\mathbf{r}(t)) = (11t^4)(t^3) \, \mathbf{i} + 3(t^3)^2 \, \mathbf{j} = 11t^7 \, \mathbf{i} + 3t^6 \, \mathbf{j}$ and $\mathbf{r}'(t) = 44t^3 \, \mathbf{i} + 3t^2 \, \mathbf{j}$. Then

$$\int_C \mathbf{F} \cdot d\mathbf{r} = \int_0^1 \mathbf{F}(\mathbf{r}(t)) \cdot \mathbf{r}'(t) \, dt = \int_0^1 (11t^7 \cdot 44t^3 + 3t^6 \cdot 3t^2) \, dt = \int_0^1 (484t^{10} + 9t^8) \, dt = \left[44t^{11} + t^9 \right]_0^1 = 45.$$

21. $\int_C \mathbf{F} \cdot d\mathbf{r} = \int_0^1 \left\langle \sin t^3, \cos(-t^2), t^4 \right\rangle \cdot \left\langle 3t^2, -2t, 1 \right\rangle dt$

$$= \int_0^1 (3t^2 \sin t^3 - 2t \cos t^2 + t^4) \, dt = \left[-\cos t^3 - \sin t^2 + \tfrac{1}{5}t^5 \right]_0^1 = \tfrac{6}{5} - \cos 1 - \sin 1$$

23. $\mathbf{F}(\mathbf{r}(t)) = (e^t)\left(e^{-t^2} \right) \mathbf{i} + \sin\left(e^{-t^2} \right) \mathbf{j} = e^{t-t^2} \, \mathbf{i} + \sin\left(e^{-t^2} \right) \mathbf{j}$, $\mathbf{r}'(t) = e^t \, \mathbf{i} - 2te^{-t^2} \, \mathbf{j}$. Then

$$\int_C \mathbf{F} \cdot d\mathbf{r} = \int_1^2 \mathbf{F}(\mathbf{r}(t)) \cdot \mathbf{r}'(t) \, dt = \int_1^2 \left[e^{t-t^2} e^t + \sin\left(e^{-t^2} \right) \cdot \left(-2te^{-t^2} \right) \right] dt$$

$$= \int_1^2 \left[e^{2t-t^2} - 2te^{-t^2} \sin\left(e^{-t^2} \right) \right] dt \approx 1.9633$$

25. $x = t^2$, $y = t^3$, $z = t^4$ so by Formula 9,

$$\int_C x \sin(y+z) \, ds = \int_0^5 (t^2) \sin(t^3 + t^4) \sqrt{(2t)^2 + (3t^2)^2 + (4t^3)^2} \, dt$$

$$= \int_0^5 t^2 \sin(t^3 + t^4) \sqrt{4t^2 + 9t^4 + 16t^6} \, dt \approx 15.0074$$

27. We graph $\mathbf{F}(x, y) = (x - y) \, \mathbf{i} + xy \, \mathbf{j}$ and the curve C. We see that most of the vectors starting on C point in roughly the same direction as C, so for these portions of C the tangential component $\mathbf{F} \cdot \mathbf{T}$ is positive. Although some vectors in the third quadrant which start on C point in roughly the opposite direction, and hence give negative tangential components, it seems reasonable that the effect of these portions of C is outweighed by the positive tangential components. Thus, we would expect $\int_C \mathbf{F} \cdot d\mathbf{r} = \int_C \mathbf{F} \cdot \mathbf{T} \, ds$ to be positive.

To verify, we evaluate $\int_C \mathbf{F} \cdot d\mathbf{r}$. The curve C can be represented by $\mathbf{r}(t) = 2\cos t \, \mathbf{i} + 2\sin t \, \mathbf{j}$, $0 \le t \le \frac{3\pi}{2}$, so $\mathbf{F}(\mathbf{r}(t)) = (2\cos t - 2\sin t) \, \mathbf{i} + 4\cos t \sin t \, \mathbf{j}$ and $\mathbf{r}'(t) = -2\sin t \, \mathbf{i} + 2\cos t \, \mathbf{j}$. Then

$$\int_C \mathbf{F} \cdot d\mathbf{r} = \int_0^{3\pi/2} \mathbf{F}(\mathbf{r}(t)) \cdot \mathbf{r}'(t) \, dt$$

$$= \int_0^{3\pi/2} [-2\sin t(2\cos t - 2\sin t) + 2\cos t(4\cos t \sin t)] \, dt$$

$$= 4 \int_0^{3\pi/2} (\sin^2 t - \sin t \cos t + 2\sin t \cos^2 t) \, dt$$

$$= 3\pi + \tfrac{2}{3} \qquad \text{[using a CAS]}$$

29. (a) $\int_C \mathbf{F} \cdot d\mathbf{r} = \int_0^1 \left\langle e^{t^2-1}, t^5 \right\rangle \cdot \left\langle 2t, 3t^2 \right\rangle dt = \int_0^1 \left(2te^{t^2-1} + 3t^7 \right) dt = \left[e^{t^2-1} + \tfrac{3}{8}t^8 \right]_0^1 = \tfrac{11}{8} - 1/e$

© 2012 Cengage Learning. All Rights Reserved. May not be scanned, copied, or duplicated, or posted to a publicly accessible website, in whole or in part.

(b) $\mathbf{r}(0) = \mathbf{0}$, $\mathbf{F}(\mathbf{r}(0)) = \langle e^{-1}, 0 \rangle$;

$\mathbf{r}\left(\frac{1}{\sqrt{2}}\right) = \left\langle \frac{1}{2}, \frac{1}{2\sqrt{2}} \right\rangle$, $\mathbf{F}\left(\mathbf{r}\left(\frac{1}{\sqrt{2}}\right)\right) = \left\langle e^{-1/2}, \frac{1}{4\sqrt{2}} \right\rangle$;

$\mathbf{r}(1) = \langle 1, 1 \rangle$, $\mathbf{F}(\mathbf{r}(1)) = \langle 1, 1 \rangle$.

In order to generate the graph with Maple, we use the `line` command in

the `plottools` package to define each of the vectors. For example,

$$\texttt{v1:=line([0,0],[exp(-1),0]):}$$

generates the vector from the vector field at the point $(0, 0)$ (but without an arrowhead) and gives it the name `v1`. To show

everything on the same screen, we use the `display` command. In Mathematica, we use `ListPlot` (with the

`PlotJoined -> True` option) to generate the vectors, and then `Show` to show everything on the same screen.

31. $x = e^{-t}\cos 4t$, $y = e^{-t}\sin 4t$, $z = e^{-t}$, $0 \le t \le 2\pi$.

Then $\dfrac{dx}{dt} = e^{-t}(-\sin 4t)(4) - e^{-t}\cos 4t = -e^{-t}(4\sin 4t + \cos 4t)$,

$\dfrac{dy}{dt} = e^{-t}(\cos 4t)(4) - e^{-t}\sin 4t = -e^{-t}(-4\cos 4t + \sin 4t)$, and $\dfrac{dz}{dt} = -e^{-t}$, so

$$\sqrt{\left(\frac{dx}{dt}\right)^2 + \left(\frac{dy}{dt}\right)^2 + \left(\frac{dz}{dt}\right)^2} = \sqrt{(-e^{-t})^2[(4\sin 4t + \cos 4t)^2 + (-4\cos 4t + \sin 4t)^2 + 1]}$$

$$= e^{-t}\sqrt{16(\sin^2 4t + \cos^2 4t) + \sin^2 4t + \cos^2 4t + 1} = 3\sqrt{2}\,e^{-t}$$

Therefore $\displaystyle\int_C x^3 y^2 z \, ds = \int_0^{2\pi} (e^{-t}\cos 4t)^3 (e^{-t}\sin 4t)^2 (e^{-t})\,(3\sqrt{2}\,e^{-t})\,dt$

$$= \int_0^{2\pi} 3\sqrt{2}\,e^{-7t}\cos^3 4t \sin^2 4t \, dt = \frac{172,704}{5,632,705}\sqrt{2}\left(1 - e^{-14\pi}\right)$$

33. We use the parametrization $x = 2\cos t$, $y = 2\sin t$, $-\frac{\pi}{2} \le t \le \frac{\pi}{2}$. Then

$ds = \sqrt{\left(\frac{dx}{dt}\right)^2 + \left(\frac{dy}{dt}\right)^2}\,dt = \sqrt{(-2\sin t)^2 + (2\cos t)^2}\,dt = 2\,dt$, so $m = \int_C k\,ds = 2k\int_{-\pi/2}^{\pi/2} dt = 2k(\pi)$,

$\overline{x} = \frac{1}{2\pi k}\int_C xk\,ds = \frac{1}{2\pi}\int_{-\pi/2}^{\pi/2}(2\cos t)2\,dt = \frac{1}{2\pi}\big[4\sin t\big]_{-\pi/2}^{\pi/2} = \frac{4}{\pi}$, $\overline{y} = \frac{1}{2\pi k}\int_C yk\,ds = \frac{1}{2\pi}\int_{-\pi/2}^{\pi/2}(2\sin t)2\,dt = 0$.

Hence $(\overline{x}, \overline{y}) = \left(\frac{4}{\pi}, 0\right)$.

35. (a) $\overline{x} = \dfrac{1}{m}\displaystyle\int_C x\rho(x, y, z)\,ds$, $\overline{y} = \dfrac{1}{m}\displaystyle\int_C y\rho(x, y, z)\,ds$, $\overline{z} = \dfrac{1}{m}\displaystyle\int_C z\rho(x, y, z)\,ds$ where $m = \int_C \rho(x, y, z)\,ds$.

(b) $m = \int_C k\,ds = k\int_0^{2\pi}\sqrt{4\sin^2 t + 4\cos^2 t + 9}\,dt = k\sqrt{13}\int_0^{2\pi} dt = 2\pi k\sqrt{13}$,

$$\overline{x} = \frac{1}{2\pi k\sqrt{13}}\int_0^{2\pi} 2k\sqrt{13}\sin t\,dt = 0, \qquad \overline{y} = \frac{1}{2\pi k\sqrt{13}}\int_0^{2\pi} 2k\sqrt{13}\cos t\,dt = 0,$$

$$\overline{z} = \frac{1}{2\pi k\sqrt{13}}\int_0^{2\pi}\left(k\sqrt{13}\right)(3t)\,dt = \frac{3}{2\pi}\left(2\pi^2\right) = 3\pi. \text{ Hence } (\overline{x}, \overline{y}, \overline{z}) = (0, 0, 3\pi).$$

© 2012 Cengage Learning. All Rights Reserved. May not be scanned, copied, or duplicated, or posted to a publicly accessible website, in whole or in part.

37. From Example 3, $\rho(x, y) = k(1 - y)$, $x = \cos t$, $y = \sin t$, and $ds = dt$, $0 \le t \le \pi$ $\Rightarrow$

$$I_x = \int_C y^2 \rho(x, y)\, ds = \int_0^\pi \sin^2 t\, [k(1 - \sin t)]\, dt = k \int_0^\pi (\sin^2 t - \sin^3 t)\, dt$$

$$= \tfrac{1}{2} k \int_0^\pi (1 - \cos 2t)\, dt - k \int_0^\pi (1 - \cos^2 t) \sin t\, dt \qquad \begin{bmatrix} \text{Let } u = \cos t,\, du = -\sin t\, dt \\ \text{in the second integral} \end{bmatrix}$$

$$= k\left[\tfrac{\pi}{2} + \int_1^{-1} (1 - u^2)\, du \right] = k\left(\tfrac{\pi}{2} - \tfrac{4}{3} \right)$$

$$I_y = \int_C x^2 \rho(x, y)\, ds = k \int_0^\pi \cos^2 t\, (1 - \sin t)\, dt = \tfrac{k}{2} \int_0^\pi (1 + \cos 2t)\, dt - k \int_0^\pi \cos^2 t \sin t\, dt$$

$$= k\left(\tfrac{\pi}{2} - \tfrac{2}{3} \right), \text{ using the same substitution as above.}$$

39. $W = \int_C \mathbf{F} \cdot d\mathbf{r} = \int_0^{2\pi} \langle t - \sin t, 3 - \cos t \rangle \cdot \langle 1 - \cos t, \sin t \rangle\, dt$

$$= \int_0^{2\pi} (t - t \cos t - \sin t + \sin t \cos t + 3 \sin t - \sin t \cos t)\, dt$$

$$= \int_0^{2\pi} (t - t \cos t + 2 \sin t)\, dt = \left[\tfrac{1}{2} t^2 - (t \sin t + \cos t) - 2 \cos t \right]_0^{2\pi} \qquad \begin{bmatrix} \text{integrate by parts} \\ \text{in the second term} \end{bmatrix}$$

$$= 2\pi^2$$

41. $\mathbf{r}(t) = \langle 2t, t, 1 - t \rangle$, $0 \le t \le 1$.

$$W = \int_C \mathbf{F} \cdot d\mathbf{r} = \int_0^1 \langle 2t - t^2, t - (1 - t)^2, 1 - t - (2t)^2 \rangle \cdot \langle 2, 1, -1 \rangle\, dt$$

$$= \int_0^1 (4t - 2t^2 + t - 1 + 2t - t^2 - 1 + t + 4t^2)\, dt = \int_0^1 (t^2 + 8t - 2)\, dt = \left[\tfrac{1}{3} t^3 + 4t^2 - 2t \right]_0^1 = \tfrac{7}{3}$$

43. (a) $\mathbf{r}(t) = at^2\, \mathbf{i} + bt^3\, \mathbf{j}$ $\Rightarrow$ $\mathbf{v}(t) = \mathbf{r}'(t) = 2at\, \mathbf{i} + 3bt^2\, \mathbf{j}$ $\Rightarrow$ $\mathbf{a}(t) = \mathbf{v}'(t) = 2a\, \mathbf{i} + 6bt\, \mathbf{j}$, and force is mass times acceleration: $\mathbf{F}(t) = m\, \mathbf{a}(t) = 2ma\, \mathbf{i} + 6mbt\, \mathbf{j}$.

(b) $W = \int_C \mathbf{F} \cdot d\mathbf{r} = \int_0^1 (2ma\, \mathbf{i} + 6mbt\, \mathbf{j}) \cdot (2at\, \mathbf{i} + 3bt^2\, \mathbf{j})\, dt = \int_0^1 (4ma^2 t + 18mb^2 t^3)\, dt$

$$= \left[2ma^2 t^2 + \tfrac{9}{2} mb^2 t^4 \right]_0^1 = 2ma^2 + \tfrac{9}{2} mb^2$$

45. Let $\mathbf{F} = 185\, \mathbf{k}$. To parametrize the staircase, let $x = 20 \cos t$, $y = 20 \sin t$, $z = \tfrac{90}{6\pi} t = \tfrac{15}{\pi} t$, $0 \le t \le 6\pi$ $\Rightarrow$

$$W = \int_C \mathbf{F} \cdot d\mathbf{r} = \int_0^{6\pi} \langle 0, 0, 185 \rangle \cdot \langle -20 \sin t, 20 \cos t, \tfrac{15}{\pi} \rangle\, dt = (185) \tfrac{15}{\pi} \int_0^{6\pi} dt = (185)(90) \approx 1.67 \times 10^4 \text{ ft-lb}$$

47. (a) $\mathbf{r}(t) = \langle \cos t, \sin t \rangle$, $0 \le t \le 2\pi$, and let $\mathbf{F} = \langle a, b \rangle$. Then

$$W = \int_C \mathbf{F} \cdot d\mathbf{r} = \int_0^{2\pi} \langle a, b \rangle \cdot \langle -\sin t, \cos t \rangle\, dt = \int_0^{2\pi} (-a \sin t + b \cos t)\, dt = \left[a \cos t + b \sin t \right]_0^{2\pi}$$

$$= a + 0 - a + 0 = 0$$

(b) Yes. $\mathbf{F}(x, y) = k\, \mathbf{x} = \langle kx, ky \rangle$ and

$$W = \int_C \mathbf{F} \cdot d\mathbf{r} = \int_0^{2\pi} \langle k \cos t, k \sin t \rangle \cdot \langle -\sin t, \cos t \rangle\, dt = \int_0^{2\pi} (-k \sin t \cos t + k \sin t \cos t)\, dt = \int_0^{2\pi} 0\, dt = 0.$$

49. Let $\mathbf{r}(t) = \langle x(t), y(t), z(t) \rangle$ and $\mathbf{v} = \langle v_1, v_2, v_3 \rangle$. Then

$$\int_C \mathbf{v} \cdot d\mathbf{r} = \int_a^b \langle v_1, v_2, v_3 \rangle \cdot \langle x'(t), y'(t), z'(t) \rangle\, dt = \int_a^b [v_1\, x'(t) + v_2\, y'(t) + v_3\, z'(t)]\, dt$$

$$= \left[v_1\, x(t) + v_2\, y(t) + v_3\, z(t) \right]_a^b = [v_1\, x(b) + v_2\, y(b) + v_3\, z(b)] - [v_1\, x(a) + v_2\, y(a) + v_3\, z(a)]$$

$$= v_1\, [x(b) - x(a)] + v_2\, [y(b) - y(a)] + v_3\, [z(b) - z(a)]$$

$$= \langle v_1, v_2, v_3 \rangle \cdot \langle x(b) - x(a), y(b) - y(a), z(b) - z(a) \rangle$$

$$= \langle v_1, v_2, v_3 \rangle \cdot [\langle x(b), y(b), z(b) \rangle - \langle x(a), y(a), z(a) \rangle] = \mathbf{v} \cdot [\mathbf{r}(b) - \mathbf{r}(a)]$$

© 2012 Cengage Learning. All Rights Reserved. May not be scanned, copied, or duplicated, or posted to a publicly accessible website, in whole or in part.

51. The work done in moving the object is $\int_C \mathbf{F} \cdot d\mathbf{r} = \int_C \mathbf{F} \cdot \mathbf{T}\, ds$. We can approximate this integral by dividing C into

7 segments of equal length $\Delta s = 2$ and approximating $\mathbf{F} \cdot \mathbf{T}$, that is, the tangential component of force, at a point (x_i^*, y_i^*) on

each segment. Since C is composed of straight line segments, $\mathbf{F} \cdot \mathbf{T}$ is the scalar projection of each force vector onto C.

If we choose (x_i^*, y_i^*) to be the point on the segment closest to the origin, then the work done is

$\int_C \mathbf{F} \cdot \mathbf{T}\, ds \approx \sum_{i=1}^{7} [\mathbf{F}(x_i^*, y_i^*) \cdot \mathbf{T}(x_i^*, y_i^*)]\, \Delta s = [2 + 2 + 2 + 2 + 1 + 1 + 1](2) = 22.$ Thus, we estimate the work done to

be approximately 22 J.

16.3   The Fundamental Theorem for Line Integrals

1. C appears to be a smooth curve, and since ∇f is continuous, we know f is differentiable. Then Theorem 2 says that the value

of $\int_C \nabla f \cdot d\mathbf{r}$ is simply the difference of the values of f at the terminal and initial points of C. From the graph, this is

$50 - 10 = 40$.

3. $\partial(2x - 3y)/\partial y = -3 = \partial(-3x + 4y - 8)/\partial x$ and the domain of $\mathbf{F}$ is $\mathbb{R}^2$ which is open and simply-connected, so by

Theorem 6 $\mathbf{F}$ is conservative. Thus, there exists a function f such that $\nabla f = \mathbf{F}$, that is, $f_x(x, y) = 2x - 3y$ and

$f_y(x, y) = -3x + 4y - 8$. But $f_x(x, y) = 2x - 3y$ implies $f(x, y) = x^2 - 3xy + g(y)$ and differentiating both sides of this

equation with respect to y gives $f_y(x, y) = -3x + g'(y)$. Thus $-3x + 4y - 8 = -3x + g'(y)$ so $g'(y) = 4y - 8$ and

$g(y) = 2y^2 - 8y + K$ where K is a constant. Hence $f(x, y) = x^2 - 3xy + 2y^2 - 8y + K$ is a potential function for $\mathbf{F}$.

5. $\partial(e^x \cos y)/\partial y = -e^x \sin y$, $\partial(e^x \sin y)/\partial x = e^x \sin y$. Since these are not equal, $\mathbf{F}$ is not conservative.

7. $\partial(ye^x + \sin y)/\partial y = e^x + \cos y = \partial(e^x + x\cos y)/\partial x$ and the domain of $\mathbf{F}$ is $\mathbb{R}^2$. Hence $\mathbf{F}$ is conservative so there

exists a function f such that $\nabla f = \mathbf{F}$. Then $f_x(x, y) = ye^x + \sin y$ implies $f(x, y) = ye^x + x\sin y + g(y)$ and

$f_y(x, y) = e^x + x\cos y + g'(y)$. But $f_y(x, y) = e^x + x\cos y$ so $g(y) = K$ and $f(x, y) = ye^x + x\sin y + K$ is a potential

function for $\mathbf{F}$.

9. $\partial(\ln y + 2xy^3)/\partial y = 1/y + 6xy^2 = \partial(3x^2y^2 + x/y)/\partial x$ and the domain of $\mathbf{F}$ is $\{(x, y) \mid y > 0\}$ which is open and simply

connected. Hence $\mathbf{F}$ is conservative so there exists a function f such that $\nabla f = \mathbf{F}$. Then $f_x(x, y) = \ln y + 2xy^3$ implies

$f(x, y) = x\ln y + x^2y^3 + g(y)$ and $f_y(x, y) = x/y + 3x^2y^2 + g'(y)$. But $f_y(x, y) = 3x^2y^2 + x/y$ so $g'(y) = 0 \Rightarrow$

$g(y) = K$ and $f(x, y) = x\ln y + x^2y^3 + K$ is a potential function for $\mathbf{F}$.

11. (a) $\mathbf{F}$ has continuous first-order partial derivatives and $\dfrac{\partial}{\partial y} 2xy = 2x = \dfrac{\partial}{\partial x} (x^2)$ on $\mathbb{R}^2$, which is open and simply-connected.

Thus, $\mathbf{F}$ is conservative by Theorem 6. Then we know that the line integral of $\mathbf{F}$ is independent of path; in particular, the

value of $\int_C \mathbf{F} \cdot d\mathbf{r}$ depends only on the endpoints of C. Since all three curves have the same initial and terminal points,

$\int_C \mathbf{F} \cdot d\mathbf{r}$ will have the same value for each curve.

© 2012 Cengage Learning. All Rights Reserved. May not be scanned, copied, or duplicated, or posted to a publicly accessible website, in whole or in part.

(b) We first find a potential function f, so that $\nabla f = \mathbf{F}$. We know $f_x(x,y) = 2xy$ and $f_y(x,y) = x^2$. Integrating

$f_x(x,y)$ with respect to x, we have $f(x,y) = x^2 y + g(y)$. Differentiating both sides with respect to y gives

$f_y(x,y) = x^2 + g'(y)$, so we must have $x^2 + g'(y) = x^2 \;\Rightarrow\; g'(y) = 0 \;\Rightarrow\; g(y) = K$, a constant.

Thus $f(x,y) = x^2 y + K$. All three curves start at $(1,2)$ and end at $(3,2)$, so by Theorem 2,

$\int_C \mathbf{F} \cdot d\mathbf{r} = f(3,2) - f(1,2) = 18 - 2 = 16$ for each curve.

13. (a) $f_x(x,y) = xy^2$ implies $f(x,y) = \frac{1}{2}x^2 y^2 + g(y)$ and $f_y(x,y) = x^2 y + g'(y)$. But $f_y(x,y) = x^2 y$ so $g'(y) = 0 \;\Rightarrow$

$g(y) = K$, a constant. We can take $K = 0$, so $f(x,y) = \frac{1}{2}x^2 y^2$.

(b) The initial point of C is $\mathbf{r}(0) = (0,1)$ and the terminal point is $\mathbf{r}(1) = (2,1)$, so

$\int_C \mathbf{F} \cdot d\mathbf{r} = f(2,1) - f(0,1) = 2 - 0 = 2$.

15. (a) $f_x(x,y,z) = yz$ implies $f(x,y,z) = xyz + g(y,z)$ and so $f_y(x,y,z) = xz + g_y(y,z)$. But $f_y(x,y,z) = xz$ so

$g_y(y,z) = 0 \;\Rightarrow\; g(y,z) = h(z)$. Thus $f(x,y,z) = xyz + h(z)$ and $f_z(x,y,z) = xy + h'(z)$. But

$f_z(x,y,z) = xy + 2z$, so $h'(z) = 2z \;\Rightarrow\; h(z) = z^2 + K$. Hence $f(x,y,z) = xyz + z^2$ (taking $K = 0$).

(b) $\int_C \mathbf{F} \cdot d\mathbf{r} = f(4,6,3) - f(1,0,-2) = 81 - 4 = 77$.

17. (a) $f_x(x,y,z) = yze^{xz}$ implies $f(x,y,z) = ye^{xz} + g(y,z)$ and so $f_y(x,y,z) = e^{xz} + g_y(y,z)$. But $f_y(x,y,z) = e^{xz}$ so

$g_y(y,z) = 0 \;\Rightarrow\; g(y,z) = h(z)$. Thus $f(x,y,z) = ye^{xz} + h(z)$ and $f_z(x,y,z) = xye^{xz} + h'(z)$. But

$f_z(x,y,z) = xye^{xz}$, so $h'(z) = 0 \;\Rightarrow\; h(z) = K$. Hence $f(x,y,z) = ye^{xz}$ (taking $K = 0$).

(b) $\mathbf{r}(0) = \langle 1, -1, 0\rangle$, $\mathbf{r}(2) = \langle 5, 3, 0\rangle$ so $\int_C \mathbf{F} \cdot d\mathbf{r} = f(5,3,0) - f(1,-1,0) = 3e^0 + e^0 = 4$.

19. The functions $2xe^{-y}$ and $2y - x^2 e^{-y}$ have continuous first-order derivatives on $\mathbb{R}^2$ and

$\dfrac{\partial}{\partial y}\left(2xe^{-y}\right) = -2xe^{-y} = \dfrac{\partial}{\partial x}\left(2y - x^2 e^{-y}\right)$, so $\mathbf{F}(x,y) = 2xe^{-y}\,\mathbf{i} + \left(2y - x^2 e^{-y}\right)\mathbf{j}$ is a conservative vector field by

Theorem 6 and hence the line integral is independent of path. Thus a potential function f exists, and $f_x(x,y) = 2xe^{-y}$

implies $f(x,y) = x^2 e^{-y} + g(y)$ and $f_y(x,y) = -x^2 e^{-y} + g'(y)$. But $f_y(x,y) = 2y - x^2 e^{-y}$ so

$g'(y) = 2y \;\Rightarrow\; g(y) = y^2 + K$. We can take $K = 0$, so $f(x,y) = x^2 e^{-y} + y^2$. Then

$\int_C 2xe^{-y}\,dx + (2y - x^2 e^{-y})\,dy = f(2,1) - f(1,0) = 4e^{-1} + 1 - 1 = 4/e$.

21. If $\mathbf{F}$ is conservative, then $\int_C \mathbf{F} \cdot d\mathbf{r}$ is independent of path. This means that the work done along all piecewise-smooth curves

that have the described initial and terminal points is the same. Your reply: It doesn't matter which curve is chosen.

23. $\mathbf{F}(x,y) = 2y^{3/2}\,\mathbf{i} + 3x\sqrt{y}\,\mathbf{j}$, $W = \int_C \mathbf{F} \cdot d\mathbf{r}$. Since $\partial(2y^{3/2})/\partial y = 3\sqrt{y} = \partial(3x\sqrt{y})/\partial x$, there exists a function f

such that $\nabla f = \mathbf{F}$. In fact, $f_x(x,y) = 2y^{3/2} \;\Rightarrow\; f(x,y) = 2xy^{3/2} + g(y) \;\Rightarrow\; f_y(x,y) = 3xy^{1/2} + g'(y)$. But

$f_y(x,y) = 3x\sqrt{y}$ so $g'(y) = 0$ or $g(y) = K$. We can take $K = 0 \;\Rightarrow\; f(x,y) = 2xy^{3/2}$. Thus

$W = \int_C \mathbf{F} \cdot d\mathbf{r} = f(2,4) - f(1,1) = 2(2)(8) - 2(1) = 30$.

© 2012 Cengage Learning. All Rights Reserved. May not be scanned, copied, or duplicated, or posted to a publicly accessible website, in whole or in part.

25. We know that if the vector field (call it **F**) is conservative, then around any closed path C, $\int_C \mathbf{F} \cdot d\mathbf{r} = 0$. But take C to be a circle centered at the origin, oriented counterclockwise. All of the field vectors that start on C are roughly in the direction of motion along C, so the integral around C will be positive. Therefore the field is not conservative.

27.

From the graph, it appears that **F** is conservative, since around all closed paths, the number and size of the field vectors pointing in directions similar to that of the path seem to be roughly the same as the number and size of the vectors pointing in the opposite direction. To check, we calculate

$$\frac{\partial}{\partial y}(\sin y) = \cos y = \frac{\partial}{\partial x}(1 + x\cos y).$$ Thus **F** is conservative, by Theorem 6.

29. Since **F** is conservative, there exists a function f such that $\mathbf{F} = \nabla f$, that is, $P = f_x$, $Q = f_y$, and $R = f_z$. Since P, Q, and R have continuous first order partial derivatives, Clairaut's Theorem says that $\partial P/\partial y = f_{xy} = f_{yx} = \partial Q/\partial x$, $\partial P/\partial z = f_{xz} = f_{zx} = \partial R/\partial x$, and $\partial Q/\partial z = f_{yz} = f_{zy} = \partial R/\partial y$.

31. $D = \{(x, y) \mid 0 < y < 3\}$ consists of those points between, but not on, the horizontal lines $y = 0$ and $y = 3$.

(a) Since D does not include any of its boundary points, it is open. More formally, at any point in D there is a disk centered at that point that lies entirely in D.

(b) Any two points chosen in D can always be joined by a path that lies entirely in D, so D is connected. (D consists of just one "piece.")

(c) D is connected and it has no holes, so it's simply-connected. (Every simple closed curve in D encloses only points that are in D.)

33. $D = \{(x, y) \mid 1 \le x^2 + y^2 \le 4, \ y \ge 0\}$ is the semiannular region in the upper half-plane between circles centered at the origin of radii 1 and 2 (including all boundary points).

(a) D includes boundary points, so it is not open. [Note that at any boundary point, $(1, 0)$ for instance, any disk centered there cannot lie entirely in D.]

(b) The region consists of one piece, so it's connected.

(c) D is connected and has no holes, so it's simply-connected.

35. (a) $P = -\dfrac{y}{x^2 + y^2}$, $\dfrac{\partial P}{\partial y} = \dfrac{y^2 - x^2}{(x^2 + y^2)^2}$ and $Q = \dfrac{x}{x^2 + y^2}$, $\dfrac{\partial Q}{\partial x} = \dfrac{y^2 - x^2}{(x^2 + y^2)^2}$. Thus $\dfrac{\partial P}{\partial y} = \dfrac{\partial Q}{\partial x}$.

(b) C_1: $x = \cos t$, $y = \sin t$, $0 \le t \le \pi$, C_2: $x = \cos t$, $y = \sin t$, $t = 2\pi$ to $t = \pi$. Then

$$\int_{C_1} \mathbf{F} \cdot d\mathbf{r} = \int_0^\pi \frac{(-\sin t)(-\sin t) + (\cos t)(\cos t)}{\cos^2 t + \sin^2 t}\, dt = \int_0^\pi dt = \pi \text{ and } \int_{C_2} \mathbf{F} \cdot d\mathbf{r} = \int_{2\pi}^\pi dt = -\pi$$

© 2012 Cengage Learning. All Rights Reserved. May not be scanned, copied, or duplicated, or posted to a publicly accessible website, in whole or in part.

Since these aren't equal, the line integral of $\mathbf{F}$ isn't independent of path. (Or notice that $\int_{C_3} \mathbf{F} \cdot d\mathbf{r} = \int_0^{2\pi} dt = 2\pi$ where C_3 is the circle $x^2 + y^2 = 1$, and apply the contrapositive of Theorem 3.) This doesn't contradict Theorem 6, since the domain of $\mathbf{F}$, which is $\mathbb{R}^2$ except the origin, isn't simply-connected.

16.4 Green's Theorem

1. (a) Parametric equations for C are $x = 2\cos t$, $y = 2\sin t$, $0 \le t \le 2\pi$. Then

$$\oint_C (x - y)\, dx + (x + y)\, dy = \int_0^{2\pi} [(2\cos t - 2\sin t)(-2\sin t) + (2\cos t + 2\sin t)(2\cos t)]\, dt$$

$$= \int_0^{2\pi} (4\sin^2 t + 4\cos^2 t)\, dt = \int_0^{2\pi} 4\, dt = 4t\Big]_0^{2\pi} = 8\pi$$

(b) Note that C as given in part (a) is a positively oriented, smooth, simple closed curve. Then by Green's Theorem,

$$\oint_C (x - y)\, dx + (x + y)\, dy = \iint_D \left[\frac{\partial}{\partial x}(x + y) - \frac{\partial}{\partial y}(x - y)\right] dA = \iint_D [1 - (-1)]\, dA = 2\iint_D dA$$

$$= 2A(D) = 2\pi(2)^2 = 8\pi$$

3. (a)

C_1: $x = t$ $\Rightarrow$ $dx = dt$, $y = 0$ $\Rightarrow$ $dy = 0\, dt$, $0 \le t \le 1$.

C_2: $x = 1$ $\Rightarrow$ $dx = 0\, dt$, $y = t$ $\Rightarrow$ $dy = dt$, $0 \le t \le 2$.

C_3: $x = 1 - t$ $\Rightarrow$ $dx = -dt$, $y = 2 - 2t$ $\Rightarrow$ $dy = -2\, dt$, $0 \le t \le 1$.

Thus

$$\oint_C xy\, dx + x^2 y^3\, dy = \oint_{C_1 + C_2 + C_3} xy\, dx + x^2 y^3\, dy$$

$$= \int_0^1 0\, dt + \int_0^2 t^3\, dt + \int_0^1 \left[-(1 - t)(2 - 2t) - 2(1 - t)^2(2 - 2t)^3\right] dt$$

$$= 0 + \left[\tfrac{1}{4}t^4\right]_0^2 + \left[\tfrac{2}{3}(1 - t)^3 + \tfrac{8}{3}(1 - t)^6\right]_0^1 = 4 - \tfrac{10}{3} = \tfrac{2}{3}$$

(b) $\oint_C xy\, dx + x^2 y^3\, dy = \iint_D \left[\frac{\partial}{\partial x}(x^2 y^3) - \frac{\partial}{\partial y}(xy)\right] dA = \int_0^1 \int_0^{2x} (2xy^3 - x)\, dy\, dx$

$$= \int_0^1 \left[\tfrac{1}{2}xy^4 - xy\right]_{y=0}^{y=2x} dx = \int_0^1 (8x^5 - 2x^2)\, dx = \tfrac{4}{3} - \tfrac{2}{3} = \tfrac{2}{3}$$

5.

The region D enclosed by C is given by $\{(x, y) \mid 0 \le x \le 2,\, x \le y \le 2x\}$, so

$$\int_C xy^2\, dx + 2x^2 y\, dy = \iint_D \left[\frac{\partial}{\partial x}(2x^2 y) - \frac{\partial}{\partial y}(xy^2)\right] dA$$

$$= \int_0^2 \int_x^{2x} (4xy - 2xy)\, dy\, dx$$

$$= \int_0^2 \left[xy^2\right]_{y=x}^{y=2x} dx$$

$$= \int_0^2 3x^3\, dx = \tfrac{3}{4}x^4\Big]_0^2 = 12$$

7. $\int_C \left(y + e^{\sqrt{x}}\right) dx + (2x + \cos y^2)\, dy = \iint_D \left[\frac{\partial}{\partial x}(2x + \cos y^2) - \frac{\partial}{\partial y}\left(y + e^{\sqrt{x}}\right)\right] dA$

$$= \int_0^1 \int_{y^2}^{\sqrt{y}} (2 - 1)\, dx\, dy = \int_0^1 (y^{1/2} - y^2)\, dy = \tfrac{1}{3}$$

© 2012 Cengage Learning. All Rights Reserved. May not be scanned, copied, or duplicated, or posted to a publicly accessible website, in whole or in part.

9. $\int_C y^3\,dx - x^3\,dy = \iint_D \left[\frac{\partial}{\partial x}\left(-x^3\right) - \frac{\partial}{\partial y}\left(y^3\right)\right] dA = \iint_D (-3x^2 - 3y^2)\,dA = \int_0^{2\pi}\int_0^2 (-3r^2)\,r\,dr\,d\theta$

$$= -3\int_0^{2\pi} d\theta \int_0^2 r^3\,dr = -3(2\pi)(4) = -24\pi$$

11. $\mathbf{F}(x,y) = \langle y\cos x - xy\sin x, xy + x\cos x\rangle$ and the region D enclosed by C is given by

$\{(x,y) \mid 0 \le x \le 2, 0 \le y \le 4 - 2x\}$. C is traversed clockwise, so $-C$ gives the positive orientation.

$\int_C \mathbf{F}\cdot d\mathbf{r} = -\int_{-C}(y\cos x - xy\sin x)\,dx + (xy + x\cos x)\,dy = -\iint_D \left[\frac{\partial}{\partial x}(xy + x\cos x) - \frac{\partial}{\partial y}(y\cos x - xy\sin x)\right]dA$

$= -\iint_D (y - x\sin x + \cos x - \cos x + x\sin x)\,dA = -\int_0^2\int_0^{4-2x} y\,dy\,dx$

$= -\int_0^2 \left[\frac{1}{2}y^2\right]_{y=0}^{y=4-2x}dx = -\int_0^2 \frac{1}{2}(4-2x)^2\,dx = -\int_0^2(8 - 8x + 2x^2)\,dx = -\left[8x - 4x^2 + \frac{2}{3}x^3\right]_0^2$

$= -\left(16 - 16 + \frac{16}{3} - 0\right) = -\frac{16}{3}$

13. $\mathbf{F}(x,y) = \langle y - \cos y, x\sin y\rangle$ and the region D enclosed by C is the disk with radius 2 centered at $(3, -4)$.

C is traversed clockwise, so $-C$ gives the positive orientation.

$$\int_C \mathbf{F}\cdot d\mathbf{r} = -\int_{-C}(y - \cos y)\,dx + (x\sin y)\,dy = -\iint_D\left[\frac{\partial}{\partial x}(x\sin y) - \frac{\partial}{\partial y}(y - \cos y)\right]dA$$

$$= -\iint_D(\sin y - 1 - \sin y)\,dA = \iint_D dA = \text{area of } D = \pi(2)^2 = 4\pi$$

15. Here $C = C_1 + C_2$ where

C_1 can be parametrized as $x = t$, $y = 1$, $-1 \le t \le 1$, and

C_2 is given by $x = -t$, $y = 2 - t^2$, $-1 \le t \le 1$.

Then the line integral is

$\oint_{C_1+C_2} y^2 e^x\,dx + x^2 e^y\,dy = \int_{-1}^1 [1\cdot e^t + t^2 e\cdot 0]\,dt$

$\qquad\qquad\qquad + \int_{-1}^1[(2-t^2)^2 e^{-t}(-1) + (-t)^2 e^{2-t^2}(-2t)]\,dt$

$= \int_{-1}^1[e^t - (2-t^2)^2 e^{-t} - 2t^3 e^{2-t^2}]\,dt = -8e + 48e^{-1}$

according to a CAS. The double integral is

$$\iint_D\left(\frac{\partial Q}{\partial x} - \frac{\partial P}{\partial y}\right)dA = \int_{-1}^1\int_1^{2-x^2}(2xe^y - 2ye^x)\,dy\,dx = -8e + 48e^{-1}, \text{ verifying Green's Theorem in this case.}$$

17. By Green's Theorem, $W = \int_C \mathbf{F}\cdot d\mathbf{r} = \int_C x(x+y)\,dx + xy^2\,dy = \iint_D(y^2 - x)\,dA$ where C is the path described in the

question and D is the triangle bounded by C. So

$$W = \int_0^1\int_0^{1-x}(y^2 - x)\,dy\,dx = \int_0^1\left[\frac{1}{3}y^3 - xy\right]_{y=0}^{y=1-x}dx = \int_0^1\left(\frac{1}{3}(1-x)^3 - x(1-x)\right)dx$$

$$= \left[-\frac{1}{12}(1-x)^4 - \frac{1}{2}x^2 + \frac{1}{3}x^3\right]_0^1 = \left(-\frac{1}{2} + \frac{1}{3}\right) - \left(-\frac{1}{12}\right) = -\frac{1}{12}$$

19. Let C_1 be the arch of the cycloid from $(0,0)$ to $(2\pi, 0)$, which corresponds to $0 \le t \le 2\pi$, and let C_2 be the segment from

$(2\pi, 0)$ to $(0,0)$, so C_2 is given by $x = 2\pi - t$, $y = 0$, $0 \le t \le 2\pi$. Then $C = C_1 \cup C_2$ is traversed clockwise, so $-C$ is

oriented positively. Thus $-C$ encloses the area under one arch of the cycloid and from (5) we have

$$A = -\oint_{-C} y\,dx = \int_{C_1} y\,dx + \int_{C_2} y\,dx = \int_0^{2\pi}(1 - \cos t)(1 - \cos t)\,dt + \int_0^{2\pi} 0\,(-dt)$$

$$= \int_0^{2\pi}(1 - 2\cos t + \cos^2 t)\,dt + 0 = \left[t - 2\sin t + \frac{1}{2}t + \frac{1}{4}\sin 2t\right]_0^{2\pi} = 3\pi$$

© 2012 Cengage Learning. All Rights Reserved. May not be scanned, copied, or duplicated, or posted to a publicly accessible website, in whole or in part.

21. (a) Using Equation 16.2.8, we write parametric equations of the line segment as $x = (1-t)x_1 + tx_2$, $y = (1-t)y_1 + ty_2$,

$0 \le t \le 1$. Then $dx = (x_2 - x_1)\,dt$ and $dy = (y_2 - y_1)\,dt$, so

$$\int_C x\,dy - y\,dx = \int_0^1 [(1-t)x_1 + tx_2](y_2 - y_1)\,dt + [(1-t)y_1 + ty_2](x_2 - x_1)\,dt$$

$$= \int_0^1 (x_1(y_2 - y_1) - y_1(x_2 - x_1) + t[(y_2 - y_1)(x_2 - x_1) - (x_2 - x_1)(y_2 - y_1)])\,dt$$

$$= \int_0^1 (x_1 y_2 - x_2 y_1)\,dt = x_1 y_2 - x_2 y_1$$

(b) We apply Green's Theorem to the path $C = C_1 \cup C_2 \cup \cdots \cup C_n$, where C_i is the line segment that joins (x_i, y_i) to

(x_{i+1}, y_{i+1}) for $i = 1, 2, \ldots, n-1$, and C_n is the line segment that joins (x_n, y_n) to (x_1, y_1). From (5),

$\frac{1}{2} \int_C x\,dy - y\,dx = \iint_D dA$, where D is the polygon bounded by C. Therefore

$$\text{area of polygon} = A(D) = \iint_D dA = \frac{1}{2} \int_C x\,dy - y\,dx$$

$$= \frac{1}{2}\left(\int_{C_1} x\,dy - y\,dx + \int_{C_2} x\,dy - y\,dx + \cdots + \int_{C_{n-1}} x\,dy - y\,dx + \int_{C_n} x\,dy - y\,dx \right)$$

To evaluate these integrals we use the formula from (a) to get

$$A(D) = \frac{1}{2}[(x_1 y_2 - x_2 y_1) + (x_2 y_3 - x_3 y_2) + \cdots + (x_{n-1} y_n - x_n y_{n-1}) + (x_n y_1 - x_1 y_n)].$$

(c) $A = \frac{1}{2}[(0 \cdot 1 - 2 \cdot 0) + (2 \cdot 3 - 1 \cdot 1) + (1 \cdot 2 - 0 \cdot 3) + (0 \cdot 1 - (-1) \cdot 2) + (-1 \cdot 0 - 0 \cdot 1)]$

$= \frac{1}{2}(0 + 5 + 2 + 2) = \frac{9}{2}$

23. We orient the quarter-circular region as shown in the figure.

$A = \frac{1}{4}\pi a^2$ so $\overline{x} = \dfrac{1}{\pi a^2/2} \oint_C x^2\,dy$ and $\overline{y} = -\dfrac{1}{\pi a^2/2} \oint_C y^2\,dx$.

Here $C = C_1 + C_2 + C_3$ where $C_1\colon x = t, \ y = 0, \ 0 \le t \le a$;

$C_2\colon x = a\cos t, \ y = a\sin t, \ 0 \le t \le \frac{\pi}{2}$; and

$C_3\colon x = 0, y = a - t, 0 \le t \le a$. Then

$\oint_C x^2\,dy = \int_{C_1} x^2\,dy + \int_{C_2} x^2\,dy + \int_{C_3} x^2\,dy = \int_0^a 0\,dt + \int_0^{\pi/2}(a\cos t)^2(a\cos t)\,dt + \int_0^a 0\,dt$

$= \int_0^{\pi/2} a^3 \cos^3 t\,dt = a^3 \int_0^{\pi/2}(1 - \sin^2 t)\cos t\,dt = a^3\left[\sin t - \frac{1}{3}\sin^3 t\right]_0^{\pi/2} = \frac{2}{3}a^3$

so $\overline{x} = \dfrac{1}{\pi a^2/2} \oint_C x^2\,dy = \dfrac{4a}{3\pi}$.

$\oint_C y^2\,dx = \int_{C_1} y^2\,dx + \int_{C_2} y^2\,dx + \int_{C_3} y^2\,dx = \int_0^a 0\,dt + \int_0^{\pi/2}(a\sin t)^2(-a\sin t)\,dt + \int_0^a 0\,dt$

$= \int_0^{\pi/2}(-a^3 \sin^3 t)\,dt = -a^3 \int_0^{\pi/2}(1 - \cos^2 t)\sin t\,dt = -a^3\left[\frac{1}{3}\cos^3 t - \cos t\right]_0^{\pi/2} = -\frac{2}{3}a^3$,

so $\overline{y} = -\dfrac{1}{\pi a^2/2} \oint_C y^2\,dx = \dfrac{4a}{3\pi}$. Thus $(\overline{x}, \overline{y}) = \left(\dfrac{4a}{3\pi}, \dfrac{4a}{3\pi}\right)$.

25. By Green's Theorem, $-\frac{1}{3}\rho \oint_C y^3\,dx = -\frac{1}{3}\rho \iint_D (-3y^2)\,dA = \iint_D y^2 \rho\,dA = I_x$ and

$\frac{1}{3}\rho \oint_C x^3\,dy = \frac{1}{3}\rho \iint_D (3x^2)\,dA = \iint_D x^2 \rho\,dA = I_y$.

27. As in Example 5, let C' be a counterclockwise-oriented circle with center the origin and radius a, where a is chosen to

be small enough so that C' lies inside C, and D the region bounded by C and C'. Here

$$P = \frac{2xy}{(x^2 + y^2)^2} \quad \Rightarrow \quad \frac{\partial P}{\partial y} = \frac{2x(x^2 + y^2)^2 - 2xy \cdot 2(x^2 + y^2) \cdot 2y}{(x^2 + y^2)^4} = \frac{2x^3 - 6xy^2}{(x^2 + y^2)^3} \text{ and}$$

© 2012 Cengage Learning. All Rights Reserved. May not be scanned, copied, or duplicated, or posted to a publicly accessible website, in whole or in part.

$$Q = \frac{y^2 - x^2}{(x^2 + y^2)^2} \quad \Rightarrow \quad \frac{\partial Q}{\partial x} = \frac{-2x(x^2 + y^2)^2 - (y^2 - x^2) \cdot 2(x^2 + y^2) \cdot 2x}{(x^2 + y^2)^4} = \frac{2x^3 - 6xy^2}{(x^2 + y^2)^3}. \text{ Thus, as in the example,}$$

$$\int_C P\,dx + Q\,dy + \int_{-C'} P\,dx + Q\,dy = \iint_D \left(\frac{\partial Q}{\partial x} - \frac{\partial P}{\partial y} \right) dA = \iint_D 0\,dA = 0$$

and $\int_C \mathbf{F} \cdot d\mathbf{r} = \int_{C'} \mathbf{F} \cdot d\mathbf{r}$. We parametrize C' as $\mathbf{r}(t) = a\cos t\,\mathbf{i} + a\sin t\,\mathbf{j}$, $0 \le t \le 2\pi$. Then

$$\int_C \mathbf{F} \cdot d\mathbf{r} = \int_{C'} \mathbf{F} \cdot d\mathbf{r} = \int_0^{2\pi} \frac{2\,(a\cos t)\,(a\sin t)\,\mathbf{i} + (a^2\sin^2 t - a^2\cos^2 t)\,\mathbf{j}}{(a^2\cos^2 t + a^2\sin^2 t)^2} \cdot \left(-a\sin t\,\mathbf{i} + a\cos t\,\mathbf{j} \right) dt$$

$$= \frac{1}{a} \int_0^{2\pi} \left(-\cos t\sin^2 t - \cos^3 t \right) dt = \frac{1}{a} \int_0^{2\pi} \left(-\cos t\sin^2 t - \cos t\,(1 - \sin^2 t) \right) dt$$

$$= -\frac{1}{a} \int_0^{2\pi} \cos t\,dt = -\frac{1}{a} \sin t \Big]_0^{2\pi} = 0$$

29. Since C is a simple closed path which doesn't pass through or enclose the origin, there exists an open region that doesn't contain the origin but does contain D. Thus $P = -y/(x^2 + y^2)$ and $Q = x/(x^2 + y^2)$ have continuous partial derivatives on this open region containing D and we can apply Green's Theorem. But by Exercise 16.3.35(a), $\partial P/\partial y = \partial Q/\partial x$, so $\oint_C \mathbf{F} \cdot d\mathbf{r} = \iint_D 0\,dA = 0$.

31. Using the first part of (5), we have that $\iint_R dx\,dy = A(R) = \int_{\partial R} x\,dy$. But $x = g(u, v)$, and $dy = \frac{\partial h}{\partial u}\,du + \frac{\partial h}{\partial v}\,dv$, and we orient ∂S by taking the positive direction to be that which corresponds, under the mapping, to the positive direction along ∂R, so

$$\int_{\partial R} x\,dy = \int_{\partial S} g(u, v) \left(\frac{\partial h}{\partial u}\,du + \frac{\partial h}{\partial v}\,dv \right) = \int_{\partial S} g(u, v) \frac{\partial h}{\partial u}\,du + g(u, v) \frac{\partial h}{\partial v}\,dv$$

$$= \pm \iint_S \left[\frac{\partial}{\partial u} \left(g(u, v) \frac{\partial h}{\partial v} \right) - \frac{\partial}{\partial v} \left(g(u, v) \frac{\partial h}{\partial u} \right) \right] dA \quad \text{[using Green's Theorem in the } uv\text{-plane]}$$

$$= \pm \iint_S \left(\frac{\partial g}{\partial u} \frac{\partial h}{\partial v} + g(u, v) \frac{\partial^2 h}{\partial u\,\partial v} - \frac{\partial g}{\partial v} \frac{\partial h}{\partial u} - g(u, v) \frac{\partial^2 h}{\partial v\,\partial u} \right) dA \quad \text{[using the Chain Rule]}$$

$$= \pm \iint_S \left(\frac{\partial x}{\partial u} \frac{\partial y}{\partial v} - \frac{\partial x}{\partial v} \frac{\partial y}{\partial u} \right) dA \quad \text{[by the equality of mixed partials]} \quad = \pm \iint_S \frac{\partial(x, y)}{\partial(u, v)}\,du\,dv$$

The sign is chosen to be positive if the orientation that we gave to ∂S corresponds to the usual positive orientation, and it is negative otherwise. In either case, since $A(R)$ is positive, the sign chosen must be the same as the sign of $\frac{\partial(x, y)}{\partial(u, v)}$.

Therefore $A(R) = \iint_R dx\,dy = \iint_S \left| \frac{\partial(x, y)}{\partial(u, v)} \right| du\,dv$.

16.5 Curl and Divergence

1. (a) $\operatorname{curl}\mathbf{F} = \nabla \times \mathbf{F} = \begin{vmatrix} \mathbf{i} & \mathbf{j} & \mathbf{k} \\ \partial/\partial x & \partial/\partial y & \partial/\partial z \\ x + yz & y + xz & z + xy \end{vmatrix}$

$$= \left[\frac{\partial}{\partial y}(z + xy) - \frac{\partial}{\partial z}(y + xz) \right] \mathbf{i} - \left[\frac{\partial}{\partial x}(z + xy) - \frac{\partial}{\partial z}(x + yz) \right] \mathbf{j} + \left[\frac{\partial}{\partial x}(y + xz) - \frac{\partial}{\partial y}(x + yz) \right] \mathbf{k}$$

$$= (x - x)\,\mathbf{i} - (y - y)\,\mathbf{j} + (z - z)\,\mathbf{k} = \mathbf{0}$$

© 2012 Cengage Learning. All Rights Reserved. May not be scanned, copied, or duplicated, or posted to a publicly accessible website, in whole or in part.

(b) $\operatorname{div}\mathbf{F} = \nabla \cdot \mathbf{F} = \dfrac{\partial}{\partial x}\,(x + yz) + \dfrac{\partial}{\partial y}\,(y + xz) + \dfrac{\partial}{\partial z}\,(z + xy) = 1 + 1 + 1 = 3$

3. (a) $\operatorname{curl}\mathbf{F} = \nabla \times \mathbf{F} = \begin{vmatrix} \mathbf{i} & \mathbf{j} & \mathbf{k} \\ \partial/\partial x & \partial/\partial y & \partial/\partial z \\ xye^z & 0 & yze^x \end{vmatrix} = (ze^x - 0)\,\mathbf{i} - (yze^x - xye^z)\,\mathbf{j} + (0 - xe^z)\,\mathbf{k}$

$\qquad = ze^x\,\mathbf{i} + (xye^z - yze^x)\,\mathbf{j} - xe^z\,\mathbf{k}$

(b) $\operatorname{div}\mathbf{F} = \nabla \cdot \mathbf{F} = \dfrac{\partial}{\partial x}\,(xye^z) + \dfrac{\partial}{\partial y}\,(0) + \dfrac{\partial}{\partial z}\,(yze^x) = ye^z + 0 + ye^x = y(e^z + e^x)$

5. (a) $\operatorname{curl}\mathbf{F} = \nabla \times \mathbf{F} = \begin{vmatrix} \mathbf{i} & \mathbf{j} & \mathbf{k} \\ \partial/\partial x & \partial/\partial y & \partial/\partial z \\ \dfrac{x}{\sqrt{x^2 + y^2 + z^2}} & \dfrac{y}{\sqrt{x^2 + y^2 + z^2}} & \dfrac{z}{\sqrt{x^2 + y^2 + z^2}} \end{vmatrix}$

$\qquad = \dfrac{1}{(x^2 + y^2 + z^2)^{3/2}}\,[(-yz + yz)\,\mathbf{i} - (-xz + xz)\,\mathbf{j} + (-xy + xy)\,\mathbf{k}] = \mathbf{0}$

(b) $\operatorname{div}\mathbf{F} = \nabla \cdot \mathbf{F} = \dfrac{\partial}{\partial x}\left(\dfrac{x}{\sqrt{x^2 + y^2 + z^2}}\right) + \dfrac{\partial}{\partial y}\left(\dfrac{y}{\sqrt{x^2 + y^2 + z^2}}\right) + \dfrac{\partial}{\partial z}\left(\dfrac{z}{\sqrt{x^2 + y^2 + z^2}}\right)$

$\qquad = \dfrac{x^2 + y^2 + z^2 - x^2}{(x^2 + y^2 + z^2)^{3/2}} + \dfrac{x^2 + y^2 + z^2 - y^2}{(x^2 + y^2 + z^2)^{3/2}} + \dfrac{x^2 + y^2 + z^2 - z^2}{(x^2 + y^2 + z^2)^{3/2}} = \dfrac{2x^2 + 2y^2 + 2z^2}{(x^2 + y^2 + z^2)^{3/2}} = \dfrac{2}{\sqrt{x^2 + y^2 + z^2}}$

7. (a) $\operatorname{curl}\mathbf{F} = \nabla \times \mathbf{F} = \begin{vmatrix} \mathbf{i} & \mathbf{j} & \mathbf{k} \\ \partial/\partial x & \partial/\partial y & \partial/\partial z \\ e^x \sin y & e^y \sin z & e^z \sin x \end{vmatrix} = (0 - e^y \cos z)\,\mathbf{i} - (e^z \cos x - 0)\,\mathbf{j} + (0 - e^x \cos y)\,\mathbf{k}$

$\qquad = \langle -e^y \cos z,\ -e^z \cos x,\ -e^x \cos y \rangle$

(b) $\operatorname{div}\mathbf{F} = \nabla \cdot \mathbf{F} = \dfrac{\partial}{\partial x}\,(e^x \sin y) + \dfrac{\partial}{\partial y}\,(e^y \sin z) + \dfrac{\partial}{\partial z}\,(e^z \sin x) = e^x \sin y + e^y \sin z + e^z \sin x$

9. If the vector field is $\mathbf{F} = P\,\mathbf{i} + Q\,\mathbf{j} + R\,\mathbf{k}$, then we know $R = 0$. In addition, the x-component of each vector of $\mathbf{F}$ is 0, so

$P = 0$, hence $\dfrac{\partial P}{\partial x} = \dfrac{\partial P}{\partial y} = \dfrac{\partial P}{\partial z} = \dfrac{\partial R}{\partial x} = \dfrac{\partial R}{\partial y} = \dfrac{\partial R}{\partial z} = 0$. Q decreases as y increases, so $\dfrac{\partial Q}{\partial y} < 0$, but Q doesn't change

in the x- or z-directions, so $\dfrac{\partial Q}{\partial x} = \dfrac{\partial Q}{\partial z} = 0$.

(a) $\operatorname{div}\mathbf{F} = \dfrac{\partial P}{\partial x} + \dfrac{\partial Q}{\partial y} + \dfrac{\partial R}{\partial z} = 0 + \dfrac{\partial Q}{\partial y} + 0 < 0$

(b) $\operatorname{curl}\mathbf{F} = \left(\dfrac{\partial R}{\partial y} - \dfrac{\partial Q}{\partial z}\right)\mathbf{i} + \left(\dfrac{\partial P}{\partial z} - \dfrac{\partial R}{\partial x}\right)\mathbf{j} + \left(\dfrac{\partial Q}{\partial x} - \dfrac{\partial P}{\partial y}\right)\mathbf{k} = (0 - 0)\,\mathbf{i} + (0 - 0)\,\mathbf{j} + (0 - 0)\,\mathbf{k} = \mathbf{0}$

11. If the vector field is $\mathbf{F} = P\,\mathbf{i} + Q\,\mathbf{j} + R\,\mathbf{k}$, then we know $R = 0$. In addition, the y-component of each vector of $\mathbf{F}$ is 0, so

$Q = 0$, hence $\dfrac{\partial Q}{\partial x} = \dfrac{\partial Q}{\partial y} = \dfrac{\partial Q}{\partial z} = \dfrac{\partial R}{\partial x} = \dfrac{\partial R}{\partial y} = \dfrac{\partial R}{\partial z} = 0$. P increases as y increases, so $\dfrac{\partial P}{\partial y} > 0$, but P doesn't change in

the x- or z-directions, so $\dfrac{\partial P}{\partial x} = \dfrac{\partial P}{\partial z} = 0$.

(a) $\operatorname{div}\mathbf{F} = \dfrac{\partial P}{\partial x} + \dfrac{\partial Q}{\partial y} + \dfrac{\partial R}{\partial z} = 0 + 0 + 0 = 0$

© 2012 Cengage Learning. All Rights Reserved. May not be scanned, copied, or duplicated, or posted to a publicly accessible website, in whole or in part.

(b) $\operatorname{curl} \mathbf{F} = \left(\dfrac{\partial R}{\partial y} - \dfrac{\partial Q}{\partial z}\right)\mathbf{i} + \left(\dfrac{\partial P}{\partial z} - \dfrac{\partial R}{\partial x}\right)\mathbf{j} + \left(\dfrac{\partial Q}{\partial x} - \dfrac{\partial P}{\partial y}\right)\mathbf{k} = (0-0)\,\mathbf{i} + (0-0)\mathbf{j} + \left(0 - \dfrac{\partial P}{\partial y}\right)\mathbf{k} = -\dfrac{\partial P}{\partial y}\mathbf{k}$

Since $\dfrac{\partial P}{\partial y} > 0$, $-\dfrac{\partial P}{\partial y}\mathbf{k}$ is a vector pointing in the negative z-direction.

13. $\operatorname{curl}\mathbf{F} = \nabla \times \mathbf{F} = \begin{vmatrix} \mathbf{i} & \mathbf{j} & \mathbf{k} \\ \partial/\partial x & \partial/\partial y & \partial/\partial z \\ y^2 z^3 & 2xyz^3 & 3xy^2 z^2 \end{vmatrix} = (6xyz^2 - 6xyz^2)\,\mathbf{i} - (3y^2 z^2 - 3y^2 z^2)\,\mathbf{j} + (2yz^3 - 2yz^3)\,\mathbf{k} = \mathbf{0}$

and $\mathbf{F}$ is defined on all of $\mathbb{R}^3$ with component functions which have continuous partial derivatives, so by Theorem 4,

$\mathbf{F}$ is conservative. Thus, there exists a function f such that $\mathbf{F} = \nabla f$. Then $f_x(x,y,z) = y^2 z^3$ implies

$f(x,y,z) = xy^2 z^3 + g(y,z)$ and $f_y(x,y,z) = 2xyz^3 + g_y(y,z)$. But $f_y(x,y,z) = 2xyz^3$, so $g(y,z) = h(z)$ and

$f(x,y,z) = xy^2 z^3 + h(z)$. Thus $f_z(x,y,z) = 3xy^2 z^2 + h'(z)$ but $f_z(x,y,z) = 3xy^2 z^2$ so $h(z) = K$, a constant.

Hence a potential function for $\mathbf{F}$ is $f(x,y,z) = xy^2 z^3 + K$.

15. $\operatorname{curl}\mathbf{F} = \nabla \times \mathbf{F} = \begin{vmatrix} \mathbf{i} & \mathbf{j} & \mathbf{k} \\ \partial/\partial x & \partial/\partial y & \partial/\partial z \\ 3xy^2 z^2 & 2x^2 yz^3 & 3x^2 y^2 z^2 \end{vmatrix}$

$= (6x^2 yz^2 - 6x^2 yz^2)\,\mathbf{i} - (6xy^2 z^2 - 6xy^2 z)\,\mathbf{j} + (4xyz^3 - 6xyz^2)\,\mathbf{k}$

$= 6xy^2 z(1-z)\,\mathbf{j} + 2xyz^2(2z - 3)\,\mathbf{k} \neq \mathbf{0}$

so $\mathbf{F}$ is not conservative.

17. $\operatorname{curl}\mathbf{F} = \nabla \times \mathbf{F} = \begin{vmatrix} \mathbf{i} & \mathbf{j} & \mathbf{k} \\ \partial/\partial x & \partial/\partial y & \partial/\partial z \\ e^{yz} & xze^{yz} & xye^{yz} \end{vmatrix}$

$= [xyze^{yz} + xe^{yz} - (xyze^{yz} + xe^{yz})]\,\mathbf{i} - (ye^{yz} - ye^{yz})\,\mathbf{j} + (ze^{yz} - ze^{yz})\,\mathbf{k} = \mathbf{0}$

$\mathbf{F}$ is defined on all of $\mathbb{R}^3$, and the partial derivatives of the component functions are continuous, so $\mathbf{F}$ is conservative. Thus

there exists a function f such that $\nabla f = \mathbf{F}$. Then $f_x(x,y,z) = e^{yz}$ implies $f(x,y,z) = xe^{yz} + g(y,z) \;\Rightarrow$

$f_y(x,y,z) = xze^{yz} + g_y(y,z)$. But $f_y(x,y,z) = xze^{yz}$, so $g(y,z) = h(z)$ and $f(x,y,z) = xe^{yz} + h(z)$.

Thus $f_z(x,y,z) = xye^{yz} + h'(z)$ but $f_z(x,y,z) = xye^{yz}$ so $h(z) = K$ and a potential function for $\mathbf{F}$ is

$f(x,y,z) = xe^{yz} + K$.

19. No. Assume there is such a $\mathbf{G}$. Then $\operatorname{div}(\operatorname{curl}\mathbf{G}) = \dfrac{\partial}{\partial x}(x\sin y) + \dfrac{\partial}{\partial y}(\cos y) + \dfrac{\partial}{\partial z}(z - xy) = \sin y - \sin y + 1 \neq 0$,

which contradicts Theorem 11.

21. $\operatorname{curl}\mathbf{F} = \begin{vmatrix} \mathbf{i} & \mathbf{j} & \mathbf{k} \\ \partial/\partial x & \partial/\partial y & \partial/\partial z \\ f(x) & g(y) & h(z) \end{vmatrix} = (0-0)\,\mathbf{i} + (0-0)\,\mathbf{j} + (0-0)\,\mathbf{k} = \mathbf{0}$. Hence $\mathbf{F} = f(x)\,\mathbf{i} + g(y)\,\mathbf{j} + h(z)\,\mathbf{k}$

is irrotational.

© 2012 Cengage Learning. All Rights Reserved. May not be scanned, copied, or duplicated, or posted to a publicly accessible website, in whole or in part.

For Exercises 23–29, let $\mathbf{F}(x, y, z) = P_1\,\mathbf{i} + Q_1\,\mathbf{j} + R_1\,\mathbf{k}$ and $\mathbf{G}(x, y, z) = P_2\,\mathbf{i} + Q_2\,\mathbf{j} + R_2\,\mathbf{k}$.

23. $\text{div}(\mathbf{F} + \mathbf{G}) = \text{div}\langle P_1 + P_2, Q_1 + Q_2, R_1 + R_2 \rangle = \dfrac{\partial(P_1 + P_2)}{\partial x} + \dfrac{\partial(Q_1 + Q_2)}{\partial y} + \dfrac{\partial(R_1 + R_2)}{\partial z}$

$$= \frac{\partial P_1}{\partial x} + \frac{\partial P_2}{\partial x} + \frac{\partial Q_1}{\partial y} + \frac{\partial Q_2}{\partial y} + \frac{\partial R_1}{\partial z} + \frac{\partial R_2}{\partial z} = \left(\frac{\partial P_1}{\partial x} + \frac{\partial Q_1}{\partial y} + \frac{\partial R_1}{\partial z} \right) + \left(\frac{\partial P_2}{\partial x} + \frac{\partial Q_2}{\partial y} + \frac{\partial R_2}{\partial z} \right)$$

$$= \text{div}\langle P_1, Q_1, R_1 \rangle + \text{div}\langle P_2, Q_2, R_2 \rangle = \text{div}\,\mathbf{F} + \text{div}\,\mathbf{G}$$

25. $\text{div}(f\mathbf{F}) = \text{div}(f \langle P_1, Q_1, R_1 \rangle) = \text{div}\langle fP_1, fQ_1, fR_1 \rangle = \dfrac{\partial(fP_1)}{\partial x} + \dfrac{\partial(fQ_1)}{\partial y} + \dfrac{\partial(fR_1)}{\partial z}$

$$= \left(f\,\frac{\partial P_1}{\partial x} + P_1\,\frac{\partial f}{\partial x} \right) + \left(f\,\frac{\partial Q_1}{\partial y} + Q_1\,\frac{\partial f}{\partial y} \right) + \left(f\,\frac{\partial R_1}{\partial z} + R_1\,\frac{\partial f}{\partial z} \right)$$

$$= f\left(\frac{\partial P_1}{\partial x} + \frac{\partial Q_1}{\partial y} + \frac{\partial R_1}{\partial z} \right) + \langle P_1, Q_1, R_1 \rangle \cdot \left\langle \frac{\partial f}{\partial x}, \frac{\partial f}{\partial y}, \frac{\partial f}{\partial z} \right\rangle = f\,\text{div}\,\mathbf{F} + \mathbf{F} \cdot \nabla f$$

27. $\text{div}(\mathbf{F} \times \mathbf{G}) = \nabla \cdot (\mathbf{F} \times \mathbf{G}) = \begin{vmatrix} \partial/\partial x & \partial/\partial y & \partial/\partial z \\ P_1 & Q_1 & R_1 \\ P_2 & Q_2 & R_2 \end{vmatrix} = \dfrac{\partial}{\partial x}\begin{vmatrix} Q_1 & R_1 \\ Q_2 & R_2 \end{vmatrix} - \dfrac{\partial}{\partial y}\begin{vmatrix} P_1 & R_1 \\ P_2 & R_2 \end{vmatrix} + \dfrac{\partial}{\partial z}\begin{vmatrix} P_1 & Q_1 \\ P_2 & Q_2 \end{vmatrix}$

$$= \left[Q_1\,\frac{\partial R_2}{\partial x} + R_2\,\frac{\partial Q_1}{\partial x} - Q_2\,\frac{\partial R_1}{\partial x} - R_1\,\frac{\partial Q_2}{\partial x} \right] - \left[P_1\,\frac{\partial R_2}{\partial y} + R_2\,\frac{\partial P_1}{\partial y} - P_2\,\frac{\partial R_1}{\partial y} - R_1\,\frac{\partial P_2}{\partial y} \right]$$

$$+ \left[P_1\,\frac{\partial Q_2}{\partial z} + Q_2\,\frac{\partial P_1}{\partial z} - P_2\,\frac{\partial Q_1}{\partial z} - Q_1\,\frac{\partial P_2}{\partial z} \right]$$

$$= \left[P_2\left(\frac{\partial R_1}{\partial y} - \frac{\partial Q_1}{\partial z} \right) + Q_2\left(\frac{\partial P_1}{\partial z} - \frac{\partial R_1}{\partial x} \right) + R_2\left(\frac{\partial Q_1}{\partial x} - \frac{\partial P_1}{\partial y} \right) \right]$$

$$- \left[P_1\left(\frac{\partial R_2}{\partial y} - \frac{\partial Q_2}{\partial z} \right) + Q_1\left(\frac{\partial P_2}{\partial z} - \frac{\partial R_2}{\partial x} \right) + R_1\left(\frac{\partial Q_2}{\partial x} - \frac{\partial P_2}{\partial y} \right) \right]$$

$$= \mathbf{G} \cdot \text{curl}\,\mathbf{F} - \mathbf{F} \cdot \text{curl}\,\mathbf{G}$$

29. $\text{curl}(\text{curl}\,\mathbf{F}) = \nabla \times (\nabla \times \mathbf{F}) = \begin{vmatrix} \mathbf{i} & \mathbf{j} & \mathbf{k} \\ \partial/\partial x & \partial/\partial y & \partial/\partial z \\ \partial R_1/\partial y - \partial Q_1/\partial z & \partial P_1/\partial z - \partial R_1/\partial x & \partial Q_1/\partial x - \partial P_1/\partial y \end{vmatrix}$

$$= \left(\frac{\partial^2 Q_1}{\partial y \partial x} - \frac{\partial^2 P_1}{\partial y^2} - \frac{\partial^2 P_1}{\partial z^2} + \frac{\partial^2 R_1}{\partial z \partial x} \right)\mathbf{i} + \left(\frac{\partial^2 R_1}{\partial z \partial y} - \frac{\partial^2 Q_1}{\partial z^2} - \frac{\partial^2 Q_1}{\partial x^2} + \frac{\partial^2 P_1}{\partial x \partial y} \right)\mathbf{j}$$

$$+ \left(\frac{\partial^2 P_1}{\partial x \partial z} - \frac{\partial^2 R_1}{\partial x^2} - \frac{\partial^2 R_1}{\partial y^2} + \frac{\partial^2 Q_1}{\partial y \partial z} \right)\mathbf{k}$$

Now let's consider $\text{grad}(\text{div}\,\mathbf{F}) - \nabla^2 \mathbf{F}$ and compare with the above.

(Note that $\nabla^2 \mathbf{F}$ is defined on page 1119 [ET 1095].)

[continued]

© 2012 Cengage Learning. All Rights Reserved. May not be scanned, copied, or duplicated, or posted to a publicly accessible website, in whole or in part.

$$\text{grad}(\text{div}\,\mathbf{F}) - \nabla^2\mathbf{F} = \left[\left(\frac{\partial^2 P_1}{\partial x^2} + \frac{\partial^2 Q_1}{\partial x\partial y} + \frac{\partial^2 R_1}{\partial x\partial z}\right)\mathbf{i} + \left(\frac{\partial^2 P_1}{\partial y\partial x} + \frac{\partial^2 Q_1}{\partial y^2} + \frac{\partial^2 R_1}{\partial y\partial z}\right)\mathbf{j} + \left(\frac{\partial^2 P_1}{\partial z\partial x} + \frac{\partial^2 Q_1}{\partial z\partial y} + \frac{\partial^2 R_1}{\partial z^2}\right)\mathbf{k}\right]$$

$$- \left[\left(\frac{\partial^2 P_1}{\partial x^2} + \frac{\partial^2 P_1}{\partial y^2} + \frac{\partial^2 P_1}{\partial z^2}\right)\mathbf{i} + \left(\frac{\partial^2 Q_1}{\partial x^2} + \frac{\partial^2 Q_1}{\partial y^2} + \frac{\partial^2 Q_1}{\partial z^2}\right)\mathbf{j}\right.$$

$$\left. + \left(\frac{\partial^2 R_1}{\partial x^2} + \frac{\partial^2 R_1}{\partial y^2} + \frac{\partial^2 R_1}{\partial z^2}\right)\mathbf{k}\right]$$

$$= \left(\frac{\partial^2 Q_1}{\partial x\partial y} + \frac{\partial^2 R_1}{\partial x\partial z} - \frac{\partial^2 P_1}{\partial y^2} - \frac{\partial^2 P_1}{\partial z^2}\right)\mathbf{i} + \left(\frac{\partial^2 P_1}{\partial y\partial x} + \frac{\partial^2 R_1}{\partial y\partial z} - \frac{\partial^2 Q_1}{\partial x^2} - \frac{\partial^2 Q_1}{\partial z^2}\right)\mathbf{j}$$

$$+ \left(\frac{\partial^2 P_1}{\partial z\partial x} + \frac{\partial^2 Q_1}{\partial z\partial y} - \frac{\partial^2 R_1}{\partial x^2} - \frac{\partial^2 R_2}{\partial y^2}\right)\mathbf{k}$$

Then applying Clairaut's Theorem to reverse the order of differentiation in the second partial derivatives as needed and comparing, we have curl curl $\mathbf{F} = \text{grad}\,\text{div}\,\mathbf{F} - \nabla^2\mathbf{F}$ as desired.

31. (a) $\nabla r = \nabla\sqrt{x^2+y^2+z^2} = \dfrac{x}{\sqrt{x^2+y^2+z^2}}\mathbf{i} + \dfrac{y}{\sqrt{x^2+y^2+z^2}}\mathbf{j} + \dfrac{z}{\sqrt{x^2+y^2+z^2}}\mathbf{k} = \dfrac{x\,\mathbf{i}+y\,\mathbf{j}+z\,\mathbf{k}}{\sqrt{x^2+y^2+z^2}} = \dfrac{\mathbf{r}}{r}$

(b) $\nabla\times\mathbf{r} = \begin{vmatrix} \mathbf{i} & \mathbf{j} & \mathbf{k} \\ \dfrac{\partial}{\partial x} & \dfrac{\partial}{\partial y} & \dfrac{\partial}{\partial z} \\ x & y & z \end{vmatrix} = \left[\dfrac{\partial}{\partial y}(z) - \dfrac{\partial}{\partial z}(y)\right]\mathbf{i} + \left[\dfrac{\partial}{\partial z}(x) - \dfrac{\partial}{\partial x}(z)\right]\mathbf{j} + \left[\dfrac{\partial}{\partial x}(y) - \dfrac{\partial}{\partial y}(x)\right]\mathbf{k} = \mathbf{0}$

(c) $\nabla\left(\dfrac{1}{r}\right) = \nabla\left(\dfrac{1}{\sqrt{x^2+y^2+z^2}}\right)$

$$= \dfrac{-\dfrac{1}{2\sqrt{x^2+y^2+z^2}}(2x)}{x^2+y^2+z^2}\mathbf{i} - \dfrac{\dfrac{1}{2\sqrt{x^2+y^2+z^2}}(2y)}{x^2+y^2+z^2}\mathbf{j} - \dfrac{\dfrac{1}{2\sqrt{x^2+y^2+z^2}}(2z)}{x^2+y^2+z^2}\mathbf{k}$$

$$= -\dfrac{x\,\mathbf{i}+y\,\mathbf{j}+z\,\mathbf{k}}{(x^2+y^2+z^2)^{3/2}} = -\dfrac{\mathbf{r}}{r^3}$$

(d) $\nabla\ln r = \nabla\ln(x^2+y^2+z^2)^{1/2} = \tfrac{1}{2}\nabla\ln(x^2+y^2+z^2)$

$$= \dfrac{x}{x^2+y^2+z^2}\mathbf{i} + \dfrac{y}{x^2+y^2+z^2}\mathbf{j} + \dfrac{z}{x^2+y^2+z^2}\mathbf{k} = \dfrac{x\,\mathbf{i}+y\,\mathbf{j}+z\,\mathbf{k}}{x^2+y^2+z^2} = \dfrac{\mathbf{r}}{r^2}$$

33. By (13), $\oint_C f(\nabla g)\cdot\mathbf{n}\,ds = \iint_D \text{div}(f\nabla g)\,dA = \iint_D[f\,\text{div}(\nabla g) + \nabla g\cdot\nabla f]\,dA$ by Exercise 25. But $\text{div}(\nabla g) = \nabla^2 g$.

Hence $\iint_D f\nabla^2 g\,dA = \oint_C f(\nabla g)\cdot\mathbf{n}\,ds - \iint_D \nabla g\cdot\nabla f\,dA$.

35. Let $f(x,y) = 1$. Then $\nabla f = \mathbf{0}$ and Green's first identity (see Exercise 33) says

$$\iint_D \nabla^2 g\,dA = \oint_C(\nabla g)\cdot\mathbf{n}\,ds - \iint_D \mathbf{0}\cdot\nabla g\,dA \quad\Rightarrow\quad \iint_D \nabla^2 g\,dA = \oint_C \nabla g\cdot\mathbf{n}\,ds. \text{ But } g \text{ is harmonic on } D, \text{ so}$$

$\nabla^2 g = 0 \quad\Rightarrow\quad \oint_C \nabla g\cdot\mathbf{n}\,ds = 0 \text{ and } \oint_C D_{\mathbf{n}}g\,ds = \oint_C(\nabla g\cdot\mathbf{n})\,ds = 0.$

37. (a) We know that $\omega = v/d$, and from the diagram $\sin\theta = d/r \quad\Rightarrow\quad v = d\omega = (\sin\theta)r\omega = |\mathbf{w}\times\mathbf{r}|$. But $\mathbf{v}$ is perpendicular to both $\mathbf{w}$ and $\mathbf{r}$, so that $\mathbf{v} = \mathbf{w}\times\mathbf{r}$.

© 2012 Cengage Learning. All Rights Reserved. May not be scanned, copied, or duplicated, or posted to a publicly accessible website, in whole or in part.

(b) From (a), $\mathbf{v} = \mathbf{w} \times \mathbf{r} = \begin{vmatrix} \mathbf{i} & \mathbf{j} & \mathbf{k} \\ 0 & 0 & \omega \\ x & y & z \end{vmatrix} = (0 \cdot z - \omega y)\,\mathbf{i} + (\omega x - 0 \cdot z)\,\mathbf{j} + (0 \cdot y - x \cdot 0)\,\mathbf{k} = -\omega y\,\mathbf{i} + \omega x\,\mathbf{j}$

(c) $\operatorname{curl} \mathbf{v} = \nabla \times \mathbf{v} = \begin{vmatrix} \mathbf{i} & \mathbf{j} & \mathbf{k} \\ \partial/\partial x & \partial/\partial y & \partial/\partial z \\ -\omega y & \omega x & 0 \end{vmatrix}$

$\qquad = \left[\dfrac{\partial}{\partial y}(0) - \dfrac{\partial}{\partial z}(\omega x) \right] \mathbf{i} + \left[\dfrac{\partial}{\partial z}(-\omega y) - \dfrac{\partial}{\partial x}(0) \right] \mathbf{j} + \left[\dfrac{\partial}{\partial x}(\omega x) - \dfrac{\partial}{\partial y}(-\omega y) \right] \mathbf{k}$

$\qquad = [\omega - (-\omega)]\,\mathbf{k} = 2\omega\,\mathbf{k} = 2\mathbf{w}$

39. For any continuous function f on $\mathbb{R}^3$, define a vector field $\mathbf{G}(x, y, z) = \langle g(x, y, z), 0, 0 \rangle$ where $g(x, y, z) = \int_0^x f(t, y, z)\,dt$.

Then $\operatorname{div} \mathbf{G} = \dfrac{\partial}{\partial x}(g(x, y, z)) + \dfrac{\partial}{\partial y}(0) + \dfrac{\partial}{\partial z}(0) = \dfrac{\partial}{\partial x} \int_0^x f(t, y, z)\,dt = f(x, y, z)$ by the Fundamental Theorem of

Calculus. Thus every continuous function f on $\mathbb{R}^3$ is the divergence of some vector field.

16.6 Parametric Surfaces and Their Areas

1. $P(7, 10, 4)$ lies on the parametric surface $\mathbf{r}(u, v) = \langle 2u + 3v, 1 + 5u - v, 2 + u + v \rangle$ if and only if there are values for u

and v where $2u + 3v = 7$, $1 + 5u - v = 10$, and $2 + u + v = 4$. But solving the first two equations simultaneously gives

$u = 2$, $v = 1$ and these values do not satisfy the third equation, so P does not lie on the surface.

$Q(5, 22, 5)$ lies on the surface if $2u + 3v = 5$, $1 + 5u - v = 22$, and $2 + u + v = 5$ for some values of u and v. Solving the

first two equations simultaneously gives $u = 4$, $v = -1$ and these values satisfy the third equation, so Q lies on the surface.

3. $\mathbf{r}(u, v) = (u + v)\,\mathbf{i} + (3 - v)\,\mathbf{j} + (1 + 4u + 5v)\,\mathbf{k} = \langle 0, 3, 1 \rangle + u\,\langle 1, 0, 4 \rangle + v\,\langle 1, -1, 5 \rangle$. From Example 3, we recognize

this as a vector equation of a plane through the point $(0, 3, 1)$ and containing vectors $\mathbf{a} = \langle 1, 0, 4 \rangle$ and $\mathbf{b} = \langle 1, -1, 5 \rangle$. If we

wish to find a more conventional equation for the plane, a normal vector to the plane is $\mathbf{a} \times \mathbf{b} = \begin{vmatrix} \mathbf{i} & \mathbf{j} & \mathbf{k} \\ 1 & 0 & 4 \\ 1 & -1 & 5 \end{vmatrix} = 4\,\mathbf{i} - \mathbf{j} - \mathbf{k}$

and an equation of the plane is $4(x - 0) - (y - 3) - (z - 1) = 0$ or $4x - y - z = -4$.

5. $\mathbf{r}(s, t) = \langle s, t, t^2 - s^2 \rangle$, so the corresponding parametric equations for the surface are $x = s$, $y = t$, $z = t^2 - s^2$. For any

point (x, y, z) on the surface, we have $z = y^2 - x^2$. With no restrictions on the parameters, the surface is $z = y^2 - x^2$, which

we recognize as a hyperbolic paraboloid.

© 2012 Cengage Learning. All Rights Reserved. May not be scanned, copied, or duplicated, or posted to a publicly accessible website, in whole or in part.

7. $\mathbf{r}(u,v) = \langle u^2, v^2, u+v \rangle$, $-1 \leq u \leq 1$, $-1 \leq v \leq 1$.

The surface has parametric equations $x = u^2$, $y = v^2$, $z = u+v$, $-1 \leq u \leq 1$, $-1 \leq v \leq 1$.

In Maple, the surface can be graphed by entering

```
plot3d([u^2,v^2,u+v],u=-1..1,v=-1..1);
```

In Mathematica we use the `ParametricPlot3D` command. If we keep u constant at u_0, $x = u_0^2$, a constant, so the corresponding grid curves must be the curves parallel to the yz-plane. If v is constant, we have $y = v_0^2$, a constant, so these grid curves are the curves parallel to the xz-plane.

9. $\mathbf{r}(u,v) = \langle u\cos v, u\sin v, u^5 \rangle$.

The surface has parametric equations $x = u\cos v$, $y = u\sin v$, $z = u^5$, $-1 \leq u \leq 1$, $0 \leq v \leq 2\pi$. Note that if $u = u_0$ is constant then $z = u_0^5$ is constant and $x = u_0\cos v$, $y = u_0\sin v$ describe a circle in x, y of radius $|u_0|$, so the corresponding grid curves are circles parallel to the xy-plane. If $v = v_0$, a constant, the parametric equations become $x = u\cos v_0$, $y = u\sin v_0$, $z = u^5$. Then $y = (\tan v_0)x$, so these are the grid curves we see that lie in vertical planes $y = kx$ through the z-axis.

11. $x = \sin v$, $y = \cos u\sin 4v$, $z = \sin 2u\sin 4v$, $0 \leq u \leq 2\pi$, $-\frac{\pi}{2} \leq v \leq \frac{\pi}{2}$.

Note that if $v = v_0$ is constant, then $x = \sin v_0$ is constant, so the corresponding grid curves must be parallel to the yz-plane. These are the vertically oriented grid curves we see, each shaped like a "figure-eight." When $u = u_0$ is held constant, the parametric equations become $x = \sin v$, $y = \cos u_0\sin 4v$, $z = \sin 2u_0\sin 4v$. Since z is a constant multiple of y, the corresponding grid curves are the curves contained in planes $z = ky$ that pass through the x-axis.

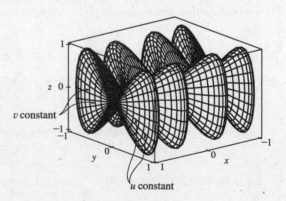

13. $\mathbf{r}(u,v) = u\cos v\,\mathbf{i} + u\sin v\,\mathbf{j} + v\,\mathbf{k}$. The parametric equations for the surface are $x = u\cos v$, $y = u\sin v$, $z = v$. We look at the grid curves first; if we fix v, then x and y parametrize a straight line in the plane $z = v$ which intersects the z-axis. If u is held constant, the projection onto the xy-plane is circular; with $z = v$, each grid curve is a helix. The surface is a spiraling ramp, graph IV.

© 2012 Cengage Learning. All Rights Reserved. May not be scanned, copied, or duplicated, or posted to a publicly accessible website, in whole or in part.

15. $\mathbf{r}(u, v) = \sin v \,\mathbf{i} + \cos u \sin 2v \,\mathbf{j} + \sin u \sin 2v \,\mathbf{k}$. Parametric equations for the surface are $x = \sin v$, $y = \cos u \sin 2v$,

$z = \sin u \sin 2v$. If $v = v_0$ is fixed, then $x = \sin v_0$ is constant, and $y = (\sin 2v_0) \cos u$ and $z = (\sin 2v_0) \sin u$ describe a

circle of radius $|\sin 2v_0|$, so each corresponding grid curve is a circle contained in the vertical plane $x = \sin v_0$ parallel to the

yz-plane. The only possible surface is graph II. The grid curves we see running lengthwise along the surface correspond to

holding u constant, in which case $y = (\cos u_0) \sin 2v$, $z = (\sin u_0) \sin 2v$ $\Rightarrow$ $z = (\tan u_0) y$, so each grid curve lies in a

plane $z = ky$ that includes the x-axis.

17. $x = \cos^3 u \cos^3 v$, $y = \sin^3 u \cos^3 v$, $z = \sin^3 v$. If $v = v_0$ is held constant then $z = \sin^3 v_0$ is constant, so the

corresponding grid curve lies in a horizontal plane. Several of the graphs exhibit horizontal grid curves, but the curves for this

surface are neither circles nor straight lines, so graph III is the only possibility. (In fact, the horizontal grid curves here are

members of the family $x = a \cos^3 u$, $y = a \sin^3 u$ and are called astroids.) The vertical grid curves we see on the surface

correspond to $u = u_0$ held constant, as then we have $x = \cos^3 u_0 \cos^3 v$, $y = \sin^3 u_0 \cos^3 v$ so the corresponding grid curve

lies in the vertical plane $y = (\tan^3 u_0) x$ through the z-axis.

19. From Example 3, parametric equations for the plane through the point $(0, 0, 0)$ that contains the vectors $\mathbf{a} = \langle 1, -1, 0 \rangle$ and

$\mathbf{b} = \langle 0, 1, -1 \rangle$ are $x = 0 + u(1) + v(0) = u$, $y = 0 + u(-1) + v(1) = v - u$, $z = 0 + u(0) + v(-1) = -v$.

21. Solving the equation for x gives $x^2 = 1 + y^2 + \frac{1}{4}z^2$ $\Rightarrow$ $x = \sqrt{1 + y^2 + \frac{1}{4}z^2}$. (We choose the positive root since we want

the part of the hyperboloid that corresponds to $x \geq 0$.) If we let y and z be the parameters, parametric equations are $y = y$,

$z = z$, $x = \sqrt{1 + y^2 + \frac{1}{4}z^2}$.

23. Since the cone intersects the sphere in the circle $x^2 + y^2 = 2$, $z = \sqrt{2}$ and we want the portion of the sphere above this, we

can parametrize the surface as $x = x$, $y = y$, $z = \sqrt{4 - x^2 - y^2}$ where $x^2 + y^2 \leq 2$.

Alternate solution: Using spherical coordinates, $x = 2 \sin \phi \cos \theta$, $y = 2 \sin \phi \sin \theta$, $z = 2 \cos \phi$ where $0 \leq \phi \leq \frac{\pi}{4}$ and

$0 \leq \theta \leq 2\pi$.

25. Parametric equations are $x = x$, $y = 4 \cos \theta$, $z = 4 \sin \theta$, $0 \leq x \leq 5$, $0 \leq \theta \leq 2\pi$.

27. The surface appears to be a portion of a circular cylinder of radius 3 with axis the x-axis. An equation of the cylinder is

$y^2 + z^2 = 9$, and we can impose the restrictions $0 \leq x \leq 5$, $y \leq 0$ to obtain the portion shown. To graph the surface on a

CAS, we can use parametric equations $x = u$, $y = 3 \cos v$, $z = 3 \sin v$ with the parameter domain $0 \leq u \leq 5$, $\frac{\pi}{2} \leq v \leq \frac{3\pi}{2}$.

Alternatively, we can regard x and z as parameters. Then parametric equations are $x = x$, $z = z$, $y = -\sqrt{9 - z^2}$, where

$0 \leq x \leq 5$ and $-3 \leq z \leq 3$.

29. Using Equations 3, we have the parametrization $x = x$, $y = e^{-x} \cos \theta$,

$z = e^{-x} \sin \theta$, $0 \leq x \leq 3$, $0 \leq \theta \leq 2\pi$.

© 2012 Cengage Learning. All Rights Reserved. May not be scanned, copied, or duplicated, or posted to a publicly accessible website, in whole or in part.

31. (a) Replacing $\cos u$ by $\sin u$ and $\sin u$ by $\cos u$ gives parametric equations

$x = (2 + \sin v)\sin u$, $y = (2 + \sin v)\cos u$, $z = u + \cos v$. From the graph, it

appears that the direction of the spiral is reversed. We can verify this observation by

noting that the projection of the spiral grid curves onto the xy-plane, given by

$x = (2 + \sin v)\sin u$, $y = (2 + \sin v)\cos u$, $z = 0$, draws a circle in the clockwise

direction for each value of v. The original equations, on the other hand, give circular

projections drawn in the counterclockwise direction. The equation for z is identical in

both surfaces, so as z increases, these grid curves spiral up in opposite directions for

the two surfaces.

(b) Replacing $\cos u$ by $\cos 2u$ and $\sin u$ by $\sin 2u$ gives parametric equations

$x = (2 + \sin v)\cos 2u$, $y = (2 + \sin v)\sin 2u$, $z = u + \cos v$. From the graph, it

appears that the number of coils in the surface doubles within the same parametric

domain. We can verify this observation by noting that the projection of the spiral grid

curves onto the xy-plane, given by $x = (2 + \sin v)\cos 2u$, $y = (2 + \sin v)\sin 2u$,

$z = 0$ (where v is constant), complete circular revolutions for $0 \le u \le \pi$ while the

original surface requires $0 \le u \le 2\pi$ for a complete revolution. Thus, the new

surface winds around twice as fast as the original surface, and since the equation for z

is identical in both surfaces, we observe twice as many circular coils in the same

z-interval.

33. $\mathbf{r}(u, v) = (u + v)\,\mathbf{i} + 3u^2\,\mathbf{j} + (u - v)\,\mathbf{k}$.

$\mathbf{r}_u = \mathbf{i} + 6u\,\mathbf{j} + \mathbf{k}$ and $\mathbf{r}_v = \mathbf{i} - \mathbf{k}$, so $\mathbf{r}_u \times \mathbf{r}_v = -6u\,\mathbf{i} + 2\,\mathbf{j} - 6u\,\mathbf{k}$. Since the point $(2, 3, 0)$ corresponds to $u = 1$, $v = 1$, a

normal vector to the surface at $(2, 3, 0)$ is $-6\,\mathbf{i} + 2\,\mathbf{j} - 6\,\mathbf{k}$, and an equation of the tangent plane is $-6x + 2y - 6z = -6$ or

$3x - y + 3z = 3$.

35. $\mathbf{r}(u, v) = u\cos v\,\mathbf{i} + u\sin v\,\mathbf{j} + v\,\mathbf{k} \;\Rightarrow\; \mathbf{r}\!\left(1, \tfrac{\pi}{3}\right) = \left(\tfrac{1}{2}, \tfrac{\sqrt{3}}{2}, \tfrac{\pi}{3}\right)$.

$\mathbf{r}_u = \cos v\,\mathbf{i} + \sin v\,\mathbf{j}$ and $\mathbf{r}_v = -u\sin v\,\mathbf{i} + u\cos v\,\mathbf{j} + \mathbf{k}$, so a normal vector to the surface at the point $\left(\tfrac{1}{2}, \tfrac{\sqrt{3}}{2}, \tfrac{\pi}{3}\right)$ is

$\mathbf{r}_u\!\left(1, \tfrac{\pi}{3}\right) \times \mathbf{r}_v\!\left(1, \tfrac{\pi}{3}\right) = \left(\tfrac{1}{2}\,\mathbf{i} + \tfrac{\sqrt{3}}{2}\,\mathbf{j}\right) \times \left(-\tfrac{\sqrt{3}}{2}\,\mathbf{i} + \tfrac{1}{2}\,\mathbf{j} + \mathbf{k}\right) = \tfrac{\sqrt{3}}{2}\,\mathbf{i} - \tfrac{1}{2}\,\mathbf{j} + \mathbf{k}$. Thus an equation of the tangent plane at

$\left(\tfrac{1}{2}, \tfrac{\sqrt{3}}{2}, \tfrac{\pi}{3}\right)$ is $\tfrac{\sqrt{3}}{2}\left(x - \tfrac{1}{2}\right) - \tfrac{1}{2}\left(y - \tfrac{\sqrt{3}}{2}\right) + 1\left(z - \tfrac{\pi}{3}\right) = 0$ or $\tfrac{\sqrt{3}}{2}x - \tfrac{1}{2}y + z = \tfrac{\pi}{3}$.

37. $\mathbf{r}(u, v) = u^2\,\mathbf{i} + 2u\sin v\,\mathbf{j} + u\cos v\,\mathbf{k} \;\Rightarrow\; \mathbf{r}(1, 0) = (1, 0, 1)$.

$\mathbf{r}_u = 2u\,\mathbf{i} + 2\sin v\,\mathbf{j} + \cos v\,\mathbf{k}$ and $\mathbf{r}_v = 2u\cos v\,\mathbf{j} - u\sin v\,\mathbf{k}$,

so a normal vector to the surface at the point $(1, 0, 1)$ is

$\mathbf{r}_u(1, 0) \times \mathbf{r}_v(1, 0) = (2\,\mathbf{i} + \mathbf{k}) \times (2\,\mathbf{j}) = -2\,\mathbf{i} + 4\,\mathbf{k}$.

Thus an equation of the tangent plane at $(1, 0, 1)$ is

$-2(x - 1) + 0(y - 0) + 4(z - 1) = 0$ or $-x + 2z = 1$.

© 2012 Cengage Learning. All Rights Reserved. May not be scanned, copied, or duplicated, or posted to a publicly accessible website, in whole or in part.

39. The surface S is given by $z = f(x,y) = 6 - 3x - 2y$ which intersects the xy-plane in the line $3x + 2y = 6$, so D is the triangular region given by $\{(x,y) \mid 0 \le x \le 2, 0 \le y \le 3 - \frac{3}{2}x\}$. By Formula 9, the surface area of S is

$$A(S) = \iint_D \sqrt{1 + \left(\frac{\partial z}{\partial x}\right)^2 + \left(\frac{\partial z}{\partial y}\right)^2}\, dA$$
$$= \iint_D \sqrt{1 + (-3)^2 + (-2)^2}\, dA = \sqrt{14} \iint_D dA = \sqrt{14}\, A(D) = \sqrt{14}\left(\frac{1}{2} \cdot 2 \cdot 3\right) = 3\sqrt{14}.$$

41. Here we can write $z = f(x,y) = \frac{1}{3} - \frac{1}{3}x - \frac{2}{3}y$ and D is the disk $x^2 + y^2 \le 3$, so by Formula 9 the area of the surface is

$$A(S) = \iint_D \sqrt{1 + \left(\frac{\partial z}{\partial x}\right)^2 + \left(\frac{\partial z}{\partial y}\right)^2}\, dA = \iint_D \sqrt{1 + \left(-\frac{1}{3}\right)^2 + \left(-\frac{2}{3}\right)^2}\, dA = \frac{\sqrt{14}}{3} \iint_D dA$$
$$= \frac{\sqrt{14}}{3} A(D) = \frac{\sqrt{14}}{3} \cdot \pi\left(\sqrt{3}\right)^2 = \sqrt{14}\,\pi$$

43. $z = f(x,y) = \frac{2}{3}(x^{3/2} + y^{3/2})$ and $D = \{(x,y) \mid 0 \le x \le 1, 0 \le y \le 1\}$. Then $f_x = x^{1/2}$, $f_y = y^{1/2}$ and

$$A(S) = \iint_D \sqrt{1 + (\sqrt{x})^2 + (\sqrt{y})^2}\, dA = \int_0^1 \int_0^1 \sqrt{1 + x + y}\, dy\, dx$$
$$= \int_0^1 \left[\frac{2}{3}(x + y + 1)^{3/2}\right]_{y=0}^{y=1} dx = \frac{2}{3} \int_0^1 \left[(x+2)^{3/2} - (x+1)^{3/2}\right] dx$$
$$= \frac{2}{3}\left[\frac{2}{5}(x+2)^{5/2} - \frac{2}{5}(x+1)^{5/2}\right]_0^1 = \frac{4}{15}(3^{5/2} - 2^{5/2} - 2^{5/2} + 1) = \frac{4}{15}(3^{5/2} - 2^{7/2} + 1)$$

45. $z = f(x,y) = xy$ with $x^2 + y^2 \le 1$, so $f_x = y$, $f_y = x$

$$A(S) = \iint_D \sqrt{1 + y^2 + x^2}\, dA = \int_0^{2\pi} \int_0^1 \sqrt{r^2 + 1}\, r\, dr\, d\theta = \int_0^{2\pi} \left[\frac{1}{3}(r^2+1)^{3/2}\right]_{r=0}^{r=1} d\theta$$
$$= \int_0^{2\pi} \frac{1}{3}(2\sqrt{2} - 1)\, d\theta = \frac{2\pi}{3}(2\sqrt{2} - 1)$$

47. A parametric representation of the surface is $x = x$, $y = 4x + z^2$, $z = z$ with $0 \le x \le 1$, $0 \le z \le 1$. Hence $\mathbf{r}_x \times \mathbf{r}_z = (\mathbf{i} + 4\mathbf{j}) \times (2z\mathbf{j} + \mathbf{k}) = 4\mathbf{i} - \mathbf{j} + 2z\mathbf{k}$.

Note: In general, if $y = f(x,z)$ then $\mathbf{r}_x \times \mathbf{r}_z = \frac{\partial f}{\partial x}\mathbf{i} - \mathbf{j} + \frac{\partial f}{\partial z}\mathbf{k}$ and $A(S) = \iint_D \sqrt{1 + \left(\frac{\partial f}{\partial x}\right)^2 + \left(\frac{\partial f}{\partial z}\right)^2}\, dA$. Then

$$A(S) = \int_0^1 \int_0^1 \sqrt{17 + 4z^2}\, dx\, dz = \int_0^1 \sqrt{17 + 4z^2}\, dz$$
$$= \frac{1}{2}\left(z\sqrt{17 + 4z^2} + \frac{17}{2}\ln\left|2z + \sqrt{4z^2 + 17}\right|\right)\Big]_0^1 = \frac{\sqrt{21}}{2} + \frac{17}{4}\left[\ln(2 + \sqrt{21}) - \ln\sqrt{17}\right]$$

49. $\mathbf{r}_u = \langle 2u, v, 0\rangle$, $\mathbf{r}_v = \langle 0, u, v\rangle$, and $\mathbf{r}_u \times \mathbf{r}_v = \langle v^2, -2uv, 2u^2\rangle$. Then

$$A(S) = \iint_D |\mathbf{r}_u \times \mathbf{r}_v|\, dA = \int_0^1 \int_0^2 \sqrt{v^4 + 4u^2v^2 + 4u^4}\, dv\, du = \int_0^1 \int_0^2 \sqrt{(v^2 + 2u^2)^2}\, dv\, du$$
$$= \int_0^1 \int_0^2 (v^2 + 2u^2)\, dv\, du = \int_0^1 \left[\frac{1}{3}v^3 + 2u^2v\right]_{v=0}^{v=2} du = \int_0^1 \left(\frac{8}{3} + 4u^2\right) du = \left[\frac{8}{3}u + \frac{4}{3}u^3\right]_0^1 = 4$$

51. From Equation 9 we have $A(S) = \iint_D \sqrt{1 + (f_x)^2 + (f_y)^2}\, dA$. But if $|f_x| \le 1$ and $|f_y| \le 1$ then $0 \le (f_x)^2 \le 1$, $0 \le (f_y)^2 \le 1 \Rightarrow 1 \le 1 + (f_x)^2 + (f_y)^2 \le 3 \Rightarrow 1 \le \sqrt{1 + (f_x)^2 + (f_y)^2} \le \sqrt{3}$. By Property 15.3.11,

$$\iint_D 1\, dA \le \iint_D \sqrt{1 + (f_x)^2 + (f_y)^2}\, dA \le \iint_D \sqrt{3}\, dA \Rightarrow A(D) \le A(S) \le \sqrt{3}\, A(D) \Rightarrow$$
$$\pi R^2 \le A(S) \le \sqrt{3}\pi R^2.$$

© 2012 Cengage Learning. All Rights Reserved. May not be scanned, copied, or duplicated, or posted to a publicly accessible website, in whole or in part.

53. $z = f(x, y) = e^{-x^2 - y^2}$ with $x^2 + y^2 \le 4$.

$$A(S) = \iint_D \sqrt{1 + \left(-2xe^{-x^2-y^2}\right)^2 + \left(-2ye^{-x^2-y^2}\right)^2} \, dA = \iint_D \sqrt{1 + 4(x^2 + y^2)e^{-2(x^2+y^2)}} \, dA$$

$$= \int_0^{2\pi} \int_0^2 \sqrt{1 + 4r^2 e^{-2r^2}} \, r \, dr \, d\theta = \int_0^{2\pi} d\theta \int_0^2 r\sqrt{1 + 4r^2 e^{-2r^2}} \, dr = 2\pi \int_0^2 r\sqrt{1 + 4r^2 e^{-2r^2}} \, dr \approx 13.9783$$

55. (a) $A(S) = \iint_D \sqrt{1 + \left(\dfrac{\partial z}{\partial x}\right)^2 + \left(\dfrac{\partial z}{\partial y}\right)^2} \, dA = \int_0^6 \int_0^4 \sqrt{1 + \dfrac{4x^2 + 4y^2}{(1 + x^2 + y^2)^4}} \, dy \, dx.$

Using the Midpoint Rule with $f(x, y) = \sqrt{1 + \dfrac{4x^2 + 4y^2}{(1 + x^2 + y^2)^4}}$, $m = 3$, $n = 2$ we have

$$A(S) \approx \sum_{i=1}^{3} \sum_{j=1}^{2} f(\overline{x}_i, \overline{y}_j) \, \Delta A = 4\left[f(1, 1) + f(1, 3) + f(3, 1) + f(3, 3) + f(5, 1) + f(5, 3)\right] \approx 24.2055$$

(b) Using a CAS we have $A(S) = \int_0^6 \int_0^4 \sqrt{1 + \dfrac{4x^2 + 4y^2}{(1 + x^2 + y^2)^4}} \, dy \, dx \approx 24.2476$. This agrees with the estimate in part (a)

to the first decimal place.

57. $z = 1 + 2x + 3y + 4y^2$, so

$$A(S) = \iint_D \sqrt{1 + \left(\dfrac{\partial z}{\partial x}\right)^2 + \left(\dfrac{\partial z}{\partial y}\right)^2} \, dA = \int_1^4 \int_0^1 \sqrt{1 + 4 + (3 + 8y)^2} \, dy \, dx = \int_1^4 \int_0^1 \sqrt{14 + 48y + 64y^2} \, dy \, dx.$$

Using a CAS, we have

$\int_1^4 \int_0^1 \sqrt{14 + 48y + 64y^2} \, dy \, dx = \dfrac{45}{8}\sqrt{14} + \dfrac{15}{16}\ln\left(11\sqrt{5} + 3\sqrt{14}\sqrt{5}\right) - \dfrac{15}{16}\ln\left(3\sqrt{5} + \sqrt{14}\sqrt{5}\right)$

or $\dfrac{45}{8}\sqrt{14} + \dfrac{15}{16}\ln\dfrac{11\sqrt{5} + 3\sqrt{70}}{3\sqrt{5} + \sqrt{70}}$.

59. (a) $x = a\sin u \cos v$, $y = b\sin u \sin v$, $z = c\cos u \;\Rightarrow$

$$\dfrac{x^2}{a^2} + \dfrac{y^2}{b^2} + \dfrac{z^2}{c^2} = (\sin u \cos v)^2 + (\sin u \sin v)^2 + (\cos u)^2$$

$$= \sin^2 u + \cos^2 u = 1$$

and since the ranges of u and v are sufficient to generate the entire graph,

the parametric equations represent an ellipsoid.

(b)

(c) From the parametric equations (with $a = 1$, $b = 2$, and $c = 3$),

we calculate $\mathbf{r}_u = \cos u \cos v \, \mathbf{i} + 2\cos u \sin v \, \mathbf{j} - 3\sin u \, \mathbf{k}$ and

$\mathbf{r}_v = -\sin u \sin v \, \mathbf{i} + 2\sin u \cos v \, \mathbf{j}$. So $\mathbf{r}_u \times \mathbf{r}_v = 6\sin^2 u \cos v \, \mathbf{i} + 3\sin^2 u \sin v \, \mathbf{j} + 2\sin u \cos u \, \mathbf{k}$, and the surface

area is given by $A(S) = \int_0^{2\pi} \int_0^{\pi} |\mathbf{r}_u \times \mathbf{r}_v| \, du \, dv = \int_0^{2\pi} \int_0^{\pi} \sqrt{36\sin^4 u \cos^2 v + 9\sin^4 u \sin^2 v + 4\cos^2 u \sin^2 u} \, du \, dv$

61. To find the region D: $z = x^2 + y^2$ implies $z + z^2 = 4z$ or $z^2 - 3z = 0$. Thus $z = 0$ or $z = 3$ are the planes where the

surfaces intersect. But $x^2 + y^2 + z^2 = 4z$ implies $x^2 + y^2 + (z - 2)^2 = 4$, so $z = 3$ intersects the upper hemisphere.

Thus $(z - 2)^2 = 4 - x^2 - y^2$ or $z = 2 + \sqrt{4 - x^2 - y^2}$. Therefore D is the region inside the circle $x^2 + y^2 + (3 - 2)^2 = 4$,

© 2012 Cengage Learning. All Rights Reserved. May not be scanned, copied, or duplicated, or posted to a publicly accessible website, in whole or in part.

that is, $D = \{(x,y) \mid x^2 + y^2 \le 3\}$.

$$A(S) = \iint_D \sqrt{1 + [(-x)(4 - x^2 - y^2)^{-1/2}]^2 + [(-y)(4 - x^2 - y^2)^{-1/2}]^2} \, dA$$

$$= \int_0^{2\pi} \int_0^{\sqrt{3}} \sqrt{1 + \frac{r^2}{4 - r^2}} \, r \, dr \, d\theta = \int_0^{2\pi} \int_0^{\sqrt{3}} \frac{2r \, dr}{\sqrt{4 - r^2}} \, d\theta = \int_0^{2\pi} \left[-2(4 - r^2)^{1/2} \right]_{r=0}^{r=\sqrt{3}} d\theta$$

$$= \int_0^{2\pi} (-2 + 4) \, d\theta = 2\theta \Big]_0^{2\pi} = 4\pi$$

63. Let $A(S_1)$ be the surface area of that portion of the surface which lies above the plane $z = 0$. Then $A(S) = 2A(S_1)$.

Following Example 10, a parametric representation of S_1 is $x = a \sin\phi \cos\theta$, $y = a \sin\phi \sin\theta$,

$z = a \cos\phi$ and $|\mathbf{r}_\phi \times \mathbf{r}_\theta| = a^2 \sin\phi$. For D, $0 \le \phi \le \frac{\pi}{2}$ and for each fixed ϕ, $\left(x - \frac{1}{2}a\right)^2 + y^2 \le \left(\frac{1}{2}a\right)^2$ or

$\left[a \sin\phi \cos\theta - \frac{1}{2}a\right]^2 + a^2 \sin^2\phi \sin^2\theta \le (a/2)^2$ implies $a^2 \sin^2\phi - a^2 \sin\phi \cos\theta \le 0$ or

$\sin\phi (\sin\phi - \cos\theta) \le 0$. But $0 \le \phi \le \frac{\pi}{2}$, so $\cos\theta \ge \sin\phi$ or $\sin\left(\frac{\pi}{2} + \theta\right) \ge \sin\phi$ or $\phi - \frac{\pi}{2} \le \theta \le \frac{\pi}{2} - \phi$.

Hence $D = \left\{(\phi, \theta) \mid 0 \le \phi \le \frac{\pi}{2}, \phi - \frac{\pi}{2} \le \theta \le \frac{\pi}{2} - \phi\right\}$. Then

$$A(S_1) = \int_0^{\pi/2} \int_{\phi - (\pi/2)}^{(\pi/2) - \phi} a^2 \sin\phi \, d\theta \, d\phi = a^2 \int_0^{\pi/2} (\pi - 2\phi) \sin\phi \, d\phi$$

$$= a^2 \left[(-\pi \cos\phi) - 2(-\phi \cos\phi + \sin\phi)\right]_0^{\pi/2} = a^2(\pi - 2)$$

Thus $A(S) = 2a^2(\pi - 2)$.

Alternate solution: Working on S_1 we could parametrize the portion of the sphere by $x = x$, $y = y$, $z = \sqrt{a^2 - x^2 - y^2}$.

Then $|\mathbf{r}_x \times \mathbf{r}_y| = \sqrt{1 + \dfrac{x^2}{a^2 - x^2 - y^2} + \dfrac{y^2}{a^2 - x^2 - y^2}} = \dfrac{a}{\sqrt{a^2 - x^2 - y^2}}$ and

$$A(S_1) = \iint_{0 \le (x - (a/2))^2 + y^2 \le (a/2)^2} \frac{a}{\sqrt{a^2 - x^2 - y^2}} \, dA = \int_{-\pi/2}^{\pi/2} \int_0^{a\cos\theta} \frac{a}{\sqrt{a^2 - r^2}} \, r \, dr \, d\theta$$

$$= \int_{-\pi/2}^{\pi/2} -a(a^2 - r^2)^{1/2} \Big]_{r=0}^{r=a\cos\theta} d\theta = \int_{-\pi/2}^{\pi/2} a^2[1 - (1 - \cos^2\theta)^{1/2}] \, d\theta$$

$$= \int_{-\pi/2}^{\pi/2} a^2(1 - |\sin\theta|) \, d\theta = 2a^2 \int_0^{\pi/2} (1 - \sin\theta) \, d\theta = 2a^2\left(\frac{\pi}{2} - 1\right)$$

Thus $A(S) = 4a^2\left(\frac{\pi}{2} - 1\right) = 2a^2(\pi - 2)$.

Notes:

(1) Perhaps working in spherical coordinates is the most obvious approach here. However, you must be careful in setting up D.

(2) In the alternate solution, you can avoid having to use $|\sin\theta|$ by working in the first octant and then multiplying by 4. However, if you set up S_1 as above and arrived at $A(S_1) = a^2\pi$, you now see your error.

© 2012 Cengage Learning. All Rights Reserved. May not be scanned, copied, or duplicated, or posted to a publicly accessible website, in whole or in part.

16.7 Surface Integrals

1. The faces of the box in the planes $x = 0$ and $x = 2$ have surface area 24 and centers $(0, 2, 3)$, $(2, 2, 3)$. The faces in $y = 0$ and $y = 4$ have surface area 12 and centers $(1, 0, 3)$, $(1, 4, 3)$, and the faces in $z = 0$ and $z = 6$ have area 8 and centers $(1, 2, 0)$, $(1, 2, 6)$. For each face we take the point P_{ij}^* to be the center of the face and $f(x, y, z) = e^{-0.1(x+y+z)}$, so by Definition 1,

$$\iint_S f(x, y, z)\, dS \approx [f(0, 2, 3)](24) + [f(2, 2, 3)](24) + [f(1, 0, 3)](12)$$
$$+ [f(1, 4, 3)](12) + [f(1, 2, 0)](8) + [f(1, 2, 6)](8)$$
$$= 24(e^{-0.5} + e^{-0.7}) + 12(e^{-0.4} + e^{-0.8}) + 8(e^{-0.3} + e^{-0.9}) \approx 49.09$$

3. We can use the xz- and yz-planes to divide H into four patches of equal size, each with surface area equal to $\frac{1}{8}$ the surface area of a sphere with radius $\sqrt{50}$, so $\Delta S = \frac{1}{8}(4)\pi(\sqrt{50})^2 = 25\pi$. Then $(\pm 3, \pm 4, 5)$ are sample points in the four patches, and using a Riemann sum as in Definition 1, we have

$$\iint_H f(x, y, z)\, dS \approx f(3, 4, 5)\,\Delta S + f(3, -4, 5)\,\Delta S + f(-3, 4, 5)\,\Delta S + f(-3, -4, 5)\,\Delta S$$
$$= (7 + 8 + 9 + 12)(25\pi) = 900\pi \approx 2827$$

5. $\mathbf{r}(u, v) = (u + v)\,\mathbf{i} + (u - v)\,\mathbf{j} + (1 + 2u + v)\,\mathbf{k}$, $0 \le u \le 2$, $0 \le v \le 1$ and

$\mathbf{r}_u \times \mathbf{r}_v = (\mathbf{i} + \mathbf{j} + 2\,\mathbf{k}) \times (\mathbf{i} - \mathbf{j} + \mathbf{k}) = 3\,\mathbf{i} + \mathbf{j} - 2\,\mathbf{k} \quad \Rightarrow \quad |\mathbf{r}_u \times \mathbf{r}_v| = \sqrt{3^2 + 1^2 + (-2)^2} = \sqrt{14}$. Then by Formula 2,

$$\iint_S (x + y + z)\, dS = \iint_D (u + v + u - v + 1 + 2u + v)\, |\mathbf{r}_u \times \mathbf{r}_v|\, dA = \int_0^1 \int_0^2 (4u + v + 1) \cdot \sqrt{14}\, du\, dv$$
$$= \sqrt{14} \int_0^1 \left[2u^2 + uv + u \right]_{u=0}^{u=2} dv = \sqrt{14} \int_0^1 (2v + 10)\, dv = \sqrt{14} \left[v^2 + 10v \right]_0^1 = 11\sqrt{14}$$

7. $\mathbf{r}(u, v) = \langle u \cos v, u \sin v, v \rangle$, $0 \le u \le 1$, $0 \le v \le \pi$ and

$\mathbf{r}_u \times \mathbf{r}_v = \langle \cos v, \sin v, 0 \rangle \times \langle -u \sin v, u \cos v, 1 \rangle = \langle \sin v, -\cos v, u \rangle \quad \Rightarrow$

$|\mathbf{r}_u \times \mathbf{r}_v| = \sqrt{\sin^2 v + \cos^2 v + u^2} = \sqrt{u^2 + 1}$. Then

$$\iint_S y\, dS = \iint_D (u \sin v)\, |\mathbf{r}_u \times \mathbf{r}_v|\, dA = \int_0^1 \int_0^\pi (u \sin v) \cdot \sqrt{u^2 + 1}\, dv\, du = \int_0^1 u\sqrt{u^2 + 1}\, du \int_0^\pi \sin v\, dv$$
$$= \left[\tfrac{1}{3}(u^2 + 1)^{3/2} \right]_0^1 \left[-\cos v \right]_0^\pi = \tfrac{1}{3}(2^{3/2} - 1) \cdot 2 = \tfrac{2}{3}(2\sqrt{2} - 1)$$

9. $z = 1 + 2x + 3y$ so $\dfrac{\partial z}{\partial x} = 2$ and $\dfrac{\partial z}{\partial y} = 3$. Then by Formula 4,

$$\iint_S x^2 yz\, dS = \iint_D x^2 yz \sqrt{\left(\frac{\partial z}{\partial x}\right)^2 + \left(\frac{\partial z}{\partial y}\right)^2 + 1}\, dA = \int_0^3 \int_0^2 x^2 y(1 + 2x + 3y) \sqrt{4 + 9 + 1}\, dy\, dx$$
$$= \sqrt{14} \int_0^3 \int_0^2 (x^2 y + 2x^3 y + 3x^2 y^2)\, dy\, dx = \sqrt{14} \int_0^3 \left[\tfrac{1}{2}x^2 y^2 + x^3 y^2 + x^2 y^3 \right]_{y=0}^{y=2} dx$$
$$= \sqrt{14} \int_0^3 (10x^2 + 4x^3)\, dx = \sqrt{14} \left[\tfrac{10}{3}x^3 + x^4 \right]_0^3 = 171\sqrt{14}$$

11. An equation of the plane through the points $(1, 0, 0)$, $(0, -2, 0)$, and $(0, 0, 4)$ is $4x - 2y + z = 4$, so S is the region in the plane $z = 4 - 4x + 2y$ over $D = \{(x, y) \mid 0 \le x \le 1, 2x - 2 \le y \le 0\}$. Thus by Formula 4,

$$\iint_S x\, dS = \iint_D x \sqrt{(-4)^2 + (2)^2 + 1}\, dA = \sqrt{21} \int_0^1 \int_{2x-2}^0 x\, dy\, dx = \sqrt{21} \int_0^1 [xy]_{y=2x-2}^{y=0}\, dx$$
$$= \sqrt{21} \int_0^1 (-2x^2 + 2x)\, dx = \sqrt{21} \left[-\tfrac{2}{3}x^3 + x^2 \right]_0^1 = \sqrt{21} \left(-\tfrac{2}{3} + 1 \right) = \tfrac{\sqrt{21}}{3}$$

© 2012 Cengage Learning. All Rights Reserved. May not be scanned, copied, or duplicated, or posted to a publicly accessible website, in whole or in part.

13. S is the portion of the cone $z^2 = x^2 + y^2$ for $1 \le z \le 3$, or equivalently, S is the part of the surface $z = \sqrt{x^2 + y^2}$ over the region $D = \{(x, y) \mid 1 \le x^2 + y^2 \le 9\}$. Thus

$$\iint_S x^2 z^2 \, dS = \iint_D x^2(x^2 + y^2)\sqrt{\left(\frac{x}{\sqrt{x^2 + y^2}}\right)^2 + \left(\frac{y}{\sqrt{x^2 + y^2}}\right)^2 + 1} \, dA$$

$$= \iint_D x^2(x^2 + y^2)\sqrt{\frac{x^2 + y^2}{x^2 + y^2} + 1} \, dA = \iint_D \sqrt{2}\, x^2(x^2 + y^2) \, dA = \sqrt{2}\int_0^{2\pi}\int_1^3 (r\cos\theta)^2 (r^2)\, r\, dr\, d\theta$$

$$= \sqrt{2}\int_0^{2\pi}\cos^2\theta\, d\theta \int_1^3 r^5\, dr = \sqrt{2}\left[\tfrac{1}{2}\theta + \tfrac{1}{4}\sin 2\theta\right]_0^{2\pi}\left[\tfrac{1}{6}r^6\right]_1^3 = \sqrt{2}\,(\pi)\cdot\tfrac{1}{6}(3^6 - 1) = \frac{364\sqrt{2}}{3}\pi$$

15. Using x and z as parameters, we have $\mathbf{r}(x, z) = x\,\mathbf{i} + (x^2 + z^2)\,\mathbf{j} + z\,\mathbf{k}$, $x^2 + z^2 \le 4$. Then

$$\mathbf{r}_x \times \mathbf{r}_z = (\mathbf{i} + 2x\,\mathbf{j}) \times (2z\,\mathbf{j} + \mathbf{k}) = 2x\,\mathbf{i} - \mathbf{j} + 2z\,\mathbf{k} \text{ and } |\mathbf{r}_x \times \mathbf{r}_z| = \sqrt{4x^2 + 1 + 4z^2} = \sqrt{1 + 4(x^2 + z^2)}. \text{ Thus}$$

$$\iint_S y\, dS = \iint_{x^2 + z^2 \le 4} (x^2 + z^2)\sqrt{1 + 4(x^2 + z^2)}\, dA = \int_0^{2\pi}\int_0^2 r^2\sqrt{1 + 4r^2}\, r\, dr\, d\theta = \int_0^{2\pi} d\theta \int_0^2 r^2\sqrt{1 + 4r^2}\, r\, dr$$

$$= 2\pi\int_0^2 r^2\sqrt{1 + 4r^2}\, r\, dr \qquad \left[\text{let } u = 1 + 4r^2 \;\Rightarrow\; r^2 = \tfrac{1}{4}(u - 1) \text{ and } \tfrac{1}{8}du = r\, dr\right]$$

$$= 2\pi\int_1^{17} \tfrac{1}{4}(u - 1)\sqrt{u}\cdot\tfrac{1}{8}du = \tfrac{1}{16}\pi\int_1^{17} (u^{3/2} - u^{1/2})\, du$$

$$= \tfrac{1}{16}\pi\left[\tfrac{2}{5}u^{5/2} - \tfrac{2}{3}u^{3/2}\right]_1^{17} = \tfrac{1}{16}\pi\left[\tfrac{2}{5}(17)^{5/2} - \tfrac{2}{3}(17)^{3/2} - \tfrac{2}{5} + \tfrac{2}{3}\right] = \frac{\pi}{60}\left(391\sqrt{17} + 1\right)$$

17. Using spherical coordinates and Example 16.6.10 we have $\mathbf{r}(\phi, \theta) = 2\sin\phi\cos\theta\,\mathbf{i} + 2\sin\phi\sin\theta\,\mathbf{j} + 2\cos\phi\,\mathbf{k}$ and

$|\mathbf{r}_\phi \times \mathbf{r}_\theta| = 4\sin\phi$. Then $\iint_S (x^2 z + y^2 z)\, dS = \int_0^{2\pi}\int_0^{\pi/2} (4\sin^2\phi)(2\cos\phi)(4\sin\phi)\, d\phi\, d\theta = 16\pi\sin^4\phi\big]_0^{\pi/2} = 16\pi.$

19. S is given by $\mathbf{r}(u, v) = u\,\mathbf{i} + \cos v\,\mathbf{j} + \sin v\,\mathbf{k}$, $0 \le u \le 3$, $0 \le v \le \pi/2$. Then

$$\mathbf{r}_u \times \mathbf{r}_v = \mathbf{i} \times (-\sin v\,\mathbf{j} + \cos v\,\mathbf{k}) = -\cos v\,\mathbf{j} - \sin v\,\mathbf{k} \text{ and } |\mathbf{r}_u \times \mathbf{r}_v| = \sqrt{\cos^2 v + \sin^2 v} = 1, \text{ so}$$

$$\iint_S (z + x^2 y)\, dS = \int_0^{\pi/2}\int_0^3 (\sin v + u^2\cos v)(1)\, du\, dv = \int_0^{\pi/2}(3\sin v + 9\cos v)\, dv$$

$$= [-3\cos v + 9\sin v]_0^{\pi/2} = 0 + 9 + 3 - 0 = 12$$

21. From Exercise 5, $\mathbf{r}(u, v) = (u + v)\,\mathbf{i} + (u - v)\,\mathbf{j} + (1 + 2u + v)\,\mathbf{k}$, $0 \le u \le 2$, $0 \le v \le 1$, and $\mathbf{r}_u \times \mathbf{r}_v = 3\,\mathbf{i} + \mathbf{j} - 2\,\mathbf{k}$. Then

$$\mathbf{F}(\mathbf{r}(u, v)) = (1 + 2u + v)e^{(u+v)(u-v)}\,\mathbf{i} - 3(1 + 2u + v)e^{(u+v)(u-v)}\,\mathbf{j} + (u + v)(u - v)\,\mathbf{k}$$

$$= (1 + 2u + v)e^{u^2 - v^2}\,\mathbf{i} - 3(1 + 2u + v)e^{u^2 - v^2}\,\mathbf{j} + (u^2 - v^2)\,\mathbf{k}$$

Because the z-component of $\mathbf{r}_u \times \mathbf{r}_v$ is negative we use $-(\mathbf{r}_u \times \mathbf{r}_v)$ in Formula 9 for the upward orientation:

$$\iint_S \mathbf{F}\cdot d\mathbf{S} = \iint_D \mathbf{F}\cdot(-(\mathbf{r}_u \times \mathbf{r}_v))\, dA = \int_0^1\int_0^2 \left[-3(1 + 2u + v)e^{u^2 - v^2} + 3(1 + 2u + v)e^{u^2 - v^2} + 2(u^2 - v^2)\right] du\, dv$$

$$= \int_0^1\int_0^2 2(u^2 - v^2)\, du\, dv = 2\int_0^1 \left[\tfrac{1}{3}u^3 - uv^2\right]_{u=0}^{u=2} dv = 2\int_0^1 \left(\tfrac{8}{3} - 2v^2\right) dv$$

$$= 2\left[\tfrac{8}{3}v - \tfrac{2}{3}v^3\right]_0^1 = 2\left(\tfrac{8}{3} - \tfrac{2}{3}\right) = 4$$

23. $\mathbf{F}(x, y, z) = xy\,\mathbf{i} + yz\,\mathbf{j} + zx\,\mathbf{k}$, $z = g(x, y) = 4 - x^2 - y^2$, and D is the square $[0, 1] \times [0, 1]$, so by Equation 10

$$\iint_S \mathbf{F}\cdot d\mathbf{S} = \iint_D [-xy(-2x) - yz(-2y) + zx]\, dA = \int_0^1\int_0^1 [2x^2 y + 2y^2(4 - x^2 - y^2) + x(4 - x^2 - y^2)]\, dy\, dx$$

$$= \int_0^1 \left(\tfrac{1}{3}x^2 + \tfrac{11}{3}x - x^3 + \tfrac{34}{15}\right) dx = \frac{713}{180}$$

© 2012 Cengage Learning. All Rights Reserved. May not be scanned, copied, or duplicated, or posted to a publicly accessible website, in whole or in part.

25. $\mathbf{F}(x, y, z) = x\,\mathbf{i} - z\,\mathbf{j} + y\,\mathbf{k}$, $z = g(x, y) = \sqrt{4 - x^2 - y^2}$ and D is the quarter disk

$\{(x, y) \mid 0 \le x \le 2, 0 \le y \le \sqrt{4 - x^2}\}$. S has downward orientation, so by Formula 10,

$$\iint_S \mathbf{F} \cdot d\mathbf{S} = -\iint_D \left[-x \cdot \tfrac{1}{2}(4 - x^2 - y^2)^{-1/2}(-2x) - (-z) \cdot \tfrac{1}{2}(4 - x^2 - y^2)^{-1/2}(-2y) + y\right] dA$$

$$= -\iint_D \left(\frac{x^2}{\sqrt{4 - x^2 - y^2}} - \sqrt{4 - x^2 - y^2} \cdot \frac{y}{\sqrt{4 - x^2 - y^2}} + y\right) dA$$

$$= -\iint_D x^2 (4 - (x^2 + y^2))^{-1/2}\, dA = -\int_0^{\pi/2} \int_0^2 (r\cos\theta)^2 (4 - r^2)^{-1/2}\, r\, dr\, d\theta$$

$$= -\int_0^{\pi/2} \cos^2\theta\, d\theta \int_0^2 r^3 (4 - r^2)^{-1/2}\, dr \qquad \left[\text{let } u = 4 - r^2 \;\Rightarrow\; r^2 = 4 - u \text{ and } -\tfrac{1}{2}\, du = r\, dr\right]$$

$$= -\int_0^{\pi/2} \left(\tfrac{1}{2} + \tfrac{1}{2}\cos 2\theta\right) d\theta \int_4^0 -\tfrac{1}{2}(4 - u)(u)^{-1/2}\, du$$

$$= -\left[\tfrac{1}{2}\theta + \tfrac{1}{4}\sin 2\theta\right]_0^{\pi/2} \left(-\tfrac{1}{2}\right)\left[8\sqrt{u} - \tfrac{2}{3}u^{3/2}\right]_4^0 = -\tfrac{\pi}{4}\left(-\tfrac{1}{2}\right)\left(-16 + \tfrac{16}{3}\right) = -\tfrac{4}{3}\pi$$

27. Let S_1 be the paraboloid $y = x^2 + z^2$, $0 \le y \le 1$ and S_2 the disk $x^2 + z^2 \le 1$, $y = 1$. Since S is a closed

surface, we use the outward orientation.

On S_1: $\mathbf{F}(\mathbf{r}(x, z)) = (x^2 + z^2)\,\mathbf{j} - z\,\mathbf{k}$ and $\mathbf{r}_x \times \mathbf{r}_z = 2x\,\mathbf{i} - \mathbf{j} + 2z\,\mathbf{k}$ (since the $\mathbf{j}$-component must be negative on S_1). Then

$$\iint_{S_1} \mathbf{F} \cdot d\mathbf{S} = \iint_{x^2 + z^2 \le 1} \left[-(x^2 + z^2) - 2z^2\right] dA = -\int_0^{2\pi} \int_0^1 (r^2 + 2r^2 \sin^2\theta)\, r\, dr\, d\theta$$

$$= -\int_0^{2\pi} \int_0^1 r^3 (1 + 2\sin^2\theta)\, dr\, d\theta = -\int_0^{2\pi} (1 + 1 - \cos 2\theta)\, d\theta \int_0^1 r^3\, dr$$

$$= -\left[2\theta - \tfrac{1}{2}\sin 2\theta\right]_0^{2\pi} \left[\tfrac{1}{4}r^4\right]_0^1 = -4\pi \cdot \tfrac{1}{4} = -\pi$$

On S_2: $\mathbf{F}(\mathbf{r}(x, z)) = \mathbf{j} - z\,\mathbf{k}$ and $\mathbf{r}_z \times \mathbf{r}_x = \mathbf{j}$. Then $\iint_{S_2} \mathbf{F} \cdot d\mathbf{S} = \iint_{x^2 + z^2 \le 1} (1)\, dA = \pi$.

Hence $\iint_S \mathbf{F} \cdot d\mathbf{S} = -\pi + \pi = 0$.

29. Here S consists of the six faces of the cube as labeled in the figure. On S_1:

$\mathbf{F} = \mathbf{i} + 2y\,\mathbf{j} + 3z\,\mathbf{k}$, $\mathbf{r}_y \times \mathbf{r}_z = \mathbf{i}$ and $\iint_{S_1} \mathbf{F} \cdot d\mathbf{S} = \int_{-1}^1 \int_{-1}^1 dy\, dz = 4$;

S_2: $\mathbf{F} = x\,\mathbf{i} + 2\,\mathbf{j} + 3z\,\mathbf{k}$, $\mathbf{r}_z \times \mathbf{r}_x = \mathbf{j}$ and $\iint_{S_2} \mathbf{F} \cdot d\mathbf{S} = \int_{-1}^1 \int_{-1}^1 2\, dx\, dz = 8$;

S_3: $\mathbf{F} = x\,\mathbf{i} + 2y\,\mathbf{j} + 3\,\mathbf{k}$, $\mathbf{r}_x \times \mathbf{r}_y = \mathbf{k}$ and $\iint_{S_3} \mathbf{F} \cdot d\mathbf{S} = \int_{-1}^1 \int_{-1}^1 3\, dx\, dy = 12$;

S_4: $\mathbf{F} = -\mathbf{i} + 2y\,\mathbf{j} + 3z\,\mathbf{k}$, $\mathbf{r}_z \times \mathbf{r}_y = -\mathbf{i}$ and $\iint_{S_4} \mathbf{F} \cdot d\mathbf{S} = 4$;

S_5: $\mathbf{F} = x\,\mathbf{i} - 2\,\mathbf{j} + 3z\,\mathbf{k}$, $\mathbf{r}_x \times \mathbf{r}_z = -\mathbf{j}$ and $\iint_{S_5} \mathbf{F} \cdot d\mathbf{S} = 8$;

S_6: $\mathbf{F} = x\,\mathbf{i} + 2y\,\mathbf{j} - 3\,\mathbf{k}$, $\mathbf{r}_y \times \mathbf{r}_x = -\mathbf{k}$ and $\iint_{S_6} \mathbf{F} \cdot d\mathbf{S} = \int_{-1}^1 \int_{-1}^1 3\, dx\, dy = 12$.

Hence $\iint_S \mathbf{F} \cdot d\mathbf{S} = \sum_{i=1}^6 \iint_{S_i} \mathbf{F} \cdot d\mathbf{S} = 48$.

31. Here S consists of four surfaces: S_1, the top surface (a portion of the circular cylinder $y^2 + z^2 = 1$); S_2, the bottom surface

(a portion of the xy-plane); S_3, the front half-disk in the plane $x = 2$, and S_4, the back half-disk in the plane $x = 0$.

On S_1: The surface is $z = \sqrt{1 - y^2}$ for $0 \le x \le 2$, $-1 \le y \le 1$ with upward orientation, so

$$\iint_{S_1} \mathbf{F} \cdot d\mathbf{S} = \int_0^2 \int_{-1}^1 \left[-x^2 (0) - y^2 \left(-\frac{y}{\sqrt{1 - y^2}} \right) + z^2 \right] dy\, dx = \int_0^2 \int_{-1}^1 \left(\frac{y^3}{\sqrt{1 - y^2}} + 1 - y^2 \right) dy\, dx$$

$$= \int_0^2 \left[-\sqrt{1 - y^2} + \tfrac{1}{3}(1 - y^2)^{3/2} + y - \tfrac{1}{3}y^3 \right]_{y=-1}^{y=1} dx = \int_0^2 \tfrac{4}{3}\, dx = \tfrac{8}{3}$$

On S_2: The surface is $z = 0$ with downward orientation, so

$$\iint_{S_2} \mathbf{F} \cdot d\mathbf{S} = \int_0^2 \int_{-1}^1 \left(-z^2 \right) dy\, dx = \int_0^2 \int_{-1}^1 (0)\, dy\, dx = 0$$

On S_3: The surface is $x = 2$ for $-1 \le y \le 1$, $0 \le z \le \sqrt{1 - y^2}$, oriented in the positive x-direction. Regarding y and z as parameters, we have $\mathbf{r}_y \times \mathbf{r}_z = \mathbf{i}$ and

$$\iint_{S_3} \mathbf{F} \cdot d\mathbf{S} = \int_{-1}^1 \int_0^{\sqrt{1-y^2}} x^2\, dz\, dy = \int_{-1}^1 \int_0^{\sqrt{1-y^2}} 4\, dz\, dy = 4A(S_3) = 2\pi$$

On S_4: The surface is $x = 0$ for $-1 \le y \le 1$, $0 \le z \le \sqrt{1 - y^2}$, oriented in the negative x-direction. Regarding y and z as parameters, we use $-(\mathbf{r}_y \times \mathbf{r}_z) = -\mathbf{i}$ and

$$\iint_{S_4} \mathbf{F} \cdot d\mathbf{S} = \int_{-1}^1 \int_0^{\sqrt{1-y^2}} x^2\, dz\, dy = \int_{-1}^1 \int_0^{\sqrt{1-y^2}} (0)\, dz\, dy = 0$$

Thus $\iint_S \mathbf{F} \cdot d\mathbf{S} = \tfrac{8}{3} + 0 + 2\pi + 0 = 2\pi + \tfrac{8}{3}$.

33. $z = xe^y \ \Rightarrow \ \partial z/\partial x = e^y$, $\partial z/\partial y = xe^y$, so by Formula 4, a CAS gives

$$\iint_S (x^2 + y^2 + z^2)\, dS = \int_0^1 \int_0^1 (x^2 + y^2 + x^2 e^{2y})\sqrt{e^{2y} + x^2 e^{2y} + 1}\, dx\, dy \approx 4.5822.$$

35. We use Formula 4 with $z = 3 - 2x^2 - y^2 \ \Rightarrow \ \partial z/\partial x = -4x$, $\partial z/\partial y = -2y$. The boundaries of the region

$3 - 2x^2 - y^2 \ge 0$ are $-\sqrt{\tfrac{3}{2}} \le x \le \sqrt{\tfrac{3}{2}}$ and $-\sqrt{3 - 2x^2} \le y \le \sqrt{3 - 2x^2}$, so we use a CAS (with precision reduced to

seven or fewer digits; otherwise the calculation may take a long time) to calculate

$$\iint_S x^2 y^2 z^2\, dS = \int_{-\sqrt{3/2}}^{\sqrt{3/2}} \int_{-\sqrt{3 - 2x^2}}^{\sqrt{3 - 2x^2}} x^2 y^2 (3 - 2x^2 - y^2)^2 \sqrt{16x^2 + 4y^2 + 1}\, dy\, dx \approx 3.4895$$

37. If S is given by $y = h(x, z)$, then S is also the level surface $f(x, y, z) = y - h(x, z) = 0$.

$\mathbf{n} = \dfrac{\nabla f(x, y, z)}{|\nabla f(x, y, z)|} = \dfrac{-h_x \mathbf{i} + \mathbf{j} - h_z \mathbf{k}}{\sqrt{h_x^2 + 1 + h_z^2}}$, and $-\mathbf{n}$ is the unit normal that points to the left. Now we proceed as in the

derivation of (10), using Formula 4 to evaluate

$$\iint_S \mathbf{F} \cdot d\mathbf{S} = \iint_S \mathbf{F} \cdot \mathbf{n}\, dS = \iint_D (P\mathbf{i} + Q\mathbf{j} + R\mathbf{k}) \frac{\dfrac{\partial h}{\partial x}\mathbf{i} - \mathbf{j} + \dfrac{\partial h}{\partial z}\mathbf{k}}{\sqrt{\left(\dfrac{\partial h}{\partial x}\right)^2 + 1 + \left(\dfrac{\partial h}{\partial z}\right)^2}} \sqrt{\left(\dfrac{\partial h}{\partial x}\right)^2 + 1 + \left(\dfrac{\partial h}{\partial z}\right)^2}\, dA$$

where D is the projection of S onto the xz-plane. Therefore $\displaystyle\iint_S \mathbf{F} \cdot d\mathbf{S} = \iint_D \left(P\frac{\partial h}{\partial x} - Q + R\frac{\partial h}{\partial z} \right) dA.$

© 2012 Cengage Learning. All Rights Reserved. May not be scanned, copied, or duplicated, or posted to a publicly accessible website, in whole or in part.

39. $m = \iint_S K\, dS = K \cdot 4\pi\left(\frac{1}{2}a^2\right) = 2\pi a^2 K$; by symmetry $M_{xz} = M_{yz} = 0$, and

$M_{xy} = \iint_S zK\, dS = K \int_0^{2\pi} \int_0^{\pi/2} (a\cos\phi)(a^2\sin\phi)\, d\phi\, d\theta = 2\pi K a^3 \left[-\frac{1}{4}\cos 2\phi\right]_0^{\pi/2} = \pi K a^3$.

Hence $(\overline{x}, \overline{y}, \overline{z}) = \left(0, 0, \frac{1}{2}a\right)$.

41. (a) $I_z = \iint_S (x^2 + y^2)\rho(x, y, z)\, dS$

(b) $I_z = \iint_S (x^2 + y^2)\left(10 - \sqrt{x^2 + y^2}\right) dS = \iint\limits_{1 \le x^2 + y^2 \le 16} (x^2 + y^2)\left(10 - \sqrt{x^2 + y^2}\right)\sqrt{2}\, dA$

$= \int_0^{2\pi} \int_1^4 \sqrt{2}\,(10r^3 - r^4)\, dr\, d\theta = 2\sqrt{2}\,\pi\left(\frac{4329}{10}\right) = \frac{4329}{5}\sqrt{2}\,\pi$

43. The rate of flow through the cylinder is the flux $\iint_S \rho\mathbf{v} \cdot \mathbf{n}\, dS = \iint_S \rho\mathbf{v} \cdot d\mathbf{S}$. We use the parametric representation

$\mathbf{r}(u, v) = 2\cos u\,\mathbf{i} + 2\sin u\,\mathbf{j} + v\,\mathbf{k}$ for S, where $0 \le u \le 2\pi$, $0 \le v \le 1$, so $\mathbf{r}_u = -2\sin u\,\mathbf{i} + 2\cos u\,\mathbf{j}$, $\mathbf{r}_v = \mathbf{k}$, and the

outward orientation is given by $\mathbf{r}_u \times \mathbf{r}_v = 2\cos u\,\mathbf{i} + 2\sin u\,\mathbf{j}$. Then

$$\iint_S \rho\mathbf{v} \cdot d\mathbf{S} = \rho \int_0^{2\pi} \int_0^1 \left(v\,\mathbf{i} + 4\sin^2 u\,\mathbf{j} + 4\cos^2 u\,\mathbf{k}\right) \cdot (2\cos u\,\mathbf{i} + 2\sin u\,\mathbf{j})\, dv\, du$$

$$= \rho \int_0^{2\pi} \int_0^1 \left(2v\cos u + 8\sin^3 u\right) dv\, du = \rho \int_0^{2\pi} \left(\cos u + 8\sin^3 u\right) du$$

$$= \rho\left[\sin u + 8\left(-\frac{1}{3}\right)(2 + \sin^2 u)\cos u\right]_0^{2\pi} = 0 \text{ kg/s}$$

45. S consists of the hemisphere S_1 given by $z = \sqrt{a^2 - x^2 - y^2}$ and the disk S_2 given by $0 \le x^2 + y^2 \le a^2$, $z = 0$.

On S_1: $\mathbf{E} = a\sin\phi\cos\theta\,\mathbf{i} + a\sin\phi\sin\theta\,\mathbf{j} + 2a\cos\phi\,\mathbf{k}$,

$\mathbf{T}_\phi \times \mathbf{T}_\theta = a^2\sin^2\phi\cos\theta\,\mathbf{i} + a^2\sin^2\phi\sin\theta\,\mathbf{j} + a^2\sin\phi\cos\phi\,\mathbf{k}$. Thus

$$\iint_{S_1} \mathbf{E} \cdot d\mathbf{S} = \int_0^{2\pi} \int_0^{\pi/2} (a^3\sin^3\phi + 2a^3\sin\phi\cos^2\phi)\, d\phi\, d\theta$$

$$= \int_0^{2\pi} \int_0^{\pi/2} (a^3\sin\phi + a^3\sin\phi\cos^2\phi)\, d\phi\, d\theta = (2\pi)a^3\left(1 + \frac{1}{3}\right) = \frac{8}{3}\pi a^3$$

On S_2: $\mathbf{E} = x\,\mathbf{i} + y\,\mathbf{j}$, and $\mathbf{r}_y \times \mathbf{r}_x = -\mathbf{k}$ so $\iint_{S_2} \mathbf{E} \cdot d\mathbf{S} = 0$. Hence the total charge is $q = \varepsilon_0 \iint_S \mathbf{E} \cdot d\mathbf{S} = \frac{8}{3}\pi a^3 \varepsilon_0$.

47. $K\nabla u = 6.5(4y\,\mathbf{j} + 4z\,\mathbf{k})$. S is given by $\mathbf{r}(x, \theta) = x\,\mathbf{i} + \sqrt{6}\cos\theta\,\mathbf{j} + \sqrt{6}\sin\theta\,\mathbf{k}$ and since we want the inward heat flow, we

use $\mathbf{r}_x \times \mathbf{r}_\theta = -\sqrt{6}\cos\theta\,\mathbf{j} - \sqrt{6}\sin\theta\,\mathbf{k}$. Then the rate of heat flow inward is given by

$\iint_S (-K\nabla u) \cdot d\mathbf{S} = \int_0^{2\pi} \int_0^4 -(6.5)(-24)\, dx\, d\theta = (2\pi)(156)(4) = 1248\pi$.

49. Let S be a sphere of radius a centered at the origin. Then $|\mathbf{r}| = a$ and $\mathbf{F}(\mathbf{r}) = c\mathbf{r}/|\mathbf{r}|^3 = (c/a^3)\,(x\,\mathbf{i} + y\,\mathbf{j} + z\,\mathbf{k})$. A

parametric representation for S is $\mathbf{r}(\phi, \theta) = a\sin\phi\cos\theta\,\mathbf{i} + a\sin\phi\sin\theta\,\mathbf{j} + a\cos\phi\,\mathbf{k}$, $0 \le \phi \le \pi$, $0 \le \theta \le 2\pi$. Then

$\mathbf{r}_\phi = a\cos\phi\cos\theta\,\mathbf{i} + a\cos\phi\sin\theta\,\mathbf{j} - a\sin\phi\,\mathbf{k}$, $\mathbf{r}_\theta = -a\sin\phi\sin\theta\,\mathbf{i} + a\sin\phi\cos\theta\,\mathbf{j}$, and the outward orientation is given

by $\mathbf{r}_\phi \times \mathbf{r}_\theta = a^2\sin^2\phi\cos\theta\,\mathbf{i} + a^2\sin^2\phi\sin\theta\,\mathbf{j} + a^2\sin\phi\cos\phi\,\mathbf{k}$. The flux of $\mathbf{F}$ across S is

$$\iint_S \mathbf{F} \cdot d\mathbf{S} = \int_0^\pi \int_0^{2\pi} \frac{c}{a^3}\,(a\sin\phi\cos\theta\,\mathbf{i} + a\sin\phi\sin\theta\,\mathbf{j} + a\cos\phi\,\mathbf{k})$$

$$\cdot (a^2\sin^2\phi\cos\theta\,\mathbf{i} + a^2\sin^2\phi\sin\theta\,\mathbf{j} + a^2\sin\phi\cos\phi\,\mathbf{k})\, d\theta\, d\phi$$

$$= \frac{c}{a^3} \int_0^\pi \int_0^{2\pi} a^3\left(\sin^3\phi + \sin\phi\cos^2\phi\right) d\theta\, d\phi = c\int_0^\pi \int_0^{2\pi} \sin\phi\, d\theta\, d\phi = 4\pi c$$

Thus the flux does not depend on the radius a.

© 2012 Cengage Learning. All Rights Reserved. May not be scanned, copied, or duplicated, or posted to a publicly accessible website, in whole or in part.

16.8   Stokes' Theorem

1. Both H and P are oriented piecewise-smooth surfaces that are bounded by the simple, closed, smooth curve $x^2 + y^2 = 4$,

$z = 0$ (which we can take to be oriented positively for both surfaces). Then H and P satisfy the hypotheses of Stokes'

Theorem, so by (3) we know $\iint_H \operatorname{curl} \mathbf{F} \cdot d\mathbf{S} = \int_C \mathbf{F} \cdot d\mathbf{r} = \iint_P \operatorname{curl} \mathbf{F} \cdot d\mathbf{S}$ (where C is the boundary curve).

3. The paraboloid $z = x^2 + y^2$ intersects the cylinder $x^2 + y^2 = 4$ in the circle $x^2 + y^2 = 4$, $z = 4$. This boundary curve C

should be oriented in the counterclockwise direction when viewed from above, so a vector equation of C is

$\mathbf{r}(t) = 2\cos t\,\mathbf{i} + 2\sin t\,\mathbf{j} + 4\,\mathbf{k}$, $0 \le t \le 2\pi$. Then $\mathbf{r}'(t) = -2\sin t\,\mathbf{i} + 2\cos t\,\mathbf{j}$,

$$\mathbf{F}(\mathbf{r}(t)) = (4\cos^2 t)(16)\,\mathbf{i} + (4\sin^2 t)(16)\,\mathbf{j} + (2\cos t)(2\sin t)(4)\,\mathbf{k} = 64\cos^2 t\,\mathbf{i} + 64\sin^2 t\,\mathbf{j} + 16\sin t\,\cos t\,\mathbf{k},$$

and by Stokes' Theorem,

$$\iint_S \operatorname{curl} \mathbf{F} \cdot d\mathbf{S} = \int_C \mathbf{F} \cdot d\mathbf{r} = \int_0^{2\pi} \mathbf{F}(\mathbf{r}(t)) \cdot \mathbf{r}'(t)\,dt = \int_0^{2\pi} (-128\cos^2 t\,\sin t + 128\sin^2 t\,\cos t + 0)\,dt$$

$$= 128\big[\tfrac{1}{3}\cos^3 t + \tfrac{1}{3}\sin^3 t\big]_0^{2\pi} = 0$$

5. C is the square in the plane $z = -1$. Rather than evaluating a line integral around C we can use Equation 3:

$\iint_{S_1} \operatorname{curl} \mathbf{F} \cdot d\mathbf{S} = \oint_C \mathbf{F} \cdot d\mathbf{r} = \iint_{S_2} \operatorname{curl} \mathbf{F} \cdot d\mathbf{S}$ where S_1 is the original cube without the bottom and S_2 is the bottom face

of the cube. $\operatorname{curl} \mathbf{F} = x^2 z\,\mathbf{i} + (xy - 2xyz)\,\mathbf{j} + (y - xz)\,\mathbf{k}$. For S_2, we choose $\mathbf{n} = \mathbf{k}$ so that C has the same orientation for

both surfaces. Then $\operatorname{curl} \mathbf{F} \cdot \mathbf{n} = y - xz = x + y$ on S_2, where $z = -1$. Thus $\iint_{S_2} \operatorname{curl} \mathbf{F} \cdot d\mathbf{S} = \int_{-1}^1 \int_{-1}^1 (x + y)\,dx\,dy = 0$

so $\iint_{S_1} \operatorname{curl} \mathbf{F} \cdot d\mathbf{S} = 0$.

7. $\operatorname{curl} \mathbf{F} = -2z\,\mathbf{i} - 2x\,\mathbf{j} - 2y\,\mathbf{k}$ and we take the surface S to be the planar region enclosed by C, so S is the portion of the plane

$x + y + z = 1$ over $D = \{(x, y) \mid 0 \le x \le 1, 0 \le y \le 1 - x\}$. Since C is oriented counterclockwise, we orient S upward.

Using Equation 16.7.10, we have $z = g(x, y) = 1 - x - y$, $P = -2z$, $Q = -2x$, $R = -2y$, and

$$\int_C \mathbf{F} \cdot d\mathbf{r} = \iint_S \operatorname{curl} \mathbf{F} \cdot d\mathbf{S} = \iint_D [-(-2z)(-1) - (-2x)(-1) + (-2y)]\,dA$$

$$= \int_0^1 \int_0^{1-x} (-2)\,dy\,dx = -2\int_0^1 (1 - x)\,dx = -1$$

9. $\operatorname{curl} \mathbf{F} = (xe^{xy} - 2x)\,\mathbf{i} - (ye^{xy} - y)\,\mathbf{j} + (2z - z)\,\mathbf{k}$ and we take S to be the disk $x^2 + y^2 \le 16$, $z = 5$. Since C is oriented

counterclockwise (from above), we orient S upward. Then $\mathbf{n} = \mathbf{k}$ and $\operatorname{curl} \mathbf{F} \cdot \mathbf{n} = 2z - z$ on S, where $z = 5$. Thus

$$\oint_C \mathbf{F} \cdot d\mathbf{r} = \iint_S \operatorname{curl} \mathbf{F} \cdot \mathbf{n}\,dS = \iint_S (2z - z)\,dS = \iint_S (10 - 5)\,dS = 5(\text{area of } S) = 5(\pi \cdot 4^2) = 80\pi$$

11. (a) The curve of intersection is an ellipse in the plane $x + y + z = 1$ with unit normal $\mathbf{n} = \frac{1}{\sqrt{3}}(\mathbf{i} + \mathbf{j} + \mathbf{k})$,

$\operatorname{curl} \mathbf{F} = x^2\,\mathbf{j} + y^2\,\mathbf{k}$, and $\operatorname{curl} \mathbf{F} \cdot \mathbf{n} = \frac{1}{\sqrt{3}}(x^2 + y^2)$. Then

$$\oint_C \mathbf{F} \cdot d\mathbf{r} = \iint_S \frac{1}{\sqrt{3}}(x^2 + y^2)\,dS = \iint_{x^2 + y^2 \le 9} (x^2 + y^2)\,dx\,dy = \int_0^{2\pi} \int_0^3 r^3\,dr\,d\theta = 2\pi\left(\tfrac{81}{4}\right) = \tfrac{81\pi}{2}$$

© 2012 Cengage Learning. All Rights Reserved. May not be scanned, copied, or duplicated, or posted to a publicly accessible website, in whole or in part.

(b)

(c) One possible parametrization is $x = 3\cos t$, $y = 3\sin t$,

$z = 1 - 3\cos t - 3\sin t$, $0 \le t \le 2\pi$.

13. The boundary curve C is the circle $x^2 + y^2 = 16$, $z = 4$ oriented in the clockwise direction as viewed from above (since S is

oriented downward). We can parametrize C by $\mathbf{r}(t) = 4\cos t\,\mathbf{i} - 4\sin t\,\mathbf{j} + 4\,\mathbf{k}$, $0 \le t \le 2\pi$, and then

$\mathbf{r}'(t) = -4\sin t\,\mathbf{i} - 4\cos t\,\mathbf{j}$. Thus $\mathbf{F}(\mathbf{r}(t)) = 4\sin t\,\mathbf{i} + 4\cos t\,\mathbf{j} - 2\,\mathbf{k}$, $\mathbf{F}(\mathbf{r}(t)) \cdot \mathbf{r}'(t) = -16\sin^2 t - 16\cos^2 t = -16$, and

$$\oint_C \mathbf{F} \cdot d\mathbf{r} = \int_0^{2\pi} \mathbf{F}(\mathbf{r}(t)) \cdot \mathbf{r}'(t)\,dt = \int_0^{2\pi} (-16)\,dt = -16\,(2\pi) = -32\pi$$

Now curl $\mathbf{F} = 2\,\mathbf{k}$, and the projection D of S on the xy-plane is the disk $x^2 + y^2 \le 16$, so by Equation 16.7.10 with

$z = g(x, y) = \sqrt{x^2 + y^2}$ [and multiplying by -1 for the downward orientation] we have

$$\iint_S \text{curl}\,\mathbf{F} \cdot d\mathbf{S} = -\iint_D (-0 - 0 + 2)\,dA = -2 \cdot A(D) = -2 \cdot \pi(4^2) = -32\pi$$

15. The boundary curve C is the circle $x^2 + z^2 = 1$, $y = 0$ oriented in the counterclockwise direction as viewed from the positive

y-axis. Then C can be described by $\mathbf{r}(t) = \cos t\,\mathbf{i} - \sin t\,\mathbf{k}$, $0 \le t \le 2\pi$, and $\mathbf{r}'(t) = -\sin t\,\mathbf{i} - \cos t\,\mathbf{k}$. Thus

$\mathbf{F}(\mathbf{r}(t)) = -\sin t\,\mathbf{j} + \cos t\,\mathbf{k}$, $\mathbf{F}(\mathbf{r}(t)) \cdot \mathbf{r}'(t) = -\cos^2 t$, and $\oint_C \mathbf{F} \cdot d\mathbf{r} = \int_0^{2\pi} (-\cos^2 t)\,dt = -\frac{1}{2}t - \frac{1}{4}\sin 2t\big]_0^{2\pi} = -\pi$.

Now curl $\mathbf{F} = -\mathbf{i} - \mathbf{j} - \mathbf{k}$, and S can be parametrized (see Example 16.6.10) by

$\mathbf{r}(\phi, \theta) = \sin\phi\cos\theta\,\mathbf{i} + \sin\phi\sin\theta\,\mathbf{j} + \cos\phi\,\mathbf{k}$, $0 \le \theta \le \pi$, $0 \le \phi \le \pi$. Then

$\mathbf{r}_\phi \times \mathbf{r}_\theta = \sin^2\phi\cos\theta\,\mathbf{i} + \sin^2\phi\sin\theta\,\mathbf{j} + \sin\phi\cos\phi\,\mathbf{k}$ and

$$\iint_S \text{curl}\,\mathbf{F} \cdot d\mathbf{S} = \iint_{x^2+z^2 \le 1} \text{curl}\,\mathbf{F} \cdot (\mathbf{r}_\phi \times \mathbf{r}_\theta)\,dA = \int_0^\pi \int_0^\pi (-\sin^2\phi\cos\theta - \sin^2\phi\sin\theta - \sin\phi\cos\phi)\,d\theta\,d\phi$$

$$= \int_0^\pi (-2\sin^2\phi - \pi\sin\phi\cos\phi)\,d\phi = \left[\tfrac{1}{2}\sin 2\phi - \phi - \tfrac{\pi}{2}\sin^2\phi\right]_0^\pi = -\pi$$

17. It is easier to use Stokes' Theorem than to compute the work directly. Let S be the planar region enclosed by the path of the

particle, so S is the portion of the plane $z = \frac{1}{2}y$ for $0 \le x \le 1$, $0 \le y \le 2$, with upward orientation.

curl $\mathbf{F} = 8y\,\mathbf{i} + 2z\,\mathbf{j} + 2y\,\mathbf{k}$ and

$$\oint_C \mathbf{F} \cdot d\mathbf{r} = \iint_S \text{curl}\,\mathbf{F} \cdot d\mathbf{S} = \iint_D \left[-8y\,(0) - 2z\,\left(\tfrac{1}{2}\right) + 2y\right]\,dA = \int_0^1 \int_0^2 \left(2y - \tfrac{1}{2}y\right)\,dy\,dx$$

$$= \int_0^1 \int_0^2 \tfrac{3}{2}y\,dy\,dx = \int_0^1 \left[\tfrac{3}{4}y^2\right]_{y=0}^{y=2}\,dx = \int_0^1 3\,dx = 3$$

19. Assume S is centered at the origin with radius a and let H_1 and H_2 be the upper and lower hemispheres, respectively, of S.

Then $\iint_S \text{curl}\,\mathbf{F} \cdot d\mathbf{S} = \iint_{H_1} \text{curl}\,\mathbf{F} \cdot d\mathbf{S} + \iint_{H_2} \text{curl}\,\mathbf{F} \cdot d\mathbf{S} = \oint_{C_1} \mathbf{F} \cdot d\mathbf{r} + \oint_{C_2} \mathbf{F} \cdot d\mathbf{r}$ by Stokes' Theorem. But C_1 is the

circle $x^2 + y^2 = a^2$ oriented in the counterclockwise direction while C_2 is the same circle oriented in the clockwise direction.

Hence $\oint_{C_2} \mathbf{F} \cdot d\mathbf{r} = -\oint_{C_1} \mathbf{F} \cdot d\mathbf{r}$ so $\iint_S \text{curl}\,\mathbf{F} \cdot d\mathbf{S} = 0$ as desired.

© 2012 Cengage Learning. All Rights Reserved. May not be scanned, copied, or duplicated, or posted to a publicly accessible website, in whole or in part.

16.9 The Divergence Theorem

1. $\operatorname{div}\mathbf{F} = 3 + x + 2x = 3 + 3x$, so

$\iiint_E \operatorname{div}\mathbf{F}\,dV = \int_0^1 \int_0^1 \int_0^1 (3x+3)\,dx\,dy\,dz = \frac{9}{2}$ (notice the triple integral is

three times the volume of the cube plus three times $\overline{x}$).

To compute $\iint_S \mathbf{F}\cdot d\mathbf{S}$, on

S_1: $\mathbf{n} = \mathbf{i}$, $\mathbf{F} = 3\mathbf{i} + y\mathbf{j} + 2z\,\mathbf{k}$, and $\iint_{S_1}\mathbf{F}\cdot d\mathbf{S} = \iint_{S_1} 3\,dS = 3$;

S_2: $\mathbf{F} = 3x\,\mathbf{i} + x\,\mathbf{j} + 2xz\,\mathbf{k}$, $\mathbf{n} = \mathbf{j}$ and $\iint_{S_2}\mathbf{F}\cdot d\mathbf{S} = \iint_{S_2} x\,dS = \frac{1}{2}$;

S_3: $\mathbf{F} = 3x\,\mathbf{i} + xy\,\mathbf{j} + 2x\,\mathbf{k}$, $\mathbf{n} = \mathbf{k}$ and $\iint_{S_3}\mathbf{F}\cdot d\mathbf{S} = \iint_{S_3} 2x\,dS = 1$;

S_4: $\mathbf{F} = \mathbf{0}$, $\iint_{S_4}\mathbf{F}\cdot d\mathbf{S} = 0$; S_5: $\mathbf{F} = 3x\,\mathbf{i} + 2x\,\mathbf{k}$, $\mathbf{n} = -\mathbf{j}$ and $\iint_{S_5}\mathbf{F}\cdot d\mathbf{S} = \iint_{S_5} 0\,dS = 0$;

S_6: $\mathbf{F} = 3x\,\mathbf{i} + xy\,\mathbf{j}$, $\mathbf{n} = -\mathbf{k}$ and $\iint_{S_6}\mathbf{F}\cdot d\mathbf{S} = \iint_{S_6} 0\,dS = 0$. Thus $\iint_S \mathbf{F}\cdot d\mathbf{S} = \frac{9}{2}$.

3. $\operatorname{div}\mathbf{F} = 0 + 1 + 0 = 1$, so $\iiint_E \operatorname{div}\mathbf{F}\,dV = \iiint_E 1\,dV = V(E) = \frac{4}{3}\pi\cdot 4^3 = \frac{256}{3}\pi$. S is a sphere of radius 4 centered at

the origin which can be parametrized by $\mathbf{r}(\phi,\theta) = \langle 4\sin\phi\cos\theta, 4\sin\phi\sin\theta, 4\cos\phi\rangle$, $0 \le \phi \le \pi$, $0 \le \theta \le 2\pi$ (similar to

Example 16.6.10). Then

$$\mathbf{r}_\phi \times \mathbf{r}_\theta = \langle 4\cos\phi\cos\theta, 4\cos\phi\sin\theta, -4\sin\phi\rangle \times \langle -4\sin\phi\sin\theta, 4\sin\phi\cos\theta, 0\rangle$$

$$= \langle 16\sin^2\phi\cos\theta, 16\sin^2\phi\sin\theta, 16\cos\phi\sin\phi\rangle$$

and $\mathbf{F}(\mathbf{r}(\phi,\theta)) = \langle 4\cos\phi, 4\sin\phi\sin\theta, 4\sin\phi\cos\theta\rangle$. Thus

$\mathbf{F}\cdot(\mathbf{r}_\phi\times\mathbf{r}_\theta) = 64\cos\phi\sin^2\phi\cos\theta + 64\sin^3\phi\sin^2\theta + 64\cos\phi\sin^2\phi\cos\theta = 128\cos\phi\sin^2\phi\cos\theta + 64\sin^3\phi\sin^2\theta$

and

$$\iint_S \mathbf{F}\cdot d\mathbf{S} = \iint_D \mathbf{F}\cdot(\mathbf{r}_\phi\times\mathbf{r}_\theta)\,dA = \int_0^{2\pi}\int_0^\pi \left(128\cos\phi\sin^2\phi\cos\theta + 64\sin^3\phi\sin^2\theta\right)d\phi\,d\theta$$

$$= \int_0^{2\pi}\left[\frac{128}{3}\sin^3\phi\cos\theta + 64\left(-\frac{1}{3}(2+\sin^2\phi)\cos\phi\right)\sin^2\theta\right]_{\phi=0}^{\phi=\pi}d\theta$$

$$= \int_0^{2\pi}\frac{256}{3}\sin^2\theta\,d\theta = \frac{256}{3}\left[\frac{1}{2}\theta - \frac{1}{4}\sin 2\theta\right]_0^{2\pi} = \frac{256}{3}\pi$$

5. $\operatorname{div}\mathbf{F} = \frac{\partial}{\partial x}(xye^z) + \frac{\partial}{\partial y}(xy^2z^3) + \frac{\partial}{\partial z}(-ye^z) = ye^z + 2xyz^3 - ye^z = 2xyz^3$, so by the Divergence Theorem,

$$\iint_S \mathbf{F}\cdot d\mathbf{S} = \iiint_E \operatorname{div}\mathbf{F}\,dV = \int_0^3\int_0^2\int_0^1 2xyz^3\,dz\,dy\,dx = 2\int_0^3 x\,dx\int_0^2 y\,dy\int_0^1 z^3\,dz$$

$$= 2\left[\frac{1}{2}x^2\right]_0^3\left[\frac{1}{2}y^2\right]_0^2\left[\frac{1}{4}z^4\right]_0^1 = 2\left(\frac{9}{2}\right)(2)\left(\frac{1}{4}\right) = \frac{9}{2}$$

7. $\operatorname{div}\mathbf{F} = 3y^2 + 0 + 3z^2$, so using cylindrical coordinates with $y = r\cos\theta$, $z = r\sin\theta$, $x = x$ we have

$$\iint_S \mathbf{F}\cdot d\mathbf{S} = \iiint_E (3y^2+3z^2)\,dV = \int_0^{2\pi}\int_0^1\int_{-1}^2 (3r^2\cos^2\theta + 3r^2\sin^2\theta)\,r\,dx\,dr\,d\theta$$

$$= 3\int_0^{2\pi}d\theta\int_0^1 r^3\,dr\int_{-1}^2 dx = 3(2\pi)\left(\frac{1}{4}\right)(3) = \frac{9\pi}{2}$$

9. $\operatorname{div}\mathbf{F} = 2x\sin y - x\sin y - x\sin y = 0$, so by the Divergence Theorem, $\iint_S \mathbf{F}\cdot d\mathbf{S} = \iiint_E 0\,dV = 0$.

11. $\operatorname{div}\mathbf{F} = y^2 + 0 + x^2 = x^2 + y^2$ so

$$\iint_S \mathbf{F}\cdot d\mathbf{S} = \iiint_E (x^2+y^2)\,dV = \int_0^{2\pi}\int_0^2\int_{r^2}^4 r^2\cdot r\,dz\,dr\,d\theta = \int_0^{2\pi}\int_0^2 r^3(4-r^2)\,dr\,d\theta$$

$$= \int_0^{2\pi}d\theta\int_0^2 (4r^3 - r^5)\,dr = 2\pi\left[r^4 - \frac{1}{6}r^6\right]_0^2 = \frac{32}{3}\pi$$

© 2012 Cengage Learning. All Rights Reserved. May not be scanned, copied, or duplicated, or posted to a publicly accessible website, in whole or in part.

13. $\mathbf{F}(x, y, z) = x\sqrt{x^2 + y^2 + z^2}\,\mathbf{i} + y\sqrt{x^2 + y^2 + z^2}\,\mathbf{j} + z\sqrt{x^2 + y^2 + z^2}\,\mathbf{k}$, so

$$\text{div }\mathbf{F} = x \cdot \tfrac{1}{2}(x^2 + y^2 + z^2)^{-1/2}(2x) + (x^2 + y^2 + z^2)^{1/2} + y \cdot \tfrac{1}{2}(x^2 + y^2 + z^2)^{-1/2}(2y) + (x^2 + y^2 + z^2)^{1/2}$$
$$+ z \cdot \tfrac{1}{2}(x^2 + y^2 + z^2)^{-1/2}(2z) + (x^2 + y^2 + z^2)^{1/2}$$
$$= (x^2 + y^2 + z^2)^{-1/2}\left[x^2 + (x^2 + y^2 + z^2) + y^2 + (x^2 + y^2 + z^2) + z^2 + (x^2 + y^2 + z^2)\right]$$
$$= \frac{4(x^2 + y^2 + z^2)}{\sqrt{x^2 + y^2 + z^2}} = 4\sqrt{x^2 + y^2 + z^2}.$$

Then
$$\iint_S \mathbf{F} \cdot d\mathbf{S} = \iiint_E 4\sqrt{x^2 + y^2 + z^2}\,dV = \int_0^{\pi/2} \int_0^{2\pi} \int_0^1 4\sqrt{\rho^2} \cdot \rho^2 \sin\phi\,d\rho\,d\theta\,d\phi$$
$$= \int_0^{\pi/2} \sin\phi\,d\phi \int_0^{2\pi} d\theta \int_0^1 4\rho^3\,d\rho = \left[-\cos\phi\right]_0^{\pi/2} \left[\theta\right]_0^{2\pi} \left[\rho^4\right]_0^1 = (1)(2\pi)(1) = 2\pi$$

15. $\iint_S \mathbf{F} \cdot d\mathbf{S} = \iiint_E \sqrt{3 - x^2}\,dV = \int_{-1}^1 \int_{-1}^1 \int_0^{2 - x^4 - y^4} \sqrt{3 - x^2}\,dz\,dy\,dx = \frac{341}{60}\sqrt{2} + \frac{81}{20}\sin^{-1}\left(\frac{\sqrt{3}}{3}\right)$

17. For S_1 we have $\mathbf{n} = -\mathbf{k}$, so $\mathbf{F} \cdot \mathbf{n} = \mathbf{F} \cdot (-\mathbf{k}) = -x^2 z - y^2 = -y^2$ (since $z = 0$ on S_1). So if D is the unit disk, we get

$\iint_{S_1} \mathbf{F} \cdot d\mathbf{S} = \iint_{S_1} \mathbf{F} \cdot \mathbf{n}\,dS = \iint_D (-y^2)\,dA = -\int_0^{2\pi} \int_0^1 r^2 (\sin^2\theta)\,r\,dr\,d\theta = -\frac{1}{4}\pi$. Now since S_2 is closed, we can use

the Divergence Theorem. Since $\text{div }\mathbf{F} = \frac{\partial}{\partial x}(z^2 x) + \frac{\partial}{\partial y}(\frac{1}{3}y^3 + \tan z) + \frac{\partial}{\partial z}(x^2 z + y^2) = z^2 + y^2 + x^2$, we use spherical

coordinates to get $\iint_{S_2} \mathbf{F} \cdot d\mathbf{S} = \iiint_E \text{div }\mathbf{F}\,dV = \int_0^{2\pi} \int_0^{\pi/2} \int_0^1 \rho^2 \cdot \rho^2 \sin\phi\,d\rho\,d\phi\,d\theta = \frac{2}{5}\pi$. Finally

$\iint_S \mathbf{F} \cdot d\mathbf{S} = \iint_{S_2} \mathbf{F} \cdot d\mathbf{S} - \iint_{S_1} \mathbf{F} \cdot d\mathbf{S} = \frac{2}{5}\pi - \left(-\frac{1}{4}\pi\right) = \frac{13}{20}\pi$.

19. The vectors that end near P_1 are longer than the vectors that start near P_1, so the net flow is inward near P_1 and $\text{div }\mathbf{F}(P_1)$ is

negative. The vectors that end near P_2 are shorter than the vectors that start near P_2, so the net flow is outward near P_2 and

$\text{div }\mathbf{F}(P_2)$ is positive.

21.

From the graph it appears that for points above the x-axis, vectors starting near a

particular point are longer than vectors ending there, so divergence is positive.

The opposite is true at points below the x-axis, where divergence is negative.

$\mathbf{F}(x, y) = \langle xy, x + y^2 \rangle \quad \Rightarrow \quad \text{div }\mathbf{F} = \frac{\partial}{\partial x}(xy) + \frac{\partial}{\partial y}(x + y^2) = y + 2y = 3y$.

Thus $\text{div }\mathbf{F} > 0$ for $y > 0$, and $\text{div }\mathbf{F} < 0$ for $y < 0$.

23. Since $\dfrac{\mathbf{x}}{|\mathbf{x}|^3} = \dfrac{x\,\mathbf{i} + y\,\mathbf{j} + z\,\mathbf{k}}{(x^2 + y^2 + z^2)^{3/2}}$ and $\dfrac{\partial}{\partial x}\left(\dfrac{x}{(x^2 + y^2 + z^2)^{3/2}}\right) = \dfrac{(x^2 + y^2 + z^2) - 3x^2}{(x^2 + y^2 + z^2)^{5/2}}$ with similar expressions

for $\dfrac{\partial}{\partial y}\left(\dfrac{y}{(x^2 + y^2 + z^2)^{3/2}}\right)$ and $\dfrac{\partial}{\partial z}\left(\dfrac{z}{(x^2 + y^2 + z^2)^{3/2}}\right)$, we have

$\text{div}\left(\dfrac{\mathbf{x}}{|\mathbf{x}|^3}\right) = \dfrac{3(x^2 + y^2 + z^2) - 3(x^2 + y^2 + z^2)}{(x^2 + y^2 + z^2)^{5/2}} = 0$, except at $(0, 0, 0)$ where it is undefined.

25. $\iint_S \mathbf{a} \cdot \mathbf{n}\,dS = \iiint_E \text{div }\mathbf{a}\,dV = 0$ since $\text{div }\mathbf{a} = 0$.

© 2012 Cengage Learning. All Rights Reserved. May not be scanned, copied, or duplicated, or posted to a publicly accessible website, in whole or in part.

27. $\iint_S \text{curl } \mathbf{F} \cdot d\mathbf{S} = \iiint_E \text{div}(\text{curl } \mathbf{F}) \, dV = 0$ by Theorem 16.5.11.

29. $\iint_S (f\nabla g) \cdot \mathbf{n} \, dS = \iiint_E \text{div}(f\nabla g) \, dV = \iiint_E (f\nabla^2 g + \nabla g \cdot \nabla f) \, dV$ by Exercise 16.5.25.

31. If $\mathbf{c} = c_1 \, \mathbf{i} + c_2 \, \mathbf{j} + c_3 \, \mathbf{k}$ is an arbitrary constant vector, we define $\mathbf{F} = f\mathbf{c} = fc_1 \, \mathbf{i} + fc_2 \, \mathbf{j} + fc_3 \, \mathbf{k}$. Then

$\text{div } \mathbf{F} = \text{div } f\mathbf{c} = \dfrac{\partial f}{\partial x} c_1 + \dfrac{\partial f}{\partial y} c_2 + \dfrac{\partial f}{\partial z} c_3 = \nabla f \cdot \mathbf{c}$ and the Divergence Theorem says $\iint_S \mathbf{F} \cdot d\mathbf{S} = \iiint_E \text{div } \mathbf{F} \, dV$ $\Rightarrow$

$\iint_S \mathbf{F} \cdot \mathbf{n} \, dS = \iiint_E \nabla f \cdot \mathbf{c} \, dV$. In particular, if $\mathbf{c} = \mathbf{i}$ then $\iint_S f\mathbf{i} \cdot \mathbf{n} \, dS = \iiint_E \nabla f \cdot \mathbf{i} \, dV$ $\Rightarrow$

$\iint_S fn_1 \, dS = \iiint_E \dfrac{\partial f}{\partial x} \, dV$ (where $\mathbf{n} = n_1 \, \mathbf{i} + n_2 \, \mathbf{j} + n_3 \, \mathbf{k}$). Similarly, if $\mathbf{c} = \mathbf{j}$ we have $\iint_S fn_2 \, dS = \iiint_E \dfrac{\partial f}{\partial y} \, dV$,

and $\mathbf{c} = \mathbf{k}$ gives $\iint_S fn_3 \, dS = \iiint_E \dfrac{\partial f}{\partial z} \, dV$. Then

$\iint_S f\mathbf{n} \, dS = \left(\iint_S fn_1 \, dS \right) \mathbf{i} + \left(\iint_S fn_2 \, dS \right) \mathbf{j} + \left(\iint_S fn_3 \, dS \right) \mathbf{k}$

$= \left(\iiint_E \dfrac{\partial f}{\partial x} \, dV \right) \mathbf{i} + \left(\iiint_E \dfrac{\partial f}{\partial y} \, dV \right) \mathbf{j} + \left(\iiint_E \dfrac{\partial f}{\partial z} \, dV \right) \mathbf{k} = \iiint_E \left(\dfrac{\partial f}{\partial x} \mathbf{i} + \dfrac{\partial f}{\partial y} \mathbf{j} + \dfrac{\partial f}{\partial z} \mathbf{k} \right) dV$

$= \iiint_E \nabla f \, dV$ as desired.

16   Review

CONCEPT CHECK

1. See Definitions 1 and 2 in Section 16.1. A vector field can represent, for example, the wind velocity at any location in space, the speed and direction of the ocean current at any location, or the force vectors of Earth's gravitational field at a location in space.

2. (a) A conservative vector field $\mathbf{F}$ is a vector field which is the gradient of some scalar function f.

(b) The function f in part (a) is called a potential function for $\mathbf{F}$, that is, $\mathbf{F} = \nabla f$.

3. (a) See Definition 16.2.2.

(b) We normally evaluate the line integral using Formula 16.2.3.

(c) The mass is $m = \int_C \rho(x, y) \, ds$, and the center of mass is $(\overline{x}, \overline{y})$ where $\overline{x} = \frac{1}{m} \int_C x\rho(x, y) \, ds$, $\overline{y} = \frac{1}{m} \int_C y\rho(x, y) \, ds$.

(d) See (5) and (6) in Section 16.2 for plane curves; we have similar definitions when C is a space curve [see the equation preceding (10) in Section 16.2].

(e) For plane curves, see Equations 16.2.7. We have similar results for space curves [see the equation preceding (10) in Section 16.2].

4. (a) See Definition 16.2.13.

(b) If $\mathbf{F}$ is a force field, $\int_C \mathbf{F} \cdot d\mathbf{r}$ represents the work done by $\mathbf{F}$ in moving a particle along the curve C.

(c) $\int_C \mathbf{F} \cdot d\mathbf{r} = \int_C P \, dx + Q \, dy + R \, dz$

5. See Theorem 16.3.2.

© 2012 Cengage Learning. All Rights Reserved. May not be scanned, copied, or duplicated, or posted to a publicly accessible website, in whole or in part.

6. (a) $\int_C \mathbf{F} \cdot d\mathbf{r}$ is independent of path if the line integral has the same value for any two curves that have the same initial and terminal points.

   (b) See Theorem 16.3.4.

7. See the statement of Green's Theorem on page 1108 [ET 1084].

8. See Equations 16.4.5.

9. (a) $\operatorname{curl} \mathbf{F} = \left(\dfrac{\partial R}{\partial y} - \dfrac{\partial Q}{\partial z}\right)\mathbf{i} + \left(\dfrac{\partial P}{\partial z} - \dfrac{\partial R}{\partial x}\right)\mathbf{j} + \left(\dfrac{\partial Q}{\partial x} - \dfrac{\partial P}{\partial y}\right)\mathbf{k} = \nabla \times \mathbf{F}$

   (b) $\operatorname{div} \mathbf{F} = \dfrac{\partial P}{\partial x} + \dfrac{\partial Q}{\partial y} + \dfrac{\partial R}{\partial z} = \nabla \cdot \mathbf{F}$

   (c) For curl $\mathbf{F}$, see the discussion accompanying Figure 1 on page 1118 [ET 1094] as well as Figure 6 and the accompanying discussion on page 1150 [ET 1126]. For div $\mathbf{F}$, see the discussion following Example 5 on page 1119 [ET 1095] as well as the discussion preceding (8) on page 1157 [ET 1133].

10. See Theorem 16.3.6; see Theorem 16.5.4.

11. (a) See (1) and (2) and the accompanying discussion in Section 16.6; See Figure 4 and the accompanying discussion on page 1124 [ET 1100].

    (b) See Definition 16.6.6.

    (c) See Equation 16.6.9.

12. (a) See (1) in Section 16.7.

    (b) We normally evaluate the surface integral using Formula 16.7.2.

    (c) See Formula 16.7.4.

    (d) The mass is $m = \iint_S \rho(x, y, z)\, dS$ and the center of mass is $(\bar{x}, \bar{y}, \bar{z})$ where $\bar{x} = \frac{1}{m} \iint_S x\rho(x, y, z)\, dS$, $\bar{y} = \frac{1}{m} \iint_S y\rho(x, y, z)\, dS$, $\bar{z} = \frac{1}{m} \iint_S z\rho(x, y, z)\, dS$.

13. (a) See Figures 6 and 7 and the accompanying discussion in Section 16.7. A Möbius strip is a nonorientable surface; see Figures 4 and 5 and the accompanying discussion on page 1139 [ET 1115].

    (b) See Definition 16.7.8.

    (c) See Formula 16.7.9.

    (d) See Formula 16.7.10.

14. See the statement of Stokes' Theorem on page 1146 [ET 1122].

15. See the statement of the Divergence Theorem on page 1153 [ET 1129].

16. In each theorem, we have an integral of a "derivative" over a region on the left side, while the right side involves the values of the original function only on the boundary of the region.

© 2012 Cengage Learning. All Rights Reserved. May not be scanned, copied, or duplicated, or posted to a publicly accessible website, in whole or in part.

TRUE-FALSE QUIZ

1. False; div $\mathbf{F}$ is a scalar field.

3. True, by Theorem 16.5.3 and the fact that div $\mathbf{0} = 0$.

5. False. See Exercise 16.3.35. (But the assertion is true if D is simply-connected; see Theorem 16.3.6.)

7. False. For example, $\operatorname{div}(y\,\mathbf{i}) = 0 = \operatorname{div}(x\,\mathbf{j})$ but $y\,\mathbf{i} \neq x\,\mathbf{j}$.

9. True. See Exercise 16.5.24.

11. True. Apply the Divergence Theorem and use the fact that div $\mathbf{F} = 0$.

EXERCISES

1. (a) Vectors starting on C point in roughly the direction opposite to C, so the tangential component $\mathbf{F} \cdot \mathbf{T}$ is negative.

Thus $\int_C \mathbf{F} \cdot d\mathbf{r} = \int_C \mathbf{F} \cdot \mathbf{T}\, ds$ is negative.

(b) The vectors that end near P are shorter than the vectors that start near P, so the net flow is outward near P and

div $\mathbf{F}(P)$ is positive.

3. $\int_C yz\cos x\, ds = \int_0^\pi (3\cos t)(3\sin t)\cos t \sqrt{(1)^2 + (-3\sin t)^2 + (3\cos t)^2}\, dt = \int_0^\pi (9\cos^2 t \sin t)\sqrt{10}\, dt$

$= 9\sqrt{10}\left(-\tfrac{1}{3}\cos^3 t\right)\Big]_0^\pi = -3\sqrt{10}\,(-2) = 6\sqrt{10}$

5. $\int_C y^3\, dx + x^2\, dy = \int_{-1}^1 \left[y^3(-2y) + (1-y^2)^2\right] dy = \int_{-1}^1 (-y^4 - 2y^2 + 1)\, dy$

$= \left[-\tfrac{1}{5}y^5 - \tfrac{2}{3}y^3 + y\right]_{-1}^1 = -\tfrac{1}{5} - \tfrac{2}{3} + 1 - \tfrac{1}{5} - \tfrac{2}{3} + 1 = \tfrac{4}{15}$

7. C: $x = 1 + 2t$ $\Rightarrow$ $dx = 2\, dt$, $y = 4t$ $\Rightarrow$ $dy = 4\, dt$, $z = -1 + 3t$ $\Rightarrow$ $dz = 3\, dt$, $0 \le t \le 1$.

$\int_C xy\, dx + y^2\, dy + yz\, dz = \int_0^1 [(1+2t)(4t)(2) + (4t)^2(4) + (4t)(-1+3t)(3)]\, dt$

$= \int_0^1 (116t^2 - 4t)\, dt = \left[\tfrac{116}{3}t^3 - 2t^2\right]_0^1 = \tfrac{116}{3} - 2 = \tfrac{110}{3}$

9. $\mathbf{F}(\mathbf{r}(t)) = e^{-t}\,\mathbf{i} + t^2(-t)\,\mathbf{j} + (t^2 + t^3)\,\mathbf{k}$, $\mathbf{r}'(t) = 2t\,\mathbf{i} + 3t^2\,\mathbf{j} - \mathbf{k}$ and

$\int_C \mathbf{F} \cdot d\mathbf{r} = \int_0^1 (2te^{-t} - 3t^5 - (t^2 + t^3))\, dt = \left[-2te^{-t} - 2e^{-t} - \tfrac{1}{2}t^6 - \tfrac{1}{3}t^3 - \tfrac{1}{4}t^4\right]_0^1 = \tfrac{11}{12} - \tfrac{4}{e}$.

11. $\tfrac{\partial}{\partial y}[(1 + xy)e^{xy}] = 2xe^{xy} + x^2ye^{xy} = \tfrac{\partial}{\partial x}\left[e^y + x^2e^{xy}\right]$ and the domain of $\mathbf{F}$ is $\mathbb{R}^2$, so $\mathbf{F}$ is conservative. Thus there

exists a function f such that $\mathbf{F} = \nabla f$. Then $f_y(x, y) = e^y + x^2e^{xy}$ implies $f(x, y) = e^y + xe^{xy} + g(x)$ and then

$f_x(x, y) = xye^{xy} + e^{xy} + g'(x) = (1 + xy)e^{xy} + g'(x)$. But $f_x(x, y) = (1 + xy)e^{xy}$, so $g'(x) = 0$ $\Rightarrow$ $g(x) = K$.

Thus $f(x, y) = e^y + xe^{xy} + K$ is a potential function for $\mathbf{F}$.

13. Since $\tfrac{\partial}{\partial y}\left(4x^3y^2 - 2xy^3\right) = 8x^3y - 6xy^2 = \tfrac{\partial}{\partial x}\left(2x^4y - 3x^2y^2 + 4y^3\right)$ and the domain of $\mathbf{F}$ is $\mathbb{R}^2$, $\mathbf{F}$ is conservative.

Furthermore $f(x, y) = x^4y^2 - x^2y^3 + y^4$ is a potential function for $\mathbf{F}$. $t = 0$ corresponds to the point $(0, 1)$ and $t = 1$

corresponds to $(1, 1)$, so $\int_C \mathbf{F} \cdot d\mathbf{r} = f(1, 1) - f(0, 1) = 1 - 1 = 0$.

© 2012 Cengage Learning. All Rights Reserved. May not be scanned, copied, or duplicated, or posted to a publicly accessible website, in whole or in part.

15. $C_1: \mathbf{r}(t) = t\,\mathbf{i} + t^2\,\mathbf{j}, -1 \le t \le 1$;

$C_2: \mathbf{r}(t) = -t\,\mathbf{i} + \mathbf{j}, -1 \le t \le 1$.

Then

$$\int_C xy^2\,dx - x^2y\,dy = \int_{-1}^1 (t^5 - 2t^5)\,dt + \int_{-1}^1 t\,dt$$

$$= \left[-\tfrac{1}{6}t^6 \right]_{-1}^1 + \left[\tfrac{1}{2}t^2 \right]_{-1}^1 = 0$$

Using Green's Theorem, we have

$$\int_C xy^2\,dx - x^2y\,dy = \iint_D \left[\frac{\partial}{\partial x}(-x^2y) - \frac{\partial}{\partial y}(xy^2) \right] dA = \iint_D (-2xy - 2xy)\,dA = \int_{-1}^1 \int_{x^2}^1 -4xy\,dy\,dx$$

$$= \int_{-1}^1 \left[-2xy^2 \right]_{y=x^2}^{y=1} dx = \int_{-1}^1 (2x^5 - 2x)\,dx = \left[\tfrac{1}{3}x^6 - x^2 \right]_{-1}^1 = 0$$

17. $\int_C x^2y\,dx - xy^2\,dy = \iint\limits_{x^2+y^2 \le 4} \left[\frac{\partial}{\partial x}(-xy^2) - \frac{\partial}{\partial y}(x^2y) \right] dA = \iint\limits_{x^2+y^2 \le 4} (-y^2 - x^2)\,dA = -\int_0^{2\pi} \int_0^2 r^3\,dr\,d\theta = -8\pi$

19. If we assume there is such a vector field $\mathbf{G}$, then $\operatorname{div}(\operatorname{curl}\mathbf{G}) = 2 + 3z - 2xz$. But $\operatorname{div}(\operatorname{curl}\mathbf{F}) = 0$ for all vector fields $\mathbf{F}$. Thus such a $\mathbf{G}$ cannot exist.

21. For any piecewise-smooth simple closed plane curve C bounding a region D, we can apply Green's Theorem to

$\mathbf{F}(x,y) = f(x)\,\mathbf{i} + g(y)\,\mathbf{j}$ to get $\int_C f(x)\,dx + g(y)\,dy = \iint_D \left[\frac{\partial}{\partial x}g(y) - \frac{\partial}{\partial y}f(x) \right] dA = \iint_D 0\,dA = 0$.

23. $\nabla^2 f = 0$ means that $\dfrac{\partial^2 f}{\partial x^2} + \dfrac{\partial^2 f}{\partial y^2} = 0$. Now if $\mathbf{F} = f_y\,\mathbf{i} - f_x\,\mathbf{j}$ and C is any closed path in D, then applying Green's

Theorem, we get

$$\int_C \mathbf{F} \cdot d\mathbf{r} = \int_C f_y\,dx - f_x\,dy = \iint_D \left[\frac{\partial}{\partial x}(-f_x) - \frac{\partial}{\partial y}(f_y) \right] dA$$

$$= -\iint_D (f_{xx} + f_{yy})\,dA = -\iint_D 0\,dA = 0$$

Therefore the line integral is independent of path, by Theorem 16.3.3.

25. $z = f(x,y) = x^2 + 2y$ with $0 \le x \le 1, 0 \le y \le 2x$. Thus

$$A(S) = \iint_D \sqrt{1 + 4x^2 + 4}\,dA = \int_0^1 \int_0^{2x} \sqrt{5 + 4x^2}\,dy\,dx = \int_0^1 2x\sqrt{5 + 4x^2}\,dx = \tfrac{1}{6}(5 + 4x^2)^{3/2} \Big]_0^1 = \tfrac{1}{6}\left(27 - 5\sqrt{5}\right).$$

27. $z = f(x,y) = x^2 + y^2$ with $0 \le x^2 + y^2 \le 4$ so $\mathbf{r}_x \times \mathbf{r}_y = -2x\,\mathbf{i} - 2y\,\mathbf{j} + \mathbf{k}$ (using upward orientation). Then

$$\iint_S z\,dS = \iint\limits_{x^2+y^2 \le 4} (x^2 + y^2)\sqrt{4x^2 + 4y^2 + 1}\,dA$$

$$= \int_0^{2\pi} \int_0^2 r^3 \sqrt{1 + 4r^2}\,dr\,d\theta = \tfrac{1}{60}\pi\left(391\sqrt{17} + 1\right)$$

(Substitute $u = 1 + 4r^2$ and use tables.)

29. Since the sphere bounds a simple solid region, the Divergence Theorem applies and

$$\iint_S \mathbf{F} \cdot d\mathbf{S} = \iiint_E \operatorname{div}\mathbf{F}\,dV = \iiint_E (z - 2)\,dV = \iiint_E z\,dV - 2\iiint_E dV$$

$$= 0 \begin{bmatrix} \text{odd function in } z \\ \text{and } E \text{ is symmetric} \end{bmatrix} - 2 \cdot V(E) = -2 \cdot \tfrac{4}{3}\pi(2)^3 = -\tfrac{64}{3}\pi$$

© 2012 Cengage Learning. All Rights Reserved. May not be scanned, copied, or duplicated, or posted to a publicly accessible website, in whole or in part.

Alternate solution: $\mathbf{F}(\mathbf{r}(\phi, \theta)) = 4 \sin \phi \cos \theta \cos \phi \, \mathbf{i} - 4 \sin \phi \sin \theta \, \mathbf{j} + 6 \sin \phi \cos \theta \, \mathbf{k}$,

$\mathbf{r}_\phi \times \mathbf{r}_\theta = 4 \sin^2 \phi \cos \theta \, \mathbf{i} + 4 \sin^2 \phi \sin \theta \, \mathbf{j} + 4 \sin \phi \cos \phi \, \mathbf{k}$, and

$\mathbf{F} \cdot (\mathbf{r}_\phi \times \mathbf{r}_\theta) = 16 \sin^3 \phi \cos^2 \theta \cos \phi - 16 \sin^3 \phi \sin^2 \theta + 24 \sin^2 \phi \cos \phi \cos \theta$. Then

$$\iint_S \mathbf{F} \cdot d\mathbf{S} = \int_0^{2\pi} \int_0^\pi (16 \sin^3 \phi \cos \phi \cos^2 \theta - 16 \sin^3 \phi \sin^2 \theta + 24 \sin^2 \phi \cos \phi \cos \theta) \, d\phi \, d\theta$$

$$= \int_0^{2\pi} \tfrac{4}{3} (-16 \sin^2 \theta) \, d\theta = -\tfrac{64}{3}\pi$$

31. Since $\operatorname{curl} \mathbf{F} = \mathbf{0}$, $\iint_S (\operatorname{curl} \mathbf{F}) \cdot d\mathbf{S} = 0$. We parametrize C: $\mathbf{r}(t) = \cos t \, \mathbf{i} + \sin t \, \mathbf{j}$, $0 \le t \le 2\pi$ and

$\oint_C \mathbf{F} \cdot d\mathbf{r} = \int_0^{2\pi} (-\cos^2 t \sin t + \sin^2 t \cos t) \, dt = \tfrac{1}{3} \cos^3 t + \tfrac{1}{3} \sin^3 t \big]_0^{2\pi} = 0$.

33. The surface is given by $x + y + z = 1$ or $z = 1 - x - y$, $0 \le x \le 1$, $0 \le y \le 1 - x$ and $\mathbf{r}_x \times \mathbf{r}_y = \mathbf{i} + \mathbf{j} + \mathbf{k}$. Then

$\oint_C \mathbf{F} \cdot d\mathbf{r} = \iint_S \operatorname{curl} \mathbf{F} \cdot d\mathbf{S} = \iint_D (-y \, \mathbf{i} - z \, \mathbf{j} - x \, \mathbf{k}) \cdot (\mathbf{i} + \mathbf{j} + \mathbf{k}) \, dA = \iint_D (-1) \, dA = -(\text{area of } D) = -\tfrac{1}{2}$.

35. $\iiint_E \operatorname{div} \mathbf{F} \, dV = \iiint\limits_{x^2 + y^2 + z^2 \le 1} 3 \, dV = 3(\text{volume of sphere}) = 4\pi$. Then

$\mathbf{F}(\mathbf{r}(\phi, \theta)) \cdot (\mathbf{r}_\phi \times \mathbf{r}_\theta) = \sin^3 \phi \cos^2 \theta + \sin^3 \phi \sin^2 \theta + \sin \phi \cos^2 \phi = \sin \phi$ and

$\iint_S \mathbf{F} \cdot d\mathbf{S} = \int_0^{2\pi} \int_0^\pi \sin \phi \, d\phi \, d\theta = (2\pi)(2) = 4\pi$.

37. Because $\operatorname{curl} \mathbf{F} = \mathbf{0}$, $\mathbf{F}$ is conservative, so there exists a function f such that $\nabla f = \mathbf{F}$. Then $f_x(x, y, z) = 3x^2yz - 3y$

implies $f(x, y, z) = x^3yz - 3xy + g(y, z) \Rightarrow f_y(x, y, z) = x^3z - 3x + g_y(y, z)$. But $f_y(x, y, z) = x^3z - 3x$, so

$g(y, z) = h(z)$ and $f(x, y, z) = x^3yz - 3xy + h(z)$. Then $f_z(x, y, z) = x^3y + h'(z)$ but $f_z(x, y, z) = x^3y + 2z$,

so $h(z) = z^2 + K$ and a potential function for $\mathbf{F}$ is $f(x, y, z) = x^3yz - 3xy + z^2$. Hence

$\int_C \mathbf{F} \cdot d\mathbf{r} = \int_C \nabla f \cdot d\mathbf{r} = f(0, 3, 0) - f(0, 0, 2) = 0 - 4 = -4$.

39. By the Divergence Theorem, $\iint_S \mathbf{F} \cdot \mathbf{n} \, dS = \iiint_E \operatorname{div} \mathbf{F} \, dV = 3(\text{volume of } E) = 3(8 - 1) = 21$.

41. Let $\mathbf{F} = \mathbf{a} \times \mathbf{r} = \langle a_1, a_2, a_3 \rangle \times \langle x, y, z \rangle = \langle a_2z - a_3y, a_3x - a_1z, a_1y - a_2x \rangle$. Then $\operatorname{curl} \mathbf{F} = \langle 2a_1, 2a_2, 2a_3 \rangle = 2\mathbf{a}$,

and $\iint_S 2\mathbf{a} \cdot d\mathbf{S} = \iint_S \operatorname{curl} \mathbf{F} \cdot d\mathbf{S} = \int_C \mathbf{F} \cdot d\mathbf{r} = \int_C (\mathbf{a} \times \mathbf{r}) \cdot d\mathbf{r}$ by Stokes' Theorem.

© 2012 Cengage Learning. All Rights Reserved. May not be scanned, copied, or duplicated, or posted to a publicly accessible website, in whole or in part.

□ PROBLEMS PLUS

1. Let S_1 be the portion of $\Omega(S)$ between $S(a)$ and S, and let ∂S_1 be its boundary. Also let S_L be the lateral surface of S_1 [that is, the surface of S_1 except S and $S(a)$]. Applying the Divergence Theorem we have $\iint_{\partial S_1} \frac{\mathbf{r} \cdot \mathbf{n}}{r^3} \, dS = \iiint_{S_1} \nabla \cdot \frac{\mathbf{r}}{r^3} \, dV$.

But

$$\nabla \cdot \frac{\mathbf{r}}{r^3} = \left\langle \frac{\partial}{\partial x}, \frac{\partial}{\partial y}, \frac{\partial}{\partial z} \right\rangle \cdot \left\langle \frac{x}{(x^2 + y^2 + z^2)^{3/2}}, \frac{y}{(x^2 + y^2 + z^2)^{3/2}}, \frac{z}{(x^2 + y^2 + z^2)^{3/2}} \right\rangle$$

$$= \frac{(x^2 + y^2 + z^2 - 3x^2) + (x^2 + y^2 + z^2 - 3y^2) + (x^2 + y^2 + z^2 - 3z^2)}{(x^2 + y^2 + z^2)^{5/2}} = 0$$

$\Rightarrow \iint_{\partial S_1} \frac{\mathbf{r} \cdot \mathbf{n}}{r^3} \, dS = \iiint_{S_1} 0 \, dV = 0$. On the other hand, notice that for the surfaces of ∂S_1 other than $S(a)$ and S,

$\mathbf{r} \cdot \mathbf{n} = 0 \Rightarrow$

$$0 = \iint_{\partial S_1} \frac{\mathbf{r} \cdot \mathbf{n}}{r^3} \, dS = \iint_{S} \frac{\mathbf{r} \cdot \mathbf{n}}{r^3} \, dS + \iint_{S(a)} \frac{\mathbf{r} \cdot \mathbf{n}}{r^3} \, dS + \iint_{S_L} \frac{\mathbf{r} \cdot \mathbf{n}}{r^3} \, dS = \iint_{S} \frac{\mathbf{r} \cdot \mathbf{n}}{r^3} \, dS + \iint_{S(a)} \frac{\mathbf{r} \cdot \mathbf{n}}{r^3} \, dS \Rightarrow$$

$\iint_{S} \frac{\mathbf{r} \cdot \mathbf{n}}{r^3} \, dS = -\iint_{S(a)} \frac{\mathbf{r} \cdot \mathbf{n}}{r^3} \, dS$. Notice that on $S(a)$, $r = a \Rightarrow \mathbf{n} = -\frac{\mathbf{r}}{r} = -\frac{\mathbf{r}}{a}$ and $\mathbf{r} \cdot \mathbf{r} = r^2 = a^2$, so

that $-\iint_{S(a)} \frac{\mathbf{r} \cdot \mathbf{n}}{r^3} \, dS = \iint_{S(a)} \frac{\mathbf{r} \cdot \mathbf{r}}{a^4} \, dS = \iint_{S(a)} \frac{a^2}{a^4} \, dS = \frac{1}{a^2} \iint_{S(a)} dS = \frac{\text{area of } S(a)}{a^2} = |\Omega(S)|$.

Therefore $|\Omega(S)| = \iint_{S} \frac{\mathbf{r} \cdot \mathbf{n}}{r^3} \, dS$.

3. The given line integral $\frac{1}{2} \int_C (bz - cy) \, dx + (cx - az) \, dy + (ay - bx) \, dz$ can be expressed as $\int_C \mathbf{F} \cdot d\mathbf{r}$ if we define the vector field $\mathbf{F}$ by $\mathbf{F}(x, y, z) = P \mathbf{i} + Q \mathbf{j} + R \mathbf{k} = \frac{1}{2}(bz - cy) \mathbf{i} + \frac{1}{2}(cx - az) \mathbf{j} + \frac{1}{2}(ay - bx) \mathbf{k}$. Then define S to be the planar interior of C, so S is an oriented, smooth surface. Stokes' Theorem says $\int_C \mathbf{F} \cdot d\mathbf{r} = \iint_S \operatorname{curl} \mathbf{F} \cdot d\mathbf{S} = \iint_S \operatorname{curl} \mathbf{F} \cdot \mathbf{n} \, dS$. Now

$$\operatorname{curl} \mathbf{F} = \left(\frac{\partial R}{\partial y} - \frac{\partial Q}{\partial z} \right) \mathbf{i} + \left(\frac{\partial P}{\partial z} - \frac{\partial R}{\partial x} \right) \mathbf{j} + \left(\frac{\partial Q}{\partial x} - \frac{\partial P}{\partial y} \right) \mathbf{k}$$

$$= \left(\tfrac{1}{2}a + \tfrac{1}{2}a \right) \mathbf{i} + \left(\tfrac{1}{2}b + \tfrac{1}{2}b \right) \mathbf{j} + \left(\tfrac{1}{2}c + \tfrac{1}{2}c \right) \mathbf{k} = a \mathbf{i} + b \mathbf{j} + c \mathbf{k} = \mathbf{n}$$

so $\operatorname{curl} \mathbf{F} \cdot \mathbf{n} = \mathbf{n} \cdot \mathbf{n} = |\mathbf{n}|^2 = 1$, hence $\iint_S \operatorname{curl} \mathbf{F} \cdot \mathbf{n} \, dS = \iint_S dS$ which is simply the surface area of S. Thus, $\int_C \mathbf{F} \cdot d\mathbf{r} = \frac{1}{2} \int_C (bz - cy) \, dx + (cx - az) \, dy + (ay - bx) \, dz$ is the plane area enclosed by C.

5. $(\mathbf{F} \cdot \nabla) \mathbf{G} = \left(P_1 \frac{\partial}{\partial x} + Q_1 \frac{\partial}{\partial y} + R_1 \frac{\partial}{\partial z} \right) (P_2 \mathbf{i} + Q_2 \mathbf{j} + R_2 \mathbf{k})$

$$= \left(P_1 \frac{\partial P_2}{\partial x} + Q_1 \frac{\partial P_2}{\partial y} + R_1 \frac{\partial P_2}{\partial z} \right) \mathbf{i} + \left(P_1 \frac{\partial Q_2}{\partial x} + Q_1 \frac{\partial Q_2}{\partial y} + R_1 \frac{\partial Q_2}{\partial z} \right) \mathbf{j}$$

$$+ \left(P_1 \frac{\partial R_2}{\partial x} + Q_1 \frac{\partial R_2}{\partial y} + R_1 \frac{\partial R_2}{\partial z} \right) \mathbf{k}$$

$$= (\mathbf{F} \cdot \nabla P_2) \mathbf{i} + (\mathbf{F} \cdot \nabla Q_2) \mathbf{j} + (\mathbf{F} \cdot \nabla R_2) \mathbf{k}.$$

© 2012 Cengage Learning. All Rights Reserved. May not be scanned, copied, or duplicated, or posted to a publicly accessible website, in whole or in part.

Similarly, $(\mathbf{G} \cdot \nabla)\mathbf{F} = (\mathbf{G} \cdot \nabla P_1)\mathbf{i} + (\mathbf{G} \cdot \nabla Q_1)\mathbf{j} + (\mathbf{G} \cdot \nabla R_1)\mathbf{k}$. Then

$$\mathbf{F} \times \operatorname{curl}\mathbf{G} = \begin{vmatrix} \mathbf{i} & \mathbf{j} & \mathbf{k} \\ P_1 & Q_1 & R_1 \\ \partial R_2/\partial y - \partial Q_2/\partial z & \partial P_2/\partial z - \partial R_2/\partial x & \partial Q_2/\partial x - \partial P_2/\partial y \end{vmatrix}$$

$$= \left(Q_1\frac{\partial Q_2}{\partial x} - Q_1\frac{\partial P_2}{\partial y} - R_1\frac{\partial P_2}{\partial z} + R_1\frac{\partial R_2}{\partial x} \right)\mathbf{i} + \left(R_1\frac{\partial R_2}{\partial y} - R_1\frac{\partial Q_2}{\partial z} - P_1\frac{\partial Q_2}{\partial x} + P_1\frac{\partial P_2}{\partial y} \right)\mathbf{j}$$

$$+ \left(P_1\frac{\partial P_2}{\partial z} - P_1\frac{\partial R_2}{\partial x} - Q_1\frac{\partial R_2}{\partial y} + Q_1\frac{\partial Q_2}{\partial z} \right)\mathbf{k}$$

and

$$\mathbf{G} \times \operatorname{curl}\mathbf{F} = \left(Q_2\frac{\partial Q_1}{\partial x} - Q_2\frac{\partial P_1}{\partial y} - R_2\frac{\partial P_1}{\partial z} + R_2\frac{\partial R_1}{\partial x} \right)\mathbf{i} + \left(R_2\frac{\partial R_1}{\partial y} - R_2\frac{\partial Q_1}{\partial z} - P_2\frac{\partial Q_1}{\partial x} + P_2\frac{\partial P_1}{\partial y} \right)\mathbf{j}$$

$$+ \left(P_2\frac{\partial P_1}{\partial z} - P_2\frac{\partial R_1}{\partial x} - Q_2\frac{\partial R_1}{\partial y} + Q_2\frac{\partial Q_1}{\partial z} \right)\mathbf{k}.$$

Then

$$(\mathbf{F} \cdot \nabla)\mathbf{G} + \mathbf{F} \times \operatorname{curl}\mathbf{G} = \left(P_1\frac{\partial P_2}{\partial x} + Q_1\frac{\partial Q_2}{\partial x} + R_1\frac{\partial R_2}{\partial x} \right)\mathbf{i} + \left(P_1\frac{\partial P_2}{\partial y} + Q_1\frac{\partial Q_2}{\partial y} + R_1\frac{\partial R_2}{\partial y} \right)\mathbf{j}$$

$$+ \left(P_1\frac{\partial P_2}{\partial z} + Q_1\frac{\partial Q_2}{\partial z} + R_1\frac{\partial R_2}{\partial z} \right)\mathbf{k}$$

and

$$(\mathbf{G} \cdot \nabla)\mathbf{F} + \mathbf{G} \times \operatorname{curl}\mathbf{F} = \left(P_2\frac{\partial P_1}{\partial x} + Q_2\frac{\partial Q_1}{\partial x} + R_2\frac{\partial R_1}{\partial x} \right)\mathbf{i} + \left(P_2\frac{\partial P_1}{\partial y} + Q_2\frac{\partial Q_1}{\partial y} + R_2\frac{\partial R_1}{\partial y} \right)\mathbf{j}$$

$$+ \left(P_2\frac{\partial P_1}{\partial z} + Q_2\frac{\partial Q_1}{\partial z} + R_2\frac{\partial R_1}{\partial z} \right)\mathbf{k}.$$

Hence

$$(\mathbf{F} \cdot \nabla)\mathbf{G} + \mathbf{F} \times \operatorname{curl}\mathbf{G} + (\mathbf{G} \cdot \nabla)\mathbf{F} + \mathbf{G} \times \operatorname{curl}\mathbf{F}$$

$$= \left[\left(P_1\frac{\partial P_2}{\partial x} + P_2\frac{\partial P_1}{\partial x} \right) + \left(Q_1\frac{\partial Q_2}{\partial x} + Q_2\frac{\partial Q_1}{\partial y} \right) + \left(R_1\frac{\partial R_2}{\partial x} + R_2\frac{\partial R_1}{\partial x} \right) \right]\mathbf{i}$$

$$+ \left[\left(P_1\frac{\partial P_2}{\partial y} + P_2\frac{\partial P_1}{\partial y} \right) + \left(Q_1\frac{\partial Q_2}{\partial y} + Q_2\frac{\partial Q_1}{\partial y} \right) + \left(R_1\frac{\partial R_2}{\partial y} + R_2\frac{\partial R_1}{\partial y} \right) \right]\mathbf{j}$$

$$+ \left[\left(P_1\frac{\partial P_2}{\partial z} + P_2\frac{\partial P_1}{\partial z} \right) + \left(Q_1\frac{\partial Q_2}{\partial z} + Q_2\frac{\partial Q_1}{\partial z} \right) + \left(R_1\frac{\partial R_2}{\partial z} + R_2\frac{\partial R_1}{\partial z} \right) \right]\mathbf{k}$$

$$= \nabla(P_1 P_2 + Q_1 Q_2 + R_1 R_2) = \nabla(\mathbf{F} \cdot \mathbf{G}).$$

© 2012 Cengage Learning. All Rights Reserved. May not be scanned, copied, or duplicated, or posted to a publicly accessible website, in whole or in part.

17 □ SECOND-ORDER DIFFERENTIAL EQUATIONS

17.1 Second-Order Linear Equations

1. The auxiliary equation is $r^2 - r - 6 = 0$ $\Rightarrow$ $(r-3)(r+2) = 0$ $\Rightarrow$ $r = 3, r = -2$. Then by (8) the general solution is $y = c_1 e^{3x} + c_2 e^{-2x}$.

3. The auxiliary equation is $r^2 + 16 = 0$ $\Rightarrow$ $r = \pm 4i$. Then by (11) the general solution is
$$y = e^{0x}(c_1 \cos 4x + c_2 \sin 4x) = c_1 \cos 4x + c_2 \sin 4x.$$

5. The auxiliary equation is $9r^2 - 12r + 4 = 0$ $\Rightarrow$ $(3r-2)^2 = 0$ $\Rightarrow$ $r = \frac{2}{3}$. Then by (10), the general solution is
$$y = c_1 e^{2x/3} + c_2 x e^{2x/3}.$$

7. The auxiliary equation is $2r^2 - r = r(2r-1) = 0$ $\Rightarrow$ $r = 0, r = \frac{1}{2}$, so $y = c_1 e^{0x} + c_2 e^{x/2} = c_1 + c_2 e^{x/2}$.

9. The auxiliary equation is $r^2 - 4r + 13 = 0$ $\Rightarrow$ $r = \dfrac{4 \pm \sqrt{-36}}{2} = 2 \pm 3i$, so $y = e^{2x}(c_1 \cos 3x + c_2 \sin 3x)$.

11. The auxiliary equation is $2r^2 + 2r - 1 = 0$ $\Rightarrow$ $r = \dfrac{-2 \pm \sqrt{12}}{4} = -\dfrac{1}{2} \pm \dfrac{\sqrt{3}}{2}$, so
$$y = c_1 e^{(-1/2 + \sqrt{3}/2)t} + c_2 e^{(-1/2 - \sqrt{3}/2)t}.$$

13. The auxiliary equation is $100r^2 + 200r + 101 = 0$ $\Rightarrow$ $r = \dfrac{-200 \pm \sqrt{-400}}{200} = -1 \pm \dfrac{1}{10}i$, so
$$P = e^{-t}\left[c_1 \cos\left(\tfrac{1}{10}t\right) + c_2 \sin\left(\tfrac{1}{10}t\right)\right].$$

15. The auxiliary equation is $5r^2 - 2r - 3 = (5r+3)(r-1) = 0$ $\Rightarrow$ $r = -\frac{3}{5}$,

$r = 1$, so the general solution is $y = c_1 e^{-3x/5} + c_2 e^x$. We graph the basic

solutions $f(x) = e^{-3x/5}, g(x) = e^x$ as well as $y = e^{-3x/5} + 2e^x$,

$y = e^{-3x/5} - e^x$, and $y = -2e^{-3x/5} - e^x$. Each solution consists of a single

continuous curve that approaches either 0 or $\pm\infty$ as $x \to \pm\infty$.

17. $r^2 - 6r + 8 = (r-4)(r-2) = 0$, so $r = 4, r = 2$ and the general solution is $y = c_1 e^{4x} + c_2 e^{2x}$. Then
$y' = 4c_1 e^{4x} + 2c_2 e^{2x}$, so $y(0) = 2$ $\Rightarrow$ $c_1 + c_2 = 2$ and $y'(0) = 2$ $\Rightarrow$ $4c_1 + 2c_2 = 2$, giving $c_1 = -1$ and $c_2 = 3$.
Thus the solution to the initial-value problem is $y = 3e^{2x} - e^{4x}$.

19. $9r^2 + 12r + 4 = (3r+2)^2 = 0$ $\Rightarrow$ $r = -\frac{2}{3}$ and the general solution is $y = c_1 e^{-2x/3} + c_2 x e^{-2x/3}$. Then $y(0) = 1$ $\Rightarrow$
$c_1 = 1$ and, since $y' = -\frac{2}{3}c_1 e^{-2x/3} + c_2 \left(1 - \frac{2}{3}x\right)e^{-2x/3}$, $y'(0) = 0$ $\Rightarrow$ $-\frac{2}{3}c_1 + c_2 = 0$, so $c_2 = \frac{2}{3}$ and the solution to
the initial-value problem is $y = e^{-2x/3} + \frac{2}{3}x e^{-2x/3}$.

© 2012 Cengage Learning. All Rights Reserved. May not be scanned, copied, or duplicated, or posted to a publicly accessible website, in whole or in part.

21. $r^2 - 6r + 10 = 0 \Rightarrow r = 3 \pm i$ and the general solution is $y = e^{3x}(c_1 \cos x + c_2 \sin x)$. Then $2 = y(0) = c_1$ and

$3 = y'(0) = c_2 + 3c_1 \Rightarrow c_2 = -3$ and the solution to the initial-value problem is $y = e^{3x}(2 \cos x - 3 \sin x)$.

23. $r^2 - r - 12 = (r - 4)(r + 3) = 0 \Rightarrow r = 4, r = -3$ and the general solution is $y = c_1 e^{4x} + c_2 e^{-3x}$. Then

$0 = y(1) = c_1 e^4 + c_2 e^{-3}$ and $1 = y'(1) = 4c_1 e^4 - 3c_2 e^{-3}$ so $c_1 = \frac{1}{7}e^{-4}$, $c_2 = -\frac{1}{7}e^3$ and the solution to the initial-value

problem is $y = \frac{1}{7}e^{-4}e^{4x} - \frac{1}{7}e^3 e^{-3x} = \frac{1}{7}e^{4x-4} - \frac{1}{7}e^{3-3x}$.

25. $r^2 + 4 = 0 \Rightarrow r = \pm 2i$ and the general solution is $y = c_1 \cos 2x + c_2 \sin 2x$. Then $5 = y(0) = c_1$ and $3 = y(\pi/4) = c_2$,

so the solution of the boundary-value problem is $y = 5 \cos 2x + 3 \sin 2x$.

27. $r^2 + 4r + 4 = (r + 2)^2 = 0 \Rightarrow r = -2$ and the general solution is $y = c_1 e^{-2x} + c_2 x e^{-2x}$. Then $2 = y(0) = c_1$ and

$0 = y(1) = c_1 e^{-2} + c_2 e^{-2}$ so $c_2 = -2$, and the solution of the boundary-value problem is $y = 2e^{-2x} - 2xe^{-2x}$.

29. $r^2 - r = r(r - 1) = 0 \Rightarrow r = 0, r = 1$ and the general solution is $y = c_1 + c_2 e^x$. Then $1 = y(0) = c_1 + c_2$

and $2 = y(1) = c_1 + c_2 e$ so $c_1 = \dfrac{e - 2}{e - 1}$, $c_2 = \dfrac{1}{e - 1}$. The solution of the boundary-value problem is $y = \dfrac{e - 2}{e - 1} + \dfrac{e^x}{e - 1}$.

31. $r^2 + 4r + 20 = 0 \Rightarrow r = -2 \pm 4i$ and the general solution is $y = e^{-2x}(c_1 \cos 4x + c_2 \sin 4x)$. But $1 = y(0) = c_1$ and

$2 = y(\pi) = c_1 e^{-2\pi} \Rightarrow c_1 = 2e^{2\pi}$, so there is no solution.

33. (a) *Case 1* $(\lambda = 0)$: $y'' + \lambda y = 0 \Rightarrow y'' = 0$ which has an auxiliary equation $r^2 = 0 \Rightarrow r = 0 \Rightarrow y = c_1 + c_2 x$

where $y(0) = 0$ and $y(L) = 0$. Thus, $0 = y(0) = c_1$ and $0 = y(L) = c_2 L \Rightarrow c_1 = c_2 = 0$. Thus $y = 0$.

Case 2 $(\lambda < 0)$: $y'' + \lambda y = 0$ has auxiliary equation $r^2 = -\lambda \Rightarrow r = \pm\sqrt{-\lambda}$ [distinct and real since $\lambda < 0$] $\Rightarrow$

$y = c_1 e^{\sqrt{-\lambda}x} + c_2 e^{-\sqrt{-\lambda}x}$ where $y(0) = 0$ and $y(L) = 0$. Thus $0 = y(0) = c_1 + c_2$ (∗) and

$0 = y(L) = c_1 e^{\sqrt{-\lambda}L} + c_2 e^{-\sqrt{-\lambda}L}$ (†).

Multiplying (∗) by $e^{\sqrt{-\lambda}L}$ and subtracting (†) gives $c_2\left(e^{\sqrt{-\lambda}L} - e^{-\sqrt{-\lambda}L}\right) = 0 \Rightarrow c_2 = 0$ and thus $c_1 = 0$ from (∗).

Thus $y = 0$ for the cases $\lambda = 0$ and $\lambda < 0$.

(b) $y'' + \lambda y = 0$ has an auxiliary equation $r^2 + \lambda = 0 \Rightarrow r = \pm i\sqrt{\lambda} \Rightarrow y = c_1 \cos \sqrt{\lambda}x + c_2 \sin \sqrt{\lambda}x$ where

$y(0) = 0$ and $y(L) = 0$. Thus, $0 = y(0) = c_1$ and $0 = y(L) = c_2 \sin \sqrt{\lambda}L$ since $c_1 = 0$. Since we cannot have a trivial

solution, $c_2 \neq 0$ and thus $\sin \sqrt{\lambda}L = 0 \Rightarrow \sqrt{\lambda}L = n\pi$ where n is an integer $\Rightarrow \lambda = n^2\pi^2/L^2$ and

$y = c_2 \sin(n\pi x/L)$ where n is an integer.

35. (a) $r^2 - 2r + 2 = 0 \Rightarrow r = 1 \pm i$ and the general solution is $y = e^x(c_1 \cos x + c_2 \sin x)$. If $y(a) = c$ and $y(b) = d$ then

$e^a(c_1 \cos a + c_2 \sin a) = c \Rightarrow c_1 \cos a + c_2 \sin a = ce^{-a}$ and $e^b(c_1 \cos b + c_2 \sin b) = d \Rightarrow$

$c_1 \cos b + c_2 \sin b = de^{-b}$. This gives a linear system in c_1 and c_2 which has a unique solution if the lines are not parallel.

If the lines are not vertical or horizontal, we have parallel lines if $\cos a = k \cos b$ and $\sin a = k \sin b$ for some nonzero

© 2012 Cengage Learning. All Rights Reserved. May not be scanned, copied, or duplicated, or posted to a publicly accessible website, in whole or in part.

constant k or $\dfrac{\cos a}{\cos b} = k = \dfrac{\sin a}{\sin b}$ $\Rightarrow$ $\dfrac{\sin a}{\cos a} = \dfrac{\sin b}{\cos b}$ $\Rightarrow$ $\tan a = \tan b$ $\Rightarrow$ $b - a = n\pi$, n any integer. (Note that none of $\cos a$, $\cos b$, $\sin a$, $\sin b$ are zero.) If the lines are both horizontal then $\cos a = \cos b = 0$ $\Rightarrow$ $b - a = n\pi$, and similarly vertical lines means $\sin a = \sin b = 0$ $\Rightarrow$ $b - a = n\pi$. Thus the system has a unique solution if $b - a \neq n\pi$.

(b) The linear system has no solution if the lines are parallel but not identical. From part (a) the lines are parallel if

$b - a = n\pi$. If the lines are not horizontal, they are identical if $ce^{-a} = kde^{-b}$ $\Rightarrow$ $\dfrac{ce^{-a}}{de^{-b}} = k = \dfrac{\cos a}{\cos b}$ $\Rightarrow$

$\dfrac{c}{d} = e^{a-b}\dfrac{\cos a}{\cos b}$. (If $d = 0$ then $c = 0$ also.) If they are horizontal then $\cos b = 0$, but $k = \dfrac{\sin a}{\sin b}$ also (and $\sin b \neq 0$) so

we require $\dfrac{c}{d} = e^{a-b}\dfrac{\sin a}{\sin b}$. Thus the system has no solution if $b - a = n\pi$ and $\dfrac{c}{d} \neq e^{a-b}\dfrac{\cos a}{\cos b}$ unless $\cos b = 0$, in

which case $\dfrac{c}{d} \neq e^{a-b}\dfrac{\sin a}{\sin b}$.

(c) The linear system has infinitely many solution if the lines are identical (and necessarily parallel). From part (b) this occurs

when $b - a = n\pi$ and $\dfrac{c}{d} = e^{a-b}\dfrac{\cos a}{\cos b}$ unless $\cos b = 0$, in which case $\dfrac{c}{d} = e^{a-b}\dfrac{\sin a}{\sin b}$.

17.2 Nonhomogeneous Linear Equations

1. The auxiliary equation is $r^2 - 2r - 3 = (r - 3)(r + 1) = 0$ $\Rightarrow$ $r = 3, r = -1$, so the complementary solution is

$y_c(x) = c_1 e^{3x} + c_2 e^{-x}$. We try the particular solution $y_p(x) = A\cos 2x + B\sin 2x$, so

$y_p' = -2A\sin 2x + 2B\cos 2x$ and $y_p'' = -4A\cos 2x - 4B\sin 2x$. Substitution into the differential equation gives

$(-4A\cos 2x - 4B\sin 2x) - 2(-2A\sin 2x + 2B\cos 2x) - 3(A\cos 2x + B\sin 2x) = \cos 2x$ $\Rightarrow$

$(-7A - 4B)\cos 2x + (4A - 7B)\sin 2x = \cos 2x$. Then $-7A - 4B = 1$ and $4A - 7B = 0$ $\Rightarrow$ $A = -\frac{7}{65}$ and

$B = -\frac{4}{65}$. Thus the general solution is $y(x) = y_c(x) + y_p(x) = c_1 e^{3x} + c_2 e^{-x} - \frac{7}{65}\cos 2x - \frac{4}{65}\sin 2x$.

3. The auxiliary equation is $r^2 + 9 = 0$ with roots $r = \pm 3i$, so the complementary solution is $y_c(x) = c_1 \cos 3x + c_2 \sin 3x$.

Try the particular solution $y_p(x) = Ae^{-2x}$, so $y_p' = -2Ae^{-2x}$ and $y_p'' = 4Ae^{-2x}$. Substitution into the differential equation

gives $4Ae^{-2x} + 9(Ae^{-2x}) = e^{-2x}$ or $13Ae^{-2x} = e^{-2x}$. Thus $13A = 1$ $\Rightarrow$ $A = \frac{1}{13}$ and the general solution is

$y(x) = y_c(x) + y_p(x) = c_1 \cos 3x + c_2 \sin 3x + \frac{1}{13}e^{-2x}$.

5. The auxiliary equation is $r^2 - 4r + 5 = 0$ with roots $r = 2 \pm i$, so the complementary solution is

$y_c(x) = e^{2x}(c_1 \cos x + c_2 \sin x)$. Try $y_p(x) = Ae^{-x}$, so $y_p' = -Ae^{-x}$ and $y_p'' = Ae^{-x}$. Substitution gives

$Ae^{-x} - 4(-Ae^{-x}) + 5(Ae^{-x}) = e^{-x}$ $\Rightarrow$ $10Ae^{-x} = e^{-x}$ $\Rightarrow$ $A = \frac{1}{10}$. Thus the general solution is

$y(x) = e^{2x}(c_1 \cos x + c_2 \sin x) + \frac{1}{10}e^{-x}$.

© 2012 Cengage Learning. All Rights Reserved. May not be scanned, copied, or duplicated, or posted to a publicly accessible website, in whole or in part.

7. The auxiliary equation is $r^2 + 1 = 0$ with roots $r = \pm i$, so the complementary solution is $y_c(x) = c_1 \cos x + c_2 \sin x$.

For $y'' + y = e^x$ try $y_{p_1}(x) = Ae^x$. Then $y'_{p_1} = y''_{p_1} = Ae^x$ and substitution gives $Ae^x + Ae^x = e^x \Rightarrow A = \frac{1}{2}$,

so $y_{p_1}(x) = \frac{1}{2}e^x$. For $y'' + y = x^3$ try $y_{p_2}(x) = Ax^3 + Bx^2 + Cx + D$. Then $y'_{p_2} = 3Ax^2 + 2Bx + C$ and

$y''_{p_2} = 6Ax + 2B$. Substituting, we have $6Ax + 2B + Ax^3 + Bx^2 + Cx + D = x^3$, so $A = 1$, $B = 0$,

$6A + C = 0 \Rightarrow C = -6$, and $2B + D = 0 \Rightarrow D = 0$. Thus $y_{p_2}(x) = x^3 - 6x$ and the general solution is

$y(x) = y_c(x) + y_{p_1}(x) + y_{p_2}(x) = c_1 \cos x + c_2 \sin x + \frac{1}{2}e^x + x^3 - 6x$. But $2 = y(0) = c_1 + \frac{1}{2} \Rightarrow$

$c_1 = \frac{3}{2}$ and $0 = y'(0) = c_2 + \frac{1}{2} - 6 \Rightarrow c_2 = \frac{11}{2}$. Thus the solution to the initial-value problem is

$y(x) = \frac{3}{2}\cos x + \frac{11}{2}\sin x + \frac{1}{2}e^x + x^3 - 6x$.

9. The auxiliary equation is $r^2 - r = 0$ with roots $r = 0$, $r = 1$ so the complementary solution is $y_c(x) = c_1 + c_2 e^x$.

Try $y_p(x) = x(Ax + B)e^x$ so that no term in y_p is a solution of the complementary equation. Then

$y'_p = (Ax^2 + (2A + B)x + B)e^x$ and $y''_p = (Ax^2 + (4A + B)x + (2A + 2B))e^x$. Substitution into the differential equation

gives $(Ax^2 + (4A + B)x + (2A + 2B))e^x - (Ax^2 + (2A + B)x + B)e^x = xe^x \Rightarrow (2Ax + (2A + B))e^x = xe^x \Rightarrow$

$A = \frac{1}{2}$, $B = -1$. Thus $y_p(x) = \left(\frac{1}{2}x^2 - x\right)e^x$ and the general solution is $y(x) = c_1 + c_2 e^x + \left(\frac{1}{2}x^2 - x\right)e^x$. But

$2 = y(0) = c_1 + c_2$ and $1 = y'(0) = c_2 - 1$, so $c_2 = 2$ and $c_1 = 0$. The solution to the initial-value problem is

$y(x) = 2e^x + \left(\frac{1}{2}x^2 - x\right)e^x = e^x\left(\frac{1}{2}x^2 - x + 2\right)$.

11. The auxiliary equation is $r^2 + 3r + 2 = (r + 1)(r + 2) = 0$, so $r = -1$, $r = -2$ and $y_c(x) = c_1 e^{-x} + c_2 e^{-2x}$.

Try $y_p = A\cos x + B\sin x \Rightarrow y'_p = -A\sin x + B\cos x$, $y''_p = -A\cos x - B\sin x$. Substituting into the differential

equation gives $(-A\cos x - B\sin x) + 3(-A\sin x + B\cos x) + 2(A\cos x + B\sin x) = \cos x$ or

$(A + 3B)\cos x + (-3A + B)\sin x = \cos x$. Then solving the equations

$A + 3B = 1$, $-3A + B = 0$ gives $A = \frac{1}{10}$, $B = \frac{3}{10}$ and the general

solution is $y(x) = c_1 e^{-x} + c_2 e^{-2x} + \frac{1}{10}\cos x + \frac{3}{10}\sin x$. The graph

shows y_p and several other solutions. Notice that all solutions are

asymptotic to y_p as $x \to \infty$. Except for y_p, all solutions approach either ∞

or $-\infty$ as $x \to -\infty$.

13. Here $y_c(x) = c_1 e^{2x} + c_2 e^{-x}$, and a trial solution is $y_p(x) = (Ax + B)e^x \cos x + (Cx + D)e^x \sin x$.

15. Here $y_c(x) = c_1 e^{2x} + c_2 e^x$. For $y'' - 3y' + 2y = e^x$ try $y_{p_1}(x) = Axe^x$ (since $y = Ae^x$ is a solution of the complementary

equation) and for $y'' - 3y' + 2y = \sin x$ try $y_{p_2}(x) = B\cos x + C\sin x$. Thus a trial solution is

$y_p(x) = y_{p_1}(x) + y_{p_2}(x) = Axe^x + B\cos x + C\sin x$.

17. Since $y_c(x) = e^{-x}(c_1 \cos 3x + c_2 \sin 3x)$ we try $y_p(x) = x(Ax^2 + Bx + C)e^{-x}\cos 3x + x(Dx^2 + Ex + F)e^{-x}\sin 3x$

(so that no term of y_p is a solution of the complementary equation).

© 2012 Cengage Learning. All Rights Reserved. May not be scanned, copied, or duplicated, or posted to a publicly accessible website, in whole or in part.

Note: Solving Equations (7) and (9) in The Method of Variation of Parameters gives

$$u_1' = -\frac{Gy_2}{a\,(y_1 y_2' - y_2 y_1')} \quad \text{and} \quad u_2' = \frac{Gy_1}{a\,(y_1 y_2' - y_2 y_1')}$$

We will use these equations rather than resolving the system in each of the remaining exercises in this section.

19. (a) Here $4r^2 + 1 = 0 \;\Rightarrow\; r = \pm\frac{1}{2}i$ and $y_c(x) = c_1 \cos\left(\frac{1}{2}x\right) + c_2 \sin\left(\frac{1}{2}x\right)$. We try a particular solution of the form

$y_p(x) = A\cos x + B\sin x \;\Rightarrow\; y_p' = -A\sin x + B\cos x$ and $y_p'' = -A\cos x - B\sin x$. Then the equation

$4y'' + y = \cos x$ becomes $4(-A\cos x - B\sin x) + (A\cos x + B\sin x) = \cos x$ or

$-3A\cos x - 3B\sin x = \cos x \;\Rightarrow\; A = -\frac{1}{3}, B = 0$. Thus, $y_p(x) = -\frac{1}{3}\cos x$ and the general solution is

$y(x) = y_c(x) + y_p(x) = c_1 \cos\left(\frac{1}{2}x\right) + c_2 \sin\left(\frac{1}{2}x\right) - \frac{1}{3}\cos x$.

(b) From (a) we know that $y_c(x) = c_1 \cos\frac{x}{2} + c_2 \sin\frac{x}{2}$. Setting $y_1 = \cos\frac{x}{2}$, $y_2 = \sin\frac{x}{2}$, we have

$y_1 y_2' - y_2 y_1' = \frac{1}{2}\cos^2\frac{x}{2} + \frac{1}{2}\sin^2\frac{x}{2} = \frac{1}{2}$. Thus $u_1' = -\dfrac{\cos x \sin\frac{x}{2}}{4\cdot\frac{1}{2}} = -\frac{1}{2}\cos\left(2\cdot\frac{x}{2}\right)\sin\frac{x}{2} = -\frac{1}{2}\left(2\cos^2\frac{x}{2} - 1\right)\sin\frac{x}{2}$

and $u_2' = \dfrac{\cos x \cos\frac{x}{2}}{4\cdot\frac{1}{2}} = \frac{1}{2}\cos\left(2\cdot\frac{x}{2}\right)\cos\frac{x}{2} = \frac{1}{2}\left(1 - 2\sin^2\frac{x}{2}\right)\cos\frac{x}{2}$. Then

$u_1(x) = \int\left(\frac{1}{2}\sin\frac{x}{2} - \cos^2\frac{x}{2}\sin\frac{x}{2}\right)dx = -\cos\frac{x}{2} + \frac{2}{3}\cos^3\frac{x}{2}$ and

$u_2(x) = \int\left(\frac{1}{2}\cos\frac{x}{2} - \sin^2\frac{x}{2}\cos\frac{x}{2}\right)dx = \sin\frac{x}{2} - \frac{2}{3}\sin^3\frac{x}{2}$. Thus

$y_p(x) = \left(-\cos\frac{x}{2} + \frac{2}{3}\cos^3\frac{x}{2}\right)\cos\frac{x}{2} + \left(\sin\frac{x}{2} - \frac{2}{3}\sin^3\frac{x}{2}\right)\sin\frac{x}{2} = -\left(\cos^2\frac{x}{2} - \sin^2\frac{x}{2}\right) + \frac{2}{3}\left(\cos^4\frac{x}{2} - \sin^4\frac{x}{2}\right)$

$= -\cos\left(2\cdot\frac{x}{2}\right) + \frac{2}{3}\left(\cos^2\frac{x}{2} + \sin^2\frac{x}{2}\right)\left(\cos^2\frac{x}{2} - \sin^2\frac{x}{2}\right) = -\cos x + \frac{2}{3}\cos x = -\frac{1}{3}\cos x$

and the general solution is $y(x) = y_c(x) + y_p(x) = c_1 \cos\frac{x}{2} + c_2 \sin\frac{x}{2} - \frac{1}{3}\cos x$.

21. (a) $r^2 - 2r + 1 = (r-1)^2 = 0 \;\Rightarrow\; r = 1$, so the complementary solution is $y_c(x) = c_1 e^x + c_2 x e^x$. A particular solution

is of the form $y_p(x) = Ae^{2x}$. Thus $4Ae^{2x} - 4Ae^{2x} + Ae^{2x} = e^{2r} \;\Rightarrow\; Ac^{2x} - e^{2x} \;\Rightarrow\; A = 1 \;\Rightarrow\; y_p(x) = e^{2x}$.

So a general solution is $y(x) = y_c(x) + y_p(x) = c_1 e^x + c_2 x e^x + e^{2x}$.

(b) From (a), $y_c(x) = c_1 e^x + c_2 x e^x$, so set $y_1 = e^x$, $y_2 = x e^x$. Then, $y_1 y_2' - y_2 y_1' = e^{2x}(1 + x) - x e^{2x} = e^{2x}$ and so

$u_1' = -x e^x \;\Rightarrow\; u_1(x) = -\int x e^x\, dx = -(x - 1)e^x$ [by parts] and $u_2' = e^x \;\Rightarrow\; u_2(x) = \int e^x\, dx = e^x$. Hence

$y_p(x) = (1 - x)e^{2x} + x e^{2x} = e^{2x}$ and the general solution is $y(x) = y_c(x) + y_p(x) = c_1 e^x + c_2 x e^x + e^{2x}$.

23. As in Example 5, $y_c(x) = c_1 \sin x + c_2 \cos x$, so set $y_1 = \sin x$, $y_2 = \cos x$. Then $y_1 y_2' - y_2 y_1' = -\sin^2 x - \cos^2 x = -1$,

so $u_1' = -\dfrac{\sec^2 x \cos x}{-1} = \sec x \;\Rightarrow\; u_1(x) = \int \sec x\, dx = \ln(\sec x + \tan x)$ for $0 < x < \frac{\pi}{2}$,

and $u_2' = \dfrac{\sec^2 x \sin x}{-1} = -\sec x \tan x \;\Rightarrow\; u_2(x) = -\sec x$. Hence

$y_p(x) = \ln(\sec x + \tan x)\cdot\sin x - \sec x \cdot \cos x = \sin x \ln(\sec x + \tan x) - 1$ and the general solution is

$y(x) = c_1 \sin x + c_2 \cos x + \sin x \ln(\sec x + \tan x) - 1$.

© 2012 Cengage Learning. All Rights Reserved. May not be scanned, copied, or duplicated, or posted to a publicly accessible website, in whole or in part.

25. $y_1 = e^x$, $y_2 = e^{2x}$ and $y_1 y_2' - y_2 y_1' = e^{3x}$. So $u_1' = \dfrac{-e^{2x}}{(1 + e^{-x})e^{3x}} = -\dfrac{e^{-x}}{1 + e^{-x}}$ and

$$u_1(x) = \int -\frac{e^{-x}}{1 + e^{-x}}\, dx = \ln(1 + e^{-x}). \quad u_2' = \frac{e^x}{(1 + e^{-x})e^{3x}} = \frac{e^x}{e^{3x} + e^{2x}} \text{ so}$$

$$u_2(x) = \int \frac{e^x}{e^{3x} + e^{2x}}\, dx = \ln\left(\frac{e^x + 1}{e^x}\right) - e^{-x} = \ln(1 + e^{-x}) - e^{-x}. \text{ Hence}$$

$y_p(x) = e^x \ln(1 + e^{-x}) + e^{2x}[\ln(1 + e^{-x}) - e^{-x}]$ and the general solution is

$y(x) = [c_1 + \ln(1 + e^{-x})]e^x + [c_2 - e^{-x} + \ln(1 + e^{-x})]e^{2x}.$

27. $r^2 - 2r + 1 = (r - 1)^2 = 0 \ \Rightarrow \ r = 1$ so $y_c(x) = c_1 e^x + c_2 x e^x$. Thus $y_1 = e^x$, $y_2 = x e^x$ and

$$y_1 y_2' - y_2 y_1' = e^x(x + 1)e^x - x e^x e^x = e^{2x}. \text{ So } u_1' = -\frac{x e^x \cdot e^x/(1 + x^2)}{e^{2x}} = -\frac{x}{1 + x^2} \ \Rightarrow$$

$$u_1 = -\int \frac{x}{1 + x^2}\, dx = -\frac{1}{2}\ln\left(1 + x^2\right), \ u_2' = \frac{e^x \cdot e^x/(1 + x^2)}{e^{2x}} = \frac{1}{1 + x^2} \ \Rightarrow \ u_2 = \int \frac{1}{1 + x^2}\, dx = \tan^{-1} x \text{ and}$$

$y_p(x) = -\frac{1}{2}e^x \ln(1 + x^2) + x e^x \tan^{-1} x$. Hence the general solution is $y(x) = e^x\left[c_1 + c_2 x - \frac{1}{2}\ln(1 + x^2) + x \tan^{-1} x\right]$.

17.3 Applications of Second-Order Differential Equations

1. By Hooke's Law $k(0.25) = 25$ so $k = 100$ is the spring constant and the differential equation is $5x'' + 100x = 0$.

The auxiliary equation is $5r^2 + 100 = 0$ with roots $r = \pm 2\sqrt{5}\, i$, so the general solution to the differential equation is

$x(t) = c_1 \cos\left(2\sqrt{5}\, t\right) + c_2 \sin\left(2\sqrt{5}\, t\right)$. We are given that $x(0) = 0.35 \ \Rightarrow \ c_1 = 0.35$ and $x'(0) = 0 \ \Rightarrow$

$2\sqrt{5}\, c_2 = 0 \ \Rightarrow \ c_2 = 0$, so the position of the mass after t seconds is $x(t) = 0.35 \cos\left(2\sqrt{5}\, t\right)$.

3. $k(0.5) = 6$ or $k = 12$ is the spring constant, so the initial-value problem is $2x'' + 14x' + 12x = 0$, $x(0) = 1$, $x'(0) = 0$.

The general solution is $x(t) = c_1 e^{-6t} + c_2 e^{-t}$. But $1 = x(0) = c_1 + c_2$ and $0 = x'(0) = -6c_1 - c_2$. Thus the position is

given by $x(t) = -\frac{1}{5}e^{-6t} + \frac{6}{5}e^{-t}$.

5. For critical damping we need $c^2 - 4mk = 0$ or $m = c^2/(4k) = 14^2/(4 \cdot 12) = \frac{49}{12}$ kg.

7. We are given $m = 1$, $k = 100$, $x(0) = -0.1$ and $x'(0) = 0$. From (3), the differential equation is $\dfrac{d^2x}{dt^2} + c\dfrac{dx}{dt} + 100x = 0$

with auxiliary equation $r^2 + cr + 100 = 0$.

If $c = 10$, we have two complex roots $r = -5 \pm 5\sqrt{3}\, i$, so the motion is underdamped and the solution is

$x = e^{-5t}\left[c_1 \cos\left(5\sqrt{3}\, t\right) + c_2 \sin\left(5\sqrt{3}\, t\right)\right]$. Then $-0.1 = x(0) = c_1$ and $0 = x'(0) = 5\sqrt{3}\, c_2 - 5c_1 \ \Rightarrow \ c_2 = -\frac{1}{10\sqrt{3}}$,

so $x = e^{-5t}\left[-0.1\cos\left(5\sqrt{3}\, t\right) - \frac{1}{10\sqrt{3}}\sin\left(5\sqrt{3}\, t\right)\right]$.

© 2012 Cengage Learning. All Rights Reserved. May not be scanned, copied, or duplicated, or posted to a publicly accessible website, in whole or in part.

If $c = 15$, we again have underdamping since the auxiliary equation has roots $r = -\frac{15}{2} \pm \frac{5\sqrt{7}}{2}i$. The general solution is

$x = e^{-15t/2}\left[c_1 \cos\left(\frac{5\sqrt{7}}{2}t\right) + c_2 \sin\left(\frac{5\sqrt{7}}{2}t\right)\right]$, so $-0.1 = x(0) = c_1$ and $0 = x'(0) = \frac{5\sqrt{7}}{2}c_2 - \frac{15}{2}c_1 \Rightarrow c_2 = -\frac{3}{10\sqrt{7}}$.

Thus $x = e^{-15t/2}\left[-0.1\cos\left(\frac{5\sqrt{7}}{2}t\right) - \frac{3}{10\sqrt{7}}\sin\left(\frac{5\sqrt{7}}{2}t\right)\right]$.

For $c = 20$, we have equal roots $r_1 = r_2 = -10$, so the oscillation is critically damped and the solution is

$x = (c_1 + c_2t)e^{-10t}$. Then $-0.1 = x(0) = c_1$ and $0 = x'(0) = -10c_1 + c_2 \Rightarrow c_2 = -1$, so $x = (-0.1 - t)e^{-10t}$.

If $c = 25$ the auxiliary equation has roots $r_1 = -5$, $r_2 = -20$, so we have overdamping and the solution is

$x = c_1e^{-5t} + c_2e^{-20t}$. Then $-0.1 = x(0) = c_1 + c_2$ and $0 = x'(0) = -5c_1 - 20c_2 \Rightarrow c_1 = -\frac{2}{15}$ and $c_2 = \frac{1}{30}$,

so $x = -\frac{2}{15}e^{-5t} + \frac{1}{30}e^{-20t}$.

If $c = 30$ we have roots $r = -15 \pm 5\sqrt{5}$, so the motion is

overdamped and the solution is $x = c_1e^{(-15+5\sqrt{5})t} + c_2e^{(-15-5\sqrt{5})t}$.

Then $-0.1 = x(0) = c_1 + c_2$ and

$0 = x'(0) = \left(-15 + 5\sqrt{5}\right)c_1 + \left(-15 - 5\sqrt{5}\right)c_2 \Rightarrow$

$c_1 = \frac{-5 - 3\sqrt{5}}{100}$ and $c_2 = \frac{-5 + 3\sqrt{5}}{100}$, so

$x = \left(\frac{-5 - 3\sqrt{5}}{100}\right)e^{(-15+5\sqrt{5})t} + \left(\frac{-5 + 3\sqrt{5}}{100}\right)e^{(-15-5\sqrt{5})t}$,

9. The differential equation is $mx'' + kx = F_0\cos\omega_0t$ and $\omega_0 \neq \omega = \sqrt{k/m}$. Here the auxiliary equation is $mr^2 + k = 0$

with roots $\pm\sqrt{k/m}\,i = \pm\omega i$ so $x_c(t) = c_1\cos\omega t + c_2\sin\omega t$. Since $\omega_0 \neq \omega$, try $x_p(t) = A\cos\omega_0t + B\sin\omega_0t$.

Then we need $(m)(-\omega_0^2)(A\cos\omega_0t + B\sin\omega_0t) + k(A\cos\omega_0t + B\sin\omega_0t) = F_0\cos\omega_0t$ or $A(k - m\omega_0^2) = F_0$ and

$B(k - m\omega_0^2) = 0$. Hence $B = 0$ and $A = \dfrac{F_0}{k - m\omega_0^2} = \dfrac{F_0}{m(\omega^2 - \omega_0^2)}$ since $\omega^2 = \dfrac{k}{m}$. Thus the motion of the mass is given

by $x(t) = c_1\cos\omega t + c_2\sin\omega t + \dfrac{F_0}{m(\omega^2 - \omega_0^2)}\cos\omega_0t$.

11. From Equation 6, $x(t) = f(t) + g(t)$ where $f(t) = c_1\cos\omega t + c_2\sin\omega t$ and $g(t) = \dfrac{F_0}{m(\omega^2 - \omega_0^2)}\cos\omega_0t$. Then f

is periodic, with period $\frac{2\pi}{\omega}$, and if $\omega \neq \omega_0$, g is periodic with period $\frac{2\pi}{\omega_0}$. If $\frac{\omega}{\omega_0}$ is a rational number, then we can say

$\frac{\omega}{\omega_0} = \frac{a}{b} \Rightarrow a = \frac{b\omega}{\omega_0}$ where a and b are non-zero integers. Then

$x\left(t + a\cdot\frac{2\pi}{\omega}\right) = f\left(t + a\cdot\frac{2\pi}{\omega}\right) + g\left(t + a\cdot\frac{2\pi}{\omega}\right) = f(t) + g\left(t + \frac{b\omega}{\omega_0}\cdot\frac{2\pi}{\omega}\right) = f(t) + g\left(t + b\cdot\frac{2\pi}{\omega_0}\right) = f(t) + g(t) = x(t)$

so $x(t)$ is periodic.

13. Here the initial-value problem for the charge is $Q'' + 20Q' + 500Q = 12$, $Q(0) = Q'(0) = 0$. Then

$Q_c(t) = e^{-10t}(c_1\cos 20t + c_2\sin 20t)$ and try $Q_p(t) = A \Rightarrow 500A = 12$ or $A = \frac{3}{125}$.

The general solution is $Q(t) = e^{-10t}(c_1\cos 20t + c_2\sin 20t) + \frac{3}{125}$. But $0 = Q(0) = c_1 + \frac{3}{125}$ and

© 2012 Cengage Learning. All Rights Reserved. May not be scanned, copied, or duplicated, or posted to a publicly accessible website, in whole or in part.

$Q'(t) = I(t) = e^{-10t}[(-10c_1 + 20c_2)\cos 20t + (-10c_2 - 20c_1)\sin 20t]$ but $0 = Q'(0) = -10c_1 + 20c_2$. Thus the charge

is $Q(t) = -\frac{1}{250}e^{-10t}(6\cos 20t + 3\sin 20t) + \frac{3}{125}$ and the current is $I(t) = e^{-10t}\left(\frac{3}{5}\right)\sin 20t$.

15. As in Exercise 13, $Q_c(t) = e^{-10t}(c_1\cos 20t + c_2\sin 20t)$ but $E(t) = 12\sin 10t$ so try

$Q_p(t) = A\cos 10t + B\sin 10t$. Substituting into the differential equation gives

$(-100A + 200B + 500A)\cos 10t + (-100B - 200A + 500B)\sin 10t = 12\sin 10t \quad\Rightarrow$

$400A + 200B = 0$ and $400B - 200A = 12$. Thus $A = -\frac{3}{250}$, $B = \frac{3}{125}$ and the general solution is

$Q(t) = e^{-10t}(c_1\cos 20t + c_2\sin 20t) - \frac{3}{250}\cos 10t + \frac{3}{125}\sin 10t$. But $0 = Q(0) = c_1 - \frac{3}{250}$ so $c_1 = \frac{3}{250}$.

Also $Q'(t) = \frac{3}{25}\sin 10t + \frac{6}{25}\cos 10t + e^{-10t}[(-10c_1 + 20c_2)\cos 20t + (-10c_2 - 20c_1)\sin 20t]$ and

$0 = Q'(0) = \frac{6}{25} - 10c_1 + 20c_2$ so $c_2 = -\frac{3}{500}$. Hence the charge is given by

$Q(t) = e^{-10t}\left[\frac{3}{250}\cos 20t - \frac{3}{500}\sin 20t\right] - \frac{3}{250}\cos 10t + \frac{3}{125}\sin 10t$.

17. $x(t) = A\cos(\omega t + \delta) \quad\Leftrightarrow\quad x(t) = A[\cos\omega t\cos\delta - \sin\omega t\sin\delta] \quad\Leftrightarrow\quad x(t) = A\left(\dfrac{c_1}{A}\cos\omega t + \dfrac{c_2}{A}\sin\omega t\right)$ where

$\cos\delta = c_1/A$ and $\sin\delta = -c_2/A \quad\Leftrightarrow\quad x(t) = c_1\cos\omega t + c_2\sin\omega t$. [Note that $\cos^2\delta + \sin^2\delta = 1 \quad\Rightarrow\quad c_1^2 + c_2^2 = A^2$.]

17.4 Series Solutions

1. Let $y(x) = \sum\limits_{n=0}^{\infty} c_n x^n$. Then $y'(x) = \sum\limits_{n=1}^{\infty} nc_n x^{n-1}$ and the given equation, $y' - y = 0$, becomes

$\sum\limits_{n=1}^{\infty} nc_n x^{n-1} - \sum\limits_{n=0}^{\infty} c_n x^n = 0$. Replacing n by $n+1$ in the first sum gives $\sum\limits_{n=0}^{\infty}(n+1)c_{n+1}x^n - \sum\limits_{n=0}^{\infty} c_n x^n = 0$, so

$\sum\limits_{n=0}^{\infty}[(n+1)c_{n+1} - c_n]x^n = 0$. Equating coefficients gives $(n+1)c_{n+1} - c_n = 0$, so the recursion relation is

$c_{n+1} = \dfrac{c_n}{n+1}$, $n = 0, 1, 2, \ldots$. Then $c_1 = c_0$, $c_2 = \frac{1}{2}c_1 = \dfrac{c_0}{2}$, $c_3 = \frac{1}{3}c_2 = \frac{1}{3}\cdot\frac{1}{2}c_0 = \dfrac{c_0}{3!}$, $c_4 = \frac{1}{4}c_3 = \dfrac{c_0}{4!}$, and

in general, $c_n = \dfrac{c_0}{n!}$. Thus, the solution is $y(x) = \sum\limits_{n=0}^{\infty} c_n x^n = \sum\limits_{n=0}^{\infty} \dfrac{c_0}{n!}x^n = c_0\sum\limits_{n=0}^{\infty}\dfrac{x^n}{n!} = c_0 e^x$.

3. Assuming $y(x) = \sum\limits_{n=0}^{\infty} c_n x^n$, we have $y'(x) = \sum\limits_{n=1}^{\infty} nc_n x^{n-1} = \sum\limits_{n=0}^{\infty}(n+1)c_{n+1}x^n$ and

$-x^2 y = -\sum\limits_{n=0}^{\infty} c_n x^{n+2} = -\sum\limits_{n=2}^{\infty} c_{n-2}x^n$. Hence, the equation $y' = x^2 y$ becomes $\sum\limits_{n=0}^{\infty}(n+1)c_{n+1}x^n - \sum\limits_{n=2}^{\infty} c_{n-2}x^n = 0$

or $c_1 + 2c_2 x + \sum\limits_{n=2}^{\infty}[(n+1)c_{n+1} - c_{n-2}]x^n = 0$. Equating coefficients gives $c_1 = c_2 = 0$ and $c_{n+1} = \dfrac{c_{n-2}}{n+1}$

for $n = 2, 3, \ldots$. But $c_1 = 0$, so $c_4 = 0$ and $c_7 = 0$ and in general $c_{3n+1} = 0$. Similarly $c_2 = 0$ so $c_{3n+2} = 0$. Finally

$c_3 = \dfrac{c_0}{3}$, $c_6 = \dfrac{c_3}{6} = \dfrac{c_0}{6\cdot 3} = \dfrac{c_0}{3^2\cdot 2!}$, $c_9 = \dfrac{c_6}{9} = \dfrac{c_0}{9\cdot 6\cdot 3} = \dfrac{c_0}{3^3\cdot 3!}, \ldots$, and $c_{3n} = \dfrac{c_0}{3^n\cdot n!}$. Thus, the solution

is $y(x) = \sum\limits_{n=0}^{\infty} c_n x^n = \sum\limits_{n=0}^{\infty} c_{3n}x^{3n} = \sum\limits_{n=0}^{\infty}\dfrac{c_0}{3^n\cdot n!}x^{3n} = c_0\sum\limits_{n=0}^{\infty}\dfrac{x^{3n}}{3^n n!} = c_0\sum\limits_{n=0}^{\infty}\dfrac{(x^3/3)^n}{n!} = c_0 e^{x^3/3}$.

© 2012 Cengage Learning. All Rights Reserved. May not be scanned, copied, or duplicated, or posted to a publicly accessible website, in whole or in part.

5. Let $y(x) = \sum_{n=0}^{\infty} c_n x^n \Rightarrow y'(x) = \sum_{n=1}^{\infty} n c_n x^{n-1}$ and $y''(x) = \sum_{n=0}^{\infty} (n+2)(n+1) c_{n+2} x^n$. The differential equation

becomes $\sum_{n=0}^{\infty} (n+2)(n+1) c_{n+2} x^n + x \sum_{n=1}^{\infty} n c_n x^{n-1} + \sum_{n=0}^{\infty} c_n x^n = 0$ or $\sum_{n=0}^{\infty} [(n+2)(n+1) c_{n+2} + n c_n + c_n] x^n = 0$

$\left[\text{since } \sum_{n=1}^{\infty} n c_n x^n = \sum_{n=0}^{\infty} n c_n x^n \right]$. Equating coefficients gives $(n+2)(n+1) c_{n+2} + (n+1) c_n = 0$, thus the

recursion relation is $c_{n+2} = \dfrac{-(n+1) c_n}{(n+2)(n+1)} = -\dfrac{c_n}{n+2}$, $n = 0, 1, 2, \ldots$. Then the even

coefficients are given by $c_2 = -\dfrac{c_0}{2}$, $c_4 = -\dfrac{c_2}{4} = \dfrac{c_0}{2 \cdot 4}$, $c_6 = -\dfrac{c_4}{6} = -\dfrac{c_0}{2 \cdot 4 \cdot 6}$, and in general,

$c_{2n} = (-1)^n \dfrac{c_0}{2 \cdot 4 \cdot \ldots \cdot 2n} = \dfrac{(-1)^n c_0}{2^n n!}$. The odd coefficients are $c_3 = -\dfrac{c_1}{3}$, $c_5 = -\dfrac{c_3}{5} = \dfrac{c_1}{3 \cdot 5}$, $c_7 = -\dfrac{c_5}{7} = -\dfrac{c_1}{3 \cdot 5 \cdot 7}$,

and in general, $c_{2n+1} = (-1)^n \dfrac{c_1}{3 \cdot 5 \cdot 7 \cdot \ldots \cdot (2n+1)} = \dfrac{(-2)^n n! \, c_1}{(2n+1)!}$. The solution is

$$y(x) = c_0 \sum_{n=0}^{\infty} \frac{(-1)^n}{2^n n!} x^{2n} + c_1 \sum_{n=0}^{\infty} \frac{(-2)^n n!}{(2n+1)!} x^{2n+1}.$$

7. Let $y(x) = \sum_{n=0}^{\infty} c_n x^n \Rightarrow y'(x) = \sum_{n=1}^{\infty} n c_n x^{n-1} = \sum_{n=0}^{\infty} (n+1) c_{n+1} x^n$ and $y''(x) = \sum_{n=0}^{\infty} (n+2)(n+1) c_{n+2} x^n$. Then

$(x-1) y''(x) = \sum_{n=0}^{\infty} (n+2)(n+1) c_{n+2} x^{n+1} - \sum_{n=0}^{\infty} (n+2)(n+1) c_{n+2} x^n = \sum_{n=1}^{\infty} n(n+1) c_{n+1} x^n - \sum_{n=0}^{\infty} (n+2)(n+1) c_{n+2} x^n$.

Since $\sum_{n=1}^{\infty} n(n+1) c_{n+1} x^n = \sum_{n=0}^{\infty} n(n+1) c_{n+1} x^n$, the differential equation becomes

$\sum_{n=0}^{\infty} n(n+1) c_{n+1} x^n - \sum_{n=0}^{\infty} (n+2)(n+1) c_{n+2} x^n + \sum_{n=0}^{\infty} (n+1) c_{n+1} x^n = 0 \Rightarrow$

$\sum_{n=0}^{\infty} [n(n+1) c_{n+1} - (n+2)(n+1) c_{n+2} + (n+1) c_{n+1}] x^n = 0$ or $\sum_{n=0}^{\infty} [(n+1)^2 c_{n+1} - (n+2)(n+1) c_{n+2}] x^n = 0$.

Equating coefficients gives $(n+1)^2 c_{n+1} - (n+2)(n+1) c_{n+2} = 0$ for $n = 0, 1, 2, \ldots$. Then the recursion relation is

$c_{n+2} = \dfrac{(n+1)^2}{(n+2)(n+1)} c_{n+1} = \dfrac{n+1}{n+2} c_{n+1}$, so given c_0 and c_1, we have $c_2 = \frac{1}{2} c_1$, $c_3 = \frac{2}{3} c_2 = \frac{1}{3} c_1$, $c_4 = \frac{3}{4} c_3 = \frac{1}{4} c_1$, and

in general $c_n = \dfrac{c_1}{n}$, $n = 1, 2, 3, \ldots$. Thus the solution is $y(x) = c_0 + c_1 \sum_{n=1}^{\infty} \dfrac{x^n}{n}$. Note that the solution can be expressed as

$c_0 - c_1 \ln(1 - x)$ for $|x| < 1$.

9. Let $y(x) = \sum_{n=0}^{\infty} c_n x^n$. Then $-x y'(x) = -x \sum_{n=1}^{\infty} n c_n x^{n-1} = -\sum_{n=1}^{\infty} n c_n x^n = -\sum_{n=0}^{\infty} n c_n x^n$,

$y''(x) = \sum_{n=0}^{\infty} (n+2)(n+1) c_{n+2} x^n$, and the equation $y'' - x y' - y = 0$ becomes

$\sum_{n=0}^{\infty} [(n+2)(n+1) c_{n+2} - n c_n - c_n] x^n = 0$. Thus, the recursion relation is

© 2012 Cengage Learning. All Rights Reserved. May not be scanned, copied, or duplicated, or posted to a publicly accessible website, in whole or in part.

$$c_{n+2} = \frac{nc_n + c_n}{(n+2)(n+1)} = \frac{c_n(n+1)}{(n+2)(n+1)} = \frac{c_n}{n+2} \text{ for } n = 0, 1, 2, \ldots. \text{ One of the given conditions is } y(0) = 1. \text{ But}$$

$$y(0) = \sum_{n=0}^{\infty} c_n(0)^n = c_0 + 0 + 0 + \cdots = c_0, \text{ so } c_0 = 1. \text{ Hence, } c_2 = \frac{c_0}{2} = \frac{1}{2}, c_4 = \frac{c_2}{4} = \frac{1}{2 \cdot 4}, c_6 = \frac{c_4}{6} = \frac{1}{2 \cdot 4 \cdot 6}, \ldots,$$

$$c_{2n} = \frac{1}{2^n n!}. \text{ The other given condition is } y'(0) = 0. \text{ But } y'(0) = \sum_{n=1}^{\infty} nc_n(0)^{n-1} = c_1 + 0 + 0 + \cdots = c_1, \text{ so } c_1 = 0.$$

By the recursion relation, $c_3 = \frac{c_1}{3} = 0, c_5 = 0, \ldots, c_{2n+1} = 0$ for $n = 0, 1, 2, \ldots.$ Thus, the solution to the initial-value

problem is $y(x) = \sum_{n=0}^{\infty} c_n x^n = \sum_{n=0}^{\infty} c_{2n} x^{2n} = \sum_{n=0}^{\infty} \frac{x^{2n}}{2^n n!} = \sum_{n=0}^{\infty} \frac{(x^2/2)^n}{n!} = e^{x^2/2}.$

11. Assuming that $y(x) = \sum_{n=0}^{\infty} c_n x^n$, we have $xy = x \sum_{n=0}^{\infty} c_n x^n = \sum_{n=0}^{\infty} c_n x^{n+1}$, $x^2 y' = x^2 \sum_{n=1}^{\infty} nc_n x^{n-1} = \sum_{n=0}^{\infty} nc_n x^{n+1}$,

$$y''(x) = \sum_{n=2}^{\infty} n(n-1)c_n x^{n-2} = \sum_{n=-1}^{\infty} (n+3)(n+2)c_{n+3} x^{n+1} \qquad \text{[replace } n \text{ with } n+3\text{]}$$

$$= 2c_2 + \sum_{n=0}^{\infty} (n+3)(n+2)c_{n+3} x^{n+1},$$

and the equation $y'' + x^2 y' + xy = 0$ becomes $2c_2 + \sum_{n=0}^{\infty} [(n+3)(n+2)c_{n+3} + nc_n + c_n] x^{n+1} = 0.$ So $c_2 = 0$ and the

recursion relation is $c_{n+3} = \frac{-nc_n - c_n}{(n+3)(n+2)} = -\frac{(n+1)c_n}{(n+3)(n+2)}, n = 0, 1, 2, \ldots.$ But $c_0 = y(0) = 0 = c_2$ and by the

recursion relation, $c_{3n} = c_{3n+2} = 0$ for $n = 0, 1, 2, \ldots.$ Also, $c_1 = y'(0) = 1$, so $c_4 = -\frac{2c_1}{4 \cdot 3} = -\frac{2}{4 \cdot 3}$,

$$c_7 = -\frac{5c_4}{7 \cdot 6} = (-1)^2 \frac{2 \cdot 5}{7 \cdot 6 \cdot 4 \cdot 3} = (-1)^2 \frac{2^2 5^2}{7!}, \ldots, c_{3n+1} = (-1)^n \frac{2^2 5^2 \cdots (3n-1)^2}{(3n+1)!}. \text{ Thus, the solution is}$$

$$y(x) = \sum_{n=0}^{\infty} c_n x^n = x + \sum_{n=1}^{\infty} \left[(-1)^n \frac{2^2 5^2 \cdots (3n-1)^2 x^{3n+1}}{(3n+1)!} \right].$$

17　Review

CONCEPT CHECK

---

1. (a) $ay'' + by' + cy = 0$ where a, b, and c are constants.

(b) $ar^2 + br + c = 0$

(c) If the auxiliary equation has two distinct real roots r_1 and r_2, the solution is $y = c_1 e^{r_1 x} + c_2 e^{r_2 x}$. If the roots are real and

　　equal, the solution is $y = c_1 e^{rx} + c_2 x e^{rx}$ where r is the common root. If the roots are complex, we can write $r_1 = \alpha + i\beta$

　　and $r_2 = \alpha - i\beta$, and the solution is $y = e^{\alpha x}(c_1 \cos \beta x + c_2 \sin \beta x).$

2. (a) An initial-value problem consists of finding a solution y of a second-order differential equation that also satisfies given

　　conditions $y(x_0) = y_0$ and $y'(x_0) = y_1$, where y_0 and y_1 are constants.

© 2012 Cengage Learning. All Rights Reserved. May not be scanned, copied, or duplicated, or posted to a publicly accessible website, in whole or in part.

(b) A boundary-value problem consists of finding a solution y of a second-order differential equation that also satisfies given boundary conditions $y(x_0) = y_0$ and $y(x_1) = y_1$.

3. (a) $ay'' + by' + cy = G(x)$ where a, b, and c are constants and G is a continuous function.

(b) The complementary equation is the related homogeneous equation $ay'' + by' + cy = 0$. If we find the general solution y_c of the complementary equation and y_p is any particular solution of the original differential equation, then the general solution of the original differential equation is $y(x) = y_p(x) + y_c(x)$.

(c) See Examples 1–5 and the associated discussion in Section 17.2.

(d) See the discussion on pages 1177–1179 [ET 1153–1155].

4. Second-order linear differential equations can be used to describe the motion of a vibrating spring or to analyze an electric circuit; see the discussion in Section 17.3.

5. See Example 1 and the preceding discussion in Section 17.4.

TRUE-FALSE QUIZ

1. True. See Theorem 17.1.3.

3. True. $\cosh x$ and $\sinh x$ are linearly independent solutions of this linear homogeneous equation.

EXERCISES

1. The auxiliary equation is $4r^2 - 1 = 0 \ \Rightarrow \ (2r+1)(2r-1) = 0 \ \Rightarrow \ r = \pm\frac{1}{2}$. Then the general solution is $y = c_1 e^{x/2} + c_2 e^{-x/2}$.

3. The auxiliary equation is $r^2 + 3 = 0 \ \Rightarrow \ r = \pm\sqrt{3}\,i$. Then the general solution is $y = c_1 \cos(\sqrt{3}\,x) + c_2 \sin(\sqrt{3}\,x)$.

5. $r^2 - 4r + 5 = 0 \ \Rightarrow \ r = 2 \pm i$, so $y_c(x) = e^{2x}(c_1 \cos x + c_2 \sin x)$. Try $y_p(x) = Ae^{2x} \ \Rightarrow \ y_p' = 2Ae^{2x}$ and $y_p'' = 4Ae^{2x}$. Substitution into the differential equation gives $4Ae^{2x} - 8Ae^{2x} + 5Ae^{2x} = e^{2x} \ \Rightarrow \ A = 1$ and the general solution is $y(x) = e^{2x}(c_1 \cos x + c_2 \sin x) + e^{2x}$.

7. $r^2 - 2r + 1 = 0 \ \Rightarrow \ r = 1$ and $y_c(x) = c_1 e^x + c_2 x e^x$. Try $y_p(x) = (Ax + B)\cos x + (Cx + D)\sin x \ \Rightarrow$ $y_p' = (C - Ax - B)\sin x + (A + Cx + D)\cos x$ and $y_p'' = (2C - B - Ax)\cos x + (-2A - D - Cx)\sin x$. Substitution gives $(-2Cx + 2C - 2A - 2D)\cos x + (2Ax - 2A + 2B - 2C)\sin x = x\cos x \ \Rightarrow \ A = 0, B = C = D = -\frac{1}{2}$. The general solution is $y(x) = c_1 e^x + c_2 x e^x - \frac{1}{2}\cos x - \frac{1}{2}(x+1)\sin x$.

9. $r^2 - r - 6 = 0 \ \Rightarrow \ r = -2, r = 3$ and $y_c(x) = c_1 e^{-2x} + c_2 e^{3x}$. For $y'' - y' - 6y = 1$, try $y_{p_1}(x) = A$. Then $y_{p_1}'(x) = y_{p_1}''(x) = 0$ and substitution into the differential equation gives $A = -\frac{1}{6}$. For $y'' - y' - 6y = e^{-2x}$ try

© 2012 Cengage Learning. All Rights Reserved. May not be scanned, copied, or duplicated, or posted to a publicly accessible website, in whole or in part.

$y_{p_2}(x) = Bxe^{-2x}$ [since $y = Be^{-2x}$ satisfies the complementary equation]. Then $y'_{p_2} = (B - 2Bx)e^{-2x}$ and

$y''_{p_2} = (4Bx - 4B)e^{-2x}$, and substitution gives $-5Be^{-2x} = e^{-2x}$ $\Rightarrow$ $B = -\frac{1}{5}$. The general solution then is

$y(x) = c_1e^{-2x} + c_2e^{3x} + y_{p_1}(x) + y_{p_2}(x) = c_1e^{-2x} + c_2e^{3x} - \frac{1}{6} - \frac{1}{5}xe^{-2x}$.

11. The auxiliary equation is $r^2 + 6r = 0$ and the general solution is $y(x) = c_1 + c_2e^{-6x} = k_1 + k_2e^{-6(x-1)}$. But

$3 = y(1) = k_1 + k_2$ and $12 = y'(1) = -6k_2$. Thus $k_2 = -2$, $k_1 = 5$ and the solution is $y(x) = 5 - 2e^{-6(x-1)}$.

13. The auxiliary equation is $r^2 - 5r + 4 = 0$ and the general solution is $y(x) = c_1e^x + c_2e^{4x}$. But $0 = y(0) = c_1 + c_2$

and $1 = y'(0) = c_1 + 4c_2$, so the solution is $y(x) = \frac{1}{3}(e^{4x} - e^x)$.

15. $r^2 + 4r + 29 = 0$ $\Rightarrow$ $r = -2 \pm 5i$ and the general solution is $y = e^{-2x}(c_1 \cos 5x + c_2 \sin 5x)$. But $1 = y(0) = c_1$ and

$-1 = y(\pi) = -c_1e^{-2\pi}$ $\Rightarrow$ $c_1 = e^{2\pi}$, so there is no solution.

17. Let $y(x) = \sum\limits_{n=0}^{\infty} c_nx^n$. Then $y''(x) = \sum\limits_{n=0}^{\infty} n(n-1)c_nx^{n-2} = \sum\limits_{n=0}^{\infty} (n+2)(n+1)c_{n+2}x^n$ and the differential equation

becomes $\sum\limits_{n=0}^{\infty} [(n+2)(n+1)c_{n+2} + (n+1)c_n]x^n = 0$. Thus the recursion relation is $c_{n+2} = -c_n/(n+2)$

for $n = 0, 1, 2, \ldots$. But $c_0 = y(0) = 0$, so $c_{2n} = 0$ for $n = 0, 1, 2, \ldots$. Also $c_1 = y'(0) = 1$, so $c_3 = -\frac{1}{3}$, $c_5 = \frac{(-1)^2}{3 \cdot 5}$,

$c_7 = \frac{(-1)^3}{3 \cdot 5 \cdot 7} = \frac{(-1)^3 2^3 3!}{7!}, \ldots, c_{2n+1} = \frac{(-1)^n 2^n n!}{(2n+1)!}$ for $n = 0, 1, 2, \ldots$. Thus the solution to the initial-value problem

is $y(x) = \sum\limits_{n=0}^{\infty} c_nx^n = \sum\limits_{n=0}^{\infty} \frac{(-1)^n 2^n n!}{(2n+1)!} x^{2n+1}$.

19. Here the initial-value problem is $2Q'' + 40Q' + 400Q = 12$, $Q(0) = 0.01$, $Q'(0) = 0$. Then

$Q_c(t) = e^{-10t}(c_1 \cos 10t + c_2 \sin 10t)$ and we try $Q_p(t) = A$. Thus the general solution is

$Q(t) = e^{-10t}(c_1 \cos 10t + c_2 \sin 10t) + \frac{3}{100}$. But $0.01 = Q'(0) = c_1 + 0.03$ and $0 = Q''(0) = -10c_1 + 10c_2$,

so $c_1 = -0.02 = c_2$. Hence the charge is given by $Q(t) = -0.02e^{-10t}(\cos 10t + \sin 10t) + 0.03$.

21. (a) Since we are assuming that the earth is a solid sphere of uniform density, we can calculate the density ρ as follows:

$\rho = \dfrac{\text{mass of earth}}{\text{volume of earth}} = \dfrac{M}{\frac{4}{3}\pi R^3}$. If V_r is the volume of the portion of the earth which lies within a distance r of the

center, then $V_r = \frac{4}{3}\pi r^3$ and $M_r = \rho V_r = \dfrac{Mr^3}{R^3}$. Thus $F_r = -\dfrac{GM_rm}{r^2} = -\dfrac{GMm}{R^3}r$.

(b) The particle is acted upon by a varying gravitational force during its motion. By Newton's Second Law of Motion,

$m\dfrac{d^2y}{dt^2} = F_y = -\dfrac{GMm}{R^3}y$, so $y''(t) = -k^2y(t)$ where $k^2 = \dfrac{GM}{R^3}$. At the surface, $-mg = F_R = -\dfrac{GMm}{R^2}$, so

$g = \dfrac{GM}{R^2}$. Therefore $k^2 = \dfrac{g}{R}$.

© 2012 Cengage Learning. All Rights Reserved. May not be scanned, copied, or duplicated, or posted to a publicly accessible website, in whole or in part.

(c) The differential equation $y'' + k^2 y = 0$ has auxiliary equation $r^2 + k^2 = 0$. (This is the r of Section 17.1, not the r measuring distance from the earth's center.) The roots of the auxiliary equation are $\pm ik$, so by (11) in Section 17.1, the general solution of our differential equation for t is $y(t) = c_1 \cos kt + c_2 \sin kt$. It follows that $y'(t) = -c_1 k \sin kt + c_2 k \cos kt$. Now $y(0) = R$ and $y'(0) = 0$, so $c_1 = R$ and $c_2 k = 0$. Thus $y(t) = R \cos kt$ and $y'(t) = -kR \sin kt$. This is simple harmonic motion (see Section 17.3) with amplitude R, frequency k, and phase angle 0. The period is $T = 2\pi/k$. $R \approx 3960$ mi $= 3960 \cdot 5280$ ft and $g = 32$ ft/s^2, so $k = \sqrt{g/R} \approx 1.24 \times 10^{-3}$ s^{-1} and $T = 2\pi/k \approx 5079$ s ≈ 85 min.

(d) $y(t) = 0 \iff \cos kt = 0 \iff kt = \frac{\pi}{2} + \pi n$ for some integer $n \implies y'(t) = -kR \sin\left(\frac{\pi}{2} + \pi n\right) = \pm kR$. Thus the particle passes through the center of the earth with speed $kR \approx 4.899$ mi/s $\approx 17{,}600$ mi/h.

© 2012 Cengage Learning. All Rights Reserved. May not be scanned, copied, or duplicated, or posted to a publicly accessible website, in whole or in part.

☐ APPENDIX

Appendix H   Complex Numbers

1. $(5 - 6i) + (3 + 2i) = (5 + 3) + (-6 + 2)i = 8 + (-4)i = 8 - 4i$

3. $(2 + 5i)(4 - i) = 2(4) + 2(-i) + (5i)(4) + (5i)(-i) = 8 - 2i + 20i - 5i^2 = 8 + 18i - 5(-1)$

$\qquad = 8 + 18i + 5 = 13 + 18i$

5. $\overline{12 + 7i} = 12 - 7i$

7. $\dfrac{1 + 4i}{3 + 2i} = \dfrac{1 + 4i}{3 + 2i} \cdot \dfrac{3 - 2i}{3 - 2i} = \dfrac{3 - 2i + 12i - 8(-1)}{3^2 + 2^2} = \dfrac{11 + 10i}{13} = \dfrac{11}{13} + \dfrac{10}{13}i$

9. $\dfrac{1}{1 + i} = \dfrac{1}{1 + i} \cdot \dfrac{1 - i}{1 - i} = \dfrac{1 - i}{1 - (-1)} = \dfrac{1 - i}{2} = \dfrac{1}{2} - \dfrac{1}{2}i$

11. $i^3 = i^2 \cdot i = (-1)i = -i$

13. $\sqrt{-25} = \sqrt{25}\,i = 5i$

15. $\overline{12 - 5i} = 12 + 15i$ and $|12 - 15i| = \sqrt{12^2 + (-5)^2} = \sqrt{144 + 25} = \sqrt{169} = 13$

17. $\overline{-4i} = \overline{0 - 4i} = 0 + 4i = 4i$ and $|-4i| = \sqrt{0^2 + (-4)^2} = \sqrt{16} = 4$

19. $4x^2 + 9 = 0 \iff 4x^2 = -9 \iff x^2 = -\frac{9}{4} \iff x = \pm\sqrt{-\frac{9}{4}} = \pm\sqrt{\frac{9}{4}}\,i = \pm\frac{3}{2}i.$

21. By the quadratic formula, $x^2 + 2x + 5 = 0 \iff x = \dfrac{-2 \pm \sqrt{2^2 - 4(1)(5)}}{2(1)} = \dfrac{-2 \pm \sqrt{-16}}{2} = \dfrac{-2 \pm 4i}{2} = -1 \pm 2i.$

23. By the quadratic formula, $z^2 + z + 2 = 0 \iff z = \dfrac{-1 \pm \sqrt{1^2 - 4(1)(2)}}{2(1)} = \dfrac{-1 \pm \sqrt{-7}}{2} = -\dfrac{1}{2} \pm \dfrac{\sqrt{7}}{2}i.$

25. For $z = -3 + 3i$, $r = \sqrt{(-3)^2 + 3^2} = 3\sqrt{2}$ and $\tan\theta = \frac{3}{-3} = -1 \implies \theta = \frac{3\pi}{4}$ (since z lies in the second quadrant).

Therefore, $-3 + 3i = 3\sqrt{2}\left(\cos\frac{3\pi}{4} + i\sin\frac{3\pi}{4}\right)$.

27. For $z = 3 + 4i$, $r = \sqrt{3^2 + 4^2} = 5$ and $\tan\theta = \frac{4}{3} \implies \theta = \tan^{-1}\left(\frac{4}{3}\right)$ (since z lies in the first quadrant). Therefore,

$3 + 4i = 5\left[\cos\left(\tan^{-1}\frac{4}{3}\right) + i\sin\left(\tan^{-1}\frac{4}{3}\right)\right].$

29. For $z = \sqrt{3} + i$, $r = \sqrt{\left(\sqrt{3}\right)^2 + 1^2} = 2$ and $\tan\theta = \frac{1}{\sqrt{3}} \implies \theta = \frac{\pi}{6} \implies z = 2\left(\cos\frac{\pi}{6} + i\sin\frac{\pi}{6}\right)$.

For $w = 1 + \sqrt{3}\,i$, $r = 2$ and $\tan\theta = \sqrt{3} \implies \theta = \frac{\pi}{3} \implies w = 2\left(\cos\frac{\pi}{3} + i\sin\frac{\pi}{3}\right)$.

Therefore, $zw = 2 \cdot 2\left[\cos\left(\frac{\pi}{6} + \frac{\pi}{3}\right) + i\sin\left(\frac{\pi}{6} + \frac{\pi}{3}\right)\right] = 4\left(\cos\frac{\pi}{2} + i\sin\frac{\pi}{2}\right)$,

$z/w = \frac{2}{2}\left[\cos\left(\frac{\pi}{6} - \frac{\pi}{3}\right) + i\sin\left(\frac{\pi}{6} - \frac{\pi}{3}\right)\right] = \cos\left(-\frac{\pi}{6}\right) + i\sin\left(-\frac{\pi}{6}\right)$, and $1 = 1 + 0i = 1(\cos 0 + i\sin 0) \implies$

© 2012 Cengage Learning. All Rights Reserved. May not be scanned, copied, or duplicated, or posted to a publicly accessible website, in whole or in part.

$1/z = \frac{1}{2}\left[\cos\left(0 - \frac{\pi}{6}\right) + i\sin\left(0 - \frac{\pi}{6}\right)\right] = \frac{1}{2}\left[\cos\left(-\frac{\pi}{6}\right) + i\sin\left(-\frac{\pi}{6}\right)\right]$. For $1/z$, we could also use the formula that precedes Example 5 to obtain $1/z = \frac{1}{2}\left(\cos\frac{\pi}{6} - i\sin\frac{\pi}{6}\right)$.

31. For $z = 2\sqrt{3} - 2i$, $r = \sqrt{\left(2\sqrt{3}\right)^2 + (-2)^2} = 4$ and $\tan\theta = \frac{-2}{2\sqrt{3}} = -\frac{1}{\sqrt{3}} \Rightarrow \theta = -\frac{\pi}{6} \Rightarrow$

$z = 4\left[\cos\left(-\frac{\pi}{6}\right) + i\sin\left(-\frac{\pi}{6}\right)\right]$. For $w = -1 + i$, $r = \sqrt{2}$, $\tan\theta = \frac{1}{-1} = -1 \Rightarrow \theta = \frac{3\pi}{4} \Rightarrow$

$w = \sqrt{2}\left(\cos\frac{3\pi}{4} + i\sin\frac{3\pi}{4}\right)$. Therefore, $zw = 4\sqrt{2}\left[\cos\left(-\frac{\pi}{6} + \frac{3\pi}{4}\right) + i\sin\left(-\frac{\pi}{6} + \frac{3\pi}{4}\right)\right] = 4\sqrt{2}\left(\cos\frac{7\pi}{12} + i\sin\frac{7\pi}{12}\right)$,

$z/w = \frac{4}{\sqrt{2}}\left[\cos\left(-\frac{\pi}{6} - \frac{3\pi}{4}\right) + i\sin\left(-\frac{\pi}{6} - \frac{3\pi}{4}\right)\right] = \frac{4}{\sqrt{2}}\left[\cos\left(-\frac{11\pi}{12}\right) + i\sin\left(-\frac{11\pi}{12}\right)\right] = 2\sqrt{2}\left(\cos\frac{13\pi}{12} + i\sin\frac{13\pi}{12}\right)$, and

$1/z = \frac{1}{4}\left[\cos\left(-\frac{\pi}{6}\right) - i\sin\left(-\frac{\pi}{6}\right)\right] = \frac{1}{4}\left(\cos\frac{\pi}{6} + i\sin\frac{\pi}{6}\right)$.

33. For $z = 1 + i$, $r = \sqrt{2}$ and $\tan\theta = \frac{1}{1} = 1 \Rightarrow \theta = \frac{\pi}{4} \Rightarrow z = \sqrt{2}\left(\cos\frac{\pi}{4} + i\sin\frac{\pi}{4}\right)$. So by De Moivre's Theorem,

$$(1 + i)^{20} = \left[\sqrt{2}\left(\cos\frac{\pi}{4} + i\sin\frac{\pi}{4}\right)\right]^{20} = (2^{1/2})^{20}\left(\cos\frac{20 \cdot \pi}{4} + i\sin\frac{20 \cdot \pi}{4}\right) = 2^{10}(\cos 5\pi + i\sin 5\pi)$$

$$= 2^{10}[-1 + i(0)] = -2^{10} = -1024$$

35. For $z = 2\sqrt{3} + 2i$, $r = \sqrt{\left(2\sqrt{3}\right)^2 + 2^2} = \sqrt{16} = 4$ and $\tan\theta = \frac{2}{2\sqrt{3}} = \frac{1}{\sqrt{3}} \Rightarrow \theta = \frac{\pi}{6} \Rightarrow z = 4\left(\cos\frac{\pi}{6} + i\sin\frac{\pi}{6}\right)$.

So by De Moivre's Theorem,

$$\left(2\sqrt{3} + 2i\right)^5 = \left[4\left(\cos\frac{\pi}{6} + i\sin\frac{\pi}{6}\right)\right]^5 = 4^5\left(\cos\frac{5\pi}{6} + i\sin\frac{5\pi}{6}\right) = 1024\left[-\frac{\sqrt{3}}{2} + \frac{1}{2}i\right] = -512\sqrt{3} + 512i.$$

37. $1 = 1 + 0i = 1(\cos 0 + i\sin 0)$. Using Equation 3 with $r = 1$, $n = 8$, and $\theta = 0$, we have

$$w_k = 1^{1/8}\left[\cos\left(\frac{0 + 2k\pi}{8}\right) + i\sin\left(\frac{0 + 2k\pi}{8}\right)\right] = \cos\frac{k\pi}{4} + i\sin\frac{k\pi}{4}, \text{ where } k = 0, 1, 2, \ldots, 7.$$

$w_0 = 1(\cos 0 + i\sin 0) = 1$, $w_1 = 1\left(\cos\frac{\pi}{4} + i\sin\frac{\pi}{4}\right) = \frac{1}{\sqrt{2}} + \frac{1}{\sqrt{2}}i$,

$w_2 = 1\left(\cos\frac{\pi}{2} + i\sin\frac{\pi}{2}\right) = i$, $w_3 = 1\left(\cos\frac{3\pi}{4} + i\sin\frac{3\pi}{4}\right) = -\frac{1}{\sqrt{2}} + \frac{1}{\sqrt{2}}i$,

$w_4 = 1(\cos\pi + i\sin\pi) = -1$, $w_5 = 1\left(\cos\frac{5\pi}{4} + i\sin\frac{5\pi}{4}\right) = -\frac{1}{\sqrt{2}} - \frac{1}{\sqrt{2}}i$,

$w_6 = 1\left(\cos\frac{3\pi}{2} + i\sin\frac{3\pi}{2}\right) = -i$, $w_7 = 1\left(\cos\frac{7\pi}{4} + i\sin\frac{7\pi}{4}\right) = \frac{1}{\sqrt{2}} - \frac{1}{\sqrt{2}}i$

39. $i = 0 + i = 1\left(\cos\frac{\pi}{2} + i\sin\frac{\pi}{2}\right)$. Using Equation 3 with $r = 1$, $n = 3$, and $\theta = \frac{\pi}{2}$, we have

$$w_k = 1^{1/3}\left[\cos\left(\frac{\frac{\pi}{2} + 2k\pi}{3}\right) + i\sin\left(\frac{\frac{\pi}{2} + 2k\pi}{3}\right)\right], \text{ where } k = 0, 1, 2.$$

$w_0 = \left(\cos\frac{\pi}{6} + i\sin\frac{\pi}{6}\right) = \frac{\sqrt{3}}{2} + \frac{1}{2}i$

$w_1 = \left(\cos\frac{5\pi}{6} + i\sin\frac{5\pi}{6}\right) = -\frac{\sqrt{3}}{2} + \frac{1}{2}i$

$w_2 = \left(\cos\frac{9\pi}{6} + i\sin\frac{9\pi}{6}\right) = -i$

41. Using Euler's formula (6) with $y = \frac{\pi}{2}$, we have $e^{i\pi/2} = \cos\frac{\pi}{2} + i\sin\frac{\pi}{2} = 0 + 1i = i$.

© 2012 Cengage Learning. All Rights Reserved. May not be scanned, copied, or duplicated, or posted to a publicly accessible website, in whole or in part.

43. Using Euler's formula (6) with $y = \dfrac{\pi}{3}$, we have $e^{i\pi/3} = \cos\dfrac{\pi}{3} + i\sin\dfrac{\pi}{3} = \dfrac{1}{2} + \dfrac{\sqrt{3}}{2}\,i$.

45. Using Equation 7 with $x = 2$ and $y = \pi$, we have $e^{2+i\pi} = e^2 e^{i\pi} = e^2(\cos\pi + i\sin\pi) = e^2(-1 + 0) = -e^2$.

47. Take $r = 1$ and $n = 3$ in De Moivre's Theorem to get

$$[1(\cos\theta + i\sin\theta)]^3 = 1^3(\cos 3\theta + i\sin 3\theta)$$

$$(\cos\theta + i\sin\theta)^3 = \cos 3\theta + i\sin 3\theta$$

$$\cos^3\theta + 3(\cos^2\theta)(i\sin\theta) + 3(\cos\theta)(i\sin\theta)^2 + (i\sin\theta)^3 = \cos 3\theta + i\sin 3\theta$$

$$\cos^3\theta + (3\cos^2\theta\,\sin\theta)i - 3\cos\theta\,\sin^2\theta - (\sin^3\theta)i = \cos 3\theta + i\sin 3\theta$$

$$(\cos^3\theta - 3\sin^2\theta\,\cos\theta) + (3\sin\theta\,\cos^2\theta - \sin^3\theta)i = \cos 3\theta + i\sin 3\theta$$

Equating real and imaginary parts gives $\cos 3\theta = \cos^3\theta - 3\sin^2\theta\,\cos\theta$ and $\sin 3\theta = 3\sin\theta\,\cos^2\theta - \sin^3\theta$.

49. $F(x) = e^{rx} = e^{(a+bi)x} = e^{ax+bxi} = e^{ax}(\cos bx + i\sin bx) = e^{ax}\cos bx + i(e^{ax}\sin bx)$ $\Rightarrow$

$$F'(x) = (e^{ax}\cos bx)' + i(e^{ax}\sin bx)'$$

$$= (ae^{ax}\cos bx - be^{ax}\sin bx) + i(ae^{ax}\sin bx + be^{ax}\cos bx)$$

$$= a[e^{ax}(\cos bx + i\sin bx)] + b[e^{ax}(-\sin bx + i\cos bx)]$$

$$= ae^{rx} + b[e^{ax}(i^2\sin bx + i\cos bx)]$$

$$= ae^{rx} + bi[e^{ax}(\cos bx + i\sin bx)] = ae^{rx} + bie^{rx} = (a+bi)e^{rx} = re^{rx}$$

© 2012 Cengage Learning. All Rights Reserved. May not be scanned, copied, or duplicated, or posted to a publicly accessible website, in whole or in part.